# Communications
# in Computer and Information Science     613

*Commenced Publication in 2007*
Founding and Former Series Editors:
Alfredo Cuzzocrea, Dominik Ślęzak, and Xiaokang Yang

More information about this series at http://www.springer.com/series/7899

Stanisław Kozielski · Dariusz Mrozek
Paweł Kasprowski · Bożena Małysiak-Mrozek
Daniel Kostrzewa (Eds.)

# Beyond Databases, Architectures and Structures

## Advanced Technologies for Data Mining and Knowledge Discovery

12th International Conference, BDAS 2016
Ustroń, Poland, May 31 – June 3, 2016
Proceedings

 Springer

*Editors*
Stanisław Kozielski
Institute of Informatics
Silesian University of Technology
Gliwice
Poland

Dariusz Mrozek
Institute of Informatics
Silesian University of Technology
Gliwice
Poland

Paweł Kasprowski
Institute of Informatics
Silesian University of Technology
Gliwice
Poland

Bożena Małysiak-Mrozek
Institute of Informatics
Silesian University of Technology
Gliwice
Poland

Daniel Kostrzewa
Institute of Informatics
Silesian University of Technology
Gliwice
Poland

ISSN 1865-0929          ISSN 1865-0937 (electronic)
Communications in Computer and Information Science
ISBN 978-3-319-34098-2          ISBN 978-3-319-34099-9 (eBook)
DOI 10.1007/978-3-319-34099-9

Library of Congress Control Number: 2016938385

This Springer imprint is published by Springer Nature
The registered company is Springer International Publishing AG Switzerland

# Preface

Collecting, processing, and analyzing data have become important branches of computer science. Many areas of our existence generate a wealth of information that must be stored in a structured manner and processed appropriately in order to gain the knowledge from the inside. Databases have become a ubiquitous way of collecting and storing data. They are used to hold data describing many areas of human life and activity, and as a consequence, they are also present in almost every IT system. Today's databases have to face the problem of data proliferation and growing variety. More efficient methods for data processing are needed more than ever. New areas of interests that deliver data require innovative algorithms for data analysis.

Beyond Databases, Architectures and Structures (BDAS) is a series of conferences located in Central Europe and very important for this geographic region. The conference intends to give the state of the art of the research that satisfies the needs of modern, widely understood database systems, architectures, models, structures, and algorithms focused on processing various types of data. The aim of the conference is to reflect the most recent developments of databases and allied techniques used for solving problems in a variety of areas related to database systems, or even go one step forward — beyond the horizon of existing databases, architectures, and data structures.

The 12th International BDAS Scientific Conference (BDAS 2016), held in Ustroń, Poland, during May 31–June 3, 2016, was a continuation of the highly successful BDAS conference series started in 2005. For many years BDAS has been attracting hundreds or even thousands of researchers and professionals working in the field of databases. Among attendees of our conference were scientists and representatives of IT companies. Several editions of BDAS were supported by our commercial, world-renowned partners, developing solutions for the database domain, such as IBM, Microsoft, Oracle, Sybase, and others. BDAS annual meetings have become an arena for exchanging information on the widely understood database systems and data-processing algorithms.

BDAS 2016 was the 12th edition of the conference, organized under the technical co-sponsorship of the IEEE Poland Section. We also continued our successful cooperation with Springer, which resulted in the publication of this book. The conference attracted more than 100 participants from 15 countries, who made this conference a successful and memorable event. There were five keynote talks given by leading scientists: Prof. Jarek Gryz from the Department of Computer Science and Engineering, York University, Toronto, Canada, had a keynote talk on "Interactive Visualization of Big Data." Prof. Abdelkader Hameurlain from Pyramid Team, Institut de Recherche en Informatique de Toulouse IRIT, Paul Sabatier University, Toulouse Cedex, France, gave an interesting lecture entitled "Big Data Management in the Cloud: Evolution or Crossroad?" Prof. Dirk Labudde from Bioinformatics group Mittweida (bigM) and Forensic Science Investigation Lab (FoSIL), University of Applied Sciences, Mittweida, Germany, gave an excellent talk entitled "From Data to Evidence: Digital

Forensic Analyses of Communication Networks." Prof. Jean-Charles Lamirel from SYNALP team, LORIA, Vandoeuvre-lès-Nancy, France, had a very enlightening speech on "Performing and Visualizing Temporal Analysis of Large Text Data Issued for Open Sources: Past and Future Methods." Prof. Zbigniew W. Raś from the Department of Computer Science, University of North Carolina, Charlotte, USA, honored us with a presentation on "Reduction of Readmissions to Hospitals Based on Actionable Knowledge Mining and Personalization." The keynote speeches and plenary sessions allowed participants to gain insight into new areas of data analysis and data processing.

BDAS is focused on all aspects of databases. It is intended to have a broad scope, including different kinds of data acquisition, processing, and storing, and this book reflects fairly well the large span of research presented at BDAS 2016. This volume consists of 57 carefully selected papers. The first four papers accompany the stunning keynote talks. The remainder of the papers are assigned to seven thematic groups:

- Artificial intelligence, data mining, and knowledge discovery
- Architectures, structures, and algorithms for efficient data processing
- Data warehousing and OLAP
- Natural language processing, ontologies, and Semantic Web
- Bioinformatics and biomedical data analysis
- Data processing tools
- Novel applications of database systems

The first group, containing eight papers, is related to various methods used in data mining and knowledge discovery. Papers assembled in this group show a wide spectrum of applications of various exploration methods, like decision rules, knowledge-based and neuro-fuzzy systems, clustering and memetic algorithms, rough sets, to solve many real problems. The second group contains 11 papers devoted to database architectures, structures, and algorithms used for efficient data processing. Papers in this group discuss hot topics of effectiveness of query execution, Big Data, testing performance of various database systems, NoSQL, scalability, task scheduling in processing data in OLTP systems, and in-memory, cloud, and probabilistic databases. The next group of papers concerns issues related to data integration, data warehousing, and OLAP. The group consists of three papers presenting research devoted to the scalability of extraction, transformation and load processes, novel data integration architectures, and spatiotemporal OLAP queries. The fourth group consists of five papers devoted to natural language processing, text mining, ontologies, and the Semantic Web. These papers discuss problems of extraction of concepts from text, mapping semantic features to words, text classification, building ontology for underutilized crops, and querying large RDF data with GPUs. The research devoted to bioinformatics and biomedical data analysis is presented in six papers gathered in the fifth group.

The next group of papers is focused on different data processing tools. It presents tools for content modeling, multitenant applications, frameworks for Big Data and biometric identification, as well as benchmarks and simulators constructed by authors.

The last group consisting of ten papers introduces novel applications for which database systems proved to be useful. Some examples include: water demand forecasting, combat identification, drug abuse extraction, hand pose recognition, or methane concentration value prediction.

We hope that the broad scope of topics related to databases covered in this proceedings volume will help the reader to understand that databases have become an important element of nearly every branch of computer science.

We would like to thank all Program Committee members and additional reviewers for their effort in reviewing the papers. Special thanks to Piotr Kuźniacki — builder and for 12 years administrator of our website: www.bdas.pl. The conference organization would not have been possible without the technical staff: Dorota Huget and Jacek Pietraszuk.

April 2016

Stanisław Kozielski
Dariusz Mrozek
Pawel Kasprowski
Bożena Małysiak-Mrozek
Daniel Kostrzewa

# Organization

## BDAS 2016 Program Committee

### Honorary Member

Lotfi A. Zadeh      University of California, Berkeley, USA

### Chair

Stanisław Kozielski      Silesian University of Technology, Poland

### Members

| | |
|---|---|
| Sansanee Auephanwiriyakul | Chiang Mai University, Thailand |
| Werner Backes | Sirrix AG Security Technologies, Bochum, Germany |
| Susmit Bagchi | Gyeongsang National University, South Korea |
| Patrick Bours | Gjøvik University College, Norway |
| George D.C. Cavalcanti | Universidade Federal de Pernambuco, Brazil |
| Andrzej Chydziński | Silesian University of Technology, Poland |
| Tadeusz Czachórski | IITiS, Polish Academy of Sciences, Poland |
| Po-Yuan Chen | China Medical University, Taichung, Taiwan, University of British Columbia, BC, Canada |
| Yixiang Chen | East China Normal University, Shanghai |
| Sebastian Deorowicz | Silesian University of Technology, Poland |
| Jack Dongarra | University of Tennessee, Knoxville, USA |
| Andrzej Drygajlo | École Polytechnique Fédérale de Lausanne, Switzerland |
| Moawia Elfaki Yahia | King Faisal University, Saudi Arabia |
| Rudolf Fleischer | German University of Technology, Oman |
| Hamido Fujita | Iwate Prefectural University, Japan |
| Krzysztof Goczyła | Gdańsk University of Technology, Poland |
| Marcin Gorawski | Silesian University of Technology, Poland |
| Jarek Gryz | York University, Ontario, Canada |
| Andrzej Grzywak | Silesian University of Technology, Poland |
| Brahim Hnich | Izmir University of Economics, Izmir, Turkey |
| Edward Hrynkiewicz | Silesian University of Technology, Poland |
| Jiewen Huang | Google Inc., Mountain View, CA, USA |
| Xiaohua Tony Hu | Drexel University, Philadelphia, PA, USA |
| Zbigniew Huzar | Wrocław University of Technology, Poland |
| Tomasz Imielinski | Rutgers University, New Brunswick, USA |
| Pawel Kasprowski | Silesian University of Technology, Poland |
| Przemysław Kazienko | Wrocław University of Technology, Poland |
| Jerzy Klamka | IITiS, Polish Academy of Sciences, Poland |

| | |
|---|---|
| Bora I. Kumova | Izmir Institute of Technology, Turkey |
| Andrzej Kwiecień | Silesian University of Technology, Poland |
| Jean-Charles Lamirel | LORIA, Nancy, France, University of Strasbourg, France |
| Sérgio Lifschitz | Pontifícia Universidade Católica do Rio de Janeiro, Brazil |
| Antoni Ligęza | AGH University of Science and Technology, Poland |
| Bożena Małysiak-Mrozek | Silesian University of Technology, Poland |
| Marco Masseroli | Politecnico di Milano, Italy |
| Zygmunt Mazur | Wrocław University of Technology, Poland |
| Yasser F.O. Mohammad | Assiut University, Egypt |
| Tadeusz Morzy | Poznań University of Technology, Poland |
| Mikhail Moshkov | King Abdullah University of Science and Technology, Saudi Arabia |
| Dariusz Mrozek | Silesian University of Technology, Poland |
| Mieczysław Muraszkiewicz | Warsaw University of Technology, Poland |
| Sergio Nesmachnow | Universidad de la República, Uruguay |
| Tadeusz Pankowski | Poznań University of Technology, Poland |
| Witold Pedrycz | University of Alberta, Canada |
| Adam Pelikant | Łódź University of Technology, Poland |
| Ewa Piętka | Silesian University of Technology, Poland |
| Bolesław Pochopień | Silesian University of Technology, Poland |
| Andrzej Polański | Silesian University of Technology, Poland |
| Hugo Proença | University of Beira Interior, Portugal |
| Riccardo Rasconi | Institute for Cognitive Sciences and Technologies, Italian National Research Council, Italy |
| Zbigniew W. Raś | University of North Carolina, Charlotte, USA |
| Marek Rejman-Greene | Centre for Applied Science and Technology in Home Office Science, UK |
| Jerzy Rutkowski | Silesian University of Technology, Poland |
| Henryk Rybiński | Warsaw University of Technology, Poland |
| Galina Setlak | Rzeszow University of Technology, Poland |
| Marek Sikora | Silesian University of Technology, Poland |
| | Institute of Innovative Technologies EMAG, Poland |
| Krzysztof Stencel | University of Warsaw, Poland |
| Przemysław Stpiczyński | Maria Curie-Skłodowska University, Poland |
| Dominik Ślęzak | University of Warsaw, Poland |
| | Infobright Inc., Poland |
| Andrzej Świerniak | Silesian University of Technology, Poland |
| Adam Świtoński | Silesian University of Technology, Poland |
| Karin Verspoor | University of Melbourne, Australia |
| Alicja Wakulicz-Deja | University of Silesia in Katowice, Poland |
| Sylwester Warecki | Intel Corporation, San Diego, California, USA |
| Tadeusz Wieczorek | Silesian University of Technology, Poland |
| Piotr Wiśniewski | Nicolaus Copernicus University, Poland |
| Konrad Wojciechowski | Silesian University of Technology, Poland |

Robert Wrembel            Poznań University of Technology, Poland
Stanisław Wrycza          University of Gdańsk, Poland
Mirosław Zaborowski       IITiS, Polish Academy of Sciences, Poland
Grzegorz Zaręba           University of Arizona, Tucson, USA
Krzysztof Zieliński       AGH University of Science and Technology, Poland
Quan Zou                  Xiamen University, People's Republic of China

## Organizing Committee

Bożena Małysiak-Mrozek     Silesian University of Technology, Poland
Dariusz Mrozek            Silesian University of Technology, Poland
Pawel Kasprowski          Silesian University of Technology, Poland
Daniel Kostrzewa          Silesian University of Technology, Poland
Piotr Kuźniacki           Silesian University of Technology, Poland

## Additional Reviewers

Augustyn Dariusz Rafał          Nurzyńska Karolina
Bach Małgorzata                 Pałka Piotr
Bajerski Piotr                  Piórkowski Adam
Brzeski Robert                  Płuciennik Ewa
Cyran Krzysztof                 Poteralski Arkadiusz
Drabowski Mieczysław            Respondek Jerzy
Duszenko Adam                   Romuk Ewa
Frączek Jacek                   Rosner Aldona
Górczyńska Sylwia               Sitek Paweł
Harężlak Katarzyna              Szwoch Mariusz
Josiński Henryk                 Szwoch Wioleta
Kawulok Michał                  Traczyk Tomasz
Kostrzewa Daniel                Tutajewicz Robert
Kozielski Michał                Werner Aleksandra
Lach Jacek                      Wyciślik Łukasz
Michalak Marcin                 Zawiślak Rafał
Momot Alina                     Zghidi Hafed
Myszkorowski Krzysztof          Zielosko Beata
Nowak-Brzezińska Agnieszka

## Sponsoring Institutions

Technical co-sponsorship of the IEEE Poland Section

# Contents

## Invited Papers

## Artifical Intelligence, Data Mining and Knowlege Discovery

## Architectures, Structures and Algorithms for Efficient Data Processing

## Data Warehousing and OLAP

## Natural Language Processing, Ontologies and Semantic Web

## Bioinformatics and Biomedical Data Analysis

## Novel Applications of Database Systems

# Invited Papers

# Interactive Visualization of Big Data

Parke Godfrey[1], Jarek Gryz[1(✉)], Piotr Lasek[1,2], and Nasim Razavi[1]

[1] York University, Toronto, Canada
{godfrey,jarek,plasek,nasim}@cse.yorku.ca
[2] Rzeszów University, Rzeszów, Poland

**Abstract.** Data becomes too big to *see*. Yet visualization is a central way people understand data. We need to learn new ways to accommodate data visualization that scales up and out for large data to enable people to explore visually their data interactively in real-time as a means to understanding it. The five V's of big data—*value, volume, variety, velocity*, and *veracity*—each highlights the challenges of this endeavor.

We present these challenges and a system, SKYDIVE, that we are developing to meet them. SKYDIVE presents an approach that tightly couples a database back-end with a visualization front-end for scaling up and out. We show how hierarchical aggregation can be used to drive this, and the powerful types of interactive visual presentations that can be supported. We are preparing for the day soon when *visualization* becomes the sixth V of big data.

## 1 Introduction

Imagine a map-view over North America showing the distribution of the population of its 476 million people[1]. One could *zoom* into regions to see greater detail, *pan* from region to region to see how the population changes. One could dictate which characteristics to see: wealth, education, age, race. One could mine this trove of census data in a natural, visual way.

But collections of data simply become too big to *see*. This is highlighted by the title of the paper, "Extreme visualization: squeezing a billion records into a million pixels," by Shneiderman in 2008 [29]. With the advent of *big data*, we now far exceed even that point of "extreme". Yet *visualization* remains one of the most potent ways to understand data. It is often a critical first step in extracting *value* from the data. There has been quite recent, consequent resurgence of interest—and calls to action [37]—on how to *scale* data visualization for big data.

There is a second, vital reason for this undertaking. Without new techniques and tools for data visualization and exploration, owners of big data apply what they have to the task. Justin Erickson, director of product management at Cloudera, has stated that a large portion of the workload in Cloudera's customers' data

---

[1] This includes U.S.A. with 319 million, Mexico with 122 million, and Canada with 35 million, as of 2013.

© Springer International Publishing Switzerland 2016
S. Kozielski et al. (Eds.): BDAS 2016, CCIS 613, pp. 3–22, 2016.
DOI: 10.1007/978-3-319-34099-9_1

centers is in direct support of visualization [9]. Visualization tools atop business intelligence and analytics applications are now dominating workloads over big data.

Better ways for interactively visualizing very large data sets, and more efficient ways to accomplish this, are needed. The issues involved in developing such approaches mirror those of big data itself. Just as the issues of big data can be typified by the five "V" 's [5,7,18,27,38], data visualization for big data can be likewise. It is common practice to use information graphics to convey value that has been discovered in data, as in presentations and annual reports [4,33]. But this is after the value had been derived by other means. We wish to derive *value* from the data *via* data visualization that enables data exploration. Interactive visualization tools then are meant to help discover value in underlying data.

Then there is sheer *volume*: the amount of data is too large to "see" at once. The data must be *reduced* before it can be visualized. As such, we do not visualize the data itself, but *synopses* of the data. Synopses can be made by *sampling* or by *aggregating* the data. But even synopses of the data themselves may be too large to see. For big data, any such approach must scale. In seminal work by Elmqvist and Fekete [8], they discuss how data may be *aggregated hierarchically* to support information visualization.

*Variety* refers to the rich types of data that appear in big data. Any technique to derive value from data must contend with these. Visualization *maps* data into visual representations. A data-visualization system then needs to provide a rich array of such *presentation models*, and *mappings* of the data into the models, for users to understand visually different projections of the data easily.

These visual models and mappings must represent the underlying data in a *truthful* way. That is, a data visualization must not "misrepresent" the data. The types of data synopses supported and used by a visualization system must serve to convey a truthful image of the underlying data. Just as the issue of *veracity* is fundamental for any big-data system, it must be *designed in* as a key component of a data-visualization system, if the system is to be worthwhile.

Lastly, visualization for big data should ultimately be able to contend with *velocity*. For many applications, new data is added rapidly. Visualization can be used to monitor how the data changes. One can extend the methodology used to summarize the data for visualization so the system can likewise visually summarize aggregate changes as new data arrives.

Of course visualizing census data and the like is not still just a dream. People are doing it. A beautiful example is provided by *The Racial Dot Map* made by the Demographics Research Group at the University of Virginia [6]. They provide a map showing "one dot per person for the entire United States," based on the U.S. 2010 Census. The dots are color coded based people's reported race, and provides one with a bird's eye view of racial diversity and *de-facto* segregation across the country. The map is interactive, allowing one to zoom and pan. It is supported by the Google Maps API, and is a materialized 1.2 million $512 \times 512$ PNG files (seven GB), and took 16 hours processing to create. While beautiful to see and explore, and eminently useful for researchers, analysts, and lay persons alike for understanding demographics in the U.S. today, the map is limited in many ways. It is *static* in that one cannot change the *view*, what is being visualized about the census data. It is a static map, after all. The map had to be generated in full before anyone could use it.

Our goal is to automate this procedure. The visualization system would provide the parts of the map at the resolutions needed in real-time on demand, rather than the map needing to be materialized in advance in entirety. The *mapping* of how the data is visually presented, and the *view* of the data, could be changed on the fly. The types of visualizations the system supports would be broad. And the *veracity* of the visualization would come guaranteed.

Our visualization system, SKYDIVE [11,13], for big data we are developing is designed to address these issues. SKYDIVE adopts the *visual information seeking mantra* of Shneiderman [28], "Overview first, zoom and filter, then details on demand." The system enables a user to *query* a spatial "cut" (a *dataset*) from the database, define a *mapping* of the "data" (inductively aggregated measures) to a *visual presentation*, and then interactively to pan over and zoom into the presentation.

SKYDIVE's architecture combines data aggregation with visual aggregation in a tightly coupled way to provide for smooth user interaction. The back-end is based on a hierarchical data structure called the *aggregate pyramid*. By interacting with the pyramid (via the database system), the front-end visualization client can quickly filter the data to move up and down in the aggregation hierarchy; thus, *skydiving* into the data.

Contributions and organization of the paper are as follows. In Sect. 2, we discuss related work, motivate what is presently missing in the state of the art, and how SKYDIVE is positioned to address this.

1. *Refining* the issues faced by all big-data systems o the specifics faced by visualization, thus
    - *illustrating* concretely the *issues* faced in designing systems for interactive data visualization for big data. (Throughout the paper.)
2. *Defining inductive aggregation* as a means to accomplish hierarchical aggregation in an *efficient* and *semantically sound* way. (Sect. 3.)
    (a) *Devising* the *aggregate pyramid* as a flexible relational data structure to support real-time, interactive visual exploration.
    (b) *Formalizing* how properly to define *inductive aggregate functions* that are well behaved
        - for *efficient computation*, and
        - to be visually *semantically sound*.
3. *Designing* and *implementing* a system *architecture* via SKYDIVE to address these issues in a comprehensive and scalable way. (Sects. 4, 5 and 6.)
    (a) *Engineering* a three-tiered architecture that exploits the memory hierarchy to support interactive visualization that enables *scale up* and *scale out*. (Sect. 4.)
    (b) *Providing* a rich array of *presentation models*, *leveraging* modern graphics. (Sect. 5.)

In Sect. 7, we discuss next steps and conclude.

## 2   Related Work

Work on information visualization dates back to the beginning of computer science itself. We cannot come close to doing justice to summarizing all work that is loosely related to our endeavor. (See the technical report [12] for a survey of interactive database visualization tools from the perspective of database support.) Herein, we focus on the recent visualization tools and methods, both academic and commercial, that are most closely related to our endeavor to support scalable, data visualization for interactive exploration.

There is also wide literature on spatial data-structures, to which we also cannot do justice here. From a data-structure point of view, there is not much novel in our aggregate pyramid. What is novel is how it is designed and tuned for use within our SKYDIVE architecture.

The following systems employ similar data structures for storing and optimizing data access by means of *aggregation*, *granulation*, and *pyramid*-like data structures to provide users with *interactive* visualizations.

**Mars and Mercury.** The MARS [20] and MERCURY [21] systems are designed for real-time support of top-$k$ spatio-temporal queries over microblogs. They employ a scalable *in-memory* index structure that is capable of digesting an incoming, high-speed, high-volume stream of microblogs. The spatio-temporal index is designed in a form of a *pyramid*, the levels (strata) of which correspond to different periods of time. Each level of the index is composed of cells (bins) containing microblogs. Within each cell, microblog and *aggregate* information are indexed temporally.

These systems have some striking resemblance to SKYDIVE's architecture, but their objectives are quite different. They hierarchically aggregate to handle large streams in time. We hierarchically aggregate spatially to allow interactive, visual exploration.

**ImMens.** The IMMENS system [19] is designed for efficient presentation of large datasets. It preprocesses aggregates in advance, storing them in the database along with the row data. Such aggregations are linked directly to their visual display for more efficient interaction at run-time. The system follows the principle that perceptual and interactive scalability should be limited by the chosen *resolution* of visualized data, not the number of records.

IMMENS is devoted to allowing the user to explore synchronized, but different, views of the (aggregated) data in real-time. They do not consider hierarchical aggregation as SKYDIVE does, nor do they accommodate the zoom and pan paradigm of SKYDIVE. The techniques and objectives of IMMENS and SKYDIVE are orthogonal, and the techniques of both could be eventually integrated.

**Tableau.** TABLEAU [30, 34, 35] is a commercial leader in data visualization. In its latest version, to reduce the computational load for processing large data sets, it employs so-named *extracts* supported by the *Tableau Data Engine*, Tableau's own in-memory, read-only column store. Extracts are filtered or sampled subsets of the original data set that are sufficient to create a truthful visualization.

This technique is used for large datasets and optimizes processing times of creating visualizations.

Still, TABLEAU is not geared for automated visualization of big-data datasets[2]. They are more geared to providing high-quality information visualizations. Interaction and exploration is provided by giving the user easy means to change the view of the data and visualization models.

**M4.** The M4 system [17] is implemented within SAP LUMIRA [16]. It addresses the problem of how much data is needed in order to render a picture truthfully, limiting the amount of data needing to be sent from a database. M4, rather than executing a query as given, relies on the parameters of the desired visualization to rewrite the query. It then develops an appropriate visualization-driven aggregation that only selects necessary data points for rendering. Thus, M4 optimizes performance at two levels: it reduces the communication costs; and it eliminates the need for data reduction at later stages. Similar ideas were implemented in the ScalaR system [3].

M4, however, presently just supports a single "presentation model", and the goal is not interactive exploration. We are striving to provide a much richer array of presentation models within SKYDIVE, and automated support of interactive exploration of visualizations via hierarchical aggregation.

## 3   Inductive Aggregation

*Inductive aggregation* involves two parts: first, how we aggregate hierarchically a very large dataset efficiently; and second, how we define aggregate functions so that they can be computed hierarchically in an efficient way, and that can be mapped to *truthful* visualizations. We define a relational data-structure that we call the *aggregate pyramid*. We want to define *principles* such that aggregate functions which obey these principles can be computed efficiently in a hierarchical manner (constituting the aggregate pyramid) and that are *veracious* in their use in visualization. We call such aggregate functions *inductive aggregate* functions.

### 3.1   The Aggregate Pyramid

The *aggregate pyramid* can be defined via a recursive SQL query as in Fig. 1. The elements used in the query are summarized in Fig. 2. Aggregate pyramids are conceptually related to *data cubes* [14]. Our raw data (`Dataset`) provides *dimensions* and *measures*. Like with data cubes, the measures are (typically) *numeric*,[3] which are to be aggregated. Unlike with data cubes, however, our dimension columns are also *numeric*, together defining a *space*. Therefore, a pyramid only makes sense for data that can be spatially plotted in a meaningful way.

---

[2] Though they are cognizant of the need, and are working toward addressing this.

[3] We shall also show ways that categorical data as measures can be accommodated.

```
-- SQL aggregate pyramid template
recursive Pyramid (c_0, ..., c_{h-1},
       divs,
       b_0, ..., b_{k-1},
       a_0, ..., a_{r-1}) as (
   -- base step
   select c_0, ..., c_{h-1},
       divs,
       integer(x_0 - low_0 * divs /
              high_0 - low_0) as b_0,
       ...,
       integer(x_{k-1} - low_{k-1} * divs /
              high_{k-1} - low_{k-1}) as b_{k-1},
       base_agg_0(...) as a_0,
       ...,
       base_agg_{r-1}(...) as a_{r-1}
   from Dataset, Params
   group by c_0, ..., c_{h-1},
       divs,
       b_0, ..., b_{k-1}
   union all
   select c_0, ..., c_{h-1},
       integer(divs / 2) as sdivs,
       integer(b_0 / 2) as sb_0,
       ...,
       integer(b_{k-1} / 2) as sb_{k-1},
       ind_agg_0(...) as sa_0,
       ...,
       ind_agg_{r-1}(...) as sa_{r-1}
   from Pyramid
   where divs > 1
   group by c_0, ..., c_{h-1},
       sdivs,
       sb_0, ..., sb_{k-1}
) -- end of the recursive definition
select c_0, ..., c_{h-1},
   -1 * integer(log(2, divs)) as stratum,
   zo(b_0, ..., b_{k-1}) as zoo,
   b_0, ..., b_{k-1},
   a_0, ..., a_{r-1}
from Pyramid
```

**Fig. 1.** SQL aggregate pyramid template.

**Dataset:** assemble a dataset of interest from the database

$c_i$'s: attributes with "categorical" data ($c_0, ..., c_{h-1}$)

$x_i$'s: the dimensions ($x_0, ..., x_{k-1}$)

$m_i$'s: the measures ($m_0, ..., m_{q-1}$)

**Params:** parameters used for constructing the pyramid; derived over dataset and/or defined by user; *returns one row*

$low_i$'s: lower grid boundaries for the pyramid

$high_i$'s: higher grid boundaries for the pyramid

$divs$: how many bin-divisions along each $x_i$ for the pyramid base; is a power of 2 (e.g., 1,024)

**Pyramid:** the constructed pyramid

$c_i$'s: constructs a pyramid per partition of $c_0, ..., c_{h-1}$

$divs$: number of bin-divisions (indicates stratum)

$b_i$'s: bin number (for each dimension)

$a_i$'s: aggregate values

**Result:** the returned aggregate pyramid

$c_i$'s: the constants

stratum: stratum of the pyramid

zoo: Z-order ordinal

$b_i$'s: bin number (for each dimension)

$a_i$'s: aggregate values

**Fig. 2.** Pyramid terminology.

**Pyramid** is constructed from input "tables" **Dataset** and **Params**. The dataset can be anything of interest extracted from the database via any SQL query (*not* shown in Fig. 1), providing $k$ dimension columns, $x_0, ..., x_{k-1}$, $q$ measure columns, $m_0, ..., m_{q-1}$, and, potentially, $h$ categorical columns, $c_0, ..., c_{h-1}$. **Params** is a one-row table that provides the *extent* for the pyramid, namely, the $low_i$'s and $high_i$'s that define the selection *window* over the dimensions, and a parameter **divs**, which specifies the initial number of *divisions* into which we will initially aggregate (to form the *base stratum* of the pyramid). It should be a power of two, which allows us to *re-aggregate* by halving **div** each time[4].

The pyramid is then constructed via a *base* step and then *inductive* steps, as per the recursive definition of **Pyramid** in Fig. 1. Let **divs** $= 2^l$ (so $l$ will be the *depth* of the pyramid). The base step does the initial aggregation (computing "base" aggregates $a_0, ..., a_{r-1}$) of the dataset's tuples as partitioned into $2^l \cdot ... \cdot 2^l$ ($2^{lk}$) *bins*. Each aggregate function **base_aggr**$_i$ can be over the input of measures, $m_0, ..., m_{q-1}$, of the dataset. Thus, the bins spatially partition the dataset's space, with the aggregated bin-tuples forming the base *stratum* (layer) of the pyramid.

Each inductive step produces the next higher stratum of the pyramid by *re-aggregating* the aggregate values of the bins of the previous stratum.

---

[4] We use the same number of divisions—power of two—along each of the dimensions, without loss of generality. It is trivial to allow for different "aspect" ratios with different numbers of divisions for different dimensions, however.

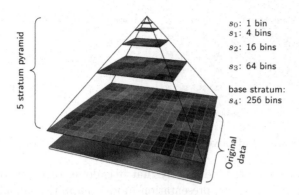

**Fig. 3.** Aggregate pyramid (with $l = 4$).

Here, each aggregate function $\mathtt{ind\_aggr}_i$ is over the input of aggregates, $a_0, \ldots, a_{r-1}$, of the immediate stratum below. This produces the "super" aggregates for the new bin, $sa_0, \ldots, sa_{r-1}$. Each "super" bin aggregates the $2^k$ underlying bins—so four bins, for $k = 2$—that it spatially encompasses. This iteratively continues until the *top* of the pyramid, the stratum consists of a single bin that aggregates the entire dataset. A pyramid over a two-dimensional dataset ($k = 2$) is pictured in Fig. 3.

We label strata from 0 at the peak of the pyramid down to $-l$ at the base of the pyramid (given we started with $\mathtt{divs} = 2^l$). Thus, the area / volume of a given bin in stratum $-i$ is $2^{-i} \cdot \ldots \cdot 2^{-i} \ (2^{-ik})$. We follow this convention to align with the picture of a pyramid, with lower strata having smaller numbers, and having the **stratum** value encode the size of bins. *Lower* strata aggregate the data at smaller granularity, thus offer a *higher resolution* picture of the underlying data[5].

Z-order, **zoo**, offers a way to index stratum for efficient access; this is discussed in Sect. 4. The $c_i$'s are columns with "categorical" data. Note they appear in each **group by**. Thus, a pyramid results per each $c_0 \times \ldots \times c_{h-1}$. For example, one might wish to aggregate separately data by *gender*, resulting in pyramids for *female* and *male*. (We consider this further in Sect. 5.3.)

## 3.2   Inductive Aggregate Functions

To compute an aggregate pyramid efficiently, we shall insist that each stratum can be computed from the immediate stratum below it. That is, the computation per stratum may not reference back to an arbitrary number of strata below or to the raw data (the dataset) each time; that would be too expensive. We call this the principle of *progressive computation*. This means our aggregate values are to be computed iteratively, stratum by stratum. We call this *inductive aggregation*. Not all aggregates can be computed inductively, and care must be taken for

---

[5] For simplicity, we shall refer to strata $s_0, \ldots, s_l$, from the top to the bottom, respectively, forgoing the minus sign when understood in context.

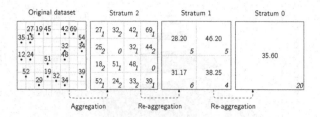

**Fig. 4.** Inductive aggregation over a sample dataset.

those that can be computed inductively that it is done correctly. To ensure this, our inductive aggregates (and presentation-model mappings) should obey given principles we call *stability principles*.

**Aggregate Functions.** Standard aggregate functions include *mean, count, sum, min,* and *max.* These are often augmented with *median, variance,* and *standard deviation.* We need to investigate what types of aggregates are suitable to support visualization, amenable to being hierarchically aggregated (and "re-aggregated") [1,8], and that are well-behaved in such a setting. We shall call these *inductive aggregate functions,* and the process *inductive aggregation.*

For our inductive construction of the aggregate pyramid, first, we aggregate the raw data (from the dataset) to construct the pyramid's *base* stratum. Next, for iteratively building each higher stratum $(i \geq -l)$, we *re-aggregate* the aggregate values of the previous stratum. An *inductive aggregate function* in this context thus consists of two functions:

- a *base-aggregate* function that aggregates the initial raw data into the base stratum; and
- a *re-aggregate* function that aggregates over the corresponding aggregate values of the corresponding sub-bins.

Consider the task of counting objects with $x, y$-coordinates. The base aggregate function will be *count,* used to count the number of objects from the data per base bin. The re-aggregate function will be *sum,* of course, which will sum the counts from the underlying bins.

Let us define these notions more precisely.

- **base aggregates**.
    - *values.* A *base aggregate value A*—or *aggregate,* for short—is a fixed-size *synopsis* (data-structure) of values (e.g., a single base-type value or a fixed-length array) that results from the application of a base-aggregate function over a set of tuples $T$.
    - *functions.* A *base aggregate function F* maps a set of tuples $T$ to a base-aggregate value (synopsis) $A$; $F(T) \rightarrow A$.
- **re-aggregates**.
    - *values.* A *re-aggregate value R* is a fixed-size synopsis that results from the application of a re-aggregate function over a set of aggregate values $\mathcal{A}$ of uniform type of the same type as $R$.

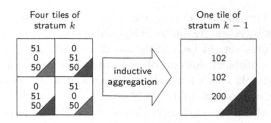

**Fig. 5.** A stability issue. (Color figure online)

- *functions.* A re-aggregate function $G$ maps a set of aggregate values $\mathcal{A}$ of a given type to (re-)aggregate value (synopsis) $R$ of the same type; $G(\mathcal{A}) \to R$.

Note that it is not an entirely trivial task to determine what types of aggregates can be handled inductively. One must be careful in their design.

Aggregate functions can thus be classified:

1. those that do not require additional information besides their corresponding values from the previous iteration;
2. those that require auxiliary aggregate values in the previous iteration; and
3. those that cannot be used inductively (as they would require referencing values from strata indefinitely below or the raw data itself).

The basic aggregate functions *sum*, *count*, *min*, and *max* are of Category 1. Others such as *average* and *expected value* can be done, but require "helper" aggregates, and so are in Category 2. Many, though, are in Category 3, and are not amenable: *mode* (returns the most frequent object from a set of objects), *percentiles*, *variance*, *standard deviation*, and *entropy*[6]. 

*Example 1.* Consider the dataset presented in Fig. 4. The input dataset contains 20 two-dimensional objects. The task is to generate the strata of the aggregate pyramid using *average* over the measure and keeping *count* of the number of objects represented. To compute aggregates of higher strata, additional information is needed; namely, the number (*count*) of objects belonging to the bin. ☐

**Stability of Aggregates.** One must design the re-aggregation functions carefully so that the aggregates of higher level strata are computed correctly, and so that they are well behaved for visualization. Towards this end, we define *stability principles* for aggregate functions as mapped to visual elements under which the mapped aggregate values are well behaved. We call this *stability* because we want the scene (the visualization) of the data that the user sees as he or she zooms and pans to be visually consistent and continuous; that is, *stable*.

Consider aggregating the count of objects up the pyramid. Say we naïvely mapped *count* to color *lightness* (*intensity*) in the heat-map image so that larger counts are dark (to represent many objects) while small counts are light (to represent few). Since bins of higher strata represent more area (and represent the *sum* of counts of their sub-bins), the count values of bins grow as one

---

[6] At least not standard versions of these.

moves up strata. This means that, as the user zooms out in the visualization, the image becomes increasingly darker. Or, conversely, as the user zooms in, the image fades out. This should be a violation of our stability principles. In this case, we can fix the problem. Instead of mapping *count*, we want to map *density* instead, which is simply count normalized by area: $density = count/area$.

Figure 5 demonstrates a stability problem that can arise which is akin to the *voter paradox* in political science and economics. In this example, the visualization intends to show the majority category per tile. There are three categories being counted; e.g., the number of votes for each of three parties within the given region. The maximum count over the three categories can be used to choose the leading category within the bin; e.g., the leading party within that region. In the visualization, the leading category (party) of the three per bin is mapped to *red*, *green*, and *blue*, respectively, to color the corresponding tile. In four adjacent tiles represented on the left (say, from stratum $i+1$), red led in two cases, and green in the other two. However, when the category counts are summed over the four for the super-bin (stratum $i$), we find that blue is the maximum category. (Blue was a very close second in all four cases.) And so the super-tile gets colored blue.

This would be jarring to a user exploring this part of the visualization. On zooming out, that part of the picture would change from being red and green to suddenly being blue. This is a failure in two ways: it breaks continuity of the visual presentation; and it somehow is not truthfully summarizing the data. When seeing the blue tile, this does not indicate that both red and green appear in significant proportion underneath. And when seeing the red and green tiles, the complete absence of blue is misleading to its significant proportion overall. (We revisit this issue in Sect. 5 to show a solution.)

We can now characterize our desirable stability principles.

1. *scale invariance.* Regardless of a stratum used for visualization, the graphical representation of data at one stratum should be *similar* to that of other strata.
2. *balance.* Values stored in aggregates should not vary significantly between strata. (E.g., *density* is better than *count*.) Values of corresponding aggregates should not grow or diminish from stratum to stratum. (A balanced aggregate function contributes to scale invariance.)
3. *veracity.* Visualization based on a given stratum must convey valid information about the underlying strata and underlying data.

## 4    Architecture

**Three-Tiered Architecture.** SKYDIVE is composed of three main components, as shown in Fig. 4:

1. the Database module (DB);
2. the Data-to-Image module (D2I); and
3. the Visualization Client (VC).

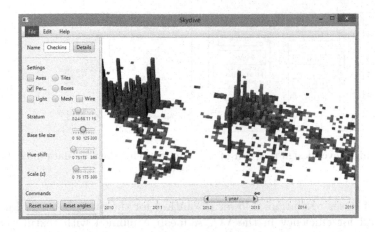

**Fig. 6.** A $2^1/_2$D bar-chart with a time-range slider.

Each is designed to exploit a different type of *computer memory*. The DB module is backed by a database system; it uses *disk* to store and manage the raw data and materialized aggregate pyramids. The D2I module always works with a small subset of the aggregated dataset—that which is to be translated to the visualization—and can store its data in *main memory* (RAM). The VC module uses the client's *graphic card's* capabilities to perform more advanced operations—such as zooming, scaling, panning, and rotation—over the graphical representation of the data.

This separation of concerns provides useful flexibility and efficiency. Each component can be implemented as a separate service, deployed on a different machine (or machines). The separation brings another useful possibility – the DB module can be easily replaced by a distributed system such as HIVE [31] or SPARK SQL [2] to run over multiple nodes for scale out in a cloud environment. This can be for example achieved by using SQL engines built on top of Hadoop [36] or other tools such as BigQuery [32], ApacheDrill [15] (an open source implementation of Google's BigQuery), Spark SQL [2], etc.

To generate a mesh, the system has to perform triangulation. In Skydive, we have implemented two methods of triangulation: the standard version, which represents a single tile by two triangles; and an extended version, which represents each tile by four rectangles with the common vertex located in the middle of a tile. A value represented by a middle vertex is computed as an average value of measure represented by four vertices of a tile. The second type of a mesh is smoother, and represents a dataset more accurately.

**Generating Textures and Meshes.** The data texture (image file) is generated by the D2I module by selecting the appropriate window out of the aggregate pyramid from the appropriate stratum, applying the visualization mapping. For the $2^1/_2$D terrain model we introduce, a mesh is additionally computed by D2I (with a UV-mapping for the texture). The mesh can be created based on a

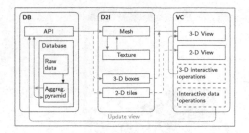

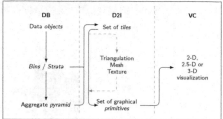

**Fig. 7.** A high level design diagram of SKYDIVE.

**Fig. 8.** The flow of data objects in SKYDIVE.

stratum of lesser resolution; a lower resolution for the mesh is better for visual appeal and for efficiency in the VC. A mesh combined with texture conveys multiple measures represented; for example, by color and height.

**User Interface.** The SKYDIVE prototype is presently implemented as a desktop application with the three modules as described above and shown in Fig. 7. The main type of window of the SKYDIVE user interface, is composed of a few simple elements: an upper menu for performing basic file operations; a left panel for tuning the current visualization; and a viewport for rotating, panning and zooming in the visualization. The VC is implemented using JavaFX, which natively supports these functions. Thus, scaling, translation, and rotation do not require to query the pyramid (the DB), as they are performed by the VC as supported by the GPU.

Other *interactive* functions do require queries to be issued from VC to the DB module. If the user wants to focus more on a certain area of a visualization, for instance, then the system must request the data from the appropriate stratum. Issuing such a request results in the loading and generating of the mesh and texture by the D2I. The VC then displays this using the GPU's graphics pipeline. The texture is mapped on the mesh using the GPU's built-in functions.

The three-tiered design influences what types of data-structures for representing data are used in each of the three modules. Figure 8 shows how data flows from one module to another. The DB deals with original data objects, generating *bins* and *strata* of the aggregate pyramid. The D2I translates the bins from the appropriate stratum to *tiles* in the visual presentation. The *presentation model* and *mapping* chosen for the visualization dictates how this translation is done. And the VC presents and manages the visualization for the user.

## 5    Presentation Models

### 5.1    Skydive's Presentation Models

A *presentation model* defines the "structure" to be visualized. A given model provides *channels* that can be used in the visual conveyance. For example, if the presentation model is a 2D heat-map, we can paint the image using the RGB colorspace; each pixel of the image having a *red*, *green*, and *blue* color channel. Considering the *colorspace* HSL—*hue*, *saturation*, and *lightness*, another common colorspace for image processing—is often advantageous for us. Other presentation models may be richer, offering more channels of conveyance. These could include an *alpha* channel in an image, for instance (how "transparent" a pixel is) and height off the plane, for instance, if our presentation model were 3D. In addition to selecting a presentation model, one also has to define the *mapping* into it. This specifies in what way the inductive aggregates are *mapped* to the channels that the model provides.

The SKYDIVE prototype presently supports three general types of presentation models.

1. 1$^1$/$_2$D  model (plot for "time series")
2. 2D models by 2D heat-map
3. 2$^1$/$_2$D  models
   (a) by 3D bar-chart
   (b) by 2$^1$/$_2$D  terrain (by mesh and UV-mapping)

**The 1$^1$/$_2$D Model.** Time-series data are ubiquitous and high volume. Thus, aggregation is needed. If incorrectly done, however, the visualization can be misleading. In [17], the authors demonstrate this. They present their M4 algorithm for aggregating time-series data to be viewed by traditional line-segment plots so that the line plots at aggregated levels remain visually similar to the line plot of the raw data. While their technique is quite suitable if one needs to maintain a traditional line-plot presentation—with which stock analysts are familiar, for example—it is wasteful when much more information about the aggregate data can be conveyed.

In this case, the data is one-dimensional (say, along time with a single measure), but in our presentation, as with line plots, we can devote two dimensions (a canvas) to the visualization. We introduce a 1$^1$/$_2$D presentation model for this. We call this "1$^1$/$_2$" as we are mapping a one-dimensional function using "$y$" axis for plotting the values. We can exploit the canvas to much greater effect, however, than line plots do, to use color to map more about the aggregated values such as *average*, *min*, and *max* into the presentation.

**The 2D Models.** It is natural to map 2D data onto a 2D canvas. The viewer then manages the visualization as an image—which we call a *data texture*—within the canvas, allowing the user to zoom and pan around it. This view can provide a *heat-map* of the data, a standard visualization technique. SKYDIVE makes it possible to explore this "heat-map" (data texture), however, dynamically in real-time.

**The 2$^1$/$_2$D Models.** With 3D rendering, we have another spatial dimension ("Z") in the rendering that we can use in addition to our data textures from the 2D models. Just as we used "$y$" in the canvas for the 1$^1$/$_2$D model, we can use "$z$" in a 3D plot to provide height over pixels of the data texture. Such a 2D *manifold* crumpled in 3D is called a *terrain*. A *manifold* is function that maps 2D coordinates to values (a "height"), we refer to these terrain models as 2$^1$/$_2$D.

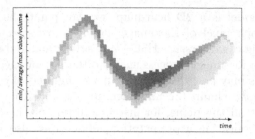

**Fig. 9.** A sample 1¹/₂D plot.

2D images are called *textures* in this context, and the mapping of textures onto the mesh surfaces is called the *UV-mapping*. This exploits modern 3D graphics rendering, which supports meshes and UV-mapping. This model offers additional channels of conveyance: *elevation* (Z); *specular*; and *normal*. A specular map determines how "reflective" a point is on the surface. As scenes in 3D have external lighting, this is quite noticeable. A normal map dictates deviations of the normals, the "perpendicular" of a point with respect to the surface. By perturbing the normals of a neighborhood (i.e., within a super-tile), that part of the surface can be made to look rough; leaving them as dictated by the mesh, the surface looks smooth.

We can use the *alpha* channel additionally, as the terrain can be floated over a flat reference plane; the bleed-through of the reference through *translucency* of the terrain is readily obvious. This means we have effectively *seven* channels of conveyance in the 2¹/₂D model, versus just the three standard ones in the 2D model.

## 5.2   Illustrations of Models and Mappings

*Example 2.* Figure 9 shows a 1¹/₂D plot. Say that it represents a stock's price for a given company over time. The stock trading price has been aggregated by time intervals. Thus, the $x$ axis represents time in intervals; say, *days*. Each 2D bar is divided into four quartiles by *hue* to indicate the *low* (*min*), 25th percentile, *average*, 75th percentile, and *high* (*max*) of the price over that period. How faint or vivid the bar is (*lightness*) is used here to indicate trading volume during the period (more vivid for heavier trading, faded for lighter trading).     □

We next demonstrate how SKYDIVE can be used to visualize data using 2D and 2¹/₂D models. The 2¹/₂D terrain model demonstrates the advantages that can be gained by using 3D rendering with the use of mesh and texture mapping.

*Example 3.* The Seattle Police Department 911 Incident Response dataset[7] contains over one million records. Each represents the police response to a 911 call within the city. Figure 10(a) shows a *density* map of the calls. *Hue* (color) denotes

---

[7] https://data.seattle.gov/.

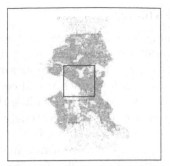

(a) Overview image.    (b) Zoomed in, terrain view.

**Fig. 10.** Visualizations of the Seattle 911 Dataset.

(a) Data texture of check-ins.    (b) Zoomed in, terrain view.

**Fig. 11.** Visualizations of the Brightkite dataset.

the number of calls made within the area represented by a "pixel" (tile). Based on the plotted heat-map, a user is not able to conclude anything more than that there are some areas of slightly higher density than their neighbors. Figure 10(b) shows a terrain view of the same data in which elevation indicates the density, and hue in this case indicates the most frequent type of a call within the cell. In this presentation, we see more detail. For example, the red pixel indicated by the arrow represents unusual activity within the magnified area.

The Brightkite Check-ins dataset[8] consists of over four millions records of geographical positions reported by users of a geo-location social service. In Fig. 11, the heat-map represents the dataset aggregated by count. Hue represents the day of week for which user activity was highest within the areas represented by the pixels. In Fig. 11(b), a zoomed-in portion of the terrain map is shown. The texture's hue again denotes the day of week with highest activity. Elevation denotes number of check-ins. We can deduce that weekends were most active days for Brightkite users in the U.S. We can additionally see the areas that were most active.                                                                                  □

---

[8] https://snap.stanford.edu/data/.

### 5.3   Cubed Pyramids and Advanced Models

Imagine we want to see how the demographics of North America changes over time. The interface of such a presentation model could show a 2D or $2^1/2$D canvas of the continent with a slider bar underneath for time. The slider bar would offer two degrees of freedom: changing the size of the slider, which would control the interval (range) of time being displayed in the map above; and sliding the slider along the bar, which would control *when*, the slice of time (interval) being shown. Figure 6 depicts this interface with a 2D bar-chart portal and a time slider bar.

To support such a presentation model, we need a richer form of hierarchical aggregation. Note that a 3D pyramid—$x,y,t$—would not suffice, because the spatial dimensions $x$ and $y$ would be aggregated together with $t$, time. Instead, we need that $x$ and $y$ are hierarchically aggregated together and that $t$ is hierarchically aggregated. Each $x,y$-bin from the $x,y$-pyramid (at every strata) needs then to be hierarchically aggregated by $t$[9]. We call this a *cubed aggregate pyramid*. Note that out definition of aggregate pyramid by SQL in Fig. 1 is designed to accommodate this. The $x,y$-pyramid can be built as depicted, with $c[0] = t$, the *time* attribute (aggregated to a base level). The output pyramid table then can be used as the input *dataset* to build a second one-dimensional ($t$) pyramid over *time*. The result will be a *cubed* pyramid over $x, y \times t$.

The ability to do this demonstrates the versatility of our aggregate pyramid, and it opens up the way for advanced presentation models. A 2D + T model would support the view in Fig. 1, and would be one way to handle *velocity*, data streaming by time. The cubed pyramid could be incrementally updated (along *time*) as more current data came in. This could provide a versatile, viable interface for monitoring data as it changes over time.

## 6   Implementation

### 6.1   Building the Pyramid

The SQL in Fig. 3 presents a declarative definition of our *aggregate pyramid* and illustrates our concept of *inductive aggregation* (as discussed in Sect. 3). Using this SQL to materialize a pyramid would likely be quite inefficient, but it can be materialized quite efficiently.

The base stratum of the pyramid needs to be computed first. The expense of this will be dictated by the size of the underlying dataset, and on whether useful indexes exist on the data to aggregate them into the spatial *bins*, as dictated by the pyramid's dimensions. The bins of the base stratum can be computed *in order*, say, in a *Z-order*,[10] with respect to the pyramid's dimensions. The tuples (bins) of the base stratum are then written sequentially to disk in Z-order[11].

---

[9] Or vice versa: the bins of the $t$-pyramid are then hierarchically aggregated by $x,y$. This is commutative.

[10] Also called Morton order [22]. This is a one-dimensional, linear ordering for any multi-dimensional data.

[11] "Bins" into which no base data aggregates—"empty bins"—are never created. These numbers in the Z-order are simply skipped over.

Each subsequent stratum is materialized from the previously materialized stratum, from the base up to the top of the pyramid. Stratum $-i$ is sequentially scanned once to write stratum $-i + 1$. As stratum $-i$'s bins are in Z-order, the bins needed to aggregate to produce the corresponding super-bin in stratum $-i + 1$ are ajacent in the sequence. And stratum $-i + 1$'s bins are produced in Z-order in this way.

Aggregate pyramids are well-behaved in size. If we choose a reasonable resolution for the base, it is no larger than the raw data-set. Call this size $B$. For 2-D then—letting $B = 2^d \times 2^d$ for some $d$, without loss of generality—by summing the number of bins per stratum, the pyramid's size is $B \sum_{i=0}^{d} 1/2^{2i} < 1\frac{1}{3}B$ (given $d + 1$ is the depth of the pyramid). It is not much larger than the original data-set itself. And it is, in fact, smaller the higher the dimensionality. For dimensionality $k$—letting $B = 2^{kd}$ for some $d$, without loss of generality—the size is $B \sum_{i=0}^{d-1} 1/2^{ki} < 1\frac{1}{k+1}B$. The reason is that we are doubling the size of a bin along *each* dimension for each higher stratum. Constrast this to a data cube, which aggregates along each dimension independently. Say we aggregate by powers of two along each dimension for a cube, as we have done for the pyramid. The cube would have $(2^{d+1} - 1)^k$ bins, thus be $2^k B$ in size.

## 6.2   Indexing the Pyramid

Once the pyramid is materialized, SKYDIVE will query against it during interactive exploration of the corresponding visualization. The pyramid is essentially, then, a *materialized view*. It is a table in our database, with its tuples, the bins, ranging over the strata representing different degrees of aggregation. The pyramid itself is not an index, in database terminology. In fact, we need to index this table in a sensible way to support SKYDIVE's operations via queries over the pyramid.

Indexing the pyramid table by *stratum, Z-order* ("zo" in the SQL in Fig. 3) as a B+-tree suffices to support well the operations we need (the pyramid "API"), as discussed in the next section[12]. This index can be created efficiently while building the pyramid, as described above.

There are alternative ways we could index the pyramid, of course. A spatial-index could be employed such as a quadtree index [24,25] or a KD-tree [26], built over the pyramid's dimensions for each stratum. We do not gain much advantage doing this, though. For our data-visualization application, the dimensionality is always quite low. B+-tree implementations in database systems have performance advantages. And we are never seeking disparit objects in space; rather, our bins *tile* our space per stratum.

## 6.3   Cubed Pyramids

*Cubed pyramids*, as introduced in Sect. 5.3, are a hybrid of the aggregrate pyramid and the data cube. We do not always need to aggregate each dimension

---

[12] This is sometimes referred to as a *linear quadtree* (for 2-D) [10].

independently, as with a data cube, nor does aggregating along all the dimensions in concert as with a data pyramid always suffice. The cubed pyramid allows us to specify which groups of dimensions to aggregate in concert, to aggregate the groups independently. This lets us specify anything along the spectrum of pyramid to cube.

The example in Sect. 5.3 was that we wanted to aggregate data spatially on $X, Y$ (2-D) and on time $(T)$. We want, however, to be able to change resolution over $X, Y$ and over $T$ independently. Thus, we need the *cube* of a 2-D $(X, Y)$ pyramid and a 1-D $(T)$ pyramid over the same dataset. Given our dataset is of size $B$ (as above),[13] the size of the $X, Y$ pyramid is $1\frac{1}{3}B$, and the size of the $T$ pyramid is $2B$. Thus, the cubed pyramid is $(1\frac{1}{3} \times 2)B = 2\frac{2}{3}B$ in size. In the extreme, that each dimension is grouped by itself, this is equivalent to a data cube.

# 7   Conclusions

The Internet is replete with beautiful information visualizations. These visualizations often take analysts and graphic designers hours, days, or weeks to design and render, with ad-hoc workflows. These visualizations also come after the fact, after the data has been understood, and people want to convey to others what the data means.

Our goal is different, to provide automated data visualization so that users can explore their data visually to understand it better. A Gartner's 2015 report [23] highlights that interactive visualization is a critical feature for today's data management and discovery tools. The demand for native access and interactive visualization over big, multi-structured and streaming data of different types is becoming central to modern business intelligence and analytics platforms. Following the vision of Wu, Battle, and Madden [37], in SKYDIVE, we seek a tight coupling of the database system back-end with a powerful visualization front-end to provide scalable and efficient real-time interactive visual exploration.

In this paper, we have presented our approach via SKYDIVE, illustrating its three-tier, modular architecture, the use of the relational concept of an aggregate pyramid to support spatial, hierarchical aggregation efficiently, and the support for rich presentation models and mappings for rendering in real-time rich, navigable visualizations. Our SKYDIVE platform provides us a number of directions for continued study.

- We look to extend the aggregate pyramid model via cubed pyramids and the like
    - to support richer presentation models and
    - to support *automated exploration* as a foundation for visual data mining.
- We plan to exploit modern graphics capabilities to provide new graphical channels for presenting data (e.g., specular and normal maps).

---

[13] The dataset is over three dimensions—$X, Y$, and $T$—so assume $B = 2^{3d}$ for some $d$, without loss of generality.

– We indend to develop a formal framework that ensures *stable* visualizations that can be used to provide guarantees of a data visualization's *veracity*.

Visualization is destined to become a core component of big data. We have shown how each of Big Data's five V's applies to the visualization of big data. Each brings its own set of challenges that we must overcome in preparation for the day soon when *visualization* becomes the sixth V of big data.

# References

1. Andrienko, N., Andrienko, G.: Exploratory analysis of spatial and temporal data: a systematic approach. Springer Science and Business Media, Heidelberg (2006)
2. Armbrust, M., Xin, R.S., Lian, C., Huai, Y., Liu, D., Bradley, J.K., Meng, X., Kaftan, T., Franklin, M.J., Ghodsi, A., et al.: Spark SQL: relational data processing in spark. In: Proceedings of SIGMOD, pp. 1383–1394. ACM (2015)
3. Battle, L., Stonebraker, M., Chang, R.: Dynamic reduction of query result sets for interactive visualizaton. In: Proceedings of the International Conference on Big Data, Santa Clara, CA, USA, pp. 1–8 (2013)
4. Bertin, J.: Semiology of Graphics. University of Wisconsin Press, Madison (1983)
5. Beyer, M.A., Laney, D.: The importance of "big data": a definition. Gartner report (2015)
6. Cable, D.: The racial dot map, demographics Research Group. Weldon Cooper Center for Public Service, University of Virginia, July 2013. www.coopercenter. org/demographics/Racial-Dot-Map
7. Dijcks, J.P.: Oracle: Big data for the enterprise. Oracle White Paper (2012)
8. Elmqvist, N., Fekete, J.D.: Hierarchical aggregation for information visualization: overview, techniques, and design guidelines. IEEE Trans. Vis. Comput. Graph. **16**(3), 439–454 (2010)
9. Erickson, J.: Private correspondence, conveyed along with permission to use by Tilmann Rabl, May 2015
10. Gargantini, I.: An effective way to represent quadtrees. Commun. ACM **25**(12), 905–910 (1982)
11. Godfrey, P., Gryz, J., Lasek, P., Razavi, N.: Skydive: an interactive data visualization engine. In: IEEE Symposium on Large Data Analytics and Visualization, Chicago, USA, October 25–26, pp. 129–130 (2015)
12. Godfrey, P., Gryz, J., Lasek, P.: Interactive visualization of large data sets. Technical report EECS-2015-03, York University, March 2015
13. Godfrey, P., Gryz, J., Lasek, P., Razavi, N.: Visualization through inductive aggregation. In: Proceedings of EDBT, March 2016
14. Gray, J., Chaudhuri, S., Bosworth, A., Layman, A., Reichart, D., Venkatrao, M., Pellow, F., Pirahesh, H.: Data cube: a relational aggregation operator generalizing group-by, cross-tab, and sub-totals. Data Min. Knowl. Disc. **1**(1), 29–53 (1997)
15. Hausenblas, M., Nadeau, J.: Apache drill: interactive ad-hoc analysis at scale. Big Data **1**(2), 100–104 (2013)
16. Jugel, U., Jerzak, Z., Hackenbroich, G., Markl, V.: Faster visual analytics through pixel-perfect aggregation. Proc. VLDB Endowment **7**(13), 1705–1708 (2014)
17. Jugel, U., Jerzak, Z., Hackenbroich, G., Markl, V.: M4: a visualization-oriented time series data aggregation. Proc. VLDB Endowment **7**(10), 797–808 (2014)
18. Laney, D.: Meta Group Res Note 6. META (2001)

19. Liu, Z., Jiang, B., Heer, J.: imMens: real-time visual querying of big data. Comput. Graph. Forum **32**(3), 421–430 (2013)
20. Magdy, A., Aly, A.M., Mokbel, M.F., Elnikety, S., He, Y., Nath, S.: Mars: real-time spatio-temporal queries on microblogs. In: ICDE, pp. 1238–1241 (2014)
21. Magdy, A., Mokbel, M.F., Elnikety, S., Nath, S., He, Y.: Mercury: a memory-constrained spatio-temporal real-time search on microblogs. In: ICDE, pp. 172–183. IEEE (2014)
22. Morton, G.M.: A Computer Oriented Geodetic Data Base and A New Technique in File Sequencing. International Business Machines Company, New York (1966)
23. Sallam, R.L., Hostmann, B., Schlegel, K., Tapadinhas, J., Parenteau, J., Oestreich, T.W.: Magic quadrant for business intelligence and analytics platforms. Gartner report (2015)
24. Samet, H.: The quadtree and related hierarchical data structures. ACM Comput. Surv. (CSUR) **16**(2), 187–260 (1984)
25. Samet, H.: Applications of Spatial Data Structures: Computer Graphics, Image Processing, and GIS. Addison-Wesley Longman Publishing Co., Inc., Boston (1990)
26. Samet, H.: Foundations of Multidimensional and Metric Data Structures. Morgan Kaufmann, San Francisco (2006)
27. Schroeck, M., Shockley, R., Smart, J., Romero-Morales, D., Tufano, P.: Analytics: The Real-World Use of Big Data. IBM Global Business Services, Somers (2012)
28. Shneiderman, B.: The eyes have it: a task by data type taxonomy for information visualizations. In: Proceedings of the 1996 IEEE Symposium on Visual Languages, pp. 336–343. IEEE (1996)
29. Shneiderman, B.: Extreme visualization: squeezing a billion records into a million pixels. In: Proceedings of the 2008 ACM SIGMOD International Conference on Management of Data, pp. 3–12. ACM (2008)
30. Stolte, C., Tang, D., Hanrahan, P.: Polaris: a system for query, analysis, and visualization of multidimensional relational databases. IEEE Trans. Vis. Comput. Graph. **8**(1), 52–65 (2002)
31. Thusoo, A., Sarma, J.S., Jain, N., Shao, Z., Chakka, P., Anthony, S., Liu, H., Wyckoff, P., Murthy, R.: Hive: a warehousing solution over a map-reduce framework. Proc. VLDB Endowment **2**(2), 1626–1629 (2009)
32. Tigani, J., Naidu, S.: Google BigQuery Analytics. John Wiley & Sons, Hoboken (2014)
33. Tufte, E.: Envisioning Information. Graphics Press, Cheshire (1990)
34. Wesley, R., Eldridge, M., Terlecki, P.T.: An analytic data engine for visualization in tableau. In: Proceedings of SIGMOD, pp. 1185–1194. ACM (2011)
35. Wesley, R.M.G., Terlecki, P.: Leveraging compression in the tableau data engine. In: Proceedings of SIGMOD, pp. 563–573. ACM (2014)
36. White, T.: Hadoop: The definitive guide. O'Reilly Media Inc, Sebastopol (2012)
37. Wu, E., Battle, L., Madden, S.R.: The case for data visualization management systems: vision paper. Proc. VLDB Endowment **7**(10), 903–906 (2014)
38. Zikopoulos, P.C., Eaton, C., DeRoos, D., Deutsch, T., Lapis, G.: Understanding Big Data. McGraw-Hill, New York (2012)

# Big Data Management in the Cloud: Evolution or Crossroad?

Abdelkader Hameurlain$^{(\boxtimes)}$ and Franck Morvan

IRIT Institut de Recherche en Informatique de Toulouse IRIT,
Paul Sabatier University, 118, Route de Narbonne, 31062 Toulouse Cedex, France
{abdelkader.hameurlain,franck.morvan}@irit.fr

**Abstract.** In this paper, we try to provide a synthetic and comprehensive state of the art concerning big data management in cloud environments. In this perspective, data management based on parallel and cloud (e.g. MapReduce) systems are overviewed, and compared by relying on meeting software requirements (e.g. data independence, software reuse), high performance, scalability, elasticity, and data availability. With respect to proposed cloud systems, we discuss evolution of their data manipulation languages and we try to learn some lessons should be exploited to ensure the viability of the next generation of large-scale data management systems for big data applications.

**Keywords:** Big data management · Data partitioning · Query processing and optimization · Parallel Relational Database Systems · High performance · Scalability · Cloud systems · Hadoop · Mapreduce · Spark · Elasticity

## 1 Introduction

Data management process dates from long ago (i.e. since 4000 BC). Between 4000 BC and 2000 AD, five generations of data management have been distinguished by [28]: "Manual processing paper and pencil; Mechanical punched card; Stored program sequential record processing; On-line navigational set processing; Nonprocedural- relational; Multimedia internetwork". In this paper, we will be interested in the penultimate generation.

A large number of datasets (structured, unstructured and semi-structured data) are produced by different sources (e.g. scientific observation, simulation, sensors, logs, social networks, finance). This large number of datasets, often referred to as Big Data and characterized by 4Vs (Volume, Variety, Velocity, and Value) [57], are distributed in large scale, heterogeneous, and produced continuously. The management of such data raises new problems and presents a real challenge such as: data modeling and storage, query processing and optimization, data replication and caching, cost models, concurrency control and transaction, data privacy and security, data streaming, monitoring services and tuning, autonomic data management (e.g. self tuning, self repairing).

© Springer International Publishing Switzerland 2016
S. Kozielski et al. (Eds.): BDAS 2016, CCIS 613, pp. 23–38, 2016.
DOI: 10.1007/978-3-319-34099-9_2

In the landscape of database management systems, data analysis (OLAP) and transaction processing systems (OLTP) are separately managed. The reasons [2] for this dichotomy are that both systems have very different functionalities, characteristics and requirements. This paper will focus only on the first class (OLAP systems).

Very recently (with respect to reference period 4000 BC), we have seen an explosion in the volume of data manipulated by applications. This phenomenon results from, on the one hand, that it is easier to collect information (e.g. log information) and, on the other hand, lower cost of storage devices. Querying and analyzing of collected data, with acceptable response time, have become essential for many companies such as web societies. Furthermore, some applications with a small amount of data need high availability of data. This is the case, for example, with data from an online game with great success. In both contexts, a uniprocessor database server can quickly become a bottleneck or result in prohibitive response times for certain queries.

To manage a huge amount of data and meet the requirements in terms of high performance (e.g. minimizing of response time) and resource availability (e.g. data source), there are two approaches: parallel database systems and cloud systems. The parallel database systems [20, 47, 62] have been an important success, both in research in the early 90s and now in industry. They have enabled many applications handling large data volumes to meet their requirements in terms of high performance and resource availability. It is recognized that parallel database systems are very expensive and require having high level skills within the company to administer the systems and databases. As for cloud systems, developed by using data processing frameworks (e.g. Hadoop MapReduce, Apache Spark) [4, 18, 29], they allow a company to reduce these costs in terms of infrastructures either by purchasing a server comprised of low-cost commodity machines or by renting services (Infrastructure-as-a Service IaaS, Platform-as-a Service PaaS, Software-as-a Service SaaS) in pay-per-use.

Currently, new tools [21, 49, 54, 65] allow to make the bridge between both approaches. These tools allow either to a MapReduce program to load data from a relational database, either to convert, through a wrapper, a file stored in an HDFS format into a relational format. This class of systems is called multistore systems [7].

In this paper we propose a synthetic and comprehensive state of the art concerning: (i) big data management in parallel database systems and cloud systems, and (ii) evolution of data manipulation languages in cloud systems. There are many synthesis papers about data management in cloud systems [2, 12, 25, 45, 55]. Agrawal et al. [2] and Chaudhuri [12] focus on the future challenges of clouds systems to meet the needs of applications. As far as the contributions of Floratou et al. [25] and Stonebraker et al. [55] they propose a performance comparison of applications with cloud systems and parallel database systems. They point out the advantages and drawbacks of each system depending on the application type.

The rest of this paper is structured as follows. In Sect. 2, data management based on parallel and cloud (e.g. MapReduce) systems are over-viewed and compared by relying on meeting software requirements (e.g. data independence, software reuse), high performance, scalability, elasticity, and data availability. Section 3 presents an overview of data manipulation languages proposed in cloud systems: (i) without relational operators, and (ii) including relational operators in data manipulation languages. In Sect. 4, with respect to proposed cloud systems, we try to learn some lessons and we discuss the evolution of their data manipulation languages. We conclude in Sect. 5.

## 2    Parallel Database Systems Versus Cloud Systems

### 2.1    Parallel Relational Database Systems

This sub-section presents an extended abstract of the paper [33]. Parallel database systems have been developed for applications processing a large volume of data. Their main objectives are to obtain high performance and resource availability. High performance can be obtained by integrating and efficiently exploiting different types of parallelism (partitioned parallelism, independent parallelism and pipeline parallelism) in relational database systems on parallel architecture models. More precisely, the objectives of parallel databases are: (i) ensuring the best cost/performance ratio compared with a mainframe solution, (ii) minimizing query response times by efficiently exploiting the different forms of parallelism and data placement approaches, (iii) improving the parallel system throughput by efficiently managing resources, and (iv) insuring scalability, which consists in holding the same performance after adding new resources and applications. A parallel relational database system is a standard relational DBMS implemented on a MIMD (Multiple Instruction Multiple Data Stream) parallel architecture. Editors of parallel servers offer three main categories of parallel RDBMS based on: (i) shared memory multiprocessor architecture, (ii) shared disks multiprocessor architecture, and (iii) shared nothing multiprocessor architecture (e.g. DBC 1012 Teradata [8,64], Tandem NonStop SQL [23], DB2 Edition [5], ORACLE Parallel Query).

Partitioning a relation is defined as distributing the tuples of the relation among several nodes (attached disks) [20]. In a parallel RDBMS, it becomes possible to: (i) improve I/O bandwidth by fully exploiting the parallelism of read operations of one or more relations (ii) apply data locality principle (operators are performed where/or very close to the data are located), and (iii) facilitate load balancing to maximize throughput. The key problem with data partitioning, also called data placement, consists in reaching and holding the best tradeoff between processing and communication [17]. Two approaches make it possible to solve the data placement problem of a set of relations in a parallel RDBMS. The first approach, called full desclustering [46], consists in distributing horizontally each base relation over all the nodes of the system. It is applied to shared memory parallel RDBMS. The second approach, called partial desclustering, consists in distributing each base relation over a subset of nodes. It is mainly found in a shared-nothing parallel RDBMS which own a large number of nodes. However, whatever the advocated data placement approach used, there are many data partitioning methods. In parallel systems, three methods are generally offered to the administrator to distribute data over nodes [20, 46]: round robin, use of a hashing function which associates a node number to one or more attribute values of a relation, and use of range partitioning given by the administrator thanks to fragmentation predicates. These partitioning methods have great influence and impact on load balancing.

The optimization phase of SQL queries consists basically of three phases: logical optimization, physical optimization and parallelization. The problem raised at physical optimization, dealing with the choice of a scheduling search strategy for join operators among enumerative strategies or randomized ones. Each one of these strategies is more or less adapted depending on the query characteristics. A query is characterized by the number of relations it refers to (i.e. size), the number of join predicates and the way they are arranged (i.e. query shape) and its nature (i.e. ad-hoc or repetitive). As far as parallelization strategies, they have been introduced concerning inter-operation parallelization phase, the key problem of optimization [13, 14, 31, 32, 34, 36, 52, 58, 68]. Indeed, two inter-operation parallelization approaches have been described in the literature [47]: the one-phase and the two-phase approaches. In the two-phase approach [13, 31, 32, 34, 36], the first phase consists in generating a query execution plan (without considering run time resources). The second phase ensures an optimal allocation of resources (memory and processors) for the previously generated plan. As for the one-phase approach [14, 52, 68], plan generation and resource allocation are packed into one single phase.

Finally, with regard to minimization of communication costs, which represents the plague of parallelism, the parallel processing of SQL queries requires the initialization of several processes on different processors with underlying data communications. The main problem to be solved by parallel execution models is to find the best processing-communication trade-off in order to maximize the system throughput and minimize the response time, while maintaining an acceptable cost optimization. In this objective, several efforts have been conducted to reduce inter-processor communication costs, and in particular to avoid the redistribution of data and minimize message transfers [23, 30, 35].

## 2.2   Cloud Systems

The rapid development of information technology and the popularity of the Internet allow the emergence of new Web applications (e.g., social networks, log and profile analysis, online document indexing). These new applications produce data that are often under the form of continuous streaming (e.g., data from sensors), with very large volume, in heterogeneous formats and distributed in a large scale. To address these data characteristics, data management community has recently proposed, with respect to traditional RDBMS, new data management systems that are more flexible in terms of data models (compared to relational model which is operational since 30 years), more cost-efficient in terms of investment (as most of these systems are open-source) and provide better availability of resources (i.e. data sources and computing resources (CPU, RAM, I/O and network bandwidth)) in terms of fault-tolerance. From this perspective, four classes of data management systems [37,56], depending on the adopted data structure type, have been designed and developed, mainly, by Google, Amazon, Microsoft, Yahoo, Facebook, IBM, and Oracle: (i) Key-Value based systems, (ii) Document based systems, (iii) Column based systems and (iv) Graph based systems for social networks.

These systems, based on a shared-nothing architecture, have been developed in Hadoop environment, using the functional programming model MapReduce and HDFS/GFS (Hadoop Distributed/Google File System). In a cloud environment, these systems are aimed to achieve high performance, maintain scalability (because they are based on massively parallel architectures), ensure elasticity (on-demand service and pay-per-use) and guarantee the fault-tolerance. High performance is based on the intensive exploitation of intra-operation parallelism of an operation (in a *map* or *reduce* operation) and independent parallelism (between multiple *map* or *reduce* programs). The elasticity paradigm [50] consists in allocating resources dynamically on demand. It extends the objective function, combined with dynamic resource allocation models, by integrating the economic model meeting the "pay-per-use" principle (tenant side), and guaranteeing a minimum profitability (provider side). Finally, in most of the proposed solutions in cloud systems fault-tolerance is managed. In fact, when a processor or process fails, only a part of the query is executed again. This is very attractive for applications that have queries that can take up to several hours as, for example, the analysis of log files.

In addition, to insure a uniform access to heterogeneous, autonomous and distributed data sources (e.g. RDBMS, NoSQL, HDFS) several mutlistore systems have been, recently, proposed. These systems, based on Mediator-Wrapper architecture [63] can be classified in three categories [7]: (i) loosely-coupled multistore systems, (ii) tightly-coupled multistore systems, based on a shared-nothing architecture, whose objective is the high performance. This approach consists in modifying the SQL engine to make the data access to HDFS transparent. Polybase [21] illustrates this approach. And (iii) hybrid/integrated multistore systems, whose objective is to query indifferently structured and unstructured data using a SQL-like declarative language. SCOPE system [67] and CoherentPaaS project [7], illustrate this approach.

## 2.3   Comparison

In this sub-section, we propose a qualitative comparison between Parallel Relational Database Systems PRDBMS and cloud systems/MapReduce by pointing out their advantages and weakness. For a quantitative comparison, we strongly suggest to authors reading the very good papers [25,38,51,55].

**Advantages and Weakness of PRDBMS.** With respect to compilation of recently published studies and experiences, the advantages of PRDBMS can be, mainly, summarized as follows: (i) logical data independence, meaning that any modification of a schema (data structure) have not any impact on application programs, (ii) regular data structure (relational schema), homogeneity and stability of parallel infrastructure (shared-nothing architecture) enable to estimate and deduce relevant annotations (Metadata, and Cost Models are used by an Optimizer-Parallelizer to generate an efficient parallel execution plan), (iii) partitioning degree for each base relation and parallelism degree estimation for each relational operator can be estimated by analytical models, (iv) declarative languages and sophisticated query Optimizer-Parallelizer (physical optimization, exploiting and integrating of partitioned, independent and dependent (pipeline) parallelisms), and (v) minimizing of communication costs by avoiding the data redistribution in some favorable cases.

However, their main weakness, in massive parallel environments, can be summarized as follows: (i) PRDBS run only on expensive servers, (ii) require very high level of skills to manage and administrate these systems, (iii) weak fault-tolerance in massive PRDBS, and (iv) hard management of Web applications (Web datasets are unstructured).

**Advantages and Weakness of Cloud Systems/MapReduce.** The main advantages of cloud systems/MapReduce are: (i) scaling very well to manage massive datasets, (ii) support the partitioned and independent parallelisms, (iii) mechanism to achieve load-balancing, and (iv) strong fault-tolerance because of HDFS characteristics and used mechanisms to data replication (a file is partitioned into fixed size chunks of 64 MB and each chunk is replicated, by default, three times).

As far as the weakness of MapReduce (initial version) two levels can be distinguished: application level and software level. In fact, application side, the developers: (i) are forced to translate their business logic to MapReduce model, (ii) have to provide efficient implementation for the *map* and *reduce* functions, and to determine the best scheduling of *map* and *reduce* operations. With regard to software side: (i) data-dependence: so, we lost the propriety of logical data independence which is a qualitative requirement of software engineering, (ii) extensive materialization of I/O (Input/Output), because each result of a *map* instance is written on the disk.

To avoid this weakness, recently, pipeline parallelism has been implemented in Tez framework which is used, recently, to improve Hive performance [24]. Also, Cloudera Impala [15] implements pipeline parallelism for all queries. However, the consequence is that the fault-tolerance or resource availability will be seriously weakened. For detailed and complete analysis of weakness of MapReduce, the most relevant work can be found in the recent survey papers [22,44].

# 3 Evolution of Data Manipulation Languages in Cloud Systems

## 3.1 Introduction

First tools, such as MapReduce [18], Bigtable [10] and PNUTS [16], were proposed to develop cloud applications. These tools allow querying data using procedural languages without relational operators. Programs written using these languages introduce a dependency between data structure and programs. Thus, when you have some modifications on the data structure it is also necessary to adapt the programs. In addition, these programs are more difficult to optimize than program writing with a declarative language. Indeed, optimization of functions defined by *"users have never been a central data management challenge in researches"* [12]. Program optimizations and their maintenance due to data structure evolution lead to important human costs. This is why the first tools were mainly used for queries which are performed only once, such as applications on logs and paper collections [26]. Recently, new tools have emerged in order to avoid the dependency between data structure and programs. Their common goal is to use the benefits of data independence and implicit (automatic) optimization programs of parallel database approaches and the advantages of scalability, fault tolerance and elasticity of cloud systems. Thus, high level of declarative languages have emerged HiveQL [60], SCOPE language [67], and CloudMdsQL [7]. This allows an automatic optimization-parallelization of queries. Moreover, new tools [21,49,54,65], based on integrated approach, allow to make the bridge between the both approaches. These tools allow either to a MapReduce program to load data from a relational database, either to convert, through a wrapper, a file stored in an HDFS format into a relational format.

In this section we propose a state of the art concerning the evolution of data manipulation languages in cloud systems, and we point out why the proposed languages have evolved. More precisely, we present an overview of data manipulation languages, first, without relational operators, and next, with relational operators in data manipulation languages.

## 3.2 Data Manipulation Languages Without Relational Operators

The first tools [10,16,18] proposed in the literature allow to manipulate massive datasets by using procedural languages. Generally, they propose relatively simple languages which permit only filter or project operation on massive datasets. These tools were, mainly, designed to serve Web applications which do not need complex queries. For example, they want to query massive datasets such as logs and click streams. However, Web applications require scalability, high performance (e.g. minimization of response time) and high availability of data.

A very popular framework for processing massive datasets is MapReduce [18]. It allows the programmer to write *map* and *reduce* functions which correspond respectively to perform grouping and reduce functions. The programming model provides a good level of abstraction. However, for some applications (e.g. applications querying a relational model) this programming model is not suitable. It can make it complex to write some programs. For example, it is not easy for a programmer to write projection, selection, or join operators over datasets with *map* and *reduce* functions [9]. It also requires valuable expertise from the programmer since users specify the physical execution plan (i.e. users implement the operators and the scheduling of operators). This physical execution plan has many chances to be sub-optimal and there does not exist an automatic optimization process for a user program. We find a first solution to these problems in Bigtable [10] and Pnuts systems [16].

Bigtable [10] is a distributed storage system which supports a simple data model that look likes a relational model. For that it relies on the file management system GFS (Google File System) [27] which provides fault-tolerance and data availability [50]. A table is a sparse, distributed, persistent multidimensional sorted *map* where data is organized into three dimensions: rows, columns and timestamps. Rows are grouped together to form the unit of distribution and load balancing. Columns are grouped to form the unit of access control. As for timestamps, they allow to differentiate different versions of the same data. Bigtable provides a basic API for creating, deleting and querying a table in a procedural language such as C++. This API provides only simple operators for iterates over subsets of data produced by a scan operator. There is no implementation of complex operators like join and minus operations. In the same way, PNUTS [16] supports only selection and projection. It presents a simplified relational data model where data are stored in hashed or ordered relation. Pnuts tables are horizontally partitioned and each partition named tablets is no bigger than 1Gbyte. Other systems like Dynamo [19] and Cassandra [42] use an even simpler data access. Dynamo is used only by Amazon's internal services. Dynamo has a simple key/value interface which offers read and write operations to a data item that is uniquely identified by a key. As far as Cassandra, the data are also partitioned using a hash function and data are accessed by a key using an API composed of three methods: insert, get and delete.

### 3.3   Data Manipulation Languages Using Relational Operators

The use of low level languages, like MapReduce, forces the users to write repetitively the same code for standard operations, like relational operators (e.g. join operator), for all new datasets. This is expensive in terms of development. Furthermore, the programs are complex to read. The bug probability is increased and an optimization process is complex. Based on these observations, the Pig Latin [26,48] and Jaql [6] languages have been proposed. These languages allow developers to work at a higher level of abstraction than MapReduce language. With Pig Latin a user can write without knowing the physical organization of the data and it introduces new operators like join. Pig Latin programs encode explicit dataflow graphs which interleave relational-style operators like join and filter with user-provided executables. A Pig Latin program is compiled in MapReduce program after four steps of transformation. Two of these steps concern the optimization process. A classical logical optimization step, where filter and project operators pushdown on the graph in order to reduce the processed data volume. With regard to the second optimization step, it optimizes the MapReduce program generated by a MapReduce compiler step. This optimization step consists in break distributive and algebraic aggregation functions into a series of *map*, *combine* and *reduce* operators. The authors [3, 26] use the Combiner/Intermediate Reduce as often as possible in order to (i) reduces the volume of data handled by the shuffle and merge operators and (ii) balance the amount of data associated for each key in order to limit the data skew for the Reduce operator. Jaql is also a procedural language where the functions are combined using '->' operator inspired from Unix pipes. A Jaql script [6] is transformed in a MapReduce program by a compiler following approximately the same optimization step as in Pig system. A difference with Pig Latin language is that the users have access to the internal system. This allows to users to develop a specific feature to solve performance problems.

A program written with Pig Latin or Jaql is complex to optimize. Indeed, the user determines a scheduling of relational operators. For an optimizer, it is difficult to determine a new optimal scheduling like in physical optimization of relational systems [43]. Furthermore, the alternative which suggests writing a program with low level languages requires a high expertise level from the user. For these reasons, Hive [59, 60], SCOPE [67] systems, and CoherentPaaS project [7] propose to use declarative languages close to SQL language. These languages allow a user to define a relation compounded of several typed columns. Each of these languages can load data from external data sources and insert query results in formats defined by users. For example, SCOPE language has been enriched by *extractor* operators in order to parse and construct rows from any kind of data sources and *outputter* operators to format the final result of a query. As for HiveQL language, the formats of an external data source or result is defined in data definition language (i.e. during create table order).

With regard to the optimization process, in Hive system, it comprises four steps: (i) a logical optimization step, (ii) a simplification step which prunes partitions and buckets that are not needed by the query, (iii) a combiner step which groups multiple joins sharing the same join attribute in order to be executed in a single MapReduce join, and (iv) a step which adds repartition operators for join, group-by and custom MapReduce operators. As for SCOPE system [67], the optimization process includes a logical optimization and chooses for each operator the best algorithm to process it (e.g. hash join or sort-merge join) and determines the scheduling of operators. With regard to avoid data reshuffling, which deteriorates the performance, a top-down approach is proposed in order that a parent operator imposes its requirements to the child operators [3, 66, 67].

## 4    Discussion

To manage a huge amount of data there are two approaches: parallel database servers and tools proposed by cloud servers. The parallel database servers have been an important success, whether in research in the early 90s and now in industry. They have enabled many applications handling large data volumes to meet their requirements in terms of high performance and resource availability. However, the use of a parallel database server is expensive for a company. Indeed, it requires the purchase of an expensive server and requires having high level skills within the company to administer servers and databases.

An alternative approach is to use tools proposed by cloud servers to manipulate massive datasets. This allows a company to reduce these costs in terms of infrastructures (Iaas) either by purchasing a server comprised of low-cost commodity machines or by renting services (PaaS and/or SaaS) in pay-per-use. An important characteristic of cloud systems is to provide a mechanism for integrated fault tolerance. This feature is important because it avoids restarting, from the beginning, a program in case of processor or process failure. As data volumes grow every day, this feature has become a critical requirement for applications. In addition, more and more applications querying only one time a dataset. For this kind of application, the loading cost of the dataset in a database server becomes prohibitive for a single query. Hence, this kind of application is not suitable for a database server, and many applications have turned to use cloud systems.

Regarding to the state of the art and previous quantitative and qualitative studies and comparisons [25,38,55] between PRDBMS and cloud systems, we can point out the following statements:

1. *Functional Complementarity* between cloud systems and parallel DB systems: in fact, the cloud systems are not intended to replace traditional RDBMS but rather to provide them with the missing features, particularly, for Web applications and Internet services. Moreover, these systems provide scalability in terms of loads adapted for Web applications. In [55], the authors have conducted a benchmark study by comparing Hadoop/MapReduce and two parallel RDBMSs. *"The results show that the DBMSs are substantially faster than the MapReduce system once the data is loaded"*. Their main conclusion of [55] is that: *"MapReduce complement DBMSs since databases are not designed for ETL (Extract-Transform-Load) tasks, a MapReduce specialty"*.
2. *Maturity*: Compared to traditional RDBMS, the cloud systems lack maturity and standardization/normalization (e.g., query languages). These systems require more experimentation and benchmarking (e.g. TPC - H, and TPC - DS) [24] with full-scale big data applications while taking into maximum consideration simultaneously their Vs which characterize them.

As far as the evolution of data manipulation languages in cloud systems, the first proposed languages allow manipulation of data stored in a cloud system, either with functional languages like MapReduce or with imperative languages like C++. These languages force the developer to: (i) modify the program if the data structure changes, and (ii) often rewrite the very similar code on different datasets. In addition, in case of performance problems, these programs are very difficult to optimize due, in particular, to their understanding which is complex. Thus, new languages like Pig Latin [26] and Jaql [6] have been proposed using relational operators such as join. As a result, user programs are more readable. Using these high level languages, the automatic optimization process has been introduced as the logic optimization, conventionally used by a RDBMS. However, these languages are still procedural and classical physical optimization process cannot be applied like in RDBMS. Therefore, Hive [60], Clydesdale [40] and SCOPE [67] systems and CoherentPaas project [7] use non-procedural (i.e. declarative) languages close to the SQL language. This helped to adapt the automatic relational optimization process proposed in the context of RDBMS (e.g. parallel RDBMS). The major drawback of this optimization process, compared to those used in RDBMS, is the scarcity of statistics stored in the meta-base. Indeed, in a cloud system, at compile time, there is no statistics on datasets (e.g. cardinality). This blinds the optimization process and impacts the choice of the optimal execution plan. Hence, a dynamic optimization process becomes necessary in order to react to sub-optimal execution plans. In this objective, [1] proposes to collect statistics at runtime, and adapts the execution plan at runtime by interfacing with a query optimizer. This proposal can be seen as an elegant adaptation of [39], proposed in a parallel database system, to a cloud system. More generally, with respect to the issue of query optimization in cloud environments, the most recent and relevant proposals are described in [11,41,53,61].

## 5    Conclusion

In this paper, we provided a synthetic and highlight state of the art concerning big data management in cloud environments. In this objective, we have tackled two major issues: (i) data management based on parallel and cloud (e.g. MapReduce) systems are over-viewed and compared by relying on meeting software requirements (e.g. data independence, software reuse), high performance, scalability, elasticity, and data availability and (ii) we mainly focused on the evolution of data manipulation languages in cloud systems. Initially, these languages were low level procedural languages as for example MapReduce. For software engineering requirements these languages evolved by introducing relational algebra operators while remaining procedural. It also allowed introducing some optimization processes used classically in RDBMS. Then, they continued their evolution for optimization needs. They became declarative to increase the opportunities of automatic optimization for user queries. The various optimization steps are very close to those used in parallel RDBMS. The main difference is due to the quasi-absence of statistics used by cost models. This blinds the optimization process and an efficient dynamic optimization becomes essential and necessary, to correct sub-optimality of sub-execution plans.

## References

1. Agarwal, S., Kandula, S., Bruno, N., Wu, M., Stoica, I., Zhou, J.: Reoptimizing data parallel computing. In: Proceedings of the 9th USENIX Symposium on Networked Systems Design and Implementation, NSDI 2012, San Jose, CA, USA, 25–27 April 2012, pp. 281–294 (2012). https://www.usenix.org/conference/nsdi12/technical-sessions/presentation/agarwal

2. Agrawal, D., El Abbadi, A., Ooi, B.C., Das, S., Elmore, A.J.: The evolving landscape of data management in the cloud. IJCSE **7**(1), 2–16 (2012). http://dx.doi.org/10.1504/IJCSE.2012.046177

3. Akbarinia, R., Liroz-Gistau, M., Agrawal, D., Valduriez, P.: An efficient solution for processing skewed mapreduce jobs. In: Chen, Q., Hameurlain, A., Toumani, F., Wagner, R., Decker, H. (eds.) DEXA 2015. LNCS, vol. 9262, pp. 417–429. Springer, Heidelberg (2015)

4. Apache Spark. https://spark.incubator.apache.org/

5. Baru, C.K., Fecteau, G., Goyal, A., Hsiao, H., Jhingran, A., Padmanabhan, S., Wilson, W.G.: An overview of DB2 parallel edition. In: Proceedings of the 1995 ACM SIGMOD International Conference on Management of Data, San Jose, California, 22–25 May 1995, pp. 460–462 (1995). http://doi.acm.org/10.1145/223784.223876

6. Beyer, K.S., Ercegovac, V., Gemulla, R., Balmin, A., Eltabakh, M.Y., Kanne, C., Özcan, F., Shekita, E.J.: JAQL: a scripting language for large scale semistructured data analysis. PVLDB **4**(12), 1272–1283 (2011). http://www.vldb.org/pvldb/vol4/p1272-beyer.pdf

7. Bondiombouy, C., Kolev, B., Levchenko, O., Valduriez, P.: Integrating big data and relational data with a functional SQL-like query language. In: Chen, Q., Hameurlain, A., Toumani, F., Wagner, R., Decker, H. (eds.) DEXA 2015. LNCS, vol. 9261, pp. 170–185. Springer, Heidelberg (2015)

8. Cariño, F., Kostamaa, P.: Exegesis of DBC/1012 and P-90 - industrial supercomputer database machines. In: Etiemble, D., Syre, J.-C. (eds.) PARLE 1992. LNCS, vol. 605, pp. 877–892. Springer, Heidelberg (1992). http://dx.doi.org/10.1007/3-540-55599-4_130

9. Chaiken, R., Jenkins, B., Larson, P., Ramsey, B., Shakib, D., Weaver, S., Zhou, J.: SCOPE: easy and efficient parallel processing of massive data sets. PVLDB **1**(2), 1265–1276 (2008). http://www.vldb.org/pvldb/1/1454166.pdf

10. Chang, F., Dean, J., Ghemawat, S., Hsieh, W.C., Wallach, D.A., Burrows, M., Chandra, T., Fikes, A., Gruber, R.E.: Bigtable: a distributed storage system for structured data. ACM Trans. Comput. Syst. **26**(2), 4 (2008). http://doi.acm.org/10.1145/1365815.1365816

11. Chang, L., Wang, Z., Ma, T., Jian, L., Ma, L., Goldshuv, A., Lonergan, L., Cohen, J., Welton, C., Sherry, G., Bhandarkar, M.: HAWQ: a massively parallel processing SQL engine in hadoop. In: International Conference on Management of Data, SIGMOD 2014, Snowbird, UT, USA, 22–27 June 2014, pp. 1223–1234 (2014). http://doi.acm.org/10.1145/2588555.2595636

12. Chaudhuri, S.: What next?: a half-dozen data management research goals for big data and the cloud. In: Proceedings of the 31st ACM SIGMOD-SIGACT-SIGART Symposium on Principles of Database Systems, PODS 2012, Scottsdale, AZ, USA, 20–24 May 2012, pp. 1–4 (2012). http://doi.acm.org/10.1145/2213556.2213558

13. Chekuri, C., Hasan, W., Motwani, R.: Scheduling problems in parallel query optimization. In: Proceedings of the Fourteenth ACM SIGACT-SIGMOD-SIGART Symposium on Principles of Database Systems, San Jose, California, USA, 22–25 May 1995, pp. 255–265 (1995). http://doi.acm.org/10.1145/212433.212471

14. Chen, M., Lo, M., Yu, P.S., Young, H.C.: Using segmented right-deep trees for the execution of pipelined hash joins. In: Proceedings of 18th International Conference on Very Large Data Bases, Vancouver, Canada, 23–27 August 1992, pp. 15–26 (1992). http://www.vldb.org/conf/1992/P015.PDF

15. Cloudera Impala. http://www.cloudera.com/content/cloudera/en/products-and-services/cdh/impala.html

16. Cooper, B.F., Ramakrishnan, R., Srivastava, U., Silberstein, A., Bohannon, P., Jacobsen, H., Puz, N., Weaver, D., Yerneni, R.: PNUTS: Yahoo!'s hosted data serving platform. PVLDB **1**(2), 1277–1288 (2008). http://www.vldb.org/pvldb/1/1454167.pdf

17. Copeland, G.P., Alexander, W., Boughter, E.E., Keller, T.W.: Data placement in bubba. In: Proceedings of the 1988 ACM SIGMOD International Conference on Management of Data, Chicago, Illinois, 1–3 June 1988, pp. 99–108 (1988). http://doi.acm.org/10.1145/50202.50213

18. Dean, J., Ghemawat, S.: Mapreduce: Simplified data processing on large clusters. In: 6th Symposium on Operating System Design and Implementation (OSDI 2004), San Francisco, California, USA, 6–8 December 2004, pp. 137–150 (2004). http://www.usenix.org/events/osdi04/tech/dean.html

19. DeCandia, G., Hastorun, D., Jampani, M., Kakulapati, G., Lakshman, A., Pilchin, A., Sivasubramanian, S., Vosshall, P., Vogels, W.: Dynamo: amazon's highly available key-value store. In: Proceedings of the 21st ACM Symposium on Operating Systems Principles 2007, SOSP 2007, Stevenson, Washington, USA, 14–17 October 2007, pp. 205–220 (2007). http://doi.acm.org/10.1145/1294261.1294281

20. DeWitt, D.J., Gray, J.: Parallel database systems: The future of high performance database systems. Commun. ACM **35**(6), 85–98 (1992). http://doi.acm.org/10.1145/129888.129894

21. DeWitt, D.J., Halverson, A., Nehme, R.V., Shankar, S., Aguilar-Saborit, J., Avanes, A., Flasza, M., Gramling, J.: Split query processing in polybase. In: Proceedings of the ACM SIGMOD International Conference on Management of Data, SIGMOD 2013, New York, NY, USA, 22–27 June 2013, pp. 1255–1266 (2013). http://doi.acm.org/10.1145/2463676.2463709

22. Doulkeridis, C., Nørvåg, K.: A survey of large-scale analytical query processing in mapreduce. VLDB J. **23**(3), 355–380 (2014). http://dx.doi.org/10.1007/s00778-013-0319-9

23. Englert, S., Glasstone, R., Hasan, W.: Parallelism and its price: a case study of nonstop SQL/MP. SIGMOD Rec. **24**(4), 61–71 (1995). http://dx.doi.org/10.1145/219713.219760

24. Floratou, A., Minhas, U.F., Özcan, F.: SQL-on-hadoop: full circle back to shared-nothing database architectures. PVLDB **7**(12), 1295–1306 (2014). http://www.vldb.org/pvldb/vol7/p1295-floratou.pdf

25. Floratou, A., Teletia, N., DeWitt, D.J., Patel, J.M., Zhang, D.: Can the elephants handle the NoSQL onslaught? PVLDB **5**(12), 1712–1723 (2012). http://vldb.org/pvldb/vol5/p1712_avriliafloratou_vldb2012.pdf

26. Gates, A., Natkovich, O., Chopra, S., Kamath, P., Narayanam, S., Olston, C., Reed, B., Srinivasan, S., Srivastava, U.: Building a highlevel dataflow system on top of mapreduce: the pig experience. PVLDB **2**(2), 1414–1425 (2009). http://www.vldb.org/pvldb/2/vldb09-1074.pdf

27. Ghemawat, S., Gobioff, H., Leung, S.: The Google file system. In: Proceedings of the 19th ACM Symposium on Operatig Systems Principles 2003, SOSP 2003, Bolton Landing, NY, USA, 19–22 October 2003, pp. 29–43 (2003). http://doi.acm.org/10.1145/945445.945450

28. Gray, J.: Evolution of data management. IEEE Comput. **29**(10), 38–46 (1996). http://dx.doi.org/10.1109/2.539719

29. Hadoop. http://hadoop.apache.org

30. Hameurlain, A., Morvan, F.: An optimization method of data communication and control for parallel execution of SQL queries. In: Proceedings of 4th International Conference on Database and Expert Systems Applications, DEXA 1993, Prague, Czech Republic, 6–8 September 1993, pp. 301–312 (1993). http://dx.doi.org/10.1007/3-540-57234-1_27

31. Hameurlain, A., Morvan, F.: A parallel scheduling method for efficient query processing. In: Proceedings of the 1993 International Conference on Parallel Processing. Algorithms & Applications, Syracuse University, NY, USA, 16–20 August 1993, vol. III, pp. 258–262 (1993). http://dx.doi.org/10.1109/ICPP.1993.31

32. Hameurlain, A., Morvan, F.: Scheduling and mapping for parallel execution of extended SQL queries. In: CIKM 1995, Proceedings of the 1995 International Conference on Information and Knowledge Management, Baltimore, Maryland, USA, 28 November–2 December 1995, pp. 197–204 (1995). http://doi.acm.org/10.1145/221270.221567

33. Hameurlain, A., Morvan, F.: Parallel relational database systems: Why, how and beyond. In: Proceedings of 7th International Conference on Database and Expert Systems Applications, DEXA 1996, Zurich, Switzerland, 9–13 September 1996, pp. 302–312 (1996). http://dx.doi.org/10.1007/BFb0034690

34. Hasan, W., Motwani, R.: Optimization algorithms for exploiting the parallelism-communication tradeoff in pipelined parallelism. In: VLDB 1994, Proceedings of 20th International Conference on Very Large Data Bases, Santiago de Chile, Chile, 12–15 September 1994, pp. 36–47 (1994), http://www.vldb.org/conf/1994/P036.PDF

35. Hasan, W., Motwani, R.: Coloring away communication in parallel query optimization. In: VLDB 1995, Proceedings of 21th International Conference on Very Large Data Bases, Zurich, Switzerland, 11–15 September 1995, pp. 239–250 (1995). http://www.vldb.org/conf/1995/P239.PDF

36. Hong, W.: Exploiting inter-operation parallelism in XPRS. In: Proceedings of the 1992 ACM SIGMOD International Conference on Management of Data, San Diego, California, 2–5 June 1992, pp. 19–28 (1992). http://doi.acm.org/10.1145/130283.130292

37. Indrawan-Santiago, M.: Database research: Are we at a crossroad? reflection on nosql. In: 15th International Conference on Network-Based Information Systems, NBiS 2012, Melbourne, Australia, 26–28 September 2012, pp. 45–51 (2012). http://dx.doi.org/10.1109/NBiS.2012.95

38. Jiang, D., Ooi, B.C., Shi, L., Wu, S.: The performance of mapreduce: an in-depth study. PVLDB 3(1), 472–483 (2010). http://www.comp.nus.edu.sg/vldb2010/proceedings/files/papers/E03.pdf

39. Kabra, N., DeWitt, D.J.: Efficient mid-query re-optimization of sub-optimal query execution plans. In: SIGMOD 1998, Proceedings of ACM SIGMOD International Conference on Management of Data, Seattle, Washington, USA, 2–4 June 1998, pp. 106–117 (1998). http://doi.acm.org/10.1145/276304.276315

40. Kaldewey, T., Shekita, E.J., Tata, S.: Clydesdale: structured data processing on mapreduce. In: Proceedings of 15th International Conference on Extending Database Technology, EDBT 2012, Berlin, Germany, 27–30 March 2012, pp. 15–25 (2012). http://doi.acm.org/10.1145/2247596.2247600

41. Karanasos, K., Balmin, A., Kutsch, M., Ozcan, F., Ercegovac, V., Xia, C., Jackson, J.: Dynamically optimizing queries over large scale data platforms. In: International Conference on Management of Data, SIGMOD 2014, Snowbird, UT, USA, 22–27 June 2014, pp. 943–954 (2014). http://doi.acm.org/10.1145/2588555.2610531

42. Lakshman, A., Malik, P.: Cassandra: a decentralized structured storage system. Opera. Syst. Rev. **44**(2), 35–40 (2010). http://doi.acm.org/10.1145/1773912. 1773922
43. Lanzelotte, R.S.G., Valduriez, P.: Extending the search strategy in a query optimizer. In: Proceedings of 17th International Conference on Very Large Data Bases, Barcelona, Catalonia, Spain, 3–6 September 1991, pp. 363–373 (1991). http://www.vldb.org/conf/1991/P363.PDF
44. Lee, K., Lee, Y., Choi, H., Chung, Y.D., Moon, B.: Parallel data processing with mapreduce: a survey. SIGMOD Rec. **40**(4), 11–20 (2011). http://doi.acm.org/10.1145/2094114.2094118
45. Li, F., Ooi, B.C., Özsu, M.T., Wu, S.: Distributed data management using mapreduce. ACM Comput. Surv. **46**(3), 31: 1–31: 42 (2014). http://doi.acm.org/10.1145/2503009
46. Livny, M., Khoshafian, S., Boral, H.: Multi-disk management algorithms. In: SIGMETRICS, pp. 69–77 (1987). http://doi.acm.org/10.1145/29903.29914
47. Lu, H., Tan, K.L., Ooi, B.C.: Query Processing in Parallel Relational Database Systems. IEEE CS Press, Los Alamitos (1994)
48. Olston, C., Reed, B., Srivastava, U., Kumar, R., Tomkins, A.: Pig Latin: a not-so-foreign language for data processing. In: Proceedings of the ACM SIGMOD International Conference on Management of Data, SIGMOD 2008, Vancouver, BC, Canada, 10–12 June 2008, pp. 1099–1110 (2008). http://doi.acm.org/10.1145/1376616.1376726
49. Oracle. http://www.oracle.com/technetwork/bdc/hadoop-loader/connectors-
50. Ozsu, M.T., Valduriez, P.: Principles of Distributed Database Systems. Springer, New York (2011)
51. Pavlo, A., Paulson, E., Rasin, A., Abadi, D.J., DeWitt, D.J., Madden, S., Stonebraker, M.: A comparison of approaches to large-scale data analysis. In: Proceedings of the ACM SIGMOD International Conference on Management of Data, SIGMOD 2009, Providence, Rhode Island, USA, 29 June–2 July 2009, pp. 165–178 (2009). http://doi.acm.org/10.1145/1559845.1559865
52. Schneider, D.A., DeWitt, D.J.: Tradeoffs in processing complex join queries via hashing in multiprocessor database machines. In: Proceedings of 16th International Conference on Very Large Data Bases, Brisbane, Queensland, Australia, 13–16 August 1990, pp. 469–480 (1990). http://www.vldb.org/conf/1990/P469.PDF
53. Soliman, M.A., Antova, L., Raghavan, V., El-Helw, A., Gu, Z., Shen, E., Caragea, G.C., Garcia-Alvarado, C., Rahman, F., Petropoulos, M., Waas, F., Narayanan, S., Krikellas, K., Baldwin, R.: Orca: a modular query optimizer architecture for big data. In: International Conference on Management of Data, SIGMOD 2014, Snowbird, UT, USA, 22–27 June 2014, pp. 337–348 (2014). http://doi.acm.org/10.1145/2588555.2595637
54. Sqoop. http://sqoop.apache.org/
55. Stonebraker, M., Abadi, D.J., DeWitt, D.J., Madden, S., Paulson, E., Pavlo, A., Rasin, A.: Mapreduce and parallel DBMSs: friends or foes? Commun. ACM **53**(1), 64–71 (2010). http://doi.acm.org/10.1145/1629175.1629197
56. Stonebraker, M., Cattell, R.: 10 rules for scalable performance in 'simple operation' datastores. Commun. ACM **54**(6), 72–80 (2011). doi:10.1145/1953122.1953144. http://doi.acm.org/10.1145/1953122.1953144
57. Stonebraker, M., Madden, S., Dubey, P.: Intel "big data" science and technology center vision and execution plan. SIGMOD Rec. **42**(1), 44–49 (2013). http://doi.acm.org/10.1145/2481528.2481537

58. Tan, K., Lu, H.: Pipeline processing of multi-way join queries in shared-memory systems. In: Proceedings of the 1993 International Conference on Parallel Processing. Architecture, Syracuse University, NY, USA, 16–20 August 1993, vol. I, pp. 345–348 (1993). http://dx.doi.org/10.1109/ICPP.1993.147
59. Thusoo, A., Sarma, J.S., Jain, N., Shao, Z., Chakka, P., Anthony, S., Liu, H., Wyckoff, P., Murthy, R.: Hive - a warehousing solution over a map-reduce framework. PVLDB **2**(2), 1626–1629 (2009)
60. Thusoo, A., Sarma, J.S., Jain, N., Shao, Z., Chakka, P., Zhang, N., Anthony, S., Liu, H., Murthy, R.: Hive - a petabyte scale data warehouse using hadoop. In: Proceedings of the 26th International Conference on Data Engineering, ICDE 2010, Long Beach, California, USA, 1–6 March 2010, pp. 996–1005 (2010). http://dx.doi.org/10.1109/ICDE.2010.5447738
61. Trummer, I., Koch, C.: Multi-objective parametric query optimization. PVLDB **8**(3), 221–232 (2014). http://www.vldb.org/pvldb/vol8/p221-trummer.pdf
62. Valduriez, P.: Parallel database systems: open problems and new issues. Distrib. Parallel Databases **1**(2), 137–165 (1993). doi:10.1007/BF01264049. http://dx.doi.org/10.1007/BF01264049
63. Wiederhold, G.: Mediators in the architecture of future information systems. IEEE Comput. **25**(3), 38–49 (1992). http://dx.doi.org/10.1109/2.121508
64. Witkowski, A., Cariño, F., Kostamaa, P.: NCR 3700 - the next-generation industrial database computer. In: Proceedings of 19th International Conference on Very Large Data Bases, Dublin, Ireland, 24–27 August 1993, pp. 230–243 (1993). http://www.vldb.org/conf/1993/P230.PDF
65. Xu, Y., Kostamaa, P., Gao, L.: Integrating hadoop and parallel DBMs. In: Proceedings of the ACM SIGMOD International Conference on Management of Data, SIGMOD 2010, Indianapolis, Indiana, USA, 6–10 June 2010, pp. 969–974 (2010). http://doi.acm.org/10.1145/1807167.1807272
66. Zha, L., Zhang, J., Liu, W., Lin, J.: An uncoupled data process and transfer model for mapreduce. In: Hameurlain, A., Küng, J., Wagner, R., Bellatreche, L., Mohania, M. (eds.) TLDKS XVII. LNCS, vol. 8970, pp. 24–44. Springer, Heidelberg (2015). http://dx.doi.org/10.1007/978-3-662-46335-2_2
67. Zhou, J., Bruno, N., Wu, M., Larson, P., Chaiken, R., Shakib, D.: SCOPE: parallel databases meet mapreduce. VLDB J. **21**(5), 611–636 (2012). http://dx.doi.org/10.1109/PDIS.1993.253066
68. Ziane, M., Zaït, M., Borla-Salamet, P.: Parallel query processing in DBS3. In: Proceedings of the 2nd International Conference on Parallel and Distributed Information Systems (PDIS 1993), Issues, Architectures, and Algorithms, San Diego, CA, USA, 20–23 January 1993, pp. 93–102 (1993). http://dx.doi.org/10.1109/PDIS.1993.253066

# Reduction of Readmissions to Hospitals Based on Actionable Knowledge Discovery and Personalization

Mamoun Almardini[1], Ayman Hajja[1], Zbigniew W. Raś[1,2(✉)], Lina Clover[3], David Olaleye[3], Youngjin Park[3], Jay Paulson[3], and Yang Xiao[3]

[1] College of Computing and Informatics, University of North Carolina, Charlotte, NC 28223, USA
{malmardi,ahajja,ras}@uncc.edu
[2] Institute of Computer Science, Warsaw University of Technology, 00-665 Warsaw, Poland
[3] SAS Institute Inc, Cary, NC 27513, USA
{Lina.Clover,David.Olaleye,Youngjin.Park,Jay.Paulson,Yang.Xiao}@sas.com

**Abstract.** In this work, we define procedure paths as the sequence of procedures that a given patient undertakes to reach a desired treatment. In addition to its value as a mean to inform the patient of his or her course of treatment, being able to identify and anticipate procedure paths for new patients is an essential task for examining and evaluating the entire course of treatments in advance, and ultimately rectifying undesired procedure paths accordingly. In this paper, we first introduce two approaches for anticipating the state of the patient that he or she will end up in after performing some procedure $p$; the state of the patient will consequently indicate the following procedure that the patient is most likely to undergo. By clustering patients into subgroups that exhibit similar properties, we improve the predictability of their procedure paths, which we evaluate by calculating the entropy to measure the level of predictability of following procedure. The clustering approach used is essentially a way of personalizing patients according to their properties. The approach used in this work is entirely novel and was designed specifically to address the twofold problem of first being able to predict following procedures for new patients with high accuracy, and secondly being able to construct such groupings in a way that allows us to identify exactly what it means to transition from one cluster to another. Then, we further devise a metric system that will evaluate the level of desirability for procedures along procedure paths, which we would subsequently map to a metric system for the extracted clusters. This will allow us to find desired transitions between patients in clusters, which would result in reducing the number of anticipated readmissions for new patients.

**Keywords:** Personalization · Side-effects · Clustering

© Springer International Publishing Switzerland 2016
S. Kozielski et al. (Eds.): BDAS 2016, CCIS 613, pp. 39–55, 2016.
DOI: 10.1007/978-3-319-34099-9_3

## 1   Introduction

Recently, expenditure on healthcare has risen rapidly in the United States. According [1], healthcare spending has been rising at twice the rate of growth of our income, for the past 40 years; the projection of the growth rate in healthcare spending is 5.8 percent during the period 2014–2024, which means that the spending will rise to 5.4 trillion by 2024. That said, the gross domestic product (GDP) growth rate is 4.7 percent (as of 2014) [2]. This increase can be attributed to several factors as listed by Price Waterhouse Coopers (PWC) research institute: over-testing, processing claims, ignoring doctors orders, ineffective use of technology, hospital readmissions, medical errors, unnecessary ER visits, and hospital acquired infections [3]. Figure 1 shows that 25 billion are spent annually on readmissions. Hospital readmissions and surgery outcomes prediction has taken a great interest recently [4–7]. Analyzing the reasons behind readmissions and reducing them can save a great amount of money. A hospital readmission is defined as a hospitalization of the patient after being discharged from the hospital. The period in average is 30 days [7].

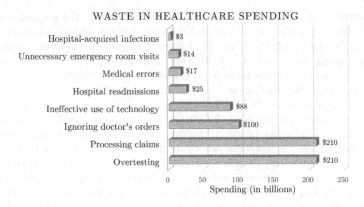

**Fig. 1.** Waste in healthcare spending as listed by Price Waterhouse Coopers (PWC) research institute [3]

One of the reasons for readmissions is negative side effects that may appear after the prescribed procedures and may not be known in advance, as a result patients may require hospital readmissions [8]. In this work, we shift our interests to the entire course of treatments for patients, and we propose ways to increase the predictability of what we call *procedure paths*, which is the sequence of procedures that a given patient undertakes to reach a desired treatment. By being able to predict the procedures that a given patient is expected to go through according to his or her diagnoses, domain experts can reevaluate the procedure path and possibly alter it to a more desired path.

## 2   HCUP Dataset Description

In this paper, we used the Florida State Inpatient Databases (SID) that is part of the Healthcare Cost and Utilization Project (HCUP) [13]. The Florida SID dataset contains records from several hospitals in the Florida State. It contains over 7.8 million visit discharges from over 3.6 million patients. The dataset is composed of five tables, namely: AHAL, CHGH, GRPS, SEVERITY, and CORE. The main table used in this work is the *Core* table. The *Core* table contains over 280 features; however, many of those features are repeated with different codification schemes. In the following experiments, we used the Clinical Classifications Software (CCS) [14] that consists of 262 diagnosis categories, and 234 procedure categories. This system is based on ICD-9-CM codes. In our experiments, we only used the features, listed in Table 1, that are relevant to the problem. Each record in the *Core* table represents a visit discharge. A patient may have several visits in the table. One of the most important features of this table is the *VisitLink* feature, which describes the patient's ID. Another important feature is the *Key*, which is the primary key of the table that identifies unique visits for the patients and links to the other tables. As mentioned earlier, a *VisitLink* might map to multiple *Key* in the database. This table reports up to 31 diagnoses per discharge as it has 31 diagnosis columns. However, patients' diagnoses are stored in a random order in this table. For example, if a particular patient visits the hospital twice with heart failure, the first visit discharge may report a heart failure diagnosis at diagnosis column number 10, and the second visit discharge may report a heart failure diagnosis at diagnosis column number 22. Furthermore, it is worth mentioning that it is often the case that patients examination returns less than 31 diagnoses. The *Core* table also contains 31 columns describing up to 31 procedures that the patient went through. Even though a patient might have gone through several procedure in a given visit, the primary procedure that occurred at the visit discharge is assumed to be the first procedure column. The *Core* table also contains a feature called *DaysToEvent*, which describes the number of days that passed between the admission to the hospital and the procedure day. This field is anonymized in order to hide the patients' identity. Furthermore, the *Core* table also contains a feature called *DIED*, that informs us on whether the patient died or survived in the hospital for a particular discharge. There are several demographic data that are reported in this table as well, such as race, age range, sex, living area, etc. Table 1 maps the features from the *Core* table to the concepts and notations used in this paper.

**Table 1.** Description of the used core table features.

| Features | Concepts |
|---|---|
| VisitLink | Patient Identifier |
| DaysToEvent | Temporal visit ordering |
| DXCCSn | $n^{th}$ Diagnosis, flexible feature |
| PRCCSn | $n^{th}$ Procedure, meta-action |
| Race, Age Range, Sex,.. | Stable features |
| DIED | Decision feature value |

## 3   Background

The idea of extracting knowledge for the purpose of guiding decision makers to make more educated decisions is certainly not an all-new concept in data mining; however, the rather new idea of extracting actionable knowledge, being-in-itself a guide that serves as a blueprint for decision makers, has only been studied recently [10,11]. The concept and motivation of actionability is based on providing another, yet more relevant, layer of knowledge to decision makers. Actionable rules specify the actions needed to be performed to transition an instance (patient in our case) from one state to another.

One main area of research that heavily involves extracting actionable patterns from mining knowledge is action rules [11]. Action rules describe the necessary transitions that need to be applied on the classification part of a rule for other desired transitions on the decision part of a rule to occur. It is often the case however, that decision makers do not have immediate control over specific transitions, instead they have control over a higher level of actions, which in the literature of action rules are referred to by the term *meta-actions* [9]. Meta-actions are defined as higher level procedures that trigger changes in flexible features of a rule either directly or indirectly according to the influence matrix [9]. For example, by extracting action rules, we may reach the conclusion that to improve a patient's condition, we would need to decrease his (or her) blood pressure. Although this may seem as an oversimplified example of action rules, it is still nonetheless essentially what action rules are meant and designed to do; that is, to provide actionable patterns of transitions. Note here however that to perform the required transition of lowering the blood pressure of the patient, we would ultimately need to perform other actions of higher-levels, called meta-actions. For this particular example, perhaps this means performing a surgery on the patient or prescribing some medication. In summary, mining actionable rules is not only the study of discovering patterns, rather it is more centered around finding ways to transition instances from one pattern to another in a way that aligns with the desires of the domain being studied.

In this work, we apply the concept of mining actionable patterns to the domain of healthcare; by identifying the level of desire for procedure paths as will be discussed in Sect. 6, we will set the foundation that will allow us to extract actionable patterns that will transition new patients from a less desired procedure path to a more desired one.

## 4   Introducing Procedure Paths and Personalization

Procedure paths are defined as the sequence of procedures that a given patient undertakes to reach a desired treatment. In other words, a procedure path is a detailed description for the course of treatments provided to an admitted patient. The length of any given procedure path is an indicator of the number of readmissions that occurred or will occur throughout the course of treatment.

For example, one procedure path for a patient could be the following: $path_x = (p_1, p_3, p_3, p_6)$, where $p$ indicates a particular procedure; according to procedure path $path_x$, the number of readmissions was 3.

In this work, we lay the foundation for predicting procedure paths by devising a system that will anticipate the following procedure (or readmission); we will also introduce a way to extract action rules that will describe transitions that will rectify the following procedure for new patients.

Although there could exist multiple metric systems for evaluating procedure paths, we decided in this work to devote our efforts on tackling the problem of reducing the number of readmissions for new patients. This means that we are interested in transitioning patients from a procedure path with more readmissions to another more desired path with less number of readmissions. That being said, our system for extracting action rules that describe transitions between procedure paths is entirely independent of the metric system used; this means that if domain experts decide to devise a new metric system by incorporating other criteria such as the cost of operation or duration in the hospital, then that would not affect our system and it would still function according to the new evaluation system.

The *procedure graph* for some procedure $p$ is defined as the tree of all possible procedure paths extracted from our dataset for patients who underwent procedure $p$ as their first procedure. The number of all procedure paths is extremely high. This high number of unique procedure paths indicates that it is not true that there exists a single universal course of treatment that patients typically follow to reach the desired state. For example, the number of patients that underwent

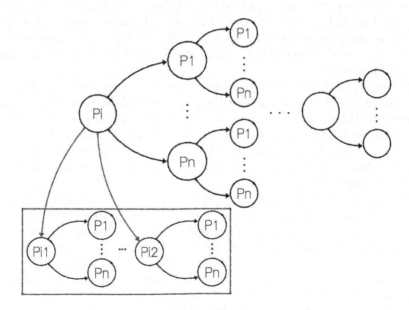

**Fig. 2.** Depiction of a procedure graph

procedure 78 (colorectal resection) is 41,753; and the number of unique procedure paths that those patients underwent is 6,774. Figure 2 shows a depiction of the *procedure graph*; $P_{(0,1)}$ is the initial procedure that patients start with; the next procedure could be any procedure from $P_{(1,1)}$ to $P_{(1,n)}$, which is determined by the resulting set of diagnoses after performing the initial procedure $P_{(0,1)}$. The first argument $x$ in the notation $P_{(x,y)}$ refers to the number (or rather level) of readmission, and the second argument $y$ refers to the procedure identifier at that level. For example, $P_{(1,2)}$ refers to the procedure with identifier 2 that occurred at the first level of readmissions (e.g. first readmission following the initial procedure). The portions of the graph that are contained in dashed boxes depict the personalization part that we introduce in the next section. The idea of personalization is to cluster patients that are scheduled to undergo procedure $P_{(0,1)}$ according to their diagnoses; as a result of this clustering, we will be able to anticipate with higher accuracy the following procedure (readmission) that the patient will undergo by identifying which cluster the new patient belongs to. In the next section, we will provide an elaborate description of the clustering approaches we propose.

## 5   Personalizing Through Diagnoses Clustering

Given a new patient, being able to anticipate the procedure path for that patient is an invaluable asset to medical doctors since it can be used as a mean to inform the patient of his or her course of treatment, and ultimately altering or amending the course of treatment accordingly. Our assumption that we use in this work, which aligns with the definition of our dataset, is that the procedure performed on the patient is determined from the set of diagnoses that the patient is diagnosed with. In addition to that, the set of diagnoses will also determine the state in which the patient ends up in; in this case, the state would be the set of diagnoses after performing the first procedure, which as result, will determine the second procedure.

Although it is theoretically possible to create a chain of predictions that will provide a complete prediction for the entire procedure path, we only examine predicting the following (next) procedure in this work for higher prediction accuracy; if the following procedure is part of a desired procedure path then no intervention is needed; otherwise, transitioning patients to another state that alters the following procedure to make it part of a more desired procedure path would be recommended if at all feasible.

To predict the procedure path, or rather the following procedure, we start by extracting knowledge from our existing dataset. The approach used to predict the following procedure is an unsupervised clustering technique based on the set of diagnoses that patients exhibit at the time of their first admission. Our assumption is that patients that exhibit similar set of diagnoses will end up with a similar set of diagnoses after the procedure; and again, by definition of our dataset, these set of diagnoses will determine the next procedure. Next, we provide two different clustering approaches; the exact matching clustering approach and the fuzzy approach.

## 5.1    Exact Matching Clustering

In the exact match clustering approach, we define a cluster by a set of diagnoses. For a given patient to belong to any cluster, he or she must have the exact same diagnoses set.

Table 2 shows some of the most common sets of exact diagnoses for patients who undertook procedure '158' (spinal fusion). From Table 2, we observe that for procedure 158, the cluster with the exact set of diagnoses {205} contains 502 patients; this means that there are 502 patients that were diagnosed with '205' and nothing else, before undertaking procedure 158. Similarly, the second row of Table 2 means that there are 128 patients that were diagnosed with both diagnostic codes '205' and '98' and nothing else, before having to undergo procedure 158.

**Table 2.** Some of the most common set of exact diagnoses for procedure 158 (spinal fusion)

| Set of diagnoses | Number of patients | Entropy |
|---|---|---|
| {205} | 502 | 3.015 |
| {205, 98} | 128 | 2.752 |
| {205, 663} | 86 | 2.543 |
| {205, 209} | 67 | 2.798 |
| {205, 211} | 51 | 2.510 |

Here is a description of each diagnostic code provided in the table:

- 205: Spondylosis; intervertebral disc disorders; other back problems
- 98: Essential hypertension
- 663: Screening and history of mental health and substance abuse codes
- 209: Other acquired deformities
- 211: Other connective tissue disease

We should also mention here that the entropy for the entire system before this personalization attempt is 3.667, which is higher than all the entropy values in the table. The weighted entropy for all exact matching clusters that have size above threshold 50 is 2.892. This implies that by applying this clustering approach, we will be able to have a higher level of predictability of which following (next) procedure is likely to be undertaken. In other words, by knowing the cluster of which a patient belongs to, we would be able to anticipate (with higher accuracy), where that patient is likely to end up after performing the first procedure.

One advantage of using the exact matching clustering is that transitions can be precisely described. For example, if we discovered that patients in cluster $c_1 = \{205\}$ tend to end up in a state that is more desired than patients in cluster $c_2 = \{205, 98\}$, then we can precisely describe the transition that needs to be done; in this case, only removing diagnostic code 98.

The fact that patients are usually admitted with other diagnoses that are often irrelevant to the main diagnosis (or diagnoses), makes this approach quite limited, which is rather evident in the frequencies (number of patients) that exhibit the most common set of exact diagnoses, compared to the number of patients that exhibit the same diagnoses but along with other diagnoses that may be irrelevant. For example, the number of patients that exhibit diagnosis code 205 along with other diagnoses is 13,096, which is substantially larger than the number of patients that only exhibit diagnosis code 205. In the next subsection, we present a new clustering approach that addresses this limitation.

## 5.2   Fuzzy Matching Clustering

In this section, we define a new and novel way for specifying the criteria that patients' properties need to satisfy for a given patient to join a cluster. Unlike the exact matching clustering approach where we define one unique set of diagnoses that needs to exactly match the patient's set of diagnoses; in this fuzzy matching clustering approach, we define three sets of diagnoses that will determine whether a patient belongs to a particular cluster or not.

The first set, which we call the *included* set, describes the set of diagnoses that any given patient needs to exhibit for that patient to belong to the cluster; the second set, which we call the *excluded* set, is the set of diagnoses that patients cannot exhibit for them to belong to that cluster; and finally the third set, which we call the *optional* set, is the set of diagnoses that patients can, but do not need to, exhibit for them to belong to that cluster.

Since the *optional* set is the complement of the *included* and *excluded* sets combined, we decided not to specify it each time we define a cluster. For example, if the entire set of diagnoses is $D = \{d_1, d_2, ..., d_{10}\}$, the *included* set of some cluster $c$ is $included(c) = \{d_1, d_2, d_5, d_7\}$, and the *excluded* set of the same cluster $c$ is $excluded(c) = \{d_6, d_8, d_9, d_{10}\}$, then the *optional* set for cluster $c$ is $optional(c) = \{d_3, d_4\}$; which is equal to $D - [included(c) \cup excluded(c)]$.

Now let's examine one real example extracted from our data set. By examining the same procedure (code 158; spinal fusion), one of the extracted clusters was: $c = \{(included : \{98\}, excluded : \{49, 138, 211\})\}$, which contained all patients that exhibited diagnostic code 98 (essential hypertension) and did not exhibit diagnostic codes 49 (diabetes mellitus without complication), 138 (esophageal disorders) and 211(other connective tissue disease).

In our approach of selecting fit diagnostic codes candidates, we started by discarding the diagnostic codes that were either always or never evident in the set of diagnoses; although this may seem counterintuitive at first, the reasoning behind doing so is rather logical. Recall again that our goal is to create clusters of patients that are similar in terms of the resulting states after applying a given procedure. Clearly, if all patients have a common diagnosis, then this diagnosis will not play any role in determining the state for which patients end up in. Accordingly, we only considered diagnostic codes that lie within a specific range; for example, when we choose the allowed range to be between 20 % and 80 %, this means that we only consider diagnoses for which the number of patients that exhibit that diagnosis is between 20 % and 80 %; this procedure is also applied to the *excluded* set, meaning that we also only consider diagnoses that were missing from 20 % to 80 % of the total number of patients. Table 3 shows different ranges that we tested, with different outcomes that aligns with our expectations.

**Table 3.** Number of clusters and the entropy for different element clusters and different ranges for procedure 158 (spinal fusion)

| | Range | | | | | |
|---|---|---|---|---|---|---|
| | 20 % to 80 % | | 10 % to 90 % | | 5 % to 95 % | |
| | # of clusters | Entropy | # of clusters | Entropy | # of clusters | Entropy |
| 1-element clusters | 14 | 5.105 | 36 | 5.109 | 66 | 5.117 |
| 2-element clusters | 37 | 5.051 | 379 | 5.028 | 1532 | 5.041 |
| 3-element clusters | 50 | 4.988 | 2097 | 4.958 | 19167 | 4.971 |
| 4-element clusters | 44 | 4.916 | 6969 | 4.905 | 155028 | 4.91 |

The main reason why this fuzzy approach is more superior than the exact matching approach is because it is not based on the assumption that each diagnosis must be relevant to the procedure. However, note that although this fuzzy approach was designed to be less strict so that it counteracts or rectifies the main disadvantage of the exact matching approach, this approach is in fact still precise enough to describe transitions rather specifically for a patient to transition from one cluster to another, as a result of the way our clusters are defined.

For example, we can achieve the transition from cluster $c_1 = \{(included : \{98\}, excluded\{53, 211\})\}$ to cluster $c_2 = \{(included : \{\}, excluded\{53, 98, 211\})\}$ by shifting diagnostic code 98 from the included set to the excluded set; which simply means applying some treatment to the patient so that he or she is no longer diagnosed with diagnostic code 98 (essential hypertension).

Here is a description of each diagnostic code provided in the two clusters from the example above:

- 53: Nutritional deficiencies
- 98: Essential hypertension
- 209: Other acquired deformities
- 211: Other connective tissue disease

The methodology used to extract all combinations of clusters is similar to the association action rules extracting approach presented in [12], we start by extracting all 1-diagnosis clusters that lie within the range specified (for both the *included* and *excluded* sets). Then we build 2-diagnosis clusters by combining all possible pairs of 1-diagnosis clusters. Next step would be to construct 3-diagnosis clusters by combining all the 2-diagnosis clusters with all the 1-diagnosis clusters, so on and so forth. Using this procedure however, the number of generated clusters will grow extremely fast. Iterating through all possible clusters for each patient to verify whether that patient belonged to a given cluster or not would be a highly inefficient implementation. Instead, we used a retrieval digital tree implementation that starts the verification process at the root of the tree, which would allow us to discard entire subtrees anytime the patient does not satisfy a node constraint (whether that constraint was an *included* constraint or an *excluded* constraint).

As mentioned earlier in this section, the main goal of clustering patients according to the diagnostic codes is to increase the predictability for the next procedure which can be measured by calculating the entropy. Constructing more combinations of the diagnostic codes enforces more personalization on the patients and as a result improves the predictability of the next procedure; the entropy decreases. Figure 3 illustrates the process of generating the clusters. The top of the tree, Level 1, shows the 1-element set clusters; which are represented by all possible diagnostic codes for a given procedure; *included sets* : $\{d1\}, \{d2\},$ and $\{d3\}$ and their negations; *excluded sets* : $\{-d1\}, \{-d2\},$ and $\{-d3\}$. The process continues and in order to generate clusters for any Level '$L_n$', we pair all elements in Level '$L_{(n-1)}$' with elements from '$L_1$'.

It is also worth mentioning here that any diagnostic code can not exist with its negation in the same cluster; for example, we can not have $\{d1\}$ and $\{-d1\}$ in the same cluster, as this means that the patients that belong to that cluster have the diagnostic code $d1$ and do not have it at the same time.

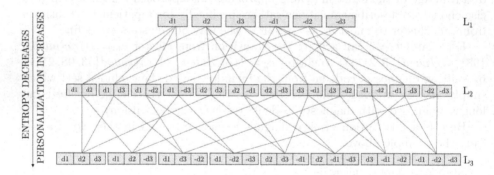

**Fig. 3.** Construction of the patients' clusters

Now let us show the benefits for clustering patients by providing a hypothetical example that mimics a real life scenario. In Fig. 4, we show a sample of some of the clusters that could be generated for the patients that are about to undertake procedure $P_{(0,1)}$ as their first procedure. The first cluster (Cluster 1) contains 60 patients; from the 60 patients that belong to Cluster 1, 10 out of which ended up undergoing procedure $P_{(1,1)}$, 45 ended up undergoing procedure $P_{(1,2)}$, only 5 patients ended up undergoing procedure $P_{(1,3)}$, and some patients didn't come back to the hospital. According to our example, this distribution of following procedures implies that if a patient exhibits the set of diagnoses that the first cluster is defined by, then that patient will most likely end up in a state that will require him or her to undergo procedure $P_{(1,2)}$. Similarly, the distribution of Cluster 2 and Cluster 3 will imply that most new patients that will belong to Cluster 2 will end up undergoing procedure $P_{(1,1)}$, and that most new patients that will belong to Cluster 3 will end up undergoing $P_{(1,3)}$. In the following section, we introduce a novel algorithm that assigns a score to each procedure in the procedure graph by taking into consideration the number of patients and the length of the procedure path. This procedures' score is later consequently used to calculate the score for each cluster to determine their level of desire, which would guide us to extracting cluster transitions.

# 6   Calculating Score

In this section, we propose a system to evaluate nodes (procedures) in a procedure graph by using the number of following anticipated readmissions as our criterion. We will propose subsequently, a score mapping system that will be used to transfer the scores of procedures in the procedure graph to clusters, which would thereafter allow us to identify feasible cluster transitions and ultimately reduce the average number of readmissions for new patients.

## 6.1   Procedure Graph

In Sect. 5, we introduced a method that will cluster patients according to their set of diagnoses; as a result, we were able to increase the predictability of the next procedure. This means that by examining the set of diagnoses for a new patient, we will be able to identify the cluster(s) that he or she belongs to, which would allow us as a result to increase the accuracy of predicting the following procedure. That being said, without having a metric system that will score the level of desirability for following procedures, we wouldn't be able to determine whether there exist(s) following procedures that are more desired than the anticipated following procedure. As we discussed in Sect. 4, our ultimate goal is to transition patients from an undesired anticipated procedure path to a more desired one; to do so however, we would need to devise a system that evaluates the level of desirability for any given procedure path.

In this section, we provide a metric system that will score each node (procedure) in the procedure graph defined in Sect. 4. There are two reasons why we decided to evaluate and score procedure graph nodes (actual procedures) rather than evaluating procedure paths. The first reason is because there is a significant number of procedure paths for any given procedure, and it would be inefficient to calculate the score for all of them; the second more important reason is the fact that by transitioning a patient from one cluster to another, we are essentially attempting to change the next procedure, which is a node in a procedure graph and not an entire procedure path.

The metric system devised in this section is based on the number of readmissions for patients. Next, we will introduce a method for calculating the score of any given node (procedure) in the procedure graph, which will represent the average number of following readmissions for patients at that particular procedure.

Let us demonstrate with an example by calculating the score for some nodes from our hypothetical procedure graph in Fig. 4. According to Fig. 4, 15 patients have undergone procedure $P_{(3,1)}$, and since there are no procedures following $P_{(3,1)}$, this means that all 15 patients have an average number of future readmission equal to 0; hence, the score of node $P_{(3,1)}$ will be equal to zero. Now let us examine node $P_{(2,1)}$; the number of patients that undergone $P_{(2,1)}$ is 35 (25 from $P_{(1,1)}$ and 10 from $P_{(1,2)}$), out of the 35, 20 patients did not come back to the hospital and 15 were readmitted to undergo procedure $P_{(3,1)}$. The score of procedure $P_{(2,1)}$ is the sum of weighted score of each possible procedure directly following the procedure $P_{(2,1)}$:

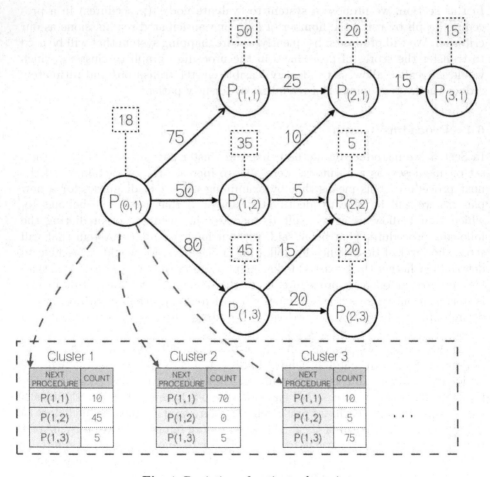

**Fig. 4.** Depiction of patients clustering

- The weight/probability of the first possibility (no further readmissions) is 20/35. The score (average number of readmissions) for patients who did not come back to the hospital is zero.
- The weight/probability of the second possibility (undergoing $P_{(3,1)}$) is 15/35, for which the score (average number of readmissions) will be 1 (which essentially reflects undergoing one more procedure) plus the score of $P_{(3,1)}$, which is zero for this example.

The score of procedure $P_{(2,1)}$ therefore becomes:

$$\text{score}(P_{(2,1)}) = \left(\frac{20}{35} * 0\right) + \left(\frac{15}{35} * \left(1 + \text{score}(P_{(3,1)})\right)\right) = \frac{15}{35} * (1 + 0) = .43$$

This essentially means that if a patient were to undergo procedure $P_{(2,1)}$, then the number of following readmission on average is .43; also, in this particular example, since we know that patients can only have one readmission ($P_{(3,1)}$), we can also state that since the score is .43, then this also means that for any patient who undertakes $P_{(1,2)}$, there will be a 43 % chance that he or she will undergo one additional readmission.

Now let us examine one more node: procedure $P_{(1,2)}$. The number of patients that underwent procedure $P_{(1,2)}$ is 50, from which we have three possibilities:

- **Possibility 1:** 35 out of 50 did not come back to the hospital.
- **Possibility 2:** 10 out of 50 were readmitted to undergo procedure $P_{(2,1)}$.
- **Possibility 3:** 5 out of 50 were readmitted to undergo procedure $P_{(2,2)}$.

To calculate the score in this case, we need to calculate the weighted score for each possible following procedure

- The weight/probability of the first possibility is 35/50; again however, the score (average number of readmissions) for patients who did not come back to the hospital is zero.
- The weight/probability of the second possibility is 10/50, for which the score (average number of readmissions) will be 1 (which essentially reflects undergoing $P_{(2,1)}$), plus the score of $P_{(2,1)}$.
- The weight/probability of the third possibility is 5/50, for which the score (average number of readmissions) will be 1 (which essentially reflects undergoing $P_{(2,2)}$), plus the score of $P_{(2,2)}$. Note here that the score of $P_{(2,2)}$ is zero since procedure $P_{(2,2)}$ was the last procedure for all patients that went through procedure $P_{(2,2)}$.

The score of procedure $P_{(1,2)}$ hence becomes:

$$\text{score}(P_{(1,2)}) = \left(\frac{35}{50} * 0\right) + \left(\frac{10}{50} * \left(1 + \text{score}(P_{(2,1)})\right)\right) + \left(\frac{5}{50} * \left(1 + \text{score}(P_{(2,2)})\right)\right)$$

$$\Rightarrow \text{score}(P_{(1,2)}) = 0 + \left(\frac{10}{50} * (1 + .43)\right) + \left(\frac{5}{50} * (1 + 0)\right) = .386$$

Which again, would mean that for patients that undergo procedure $P_{(1,2)}$, the number of following readmission on average is .386; this however does not mean that there is a 39 % chance that the patients will undergo additional readmissions, since a single patient may undergo two readmissions.

We define the procedure score recurrence function as:

$$\text{score}(P_x) = \begin{cases} \sum_{k=1}^{n} \frac{|P_k|}{|P_x|} * (1 + \text{score}(P_k)) & \text{if } n \geq 1 \\ 0, & \text{otherwise} \end{cases}$$

where $n$ denotes the number of procedures directly following procedure $P_x$, $P_k$ denotes the $kth$ procedure right after $P_x$, and $|P_k|$ denotes the number of patients that underwent the $kth$ procedure.

Figure 5 shows the score for each node in our procedure graph.

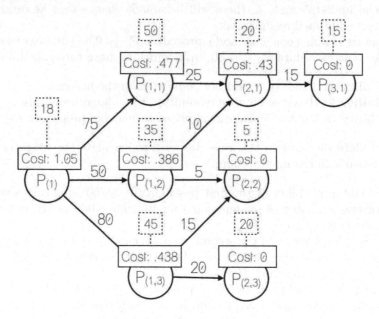

**Fig. 5.** Procedure graph with procedure scores

## 6.2   Calculating the Score for Clusters

By following the graph clustering approach, proposed in Sect. 5, we can identify the cluster(s) that the patient belongs to by only examining the set of diagnoses characterizing that patient. In this section, we will use the procedure graph metric system devised in the previous section to introduce a mapping between the scores of nodes in our procedure graph to the scores of clusters.

Since clusters contain patients that undergo the same procedure but the following procedures may differ for them, the score of a cluster is therefore defined as the sum of the weighted score of the procedures directly following that cluster. So, we define the score or cluster $C_x$ as:

$$\text{score}(C_x) = \sum_{k=1}^{m} \frac{|P_{(x,k)}|}{m} * \text{score}(P_k)$$

where $m$ denotes the total number of patients in Cluster $C_x$, and $|P_{(x,k)}|$ denotes the number of patients that underwent the $kth$ next procedure for Cluster $C_x$. Clearly, $C_x = \bigcup\{P_k : 1 \le k \le n\}$.

For example, according to Fig. 4 shown in Sect. 5, the score for each cluster is calculated as follows:

– score(Cluster 1) = $\frac{10}{60} * .477 + \frac{45}{60} * .386 + \frac{5}{60} * .438 = .406$

– score(Cluster 2) = $\frac{70}{75} * .477 + \frac{0}{75} * .386 + \frac{5}{75} * .438 = .474$

– score(Cluster 3) = $\frac{10}{90} * .477 + \frac{5}{90} * .386 + \frac{75}{90} * .438 = .439$

According to the scores above, the most desired cluster would be Cluster 1. By transitioning patients from Cluster 2 (least desired) to Cluster 1 (most desired), we would be able to reduce the number of following readmissions by almost 7 %; meaning that we would decrease the total number of following readmissions for 100 patients by 7.

# 7    Results of Cluster Transitions from Our Dataset

In this section, we provide some of the results obtained after extracting clusters of patients from the Florida State Inpatient Databases (SID), followed by calculating their scores according to the definitions presented above. After calculating the scores for the extracted clusters, any clinically feasible transition from a cluster with a higher score to another cluster with a lower score would essentially be considered a valid action rule (applicable set of actions) that will help reduce the average number of following readmissions.

Recall that as the number of elements (diagnostic codes) defined in our clusters increases, the number of actual clusters extracted will also increase, as shown in Table 3. Also, as the number of clusters increases, the number of transitions between clusters will therefore increase as well. The number of cluster transitions for $n$ clusters will be roughly $n(n-1)/2$, since the patients in the cluster with the highest score can transition to all other clusters ($n-1$ transitions), and the patients in the cluster with the second highest score can transition to all other clusters with lower score ($n-2$ transitions). Clearly, most of these transitions are not clinically valid; nevertheless, we believe that the number of clinically valid transitions is still substantial.

In this section, we examine few 3-element clusters extracted from procedure 44 (coronary artery bypass graft) using the range 10 % to 90 %. Table 4 shows three clusters extracted from patients who undertook procedure 44, which is the most common type of open-heart surgery, as their first procedure.

**Table 4.** A sample of three clusters with their scores and their included and excluded sets for procedure 44 (coronary artery bypass graft)

| Name | # of Patients in Cluster | Included Set | Excluded Set | Score |
|---|---|---|---|---|
| Cluster 1 | 250 | {} | {-53, -59, -238} | 0.058 |
| Cluster 2 | 258 | {59, 49} | {-257} | 0.073 |
| Cluster 3 | 257 | {238} | {-108, -663} | 0.078 |

Here is a description of the diagnostic codes shown in Table 4:

- 49: Diabetes mellitus without complication
- 53: Nutritional deficiencies
- 59: Deficiency in the red blood cells
- 108: Congestive heart failure; nonhypertensive
- 238: Complications of surgical procedures or medical care
- 663: Screening and history of mental health and substance abuse codes

According to all extracted clusters, Cluster 1 (which has score 0.058 as shown in Table 4) is the best cluster for procedure 44 (amongst all 13,140 extracted clusters); this means that if a new patient were to belong to Cluster 1, then there is no transition that could reduce the anticipated number of readmissions for that patient. However, if a new patient belongs to any other cluster, then there is at least one transition that would reduce the anticipated number of following readmissions.

By examining Table 4, we can infer that a transition from Cluster 2 to Cluster 1 will reduce the number of following readmissions on average by 1.5 %; this means that by treating the diagnostic code 59 (shifting it from the included set to the excluded set) and making sure that the patients do not have diagnostic codes 53 and 238 before performing procedure 44, we would decrease the number of following readmissions by 3.87 for the 258 patients in Cluster 2.

Similarly, by transitioning patients from Cluster 3 to Cluster 1 (by treating the diagnostic code 238 and making sure that the patients do not have diagnostic codes 53 and 59 before performing procedure 44), we would decrease the number of following readmissions by 5.14 for the 257 patients in Cluster 3.

## 8    Conclusion

Predicting all possible paths that a new patient may undertake during his or her stay at a hospital is of great help to physicians; since it allows them to choose the best treatment in order to achieve the desired outcomes. In this research, we examined the problem of predicting the following procedure by proposing two novel approaches to personalize patients by clustering them into subgroups that exhibit similar properties. To evaluate our personalizing approach we calculated the entropy and showed that by personalization, the accuracy of predicting following procedures indeed increases. We then introduced and devised a metric system that evaluated nodes (procedures) in the procedure graph, followed by a mapping that will transfer the scores to the extracted clusters by calculating the entropy to measure the predictability of the next procedure.

# References

1. Goodman, J.C.: Priceless: Curing the Healthcare Crisis. Independent Institute, Oakland (2012)
2. Keehan, S.P., et al.: National health expenditure projections, 2014–24: spending growth faster than recent trends. Health Aff. **34**(8), 1407–1417 (2015)
3. Pricewaterhouse Coopers: The Price of Excess: Identifying Waste in Healthcare Spending (2006)
4. Touati, H., Raś, Z.W., Studnicki, J., Wieczorkowska, A.A.: Mining surgical meta-actions effects with variable diagnoses' number. In: Andreasen, T., Christiansen, H., Cubero, J.-C., Raś, Z.W. (eds.) ISMIS 2014. LNCS, vol. 8502, pp. 254–263. Springer, Heidelberg (2014)
5. Lally, A., Bachi, S., Barborak, M.A., Buchanan, D.W., Chu-Carroll, J., Ferrucci, D.A., Glass, M.R., et al.: WatsonPaths: Scenario-based Question Answering and Inference over Unstructured Information. Technical Report Research Report RC25489. IBM Research (2014)
6. Tremblay, M.C., Berndt, D.J., Studnicki, J.: Feature selection for predicting surgical outcomes. In: Proceedings of the 39th Annual Hawaii International Conference System Sciences, HICSS 2006, vol. 5. IEEE (2006)
7. Silow-Carroll, S., Edwards, J.N., Lashbrook, A.: Reducing hospital readmissions: lessons from top-performing hospitals. Care Manage. **17**(5), 14 (2011)
8. Hajja, A., Touati, H., Raś, Z.W., Studnicki, J., Wieczorkowska, A.A.: Predicting negative side effects of surgeries through clustering. In: Appice, A., Ceci, M., Loglisci, C., Manco, G., Masciari, E., Ras, Z.W. (eds.) NFMCP 2014. LNCS, vol. 8983, pp. 41–55. Springer, Heidelberg (2015)
9. Raś, Z.W., Dardzińska, A.: Action rules discovery based on tree classifiers and meta-actions. In: Rauch, J., Raś, Z.W., Berka, P., Elomaa, T. (eds.) ISMIS 2009. LNCS, vol. 5722, pp. 66–75. Springer, Heidelberg (2009)
10. Touati, H., Raś, Z.W., Studnicki, J., Wieczorkowska, A.: Side effects analysis based on action sets for medical treatments. In: Proceedings of the Third ECML-PKDD Workshop on New Frontiers in Mining Complex Patterns, Nancy, France, pp. 172–183, 15–19 September 2014
11. Raś, Z.W., Wieczorkowska, A.A.: Action-rules: how to increase profit of a company. In: Zighed, D.A., Komorowski, J., Żytkow, J.M. (eds.) PKDD 2000. LNCS (LNAI), vol. 1910, pp. 587–592. Springer, Heidelberg (2000)
12. Raś, Z.W., Dardzińska, A., Tsay, L.S., Wasyluk, H.: Association action rules. In: IEEE/ICDM Workshop on Mining Complex Data (MCD 2008), Pisa, Italy. ICDM Workshops Proceedings, pp. 283–290. IEEE Computer Society (2008)
13. HCUP-US: Overview Of The State Inpatient Databases (SID). Web. 12 February 2016
14. HCUP-US: Clinical Classifications Software (CCS). Web. 12 February 2016

# Performing and Visualizing Temporal Analysis of Large Text Data Issued for Open Sources: Past and Future Methods

Jean-Charles Lamirel[1]([⊠]), Nicolas Dugué[1], and Pascal Cuxac[2]

[1] Equipe Synalp, Bâtiment B, 54506 Vandoeuvre-lès-Nancy, France
{lamirel,Nicolas.Dugue}@loria.fr
[2] CNRS-Inist, Vandoeuvre-lès-Nancy, France
Pascal.Cuxac@inist.fr

**Abstract.** In this paper we first propose a state of the art on the methods for the visualization and for the interpretation of textual data, and in particular of scientific data. We then shortly present our contributions to this field in the form of original methods for the automatic classification of documents and easy interpretation of their content through characteristic keywords and classes created by our algorithms. In a second step, we focus our analysis on the data evolving over time. We detail our diachronic approach, especially suitable for the detection and for visualization of topic changes. This allows us to conclude with Diachronic'Explorer, our upcoming visualization tool for visual exploration of evolutionary data.

**Keywords:** Visualization · Diachrony · Clustering · Feature selection · Open data

## 1 Introduction

Databases of scientific literature and patents provide volumes of significant data for the study of scientific production. These data are also very rich and so complex. Indeed, the textual content of the publications, keywords used for archiving in these databases, the citations they contain and affiliations of the authors are as much information that it is possible to exploit for studying corpora of publications.

These corpora are therefore a boon for the analysis of scientific and technical information. In this article, we focus in particular on one of the main objectives, which is to identify and describe major changes related to developments in the science in a textual and visual way. In the activity of researchers, monitoring the development of transversal themes as well as detection of emerging themes or bridges between themes allows them to ensure of the innovative character of their area of research.

Furthermore, in managing the financing of research by the European Commission (EC) the detection of emerging issues is fundamental, as shown in the following examples. The NEST (New and Emerging Science and Technology)

© Springer International Publishing Switzerland 2016
S. Kozielski et al. (Eds.): BDAS 2016, CCIS 613, pp. 56–76, 2016.
DOI: 10.1007/978-3-319-34099-9_4

program was a specific EC program in FP6. Its objective was to encourage a visionary and unconventional research at the frontiers of knowledge and at the interface of disciplines. To organize this program, the EC launched a call for "support actions" to follow and evaluate the projects but also to identify future research opportunities. Similarly, alongside the thematic in ICT (Information & Communication Technologies) program, the European Commission has set up, in the 7th Framework Program (FP7), the FET program (Future and emerging technologies) to promote research in the long term, or high risk, but with potentially strong impact from a societal or industrial point of view[1].

The detection of emerging technologies remains a complex process, and is therefore subject to studies in a broad spectrum of areas ranging from marketing to bibliometrics.

The selection tree proposed by [1] gives a good image of all forecasting methods that can be applied, in particular for the detection of these emerging technologies. It illustrates very well the dichotomy between quantitative methods and those based on expertise and shows the great diversity of existing approaches: Delphi and Nominal Group technique, methods based on the confrontation of the opinion of experts, scenario methods designed to scan different possible futures [20], until the methods combining the knowledge of the experts of a field and statistical techniques, allowing the identification of trends affecting causal factors.

The size and the complexity of the data that can be exploited to study the resulting databases of scientific publications require the development of quantitative methods for the detection of emerging topics by bibliometric methods, applying relatively simple statistical techniques as growth curves, or more sophisticated ones, such as automatic classification and analysis of networks [4,13,19,21]. Another concern is also to provide tools capable of producing outputs exploitable by the end user. These outputs should be descriptive, intelligible and viewable.

Therefore, we separate our analysis into two parts. In the first part, we describe the quantitative and automatic methods that allow the extraction of relevant information from a corpus. In particular, these techniques offer to detect characteristic keywords from documents, or underlying topics - and keywords that describe them - referred in the documents. We also discuss of the visual exploitation which may be made of these methods. In a second part, we detail the methods for studying topic changes within a corpus whose data are anchored in time. We insist in particular on diachronic analysis methods, particularly effective to track these changes in a step by step and in a synthetic way.

Finally, in a last part, we detail Diachronic'Explorer, our open source tool for the production and viewing of diachronic analysis results. We show the effectiveness of the extraction methods that we offer through complementary and dynamic visualizations using recent technologies.

## 2    Exploitation of Textual Data and Visualization

Topic identification is a technique which consists in understanding the meaning of the content of the documents of a corpus in an unsupervised way (without prior

---

[1] See URL: http://cordis.europa.eu/fp7/ict/programme/fet_en.html.

knowledge on the corpus and without human intervention) by isolating the topics underlying this content. These topics are usually represented by coordinated phrases and are often ranked in order of importance in the documents. As shown in [5], many techniques can be applied for topic identification and they might exploit research issued from several different communities, such as data mining, computational linguistics and information retrieval. We present hereafter two different types of identification techniques and their related visualization tools: the first is a widely used state of the art technique, the second is an alternative technique that we propose.

## 2.1    LDA

**The Method.** The LDA method is a probabilistic method for topic extraction who considers that the underlying topics of a corpus of documents can be characterized by multinomial distributions of words present in the documents [2]. According to this principle, each document is then considered as a composition of the topics extracted of the studied corpus. Figure 1 presents a list of topics produced by LDA, and their manifestation in a document. LDA uses a Dirichlet law to allow a careful choice of the parameters of the multinomial distributions. In practice, the extraction of these parameters is however complex and costly regarding computation time. It requires to exploit expectation maximization algorithms [7], which are prone to produce sub-optimal solutions, and in particular trivial or general results that are not usable in many cases, as in the context of the diachronic analysis (Sect. 3). This last type of analysis, which aims at comparing topics evolving over time, indeed requires working with accurate topic descriptions to isolate changes. Finally, the importance of the words in the documents can only be estimated in an indirect way by LDA and the method does not work on isolated documents. We present later a method based on feature maximization metric we have developed that does not have these drawbacks.

**Visualization Using LDA.** In LDAExplore [9], the authors use a Treemap (Fig. 2) to represent the distribution of the importance of keywords in topics learned by the LDA. The representation of the topics in each document is also displayed, but in the form of curves where each point x-coordinate is a topic, and its weight is on the ordinate. Guille and Morales offer as a complete library for topic modelling, including LDA, which can also produce visualizations in the form of word clouds or histograms [11].

## 2.2    Feature Maximization for Feature Selection

**The Method.** To introduce the feature maximization metric and process [17], we first use an example. We then explain its use in our application framework. In Fig. 3, we present sample data collected from a panel of $Men$ ($M$) and $Women$ ($W$) described by three features ($Nose\_Size$ ($N$), $Hair\_Length$ ($C$), $Shoes\_Size$ ($S$)). The problem of supervised classification in computer science is to learn to discriminate the class of $Men$ of the class of $Women$ automatically by using

| "Arts" | "Budget" | "Children" | "Education" |
|--------|----------|------------|-------------|
| NEW | MILLION | CHILDREN | SCHOOL |
| FILM | TAX | WOMEN | STUDENTS |
| SHOW | PROGRAM | PEOPLE | SCHOOLS |
| MUSIC | BUDGET | CHILD | EDUCATION |
| MOVIE | BILLION | YEARS | TEACHERS |
| PLAY | FEDERAL | FAMILIES | HIGH |
| MUSICAL | YEAR | WORK | PUBLIC |
| BEST | SPENDING | PARENTS | TEACHER |
| ACTOR | NEW | SAYS | BENNETT |
| FIRST | STATE | FAMILY | MANIGAT |
| YORK | PLAN | WELFARE | NAMPHY |
| OPERA | MONEY | MEN | STATE |
| THEATER | PROGRAMS | PERCENT | PRESIDENT |
| ACTRESS | GOVERNMENT | CARE | ELEMENTARY |
| LOVE | CONGRESS | LIFE | HAITI |

The William Randolph Hearst Foundation will give $1.25 million to Lincoln Center, Metropolitan Opera Co., New York Philharmonic and Juilliard School. "Our board felt that we had a real opportunity to make a mark on the future of the performing arts with these grants an act every bit as important as our traditional areas of support in health, medical research, education and the social services." Hearst Foundation President Randolph A. Hearst said Monday in announcing the grants. Lincoln Center's share will be $200,000 for its new building, which will house young artists and provide new public facilities. The Metropolitan Opera Co. and New York Philharmonic will receive $400,000 each. The Juilliard School, where music and the performing arts are taught, will get $250,000. The Hearst Foundation, a leading supporter of the Lincoln Center Consolidated Corporate Fund, will make its usual annual $100,000 donation, too.

**Fig. 1.** Results of LDA from [2].

these features. To achieve this, it is worthwhile for the algorithms to exploit the features that best separate the *Men* from the *Women*.

The process of feature maximization is comparable to a feature selection process. This process is based on the feature F-measure. The feature F-measure $FF_c(f)$ of a feature $f$ associated with a cluster $c$ ($M$ or $W$ in our example) is defined as the harmonic mean of the feature recall $FR_c(f)$ and of the feature predominance $FP_c(f)$, themselves defined as follows:

$$FR_c(f) = \frac{\Sigma_{d\in c'}W_d^f}{\Sigma_{c'\in C}\Sigma_{d\in c'}W_d^f} \quad FP_c(f) = \frac{\Sigma_{d\in c}W_d^f}{\Sigma_{f'\in F_c, d\in c}W_d^{f'}} \tag{1}$$

with

$$FF_c(f) = 2\left(\frac{FR_c(f) \times FP_c(f)}{FR_c(f) + FP_c(f)}\right) \tag{2}$$

where $W_d^f$ represents the weight[2] of the feature $f$ for the data $d$ and $F_c$ represents all the features present in the dataset associated with the class $c$.

The feature selection process based on feature maximization can thus be defined as a parameter-free process in which a class feature is characterized

---

[2] The choice of the weighting scheme is not really constrained by the approach instead of producing positive values. Such scheme is supposed to figure out the significance (i.e. semantic and importance) of the feature for the data.

Feature Recall is a scale independent measure but feature Predominance is not. We have however shown experimentally in (Lamirel et al., 2014a) that the F-measure which is a combination of these two measures is only weakly influenced by feature scaling. Nevertheless, to guaranty full scale independent behavior for this measure, data must be standardized.

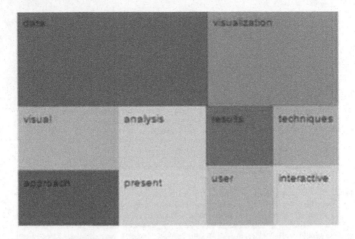

**Fig. 2.** Treemap based on LDA from [9].

by using both his ability to discriminate the class to which it relates ($FR_c(f)$ index) and its ability to faithfully represent the data of this class ($FP_c(f)$ index). Figure 4 presents how does operate the calculation of the feature F-measure of the *Shoes_Size* feature to the *Men* class.

Once the capacity to discriminate a class ($FR_c(f)$ index) and to faithfully represent the data of a class ($FP_c(f)$ index) calculated for each feature (Fig. 5), the further step consists in automatically selecting the most relevant features for distinguishing classes. The set $S_c$ of features that are characteristic of a given class $c$ belonging to the group of classes $C$ is represented as:

$$S_c = \{f \in F_c \mid FF_c(f) > \overline{FF}(f) \text{ and } FF_c(f) > \overline{FF}_D\} \text{ where} \tag{3}$$

$$\overline{FF}(f) = \Sigma_{c' \in C} \frac{FF_{c'}(f)}{|C_{/f}|} \text{ and } \overline{FF}_D = \Sigma_{f \in F} \frac{\overline{FF}(f)}{|F|} \tag{4}$$

where $C_{/f}$ represents the subset of $C$ in which the $f$ feature is represented.

Finally, the set of all selected features $S_C$ is the subset of $F$ defined by:

$$S_C = \cup_{c \in C} S_C. \tag{5}$$

The features that are considered relevant for a given class are the features whose representations are better in that class than their average representation in all classes, and also better than the average representation of all features, as regards to feature F-measure. Thus, features whose feature F-measure is always lower than the overall average are eliminated, and the variable *Nose_Size* is therefore suppressed in our example ($0.3 < 0.38$ and $0.24 < 0.38$).

A complementary step to estimate the contrast may be operated in addition to the first stage of selection. The role of this step is to estimate the information gain produced by a feature on a class. It is proportional to the ratio between

| Shoes Size | Hair Length | Nose Size | Class |
|:---:|:---:|:---:|:---:|
| 9 | 5 | 5 | M |
| 9 | 10 | 5 | M |
| 9 | 20 | 6 | M |
| 5 | 15 | 5 | W |
| 6 | 25 | 6 | W |
| 5 | 25 | 5 | W |

**Fig. 3.** Sample data for supervised classification in *Men/Women* classes.

| Shoes Size | Hair Length | Nose Size | Class |
|:---:|:---:|:---:|:---:|
| 9 | 5 | 5 | M |
| 9 | 10 | 5 | M |
| 9 | 20 | 6 | M |
| 5 | 15 | 5 | W |
| 6 | 25 | 6 | W |
| 5 | 25 | 5 | W |

$$FR(S,M) = 27/43 = 0.62$$

$$FP(S,M) = 27/78 = 0.35$$

$$FF(S,M) = \frac{2(FR(S,M) \times FP(S,M))}{FR(S,M) + FP(S,M)}$$
$$= 0.48$$

**Fig. 4.** Sample data and computation of feature F-measure for *the Shoes_Size* feature.

the value of the F-measure of a feature in the class and the average value of the F-measure of that feature in all classes. For a feature $f$ belonging to the group of selected features $S_c$ from a class $c$, the gain $G_c(f)$ is expressed as:

$$G_c(f) = FF_c(f)/\overline{FF}(f) \tag{6}$$

Finally, active, or descriptive, features of a class are those for which the contrast is greater than 1 in those above. Thus, the selected features are considered active in the classes in which the feature F-measure is higher than the marginal average:

- *Shoes_Size* is active in the *Men* class (0.48 > 0.35),
- *Hair_Length* is active in the *Women* class (0.66 > 0.53).

Contrast ratio highlights the degree of activity/passivity of the features selected compared to their average F-measure. Figure 6 shows how the contrast

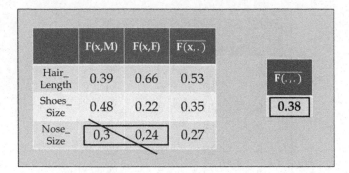

**Fig. 5.** Feature F-measure of the feature and related marginal average.

is calculated on the presented example. In this context, the contrast may thus be considered as a function that will virtually have the following effects:

– Increase the *length* of *women'shair*,
– Increase the *size* of the *men'sshoes*,
– Reduce the *length* of the *men'shair*,
– Decrease the *size* of *women'sshoes*.

Preliminary cluster labelling experiments showed that feature maximization metric has discrimination capabilities similar to Chi-squared metric, while with generalization capabilities very appreciably higher [18]. Moreover, this technique proved to have very low computation time, unlike LDA. It often has a dual function in learning and visualization, as shown by experiments in [6] or in [17]. In the classification context, it can thus optimize performance of the classifiers, while producing class profiles exploitable for the visualization of the content of the classes.

Figure 7 shows an application of the method on textual data with the goal of establishing discriminative profiles of Chirac and Mitterrand presidents using data of the DEFT 2006 corpus containing around 80000 extracts of their speeches [15]. It shows in particular that contrast allows to quantify the influence of features in classes (typical terms of each speaker). Extracted features and their contrast can thus act as profiles of classes for a classifier, as well as indicators of content or meaning for an analyst.

This first example shows how to use feature maximization in the general framework of text datasets. We show more specifically in Sect. 3 how our feature maximization method can be applied for the more precise and difficult task of topics changes detection in corpora of scientific papers and describe in Sect. 4 advanced visualizations of that changes. But, to clearly illustrate the potential and the scope of the method, we firstly give hereafter a more specific example, related to synthetic visualization and automatic summary of the individual content of the documents from this method.

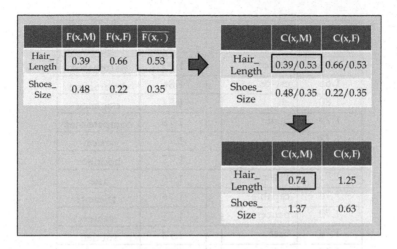

**Fig. 6.** Principle of computation of contrast on selected features and obtained results.

**Visualization Using Feature Maximization.** For the synthetic representation of the content in a single document, we propose an original method based on competition between blocks of text, coupled with the feature maximization metric. This approach allows to overcome the lack of metadata to describe the texts. Furthermore, it has the advantage of being independent of the language, to function without external knowledge source and parameters, and is likely to have multiple applications: generation of metadata or input data for the clustering, generation of automatic summaries and explanations of several levels of generality. It consists here in performing indexing of full text papers taking benefit of their structure. As regards to this approach, each part of the paper can be seen as a class, the paper itself being a classification: for example, the exploited classes might be:"introduction, methodology, state of the art, results, conclusion... ".

We will illustrate our point with this scientific paper (randomly selected in the ISTEX[3] reservoir): Hauk P, Friedl K, Kaufmehl K, Urbanek R, Forster J.: Subsequent insect stings in children with hypersensitivity to Hymenoptera. J Pediatr. 1995 Feb; 126 (2): 185–90.

This paper includes the following major parts: Introduction / Methods / Results / Discussion (Fig. 8).

---

[3] The ISTEX project (Initiative d'Excellence pour l'Information Scientifique et Technique) fits in the "Investment for the future" program, initiated by the French Ministry of Higher Education and Research (MESR), whose ambition is to strengthen research and French higher education on the world level. The ISTEX project main objective is to offer to the whole of the community of higher education and research, online access to the retrospective collections of scientific literature in all disciplines by engaging a national policy of massive acquisition of documentation: archives of journals, databases, corpus of texts.
Reference: http://www.istex.fr.

| Mitterrand | | Chirac | |
|:---:|:---:|:---:|:---:|
| Contraste | Terme | Contraste | Terme |
| 1.88 | douze | 1.93 | partenariat |
| 1.85 | est-ce | 1.86 | dynamisme |
| 1.80 | eh | 1.81 | exigence |
| 1.79 | quoi | 1.78 | compatriotes |
| 1.78 | - | 1.77 | vision |
| 1.76 | gens | 1.77 | honneur |
| 1.75 | assez | 1.76 | asie |
| 1.74 | capables | 1.76 | efficacité |
| 1.72 | penser | 1.75 | saluer |
| 1.70 | bref | 1.74 | soutien |
| 1.69 | puisque | 1.74 | renforcer |
| 1.67 | on | 1.72 | concitoyens |
| 1.66 | étais | 1.71 | réforme |
| 1.62 | parle | 1.70 | devons |
| 1.62 | fallait | 1.70 | engagement |
| 1.60 | simplement | 1.69 | estime |

**Fig. 7.** Most contrasted features (terms) in *Mitterrand* and *Chirac* speeches (extract).

After extraction of the terms by a conventional PoS tagging method, the feature maximization method described Sect. 2.2 allows to obtain a list of specific terms for each part of the paper, weighted by their importance. From that, it is possible to build up a vectorial (i.e. Bag-of-Words) representation of the paper, or alternatively, to build up a weighted graph (*paper parts/selected terms*) that will illustrate clearly the scientific contents of the text (Fig. 9).

If we follow an approach of automatic summarization [10], each selected term being weighted for each identified part of the paper, it is easy to balance the sentences containing these terms by adding their weight. We furthermore assign an additional weight to terms that are also part of the title of the paper. The curve of the weights of the sentences thus calculated for each of them always shows a plateau (Fig. 10); then, we choose to keep the sentences whose weight is greater or equal to this level and reorder them by rank of appearance in the text.

We then have a summary obtained by extraction of meaningful sentences of text that has generally small size (less than 12 sentences in all our experiments). For the paper used as an example, the summary is described on Fig. 11.

## Subsequent insect stings in children with hypersensitivity to Hymenoptera

Pia Hauk, MD, Katrin Friedl, Klaus Kaufmehl, MD, Radvan Urbanek, MD, and Johannes Forster, MD

From University Children's Hospitals, Freiburg, Germany, and Vienna, Austria

To investigate the risk of life-threatening reactions to future stings, we sequentially challenged 113 children (aged 2 to 17 years) allergic to insect stings with a sting by the relevant insect. The time interval between the challenges varied from 2 to 6 weeks. The history of the index stings was a large local reaction (LR) in 16% and a systemic reaction (SR) in 84% of the test subjects. On the first challenge, 76% had a normal LR, 11% a large LR, and 43% an SR. On the second challenge, 78% of the children had a normal LR, 5% a large LR, and 17% an SR. Thirty-nine of the untreated children were exposed to a field sting during the subsequent 3-year follow-up period. In comparison with other diagnostic evaluations such as skin-prick tests, determinations of specific IgE and IgG antibodies, and single-sting exposure, the dual sting challenge scheme appears to be the best predictor of reactions to subsequent stings. It also appears to be helpful in selecting patients with an uncertain sensitization status for venom immunotherapy. (J PEDIATR 1995;126:185-90)

In childhood, allergy to Hymenoptera venom is mainly caused by stings of honeybees and wasps. In Europe, yellow jackets are known as "wasps," whereas in the United States, Polistes wasps are known as "wasps."[1] Between 0.4% and 4% of the population have systemic allergic reactions to insect stings.[2-4] The incidence of systemic reactions to subsequent stings is lower in children and adolescents than in adults.[3-8] Prospective observations of the natural course of insect allergy show that adults have a risk of 27% to 57%[3, 9-11] of having repeated systemic allergic reactions, in comparison with a risk of 10% to 20% in children.[4-6, 8] Therefore venom immunotherapy should be indicated less frequently in children.[8] In vitro assays and risk scores provide only limited help in identifying those patients at risk of having further life-threatening allergic reactions. Numerous studies[12-15] have been unsuccessful in showing a correlation between the standard diagnostic methods—mainly skin-prick tests and measurements of specific IgE and IgG

Submitted for publication April 15, 1994; accepted Aug. 10, 1994.
Reprint requests: Johannes Forster, MD, University Children's Hospital, Mathildenstr. 1, D-79106 Freiburg, Germany.
Copyright © 1995 by Mosby–Year Book, Inc.
0022-3476/95/$3.00 + 0   9/20/59779

antibodies—and the reactions to subsequent insect stings. Treatment recommendations based only on those criteria typically lead to an overestimation of the number of children who require venom immunotherapy.[6, 8, 16]

Although single diagnostic sting challenges give additional information, there is increasing concern about the possible booster effect. From the natural history of bee venom allergy, we know that one sting followed by another

See commentary, p. 257.

| AU | Arbitrary unit(s) |
| LR | Local reaction |
| SR | Systemic reaction |

2 to 4 weeks later will result in the highest incidence of systemic reactions. We tried to mimic this naturally occurring event by subjecting test subjects to sequential sting challenges to detect the group of patients at highest risk. Those who did not react and therefore were not assigned to receive venom immunotherapy were followed for up to 3 years for life-threatening events after natural stings.

**Fig. 8.** First page of the selected paper.

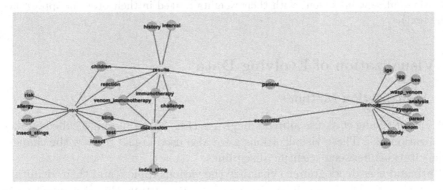

**Fig. 9.** Graphical representation of the content of a scientific paper with the use of its structure.

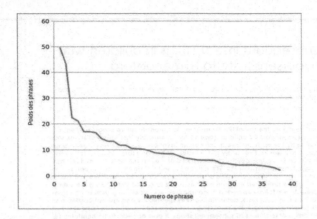

**Fig. 10.** Distribution of the weights of the sentences.

Although there were more severe reactions in the group of children who required immunotherapy according to our assessment, no significant correlation could be detected between the reactions to the index sting and to the challenge stings, or between the reactions to the index sting and to the field sting. 98.2682

Considering the previous reaction to the index sting and the results of skin-prick tests and venom-specific IgE measurements as criteria for the recommendation of venom immunotherapy, 41% of the scored bee venom- and wasp venom-allergic children would have been assigned to this treatment, but only 9% received venom immunotherapy as a result of the clinical reaction to the second challenge.    98.6776

Although this is not a 100% safety record, we believe that the sequential insect sting challenge performed in the hospital represents the safest and most informative method of eliminating unnecessary venom immunotherapy in children having mild to moderate SRs to an index sting.    137.1604

On the basis of the data presented, we suggest the following diagnostic and therapeutic procedures for children up to 16 years of age: Sensitized patients, identified by a positive skin-prick test result or specific IgE finding, who had only a large LR to the index sting, need neither a challenge sting nor venom immunotherapy. 104.307

**Fig. 11.** Automatic summary of the paper of Fig. 8 produced by sentence extraction (most relevant sentences along with their weights ranked in their order of appearance in the text).

## 3    Visualization of Evolving Data

### 3.1    Visualization Methods

In Neviewer, Wang et al. use alluvial diagrams (Fig. 12), sometimes called Sankey visualization [28]. These visualizations were also used in [27] to view the changes in the citations between scientific disciplines.

Ratinaud uses dendogram to visualize the various topics (and their vicinities) mentioned on Twitter with the hashtag #mariagepourtous [26]. It is nonetheless Treemaps (Sect. 2, Fig. 2) that enable him to show the progression or regression of topics in time. On their own side, Osborne and Motta use graphics with stacked areas to follow the evolution of the amount of publications grouped into topics across time [22].

These methods have all of the interesting benefits and we show their complementary exploitation use in Sect. 4.

## 3.2   Diachronic Analysis

Diachronic analysis, which consists in comparing data or results by time step, is extensively used. In linguistics, Perea uses this technique to follow the evolution of the Catalan language through time [24]. Cardon et al. study the evolution of blogs and their importance on the web in a diachronic way [3]. Similarly, the activity of bloggers and the evolution of their interests are studied in a diachronic way by [12]. The work of Wang et al., more connected to our field of applications, analyzes the thematic evolution of the research in such a way [28].

In our case, we develop a parameter-free method, directly exploitable by the user and based on feature maximization [17] to identify and describe the topics of a corpus. This approach allows identification of keywords issued from the full text content of documents which are characteristic of each topics, conversely to methods based on keywords indexed by the publication databases [23,28]. Furthermore, the absence of parameters in the process allows to completely automate the task of indexing the documents passing through the detection of topics and up to the visualization of their evolution. Figure 13 shows the progress of the complete method up to visualization. The whole process is also detailed hereafter:

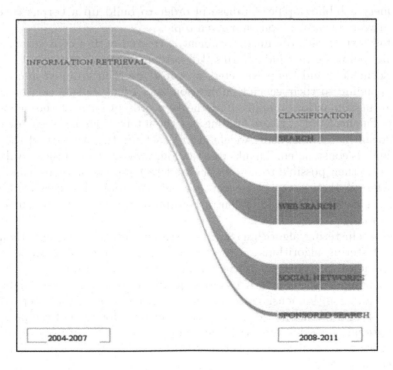

**Fig. 12.** Alluvial diagram from Neviewer [28].

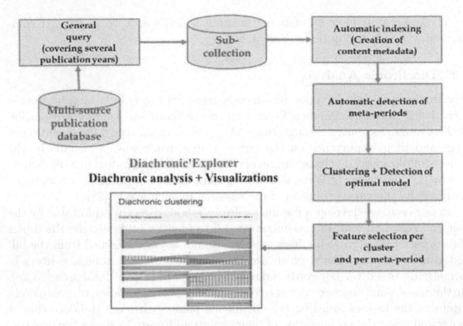

**Fig. 13.** Detailed diachronic analysis process (incl. Diachronic Explorer tool).

1. We query a bibliographic database in order to build up a corpus covering several years of publication on a given topic.
2. The full text of each obtained document is treated with a conventional PoS tagging tool to extract index terms (keywords).
3. The documents and their extracted keywords are then grouped into classes corresponding to their year of publication and a feature selection based on feature maximization is applied on the keywords of each of the document group. Furthermore, a graph figuring out the interactions between keywords and document groups, using weighted links setting the strength of the relationships, is constructed. Thanks to a random walk algorithm (here Walktrap [25]), it is then possible to automatically detect groups of years (time periods) who will then serve as time steps for the diachronic clustering algorithm. Figure 14 gives an example of contrast graph as well as resulting cutting in time periods.
4. A neural clustering algorithm [8], more stable and more efficient than the usual clustering algorithms on the textual data, is applied multiple times, with standard parameters, on the data of each time period by varying the desired number of clusters. Clustering quality criteria that are reliable for textual and multidimensional data are exploited in a further step [16] to isolate an optimal model (ideal number of clusters) for each of the periods.
5. The optimal clustering models of each period are post processed separately using feature maximization so as to extract the salient features of each cluster in each time period.

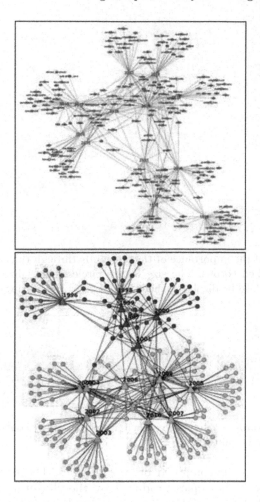

**Fig. 14.** Contrast graph per year (up) and cutting into time periods (down).

6. The feature maximization results, as well as the overall clustering results, are transmitted to the diachronic module of the Diachronic'Explorer tool presented in Sect. 4. This tool implements both diachronic analysis functions, based on unsupervised Bayesian reasoning [14], in order to detect thematic connections and differences between the time periods, as well as many functions of visualization of the results.

## 4    Diachronic'Explorer Tool

We now introduce Diachronic'Explorer[4], our open source tool for the production and visual exploitation of diachronic analysis results.

---

[4] A demo version of the tool can be found at URL: http://github.com/nicolasdugue/istex-demonstrateur.

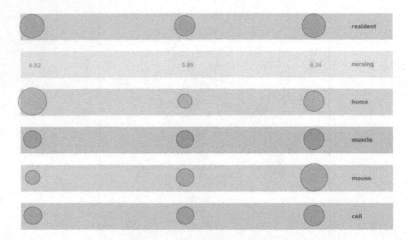

**Fig. 15.** Evolution of the importance of the keywords through the curse of the 3 time periods for the studied corpus. The size of the circles materialize this importance. Contrast values can also be displayed (here for the keyword nursing).

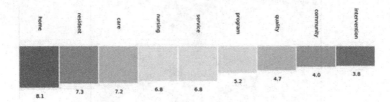

**Fig. 16.** List of keywords associated to a cluster and ranked by contrast values. Colors materialize the different values.

Diachronic'Explorer is composed of two modules. The first module allows to use descriptions of the topics produced by the clustering and the feature maximization (Sect. 3.2) to track these topics over time and detect the changes and similarities between time periods. The second module is dedicated to the visualization of the results produced by the first. Designed in the form of web platform using modern technology, the tool offers the possibility to explore the corpus through various complementary visualizations each representing a different level of granularity in the exploration of this corpus.

We will detail below how the tool can be used from the finest-grained level to more synthetic visualization of the corpus and its evolutions. For that purpose, we will take, as example corpus, the evaluation corpus operated by the French ISTEX project. This corpus comprises 9779 records related to research conducted in the general field of Gerontology/Geriatrics between 1996 and 2010. The result of the analysis steps described in Sect. 3 on this corpus is a division into three time periods.

First of all, the tool allows to study keywords that are particularly characteristics of the topics and to see the evolution of their importance in time. Figure 15 shows for example the main keywords and their evolution in the above-mentioned

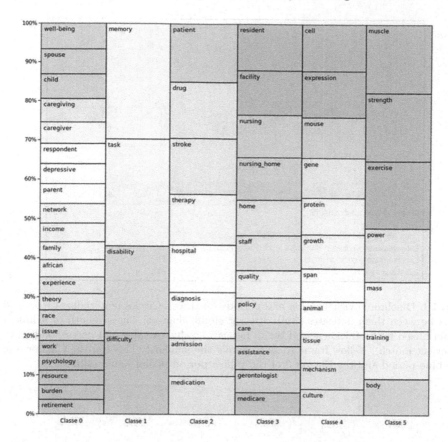

**Fig. 17.** Topics of a period described by their associated keywords. Each topic is a column and the height of a cell materializes the importance of the keyword related to this cell. The color represents the intensity of its contrast.

corpus. The size of a circle is proportional to the importance of the related keyword in its topic. This size is therefore conditioned by the contrast value produced by the feature maximization process (Sect. 3.2). Figure 16 shows the description of a topic (cluster) through keywords that are representative. The size of the rectangles is also proportional to the value of contrast.

Taking some distance from the corpus, with the Fig. 17, the topics of a period can be observed in a global way. In this figure, each column represents a topic, and the cells in the column are the keywords that describe this topic. The size of these cells is conditioned by the relative importance of the related keywords, in terms of contrast, in the topic. The colors represent the intensity of the contrast of the considered keywords.

Figure 18 allows take some more distance, and to consider interactions between couples of periods. This visualization provides a detailed representation of the diachronic analysis between two periods. The blue rectangle represents the period prior to that represented by the yellow rectangle. The circles

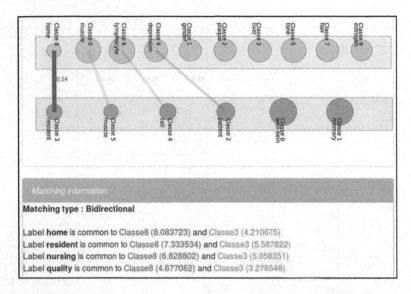

**Fig. 18.** Diachronic visualization of 2 periods (extract). Circles materialize topics and links between them indicates similarities or slight changes. The strength of a link is materialized by its thickness and by an indicator value (a value of 1 corresponding to a perfect match). Yellow frame (down) details the similarity between the Cluster 8 of the blue period and the Cluster 3 of the yellow period (Color figure online).

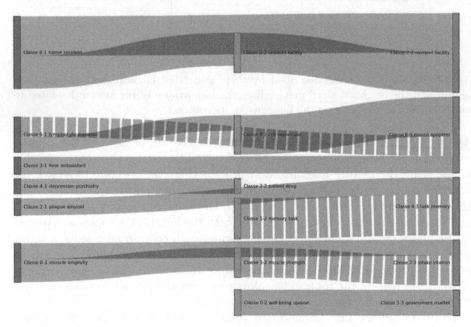

**Fig. 19.** Diachronic visualization of all the periods of a corpus. Each color represents one period and vertical rectangles materialize topics. In grey color, one can observe the links between the topics of the different periods (Color figure online).

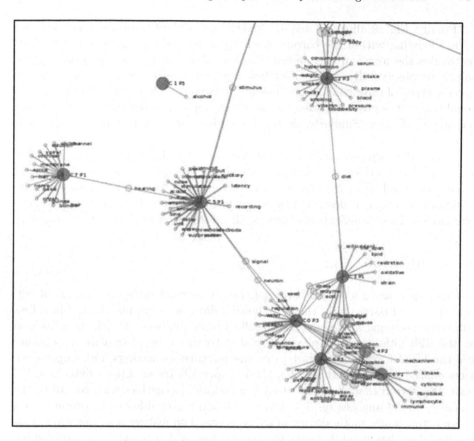

**Fig. 20.** Diachronic visualization of all the periods of a corpus in the form of a contrast grap (extract). Each color represents one specific period and big circle materialize the topics. In orange color one can observe the keywords related to the topics (Color figure online).

represent topics whose size is determined by the quantity of documents they contain. The label of each circle is currently the most characteristic keyword (i.e. the more contrasted one) of the period (for more details, it is possible to select several labels). In addition, the links between circles represent inter-period topic links. They are characterized by a force, which depends on both the number of keywords shared between the topics of the two periods and the contrast of those shared keywords[5]. This force provides the thickness of the link. The yellow rectangle below details similarities for each link between two topics. The dissimilarities between topics can be displayed as well, but there are not shown in the figure.

---

[5] The principle of computation of the strength of a topic link between periods is explained with more details in [14].

Finally, Fig. 19 allows us acquire knowledge of all of the topic links between period existing within the corpus. Each color represents a period and vertical rectangles the topics of this period who have links with other periods. In grey color, we observe the inter-period link between topics. Dotted links indicate specific types of topic connections: one of the two topics has a broader descriptive vocabulary. It is possible to see the details of topics passing the mouse over the rectangles of color. Similarly, details on topics links are available on the grey areas.

To see the corpus content in its entirety, we offer also in Diachronic'Explorer an original method of visualization that shows the information in the form of a contrast graph (Fig. 20). The big circles represent topics, the small ones, the keywords. If a topic is described by a keyword then a link is present between the two circles. This visualization shows so all information in a condensed manner.

# 5   Conclusion

In this paper, using a detailed state of the art on work done for analysis of textual data, and particularly that of scientific data, we have highlighted in a first time, the strategic importance of the diachronic analysis of such data, as well as the difficulties and complexities related to this type of treatment, whether it's the lack of available metadata or the parameters settings and scope problems related to usual methods of analysis, especially those of topic detection. We have also shown that there are many interesting alternatives with regard to the visualization of analysis results. This discussion has enabled us to propose, as a second step, a new methodology of analysis based on feature maximization. This methodology has many benefits to the existing, which be without parameters, to be applicable at different scale levels, from the corpus to the document, with ease of calculation, and finally, to have strong capabilities of synthesis, which makes her results easily interpretable, even for complex problems and large corpora. All these decisive advantages have in particular helped us to create the Diachronic Explorer tool, which, by integrating the unsupervised Bayesian reasoning and feature maximization with many methods of visualization, provides effective solutions to deal with a problem as complex as that of the detection of changes, from the full text, within a large corpus of scientific publications whose topics evolve with time. This indirectly shows that this type of methodology, due to its great flexibility, has a field of application of much wider range than that presented in this paper. Its synthesis capabilities make it indispensable, especially upstream visualization methods, when the representation of the data itself is complex.

# References

1. Amstrong, J., Green, K., Graefe, A.: Forescating principles. In: Encyclopedia of Statistical Sciences (2011)

2. Blei, D., Ng, A.Y., Jordan, M.: Latent dirichlet allocation. J. Mach. Learn. Res. **3**, 993–1022 (2003)
3. Cardon, D., Fouetillou, G., Roth, C.: Two paths of glory-structural positions and trajectories of websites within their topical territory. In: Fifth International AAAI Conference on Weblogs and Social Media - ICWSM 2011 (2011)
4. Daim, T., Rueda, G., Martin, H., Gerdsri, P.: Forecasting emerging technologies: Use of bibliometrics and patent analysis. Technol. Forecast. Soc. Chang. **73**, 981–1012 (2006)
5. Dermouche, M., Velcin, J., Loudcher, S., Khouas, L.: Une nouvelle mesure pour lévaluation des méthodes dextraction de thématiques : la vraisemblance généralisée. In: Actes des 13ièmes Journées Francophones dExtraction et de Gestion des Connaissances (EGC 2013), Toulouse, France, pp. 317–328 (2013)
6. Falk, I., Lamirel, J.C., Gardent, C.: Classifying french verbs using french and english lexical resources. In: International Conference on Computational Linguistic (ACL 2012), Jeju Island, Korea (2012)
7. Francesiaz, T., Graille, R., Metahri, B.: Introduction aux modèles probabilistes utilisés en fouille de données (2015)
8. Fritzke, B.: A growing neural gas network learns topologies. In: Advances in Neural Information Processing Systems, vol. 7, pp. 625–632 (1995)
9. Ganesan, A., Brantley, K., Pan, S., Chen, J.: Ldaexplore: Visualizing topic models generated using latent dirichlet allocation. arXiv preprint arxiv:1507.06593 (2015)
10. Goldstein, J., Mittal, V., Carbonell, J., Kantrowitz, M.: Multi-document summarization by sentence extraction. In: Workshop on Automatic summarization, NAACL-ANLP, pp. 40–48 (2000)
11. Guille, A., Soriano-Morales, E.P.: Tom: A library for topic modeling and browsing. In: Actes des 16ièmes Journées Francophones dExtraction et de Gestion des Connaissances (EGC 2016), Reims, France, pp. 451–456 (2016)
12. Itoh, M., Yoshinaga, N., Toyoda, M., Kitsuregawa, M.: Analysis and visualization of temporal changes in bloggers' activities and interests. In: 2012 IEEE Pacific Visualization Symposium (PacificVis), pp. 57–64 (2012)
13. Kajikawa, Y., Yoshikawa, J., Takeda, Y., Matushima, K.: Tracking emerging technologies in energy research: Toward a roadmap for sustainable energy. Technol. Forecast. Soc. Chang. **75**, 771–782 (2008)
14. Lamirel, J.C.: A new diachronic methodology for automatizing the analysis of research topics dynamics: an example of application on optoelectronics research. Scientometrics **93**(1), 151–166 (2012)
15. Lamirel, J.C., Cuxac, P.: Une nouvelle méthode statistique pour la classification robuste des données textuelles : le cas mitterand-chirac. In: JADT, Paris, France (2014)
16. Lamirel, J.C., Cuxac, P.: New quality indexes for optimal clustering model identification with high dimensional data. In: Proceedings of ICDM-HDM15 - International Workshop on High Dimensional Data Mining, Atlantic City, USA (2015)
17. Lamirel, J.C., Cuxac, P., Chivukula, A., Hajlaoui, K.: Optimizing text classification through efficient feature selection based on quality metric. J. Intell. Inf. Syst. **2013**, 1–18 (2014). Special issue on PAKDD-QIMIE 2013
18. Lamirel, J.C., Ta, A., Attik, M.: Novel labeling strategies for hierarchical representation of multidimensional data analysis results. In: Proceedings of IASTED International Conference on Artificial Intelligence and Applications (AIA), Innsbruck, Austria (2008)

19. Mogoutov, A., Kahane, N.: Data search strategy for science and technology emergence: A scalable and evolutionary query for nanotechnology tracking. Res. Policy **36**, 893–903 (2007)

20. Monti, R., Roubelat, F.: La boîte outils de prospective stratégique et la prospective de dfense: rétrospective et perspectives. In: Actes des Entretiens Science & Défense, Paris (1998)

21. Noyons, E.: Science maps within a science policy context. In: Moed, H.F., Glänzel, W., Schmoch, U. (eds.) Handbook of Quantitative Science and Technology Research, pp. 237–255. Springer, Heidelberg (2004)

22. Osborne, F., Motta, E.: Understanding research dynamics. In: Presutti, V., Stankovic, M., Cambria, E., Cantador, I., Di Iorio, A., Di Noia, T., Lange, C., Reforgiato Recupero, D., Tordai, A. (eds.) SemWebEval 2014. CCIS, vol. 475, pp. 101–107. Springer, Heidelberg (2014)

23. Osborne, F., Scavo, G., Motta, E.: Identifying diachronic topic-based research communities by clustering shared research trajectories. In: Presutti, V., d'Amato, C., Gandon, F., d'Aquin, M., Staab, S., Tordai, A. (eds.) ESWC 2014. LNCS, vol. 8465, pp. 114–129. Springer, Heidelberg (2014)

24. Perea, M.P.: Dynamic cartography with diachronic data: dialectal stratigraphy. Literary Linguist. Comput. **28**(1), 147–156 (2013)

25. Pons, P., Latapy, M.: Computing communities in large networks using random walks. In: Yolum, I., Güngör, T., Gürgen, F., Özturan, C. (eds.) ISCIS 2005. LNCS, vol. 3733, pp. 284–293. Springer, Heidelberg (2005)

26. Ratinaud, P.: Visualisation chronologique des analyses alceste: application twitter avec l'exemple du hashtag # mariagepourtous (2014)

27. Rosvall, M., Bergstrom, C.T.: Mapping change in large networks. PloS ONE **5**(1), e8694 (2010)

28. Wang, X., Cheng, Q., Lu, W.: Analyzing evolution of research topics with neviewer: a new method based on dynamic co-word networks. Scientometrics **101**(2), 1253–1271 (2014)

# Artifical Intelligence, Data Mining
# and Knowlege Discovery

# Influence of Outliers Introduction on Predictive Models Quality

Mateusz Kalisch[1], Marcin Michalak[2,3(✉)], Marek Sikora[2,3], Łukasz Wróbel[3], and Piotr Przystałka[1]

[1] Institute of Fundamentals of Machinery Design, Silesian University of Technology, ul. Akademicka 16, 44-100 Gliwice, Poland
{Mateusz.Kalisch,Piotr.Przystalka}@polsl.pl
[2] Institute of Informatics, Silesian University of Technology, ul. Akademicka 16, 44-100 Gliwice, Poland
{Marcin.Michalak,Marek.Sikora}@polsl.pl
[3] Institute of Innovative Technologies EMAG, ul. Leopolda 31, 40-186 Katowice, Poland
Lukasz.Wrobel@ibemag.pl

**Abstract.** The paper presents results of the research related to influence of the level of outliers in the data (train and test data considered separately) on the quality of a model prediction in a classification task. The set of 100 semi–artificial time series was taken into consideration, which independent variables was close to real ones, observed in a underground coal mining environment and dependent variable was generated with the decision tree. For every considered method (decision trees, naive bayes, logistic regression and kNN) a reference model was built (no outliers in the data) which quality was compared with the quality of two models: Out–Out (outliers in train and test data) and Non-out–Out (outliers only in test data). 50 levels of outliers in the data were considered, from 1 % to 50 %. Statistical comparison of models was done on the basis of sign test.

**Keywords:** Data analysis · Classification · Outlier detection · Time series

## 1 Introduction

Decision support systems can be considered as a higher level of abstraction of monitoring and diagnostic systems. In the past, a simple diagnostic systems took into consideration only the current values of monitored variables and compared them with a finite and constant historic data. Nowadays, it becomes more popular to analyse the stream data and refer them to a history which changes with time. This kind of on-line analysis requires also an on-line application of data preparation techniques (replacing missing values, detection of outliers, preprocessing).

Our point of interest is a decision support system dedicated for the underground coal mining industry. The input data for this system are multivariate

© Springer International Publishing Switzerland 2016
S. Kozielski et al. (Eds.): BDAS 2016, CCIS 613, pp. 79–93, 2016.
DOI: 10.1007/978-3-319-34099-9_5

time series. Partially due to the nature of these data – a specific machine operating, measurement devices exploitation requirements – and partially due to the specific environment conditions, these data contain outliers and missing values.

The goal of this paper is to find out how different algorithms are resistant for the level of outliers occurrence in the data, that are expected as the input data for the developed decision support system. How many percents outliers in the data are critical for the classifier to worse the classification quality? How many percent of outliers in the data does the prediction model quality remain on the same level? Is there any difference between model trained on outliered or cleared data? Answers for these question will help to develop the system in such a way that it will become more visible whether detect outliers in the newcoming data or not? The possibility of not detecting outliers will shorten the phase of new data gathering.

The analysis of outliers occurrence level and its influence of training and testing prediction models was performed on the data that are typical for the underground coal mining. Five real time series were used as independent variables while the sixth one was calculated with the artificial decision tree. One hundred of noised replication of this real time series were tested with the introduction of outliers changing from 0 % to 50 % (with the step of 1 %). Introduction of outliers took the place in one and in multiple dimensions. Due to the unbalanced classes a balanced accuracy was taken into consideration as the classification quality measure.

The paper is organised as follows: it starts from the brief comparison of different definitions of outliers, then a review of outlier detection methods is presented, afterwards experiments on semi-artificial data are described. The paper ends with some final words and conclusions. Also perspectives of further works are presented.

## 2    Background

The Grubbs' definition of an outlier [11] states that "an outlying observation is one that appears to deviate markedly from other members of the sample in which it occurs". Grubbs also points two possible causes of this occurence: an extreme manifestation of the random variability inherent in the data or the result of the gross deviation from prescribed experimental procedure or an error in a data acquisition. Following the Hawkins definition "An outlier is an observation which deviates so much from the other observations as to arouse suspicions that it was generated by a different mechanism" [13]. Another definition by Barnett and Lewis [3] is that outlier is an observation (or subset of observations) which appears to be inconsistent with the remainder of that set of data. Weisberg claims that outlier is a case that does not follow the same model as the rest of the data [28]. An incorrect class labelling based definition of an outlier can be found in [16]: an object from class A is incorrectly assigned to the class B, so from the B class point of view it can be treated as an outlying observation. Aggarwal and Yu in their paper [1] distinguish two types of outliers: observations lying outside the data but interpretable as the noise and observations lying outside of the data and the noise.

We can distinguish three main types of outliers: point outlier, context outlier and a collective outlier. A point outlier is a single observation which differs from the data in its neighbourhood and is the most intuitive interpretation of definitions presented above. A context outlier deals with the situation when the value of the point seems correct (does not exceed the typical range of values) but its occurrence in time is not typical. It can be observed especially in time series analysis. Collective outliers are also typical for time series as they represent the situation when a subset of following observations do not behave typically.

## 3   Related Works

A problem of outliers in the data is very common in many fields of application of data analysis [2,12,19,22]. Three main approach of outlier analysis can be found in a literature [14]:

– with no prior knowledge about the data, which is similar to unsupervised clustering,
– modelling the known normality and known abnormality,
– modelling the known normality with a very few cases of abnormality [9,15].

Generally, two groups of outliers detection algorithms can be mentioned: statistical methods and method based on the spatial proximity.

For the one-dimensional data, that has a normal distribution, a simple $3\sigma$ criterion can be applied: a value that differs from the mean value more than three standard deviations is considered as an outlying one. For this type of data the appearance of outlier can be also checked with the Grubb's test [10].

Spatial proximity based methods of outlier detection generally takes into the consideration one of two distant measures: Euclidean or Mahalanobis. Euclidean distance is intuitively a distance between $k-$dimensional objects $x$ and $y$ while Mahalanobis distance is the distance between the object $x$ and a set of objects, described with a centroid $\mu$ ($C$ is a covariance matrix for the set of objects, described by a centroid $\mu$).

In a paper [21] a k-NN based approach for outliers is presented: "A point in a data set is an outlier with respect to parameters $k$ and $d$ if no more than $k$ points in the data set are at a distance of $d$ or less from $p$."

A similar idea is presented in [18]: if $p$ of the $k$ nearest neighbours of a data point ($p < k$) are closer than a specified threshold $D$ than the point is considered as a normal one (not an outlier). A simplified version bases only on comparing the distance of the $m-$th closest neighbour with the threshold $D$ (elimination of the $k$ parameter) presented in [7].

A well known clustering algorithm DBSCAN [8] can also be used for the proximity based detection of outliers. On the basis of the $\varepsilon-$reachability and $\varepsilon-$connectivity it is possible to define a cluster as a subset of points, satisfying a density criterion. Points that do not belong to any cluster are considered as a noise or — from the outliers detection point of view — as outliers. The problem of a data with regions of different points density was solved among others in [5,6].

Another approach for finding outliers is called Minimum Volume Ellipsoid Estimation (MVE) [23]. It consists on finding the smallest possible permissible ellipsoid volume containing an assumed (usually 50 %) fraction of the data points. Points not covered by the ellipsoid volume are considered as outliers.

Support Vector Machine and its modifications are also used in a problem of outlier detection. A modification called one-class SVM [25] tries to separate all data points from the rest of the space. If the soft criterion will be applied some points will be excluded from the selection and considered as outliers. In a paper [20] a Support Vector Regression based outlier detection algorithm is presented. On the basis of the temporal Support Vector Regression model a new observation is considered as an outlier if it exceeds an assumed margin.

A regression based method of finding outliers is also proposed by Torr and Murray in [26]. Starting from the whole set of a data the regression model is built. Then, iteratively a point with the highest influence on the model is being statistically tested whether it is an outlier (a null hypothesis) or not (an alternative hypothesis). After the point removal (what means that the point really is an outlier) the regression model is rebuilt.

The Tukey's proposal of point half–space depth [27] also leads to the definition of outliers. Points on the convex hull of the full data space have a depth equal to 1. Points on the convex hull of the data after removing all points with the depth equal to one have a depth equal two. The procedure is performed as long as there is no more data points. Then points having a depth less or equal a given number $k$ are treated as outliers. This approach was successfully applied in [17,24].

All mentioned approaches base on the assumption that data contain only numerical attributes. A problem arises when categorical (nominal) attributes are present in the data and a typical definition of distance (difference) between values should be redefined. In a paper [4] an overview of 14 measures of similarity (which inverse can represent an attribute distance) is presented.

## 4    Experiments

The main motivation of research, which results are presented in this paper, was to check influence of the increasing level of outliers in the data on the prediction task quality. This led to the statement of two basic questions, which are also more detailed tasks of the data analysis:

- How does the outliers content level in the data influence the prediction ability?
- How many outliers can be in the test data (new data) that it does not influence on the prediction accuracy significantly?

We performed experiments on time series, very close to the real time series, predicting a binary dependent variable. The original data were noised and introduced an increasing level of outliers. A more detailed description of the analysed data and outliers introduction procedure will be presented in the following two subsections.

We also planned two models of building–testing schemes. The first scheme was based on the train and test set containing outliers (called "Out–Out", from "Outliered train – Outliered test") – the level of outliers was the same in the both of them. The second scheme was based on non-outliered train set and outliered test set (called "Non-out–Out"). It is very important to stress that at any part of processing the data, building models and their application none outlier detection/removing procedure was applied. As the reference model for the mentioned scheme a model built on non-outliered train data and tested on non-outliered train data was considered.

Due to the inbalanced classes in the data the balanced accuracy was used as the criterion of a prediction quality. In the further reading notions "accuracy" and "quality" will be used instead of "balanced accuracy" alternatively.

## 4.1 Semi-artificial Time Series Description

For the experiments of the outliers introduction influence a semi-artificial time series was developed. It is a semi-artificial as its independent variables are real observations which comes from underground atmosphere monitoring system installed at one of Polish coal mines. These five variables are as follows:

– $AN$: air flow (in [m/s]),
– $MM$: methan concentration (in [% CH$_4$]),
– $TP1, TP2$: air temperature (in [°C]),
– $BA$: air pressure (in [hPa]).

The data comes from the over 100 h of observations. All data were aggregated into 60 s intervals (originally, raw data were gathered every 2–3 s). Depending from the measuring device a different method of aggregation was applied: a minimum (anemometer), a maximum (methanometer) or an average (thermometers and barometer). An aggregated data were also smoothed with the 24 prevoius observation window. Time series are presented on Fig. 1 and a brief statistical description of its variables is presented in Table 1.

**Table 1.** A brief statistical description of the real independent time series.

|        | AN    | MM     | TP1   | TP2   | BA   |
|--------|-------|--------|-------|-------|------|
| $min$  | 0.100 | 0.1000 | 27.21 | 23.90 | 1090 |
| $Q_1$  | 1.300 | 0.3000 | 28.33 | 24.68 | 1101 |
| $Q_2$  | 1.400 | 0.4000 | 28.80 | 25.31 | 1105 |
| $mean$ | 1.374 | 0.5048 | 28.91 | 25.47 | 1106 |
| $Q_3$  | 1.500 | 0.6000 | 29.39 | 26.06 | 1109 |
| $max$  | 1.800 | 3.0000 | 31.10 | 27.80 | 1131 |

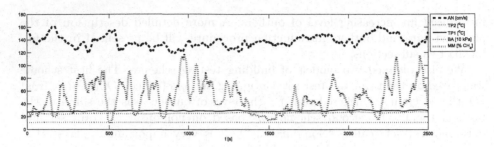

**Fig. 1.** Five independent time series.

On the basis of this five real variables a set of 100 random replications were generated. Each replication contained noised five real variables (a noise with the normal distibution: mean equal to 0 and standard deviation of difference between raw and smoothed values; in case of exceeding the [min, max] range an appropriate boundary value was set) and a artificial dependent variable. The value of the dependent variable was generated with the following decision tree formula:

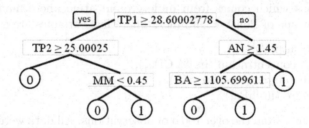

**Fig. 2.** A decision tree dependence between variables.

The decision tree, presented above, was created manually. Starting from the structure the conditional variables were selected. For each variable the threshold of the condition was choosen as a close to the median of a remaining values: for TP1 close to the median of all of the data, for AN close to the median of objects that did not fulfill the TP1 criterion. Values of the dependent variable in the final nodes were also selected arbitrally but in order to assure maximal balance of the classes.

The introduced dependent variable has no rational interpretation.

## 4.2 Outliers Introduction Procedure

The original data and 100 of its randomly noised reproductions did not include outliers. So, to observe how the level of outliers in the data influences the predictor quality a procedure of outliers introduction should be prepared. For this purpose it was necessary to define different types of outliers, at least due to their duration (time aspect), range (variables aspect). In our approach a two criteria of outliers were defined:

- duration:
  - short time outlier (point outlier): several following observation of variable are outliers, the duration of the outlying observation is randomly chosen from the uniform distribution on an interval $[1, 4]$;
  - long time outlier (interval outlier): a longer period of outliers in the data, the duration is also randomly chosen from the uniform distribution on an interval $[t_{min}, t_{max}]$.
- scope:
  - one-dimensional: only one variable in the vector is an outlier; the probability of each variable selection is unique;
  - multi-dimensional: a subset of variables in the vector is an outlier; to assure a unique distribution of a scope in the first step a number of variables is chosen randomly and in the second step a random selection of a specified number of variables is performed;

A specific type of outlier for the methane concentration variable $MM$ is dealt with a common calibration process: this process consists of introducing a sample of a gas with a very high methane concentration to observe the methanometer measures. This causes a linear increase in time. This insertion of outliers generally takes from 25 to 40 s and the methane concentration reaches 30 % and decreases to the initial level. The calibration insertion is made in every third time.

An additional information about outlier is its "direction" – whether the outlying value is smaller than a minimal typical variable values or it is higher than a typical maximal variable values.

In this research as a outlier vector a vector with at least one variable which value is an outlier is considered. We also assumed that the number (or fraction) of outliers in the data describes the number (or fraction) of outlier vectors in the total number of vectors. A vector with several outlier values is treat as a single outlier.

The algorithm of outliers introduction has several parameters:

- a percentage of outliers in the data;
- a percentage of interval outliers in the total number of outliers;

The outlier introduction procedure consists of two main loops. First loop is responsible for introducing interval outliers as long as their total number reaches the demanded level of this type outliers. A dimensionality of each interval outlier, the initial point (initial vector of data) and the duration are randomly chosen. Second loop performs similar operations but the duration of outlier observation is limited to 4 (a short time outlier).

## 4.3   Results

The influence of outliers level on prediction accuracy was checked on several classifiers, starting from decision trees, through naive Bayes, $kNN$ and logistic regression. It is important to remind that not the prediction accuracy but the influence of outliers level was the goal of the research. We put much more attention into the change of prediction accuracy with the increase of the outliers level for the same classifier than for the comparison of methods between themselves.

Decision trees were analysed as the first due to the nature of the data as its decision variable was derived with a given decision tree (see Fig. 2).

On the Fig. 3(a) the result of increasing level of outliers in the both of data (Out–Out model) is presented. Five series – from the upper to the lower – represent the maximal accuracy, the third, second (median) and first quartile and the minimal accuracy from the set of 100 experiments.

The X axis represents the level of outliers introduction in the data in %. As it is expected with the increase of the outlier level the decrease of the model accuracy is observed. As the best models still achieve almost 95 % accuracy the median accuracy decrease to the level of over 75 %.

The analysis of statistical significance of models accuracy decrease a sign test was used. The number of wins of $N$ experiments, which allows to reject the null hypothesis (algorithms give comparable results), should be higher than $0.5N + 1.96N^{-0.5}$. For the presented comparison of results on $N = 100$ sets the critical value becomes 59.8.

On the Fig. 3(b) the series representing the number of wins of the model derived from the non-outliered data over the model derived from the data with specified level of outliers. Starting point of the series (0,0) can be interpreted as the comparison between the non-outliered model with itself. We can notice that even very small introduction of outliers (1 % in the train and test data) influences on the model accuracy statistically significantly.

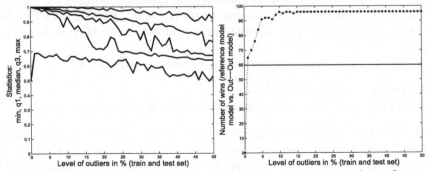

(a) Decrease of the decision tree models accuracy with the increase of outliers level for a Out–Out scheme.

(b) Increase of a statistical significance between reference model and Out–Out models.

**Fig. 3.** Results of outliers introduction for decision tree model in an Out–Out scheme.

From the other hand it may be interesting how the prediction accuracy of the model, derived from the non-outliered data, decreases as the level of outliers in the new data increases. On the Fig. 4(a) the analogical chart is presented which shows the decrease of prediction accuracy statistics as the model from non-outliered data is applied for the data with an increasing level of introduced outliers.

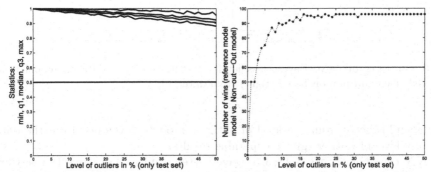

(a) Decrease of the decision tree models accuracy with the increase of outliers level for a Non-Out–Out scheme.

(b) Increase of a statistical significance between reference model and Non-out–Out models.

**Fig. 4.** Results of outliers introduction for decision tree model in an Out–Out scheme.

On the Fig. 4(a) (and partially on a Fig. 3(a)) we can observe a situation, when the prediction accuracy remains on 0.5 level. It occurs when it was difficult to create a proper decision tree with default parameters: the final tree contained only one leaf, representing all data in a training set. As the balanced accuracy is considered as the quality measure it corresponds to evaluate the tree with the 0.5 value.

We can also observe the decrease of models statistics but what is interesting, statistics decrease much slower than in the previous experiments. For the test data with a 50 % of outliers over 50 % models achieved more than 90 % of accuracy – for the model derived from data with 50 % of outliers almost all model gave more than 90 % of accuracy.

On the Fig. 4(b) the increase of a statistically significance between results of model coming from non-outliered data tested on non-outliered and outliered data is presented. In this case results become significantly worse when the level of outliers achieves 3 % in the test data.

The next considered classification algorithm was a naive Bayes. Statistics of models are presented on the Fig. 5.

In the case of this method it occurred that any of model with outliers became statistically worse than a reference one – no matter whether it contained outliers in the train set or not (even the 1 % of the outliers in the data worsen the quality significantly). What is more important, the worsening of classification result does

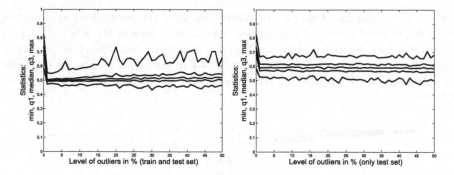

**Fig. 5.** Statistics of naive Bayes classifier for testing outliered data on model from outliered (left) and non-outliered (right) train data.

not depend strictly from the level of outliers: models developed from the data with any level of outliers give comparable results.

$kNN$ classifier with $k \in \{1, 3, 5\}$ gave better results than naive Bayes. Corresponding figures are presented on Figs. 6, 7 and 8.

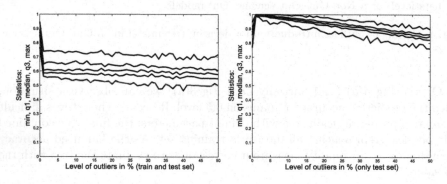

**Fig. 6.** Statistics of 1nn classifier for testing outliered data on model from outliered (left) and non-outliered (right) train data.

The last considered method of classification was a logistic regression, which results for Out–Out and Non-out–Out schemes are presented on Fig. 9. As in case of the first type of models results are rather expected: as the level of outliers in train and test set increase the quality of models decreases. The minimal critical level of outliers, which generates statistically worse models is again 1 %. But situation looks completely different when the Non-out–Out model is being considered. Models developed on non-outliered data behave better on outliered data than on the non-outliered.It is clearly visible on the left chart of the Fig. 9. As the level of outliers increase the quality of the model also increase. The reference model becomes better from the outliered one for 48 % of outliers in the data.

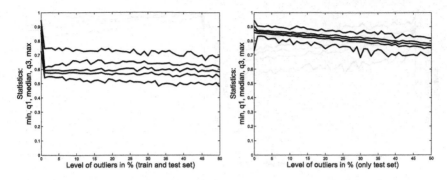

**Fig. 7.** Statistics of 3nn classifier for testing outliered data on model from outliered (left) and non-outliered (right) train data.

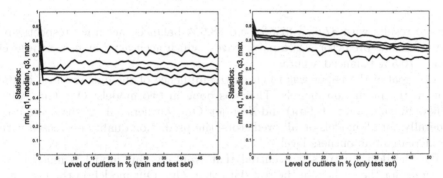

**Fig. 8.** Statistics of 5nn classifier for testing outliered data on model from outliered (left) and non-outliered (right) train data.

Up to 36 % of outliers in the data the outliered test data model give statistically better results than a reference one.

In this case also occurred that application of outliered data can worse the prediction quality significantly. In case of Out–Out models even 1 % of outliers in the data cause a statistically worse models. In case of Non-out–Out model: comparably. Only for $k = 1$ a minimal level of outliers in the test data, which influences on models qualities significantly, become 3 %.

## 4.4   Discussion

Four methods of time series prediction were analysed: decision trees, kNN, naive Bayes and logistic regression. Despite the comparison of quality of different methods was not the goal of the analysis, it confirmed that decision trees gave best results on the decision tree based data (in order to a reference model). Excluding 4 worst models (with accuracy 0.5) the rest 96 of them have at least 97.5 % accuracy. Logistic regression should be admitted as the second best predictor with an average accuracy 90.6 %. $kNN$ results, with $k \in \{1, 3, 5\}$ gave worse

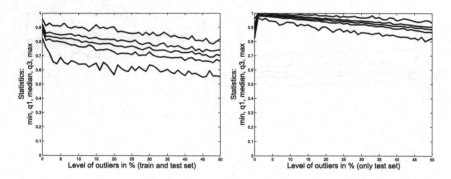

**Fig. 9.** Statistics of logistic regression classifier for testing outliered data on model from outliered (left) and non-outliered (right) train data.

average results with 87.6 %, 87.8 % and 88.9 % balanced accuracy respectively. The worst one predictor was a naive bayes, which accuracy balances at a level of only 70.5 % balanced accuracy.

The goal of the paper was to check the influence of the outliers level in the data on the prediction models. This was done in two models: Out–Out (with outliers in train and test data) and Non-out–Out (outliers only in the test data). Generally, for all models on all predictors, the prediction quality decreased with the increase of an outliers level.

For the decision trees it occurred that Non-out–Out model is much more resistance for the outliers in the test data than Out–Out model: in the first case even the test data with 50 % of outliers were classified with a balanced accuracy not smaller than 90 % in more than 50 % of models. Considering the Out–Out model, it assures comparable model qualities only for the outliers level at 3 %.

Statistical analysis on numbers of wins of a reference model with Out–Out and Non-out–Out models leads to the conclusion, that for the first one the statistically significant level of outliers in the data is 1 % and 3 % respectively.

The analysis of naive bayes classifier results leads to a comparable conclusions: Non-out–Out model is more resistant for the outliers in the data. Any level of outliers decreases the classification quality very significantly, but the it can be stated that the strength of this influence does not depend on the level of outliers.

For all considered $k$'s in $kNN$ models behave also comparable: the classification accuracy decrease significantly from the 1 % of outliers in the data for an Out–Out model and then oscillates around the 0.6 level what is much less than reference accuracies 87.6 %, 87.8 % and 88.9 %. What is very interesting for a $1NN$ Non-out–Out model – introduction of outliers into the test data improves the classification accuracy. For the 1 % of outliers accuracy of 100 models is from the range [98.8 %, 100.0 %]. The further outliers introduction decreases the range limits. Finally, at a level of 43 % in the data a reference model become statistically better than a Non-out–Out one.

The last method – a logistic regression – seems to behave similar as $1NN$ classifier due to the level of outliers in the data. Model trained on clear data applied on outliered test data gives better results than a reference one gives. Also the high level of outliers in the data does not worsen its quality. The critical level of outliers in this model is 48 %.

From the other hand it is interesting to determine from which level of outliers occurrence should we pay so much attention to outliers removing? In the Table 2 which caused the prediction level on train data statistically worse than on a non-outliered train data – is presented.

Models built on the basis of the data cleared from outliers are more resistant on outliers in the test data. However, it could be expected that models could learn the outliers as a correct observations and provide a new knowledge for classifying newly coming outlier observations.

It is also very important to stress that this aspect of analysis leads to the statement that – depending on the predictor – a certain level of outliers in the newly coming data does not require the procedure of data cleaning what may become a very important observation for application of prediction methods in real-time (or close to real-time) prediction tasks.

## 5    Conclusions and Further Works

Most important, and partially expected, conclusion from the carried out experiments is that outliers in the data influences on the quality of built models. The statistically important levels of outliers in the data for several predictors and two schemes of train/test model are presented in the Table 2.

**Table 2.** Minimal outlier level in data which build statistically worse model than built on non-outliered data.

| Predictor | Decision | Naive | $kNN$ | | | Logistic |
|---|---|---|---|---|---|---|
| model | trees | bayes | 1 | 3 | 5 | regression |
| out–out | 1 | 1 | 1 | 1 | 1 | 1 |
| non-out–out | 3 | 1 | 43 | 1 | 1 | 48 |

This influence becomes especially visible when model is derived from the data that also contains outliers (none clearing data technique was applied), what is represented in the first row of the Table 2. It means that detection of outliers is almost necessary in a situation when we expect outliers in the new data, even if they will appear as a one observation for a hundred.

Generally, elimination of outliers in the train data may give us a model more resistant for the outliers in the test data. Its accuracy may decrease with the increase of the outliers level, but the range of results quality remains on a comparable level. For the analysed data a decision tree model gives statistically not worse results as long as the level of outliers in a new data does not approach to 3 %. This remark does not have to be an obvious one as one may expect that a small level of outliers in the train data will help to classify test objects which also contain some outliers.

A completely unexpected behavior of two methods ($1NN$ and logistic regression) was observed: introduction outliers only for a test data improves classification accuracy significantly, even if the level of outliers in the data was increasing.

Concluding, it gives better results to prepare a good quality train data (outliers removing is being considered). An application of outlier detection for the new-coming data could depend on the expected level of outliers – if it is expected to be small it may occur that the time cost is significantly higher than an accuracy decrease. Our further works will focus on comparison of existing techniques of outlier detection and development of new algorithms of outlier detection for a stream data.

**Acknowledgments.** This work was partially supported by Polish National Centre for Research and Development (NCBiR) grant PBS2/B9/20/2013 in frame of Applied Research Programmes. The infrastructure was supported by "PL-LAB2020" project, contract POIG.02.03.01-00-104/13-00.

# References

1. Aggarwal, C.C., Yu, P.S.: Outlier detection for high dimensional data. SIGMOD Rec. **30**(2), 37–46 (2001). http://doi.acm.org/10.1145/376284.375668
2. Ahmed, B., Thesen, T., Blackmon, K.E., Zhao, Y., Devinsky, O., Kuzniecky, R., Brodley, C.E.: Hierarchical conditional random fields for outlier detection: an application to detecting epileptogenic cortical malformations. In: Proceedings of the 31st International Conference on Machine Learning, Beijing, China (2014)
3. Barnett, V., Lewis, T.: Outliers in Statistical Data, 3rd edn. Wiley, Chichester (1994)
4. Boriah, S., Chandola, V., Kumar, V.: Similarity measures for categorical data: a comparative evaluation. In: Proceedings of the SIAM Internation Conference on Data Mining, pp. 243–254 (2008)
5. Breunig, M.M., Kriegel, H.P., Ng, R.T., Sander, J.: Lof: identifying density-based local outliers. In: Proceedings of the 2000 ACM SIGMOD International Conference on Management of Data, pp. 93–104. ACM, New York (2000)
6. Breunig, M.M., Kriegel, H.-P., Ng, R.T., Sander, J.: OPTICS-OF: identifying local outliers. In: Żytkow, J.M., Rauch, J. (eds.) PKDD 1999. LNCS (LNAI), vol. 1704, pp. 262–270. Springer, Heidelberg (1999)
7. Byers, S., Raftery, A.E.: Nearest-neighbor clutter removal for estimating features in spatial point processes. J. Am. Stat. Assoc. **93**(442), 577–584 (1998)
8. Ester, M., Kriegel, H.P., Sander, J., Xu, X.: A density-based algorithm for discovering clusters in large spatial databases with noise. In: Proceedings of the Second International Conference on Knowledge Discovery and Data Mining, pp. 226–231 (1996)
9. Fawcett, T., Provost, F.: Activity monitoring: noticing interesting changes in behavior. In: Proceedings of the Fifth ACM SIGKDD International Conference on Knowledge Discovery and Data Mining, KDD 1999, pp. 53–62. ACM, New York (1999)

10. Grubbs, F.E.: Sample criteria for testing outlying observations. Ann. Math. Stat. **21**(1), 27–58 (1950)
11. Grubbs, F.E.: Procedures for detecting outlying observations in samples. Technometrics **11**(1), 1–21 (1969)
12. Gupta, M., Gao, J., Aggarwal, C., Han, J.: Outlier detection for temporal data: a survey. IEEE Trans. Knowl. Data Eng. **26**(9), 2250–2267 (2014)
13. Hawkins, D.M.: Identification of Outliers. Monographs on Applied Probability and Statistics. Springer, Netherlands (1980)
14. Hodge, V., Austin, J.: A survey of outlier detection methodologies. Artif. Intell. Rev. **22**(2), 85–126 (2004)
15. Japkowicz, N., Myers, C., Gluck, M.: A novelty detection approach to classification. In: Proceedings 14th International Joint Conference Artificial Intelligence, pp. 518–523 (1995)
16. John, G.H.: Robust decision trees: removing outliers from databases. In: Knowledge Discovery and Data Mining, pp. 174–179. AAAI Press (1995)
17. Johnson, T., Kwok, I., Ng, R.T.: Fast computation of 2-dimensional depth contours. In: Agrawal, R., Stolorz, P.E., Piatetsky-Shapiro, G. (eds.) Internation Conference on Knowledge Discovery and Data Mining (KDD), pp. 224–228. AAAI Press (1998)
18. Knorr, E.M., Ng, R.T.: Algorithms for mining distance-based outliers in large datasets. In: Proceedings of the 24rd International Conference on Very Large Data Bases, VLDB 1998, pp. 392–403. Morgan Kaufmann Publishers Inc., San Francisco (1998). http://dl.acm.org/citation.cfm?id=645924.671334
19. Kuna, H., Garcia-Martinez, R., Villatoro, F.: Outlier detection in audit logs for application systems. Inf. Syst. **44**, 22–33 (2014)
20. Ma, J., Perkins, S.: Online novelty detection on temporal sequences. In: Proceedings of the Ninth ACM SIGKDD International Conference on Knowledge Discovery and Data Mining, KDD 2003, pp. 613–618. ACM, New York (2003)
21. Ramaswamy, S., Rastogi, R., Shim, K.: Efficient algorithms for mining outliers from large data sets. SIGMOD Rec. **29**(2), 427–438 (2000)
22. Ritter, G., Gallegos, M.T.: Outliers in statistical pattern recognition and an application to automatic chromosome classification. Pattern Recogn. Lett. **18**(6), 525–539 (1997)
23. Rousseeuw, P.J.: Multivariate estimation with high breakdown point. In: Grossmann, W., Pflug, G., Vincze, I., Wertz, W. (eds.) Mathematical Statistics and Applications, vol. B, pp. 283–297. Reidel, Dordrecht (1985)
24. Ruts, I., Rousseeuw, P.J.: Computing depth contours of bivariate point clouds. Comput. Stat. Data Anal. **23**(1), 153–168 (1996)
25. Schölkopf, B., Williamson, R.C., Smola, A.J., Shawe-Taylor, J., Platt, J.C.: Support vector method for novelty detection. In: Solla, S., Leen, T., Müller, K. (eds.) Advances in Neural Information Processing Systems 12, pp. 582–588. MIT Press (2000)
26. Torr, P.H.S., Murray, D.W.: Outlier detection and motion segmentation, vol. 2059, pp. 432–443 (1993)
27. Tukey, J.: Exploratory Data Analysis. Addison-Wesley Publishing Company, Reading (1977)
28. Weisberg, S.: Applied Linear Regression. Wiley Series in Probability and Statistics, 3rd edn. Wiley & Sons, Hoboken (2005)

# Mining Rule-Based Knowledge Bases

Agnieszka Nowak-Brzezińska[✉]

Department of Computer Science, Institute of Computer Science,
Silesian University, Sosnowiec, Poland
agnieszka.nowak@us.edu.pl

**Abstract.** Rule-based knowledge bases are constantly increasing in volume, thus the knowledge stored as a set of rules is getting progressively more complex and when rules are not organized into any structure, the system is inefficient. In the author's opinion, modification of both the knowledge base structure and inference algorithms lead to improve the efficiency of the inference process. Rules partition enables reducing significantly the percentage of the knowledge base analysed during the inference process. The form of the group's representative plays an important role in the efficiency of the inference process. The good performance of this approach is shown through an extensive experimental study carried out on a collection of real knoswledge bases.

**Keywords:** Rule-based knowledge base · Inference process · Rules partition · Similarity of rules

## 1 Introduction

For the last twenty years, there has been an enormous interest in integrating database and knowledge-based system technologies to create an infrastructure for modern advanced applications. The result of it is a knowledge base ($KB$) system which consists of database systems extended with some kind of knowledge, usually expressed in the form of *rules*[1] - logical statements (implications) of the "*if* condition$_1$ & ... & condition$_n$ *then* conclusion" type. The knowledge of experts, expressed in such a natural way, makes rules easily understood by people not involved in the expert system building. $KB$s are constantly increasing in volume, thus the knowledge stored as a set of rules is getting progressively more complex and when rules are not organized into any structure, the system is inefficient. There is a growing research interest in managing large sets of rules using the clustering approach as well as joining and reducing rules. Interesting and effective approaches to this problem can be found in [5,9], where known systems like XTT2 or D3RJ are described. The goal of these systems is to aggregate the rules built from the same conditional attributes, or the rules that lie close to each other

---

[1] Rules have been extensively used in knowledge representation and reasoning. It is very space efficient: only a relatively small number of facts needs to be stored in the $KB$ and the rest can be derived by the inference rules.

© Springer International Publishing Switzerland 2016
S. Kozielski et al. (Eds.): BDAS 2016, CCIS 613, pp. 94–108, 2016.
DOI: 10.1007/978-3-319-34099-9_6

(if their joint causes no deterioration in accuracy of the obtained rule). In most of these tools, a global set of rules is partitioned by the system designer into several parts in an arbitrary way. This paper presents a different idea, in which rules are divided into a number of groups based on similar premises. The process is called *rules partition*[2] and is conducted by the so-called *partition strategies*. In the author's opinion, rules partition leads to the improvement of the inference process efficiency, because instead of searching within the whole set of rules (as in case of traditional inference processes), only representatives of groups are compared with the set of facts and/or hypothesis to be proven. The most relevant group of rules is selected and the exhaustive searching is done only within a given group [10]. The final efficiency depends on different representation of rules partition. In this article, the author presents three different methods for representatives of the created rules partition. Given a set of rules, new knowledge may be derived using a standard forward chaining inference process, which can be described as follows: each cycle of deductions starts with matching the condition part of each rule with known facts. If at least one rule matches the facts just asserted into the rule base, it is fired. The process stops if there is no new assertion at the end of the cycle.

## 1.1 Proposed Approach

From the point of view of the efficiency of the forward inference process, grouping similar rules seems to be the most optimal solution. It is important to strongly emphasize the role of representatives of groups. It is not enough to combine rules into groups when they provide high capacity to find the wanted information in them. Therefore, an equally important task is to create representatives of the groups. Later, only these representatives are analysed. Among them, the representatives that most closely match the input data are selected. Thus, after the clustering process (which provides groups of the most similar rules), the creation of appropriate quality representatives for the groups takes place. Using different methods for representing rules partition, it is possible to manipulate the size of the result set of rules activated in the inference process. It all controls the efficiency of inference, which in turn influences the quality of the knowledge extracted from $KBs$. The main goal of the *rules partition* idea is to get the same knowledge as would be achieved by using a classical inference algorithm, even if only selected groups of rules are analysed instead of every single rule separately. The results of the experiment show the ability to find all the candidates that can be fired in the inference process, what results in limiting the review to only the representatives of the groups of rules. The idea of rules partition is implemented in the kbExplorer system[3], which is not limited to the inference optimization. The practical goal of the project is to create an expert system shell that allows for

---

[2] The idea is new but it is based on the author's previous research, where the idea of *clustering rules* as well as creating the so-called *decision units* was introduced [10,11].

[3] http://kbexplorer.ii.us.edu.pl/.

flexible switching between different inference methods based on the knowledge engineer's preferences[4].

The rest of the paper is organized as follows: in Sect. 2 the description of the rules partition idea for rules in $KB$s is included. It contains the definition of rules partition, the partition strategy (similarity-based) with the pseudocode of the hierarchical clustering algorithm with modification. Section 3 presents the definition of the representatives' forms for rules partition. The experiments with the analysis of their results are considered in Sect. 4. Section 5 contains the summary.

# 2 Rules Partition Idea

The initial and fundamental step in data analysis is clustering. Broadly speaking, clustering algorithms can be divided into two types - partitional and hierarchical. The agglomerative hierarchical clustering algorithm, proposed by the author, is based on the similarity approach. It starts with each rule in a separate group and successively combines them until there is only one group remaining or a specified termination condition is satisfied. In this research, a termination condition is given as a moment, in which there is no similarity between combined groups or the similarity is no longer at least equal to some given minimal value. Taking into account the fact that the clustered objects are rules (which are much more difficult to analyse than standard records in a database or rows in a decision table because they are composed of a number of premises and the conclusion), we must be aware that different attributes may exist for these rules and that even if some attributes are common, they still may differ in the number of premises. It is also possible that two rules have the same attributes in the premises and/or conclusions (and the same number of premises) but different values of these attributes. Then, it is much more difficult to establish their similarity. The next subsection presents the basic concepts and notations of $KB$ partitioning.

## 2.1 The Definition of Rules Partition

The proposed idea assumes division of a $KB$ into coherent subgroups of rules. Therefore, this section presents the basic concepts and notations of $KB$ partitioning. We assume that a $KB$ is a single set $\mathcal{R} = \{r_1, \ldots, r_i, \ldots, r_n\}$ without any order of $n$ rules. Each rule $r_i \in \mathcal{R}$ is stored as a Horn clause defined as: $r_i : p_1 \wedge p_2 \wedge \ldots \wedge p_m \to c$, where $p_s$ is $s$-th literal (a pair of an attribute and its value) $(a, v_i^a)$ $(s = 1, 2, \ldots, m)$. Attribute $a \in A$ may be a conclusion of rule $r_i$ as well as a part of the premises. For every $KB$ with $n$ rules, the number of

---

[4] The user has the possibility of creating $KB$s using a special creator or by importing a $KB$ from a given data source. The format of the $KB$s enables working with a rule set generated automatically based on the rough set theory as well as with rules given apriori by the domain expert. The $KB$ can have one of the following file formats: XML, RSES, TXT. It is possible to define attributes of any type: nominal, discrete or continuous. There are no limits for the number of rules, attributes, facts or the length of the rule.

possible subsets is $2^n$. Any arbitrarily created subset of rules $R \in 2^{\mathcal{R}}$ is called *partition of rules* $(PR)$ and it can be generated by one of many possible *partition strategies* $(PS)$.

In mathematical meaning, a partition of rules is a collection of subsets $\{R_j\}_{j \in J}$ of $R$ such that: $R = \bigcup_{j \in J} R_j$ and if $j, l \in J$ and $j \neq l$ then $R_j \cap R_l = \emptyset$. The subsets of rules are non-empty and every rule is included in one and only one of the subsets. The partition of rules $PR \subseteq 2^{\mathcal{R}} : PR = \{R_1, R_2, \ldots, R_k\}$, where: $k$ - the number of groups of rules creating the partition $PR$, $R_j$ - $j$-th group of rules, $R \in 2^{\mathcal{R}}$ and $j = 1, \ldots, k$, is generated by the partitioning strategy $(PS)$.

## 2.2  Partition Strategies

*Partition strategy* is a general concept. It is possible to point out a number of different approaches for creating groups of rules. The most general division of $PS$ talks about two types of strategies: *simple* and *complex*. Simple strategies[5] allocate every rule $r_i$ to the proper group $R_j$, according to the value of the function $mc(r_i, R_j)$. The example is the strategy of finding a pair of rules that are the most similar in a given context, i.e. premises of rules. *Complex strategies* usually do not generate the final partition in a single step. It is defined by *a sequence of simple strategies* or *a combination of them*, or by *iteration* of a single simple strategy. An example is the strategy of creating *similarity based partition*, which stems from the method of *cluster analysis* used for rules clustering. It uses a simple strategy, which finds pairs of the most similar rules many times. The process terminates if the similarity is no longer at least $T$[6]. In effect, we get $k$ groups of rules $R_1, R_2, \ldots, R_l, \ldots, R_k$ such that $\bigwedge_{r_i, r_f \in R_l} sim(r_i, r_f) \geq T$. Each group $R_l$ contains rules for which mutual similarity is at least equal to $T$. It is possible to obtain many different rules partitions. In this strategy, rules that form the same group are said to have the same or similar premises. It uses the similarity function $sim(r_i, r_f)$: $\mathcal{R} \times \mathcal{R} \rightarrow [0..1]$ which can be defined in a variety of ways. The larger the value of a similarity coefficient, the more similar the objects are, the smaller the values are, the more dissimilar the objects are. There are many possible similarity coefficients [13]. In accordance with the fact that the research is based on rules in $KBs$ (very specific knowledge representation), the author uses the similarity measure, based on the so-called $SMC$ coefficient, which in its simplest form is the ratio of the number of the attributes common to all the attributes. Rules may have different attributes in the premises part, thus the following modification is proposed. The modified $SMC$ coefficient is the ratio of the number of common premises in two compared objects (rules or groups of rules) to all possible premises for these two objects, and it can be defined as:

---

[5] It allows for partitioning the rules using the algorithm with time complexity not higher than $O(nk)$, where $n = |R|$ and $k = |PR|$. Simple strategies create final partition $PR$ by a single search of rules set $R$ according to the value of $mc(r_i, R_j)$ function described above. For complex strategies, time complexity is rather higher than any simple partition strategy.

[6] Let us assume that threshold value $0 \leq T \leq 1$ exists.

$$sim(r_i, r_f) = \frac{|cond(r_i) \cap cond(r_f)|}{|cond(r_i) \cup cond(r_f)|}.$$

In this formula $cond(r_i)$ denotes the conditional part of rule $r_i$ and $concl(r_i)$ its conclusion. The value of $sim(r_i, r_f)$ is equal to 1 if rules are formed by the same literals and 0 when they do not have any common literals. At the beginning of the clustering process, there are only single rules to combine in groups, but when two rules $r_i, r_f$ create a group $R_j$ we need to know how to calculate the similarity between two groups $R_{j_1}, R_{j_2}$ or between a group $R_j$ and a rule $r_i$. The similarity measure presented above works properly for single rules as well as for groups' representatives. If groups' representatives are formulated from the conditional part of rules[7] then having representatives of groups $R_{j_1}$ and $R_{j_2}$ (which has a form of a set of pairs $(a, v)$ like premises of a single rule) the similarity is calculated in the same way.

## 2.3  Hierarchical Rules Clustering Algorithm — the Study of the Proposed Approach

Clustering algorithms proposed by the author look for two the most similar rules (related to their premises) and combine them iteratively. Very often there are $KBs$ in which many rules are totally dissimilar (do not have at least one common premise). That is why it was natural to form a termination condition as a moment in which there are no rules or rule groups that are similar somehow. Usually, the resultant structure contains both some number of groups of rules and a set of small groups of rules with only one (singular) rule inside. At the beginning, each object is considered a cluster itself (or one may say that each object is placed within a cluster that consists only of that object) [4]. During each iteration step of agglomerative hierarchical clustering ($AHC$) algorithms, clusters are merged with other clusters. The process is repeated until the similarity between joined groups is at least equal to the given threshold (in $mAHC$) or until there is any (even very small) similarity between joined groups ($AHC$). In comparison to the basic $AHC$, known from the literature, objects are combined until some (predefined) similarity between them is seen. It means that very often there is some number of groups of rules at the end, instead of one group containing many subgroups, as it is in case of the classical $AHC$. The problem with the classical $AHC$ is that the group created as the root of the dendrogram is very often inadequate for the group's content. In fact, it is not a good representative of the subgroups' rules, because, depending on the method used to create the representative, it contains either too little information about the group or too much information corresponds only to the selected fragment of the whole set rather than to the entire knowledge base. Based on the idea of rules partition (described in detail in Sect. 2.1) and using the complex partition strategy (called *similarity-based partition strategy*), we get a set of groups of rules with similar premises. The pseudocode of the proposed algorithm is presented below.

---

[7] Thus, conjunction of pairs $(a_i, v_{1i})(a_2, v_{2i}) \ldots (a_s, v_{si}))$ may be both a conditional part of one rule and a representative of some group of rules.

**Similarity-Based Partition Strategy:** The algorithm joins similar rules (groups of rules) iteratively, until it reaches the similarity between the groups of rules defined by given threshold $T$. In the first step of the algorithm, a specific case of partition, in which every rule forms a separate group, is created. The pseudocode is suitable for both the $AHC$ and the $mAHC$ algorithms because the similarity threshold is given as an argument in the *createPartition* procedure, and further checked in the *findTwoMostSimilarGroups* procedure.

**Require:** $\mathcal{R}$, *sim*, $T$;
**Ensure:** $PR = \{R_1, R_2, \ldots, R_k\}$;
   **procedure** *createPartitions*( $\mathcal{R}$, **var** $PR$, *sim*, $T$ )
   **var** $R_i$, $R_j$;
   **begin**
   $PR = createSingletonGroups(\mathcal{R})$;
   $R_i = R_j = \{\}$;
   **while** *findTwoMostSimilarGroups*($sim$, $R_i$, $R_j$, $PR$) $\geqslant T$ **do**
     $PR = rearrangeGroups(R_i, R_j, PR)$;
   **end while**
   **end procedure**

## 2.4   A Role of Rules Partition Idea in the Exploration of $KBs$

$KBS$ (knowledge-based system) provides expert conclusions about specialized areas, generally obtained through the inference algorithms, which are responsible for generating new knowledge stored in rules and a set of facts ($F$). The inference process can be divided into three stages: (1) matching, (2) choosing and (3) execution. In the matching phase, the premises of each and every rule are matched with $F$, and selected rules form the conflict set[8]. Next (in the choosing phase), only one rule is selected from the conflict set. Finally, the previously selected rule is activated and the set of facts is updated (execution step). From the two methods of inference, *forward chaining* and *backward chaining*, only the forward inference will be examined in this paper. The goal of the *forward inference* process is to find a set of rules with given data included in the premises part of a particular rule. The shorter the time needed to perform such a process, the better. The most time consuming procedure searches rules and finds those relevant to the given data (goal/facts) [11]. In a large $KBS$, the process of activating multiple rules and rapid growth of the set of facts can slow down the

---

[8] When more than one rule matches the working memory (it is called *conflict set*) and only one has to be selected, it is possible to use the following strategies: *random, textual order, recency, specificity* and *refractoriness* [1]. In the research, the author uses the *textual order*, the *recency* and the modified *specificity* strategies. The *textual order* fires the first matching rule, while the *recency* fires the rule which uses the data added most recently to the working memory and the *specificity* (*complexity*) fires the rule with the most conditions attached, which means that rules with a greater number of conditions or fewer variables are more specific and should be applied earlier because they use more data and can be used for special cases or exceptions to general rules.

inference process significantly. Therefore, some improvements are needed, like the `Rete` algorithm or its modifications (`Treat`,`Gator` or `Leaps`), but they do not work well with big $KBs$ [2,3,8].

**Forward Inference for Partition of Rules:** The forward inference algorithm, based on the rules partition, reduces search space by choosing rules only from a particular group of rules. In every step, the inference engine matches facts to groups' representatives and finds a group with the greater similarity value. Premises of each and every rule are analysed to match them with the current set of facts $(F)$ and the subset of the conflict set, consisting of the rules that are actually activated, is chosen. Finally, the previously selected rules are analysed and the set of facts is updated. To find a relevant group of rules, during the inference process, the maximal value of similarity between the set of facts $F$ (and/or the hypothesis to be proven) and the representatives of each group of rules, $Profile(R_j)$ is searched[9]. The more facts with hypothesis are included in the profile of group $R_j$, the greater the value of similarity. At the end, rules from the most promising group are activated. The most important step, the selectBestFactMatchingGroup procedure, is responsible for finding the most relevant group of rules. The pseudocode of the algorithm is following[10]:

**Require:** $\mathcal{PR}$, $\mathcal{F}$;
**Ensure:** $\mathcal{F}$;
  **procedure** *forwardInference*( $\mathcal{PR}$, **var** $\mathcal{F}$ )
  **var** R,r;
  **begin**
  R := *selectBestFactMatchingGroup*( $\mathcal{PR}$, $\mathcal{F}$ );
  **while** $R \neq \emptyset$ **do**
    r:=*selectRule*( R, strategy);
    *activeRule*( r );
    *excludeRule*( R, r );
    R := *selectBestFactMatchingGroup*( $\mathcal{PR}$, $\mathcal{F}$ );
  **end while**
  **end procedure**

If the result of the selection is not empty $(R \neq \emptyset)$, the *selectRule* procedure is commenced. It finds the rule in which the premises part is fully covered by the facts. If there is more than one rule, the strategy of the conflict set problem plays a significant role in final selection of one rule. If that happens, the conflict

---

[9] $sim(F, R_j) = \frac{|F \cap Profile(R_j)|}{|F \cup Profile(R_j)|}$. The value of $sim(F, Profile(R_j))$ is equal to 0 if there is no such fact $f_i$ (or the hypothesis to prove) which is included in the representative of any group $R_j$. It is equal to 1 if all facts (and/or hypothesis) are included in $Profile(R_j)$ of group $R_j$.

[10] The inputs are: $PR$ - groups of rules with the representatives and $F$ - the set of facts. The output is $F$ the set of facts, including possible new facts obtained through the inference. The algorithm uses temporary variable $R$, which is the set of rules that is the result of the previous selection.

set strategy[11] enables selecting one rule. Activating one rule means its blocking for further searching (*excludeRule*($R$, $r$);). The *activeRule* procedure adds a new fact (the conclusion of the activated rule) to the $KB$. The *excludeRule* function temporarily removes the activated rule from further searching. Afterwards, the *selectBestFactMatchingGroup* procedure is called again. The whole presented process is repeated in a while loop until the selection becomes empty. Among the advantages are: reducing the time necessary to search within the whole $KB$ in order to find rules to activate and achieving additional information about fraction of rules that match the input knowledge (the set of facts). It is worth mentioning that such a modification enables firing not only certain rules but also the approximate rules for which the similarity with the given set of facts is at least equal to $T$ or is greater than 0.

To sum up the state-of-the-art of the inference algorithms above, it can be said that there are many possibilities for improving the efficiency of that process, such as dividing the $KB$ into the groups of rules similar in some context and then, during the inference process, activating only necessary rules. The time of the inference process may be reduced significantly by dealing with the structure of groups of rules that are similar in some way. It is also important to improve the inference algorithms. This improvement is possible, because the classical inference algorithm is based on the assumption that rules in a KB are not grouped and arranged in any way, and thus it is necessary to search every single rule to make sure that every possible (relevant) rule has been found and activated. Now, if we have a group of similar rules, and the representatives for them, we may modify the inference algorithm. It is no longer necessary to search the entire KB and analyze every single rule. It is enough to search only the groups' representatives and select the most appropriate one. This can significantly reduce the time of the inference (sometimes time is reduced by up to 90 percent) in comparison to the classical version of the algorithm.

## 3    Rules Partition Representations

Groups' representatives enables generalizing the information stored in a given group of rules. It may be important information for knowledge engineers in the context of $KBs$ exploration (for deeper exploration of the domain knowledge). Combining some number of rules into one group and forming a representative for this group allows for generalizing the description of it, which may be helpful in further analysis. If all the groups of rules have representatives, it means that instead of searching every single rule in the entire $KB$ we may only analyze the representatives of the groups and select the group or groups the most relevant

---

[11] In this paper, the following strategies are used: FR (*first rule*) — which fires the first rule in the conflict set (so it is relevant to the *textual order* strategy), LR (*last rule*) — which selects the rules recently added to the conflict set (so it works similarly to the *recently* strategy), SR (*shortest rule*) — which selects the rule with the smallest number of conditions, LOR (*longest rule*) — which chooses rules with the greatest number of conditions (so it is relevant to the *specificity* strategy).

to the information we are looking for. This will be possible if we choose the correct method of creating the representative. However, the method of groups' representatives is not sufficient to achieve optimal results.

## 3.1 Study of the Methods for Creating the Rules Groups' Representatives

Cluster analysis methods are meant to reflect human categorization, however standard clustering methods split observations into a set of clusters without providing any generalized descriptions of the clusters. The potential characteristic of clusters that may be desired, and that can be checked using the available data is the following: within-cluster dissimilarities should be small while between-cluster dissimilarities should be large. It should be possible to characterize the clusters using a small number of variables. Representing objects by centroids well may require some clusters with little or no gap between them. Stability is often easier to achieve with few clusters; but more clusters may be required in situations where they need to be very homogeneous. One of the most popular methods for determining the most representative elements is to optimize descriptions during the rule generation process [7]. Such a method employs a rule quality criterion, defined by the user, that specifies a trade off between completeness and consistency of a rule. At each stage of rule learning, candidate hypotheses are overgeneralized (introducing inconsistency, but increasing rule coverage), and then evaluated using the rule quality criterion. As outlined in the [6], there are three main ways to represent clusters:

1. Use the centroid of each cluster to represent the cluster. This is the most popular way. The centroid tells us where the center of the cluster is. One may also compute the radius and standard deviation of the cluster to determine the spread in each dimension,
2. Use the classification models to represent clusters. In this method, we treat each cluster as a class. That is, all the data points in a cluster are regarded as having the same label. We then run a supervised learning algorithm on the data to find a classification model. For example, we may use the decision tree learning to distinguish the clusters. The resulting tree or set of rules provide an understandable representation of the clusters,
3. Use frequent values in each cluster to represent it. This method is mainly for clustering of categorical data. It is also the key method used in text clustering, where a small set of frequent words in each cluster is selected to represent the cluster. The centroid of a cluster represents the average (or center of gravity) of all the documents in the cluster.

The description of the group can contain values of attributes that are the same for every member of the group. We may leave these values of attributes that are different from other attributes. This type of a group's description is called the generalization. In accordance with the principle of Occam's razor and the principle of minimum description length (MDL), the aim is to find the shortest possible descriptions of the groups' rules, because they are the easiest to interpret [12].

It is sometimes claimed that the MDL prefers simpler classifiers over complex ones. Briefly stated, the approach looks for finding irreducible, smallest representation of all members of a category. The most general and most detailed descriptions represent two extreme cases that are undesirable. Both overgeneralization and excessive specialization are not optimal solutions. The principle of the MDL slightly prefers overgeneralization. The more premises are common for all the rules in a given group, the more probable it is that the inference succeeds. The less premises are common for all the rules in a given group, the more probable it is that its representative is empty and the group is excluded from the searching process. Section 3.2 presents three different approaches to rules partition's representation.

## 3.2   Partition's Representatives — *Profiles*

Representatives of the rules groups influence the quality of rules partition (cohesion and separation of the created groups) and the efficiency of the inference process. A lot depends on the form of the representative of a given group. If it does not contain the proper information, such a group may not be found at all. All the literature-based approaches and conditions for clusters' representatives were taken into account by the author. Based on the most popular and effective solutions, the following three were used in the research:

- The conjunction of the premises of all the rules included in a given group $R_j$ (here denoted as $Profile_{general}(R_j)$),
- The set of all the premises of rules that form group $R_j$ ($Profile_{specialized}(R_j)$),
- The conjunction of the selected premises of all the rules included in a given group $R_j$ (here denoted as $Profile_{weighted}(R_j)$).

The first method forms a representative for a group of rules as a general description containing only features (premises) appearing in every rule in a given group. It is defined as:

$$Profile_{general}(R_j) = \bigcup\{p_s : \forall_{r_i \in R_j} \quad p_s \in cond(r_i)\},$$

and it contains all the premises that definitely form all the rules in this group. The second approach creates a representative as an extended description, which contains every feature appearing at least once in any of the rules in a given group. It can be defined in the following way:

$$Profile_{specialized}(R_j) = \bigcup\{p_s : \exists_{r_i \in R_j} \quad p_s \in cond(r_i)\}.$$

The last method builds a representative as a set of frequent premises. It will be set containing the premises for which the weights was at least equal to a given threshold. Its definition is as follows:

$$Profile_{weighted}(R_j) = \bigcup\{p_s : \forall_{r_i \in R_j} \quad \frac{n_{p_s R_j}}{n_{R_j}} \geq T \quad p_s \in cond(r_i)\}$$

It can be any value from the range $(0, 1)$. It is specified by the user. The weights are calculated in the following way: for every group $R_j$ a number of rules in it is calculated $(n_{R_j})$ as well as a number of premises (for every premise $p_s$ its support is calculated as a number of rules in a group $R_j$ containing a premise $p_s$ - $n_{p_s R_j}$). Then a weight for a premise $p_s$ is a ratio of $n_{p_s R_j}$ and $n_{R_j}$, expressed in a percentage scale. It means that if a threshold is specified as the default value (50), we expect that a representative is composed by premises which appear in at least 50 % of the rules in a given group $R_j$. During the experiments, 5 different threshold values were analyzed: 10 %, 25 %, 50 %, 75 %, and 100 %. It is possible to say that the third method is a compromise between these two approaches presented above. It assumes that a group's representative should contain only premises with proper weights. Rules premises have got weights that are values determined as a ratio of the number of rules in the entire $KB$, in which a given premise $p_s$ appears, to the number of all the rules in the $KB$. It means that the more rules with a given premise, the higher the weight of it. It gives the same result as the first method.

# 4    Experiments

The main goal of the experiments is to compare the three different methods for building groups' representatives in accordance with their influence on the inference efficiency (for 17 different $KBs$). Intuitively, we may say that reducing the groups' descriptions may cause a problem with finding a relevant group of rules which is activated during the inference process. Thus, it can distort the inference efficiency. In the opposite case, when we extend the groups' descriptions, all the relevant groups should be found. Let us start to analyze the performance of the proposed approach by presenting an example of rules partition's representatives for a simple $KB$ containing the following 5 rules:

$$r_1 : (a, 1) \wedge (b, 1) \wedge (c, 1) \rightarrow (dec, A) \quad r_2 : (a, 1) \rightarrow (dec, B)$$
$$r_3 : (d, 1) \rightarrow (dec, A) \qquad\qquad r_4 : (d, 1) \wedge (e, 1) \rightarrow (dec, C)$$
$$r_5 : (a, 1) \wedge (b, 1) \rightarrow (dec, B)$$

When we are interested in finding only rules for which similarity is not less than 0.5, the partition will be as follows: $\{PR\}^* = \{\{r_1, r_2, r_5\}, \{r_3, r_4\}\}$. We got two groups: $R_1$ containing rules $r_3, r_4$ and $R_2$ containing rules $r_1, r_2, r_5$, The different representatives for group $R_2$ for the three methods introduced above are the following:

$$Profile_{general}(R_2) = \{(a, 1)\}$$

$$Profile_{specialized}(R_2) = \{(a, 1), (b, 1), (c, 1), (d, 1)\}$$

and the last one assuming that the threshold value is equal to 50 %:

$$Profile_{frequent}(R_2) = \{(a, 1), (b, 1)\}$$

## 4.1 The Results of the Experiments

The first step of the experiments was to compare the efficiency of using the approach of rules partition in accordance with use two different clustering algorithms: $AHC$ and $mAHC$ (and to compare it with the classic $KB$, without rules partition). The results are presented in Table 1. It contains the following information: [KB] - the knowledge base, the number of rules in the $KB$ ($N_{Rules}$), the number of groups using the $AHC$ algorithm ($G\_AHC$) and the number of groups of rules created using the $mAHC$ algorithm ($G\_mAHC$). In the brackets, there are percentages of the $KB$ actually searched during the inference process in comparison to the size of the $KB$ searched normally for the $AHC$ and the $mAHC$ algorithms.

**Table 1.** The % of the KB analyzed during the inference for every KB

| KB | $N_{Rules}$ | Number of groups (% of KB analyzed) | |
|---|---|---|---|
| | | $G\_AHC$ | $G\_mAHC$ |
| 1-3 | 10 | 6 (60, 00 %) | 8 (80, 00 %) |
| 4-6 | 17 | 9 (52, 94 %) | 11 (64, 71 %) |
| 7 - 9 | 199 | 9 (4, 52 %) | 72 (36, 18 %) |
| 10-12 | 416 | 13 (3, 13 %) | 73(17, 55 %) |
| 13-15 | 1199 | 14 (1, 17 %) | 230 (19, 18 %) |
| 16 | 13 | 3 (23, 08 %) | 4 (30, 77 %) |
| 17 | 10 | 10 (100 %) | 10(100 %) |

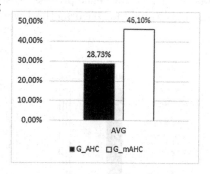

**Fig. 1.** The average % of the KB analyzed during the inference

An interpretation of the first row is as follows: the first three $KB$s consists of 10 rules, and when we cluster the rules in them, during the inference process it is enough to search 60 % or 80 % instead of the entire KB in case of the classic inference process. What is more impressive, in case of $KB$s with 1199 of rules, it is necessary to search only less than 2 % (in case of $AHC$) or less than 20 % of the $KB$. There is a trend that the more rules, the better while using the modified version of the classic forward inference algorithm in comparison to the classical deduction. The more rules in the $KB$, the longer the time of inference. The reason is the necessity of matching every rule with a given set of facts. Last but not least, the partition-based inference (in contrast to the classical approach) is characterized by the following rule: the more rules, the faster the inference process. The time of rules partitioning is significantly long in comparison to the time of inference. Partition of rules should be stored as a static structure (assuming that it is rebuilt whenever the KB is modified). Figure 1 presents the average effectiveness of the inference process using the $AHC$ and the $mAHC$ algorithms. We can observe that while using the $AHC$ algorithm, we search less of the $KB$ in comparison to the $mAHC$. It is because the $mAHC$ usually cuts the clustering when the number of groups is big, but the similarity is high. More groups to search increases the % of the $KB$ necessary to search during the inference process. The second (main) goal of the experiments shows the differences between the three different methods for groups of rules' representation proposed in Sect. 3.2. The results are presented in Fig. 2, which shows the results of the classic forward inference algorithm (in Fig. 2 it is noticed as $FI\_classic$) for 17 different $KBs$ compared to the results of using the rules partition-based forward inference algorithm, created by different similarity-based rules partition strategies: $AHC$ or $mAHC^{12}$ in accordance with the methods of rules partition's representatives: $Profile_{general}(R_j)$, $Profile_{specialized}(R_j)$ and $Profile_{weighted}(R_j)^{13}$. The results are the average values of the number of new facts (expressed in percentages) generated from rules, during the forward inference process in comparison to the number of new facts generated during the classic forward inference process. Tests on 17 different $KBs$ were performed, the results obtained for each method were averaged and presented in Fig. 2.

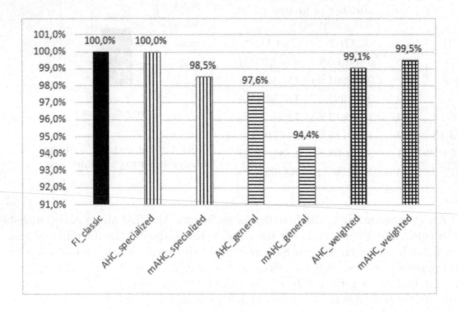

**Fig. 2.** The average effectiveness of inference for the analyzed methods

Looking at Fig. 2 it is possible to see that only for the $AHC$ algorithm with specialized-based representatives ($Profile_{specialized}$) the recall is the same (100 %) as for the classic forward inference algorithm without the rules partition.

---

[12] The $mAHC$ approach for rules partitioning with using four different thresholds of similarity: $k = 0, k = 0.25, k = 0.5, k = 1.0$.

[13] In Fig. 2 they are noticed as $AHC\_general$, $mAHC\_general$, $AHC\_specialized$, $mAHC\_specialized$, $AHC\_weighted$ or $mAHC\_weighted$.

When using the $mAHC$ algorithm (with specialized-based representatives) the recall deteriorates (98.5 %). Quite good results are achieved when the weighted-based approach is used: for the $AHC$ algorithm the average efficiency of the inference process is equal to 99.1 % while for $mAHC$ it is equal to 99.5 %. With the general-based representatives ($Profile_{general}$) the recall is the smallest in comparison to all the other methods (when using the $mAHC$ partition algorithm it is even equal to 94.4 %). Generally, we may say that unfortunately the $mAHC$ algorithm (for rules partitioning) sometimes, in the matching process, loses some rules that are expected to be found. There is a tendency: using the general-based representatives we lose more rules than when using the specialized-based representatives. It seems that the specialized-based approach lets us find much more relevant rules than the general-based method. The compromise is the third method, weighted-based, when the user may specify the size of the representatives. The general-based profile provides a smaller number of generated facts, often ending in failure, while both the specialized-based and weighted-based profiles generate a number of facts comparable to the classic approach.

## 5   Summary

Exploration of rule-based $KBs$ involves clustering similar rules, discovering similarities in groups of rules or identifying previously unknown relationships between rules or groups of rules as well as extracting new knowledge from the knowledge already known and stored in the $KB$ in the form of rules or facts using the inference algorithms. In this paper, the author has considered modifying the structure of large $KBs$ using the rules partition idea and introduced the forward inference algorithm for the structure of these specific rules. From many studied hierarchical clustering algorithms, two were analyzed in this paper: $AHC$ based on the similarity analysis and its modification $mAHC$ with a different $k$ threshold as a minimal similarity value necessary to combine rules or groups of rules together. Knowing that the efficiency of the forward inference depends strongly on the matching results, three different methods for groups' representatives were proposed: the general-based, specialized-based and weighted-based. The first part of the experiments show that the rules partition idea enables decreasing significantly the number of rules necessary to perform analysis in comparison to the number of rules which is analysed in case of the classic forward inference process when rules in the $KB$ are not grouped. The second part of the experiments shows that using different representative's forms for groups of rules enables managing the value of the recall parameter.

**Acknowledgement.** This work is a part of the project "Exploration of rule knowledge bases" founded by the Polish National Science Centre (NCN: 2011/03/D/ST6/03027).

# References

1. Akerkar, R., Sajja, P.: Knowledge-Based Systems. Jones & Bartlett Learning, Sudbury (2010)
2. Forgy, C.L.: Rete: A fast algorithm for the many pattern many object pattern match problem. Artif. Intell. **19**, 17–37 (1981)
3. Hanson, E., Hasan, M.S.: Gator: An optimized discrimination network for active database rule condition testing. Tech. rep. (1993)
4. Jain, A.K., Dubes, R.C.: Algorithms for Clustering Data. Prentice Hall, New Jersey (1988)
5. Latkowski, R., Mikołajczyk, M.: Data decomposition and decision rule joining for classification of data with missing values. In: Tsumoto, S., Słowiński, R., Komorowski, J., Grzymała-Busse, J.W. (eds.) RSCTC 2004. LNCS (LNAI), vol. 3066, pp. 254–263. Springer, Heidelberg (2004)
6. Markov, Z., Larose, D.T.: Data Mining the Web: Uncovering Patterns in Web Content, Structure, and Usage. Wiley, Hoboken (2007)
7. Michalski, R.S., Larson, J.B.: Selection of Most Representative Training Examples and Incremental Generation of vl Hypotheses. The Underlying Methodology and the Description of the Programs esel and aq11. University of Illinois, Department of Computer Science, Urbana (1978)
8. Miranker, D.P.: Treat: A better match algorithm for ai production systems. Department of Computer Sciences, University of Texas at Austin, Technical report (1987)
9. Nalepa, G.J., Ligęza, A., Kaczor, K.: Overview of knowledge formalization with XTT2 rules. In: Bassiliades, N., Governatori, G., Paschke, A. (eds.) RuleML 2011 - Europe. LNCS, vol. 6826, pp. 329–336. Springer, Heidelberg (2011)
10. Nowak-Brzezińska, A., Simiński, R.: Knowledge mining approach for optimization of inference processes in rule knowledge bases. In: Herrero, P., Panetto, H., Meersman, R., Dillon, T. (eds.) OTM-WS 2012. LNCS, vol. 7567, pp. 534–537. Springer, Heidelberg (2012)
11. Pedrycz, W., Cholewa, W.: Expert Systems. Silesian University of Technology, Section of Scientific Publications, Poland (1987). [in polish]
12. Rissanen, J.: Paper: Modeling by shortest data description. Automatica 14(5), 465–471, September 1978. http://dx.doi.org/10.1016/0005-1098(78)90005-5
13. Sarker, B.R.: The resemblance coefficients in group technology: A survey and comparative study of relational metrics. Comput. Ind. Eng. **30**(1), 103–116 (1996). Elsevier Science Ltd., Printed in Great Britain

# Two Methods of Combining Classifiers, Which are Based on Decision Templates and Theory of Evidence, in a Dispersed Decision-Making System

Małgorzata Przybyła-Kasperek[(✉)]

Institute of Computer Science, University of Silesia,
Będzińska 39, 41-200 Sosnowiec, Poland
malgorzata.przybyla-kasperek@us.edu.pl
http://www.us.edu.pl

**Abstract.** Issues that are related to decision making that is based on dispersed knowledge are discussed in the paper. The main aim of the paper is to compare the results obtained using two different methods of conflict analysis in a dispersed decision-making system. The conflict analysis methods, used in the article, are discussed in the paper of Kuncheva et al. [5] and in the paper of Rogova [16]. These methods are used if the individual classifiers generate vectors that represent the probability distributions over different decision. Both methods belong to the class-indifferent group, i.e. methods that use all of decision profile matrices to calculate the support for each class. Also, both methods require training. These methods were used in a dispersed decision-making system which was proposed in the paper [12].

**Keywords:** Decision-making system · Dispersed knowledge · Conflict analysis · Decision templates · Theory of evidence

## 1 Introduction

In the paper, issues that are related to decision making that is based on dispersed knowledge are considered. We use the knowledge that is stored in many different knowledge bases, which is a difficult and challenging issue. In this paper an approach proposed in the article [12] is used. The aim of the paper is to investigate the use of two selected conflict analysis methods in the system with dynamically generated clusters. Two methods known from the literature [5,16] were used to conflict analysis: the method that is based on decision templates and the method that is based on theory of evidence.

The issue of making decisions based on dispersed knowledge is widely discussed in the literature. For example, this issue is discussed in the multiple model approach [3,4]. In the paper [6], an overview of methodological frames and software systems that have been developed for collaborative decision making is presented. The paper [2] systematises and evaluates several existing mechanisms and techniques for group decision-making that limit complexity.

© Springer International Publishing Switzerland 2016
S. Kozielski et al. (Eds.): BDAS 2016, CCIS 613, pp. 109–119, 2016.
DOI: 10.1007/978-3-319-34099-9_7

## 2    A Brief Overview of a Dispersed Decision-Making System

The concept of a dispersed decision-making system is being considered by the author in the papers [11–15,17]. In the paper [12] a system with a dynamic structure has been proposed. During the construction of this system's structure a negotiation stage is used. The main assumptions, notations and definitions of the system are described below. We assume that the knowledge is available in a dispersed form, which means in a form of several decision tables. Each local knowledge base is managed by one agent, which is called a resource agent. We call $ag$ in $Ag = \{ag_1, \ldots, ag_n\}$ a resource agent if he has access to resources represented by a decision table (see [8]) $D_{ag} := (U_{ag}, A_{ag}, d_{ag})$, where $U_{ag}$ is the universe; $A_{ag}$ is a set of conditional attributes, $V_{ag}^a$ is a set of attribute $a$ values; $d_{ag}$ is a decision attribute. We assume that the same decision attribute occurs in all decision tables. We want to designate homogeneous groups of resource agents. The agents who agree on the classification for a test object into the decision classes will be combined in the group. It is realized in two steps. At first initial coalitions are created. Then the negotiation stage is implemented. These two steps are based on the test object classification carried out by the resource agents. For each agent $ag_i \in Ag$ the classification is represented as a vector of values $[\bar{\mu}_{i,1}(x), \ldots, \bar{\mu}_{i,c}(x)]$, whose dimension is equal to the number of decision classes $c = card\{V^d\}$, where $V^d = \bigcup_{ag \in Ag} V_{ag}^d$. This vector will be defined on the basis of certain relevant objects. That is $m_1$ objects from each decision class of the decision tables of agents that carry the greatest similarity to the test object. The value of the parameter $m_1$ is selected experimentally. The value $\bar{\mu}_{i,j}(x)$ is defined as follows:

$$\bar{\mu}_{i,j}(x) = \frac{\sum_{y \in U_{ag_i}^{rel} \cap X_{ag_i}^{v_j}} s(x,y)}{card\{U_{ag_i}^{rel} \cap X_{ag_i}^{v_j}\}}, i \in \{1, \ldots, n\}, j \in \{1, \ldots, c\}, \tag{1}$$

where $U_{ag_i}^{rel}$ is the subset of relevant objects selected from the decision table $D_{ag_i}$ of a resource agent $ag_i$ and $X_{ag_i}^{v_j} = \{x \in U_{ag_i} : d_{ag_i}(x) = v_j\}$ is the decision class of the decision table of resource agent $ag_i$; $s(x,y)$ is the measure of similarity between objects $x$ and $y$. In the experimental part of this paper the Gower similarity measure [13] was used. This measure enables the analysis of data sets that have qualitative, quantitative and binary attributes. On the basis of the vector of values defined above, for each resource agent $ag_i, i \in \{1, \ldots, n\}$ a vector of the rank $[\bar{r}_{i,1}(x), \ldots, \bar{r}_{i,c}(x)]$ is specified. Rank 1 is assigned to the values of the decision attribute that are taken with the maximum level of certainty. In order to create clusters of agents, relations between the agents are defined. These definitions were based on the papers of Pawlak [7,9]. Relations between agents are defined by their views on the classification of the test object $x$ to the decision class. We define the distance between agents $\rho^x$ for the test object $x$: $\rho^x : Ag \times Ag \to [0,1]$ (for more details see [12]).

**Definition 1.** *Let $p$ be a real number, which belongs to the interval $[0, 0.5)$. We say that agents $ag_i, ag_k \in Ag$ are in a friendship relation due to the object $x$, which is written $R^+(ag_i, ag_k)$, if and only if $\rho^x(ag_i, ag_k) < 0.5 - p$. Agents $ag_i, ag_k \in Ag$ are in a conflict relation due to the object $x$, which is written $R^-(ag_i, ag_k)$, if and only if $\rho^x(ag_i, ag_k) > 0.5 + p$. Agents $ag_i, ag_k \in Ag$ are in a neutrality relation due to the object $x$, which is written $R^0(ag_i, ag_k)$, if and only if $0.5 - p \leq \rho^x(ag_i, ag_k) \leq 0.5 + p$.*

The initial cluster due to the classification of object $x$ is the maximum, due to the inclusion relation, subset of resource agents $X \subseteq Ag$ such that $\forall_{ag_i, ag_k \in X} \ R^+(ag_i, ag_k)$. In the second stage of clustering, limitations imposed on compatibility of agents are relaxed. During the negotiation stage, the intensity of the conflict between the two groups of agents is determined by using the generalized distance $\rho_G^x$; $\rho_G^x : 2^{Ag} \times 2^{Ag} \rightarrow [0, \infty]$ (for more details see [12]). For each agent $ag$ that has not been included to any initial clusters, the generalized distance value is determined for this agent and all initial clusters, with which the agent $ag$ is not in a conflict relation and for this agent and other agents without coalition, with which the agent $ag$ is not in a conflict relation. Then the agent $ag$ is included to all initial clusters, for which the generalized distance does not exceed a certain threshold, which is set by the system's user. Also agents without coalition, for which the value of the generalized distance function does not exceed the threshold, are combined into a new cluster. After completion of the second stage of the process of clustering we get the final form of clusters. For each cluster, a superordinate agent is defined, which is called a synthesis agent, $as_j$, where $j$- number of cluster. $As_x$ is a finite set of synthesis agents defined for the clusters that are dynamically generated for test object $x$. By a dispersed decision-making system with dynamically generated clusters we mean $WSD_{Ag}^{dyn} = \langle Ag, \{D_{ag} : ag \in Ag\}, \{As_x : x \text{ is a classified object}\}, \{\delta_x : x \text{ is a classified object}\}\rangle$ where $Ag$ is a finite set of resource agents; $\{D_{ag} : ag \in Ag\}$ is a set of decision tables of resource agents; $As_x$ is a set of synthesis agents, $\delta_x : As_x \rightarrow 2^{Ag}$ is a injective function that each synthesis agent assigns a cluster. Next, an approximated method of the aggregation of decision tables have been used to generate decision tables for synthesis agents (see [11–13] for more details). Based on these aggregated decision tables global decisions are taken using the methods of conflict analysis.

## 3   Methods of Conflict Analysis

In this article, we use two different methods of conflict analysis: the method that is based on decision templates and the method that is based on theory of evidence. The first method was proposed by Kuncheva et al. in the paper [5] and the second method was proposed by Rogova in the paper [16]. Both methods are quite sophisticated, computationally complex and require training. These methods are used if the individual classifiers generate vectors of probabilities instead of unique class choices. Below the methods will be discussed, then the proposed modification introduced in order to apply to the dispersed decision-making system will be presented. We assume that each $j$-th base classifier (agent) generates a vector of probabilities $[\mu_{j,1}(x), \ldots, \mu_{j,c}(x)]$ for object $x$, where $c$ is the number of all of the decision classes. The following conditions are fulfilled $\forall_{i \in \{1,\ldots,c\}} \mu_{j,i}(x) \in [0,1]$, and $\sum_{i=1}^{c} \mu_{j,i}(x) > 0$. In the paper [5], it was proposed that the classifier outputs can be organized in a decision profile (DP) as the matrix. The decision profile is a matrix with dimensions $L \times c$, where $L$ is the number of base classifiers and $c$ is the number of all of the decision classes. The decision profile is defined as follows

$$DP(x) = \begin{bmatrix} \mu_{1,1}(x) \cdots \mu_{1,i}(x) \cdots \mu_{1,c}(x) \\ \cdots \\ \mu_{L,1}(x) \cdots \mu_{L,i}(x) \cdots \mu_{L,c}(x) \end{bmatrix}$$

The $j$-th row of the matrix saves the output of $j$-th base classifier and the $i$-th column of the matrix saves support from base classifiers for decision class $i$.

Some methods calculate the support for class $i$ using only the $i$-th column of $DP(x)$. But the methods that are considered in this paper use all of $DP(x)$ to calculate the support for each class. Such fusion methods will be called class-indifferent [5]. On the basis of decision profiles, that are constructed for the objects from the training set, decision templates of each class are defined. The decision template $DT_i$ for class $i$ is the average of the decision profiles of the objects of the training set labeled in class $i$, thus $DT_i = \frac{1}{card\{Z_i\}} \sum_{x \in Z_i} DP(x)$, where $Z_i$ is a set of objects from the training set that belong to the class $i$.

**Method that is Based on Decision Templates.** The training process in this method consists in determining decision templates of each decision class. For the test object $x$, at first the decision profile of object $x$ will be built, and then the value of a certain similarity measure between the $DP(x)$ and decision templates $DT_i$ of each class $i \in \{1, \ldots, c\}$ will be calculated. The higher the similarity between the decision profile of the object $x$ and the decision template for class $i$, the higher the support for that class.

**Method that is Based on Theory of Evidence.** In this method as with the decision templates method, the decision templates, $DT_i, i \in \{1, \ldots, c\}$ are designated from the data. Then, instead of calculating the similarity between the decision template $DT_i$ and the decision profile $DP(x)$, in this method the Dempster-Shafer theory is used and belief is calculated. The following steps are performed in the Dempster-Shafer algorithm:

1. Let $DT_i(m, \cdot)$ denote the $m$-th row of the decision template for class $i$ and $DP_{m,\cdot}(x)$ denote the $m$-th row of the decision profile for the object $x$. The proximity between the prediction of the $m$-th base classifier $DP_{m,\cdot}(x)$ and the $m$-th row of the decision template for every class $i \in \{1, \ldots, c\}$ and for every classifier $m \in \{1, \ldots, L\}$ is calculated $\phi_{i,m}(x) = \frac{(1+\|DT_i(m,\cdot)-DP_{m,\cdot}(x)\|^2)^{-1}}{\sum_{k=1}^{c}(1+\|DT_k(m,\cdot)-DP_{m,\cdot}(x)\|^2)^{-1}}$ where $\|\cdot\|$ is norm, in the paper the Euclidean norm was applied.

2. For every class $i \in \{1, \ldots, c\}$ and for every classifier $m \in \{1, \ldots, L\}$ the following belief degrees is calculated $Bel_i(DP_{m,\cdot}(x)) = \frac{\phi_{i,m}(x)\prod_{k \neq i}(1-\phi_{k,m}(x))}{1-\phi_{i,m}(x)[1-\prod_{k\neq i}(1-\phi_{k,m}(x))]}$.

3. The Dempster-Shafer membership degrees for every class $i \in \{1, \ldots, c\}$ is calculated $\mu_i(x) = K \prod_{m=1}^{L} Bel_i(DP_{m,\cdot}(x))$ where $K$ is a constant which ensure that $\mu_i(x) \leq 1$.

In order to apply these methods in the dispersed decision-making system, some modifications were adopted. The beginning of the calculation is the same for both methods. At first the decision profiles of the resource agents for training objects are calculated

$$DP^{Ag}(x) = \begin{bmatrix} \bar{\mu}_{1,1}(x) & \cdots & \bar{\mu}_{1,i}(x) & \cdots & \bar{\mu}_{1,c}(x) \\ \cdots & & & & \\ \bar{\mu}_{card\{Ag\},1}(x) & \cdots & \bar{\mu}_{card\{Ag\},i}(x) & \cdots & \bar{\mu}_{card\{Ag\},c}(x) \end{bmatrix}$$

the values $\bar{\mu}$ are defined in Formula 1. Based on the decision profiles of the resource agents, the decision templates of the resource agents are determined $DT_i^{Ag} = \frac{1}{card\{Z_i\}} \sum_{x \in Z_i} DP^{Ag}(x)$, where $Z_i$ is a set of objects from the training set that belong to the class $i$. For the test object $x$, the clusters and the system structure are generated as described in Sect. 2. In both discussed methods of conflict analysis, the training process consists in determining the decision templates of the synthesis agents for each class $DT_i^{As_x}$, $i \in \{1, \ldots, c\}$. The decision template is a matrix with dimensions $card\{As_x\} \times c$. The decision templates of the synthesis agents is determined based on the decision templates of the resource agents in the following way. The $j$-th row of the decision template should saves the output of $j$-th synthesis agent. The $j$-th row of the decision template is calculated as the average of the rows of decision templates of the resource agents that correspond to the resource agents that belong to the cluster subordinate to $j$-th synthesis agent

$$DT_i^{As_x} = \begin{bmatrix} \frac{\sum_{agp \in \delta_x(as_1)} DT_i^{Ag}(p,1)}{card\{\delta_x(as_1)\}} & \cdots & \frac{\sum_{agp \in \delta_x(as_1)} DT_i^{Ag}(p,c)}{card\{\delta_x(as_1)\}} \\ \cdots & & \\ \frac{\sum_{agp \in \delta_x(as_k)} DT_i^{Ag}(p,1)}{card\{\delta_x(as_k)\}} & \cdots & \frac{\sum_{agp \in \delta_x(as_k)} DT_i^{Ag}(p,c)}{card\{\delta_x(as_k)\}} \end{bmatrix}$$

where $As_x = \{as_1, \ldots, as_k\}$ and $DT_i^{Ag}(p,l)$ is an element at the $p$-th row and the $l$-th column of the matrix $DT_i^{Ag}$. Then the decision profile of the test object $DP(x)$ is defined in the following way. On the basis of each aggregated decision table of synthesis agent a vector of probabilities is generated. A $c$-dimensional vector of values $[\mu_{j,1}(x), \ldots, \mu_{j,c}(x)]$ is generated for each $j$-th cluster. This vector will be defined on the basis of relevant objects. From each aggregated decision table and from each decision class, the smallest set containing at least $m_2$ objects for which the values of conditional attributes bear the greatest similarity to the test object is chosen. The value of the parameter $m_2$ is selected experimentally. The value $\mu_{j,i}(x)$ is equal to the average value of the Gower similarity of the test object to the relevant objects form $j$-th aggregated decision table, belonging to the decision class $v_i$. In this way, for each cluster the vector of probabilities is generated. The decision profile of the test object is defined as follows

$$DP(x) = \begin{bmatrix} \mu_{1,1}(x) & \cdots & \mu_{1,i}(x) & \cdots & \mu_{1,c}(x) \\ \cdots & & & & \\ \mu_{card\{As_x\},1}(x) & \cdots & \mu_{card\{As_x\},i}(x) & \cdots & \mu_{card\{As_x\},c}(x) \end{bmatrix}$$

From this moment, both methods have different course.

The next step in the method that is based on decision templates is to calculate the similarity measure between the decision profile of the test object and the decision templates $DT_i^{As_x}$ of each class $i \in \{1, \ldots, c\}$. In the paper four different similarity measures were used:

1. The similarity measure that uses the normalized Euclidean distance
$$s(DP(x), DT_i^{As_x}) = 1 - \frac{1}{card\{As_x\} \cdot c} \sum_{m=1}^{card\{As_x\}} \sum_{l=1}^{c} \left( DP_{m,l}(x) - DT_i^{As_x}(m,l) \right)^2,$$
where $DP_{m,l}(x)$ and $DT_i^{As_x}(m,l)$ is an element at the $m$-th row and the $l$-th column of the matrix $DP(x)$ or $DT_i^{As_x}$ respectively.

2. The similarity measure that uses the symmetric difference defined by the Hamming distance
$$s(DP(x), DT_i^{As_x}) = 1 - \frac{1}{card\{As_x\} \cdot c} \sum_{m=1}^{card\{As_x\}} \sum_{l=1}^{c} |DP_{m,l}(x) - DT_i^{As_x}(m,l)|$$

3. The Jaccard similarity coefficient
$$s(DP(x), DT_i^{As_x}) = \frac{\sum_{m=1}^{card\{As_x\}} \sum_{l=1}^{c} \min\{DP_{m,l}(x), DT_i^{As_x}(m,l)\}}{\sum_{m=1}^{card\{As_x\}} \sum_{l=1}^{c} \max\{DP_{m,l}(x), DT_i^{As_x}(m,l)\}}$$

4. The similarity measure that uses the symmetric difference
$$s(DP(x), DT_i^{As_x}) = 1 - \frac{1}{card\{As_x\} \cdot c} \cdot \sum_{m=1}^{card\{As_x\}} \sum_{l=1}^{c} \max\{\min\{DP_{m,l}(x), 1 - DT_i^{As_x}(m,l)\}, \min\{1 - DP_{m,l}(x), DT_i^{As_x}(m,l)\}\}$$

All these measures were also used by the authors of the method based on decision templates in the paper [5]. The global decision is defined by selecting the decision that has the maximum value of similarity.

The next step in the method that is based on theory of evidence is to calculate the steps in the Dempster-Shafer algorithm. For this purpose, we use the decision templates of the synthesis agents $DT_i^{As_x}$, $i \in \{1, \ldots, c\}$ and the decision profile of the test object $DP(x)$. The global decision is defined by selecting the decision that has the maximum value of the Dempster-Shafer membership degrees.

## 4   Experiments

The aim of the experiments is to compare the quality of the classification made by the dispersed decision-making system using two different methods of conflict analysis. The method that is based on decision templates and the method that is based on theory of evidence were considered. These methods are used when vectors of probabilities are generated by the base classifiers. For the experiments the following data, which are in the UCI repository [1], were used: Soybean data set and Vehicle Silhouettes data set. A numerical summary of the data sets is as follows: Soybean: # The training set - 307; # The test set - 376; # Conditional - 35; # Decision - 19; Vehicle Silhouettes: # The training set - 592; #

The test set - 254; # Conditional - 18; # Decision - 4. In order to determine the efficiency of inference of the dispersed decision-making system with respect to the analyzed data, the Vehicle Silhouettes data set was divided into two disjoint subsets: a training set and a test set. The Soybean data set is available on the UCI repository website in a divided form: a training and a test set. Because the available data sets are not in the dispersed form, in order to test the dispersed decision-making system the training set was divided into a set of decision tables. For each of the data sets used, the decision-making system with five different versions (with 3, 5, 7, 9 and 11 decision tables) were considered. For these systems, we use the following designations: $WSD_{Ag1}^{dyn}$ - 3 decision tables; $WSD_{Ag2}^{dyn}$ - 5 decision tables; $WSD_{Ag3}^{dyn}$ - 7 decision tables; $WSD_{Ag4}^{dyn}$ - 9 decision tables; $WSD_{Ag5}^{dyn}$ - 11 decision tables. In the description of the results of experiments for clarity some designations for algorithms and parameters have been adopted: $m_1$ - parameter which determines the number of relevant objects that are selected from each decision class of the decision table and are then used in the process of cluster generation; $p$ - parameter which occurs in the definition of friendship, conflict and neutrality relations; $A(m)$ - the approximated method of the aggregation of decision tables; $C(m_2)$ - the method of conflict analysis (the method that is based on decision templates and the method that is based on theory of evidence), with parameter which determines the number of relevant objects that are used to generate the decision profile of the test object. The measure of determining the quality of the classification is the error rate, that is equal to a fraction of wrong predictions. Note that the system generates unambiguous decisions. The process of parameters optimization was carried out as follows. A series of tests for different parameter values were performed: $m_1 \in \{1, 6, 11\}$, $m, m_2 \in \{1, \ldots, 10\}$ and $p \in \{0.05, 0.15\}$. Thus, for each of the considered dispersed systems, 600 tests were conducted ($1000 = 3 \cdot 10 \cdot 10 \cdot 2$). From all of the obtained results, one was selected that guaranteed a minimum value of error rate. In tables presented below the best results, obtained for optimal values of the parameters, are given. The results of the experiments with the Soybean data set are presented in Table 1. In the table the following information is given: the name of dispersed decision-making system (System); the selected, optimal parameter values (Parameters); the algorithm's symbol (Algorithm); the error rate $e$; the time $t$ needed to analyze a test set expressed in minutes. In the table below the best results are bolded. As can be seen, for the Soybean data set, the best results were achieved for the method that is based on theory of evidence and for the method that is based on decision templates with the similarity measure that uses the symmetric difference defined by the Hamming distance. An important property of the methods considered in this paper, is that they generate unambiguous decisions without any ties. In the paper [10] experiments with the Soybean data set were also described. These results were obtained by using a dispersed decision-making system, wherein the method of forming the system structure is the same as that presented in this paper. However, in this approach different methods of conflict analysis were considered. When we compare the results from this paper with the results presented in the paper [10] it can be concluded that the simple fusion methods give surprisingly good results. For example, simple methods like the sum rule and the product rule give better results than the complex method discussed in this paper.

**Table 1.** Summary of experiments results with the Soybean data set

| Decision templates and the normalized Euclidean distance | | | | |
|---|---|---|---|---|
| System | Parameters | Algorytm | $e$ | $t$ |
| $WSD_{Ag1}^{dyn}$ | $m_1 = 1, p = 0.15$ | $A(3)C(1)$ | 0.335 | 0.08 |
| $WSD_{Ag2}^{dyn}$ | $m_1 = 1, p = 0.05$ | $A(2)C(4)$ | 0.226 | 0.09 |
| $WSD_{Ag3}^{dyn}$ | $m_1 = 6, p = 0.05$ | $A(4)C(4)$ | 0.176 | 0.15 |
| $WSD_{Ag4}^{dyn}$ | $m_1 = 6, p = 0.15$ | $A(1)C(1)$ | 0.202 | 0.23 |
| $WSD_{Ag5}^{dyn}$ | $m_1 = 6, p = 0.05$ | $A(1)C(1)$ | 0.218 | 2.36 |
| Decision templates and the Hamming distance | | | | |
| System | Parameters | Algorytm | $e$ | $t$ |
| $WSD_{Ag1}^{dyn}$ | $m_1 = 1, p = 0.15$ | $A(1)C(1)$ | **0.290** | 0.05 |
| $WSD_{Ag2}^{dyn}$ | $m_1 = 11, p = 0.05$ | $A(1)C(1)$ | 0.229 | 0.07 |
| $WSD_{Ag3}^{dyn}$ | $m_1 = 6, p = 0.05$ | $A(1)C(2)$ | 0.186 | 0.11 |
| $WSD_{Ag4}^{dyn}$ | $m_1 = 11, p = 0.15$ | $A(1)C(1)$ | 0.221 | 0.24 |
| $WSD_{Ag5}^{dyn}$ | $m_1 = 11, p = 0.05$ | $A(3)C(1)$ | **0.210** | 2.38 |
| Decision templates and the Jaccard similarity coefficient | | | | |
| System | Parameters | Algorytm | $e$ | $t$ |
| $WSD_{Ag1}^{dyn}$ | $m_1 = 6, p = 0.05$ | $A(1)C(1)$ | 0.295 | 0.05 |
| $WSD_{Ag2}^{dyn}$ | $m_1 = 1, p = 0.05$ | $A(1)C(1)$ | 0.231 | 0.06 |
| $WSD_{Ag3}^{dyn}$ | $m_1 = 6, p = 0.15$ | $A(1)C(1)$ | 0.197 | 0.10 |
| $WSD_{Ag4}^{dyn}$ | $m_1 = 6, p = 0.15$ | $A(1)C(1)$ | 0.210 | 0.23 |
| $WSD_{Ag5}^{dyn}$ | $m_1 = 11, p = 0.05$ | $A(1)C(1)$ | 0.229 | 2.36 |
| Decision templates and the symmetric difference | | | | |
| System | Parameters | Algorytm | $e$ | $t$ |
| $WSD_{Ag1}^{dyn}$ | $m_1 = 6, p = 0.05$ | $A(3)C(3)$ | 0.391 | 0.07 |
| $WSD_{Ag2}^{dyn}$ | $m_1 = 11, p = 0.05$ | $A(5)C(5)$ | 0.399 | 0.11 |
| $WSD_{Ag3}^{dyn}$ | $m_1 = 6, p = 0.05$ | $A(2)C(3)$ | 0.301 | 0.13 |
| $WSD_{Ag4}^{dyn}$ | $m_1 = 1, p = 0.05$ | $A(3)C(1)$ | 0.375 | 0.41 |
| $WSD_{Ag5}^{dyn}$ | $m_1 = 11, p = 0.05$ | $A(4)C(7)$ | 0.327 | 2.44 |
| Theory of evidence | | | | |
| System | Parameters | Algorytm | $e$ | $t$ |
| $WSD_{Ag1}^{dyn}$ | $m_1 = 6, p = 0.15$ | $A(1)C(1)$ | 0.293 | 0.06 |
| $WSD_{Ag2}^{dyn}$ | $m_1 = 1, p = 0.05$ | $A(1)C(2)$ | **0.194** | 0.06 |
| $WSD_{Ag3}^{dyn}$ | $m_1 = 11, p = 0.05$ | $A(1)C(1)$ | **0.157** | 0.10 |
| $WSD_{Ag4}^{dyn}$ | $m_1 = 11, p = 0.15$ | $A(1)C(1)$ | **0.178** | 0.24 |
| $WSD_{Ag5}^{dyn}$ | $m_1 = 11, p = 0.05$ | $A(1)C(1)$ | 0.221 | 2.35 |

**Table 2.** Summary of experiments results with the Vehicle Silhouettes data set

| Decision templates and the normalized Euclidean distance | | | | |
|---|---|---|---|---|
| System | Parameters | Algorytm | $e$ | $t$ |
| $WSD_{Ag1}^{dyn}$ | $m_1 = 1, p = 0.05$ | $A(1)C(1)$ | 0.331 | 0.12 |
| $WSD_{Ag2}^{dyn}$ | $m_1 = 6, p = 0.05$ | $A(9)C(1)$ | 0.480 | 0.33 |
| $WSD_{Ag3}^{dyn}$ | $m_1 = 1, p = 0.05$ | $A(1)C(1)$ | 0.406 | 0.16 |
| $WSD_{Ag4}^{dyn}$ | $m_1 = 11, p = 0.05$ | $A(7)C(1)$ | 0.496 | 4.15 |
| $WSD_{Ag5}^{dyn}$ | $m_1 = 11, p = 0.05$ | $A(3)C(2)$ | 0.547 | 2.33 |
| Decision templates and the Hamming distance | | | | |
| System | Parameters | Algorytm | $e$ | $t$ |
| $WSD_{Ag1}^{dyn}$ | $m_1 = 1, p = 0.05$ | $A(2)C(1)$ | **0.323** | 0.11 |
| $WSD_{Ag2}^{dyn}$ | $m_1 = 6, p = 0.05$ | $A(10)C(1)$ | **0.433** | 0.38 |
| $WSD_{Ag3}^{dyn}$ | $m_1 = 1, p = 0.05$ | $A(8)C(1)$ | **0.335** | 0.36 |
| $WSD_{Ag4}^{dyn}$ | $m_1 = 6, p = 0.05$ | $A(5)C(1)$ | **0.445** | 0.59 |
| $WSD_{Ag5}^{dyn}$ | $m_1 = 11, p = 0.05$ | $A(5)C(1)$ | 0.472 | 4.41 |
| Decision templates and the Jaccard similarity coefficient | | | | |
| System | Parameters | Algorytm | $e$ | $t$ |
| $WSD_{Ag1}^{dyn}$ | $m_1 = 1, p = 0.05$ | $A(2)C(1)$ | 0.327 | 0.12 |
| $WSD_{Ag2}^{dyn}$ | $m_1 = 6, p = 0.05$ | $A(10)C(1)$ | **0.433** | 0.38 |
| $WSD_{Ag3}^{dyn}$ | $m_1 = 1, p = 0.05$ | $A(8)C(1)$ | 0.339 | 0.36 |
| $WSD_{Ag4}^{dyn}$ | $m_1 = 1, p = 0.05$ | $A(4)C(1)$ | 0.449 | 0.30 |
| $WSD_{Ag5}^{dyn}$ | $m_1 = 11, p = 0.05$ | $A(5)C(1)$ | **0.469** | 4.42 |
| Decision templates and the symmetric difference | | | | |
| System | Parameters | Algorytm | $e$ | $t$ |
| $WSD_{Ag1}^{dyn}$ | $m_1 = 1, p = 0.05$ | $A(1)C(2)$ | 0.472 | 0.12 |
| $WSD_{Ag2}^{dyn}$ | $m_1 = 6, p = 0.05$ | $A(8)C(1)$ | 0.547 | 0.28 |
| $WSD_{Ag3}^{dyn}$ | $m_1 = 11, p = 0.05$ | $A(7)C(1)$ | 0.457 | 0.48 |
| $WSD_{Ag4}^{dyn}$ | $m_1 = 1, p = 0.05$ | $A(5)C(1)$ | 0.508 | 0.39 |
| $WSD_{Ag5}^{dyn}$ | $m_1 = 6, p = 0.05$ | $A(2)C(1)$ | 0.551 | 2.27 |
| Theory of evidence | | | | |
| System | Parameters | Algorytm | $e$ | $t$ |
| $WSD_{Ag1}^{dyn}$ | $m_1 = 1, p = 0.05$ | $A(1)C(1)$ | 0.331 | 0.12 |
| $WSD_{Ag2}^{dyn}$ | $m_1 = 6, p = 0.05$ | $A(9)C(1)$ | 0.480 | 0.32 |
| $WSD_{Ag3}^{dyn}$ | $m_1 = 1, p = 0.05$ | $A(1)C(1)$ | 0.406 | 0.16 |
| $WSD_{Ag4}^{dyn}$ | $m_1 = 6, p = 0.05$ | $A(6)C(1)$ | 0.500 | 1.48 |
| $WSD_{Ag5}^{dyn}$ | $m_1 = 11, p = 0.05$ | $A(3)C(2)$ | 0.543 | 2.33 |

The results of the experiments with the Vehicle data set are presented in Table 2. In the table the best results are bolded. In the case of the Vehicle Silhouettes data set, the system with the method that is based on decision templates and the similarity measure that uses the symmetric difference defined by the Hamming distance generates the best results. For this data set also methods like the sum rule and the product rule (considered in the paper [10]) generate better results. As Kuncheva et al. noticed in the paper [5] "it is somewhat surprising to see how well the simple aggregation rules, with no second-level training, compete with the more sophisticated ones. One problem with simple aggregation is, that although they have good overall performance, it is not clear which one is good for a particular data set." In addition it is noted that the advantages of methods, which are based on decision templates, are that they do not rely on questionable assumptions (as attributes independence) and are less likely to overtrain.

## 5   Conclusion

In this article, two different methods of conflict analysis were used in the dispersed decision-making system: the method that is based on decision templates and the method that is based on theory of evidence. In the first method, four different similarity measures were applied: the similarity measure that uses the normalized Euclidean distance, the similarity measure that uses the symmetric difference defined by the Hamming distance, the Jaccard similarity coefficient and the similarity measure that uses the symmetric difference. In the experiments, which are presented, dispersed data were used: Soybean data set and Vehicle Silhouettes data set. Based on the presented results of experiments it can be concluded that the method that is based on decision templates and the similarity measure that uses the symmetric difference defined by the Hamming distance generates the best results.

## References

1. http://mlr.cs.umass.edu/ml/datasets.html
2. Bregar, A.: Towards a framework for the measurement and reduction of user-perceivable complexity of group decision-making methods. IJDSST **6**(2), 21–45 (2014)
3. Gatnar, E.: Multiple-Model Approach to Classification and Regression. PWN, Warsaw (2008)
4. Kuncheva, L.I.: Combining Pattern Classifiers Methods and Algorithms. Wiley, Chichester (2004)
5. Kuncheva, L.I., Bezdek, J.C., Duin, R.P.W.: Decision templates for multiple classifier fusion: an experimental comparison. Pattern Recogn. **34**(2), 299–314 (2001)
6. Matsatsinis, N., Samaras, A.P.: MCDA and preference disaggregation in group decision support systems. EJOR **130**(2), 414–429 (2001)
7. Pawlak, Z.: On conflicts. Int. J. Man-Mach. Stud. **21**(2), 127–134 (1984)
8. Pawlak, Z.: Rough Sets: Theoretical Aspects of Reasoning about Data. Kluwer, Dordrecht (1991)

9. Pawlak, Z.: An inquiry into anatomy of conflicts. Inf. Sci. **109**(1–4), 65–78 (1998)
10. Przybyła-Kasperek, M.: Selected methods of combining classifiers, when predictions are stored in probability vectors, in a dispersed decision-making system. In: 24th International Workshop, CS&P, pp. 211–222 (2015)
11. Przybyła-Kasperek, M., Wakulicz-Deja, A.: Application of reduction of the set of conditional attributes in the process of global decision-making. Fundam. Inform. **122**(4), 327–355 (2013)
12. Przybyła-Kasperek, M., Wakulicz-Deja, A.: A dispersed decision-making system - the use of negotiations during the dynamic generation of a system's structure. Inf. Sci. **288**, 194–219 (2014)
13. Przybyła-Kasperek, M., Wakulicz-Deja, A.: Global decision-making system with dynamically generated clusters. Inf. Sci. **270**, 172–191 (2014)
14. Przybyła-Kasperek, M.: Global decisions taking process, including the stage of negotiation, on the basis of dispersed medical data. In: Kozielski, S., Mrozek, D., Kasprowski, P., Małysiak-Mrozek, B., Kostrzewa, D. (eds.) BDAS 2014. CCIS, vol. 424, pp. 290–299. Springer, Heidelberg (2014)
15. Przybyła-Kasperek, M.: Application of the Shapley-Shubik power index in the process of decision making on the basis of dispersed medical data. In: Kozielski, S., Mrozek, D., Kasprowski, P., Małysiak-Mrozek, B., Kostrzewa, D. (eds.) BDAS 2015. CCIS, vol. 521, pp. 277–287. Springer, Heidelberg (2015)
16. Rogova, G.L.: Combining the results of several neural network classifiers. Neural Netw. **7**(5), 777–781 (1994)
17. Wakulicz-Deja, A., Przybyla-Kasperek, M.: Application of the method of editing and condensing in the process of global decision-making. Fundam. Inform. **106**(1), 93–117 (2011)

# Methods for Selecting Nodes for Maximal Spread of Influence in Recommendation Services

Bogdan Gliwa and Anna Zygmunt[✉]

AGH University of Science and Technology, Al. Mickiewicza 30,
30-059 Kraków, Poland
{bgliwa,azygmunt}@agh.edu.pl

**Abstract.** Social network analysis is a tool to assess social interactions between people e.g. in the Internet. One of the most active areas in this field are modeling influence of users and finding influential users. These areas have many applications, e.g., in marketing, business or politics. Several models of influence have been described in literature, but there is no single model that best describes the process of spreading entities (e.g. information, behaviour) through the network. Interesting and practical problem is how to choose a small number of users that will guarantee maximal spread of entities over the whole network (*influence maximization problem*). In this paper we studied this problem using various centrality metrics with different models of influence propagation. Experiments were conducted on three, real-world datasets regarding the domain of recommendation services.

**Keywords:** Influence · Influence maximization problem · Influence models · Centrality measures · Social networks · Social media

## 1 Introduction

Social network analysis (SNA) is an active and attractive area of research in computer science. It describes social relationships of users in the form of a network. Such a network, however, is not homogeneous and we can distinguish some parts (called groups) that are more dense than others. One of the directions of research is analysis of groups, their structure and dynamics. Another important aspect of analysis is the position of users in a network which may be described using centrality metrics. To the most interesting and promising directions of research we can also include finding influential users and analysis of influence that social network's users have on each other. Such studies are widely used mainly in business, marketing, but also in epidemiology and politics. In today's viral marketing, potential customers will pass each other information about the company, services or products. In epidemiology, it is important to detect sources of an epidemic and identifying trends.

Recommendation services are more and more popular among the Internet's users. People use them to express their opinions about products or services.

© Springer International Publishing Switzerland 2016
S. Kozielski et al. (Eds.): BDAS 2016, CCIS 613, pp. 120–134, 2016.
DOI: 10.1007/978-3-319-34099-9_8

Such portals help them to make a decision about buying a product or using a service. Consumers often take into consideration the recommendations and opinions of trusted people so the use of algorithms that determine influentials can help, among others, in conducting a successful advertising campaign, reducing the evolution of the disease or predicting trends of popularity of a political party, a piece of music or a film. There are many interesting questions: why one user is more influential than other, what conditions must arise in the neighborhood that the user will become influential, if one is influential, it means that it has good network position or maybe some individual characteristics are important? Several models of influence have been proposed (e.g., IC, LT, LTC). In the context of these works, an interesting direction seems to be how to choose a small number of users (within predefined budget for given number of people) that will guarantee maximal spread over the whole network, which is called *influence maximization problem.*

The majority of work regarding the topic of influence in social networks has been dedicated to studying three different aspects of influence analysis: (1) influence modeling, (2) finding influential users, (3) influence maximization problem. In [16] we compared various influence models and proposed our new algorithm which takes into account users behaviours and their expertise.

In this paper we study the problem of influence maximization. We present an approach based on integrating various influence models (IC, ICN, LT, LTC, LTB) with different centrality measures (degrees, closeness, PageRank, eigenvector, betweenness). We study this in three different recommendation services: *Last.fm* (music recommendations), *Flixster* (movies recommendations) and *Yelp* (focused on reviews on local businesses, e.g., restaurants).

The remainder of this paper is organized as follows: Sect. 2 presents an overview of research in the area of the influence maximization problem. Section 3 gives a description of concept and methods. Section 4 presents experimental evaluation, and Sect. 5 provides some final conclusions and presents future directions.

## 2    Related Work

Presenting interactions between users of the given social media site in the form of social network allows discovering important users as well as observing how much influenced they are. Understanding and modeling the spread of influence is an important topic in social network analysis and therefore attracts many researchers to this area.

Spread of an entity (e.g., information or products) in the social network can be seen as cascading actions taken by adopting users (e.g., through commenting, evaluation or purchase the entity). Typically, a group of the initiators adopt the entity, then *infect* their neighbors (e.g., by sending messages, recommendations), which in turn may also decide on the entity adoption continuing its spread. Analysis of this process relates to a study of the structure of social networks, as well as the ability and the susceptibility of the individual users during the adoption of the entity [6].

Influence modeling is often seen as a variation of an epidemic spread problem. Many models have been proposed, namely Susceptible-Infected-Recovered (SIR [11]) model and its variations (e.g., SIS, SIRS, SEIS). These approaches assume an implicit network and do not include the chronological sequence of infections. Therefore, they are only able to predict general trends such as the percentage of population being infected or cured.

Kempe et al. in [10] investigated the *Linear Threshold Model (LT)* and *Independent Cascade (IC)* to simulate exact propagation of information in social networks. The *IC* model may be considered as a push model, because active nodes independently try to propagate information to their neighborhoods. In contrast, the *LT* model may be described as a pull model, where inactive nodes become activated when the majority of their neighbors (with respect to a certain threshold) have already accepted the information. Both models have remained a strong foundation for successive studies. The *Linear Threshold Colors (LT-C)* model was introduced to capture product adoption rather than influence [1]. Each node after activation can end up in one of three states/colors: *Adopt, Promote* or *Inhibit*. The two latter states correspond to the users who report positively (*Promote*) or discourage the product without actually adopting it (*Inhibit*). In [21] *Linear Threshold Value (LT-V)* assumes a distinction between users who cannot adopt certain product due to, e.g., a high price, and those who can. In [2], Chen et al. defined the *Independent Cascade Model with Negative Opinions (IC-N)* that allows for modeling influence of both positive and negative opinions. New model, taking into account (apart from user's connections) both users behaviours and their expertise was proposed in [16].

In our previous experiment described in [16] we showed that the accuracy of particular model highly depends on the kind of the entity that is being spread through the network and parameters that are associated with it (e.g., price, knowledge, accessibility).

The field of research dedicated to influence maximization problem was formalized by Kempe et al. in [10]. The idea is to find minimal set of nodes (within given budget), that will eventually result in maximal spread of information over the whole network. The problem is NP-hard under IC (Independent Cascade) and LT (Linear Threshold) model.

There are several approaches to solve influence maximization problem. In the simplest one, the centrality measure is chosen (e.g., degree centrality [20], closeness centrality [13], betweenness centrality [5], eigenvector centrality [18], PageRank [17]) together with selected model of propagation. The measure is then calculated for each node and those nodes with the highest value are selected. Centrality measures are based only on the structure of graph and nodes position. However, several efficient algorithms for effective implementation of the calculations [12] have been proposed.

Kempe in [10] proposed to use greedy algorithm and Monte Carlo simulation. Initial nodes are chosen in such way that in subsequent time steps they activate the greatest number of their neighbors. Process of neighbors activation is random, so selecting the appropriate node in each step is nondeterministic.

Therefore, the node activation process is repeated in several hundred iterations of the Monte Carlo simulation, checking thus the number of neighbors that a node is able to activate in each iteration. The node with the highest average number of active neighbors in all iterations is chosen [18].

Goyal in [7] presented an approach that completely omitted models of propagation. To determine the initial set of nodes, the method used only traces of the previous propagating information in the network. Method did not model neither the network nor actions taken by users. Narayanam proposed in [15] approach based on game theory. Presented algorithm *SPIN* (Sharply Value-Based Influential Nodes) turned out to be more time efficient than greedy algorithms.

Chen et al. in [3] presented new algorithms for influence maximization that employed community structure to improve nodes' selection. There are also some approaches to influence maximization in dynamic social networks (when the network changes over time) [22].

## 3   Concept and Methods

In this section we describe basic definitions used in the task of influence maximization problem and later, we provide descriptions of influence propagation models used in the experiments.

### 3.1   Basic Definitions

To describe different models of influence propagation, we introduced the following definitions:

– a *social network* is a directed weighted graph denoted $G = (V, E, W)$ where $V$ is a set of nodes representing users of the network, $E$ is a set of directed edges representing social relationship between the users, and $W$ is a set of weights for the edges. Notation $V_{in}(v)$ stands for all in-neighbors of the node $v$ and $V_{out}(v)$ denotes all out-neighbors of the node $v$. The weight $w_e \in W$ of the directed edge $(u, v)$ between the nodes $u$ and $v$ is an indicator of the influence strength and its weight implies probability of $u$ influencing $v$. The edge weight is calculated as stated in Eq. 1

$$w_{u,v} = \frac{r_{v,u}}{\sum_{n \in V_{in}(v)} r_{v,n}}, w_{u,v} \in [0; 1], \tag{1}$$

where $r_{a,b}$ means the number of entities that user $a$ rated after user $b$. This formulation implies the following property:

$$\sum_{n \in V_{in}(u)} w_{n,u} = 1. \tag{2}$$

– an *entity* represents a thing that is being spread through the network, e.g., a movie, an artist, or a product. Each user may evaluate the entity with appropriate numeric rating/evaluation.

– *a node activation* is an action of accepting the entity by a given node. Such an action depends on a particular social service and may correspond to: liking, tagging, sharing, evaluating or commenting the entity. Time variable must be associated with each activation to let us order the sequence of nodes' activations. $A_e$ denotes a set of active nodes during propagation of the entity $e$.

– *a seed set* is an initial set of active nodes that begins the propagation in the network. In the task of influence maximization a seed set is chosen by the order of nodes' ranks that can be determined using different metrics.

– *a coverage* is a final set of activated nodes at the end of simulation (when the propagation of entities is finished). Intuitively, it is a spread or a range of the entity in the network.

Influence maximization task involves the selection of a method of ranking nodes to achieve the maximum coverage using given size of seed set. In experiments we repeated each simulation for every seed size in a given model 100 times and all figures present mean values of a coverage.

## 3.2   IC Model

The *Independent Cascade Model* (denoted *IC*) is a one of the basic models used in the influence modeling which was studied by Kempe in [10]. Each edge has assigned activation probability corresponding to a weight $w$ described in Sect. 3.1. Each node can be either in one of two states – *Active* or *Inactive*. The process of nodes activation starts with a seed $A_e^0$ and during the process continues to activate further nodes as follows: when a node $u$ becomes activated in step $t$, it gets only single chance to activate each of its inactive neighbors, denoted $n$ with a probability $p = w_{u,n}$. Once the node is activated, it cannot be deactivated. The process runs until no more activations are possible.

## 3.3   ICN Model

The *Independent Cascade Model with Negative Opinions* (denoted *ICN*) [2] is an extension of *IC* model that allows for modeling of both positive and negative opinions. Each node has one of three states – *neutral, positive* or *negative*. The node is treated as activated in a timestep $t$ if it is positive or negative in a timestep $t$ and *neutral* in the previous timestep. *ICN* model has a parameter $q$ (*quality factor*) describing the probability of staying *positive* when was activated by *positive* neighbour. Initially, all nodes from seed set are either *positive* (with probability $q$) or *negative* (with probability $1 - q$). In each timestep, they try to activate their inactive neighbours. If *negative* node infects one of its neighbour, then such neighbour becomes also *negative*, but in the case when infecting node is *positive* then the neighbour may become either *positive* (with probability $q$) or *negative* (with probability $1 - q$). The process of activation stops when no more activations are possible.

In experiments we set value of probability $q$ to value 0.7.

### 3.4   LT Model

The *Linear Threshold Model* (denoted *LT*) [10] assumes directed, weighted graph $G$. Nodes can also be either *Active* or *Inactive*. Each node $v \in V$ selects random, chosen with uniform distribution, threshold number $\theta_v$. At the beginning, only a seed set is active. With each timestep, active nodes $(A_e)$ try to activate their inactive neighbors. If the sum of weights of all active in-neighbor's edges is greater than or equal to the node threshold, the node becomes activated. The active nodes cannot become inactive and, again, the process ends when no more activations are possible.

### 3.5   LTC Model

The *Linear Threshold with Colors Model* (denoted *LTC*) [1] introduced three additional states to differentiate the level of entities adoption (as described in Fig. 1).

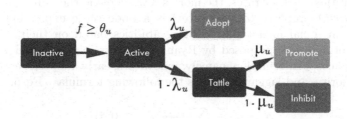

**Fig. 1.** Node states in LTC model (Color figure online).

Except new states, the main difference between *LT* model and *LTC* model is the condition of activation. Nodes activate its neighbours with similar rules as specified in the *LT* model, but an activation condition differs significantly. Activation function $f$ for a node $u$ during propagation of an entity $e$ is defined by Eq. 3, where $s_{min}$ and $s_{max}$ are the lowest and the highest ranking scores respectively, and $s_{n,e}$ is the rating score that the neighbor user $n$ assigned to the entity $e$.

$$f(u,e) = \frac{\sum_{n \in V_{in}(u), n \in A_e} w_{n,u}(s_{n,e} - s_{min})}{s_{max} - s_{min}}, f \in [0;1]. \qquad (3)$$

Once the node changes its state to *Active*, it again picks a random value with uniform distribution from $[0;1]$. If the value is grater than $\lambda$, the node fully adopts the entity (green). If not, it goes to *Tattle* state which indicates users that cannot afford the entity or do not want to buy the entity, but are still expressing opinions about it. In the next random process, the node has a $\mu$ probability to eventually get into *Promote* (blue) state and $1 - \mu$ probability to completely inhibit the entity and express negative opinion (red).

## 3.6   LTB Model

*Linear Threshold Behavioral Model* (denoted *LTB*) [16] is based on *LT* model, but enhances the activation condition from *LTC* model. It incorporates into the influence propagation model additional parts connected with personality (*Activity* parameter) and expertise (*Expertise* parameter) to fulfill the criteria of influence defined by two sociologists, Katz and Lazarsfeld in [9] (also confirmed in the context of online social networks by [4]).

The *Expertise* property is a node score that tries to describe the competence of the node in a given field of interest. General expertise $\gamma_u$ of the node $u$ in a social network is measured as a fraction of the number of entities rated by the user over the number of all entities (as described in Eq. 4):

$$\gamma_u = \frac{r_u}{\sum_{v \in V} r_v}, \gamma_u \in [0; 1], \tag{4}$$

where $r_u$ denotes the number of all ratings issued by a node $u$. We assume that the more entities the user rates, the more knowledgeable they are.

The *Activity* property ($\alpha$ parameter) is a node score expressing the user engagement in social media and adopting abilities based on their personality (categories of adopters proposed by Ryan and Gross in [19] linked with their eagerness to accept information, namely: innovators, early adopters, early majority, late majority, and laggards). It has the following formula – Eq. 5:

$$\alpha_u = \frac{1}{|R_u|} \sum_{e \in R_u} \frac{i_{e,u}}{|I_e|}, \alpha_u \in [0; 1], \tag{5}$$

where: $R_u$ is a set of entities rated by $u$; $I_e$ is a sequence of users who rated entity $e$, ordered ascendingly by rating date; $i_{e,u}$ is an index of $u$ in the sequence $I_e$. Top 2.5 % of the user with a lower $\alpha$ parameter may be called *Innovators*, the next 13.5 % - *Early Adopters*, etc. In other words, the lower $\alpha$ parameter is, the more eager the node is to accept the entity.

Activation condition (denoted as $Influence(u, e)$) to compute total influence exerted on the node $u \in N$ by its active neighbors while evaluating the entity $e$ has the following form:

$$Influence(u, e) = \frac{\sum_{n \in V_{in}(u), n \in A_e} w_{n,u}(s_{n,e} - s_{min})\gamma_n}{s_{max} - s_{min}}. \tag{6}$$

Similar to the *LTC* model, $s_{n,e}$ is the rating score that user $n$ assigned to the entity $e$; and $s_{min}$, $s_{max}$ are the lowest and the highest ranking scores respectively. Such function guarantees, that the more experienced the user is, the more power they have to influence others. In addition, higher neighbors' ratings of a particular entity increase overall influence.

The nodes' states are the same as in the *LT* model, that is: *Inactive* and *Active*. Each node $u \in V$ picks random threshold number $\theta_u$ between 0 and $\alpha_u$ - the more innovative the user is, the more likely they are to adopt the new entity. In each timestep, each of the inactive nodes computes total value of *Influence* exerted on it by its neighbors. Once the sum exceed threshold $\theta_u$, the node $u$ becomes *Active*. The propagation continues until no more inactive nodes are left.

# 4  Results

In this section we provide the description of datasets, then we present experiments on full dataset. Due to the expensive computational complexity of two centrality metrics – closeness and betweenness centrality, they were not calculated for *Flixter* and *Yelp* dataset. We decided to reduce graphs generated from these datasets and we demonstrate results for one chosen model of influence propagation (IC).

## 4.1  Description of Datasets

Experiments were conducted on three real world data sets from the domain of social services:

**Last.fm**[1] - an online music service with songs ratings. We utilized a dataset that was released for the Workshop on Information Heterogenity and Fusion in Recommender System (HetRec 2011)[2]. The dataset consists of 1892 users with 25.4 K relations among them. Listening history is represented in the form of the number of times a user listened to a song performed by an artist. The dataset also includes tagging history with date timestamps, but does not provide explicit user ratings, so it is calculated based on the frequency of playing songs of artists by users. Exact way of assigning scores for songs was described in [16]. In *Last.fm*, an entity symbolize an artist, and node activation indicates an action of tagging or listening to an artist for the first time.

**Flixster**[3] - a social network for movies aimed at sharing opinions, making reviews and rating movies. *Flixster* dataset was published by Jamali et al. [8] and contains 1 M users, 14 M undirected friendships relations between users, and 8.2 M ratings. In this case, an entity is a movie, and node activation represents an action of rating a movie.

**Yelp**[4] - a social network focused on enabling users to publish opinions on companies such as restaurants, cafes etc. The dataset was published as a part of Yelp Challange[5]. It consists of 1.6 M reviews published by 366 K users with 2.9 M social connections for 61 K business companies. In this dataset, the following definitions are assumed: an entity describes a business company and node activation means an action reviewing a company by a user.

In all datasets, nodes that did not rate any entity or do not have any neighbours, were excluded.

---

[1] http://www.last.fm.
[2] http://ir.ii.uam.es/hetrec2011.
[3] http://www.flixster.com.
[4] http://www.yelp.com.
[5] http://www.yelp.com/dataset_challenge.

## 4.2   Influence Maximization

**LastFm Dataset.** LastFm is the smallest dataset among tested ones. Figure 2a and b presents results (which are very similar to each other) for models IC and ICN, respectively. Greedy algorithm provides the best coverage. Good results are also obtained using PageRank and out-degree metrics. Eigenvector and betweenness gave worse coverage. Interesting behaviour can be observed for closeness centrality i.e. for some seed sizes the coverage is very small (even worse than random selection of nodes), but later, it has sudden increase. It may be explained by the fact that closeness metrics find nodes that have central position in network (nodes with the smallest total distances from all other nodes) and therefore, are close to each other (often may be neighbours), so adding more nodes with the highest values of this metrics does not improve the coverage until we add a node that has higher values of other metrics.

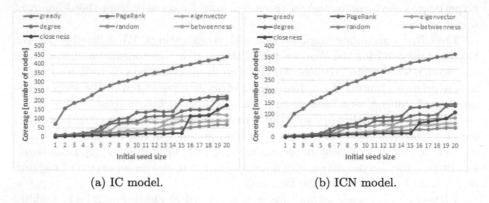

(a) IC model.                    (b) ICN model.

**Fig. 2.** LastFm - models based on IC.

Figure 3a shows results for LT model. The tendency is similar to the previous cases (but eigenvector centrality gives comparable results to out-degree). In LTC model (Fig. 3b) the difference between various centrality metrics is small (the tendency is also similar). In LTB model (Fig. 3c) the centrality metrics do not improve at all the total coverage, they may be even worse than random selection of nodes. It is also worth noticing that models LTC have 10 times less coverage than LT model, and LTB - 20 times.

Table 1a shows calculation times for different centrality measures. We provided such comparison only for LastFm, because only in this dataset we could calculate all tested centrality metrics. We can spot that out-degree, eigenvector and PageRank are very fast, but betweenness and closeness very slow. Greedy algorithm gives best results in every case, but is also very expensive computationally – in Table 1b we can see mean times for choosing a single node. The difference of times between different models is very small (the largest value is for LTC model).

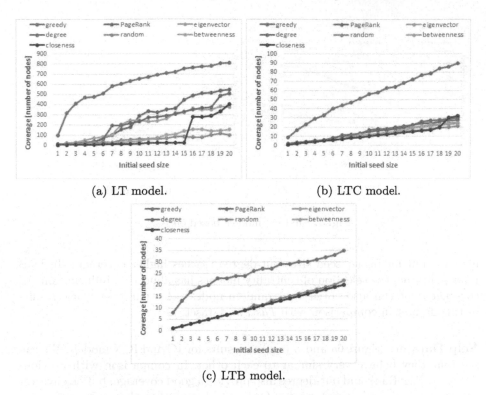

(a) LT model.                                    (b) LTC model.

(c) LTB model.

**Fig. 3.** LastFm - models based on LT.

**Table 1.** Calculation times for LastFm dataset.

(a) Centrality metrics (times for all nodes)

(b) Greedy algorithm in different models (times for choosing a single node)

| Algorithm of metrics | Time [s] | Propagation model | Mean time [s] |
| --- | --- | --- | --- |
| Out-degree | 0.008 | IC | 0.014 |
| Eigenvector | 0.198 | ICN | 0.013 |
| PageRank | 0.136 | LT | 0.037 |
| Betweenness | 73.504 | LTC | 0.095 |
| Closeness | 2 262.865 | LTB | 0.014 |

**Flixter Dataset.** In IC and IC models (Fig. 4a, b) we can observe analogous behaviour as it was presented in LastFm dataset - among centrality metrics, the best results are for PageRank, little worse results for out-degree and eigenvector centrality metrics.

For LT model (Fig. 5a) the situation is little different – eigenvector centrality provides better results than PageRank. Good coverage is also obtained using out-degree. In LTC model (Fig. 5b) PageRank has the best results among centrality

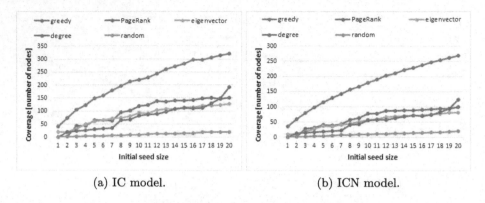

(a) IC model.             (b) ICN model.

**Fig. 4.** Flixter - models based on IC.

metrics (but for bigger seed sizes out-degree reaches higher coverage). In LTB model (Fig. 5c) the selection of centrality metrics has almost no influence on the coverage. One can also notice that random nodes' selection gives worse results in this dataset in comparison with *LastFm* dataset.

**Yelp Dataset.** Figure 6a and b present results for IC and ICN models. We can see that they behave very similar to each other. In comparison with previous datasets, PageRank and out-degree also provide a good coverage, but eigenvector centrality gave much worse results. One can observe that the coverage for this dataset is 10 times bigger than it was in previous datasets.

For LT and LTC models (Fig. 7a and b, respectively) the results are very similar to presented ones for IC and ICN models. In LTB model (Fig. 7c) centrality metrics have some effect on the final value of coverage (bigger than in previous datasets). The results obtained using centrality metrics are closer (than in previous datasets) to the coverage of greedy algorithm, and even in some cases greedy algorithm produced worse results than results from centrality measures (e.g. in LT model for selection only one node). One can notice that random selection of nodes gives bad results.

### 4.3   Influence Maximization on Reduced Graphs

All centrality metrics were only calculated on *LastFm* dataset, but for *Flixter* and *Yelp* it was not possible due to high computation complexity. Therefore, we decided to reduce graphs from these datasets. The reduction was conducted using *RandomWalk* algorithm [14]. In this subsection we provide results only for IC model – Fig. 8a and b. After the reduction, the number of nodes in the graph generated from Flixter dataset shrinked 6 times and in analogous graph from Yelp – 8 times, so it changes graphs significantly and conclusions drawn from these reduced graphs may be different than ones from full graphs. In both datasets the results obtained using PageRank, out-degree and eigenvector centrality are very similar. We can observe that betweenness produced very small

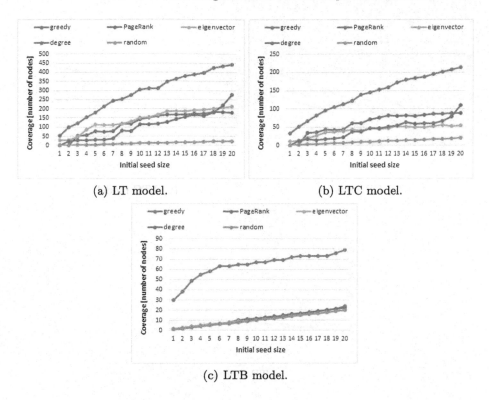

(a) LT model.

(b) LTC model.

(c) LTB model.

**Fig. 5.** Flixter - models based on LT.

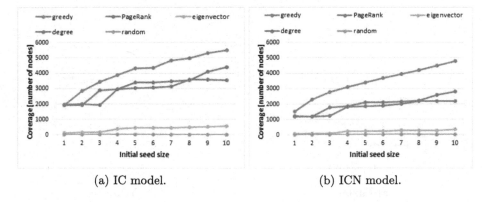

(a) IC model.

(b) ICN model.

**Fig. 6.** Yelp - models based on IC.

coverage. Regarding the centrality metrics, the behaviour is similar to observed one in the case of *LastFm* i.e. the selection using this metric for some high ranked nodes gives poor results, but at some point there is a sudden increase, and in the case of reduced *Yelp* dataset its coverage was higher than the coverage of other centrality metrics.

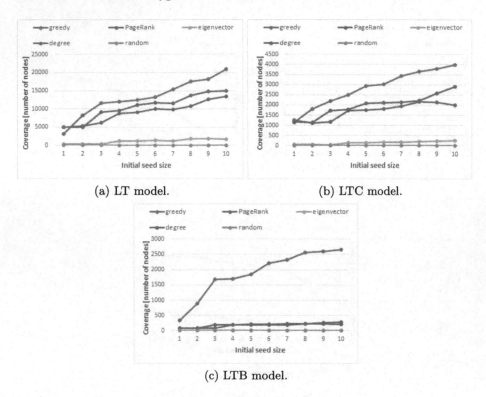

(a) LT model.

(b) LTC model.

(c) LTB model.

**Fig. 7.** Yelp - models based on LT.

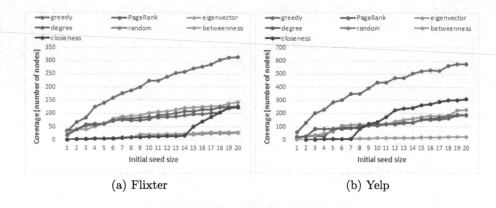

(a) Flixter

(b) Yelp

**Fig. 8.** Reduced graphs - IC models.

## 5    Conclusion and Future Work

In this paper we studied using various centrality metrics in the influence maximization problem. We considered different models of influence propagation.

In all cases, the best results were achieved using greedy algorithm, which may be time-consuming when we select many nodes. Therefore, selection of nodes based on their ranking from centrality metrics may be a good option. Generally, very good results are provided when we select nodes on the basis of ranking from PageRank. Using out-degree is also a good alternative. Betweenness and closeness produced significantly worse results. Moreover, they are also very expensive computationally and rather should not be considered for nodes selection.

Analyzing the results of choosing different methods of nodes selection in various propagation models, we can notice similar behavior and usage of centrality metrics gave much better results than random selection of nodes. In LTB model, such difference was the smallest (only in the case of *Yelp* dataset there is significant gain from choosing nodes based on the centrality metrics instead of choosing them randomly).

Future work will cover comparison with other methods of node selection. We are also planning to conduct comparisons of such methods on different datasets. Finally, we would like to repeat experiments with taking into account dynamics of networks (respecting the fact that edges can change in time).

**Acknowledgements.** The research reported in the paper was partially supported by the grant No. DOBR-BIO4/060/13423/2013 from the Polish National Centre for Research and Development.

The authors thank to Bartosz Niemczura, a student of Computer Science of AGH-UST, for his collaboration.

# References

1. Bhagat, S., Goyal, A., Lakshmanan, L.V.: Maximizing product adoption in social networks. In: Proceedings of the Fifth ACM International Conference on Web Search and Data Mining, pp. 603–612. ACM, New York (2012)
2. Chen, W., Collins, A., Cummings, R., Ke, T., Liu, Z., Rincón, D., Sun, X., Wang, Y., Wei, W., Yuan, Y.: Influence maximization in social networks when negative opinions may emerge and propagate. In: Proceedings of the Eleventh International Conference on Data Mining, SDM 2011, Mesa, Arizona, USA, 28–30 April 2011, pp. 379–390 (2011)
3. Chen, Y.C., Peng, W.C., Lee, S.Y.: Efficient algorithms for influence maximization in social networks. Knowl. Inf. Syst. **33**(3), 577–601 (2012). http://dx.org/10.1007/s10115-012-0540-7
4. Eccleston, D., Griseri, L.: How does web 2.0 stretch traditional influencing patterns. Int. J. Mark. Res. **50**(5), 591–616 (2008)
5. Freeman, L.C.: Centrality in social networks conceptual clarification. Soc. Netw. **1**(3), 215–239 (1979)
6. Godes, D., Mayzlin, D.: Firm-created word-of-mouth communication: evidence from a field test. Mark. Sci. **28**(4), 721–739 (2009)
7. Goyal, A., Bonchi, F., Lakshmanan, L.V.S.: A data-based approach to social influence maximization. Proc. VLDB Endow. **5**(1), 73–84 (2011)
8. Jamali, M., Ester, M.: A matrix factorization technique with trust propagation for recommendation in social networks. In: Proceedings of the Fourth ACM Conference on Recommender Systems, pp. 135–142. ACM, New York (2010)

9.  Katz, E., Lazarsfeld, P.: Personal Influence: The Part Played by People in the Flow of Mass Communications. Transaction Publishers, New Brunswick (2005)
10. Kempe, D., Kleinberg, J., Tardos, É.: Maximizing the spread of influence through a social network. In: Proceedings of the Ninth ACM SIGKDD International Conference on Knowledge Discovery and Data Mining, pp. 137–146. ACM (2003)
11. Kermack, W.O., McKendrick, A.G.: Contributions to the mathematical theory of epidemics-II. The problem of endemicity. Proc. Roy. Soc. London Math. Phys. Eng. Sci. **138**(834), 55–83 (1932)
12. Kundu, S., Murthy, C.A., Pal, S.K.: A new centrality measure for influence maximization in social networks. In: Kuznetsov, S.O., Mandal, D.P., Kundu, M.K., Pal, S.K. (eds.) PReMI 2011. LNCS, vol. 6744, pp. 242–247. Springer, Heidelberg (2011)
13. Leavitt, H.J.: Some effects of certain communication patterns on group performance. J. Abnorm. Soc. Psychol. **46**(1), 38 (1951)
14. Lovasz, L.: Random walks on graphs: A survey (1993)
15. Narayanam, R., Narahari, Y.: A shapley value-based approach to discover influential nodes in social networks. IEEE Trans. Autom. Sci. Eng. **8**(1), 130–147 (2011)
16. Niemczura, B., Gliwa, B., Zygmunt, A.: Linear threshold behavioral model for the spread of influence in recommendation services. In: The Second European Network Intelligence Conference, ENIC 2015, pp. 98–105 (2015)
17. Page, L., Brin, S., Motwani, R., Winograd, T.: The pagerank citation ranking: Bringing order to the web (1999)
18. Zafarani, R., Abbasi, M.A., Liu, H.: Influence and homophily. In: Zafarani, R., Abbasi, M.A., Liu, H. (eds.) Social Media Mining. Cambridge University Press (2014)
19. Ryan, B., Gross, N.C.: Acceptance and diffusion of hybrid corn seed in two Iowa communities, vol. 372. Agricultural Experiment Station, Iowa State College of Agriculture and Mechanic Arts (1950)
20. Shaw, M.E.: Group structure and the behavior of individuals in small groups. J. Psychol. **38**(1), 139–149 (1954)
21. Wei, L., Laks, V.S.: Profit maximization over social networks. CoRR abs/1210.4211 (2012)
22. Zhuang, H., Sun, Y., Tang, J., Zhang, J., Sun, X.: Influence maximization in dynamic social networks. In: 2013 IEEE 13th International Conference on Data Mining (ICDM), pp. 1313–1318, December 2013

# Memetic Neuro-Fuzzy System
# with Differential Optimisation

Krzysztof Siminski[✉]

Institute of Informatics, Silesian University of Technology,
ul. Akademicka 16, 44-100 Gliwice, Poland
Krzysztof.Siminski@polsl.pl

**Abstract.** Neuro-fuzzy systems are capable of tuning theirs parameters
on presented data. Both global and local techniques can be used. The
paper presents a hybrid memetic approach where local (gradient descent)
and global (differential evolution) approach are combined to tune parameters of a neuro-fuzzy system. Application of the memetic approach
results in lower error rates than either gradient descent optimisation or
differential evolution alone. The results of experiments on benchmark
datasets have been statistically verified.

**Keywords:** Neuro-fuzzy system · Memetic algorithm · Differential evolution

## 1 Introduction

Neuro-fuzzy systems are successful in modelling of imprecise data. They are
able to extract data from presented data. It is done in two stages: identification of model structure and identification of values of parameters (tuning of
parameters). The first method of tuning of neuro-fuzzy systems were gradient
techniques. These are local optimisation techniques. They can stop in a local
minimum. The alternatives are global techniques as simulated annealing [6],
genetic algorithms [2,3].

Memetic algorithms are an implementation of the Lamarckian evolutionary
theory. This theory claims that acquired features can be inherited by offsprings.
It turned to be false in biology, but it can be successfully applied in optimisation
tasks. This technique joins an evolutionary approach with a local optimisation
(acquired features). The memetic algorithms are widely used in machine learning [7], molecular optimisation, automatic control, pattern recognition, robotics,
image processing [5], neuro-fuzzy systems [11].

The novelty of the paper is an application of memetic optimisation (gradient
descent local optimisation with differential evolutionary global optimisation) to
tuning parameters of a neuro-fuzzy system with logical interpretation of fuzzy
rules.

S. Kozielski et al. (Eds.): BDAS 2016, CCIS 613, pp. 135–145, 2016.
DOI: 10.1007/978-3-319-34099-9_9

## 2   Memetic Fuzzy Inference System

The memetic fuzzy inference system is based on fuzzy systems with logical interpretation of fuzzy rules. The general pattern of the training is presented in Algorithm 1.

```
1  procedure train (X, g, t, Q, λ, L, f, λ)
2  /* X : train dataset */
3  /* g : number of generations */
4  /* Q : number of individuals */
5  /* t : number of tuning iterations */
6  /* λ : parameter */
7  /* L : number of rules */
8  /* f : quality function */
9  /* λ : difference importance parameter (Eq. (8)) */
10
11     /* create random population Q of individuals (models) */
12     Q := create_individuals (Q, L);
13
14     foreach q in Q      /* tune each individual ...   */
15         tune(q, X, t);  /* ... with gradient descent method */
16
17     /* differential evolution: */
18     while g > 0 do
19         pick four individuals q_1 ≠ q_2 ≠ q_3 ≠ q_4;
20         q_n := λ(q_2 - q_1) + q_3;  /* new individual */
21         tune(q_n, X, t);   /* tune new individual with gradient
                                  descent method */
22
23         /* check quality with f function */
24         if f(q_n) ≤ f(q_4) then  /* replace q_4 with q_n */
25             Q := Q \ {q_4};
26             Q := Q ∪ {q_n};
27         end if;
28         g := g - 1;
29     end while;
30
31     /* find the best individual in population Q */
32     find q_b with minimal f(q_b);
33
34     return q_b;   /* the best individual */
35 end procedure;
```

**Algorithm** 1: Pseudocode of the memetic training procedure.

## 2.1   Neuro-Fuzzy System

The neuro-fuzzy system we use in our experiments is ANNBFIS (Artificial Neural Network Based Fuzzy Inference System) [4]. It is a multiple input single output (MISO) system. The crucial part of the systems is fuzzy rule base. Each fuzzy rule is a fuzzy logical implication.

Rule base $\mathbb{L}$ contains fuzzy rules $l$ (fuzzy implications)

$$l : \mathbf{x} \text{ is } \mathfrak{a} \rightsquigarrow y \text{ is } \mathfrak{b}, \tag{1}$$

where $\mathbf{x} = [x_1, x_2, \ldots, x_D]^\mathrm{T}$ and $y$ are linguistic variables, $\mathfrak{a}$ and $\mathfrak{b}$ are fuzzy linguistic values. The squiggle arrow ($\rightsquigarrow$) stands for a fuzzy implication.

The rule's premise is built with a fuzzy set $\mathbb{A}$ in $D$-dimensional space. In each dimension $d$ the set $\mathbb{A}$ is described with the Gaussian membership function:

$$\mu_d(x_d) = \exp\left(-\frac{(x_d - v_d)^2}{2s_d^2}\right), \tag{2}$$

where $v_d$ is the core location for the $d$-th attribute and $s_d$ is this attribute's deviation. The membership of the variable $\mathbf{x}$ to the fuzzy set $\mathbb{A}^{(l)}$ (in the $l$th rule is defined as a T-norm of membership values of all attributes:

$$\mu_{\mathbb{A}^{(l)}}(\mathbf{x}) = \mu_{d_1^{(l)}}(x_1) \star \cdots \star \mu_{d_D^{(l)}}(x_D) = \bigstar_{d \in \mathbb{D}} \mu_{d^{(l)}}(x_d), \tag{3}$$

where $\star$ denotes the T-norm and $\mathbb{D}$ stands for the set of attributes (in ANNBFIS the product T-norm is used).

The term $\mathfrak{b}$ (in the rule's consequence) is represented by an normal isosceles triangle fuzzy set $\mathbb{B}$ with the base width $w$. The localisation of the core of the triangle membership function is a linear function of attribute values:

$$y^{(l)}(\mathbf{x}) = \left(\mathbf{p}^{(l)}\right)^\mathrm{T} \cdot [1, \mathbf{x}^\mathrm{T}]^\mathrm{T} = \left[p_0^{(l)}, p_1^{(l)}, \ldots, p_D^{(l)}\right] \cdot [1, x_1, \ldots, x_D]^\mathrm{T}, \tag{4}$$

where $\mathbf{p}^{(l)}$ is the parameter vector of the consequence for $l$th rule. The localisation of the fuzzy sets depends on the values of the input vector $\mathbf{x}$.

The output of the rule is a value of the fuzzy implication:

$$\mu_{\mathbb{B}'^{(l)}}(\mathbf{x}) = \mu_{\mathbb{A}^{(l)}}(\mathbf{x}) \rightsquigarrow \mu_{\mathbb{B}^{(l)}}(\mathbf{x}). \tag{5}$$

The answer of the system is the aggregation of all rules:

$$\mu_{\mathbb{B}'}(\mathbf{x}) = \bigoplus_{l=1}^{L} \mu_{\mathbb{B}'^{(l)}}(\mathbf{x}). \tag{6}$$

In order to get a crisp answer $y_0$ the fuzzy set $\mathbb{B}'$ is defuzzified with modified indexed centre of gravity MICOG method [4]. The shape of the fuzzy set $\mathbb{B}'$

is usually quite complicated what causes expensive aggregation and defuzzyfication, but it has been proved [4] that the crisp system output can be expressed as:

$$y_0 = \frac{\sum_{l=1}^{L} g^{(l)}(\mathbf{x}) y^{(l)}(\mathbf{x})}{\sum_{l=1}^{L} g^{(l)}(\mathbf{x})}. \tag{7}$$

The function $g$ depends on the fuzzy implication. The Artificial Neural Network Based Fuzzy Interence System (ANNBFIS) system [4] uses Reichenbach implication [9].

In local search optimisation parameters of the premises are tuned with backpropagation gradient descent method. The parameters of the linear models in conclusions are calculated with least square linear regression. The widths of the supports $w$ of sets in consequences are tuned with backpropagation gradient descent method.

## 2.2 Differential Evolution Algorithm

Differential evolution algorithm [12] has strong global search capability, fast convergence, and user friendliness [8].

The idea of the algorithm is presented in Fig. 1. First the population $\mathbb{Q}$ of $Q$ individuals is created at random. Each individual is a fuzzy model (with fuzzy rules), as described in Sect. 2.1. Then in loop four different individuals $q_1 \neq q_2 \neq q_3 \neq q_4$ are picked from the population. New individual $q_n$ is calculated as

$$q_n = \lambda(q_2 - q_1) + q_3, \tag{8}$$

where $\lambda > 0$. If the quality function $f$ of the new individual $q_n$ is lower than $f(q_4)$, the new individual $q_n$ substitutes individual $q_4$ in the population $\mathbb{Q}$. The calculation of new individual $q_n$ in two-dimensional task is presented in Fig. 1. The $f$ function used in the proposed system is defined in Sect. 3.2.

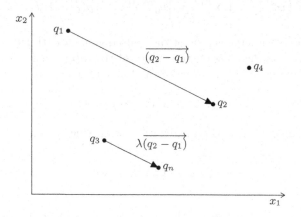

**Fig. 1.** Calculation of a new individual $q_n$ in differential evolution optimisation technique for two-dimensional space.

The original differential evolution method implements also a crossing-over operator. We do not use it in our case, because the crossing-over of two fuzzy model severely decreases the quality of models. The creation of a new individual (model) with the formula (8) keeps the meaning of model's parameters.

## 3 Experiments

The memetic neufo-fuzzy system first identifies the structure of the model. In experiments fuzzy models have $L = 6$ rules. In our experiments we do not aim at finding the best number of rules in the fuzzy model. We use constant number of rules. The number of rules in neuro-fuzzy systems is an open question. Commonly the number or rules is low to keep the fuzzy model interpretable by humans [10].

In preliminary experiments the parameter $\lambda$ was calibrated and in later on the value $\lambda = 0.5$ was used.

The reference system is an Artificial Neural Network Based Fuzzy Inference System (ANNBFIS) – fuzzy inference system with logical interpretation of rules. The ANNBFIS systems has the same number of fuzzy rules $L = 6$, 100 of iterations in fuzzy model identification and 100 iterations of gradient descent tuning. The experiments were conducted as a 5-fold cross validation.

### 3.1 Data Sets

In experiments we use data sets from KEEL repository [1]: 'treasury', 'friedman', and 'laser'. For the brevity of the paper only the essential information on the datasets is provided.

The 'treasury' dataset contains the week economic information of USA from 1980-04-01 to 2000-04-02. The target attribute is the prediction in one month rate. The data set holds 1049 data item with 15 attributes. The 'laser' dataset holds 993 values recorded from a far-infrared-laser in a chaotic state (with 4 attributes). The 'friedman' dataset is a synthetic benchmark dataset proposed by Friedman in 1991. The values of five attributes are taken independently at random from interval $[0.0, 1.0]$ with uniform distribution. The target attribute $y$ is calculated with a formula

$$y = 10(\sin(\pi x_1 x_2)) + 20(x_3 - 0.5)^2 + 10x_4 + 5x_5 + N(0, 1), \qquad (9)$$

where $N(0, 1)$ stands for Gaussian random noise. The dataset holds 1200 data instances.

The datasets were downloaded from the KEEL repository. The repository provides datasets split into subsets for 5-fold cross validation. In experiment we have tested fuzzy models with $L = 6$ rules. We have tested systems with $Q \in \{10, 20, 50, 100\}$ individuals, $g \in \{200, 500, 1000, 2000\}$ generations, and $t \in \{0, 10, 20, 50\}$ iterations of local optimisation. Value $t = 0$ iterations of local optimisation means that the models were only optimised with differential evolution without any local gradient optimisation. The reference system was the ANNBFIS system with gradient local optimisation. The systems were tested in two paradigms:

– DA (data approximation) – system trained and tested with the same data,
– KG (knowledge generalisation) – system tested with unseen data.

**Table 1.** The results elaborated by the memetic DE NFS systems for 'friedman' data set. The table holds the averages and standard deviations of the root mean square error (RMSE) for DA and KG paradigms. The notation $\mu \pm \sigma$ stand for average and standard deviation, $g$ stands for number of generations, $t$ stands for number of tuning iterations in memetic local search, symbol $(*)$ denotes statistically significantly better results than those elaborated by ANNBFIS.

| $g$ | $t$ | number of individuals $Q$ | | | |
|---|---|---|---|---|---|
| | | 10 | 20 | 50 | 100 |
| DA: data approximation | | | | | |
| 200 | 0 | 1.174224 ± 0.3648 | 1.6147 ± 0.3106 | 3.5085 ± 0.6952 | 3.4006 ± 0.5622 |
| | 10 | 0.419060 ± 0.0119 | 0.4198 ± 0.0349 | 0.4348 ± 0.0154 | 0.4405 ± 0.0176 |
| | 20 | 0.421156 ± 0.0214 | 0.4372 ± 0.0363 | 0.4173 ± 0.0155 | 0.4354 ± 0.0151 |
| | 50 | 0.412756 ± 0.0330 | 0.4078 ± 0.0087 | 0.4212 ± 0.0124 | 0.4310 ± 0.0086 |
| 500 | 0 | 1.218500 ± 0.3226 | 0.7621 ± 0.1191 | 2.4273 ± 0.5621 | 3.2011 ± 0.5414 |
| | 10 | 0.410281 ± 0.0209 | 0.3946 ± 0.0100(*) | 0.4293 ± 0.0049 | 0.4149 ± 0.0287 |
| | 20 | 0.433408 ± 0.0079 | 0.4358 ± 0.0137 | 0.4113 ± 0.0290 | 0.4337 ± 0.0082 |
| | 50 | 0.414046 ± 0.0200 | 0.4052 ± 0.0151 | 0.4088 ± 0.0238 | 0.4168 ± 0.0123 |
| 1000 | 0 | 1.089926 ± 0.2299 | 0.9340 ± 0.2085 | 1.0909 ± 0.3314 | 2.3762 ± 0.4333 |
| | 10 | 0.410083 ± 0.0237 | 0.4119 ± 0.0241 | 0.4216 ± 0.0187 | 0.4142 ± 0.0164 |
| | 20 | 0.372967 ± 0.0333(*) | 0.4087 ± 0.0253 | 0.4111 ± 0.0168 | 0.4221 ± 0.0266 |
| | 50 | 0.386073 ± 0.0448(*) | 0.3812 ± 0.0284(*) | 0.4101 ± 0.0145 | 0.4005 ± 0.0113(*) |
| 2000 | 0 | 1.120406 ± 0.2691 | 0.6259 ± 0.1113 | 0.6078 ± 0.0794 | 1.8690 ± 0.3232 |
| | 10 | 0.393934 ± 0.0230(*) | 0.4023 ± 0.0174(*) | 0.4000 ± 0.0104(*) | 0.4147 ± 0.0168 |
| | 20 | 0.391021 ± 0.0208(*) | 0.3975 ± 0.0297(*) | 0.4032 ± 0.0364 | 0.4064 ± 0.0088 |
| | 50 | 0.381450 ± 0.0300(*) | 0.3820 ± 0.0286(*) | 0.4145 ± 0.0096 | 0.4025 ± 0.0132(*) |
| KG: knowledge generalisation | | | | | |
| 200 | 0 | 1.185371 ± 0.4019 | 1.5517 ± 0.2705 | 3.5604 ± 0.7008 | 3.4588 ± 0.5761 |
| | 10 | 0.446111 ± 0.0479 | 0.4449 ± 0.0137 | 0.4457 ± 0.0370 | 0.4718 ± 0.0520 |
| | 20 | 0.435541 ± 0.0548 | 0.4358 ± 0.0066 | 0.4139 ± 0.0311(*) | 0.4670 ± 0.0491 |
| | 50 | 0.426995 ± 0.0359(*) | 0.4293 ± 0.0432 | 0.4419 ± 0.0384 | 0.4317 ± 0.0297 |
| 500 | 0 | 1.190205 ± 0.3274 | 0.7631 ± 0.1370 | 2.3769 ± 0.5047 | 3.1710 ± 0.5541 |
| | 10 | 0.434512 ± 0.0413 | 0.4062 ± 0.0442(*) | 0.4369 ± 0.0500 | 0.4394 ± 0.0528 |
| | 20 | 0.445498 ± 0.0375 | 0.4573 ± 0.0266 | 0.4238 ± 0.0475(*) | 0.4476 ± 0.0512 |
| | 50 | 0.427509 ± 0.0498 | 0.4256 ± 0.0454(*) | 0.4196 ± 0.0267(*) | 0.4426 ± 0.0432 |
| 1000 | 0 | 1.087825 ± 0.2283 | 0.9613 ± 0.2277 | 1.0864 ± 0.3266 | 2.4366 ± 0.5110 |
| | 10 | 0.413079 ± 0.0231(*) | 0.4194 ± 0.0369(*) | 0.4288 ± 0.0315(*) | 0.4377 ± 0.0345 |
| | 20 | 0.387021 ± 0.0720(*) | 0.4188 ± 0.0462(*) | 0.4212 ± 0.0273(*) | 0.4444 ± 0.0546 |
| | 50 | 0.400426 ± 0.0443(*) | 0.4071 ± 0.0365(*) | 0.4323 ± 0.0379 | 0.4046 ± 0.0415(*) |
| 2000 | 0 | 1.104279 ± 0.2439 | 0.6365 ± 0.1077 | 0.6161 ± 0.1047 | 1.8470 ± 0.2598 |
| | 10 | 0.401054 ± 0.0557(*) | 0.4164 ± 0.0396(*) | 0.4173 ± 0.0488(*) | 0.4352 ± 0.0299 |
| | 20 | 0.405673 ± 0.0528(*) | 0.4144 ± 0.0408(*) | 0.4112 ± 0.0229(*) | 0.4306 ± 0.0352 |
| | 50 | 0.401518 ± 0.0528(*) | 0.3917 ± 0.0398(*) | 0.4372 ± 0.0408 | 0.4257 ± 0.0400(*) |

**Table 2.** The results elaborated by the memetic DE NFS systems for 'laser' data set. The table holds the averages and standard deviations of the root mean square error (RMSE) for DA and KG paradigms. The notation $\mu \pm \sigma$ stand for average and standard deviation, $g$ stands for number of generations, $t$ stands for number of tuning iterations in memetic local search, symbol $(*)$ denotes statistically significantly better results than those elaborated by ANNBFIS.

| $g$ | $t$ | number of individuals $Q$ | | | |
|---|---|---|---|---|---|
| | | 10 | 20 | 50 | 100 |
| DA: data approximation | | | | | |
| 200 | 0 | $0.8683 \pm 0.1752$ | $1.1593 \pm 0.3339$ | $2.0604 \pm 0.2096$ | $2.5442 \pm 0.3576$ |
| | 10 | $0.2884 \pm 0.0335(*)$ | $0.2685 \pm 0.0282(*)$ | $0.2642 \pm 0.0149(*)$ | $0.2811 \pm 0.0175(*)$ |
| | 20 | $0.2554 \pm 0.0420(*)$ | $0.2494 \pm 0.0222(*)$ | $0.2590 \pm 0.0426(*)$ | $0.2592 \pm 0.0160(*)$ |
| | 50 | $0.2305 \pm 0.0209(*)$ | $0.2502 \pm 0.0314(*)$ | $0.2383 \pm 0.0228(*)$ | $0.2205 \pm 0.0245(*)$ |
| 500 | 0 | $0.8461 \pm 0.2552$ | $0.7553 \pm 0.2059$ | $1.5823 \pm 0.2667$ | $2.2960 \pm 0.1968$ |
| | 10 | $0.2267 \pm 0.0358(*)$ | $0.2661 \pm 0.0241(*)$ | $0.2519 \pm 0.0247(*)$ | $0.2394 \pm 0.0104(*)$ |
| | 20 | $0.2466 \pm 0.0389(*)$ | $0.2406 \pm 0.0167(*)$ | $0.2368 \pm 0.0282(*)$ | $0.2409 \pm 0.0067(*)$ |
| | 50 | $0.2286 \pm 0.0417(*)$ | $0.2309 \pm 0.0187(*)$ | $0.2340 \pm 0.0232(*)$ | $0.2367 \pm 0.0125(*)$ |
| 1000 | 0 | $0.9775 \pm 0.3364$ | $0.6173 \pm 0.0993$ | $0.8467 \pm 0.3157$ | $1.7858 \pm 0.2810$ |
| | 10 | $0.2283 \pm 0.0369(*)$ | $0.2346 \pm 0.0269(*)$ | $0.2377 \pm 0.0222(*)$ | $0.2411 \pm 0.0130(*)$ |
| | 20 | $0.2287 \pm 0.0340(*)$ | $0.2193 \pm 0.0126(*)$ | $0.2259 \pm 0.0142(*)$ | $0.2351 \pm 0.0177(*)$ |
| | 50 | $0.2403 \pm 0.0358(*)$ | $0.2276 \pm 0.0182(*)$ | $0.2381 \pm 0.0136(*)$ | $0.2265 \pm 0.0146(*)$ |
| 2000 | 0 | $0.9493 \pm 0.2393$ | $0.6284 \pm 0.0812$ | $0.5230 \pm 0.0160$ | $1.1377 \pm 0.3183$ |
| | 10 | $0.1940 \pm 0.0246(*)$ | $0.2059 \pm 0.0184(*)$ | $0.2330 \pm 0.0108(*)$ | $0.2287 \pm 0.0114(*)$ |
| | 20 | $0.2500 \pm 0.0461(*)$ | $0.2115 \pm 0.0137(*)$ | $0.2348 \pm 0.0190(*)$ | $0.2282 \pm 0.0069(*)$ |
| | 50 | $0.2263 \pm 0.0617(*)$ | $0.1977 \pm 0.0125(*)$ | $0.2230 \pm 0.0154(*)$ | $0.2215 \pm 0.0040(*)$ |
| KG: knowledge generalisation | | | | | |
| 200 | 0 | $0.8574 \pm 0.1774$ | $1.1556 \pm 0.2966$ | $2.0989 \pm 0.2819$ | $2.6264 \pm 0.3564$ |
| | 10 | $0.3174 \pm 0.0946$ | $0.3067 \pm 0.0929$ | $0.3027 \pm 0.0308$ | $0.3104 \pm 0.0170$ |
| | 20 | $0.2549 \pm 0.0319(*)$ | $0.2588 \pm 0.0454(*)$ | $0.3003 \pm 0.0698$ | $0.2620 \pm 0.0369(*)$ |
| | 50 | $0.2361 \pm 0.0173(*)$ | $0.2786 \pm 0.0566$ | $0.2549 \pm 0.0322(*)$ | $0.2700 \pm 0.0463$ |
| 500 | 0 | $0.9057 \pm 0.3092$ | $0.8103 \pm 0.2742$ | $1.5278 \pm 0.2652$ | $2.5053 \pm 0.2507$ |
| | 10 | $0.2460 \pm 0.0260(*)$ | $0.3176 \pm 0.0619$ | $0.2593 \pm 0.0164(*)$ | $0.2731 \pm 0.0306$ |
| | 20 | $0.2819 \pm 0.0362$ | $0.2755 \pm 0.0462$ | $0.2544 \pm 0.0292(*)$ | $0.2586 \pm 0.0292(*)$ |
| | 50 | $0.2447 \pm 0.0527(*)$ | $0.2477 \pm 0.0365(*)$ | $0.2493 \pm 0.0149(*)$ | $0.2584 \pm 0.0290(*)$ |
| 1000 | 0 | $1.0128 \pm 0.3160$ | $0.5969 \pm 0.0471$ | $0.9212 \pm 0.5094$ | $1.7704 \pm 0.2916$ |
| | 10 | $0.2441 \pm 0.0551(*)$ | $0.2654 \pm 0.0229(*)$ | $0.2608 \pm 0.0352(*)$ | $0.2574 \pm 0.0241(*)$ |
| | 20 | $0.2573 \pm 0.0471(*)$ | $0.2571 \pm 0.0362(*)$ | $0.2423 \pm 0.0308(*)$ | $0.2630 \pm 0.0474$ |
| | 50 | $0.2640 \pm 0.0520$ | $0.2566 \pm 0.0339(*)$ | $0.2597 \pm 0.0303(*)$ | $0.2500 \pm 0.0235(*)$ |
| 2000 | 0 | $0.9645 \pm 0.2158$ | $0.6377 \pm 0.0892$ | $0.5228 \pm 0.0587$ | $1.2046 \pm 0.3401$ |
| | 10 | $0.1956 \pm 0.0449(*)$ | $0.2351 \pm 0.0357(*)$ | $0.2666 \pm 0.0343$ | $0.2543 \pm 0.0299(*)$ |
| | 20 | $0.2629 \pm 0.0504$ | $0.2524 \pm 0.0430(*)$ | $0.2526 \pm 0.0509(*)$ | $0.2770 \pm 0.0471$ |
| | 50 | $0.2575 \pm 0.0450(*)$ | $0.2213 \pm 0.0388(*)$ | $0.2493 \pm 0.0255(*)$ | $0.2396 \pm 0.0319(*)$ |

## 3.2   Results

The results are gathered in Tables 1 ('friedman'), 2 ('laser'), and 4 ('treasury'). The symbol ($*$) denotes statistically significantly better results elaborated by memetic approach than by ANNBFIS (Table 3). The tables present mean values and standard deviations of root mean square error (RMSE) for the data set $\mathbb{X}$:

$$E_{\mathrm{RMSE}}(\mathbb{X}) = \sqrt{\frac{1}{X}\sum_{i=1}^{X}\left[y_0\left(\mathbf{x}_i\right) - y\left(\mathbf{x}_i\right)\right]^2}, \tag{10}$$

where $X$ stands for a number of data items in a dataset, $y_0\left(\mathbf{x}_i\right)$ is the answer elaborated by the system for the $i$th data item (cf Eq. 7), and $y\left(\mathbf{x}_i\right)$ is expected value for this data item. This error measure for the data set is used as a quality function $f$ of the fuzzy model in the `train` procedure (Algorithm 1). Some other error measures can be used instead as mean absolute error or mean percentage error.

**Table 3.** Results elaborated by the reference system ANNBFIS.

| dataset | DA | KG |
|---|---|---|
| 'friedman' | $0.61219 \pm 0.19399$ | $0.63043 \pm 0.18634$ |
| 'laser' | $0.41060 \pm 0.03553$ | $0.41111 \pm 0.13418$ |
| 'treasury' | $0.21934 \pm 0.10572$ | $0.23934 \pm 0.13257$ |

In statistical testing we use the Cochran-Cox test. We assume that the averages of RMSE elaborated by ANNBFIS (subscript $_1$) and proposed approach – ANNBFIS with DE and gradient optimisation (subscript $_2$) have normal distributions $N(\mu_1, \sigma_1)$ i $N(\mu_2, \sigma_2)$ with unknown standard deviations $\sigma_1$ and $\sigma_2$. The null hypothesis states that the averages are equal: $H_0 : \mu_1 = \mu_2$. The one-tailed alternative hypothesis $H_1$ states that the error elaborated by ANNBFIS is greater: $H_1 : \mu_1 > \mu_2$. The significance level $1 - \alpha = 0.95$ leads to critical interval $[2.132, +\infty)$.

The proposed approach has been also compared with memetic fuzzy systems described in [11], where Big-Bang-Big-Crunch global technique is used. The experiments were repeated with memetic NFS with BBBC optimisation, the results are gathered in Table 5. The results show that for the 'laser' and 'treasury' dataset the proposed approach can outperform the BBBC memetic system. Otherwise the latter technique elaborated lower error for the 'friedman' dataset.

**Table 4.** The results elaborated by the memetic DE NFS systems for 'treasury' data set. The table holds the averages and standard deviations of the root mean square error (RMSE) for DA and KG paradigms. The notation $\mu \pm \sigma$ stands for average and standard deviation, $g$ stands for number of generations, $t$ stands for number of tuning iterations in memetic local search, symbol $(*)$ denotes statistically significantly better results than those elaborated by ANNBFIS.

| $g$ | $t$ | number of individuals $Q$ | | | |
| | | 10 | 20 | 50 | 100 |
|---|---|---|---|---|---|
| DA: data approximation | | | | | |
| 200 | 0 | $0.9400 \pm 0.3727$ | $2.3483 \pm 0.5730$ | $3.0247 \pm 0.3105$ | $3.2118 \pm 0.6580$ |
| | 10 | $0.0620 \pm 0.0059(*)$ | $0.0636 \pm 0.0048(*)$ | $0.0627 \pm 0.0041(*)$ | $0.0640 \pm 0.0053(*)$ |
| | 20 | $0.0587 \pm 0.0077(*)$ | $0.0598 \pm 0.0020(*)$ | $0.0603 \pm 0.0042(*)$ | $0.0645 \pm 0.0034(*)$ |
| | 50 | $0.0526 \pm 0.0024(*)$ | $0.0591 \pm 0.0059(*)$ | $0.0611 \pm 0.0040(*)$ | $0.0632 \pm 0.0038(*)$ |
| 500 | 0 | $0.7849 \pm 0.1785$ | $0.7751 \pm 0.5126$ | $2.8234 \pm 0.3567$ | $2.9969 \pm 0.3208$ |
| | 10 | $0.0592 \pm 0.0067(*)$ | $0.0601 \pm 0.0032(*)$ | $0.0602 \pm 0.0019(*)$ | $0.0635 \pm 0.0020(*)$ |
| | 20 | $0.0571 \pm 0.0010(*)$ | $0.0587 \pm 0.0030(*)$ | $0.0577 \pm 0.0028(*)$ | $0.0590 \pm 0.0033(*)$ |
| | 50 | $0.0598 \pm 0.0096(*)$ | $0.0607 \pm 0.0060(*)$ | $0.0590 \pm 0.0047(*)$ | $0.0596 \pm 0.0056(*)$ |
| 1000 | 0 | $1.0996 \pm 0.4714$ | $0.5436 \pm 0.2964$ | $2.1330 \pm 0.4014$ | $2.7258 \pm 0.4465$ |
| | 10 | $0.0603 \pm 0.0056(*)$ | $0.0605 \pm 0.0051(*)$ | $0.0607 \pm 0.0054(*)$ | $0.0621 \pm 0.0053(*)$ |
| | 20 | $0.0619 \pm 0.0045(*)$ | $0.0562 \pm 0.0033(*)$ | $0.0613 \pm 0.0049(*)$ | $0.0627 \pm 0.0041(*)$ |
| | 50 | $0.0523 \pm 0.0044(*)$ | $0.0565 \pm 0.0037(*)$ | $0.0586 \pm 0.0033(*)$ | $0.0587 \pm 0.0041(*)$ |
| 2000 | 0 | $1.1767 \pm 0.3603$ | $0.5948 \pm 0.1607$ | $0.6687 \pm 0.4512$ | $2.3063 \pm 0.2248$ |
| | 10 | $0.0591 \pm 0.0042(*)$ | $0.0593 \pm 0.0046(*)$ | $0.0591 \pm 0.0039(*)$ | $0.0612 \pm 0.0028(*)$ |
| | 20 | $0.0561 \pm 0.0060(*)$ | $0.0549 \pm 0.0056(*)$ | $0.0603 \pm 0.0059(*)$ | $0.0622 \pm 0.0040(*)$ |
| | 50 | $0.0553 \pm 0.0042(*)$ | $0.0549 \pm 0.0058(*)$ | $0.0577 \pm 0.0027(*)$ | $0.0609 \pm 0.0024(*)$ |
| KG: knowledge generalisation | | | | | |
| 200 | 0 | $0.8976 \pm 0.3636$ | $2.3678 \pm 0.6788$ | $2.9479 \pm 0.2755$ | $3.2405 \pm 0.6381$ |
| | 10 | $0.0733 \pm 0.0145(*)$ | $0.0811 \pm 0.0124(*)$ | $0.0720 \pm 0.0105(*)$ | $0.0753 \pm 0.0135(*)$ |
| | 20 | $0.0860 \pm 0.0163(*)$ | $0.0753 \pm 0.0132(*)$ | $0.0793 \pm 0.0107(*)$ | $0.0850 \pm 0.0256(*)$ |
| | 50 | $0.0939 \pm 0.0194(*)$ | $0.1186 \pm 0.0547$ | $0.0899 \pm 0.0188(*)$ | $0.0754 \pm 0.0104(*)$ |
| 500 | 0 | $0.8127 \pm 0.2164$ | $0.8059 \pm 0.5322$ | $2.8782 \pm 0.3346$ | $2.9342 \pm 0.3114$ |
| | 10 | $0.0802 \pm 0.0195(*)$ | $0.0841 \pm 0.0189(*)$ | $0.0867 \pm 0.0205(*)$ | $0.0814 \pm 0.0166(*)$ |
| | 20 | $0.0808 \pm 0.0122(*)$ | $0.0840 \pm 0.0204(*)$ | $0.0860 \pm 0.0229(*)$ | $0.1454 \pm 0.0875$ |
| | 50 | $0.0830 \pm 0.0159(*)$ | $0.0902 \pm 0.0212(*)$ | $0.0747 \pm 0.0117(*)$ | $0.0879 \pm 0.0237(*)$ |
| 1000 | 0 | $1.0943 \pm 0.4712$ | $0.5733 \pm 0.3437$ | $2.1357 \pm 0.3868$ | $2.7721 \pm 0.4732$ |
| | 10 | $0.0786 \pm 0.0107(*)$ | $0.0875 \pm 0.0168(*)$ | $0.0825 \pm 0.0272(*)$ | $0.0762 \pm 0.0151(*)$ |
| | 20 | $0.0897 \pm 0.0215(*)$ | $0.0977 \pm 0.0405$ | $0.0867 \pm 0.0168(*)$ | $0.0789 \pm 0.0137(*)$ |
| | 50 | $0.1130 \pm 0.0304$ | $0.0916 \pm 0.0138(*)$ | $0.0727 \pm 0.0083(*)$ | $0.0800 \pm 0.0262(*)$ |
| 2000 | 0 | $1.1770 \pm 0.3567$ | $0.6146 \pm 0.1431$ | $0.6679 \pm 0.4506$ | $2.3037 \pm 0.2585$ |
| | 10 | $0.0804 \pm 0.0148(*)$ | $0.0867 \pm 0.0215(*)$ | $0.0908 \pm 0.0231(*)$ | $0.0898 \pm 0.0119(*)$ |
| | 20 | $0.0933 \pm 0.0192(*)$ | $0.0845 \pm 0.0147(*)$ | $0.0833 \pm 0.0186(*)$ | $0.0860 \pm 0.0151(*)$ |
| | 50 | $0.0923 \pm 0.0186(*)$ | $0.0911 \pm 0.0172(*)$ | $0.0877 \pm 0.0224(*)$ | $0.0954 \pm 0.0298$ |

**Table 5.** Comparison of memetic neuro-fuzzy systems with differential (DE) and Big-Bang-Big-Crunch (BBBC) optimisation.

| dataset | DE | BBBC |
|---|---|---|
| 'friedman' | $0.4011 \pm 0.0557$ | $0.2722 \pm 0.0259$ |
| 'laser' | $0.2213 \pm 0.0388$ | $0.2980 \pm 0.0300$ |
| 'treasury' | $0.0720 \pm 0.0105$ | $0.1075 \pm 0.0101$ |

## 4    Conclusions

Neuro-fuzzy system tuned only with differential evolutionary approach is not able to achieve results lower than local gradient optimisation. Application of both evolutionary approach (DE algorithm) and local search (gradient method) enables elaborating of lower errors than either evolutionary approach or local search. The results elaborated with a memetic approach have lower standard deviation than results elaborated by ANNBFIS (only local search) or the system with only evolutionary approach. The important conclusion from the executed experiments is that the number of generations is more important than the number of individuals in the population. When the population is too big and number of generations too low, most individuals do not take part in optimisation procedure. When the population is small and number of generation high, the chance that a certain individual is involved into optimisation is higher. Thus the proper balance between number of generations and number of individuals should not be neglected.

## References

1. Alcalá-Fdez, J., Fernandez, A., Luengo, J., Derrac, J., García, S., Sánchez, L., Herrera, F.: KEEL data-mining software tool: data set repository, integration of algorithms and experimental analysis framework. J. Multiple-Valued Logic Soft Comput. **17**(2–3), 255–287 (2011)
2. Cordón, O., Herrera, F.: Identification of linguistic fuzzy models by means of genetic algorithms. In: Hellendoorn, H., Driankov, D. (eds.) Fuzzy Model Identification, pp. 215–250. Springer, Heidelberg (1997). http://dx.doi.org/10.1007/978-3-642-60767-7_7
3. Cordón, O., Herrera, F.: A three-stage evolutionary process for learning descriptive and approximate fuzzy-logic-controller knowledge bases from examples. Int. J. Approximate Reasoning **17**(4), 369–407 (1997). Genetic Fuzzy Systems for Control and Robotics
4. Czogała, E., Łeski, J.: Fuzzy and Neuro-Fuzzy Intelligent Systems. STUDFUZZ. Physica-Verlag, A Springer-Verlag Company, Heidelberg, New York (2000)
5. Di Gesù, V., Lo Bosco, G., Millonzi, F., Valenti, C.: A memetic algorithm for binary image reconstruction. In: Brimkov, V.E., Barneva, R.P., Hauptman, H.A. (eds.) IWCIA 2008. LNCS, vol. 4958, pp. 384–395. Springer, Heidelberg (2008)

6. Łeski, J., Czogała, E.: A neuro-fuzzy inference system optimized by deterministic annealing. In: Hampel, R., Wagenknecht, M., Chaker, N. (eds.) Fuzzy Control. Advances in Soft Computing, vol. 6, pp. 287–293. Physica-Verlag HD (2000). http://dx.doi.org/10.1007/978-3-7908-1841-3_25

7. Nalepa, J., Kawulok, M.: A memetic algorithm to select training data for support vector machines. In: Proceedings of the 2014 conference on Genetic and evolutionary computation, pp. 573–580 (2014)

8. Qing, A., Lee, C.K. (eds.): Differential Evolution in Electromagnetics. ALO, vol. 4. Springer, Heidelberg (2010)

9. Reichenbach, H.: Erkenntnis. Wahrscheinlichkeitslogik **5**, 37–43 (1935)

10. Saaty, T.L., Ozdemir, M.S.: Why the magic number seven plus or minus two. Math. Comput. Model. **38**(3–4), 233–244 (2003)

11. Siminski, K.: Memetic neuro-fuzzy system with Big-Bang-Big-Crunch optimisation. In: Gruca, A., Brachman, A., Kozielski, S., Czachórski, T. (eds.) Man-Machine Interactions 4. AISC, pp. 583–592. Springer International Publishing, New York (2016)

12. Storn, R., Price, K.V.: Differential evolution - a simple and efficient adaptive scheme for global optimization over continuous spaces. Technical report, International Computer Science Insitute (1995)

# New Rough-Neuro-Fuzzy Approach
# for Regression Task in Incomplete Data

Krzysztof Siminski[✉]

Institute of Informatics, Silesian University of Technology,
ul. Akademicka 16, 44-100 Gliwice, Poland
Krzysztof.Siminski@polsl.pl

**Abstract.** A fuzzy rule base is a crucial part of neuro-fuzzy systems. Data items presented to a neuro-fuzzy system activate rules in a rule base. For incomplete data the firing strength of the rules cannot be calculated. Some neuro-fuzzy systems impute the missing firing strength. This approach has been successfully applied. Unfortunately in some cases the imputed firing strength values are very low for all rules and data items are poorly recognized by the system. That may deteriorate the quality and reliability of elaborated results.

The paper presents a new method for handling missing values in neuro-fuzzy systems in a regression task. The new approach introduces a new imputation technique (*imputation with group centres*) to avoid very low firing strength for incomplete data items. It outperforms previous method (elaborates lower error rates), avoids numerical problems with very low firing strengths in all fuzzy rules of the system. The proposed systems elaborated interval answer without Karnik-Mendel algorithm. The paper is accompanied by numerical examples and statistical verification on real life data sets.

**Keywords:** Incomplete data · Missing values · Neuro-fuzzy system · Rough fuzzy clustering

## 1 Introduction

Errors in data acquisition, random noise, impossible values etc. lead to incompleteness of data. Sometimes data are compiled from various sources and may miss some values. In certain cases it is practically impossible to collect all the data. The incomplete data may hold important information and its analysis is a challenging task.

Generally three approaches are used to handle the problem of missing values: (1) imputation of missing values [16,22,24] (2) marginalisation of incomplete data items [23] or incomplete data attributes [2]; (3) handling the imprecise information with rough sets [5,11,15,20,21]. The marginalisation is the simplest approach – leaves only complete data. The marginalisation of incomplete attributes leads to a reduction of dimensionality of the task domain. Although

© Springer International Publishing Switzerland 2016
S. Kozielski et al. (Eds.): BDAS 2016, CCIS 613, pp. 146–156, 2016.
DOI: 10.1007/978-3-319-34099-9_10

the imputation is more complicated than marginalisation, it is used more frequently [6]. Many techniques of imputation have been proposed (imputation with constants, averages, medians or sophisticatedly elaborated values). The comparison of imputation techniques for clustering of missing data is provided in [10]. The analysis of techniques used to handle the incomplete data for neuro-fuzzy systems is provided in [17]. The essential problem with the imputation is the fact that imputed values cannot be fully trusted, they may have no meaning in the data set. The rough set approach saves the distinction between original and imputed values [20].

The neuro-fuzzy systems provide an efficient technique for data mining. They can elaborate precise models with good generalisation ability. The neuro-fuzzy system for handling incomplete data have been proposed for classification [8,11,13] and for regression [19]. The systems described in [13] (for classification) and [19] (for regression) apply the minimal and maximal firing strengths in case of missing values (imputation of firing strengths with minimal and maximal values). The practical application of such an approach reveals in some cases such low values ($< 10^{-6}$) of firing strengths of all rules. It causes numerical problems in elaboration of the answer. This leads to poor answers of the systems. The objective of the paper is a new neuro-fuzzy system for incomplete data that avoids this undesired phenomenon. The higher recognition of data items by the rules makes the fuzzy model more reliable. In this proposition the fuzzy model better recognises the data items and minimal values of firing strengths do not drop to very low values. The paper presents the extension of the system proposed in [19] and in some sense also the extension of the systems for classification [11,13]. The system described in [19] applies imputation of firing strength for missing attributes and uses gradient analysis for elaboration of the final answer. The proposed system uses an *imputation with group centres* (cf. Sect. 3.1) for missing values, what prevents low firing values. This is the complete system: it creates the fuzzy model with both complete and incomplete data, and elaborates the results both for complete and incomplete data. The system applies marginalisation, imputation (twice) and rough sets for handling missing data. The system elaborates rough answers for the incomplete data and keeps the distinction between original and imputed data items.

In the paper the blackboard bold characters ($\mathbb{A}$) are used to denote the sets, bolds ($\mathbf{a}$) – matrices and vectors, uppercase italics ($A$) – the cardinality of sets, lowercase italics ($a$) – scalars and set elements.

## 2   Neuro-Fuzzy System with Parametrised Consequences

The system with parametrised consequences (ANNBFIS) [3] is a MISO (Multiple Input Single Output) system. The rule base $\mathbb{L}$ contains fuzzy rules $l$ (fuzzy implications). The premises are built with Gaussian fuzzy sets. The consequences are composed of isosceles triangular fuzzy sets, the localisation of each fuzzy set is a linear combination of input values. The implication in rules is a genuine implication (logical type implication fuzzy system). The creation of the fuzzy model (elaboration of fuzzy rule base) is done in three steps: clustering of the input domain, extraction of rules' premises, and tuning of the rules (this step is also responsible for creation of rules consequences) [3].

## 3  New Approach to Handle Incomplete Data

The premises of the fuzzy rules are elaborated with the scatter domain partition. The data are clustered and each cluster is transformed into a premise of a fuzzy rule. Thus the number of clusters $C$ and number of rules $L$ are equal: $C = L$. For clustering of incomplete data the rough fuzzy clustering technique [20] is used. This algorithm requires the preprocessing of data. The marginalisation of incomplete data vectors creates the data set $\underline{\mathbb{X}}$. This data sets holds only complete original data item. If the tuple lacks $n$ values, it is substituted with $k^n$ tuples with all combination of imputed values (these are the mean values $m$ of missing attribute calculated from values existing in other tuples, $m + \sigma$, where $\sigma$ is the standard deviation of the attribute, $m - \sigma$, thus $k = 3$). The maximum and minimum values are not used here, because the extreme values may be outliers and one extreme value can substantially influence the clustering process. If the tuple with missing values is substituted with $n$ imputed tuples, each of these imputed tuples is assigned a weight $\eta = 1/n$. The weight is treated as a condition in the clustering procedure. The complete data items and imputed items constitute the data set $\overline{\mathbb{X}}$. The $\underline{\mathbb{X}}$ is a subset of $\overline{\mathbb{X}}$, when the original data set lacks no values both preprocessed data sets are equal: $\underline{\mathbb{X}} \subseteq \overline{\mathbb{X}}$.

In the system the hybrid fuzzy $c$-means (FCM) algorithm is used [20]. The clustering procedure uses both $\underline{\mathbb{X}}$ and $\overline{\mathbb{X}}$ data sets. The "upper" (elaborated with $\overline{\mathbb{X}}$ data set) and "lower" (elaborated with $\underline{\mathbb{X}}$) clusters are joined into pairs with common cluster centres. The "lower" data set holds only original (not imputed) data item. This data set is treated as more reliable (although is does not have to be so) and determined the localisation of the cluster centres. This requires presence of complete data tuples in the data set. There should be more complete data tuples than clusters.

The results of clustering are transformed into the premises of the rules. The number of rules $L$ equals the number of clusters $C$. The cores of the premises of rules equal the centres of the clusters $\mathbf{c} = \underline{\mathbf{c}} = \overline{\mathbf{c}} = \mathbf{v}$.

The premises of the rules for each attribute are composed of twin Gaussian functions with the same core. The values of the membership functions created with the data set $\underline{\mathbb{X}}$ are not greater than the values of the membership functions created with the data set $\overline{\mathbb{X}}$. The second function (calculated with original complete data and imputed data) constitutes the "cloud" around the first function (calculated with only complete data). The form of membership functions in premises of the rules can be described as interval type-2 fuzzy set with uncertain fuzziness.

The clusters enable the initialisation of the premises of the rules. This stage has two aims: (1) fitting of the model to the presented data and (2) calculation of the premises.

The premises of the rules and the widths of supports of fuzzy sets in consequences are tuned with gradient method, whereas the linear coefficients of the consequences are elaborated with linear regression (least mean square method).

Similarly as during clustering the centres of clusters are tuned basing on the "lower" data set. Other "lower" parameters are tuned with $\underline{\mathbb{X}}$ and respectively the "upper" parameters are tuned with $\overline{\mathbb{X}}$ data sets.

The consequences of the rules are calculated with $\underline{\mathbb{X}}$ data. In [19] the consequences in "upper" rules are tuned with $\overline{\mathbb{X}}$ and "lower" with $\underline{\mathbb{X}}$. The experiments show that in our system it is better to use only $\underline{\mathbb{X}}$ for tuning.

### 3.1   Elaboration of the Answer

The models can be created both with complete and incomplete data. The system is able to elaborate the answer for complete and incomplete data. Thus there are four variants: CC, CI, IC, and II.

'**CC**' When the model is created with the complete data set and the system should elaborate the answer for the complete data, the system reduces to the ANNBFIS, the base system. The data sets $\mathbb{X} = \underline{\mathbb{X}} = \overline{\mathbb{X}}$, the twin functions in the premise of the rule are the same, because $\underline{\mathbf{s}} = \overline{\mathbf{s}}$. We label this situation with 'CC' (complete train and complete test sets).

'**IC**' When the model is created with the incomplete each premise has twin membership functions. Each rules is fired in two ways. Thus for each complete data item the systems elaborated two answers: "lower" and "upper". We label this situation with 'IC' (incomplete train and complete test sets).

'**CI**' In the third situation the fuzzy model is created with the complete data set and then tested with the incomplete data. The premises of the rules have the twin functions that are represented with the same function (as in 'CC'). When the system is supposed to give an answer for an incomplete data item, the problem of missing values has to be solved. The paper [4] describes the system where missing attributes are fired with maximal (equal 1) strength. Systems [14,19] implement more sophisticated approach: the firing strength of an missing attribute is substituted with the highest and lowest value of firing strength of this attribute. Unfortunately this approach sometimes generates a serious problem: when the minimal firing strength of the attribute is very close to zero, it makes the firing strength of the rule's premise very low. The firing strengths of the rules are substituted independently and it is quite possible that lower firing strengths of all rules are very low. This may cause numerical problems and severely distort the aggregated answer of the system.

Here a different approach to handle incomplete data is proposed. To avoid very low firing strength the maximal and minimal firing strengths for the data item are not used. The missing attribute's value in the data tuple is imputed with the values of this attribute building the cores of the fuzzy set in the rule premises. These cores equal the cluster (rules) centres. We call it *imputation with group centres*. Each missing attribute is imputed with $C = L$ values. Let's assume there are $C$ clusters with centres $\mathbf{v}_1, \mathbf{v}_2, \ldots, \mathbf{v}_C$. If the data item lacks the value of the $d$th attribute, it is replaced with $C$ data items that have the same values of the attributes as the item in question and the missing value is imputed with $v_{1d}, v_{2d}, \ldots, v_{Cd}$ values. The Fig. 1 presents an example with $C = 3$ cluster with centres: $(x_a, y_a)$, $(x_b, y_b)$ and $(x_c, y_c)$. The data tuple $(x_1, ?)$ misses the second attribute. The missing values is substituted with $y_a, y_b$ and $y_c$, thus the data tuple is substituted with three data items: $(x_1, y_a)$, $(x_1, y_b)$ and $(x_1, y_c)$. The new data items are the closest possible items to all existing clusters centres. This approach prevents from very low values of firing strengths that may cause severe problems in system described in [19]. This approach does not need the Karnik-Mendel algorithm [7] for elaboration of the answer. For each data item the system elaborates two answers (for two firing strengths). Then from the results the first and third quadrilles (to avoid outliers) are taken as 'lower' and 'upper' limits for interval answer. We label this situation with 'CI' (complete train and incomplete test sets).

'II' The last situation is when the system is both trained and tested with incomplete data. Then each premise has twin membership function and the missing values in test data item are imputed with cluster centres. This situation joins the situations 'IC' and 'CI'. We label this situation with 'II' (incomplete train and incomplete test sets).

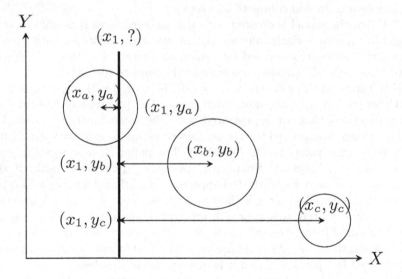

**Fig. 1.** A two dimensional example of *imputation with group centres*. The data item $(x_1, ?)$ lacks the attribute value marked with '?'. The clusters have centres: $(x_a, y_a)$, $(x_b, y_b)$ and $(x_c, y_c)$. The fuzzification of cluster is symbolically rendered with circles. The missing value is imputed with attribute's values from all clusters, so the incomplete tuples is imputed with three values: $(x_1, y_a)$, $(x_1, y_b)$ and $(x_1, y_c)$.

## 4   Experiments

The experiments were conducted on real life data. The 'Gas Furnace' is a popular real life data set depicting the concentration of methane ($x$) and carbon dioxide ($y$) in the gas furnace [1]. The data sets contains 290 tuples organised according to the template $[y_{n-1}, \ldots, y_{n-4}, x_{n-1}, \ldots, x_{n-6}, y_n]$. The last item in the data tuple is the decision attribute.

The concentration of leukocytes in blood ('Mackey-Glass' data set) is modelled with the Mackey-Glass equation $\frac{dx(t)}{dt} = \frac{ax(t-\tau)}{1+(x(t-\tau))^{10}} - bx(t)$, where $x$ is concentration of leukocytes, $a = 0.2$, $b = 0.1$ and $\tau = 17$ are constants [9]. The equation was solved with condition $x(0) = 0.1$ with the Runge-Kutt method with step $k = 0.1$ [3]. The data series was the base for creation of tuples with the template $[x(t), x(y-6), x(t-12), x(t-18), x(k+6)]$. The 'Methane' data set contains the real life measurements of air parameters in a coal mine in Upper Silesia (Poland) [18]. The parameters (measured in 10 s intervals) are: AN31 – the flow of air in the shaft, AN32 – the flow of air in the adjacent shaft, MM32 – concentration of methane ($CH_4$), production of coal, the day of week. To the tuples the 10-min sums of measurements of AN31, AN32, MM32 are added as dynamic attributes. The task is to predict the concentration of the methane in 10 min.

**Table 1.** The comparison of errors elaborated by old and new rough neuro-fuzzy system for 'Gas Furnace' data set. The symbol $E_\triangle$ represents the error measure defined with formula 2 (total value and its two components), the symbol $E_l$ stands for maximal linear error (defined with formula 3), 'norm' – normalisation, 'DA' – data approximation, 'KG' – knowledge approximation. The statistically verified lower error of the new system is denoted with '✔', the lack of statistical significance is marked with '✘'. The abbreviation 'CC' stands for complete train, complete test data sets; 'CI' – complete train and incomplete test sets; 'IC' – incomplete train and complete test data sets; 'II' – incomplete train and incomplete test data sets.

| | | | non normalised data | | | | | | normalised data | | | |
| | | | $E_\triangle$ | | | | | | $E_\triangle$ | | | |
| P | T | | total | interval | deviation | $E_l$ | P | T | | total | interval | deviation | $E_l$ |
|---|---|---|---|---|---|---|---|---|---|---|---|---|---|
| DA | CC | old | 0.1601 | 0.0000 | 0.1601 | 0.1601 | DA | CC | old | 0.0530 | 0.0000 | 0.0530 | 0.0530 |
| | | new | 0.1601 | 0.0000 | 0.1601 | 0.1601 | | | new | 0.0532 | 0.0000 | 0.0532 | 0.0532 |
| | CI | old | 8.3182 | 8.3176 | 0.0933 | 6.0796 | | CI | old | 2.6803 | 2.6801 | 0.0369 | 1.9412 |
| | | new | 2.1162 ✔ | 1.9005 ✔ | 0.9063 ✘ | 2.2534 ✔ | | | new | 0.6800 ✔ | 0.6162 ✔ | 0.2766 ✘ | 0.6817 ✔ |
| | IC | old | 1.1281 | 1.1260 | 0.0677 | 0.8543 | | IC | old | 0.2932 | 0.2909 | 0.0331 | 0.2609 |
| | | new | 0.4790 ✔ | 0.4527 ✔ | 0.1511 ✘ | 0.4610 ✔ | | | new | 0.1861 ✔ | 0.1750 ✔ | 0.0606 ✘ | 0.1858 ✔ |
| | II | old | 10.1281 | 10.1279 | 0.0461 | 8.0685 | | II | old | 3.6855 | 3.6854 | 0.0088 | 3.0167 |
| | | new | 3.3473 ✔ | 3.2429 ✔ | 0.7868 ✘ | 3.2833 ✔ | | | new | 1.0043 ✔ | 0.9747 ✔ | 0.2367 ✘ | 0.9959 ✔ |
| KG | CC | old | 0.5429 | 0.0000 | 0.5429 | 0.5429 | KG | CC | old | 0.3363 | 0.0000 | 0.3363 | 0.3363 |
| | | new | 0.5429 | 0.0000 | 0.5429 | 0.5429 | | | new | 0.3366 | 0.0000 | 0.3366 | 0.3366 |
| | CI | old | 7.3581 | 7.3512 | 0.3163 | 5.0630 | | CI | old | 2.6630 | 2.6555 | 0.1955 | 1.9854 |
| | | new | 2.1084 ✔ | 1.7103 ✔ | 1.1933 ✘ | 2.2023 ✔ | | | new | 0.7410 ✔ | 0.5754 ✔ | 0.4532 ✘ | 0.8202 ✔ |
| | IC | old | 1.7179 | 1.7041 | 0.2006 | 1.3801 | | IC | old | 0.5837 | 0.5688 | 0.1084 | 0.5095 |
| | | new | 0.8461 ✔ | 0.7105 ✔ | 0.4329 ✘ | 0.9424 ✔ | | | new | 0.4466 ✔ | 0.4002 ✔ | 0.1749 ✘ | 0.4542 ✔ |
| | II | old | 9.9187 | 9.9155 | 0.2444 | 7.9197 | | II | old | 4.1890 | 4.1873 | 0.0920 | 3.4526 |
| | | new | 3.1329 ✔ | 2.9556 ✔ | 0.9596 ✘ | 3.2267 ✔ | | | new | 1.1192 ✔ | 1.0667 ✔ | 0.3296 ✘ | 1.1510 ✔ |

The system was tested in two paradigms: (1) data approximation (DA): the system was trained and tested on the same data set; (2) knowledge generalisation (KG): the train and test data sets are disjoint. For the knowledge generalisation (KG) the data sets described above were divided into train and test sets:

- 'Mackey-Glass' – train data sets: tuples 1–200, test data sets: tuples 201–500
- 'Methane' – train data sets: tuples 1–511, test data sets: tuples 512–1022
- 'Gas Furnace' – train data sets: tuples 1–145, test data sets: tuples 146–290

For the data approximation the train data sets are used both for training and testing. The experiments are conducted both on original (non-normalised) and normalised data.

In the experiments the neuro-fuzzy system with $L = 6$ rules in the rule base is used. Our aim is to test new method of handling missing data, not to discuss the optimal number of rules in the fuzzy systems.

## 4.1   Measure of Error

In the ANNBFIS system the RMSE (root mean square error) is calculated as $E = \sqrt{\frac{1}{X} \sum_{\mathbf{x} \in \mathbb{X}} [y_0(\mathbf{x}) - y(\mathbf{x})]^2}$, where $y_0(\mathbf{x})$ is the answer of the ANNBFIS system for the $\mathbf{x}$ data tuple, $y(\mathbf{x})$ is the original expected value of the decision attribute of the tuple $\mathbf{x}$. For a rough neuro-fuzzy system the above formula should be modified. The system elaborates two answers $\overline{y}_0(\mathbf{x})$ and $\underline{y}_0(\mathbf{x})$ for each tuple $\mathbf{x}$. Instead of one value the system returns the interval $\left[\underline{y}_0(\mathbf{x}), \overline{y}_0(\mathbf{x})\right]$. The deviation $\Delta y(\mathbf{x})$ of the original value from the returned interval is determined as

$$\Delta y(\mathbf{x}) = \begin{cases} y(\mathbf{x}) - \overline{y}_0(\mathbf{x}), & \text{if } y(\mathbf{x}) > \overline{y}_0(\mathbf{x}), \\ 0, & \text{if } y(\mathbf{x}) \in \left[\underline{y}_0(\mathbf{x}), \overline{y}_0(\mathbf{x})\right], \\ \underline{y}_0(\mathbf{x}) - y(\mathbf{x}), & \text{if } y(\mathbf{x}) < \underline{y}_0(\mathbf{x}). \end{cases} \tag{1}$$

Such a definition of the deviation would prise models that elaborate very wide intervals in an answer. This is why the length of the interval is also taken into account in measuring the system answer (triangle measure) [19]

$$E_\triangle (\mathbb{X}) = \sqrt{\frac{1}{X} \sum_{\mathbf{x} \in \mathbb{X}} \left( [\Delta y(\mathbf{x})]^2 + \left[\overline{y}_0(\mathbf{x}) - \underline{y}_0(\mathbf{x})\right]^2 \right)}. \tag{2}$$

Slightly different error measure – maximal linear error – is also used here. This measure takes into regard the maximum distance of the expected value from the ends of the elaborated interval [21]:

$$E_l (\mathbb{X}) = \sqrt{\frac{1}{X} \sum_{\mathbf{x} \in \mathbb{X}} \left[ \max \left( \left| y(\mathbf{x}) - \underline{y}_0(\mathbf{x}) \right|, \left| y(\mathbf{x}) - \overline{y}_0(\mathbf{x}) \right| \right) \right]^2}. \tag{3}$$

## 4.2    Results

In the experiments the system is compared with a system described in [19]. This system imputes the firing strength of missing attributes with maximal and minimal values. This may cause that all rules are fired very weakly what may deteriorate the quality of the answers of the system. To the best of our knowledge the system cited above and one described in this paper are the only complete neuro-fuzzy systems that elaborate the interval answer for a regression task in incomplete data. To minimise the influence of randomness on the elaborated results the experiments for incomplete data were repeated $N = 20$ times with 10 % of values missing completely at random. The results are verified with statistical dependent $t$-test for paired samples. The null hypothesis $H_0$ states that there is no statistical difference between both methods, the alternative one-tailed hypothesis $H_1$ states that new method elaborates lower error rates than method described in [19]. The results are tested with statistical significance $1 - \alpha = 0.95$. The results of our experiments are presented in Table 1 (for 'Gas Furnace'), 2 (for 'Methane') and 3 (for 'Mackey-Glass'). The error $E_\triangle$ is defined with Eq. (2), and $E_l$ with Eq. (3). The column 'total' holds $E_\triangle$ error, and columns 'deviation' and 'interval' hold the components of $E_\triangle$: $\sqrt{\frac{1}{X}\sum_{\mathbf{x}\in\mathbb{X}}(\triangle y(\mathbf{x}))^2}$ and

$\sqrt{\frac{1}{X}\sum_{\mathbf{x}\in\mathbb{X}}\left[\overline{y}_0(\mathbf{x}) - \underline{y}_0(\mathbf{x})\right]^2}$ respectively.

**Table 2.** The comparison of errors elaborated by old and new rough neuro-fuzzy system for 'Methane' data set. The symbols are explained in the caption of Fig. 1.

| P | T | | total | interval | deviation | $E_l$ | P | T | | total | interval | deviation | $E_l$ |
|---|---|---|---|---|---|---|---|---|---|---|---|---|---|
| | | | non normalised data $E_\triangle$ | | | | | | | normalised data $E_\triangle$ | | | |
| DA | CC | old | 0.0919 | 0.0000 | 0.0919 | 0.0919 | DA | CC | | 0.3766 | 0.0000 | 0.3766 | 0.3766 |
| | | new | 0.0919 | 0.0000 | 0.0919 | 0.0919 | | | | 0.3751 | 0.0000 | 0.3751 | 0.3751 |
| | CI | old | 1.7841 | 1.7664 | 0.0683 | 0.5717 | | CI | old | 1.9251 | 1.9020 | 0.2929 | 1.5800 |
| | | new | 0.1569 ✔ | 0.1205 ✔ | 0.1000 ✖ | 0.1693 ✔ | | | new | 0.8049 ✔ | 0.5702 ✔ | 0.5580 ✖ | 0.8756 ✔ |
| | IC | old | 0.7906 | 0.7885 | 0.0452 | 0.7163 | | IC | old | 1.1074 | 1.0713 | 0.2680 | 1.0326 |
| | | new | 0.1332 ✔ | 0.1032 ✔ | 0.0835 ✖ | 0.1449 ✔ | | | new | 0.5450 ✔ | 0.3913 ✔ | 0.3575 ✖ | 0.6064 ✔ |
| | II | old | 2.1796 | 2.1714 | 0.0386 | 1.0594 | | II | old | 2.4739 | 2.4630 | 0.2237 | 1.9526 |
| | | new | 0.1743 ✔ | 0.1486 ✔ | 0.0905 ✖ | 0.1886 ✔ | | | new | 0.6016 ✔ | 0.4705 ✔ | 0.3707 ✖ | 0.6946 ✔ |
| KG | CC | old | 0.2274 | 0.0000 | 0.2274 | 0.2274 | KG | CC | old | 0.4841 | 0.0000 | 0.4841 | 0.4841 |
| | | new | 0.2849 | 0.0000 | 0.2849 | 0.2849 | | | new | 0.4077 | 0.0000 | 0.4077 | 0.4077 |
| | CI | old | 1.5074 | 1.6913 | 0.1934 | 0.6514 | | CI | old | 1.9341 | 1.9021 | 0.3396 | 1.5953 |
| | | new | 0.2848 ✔ | 0.1385 ✔ | 0.2479 ✖ | 0.3075 ✔ | | | new | 0.5852 ✔ | 0.4034 ✔ | 0.4159 ✖ | 0.6097 ✔ |
| | IC | old | 1.0084 | 1.0065 | 0.0525 | 0.8965 | | IC | old | 1.1466 | 1.1307 | 0.1714 | 0.9754 |
| | | new | 0.2766 ✔ | 0.1835 ✔ | 0.1851 ✖ | 0.3149 ✔ | | | new | 0.5875 ✔ | 0.4999 ✔ | 0.2808 ✖ | 0.6211 ✔ |
| | II | old | 2.3984 | 2.3950 | 0.0517 | 1.2930 | | II | old | 2.8685 | 2.8635 | 0.1491 | 2.2550 |
| | | new | 0.3097 ✔ | 0.2285 ✔ | 0.2034 ✖ | 0.3500 ✔ | | | new | 0.6583 ✔ | 0.5747 ✔ | 0.3054 ✖ | 0.6953 ✔ |

The experiments show also the concordance of the two error measures: triangle measure ($E_\triangle$) (formula 2) and max linear error ($E_l$) (formula 3). The latter seems to be more natural and the former is used in [19]. For comparison of two systems both measures were used.

**Table 3.** The comparison of errors elaborated by old and new rough neuro-fuzzy system for 'Mackey-Glass' data set. The symbols are explained in the caption of Fig. 1.

| | | | non normalised data $E_\triangle$ | | | | | | normalised data $E_\triangle$ | | | |
|----|----|-----|-------|----------|-----------|--------|----|----|-------|----------|-----------|--------|
| P | T | | total | interval | deviation | $E_l$ | P | T | total | interval | deviation | $E_l$ |
| DA | CC | old | 0.000 | 0.000 | 0.000 | 0.000 | DA | CC | 0.001 | 0.000 | 0.001 | 0.001 |
| | | new | 0.000 | 0.000 | 0.000 | 0.000 | | | 0.002 | 0.000 | 0.002 | 0.002 |
| | CI | old | 18.645 | 18.645 | 0.000 | 14.675 | | CI | 85.099 | 85.088 | 1.188 | 69.362 |
| | | new | 6.782 ✔ | 6.001 ✔ | 3.071 ✘ | 7.626 ✔ | | | 34.739 ✔ | 31.050 ✔ | 15.152 ✘ | 37.583 ✔ |
| | IC | old | 0.099 | 0.099 | 0.558 | 0.096 | | IC | 1.119 | 1.116 | 0.061 | 0.988 |
| | | new | 0.002 ✔ | 0.002 ✔ | 0.000 ✘ | 0.002 ✔ | | | 0.754 ✔ | 0.695 ✔ | 0.248 ✘ | 0.818 ✔ |
| | II | old | 7.533 | 7.533 | 2.284 | 5.886 | | II | 33.766 | 33.766 | 0.014 | 32.030 |
| | | new | 2.967 ✔ | 2.926 ✔ | 0.435 ✔ | 2.965 ✔ | | | 14.377 ✔ | 14.061 ✔ | 2.715 ✘ | 14.65 ✔ |
| KG | CC | old | 0.000 | 0.000 | 0.000 | 0.000 | KG | CC | 0.002 | 0.000 | 0.002 | 0.002 |
| | | new | 0.000 | 0.000 | 0.000 | 0.000 | | | 0.002 | 0.000 | 0.002 | 0.002 |
| | CI | old | 18.121 | 18.121 | 0.002 | 14.160 | | CI | 82.960 | 82.955 | 0.949 | 67.435 |
| | | new | 6.708 ✔ | 6.138 ✔ | 2.620 ✘ | 7.244 ✔ | | | 33.044 ✔ | 29.156 ✔ | 15.210 ✔ | 36.006 ✔ |
| | IC | old | 0.098 | 0.098 | 0.000 | 0.094 | | IC | 1.096 | 1.093 | 0.058 | 0.947 |
| | | new | 0.002 ✔ | 0.002 ✔ | 0.000 ✘ | 0.002 ✔ | | | 0.826 ✔ | 0.770 ✔ | 0.219 ✘ | 0.872 ✘ |
| | II | old | 6.780 | 6.780 | 2.978 | 5.351 | | II | 33.398 | 33.398 | 0.039 | 31.720 |
| | | new | 2.903 ✔ | 2.860 ✔ | 0.448 ✔ | 2.930 ✔ | | | 14.161 ✔ | 13.795 ✔ | 2.985 ✘ | 14.718 ✔ |

For the 'Mackey-Glass' and 'Gas Furnace' data sets the results reveal an interesting feature: The errors for the paradigms 'CC', 'CI', 'IC', and 'II' can be ordered in the following way: $E_{CC} < E_{IC} < E_{II} < E_{CI}$. The lowest error is for the 'CC' paradigm (complete train and complete test sets), in other data sets the incompleteness of data manifests in a higher error. The interesting relation is $E_{II} < E_{CI}$, what in symbolic way means that the error for the 'II' paradigm is lower than for the 'CI' paradigm. This shows that when the system is supposed to elaborate the answer for the incomplete test data, it is more advantageous to create the fuzzy model (for the neuro-fuzzy system) with the incomplete train data. This is valid both for data approximation (DA) and knowledge generalisation (KG).

## 5    Conclusions

The imputation of firing strength for missing attributes is applied in some neuro-fuzzy systems [12,13,19]. The firing strengths are commonly imputed with maximum and minimum values. Unfortunately this may cause very low firing strength of all rules. This may severely distort the answers of the system, because the rule base of the system hardly recognized the data items and the results cannot be fully trusted.

The paper presents a new approach for elaboration of an answer for incomplete data with neuro-fuzzy systems. In the system the values of missing attributes are imputed with cores of rules (*imputation with group centres*). This decreases the danger of very low firing strengths in all rules. So the fuzzy rule base better recognizes presented data items and the answers are more reliable. The presented system is a complete one. It can elaborate the fuzzy rule base both for complete and incomplete data. It can also calculate answer for complete and incomplete data. For incomplete data the system gives an interval answer.

The numerical examples with statistical verification show that the new approach can elaborate lower errors for a regression task in incomplete data.

# References

1. Box, G.E.P., Jenkins, G.: Time Series Analysis, Forecasting and control. Holden Day, Incorporated, Oakland (1970)
2. Cooke, M., Green, P., Josifovski, L., Vizinho, A.: Robust automatic speech recognition with missing and unreliable acoustic data. Speech Commun. **34**, 267–285 (2001)
3. Czogała, E., Łęski, J.: Fuzzy and Neuro-Fuzzy Intelligent Systems. Series in Fuzziness and Soft Computing. Physica-Verlag, Springer, Heidelberg, New York (2000)
4. Gabriel, T.R., Berthold, M.R.: Missing values in fuzzy rule induction. In: SMC, pp. 1473–1476 (2005)
5. Grzymala-Busse, J.W.: A rough set approach to data with missing attribute values. In: Wang, G.-Y., Peters, J.F., Skowron, A., Yao, Y. (eds.) RSKT 2006. LNCS (LNAI), vol. 4062, pp. 58–67. Springer, Heidelberg (2006)
6. Himmelspach, L., Conrad, S.: Fuzzy clustering of incomplete data based on cluster dispersion. In: Hüllermeier, E., Kruse, R., Hoffmann, F. (eds.) IPMU 2010. LNCS, vol. 6178, pp. 59–68. Springer, Heidelberg (2010)
7. Karnik, N.N., Mendel, J.M.: Centroid of a type-2 fuzzy set. Inf. Sci. **132**, 195–220 (2001)
8. Korytkowski, M., Nowicki, R., Scherer, R., Rutkowski, L.: Ensemble of rough-neuro-fuzzy systems for classification with missing features. In: IEEE International Conference on Fuzzy Systems, FUZZ-IEEE 2008 (IEEE World Congress on Computational Intelligence), Hong Kong, pp. 1745–1750, June 2008
9. Mackey, M.C., Glass, L.: Oscillation and Chaos in physiological control systems. Science **197**(4300), 287–289 (1977)
10. Matyja, A., Simiński, K.: Comparison of algorithms for clustering incomplete data. Found. Comput. Decis. Sci. **39**(2), 107–127 (2014)
11. Nowicki, R.: Rough-neuro-fuzzy system with MICOG defuzzification. In: 2006 IEEE International Conference on Fuzzy Systems, Vancouver, Canada, pp. 1958–1965 (2006)
12. Nowicki, R.: On combining neuro-fuzzy architectures with the rough set theory to solve classification problems with incomplete data. IEEE Trans. Knowl. Data Eng. **20**(9), 1239–1253 (2008)
13. Nowicki, R.K.: Rough-neuro-fuzzy structures for classification with missing data. IEEE Trans. Syst. Man Cybern. Part B: Cybern. **39**(6), 1334–1347 (2009)
14. Nowicki, R.K.: On classification with missing data using rough-neuro-fuzzy systems. Int. J. Appl. Math. Comput. Sci. **20**(1), 55–67 (2010)
15. Nowicki, R.K., Korytkowski, M., Scherer, R., Nowak, B.A.: Design methodology for rough-neuro-fuzzy classification with missing data. In: 2015 IEEE Symposium Series on Computational Intelligence, pp. 1650–1657 (2015)
16. Renz, C., Rajapakse, J.C., Razvi, K., Liang, S.K.C.: Ovarian cancer classification with missing data. In: Proceedings of the 9th International Conference on Neural Information Processing, ICONIP 2002, vol. 2, pp. 809–813, Singapore (2002)

17. Sikora, M., Simiński, K.: Comparison of incomplete data handling techniques for neuro-fuzzy systems. Comput. Sci. **15**(4), 441–458 (2014)
18. Sikora, M., Krzystanek, Z., Bojko, B., Śpiechowicz, K.: Application of a hybrid method of machine learning for description and on-line estimation of methane hazard in mine workings. J. Min. Sci. **47**(4), 493–505 (2011)
19. Simiński, K.: Neuro-rough-fuzzy approach for regression modelling from missing data. Int. J. Appl. Math. Comput. Sci. **22**(2), 461–476 (2012)
20. Simiński, K.: Clustering with missing values. Fundamenta Informaticae **123**(3), 331–350 (2013)
21. Simiński, K.: Rough subspace neuro-fuzzy system. Fuzzy Sets Syst. **269**, 30–46 (2015). http://www.sciencedirect.com/science/article/pii/S0165011414003108
22. Siminski, K.: Imputation of missing values by inversion of fuzzy neuro-system. In: Gruca, A., Brachman, S., Czachórski, T. (eds.) Man-Machine Interactions 4. AISC, pp. 573–582. Springer International Publishing, Heidelberg (2016)
23. Troyanskaya, O., Cantor, M., Sherlock, G., Brown, P., Hastie, T., Tibshirani, R., Botstein, D., Altman, R.B.: Missing value estimation methods for DNA microarrays. Bioinformatics **17**(6), 520–525 (2001)
24. Zhang, S.: Shell-neighbor method and its application in missing data imputation. Appl. Intell. **35**(1), 123–133 (2011)

# Improvement of Precision of Neuro-Fuzzy System by Increase of Activation of Rules

Krzysztof Siminski[✉]

Institute of Informatics, Silesian University of Technology,
ul. Akademicka 16, 44-100 Gliwice, Poland
Krzysztof.Siminski@polsl.pl

**Abstract.** Neuro-fuzzy systems have proved to be a powerful tool for data approximation and generalization. A rule base is a crucial part of a neuro-fuzzy system. The data items activate the rules and their answers are aggregated into a final answer. The experiments reveal that sometimes the activation of all rules in a rule base is very low. It means the system recognizes the data items very poorly. The paper presents a modification of the neuro-fuzzy system: the tuning procedure has two objectives: minimizing of the error of the system and maximizing of the activation of rules. The higher activation (better recognition of the data items) makes the model more reliable. The increase of the activation of rules may also decrease the error rate for the model. The paper is accompanied by the numerical examples.

**Keywords:** Neuro-fuzzy system · Activation of rules

## 1 Introduction

Neuro-fuzzy systems are a good tool for classification and approximation of data. They can handle imprecision of data by fuzzy approach and are able to identify a fuzzy model and then tune its parameters to better fit the presented data. This makes them a good tool for data analysis. A model created by neuro-fuzzy models has a form of a set of fuzzy rules and is more interpretable by humans than knowledge disseminated in weights of synapses in artificial neural networks. Although there are some attempts [21] at interpretation of weights in neurons of artificial neural networks, the knowledge is hard to extract. The premises and consequences of the rules in neuro-fuzzy systems have human friendly form of IF-THEN rules. The triangle, trapezoidal or Gaussian membership functions in premises of the rules enable the interpretation of the attributes' values with linguistic terms. This also enables incorporation of an expert's knowledge into the system. Also the triangle fuzzy sets in consequences (Mamdani type [10]), singletons (Sugeno type [18]), or parameterized consequences (Czogała-Łęski type [4]) alleviate the interpretation of the rules.

The main aim in the research of neuro-fuzzy systems is the increase in precision with satisfactory interpretability of the fuzzy model. Many techniques of

© Springer International Publishing Switzerland 2016
S. Kozielski et al. (Eds.): BDAS 2016, CCIS 613, pp. 157–167, 2016.
DOI: 10.1007/978-3-319-34099-9_11

creation of the rule base have been proposed: grid partition (the simplest technique) [11,12], scatter partition (the most popular), and hierarchical (the most precise) [2,7,14].

The paper presents two modification of the ANNBFIS system [4]. It is a neuro-fuzzy system with logical interpretation of fuzzy rules. Each fuzzy rule in the rule base is a logical implication. The premises are built with Gaussian function and consequences of rules are symmetrical triangular fuzzy sets (as in Mamdani-Assilan approach) whose localizations are linear combinations of input values (as in Takagi-Sugeno-Kang approach). This system has shown its precision in data modeling and has many modifications [3,8,15–17].

Unfortunately the application of neuro-fuzzy systems shows that in some cases there is a problem with elaboration of answer because of very low firing strengths of all rules in the fuzzy rule base. This proves that the fuzzy rule base is not properly created. The aim of this paper is a tuning algorithm for neuro-fuzzy system that can both lower the errors elaborated by the system and increase the recognition of the data examples by the system. The latter aim can be treated as an indirect increase of interpretability of the model. The model is more reliable, because it better recognizes the presented data.

## 2    Proposed Approach

The proposed approach is based on the ANNBFIS neuro-fuzzy system. For the brevity of the paper only essential information on this system is provided. For detail please consult [4].

ANNBFIS is a neuro-fuzzy system with parametrized consequences [4,9] and logical interpretation of fuzzy implications (fuzzy rules). The premises of the rules are defined with Gaussian type-1 fuzzy sets. The membership function $\mu_d$ for each descriptor $x_d$ in a rule is defined as $\mu_d(x_d) = \exp\left(-\frac{(x_d - v_d)^2}{2s_d^2}\right)$. The T-norm (in ANNBFIS: product) of memberships of all descriptor define the firing strength of the premise (and rule). The consequence of the rule is represented by an isosceles triangular normal fuzzy set $\mathbb{B}$ with the support $w$. The localisation of the core of the triangular membership function is a linear combination of input attribute values. The value of the rule is a fuzzy value of a fuzzy implication (logical interpretation of fuzzy rule). The answers $\mu_{\mathbb{B}'(l)}$ of all $L$ rules are then aggregated into one fuzzy answer of the system. The crisp answer is elaborated with MICOG method [4]. The input domain is split with scatter partition – FCM clustering. In the ANNBFIS system the parameters of the model are tuned with a gradient method. The only exception are the parameters $\mathbf{p}$ – they are calculated with a system of linear equations.

We propose two modifications of the original ANNBFIS system: tuning of firing strengths and two phase domain partition.

### 2.1    Tuning of Firing Strengths

The first technique (labeled FIRING) modifies the tuning procedure of the ANNBFIS system. Tuning of the system's parameters has two aims:

```
1 procedure ANNBFIS_FIRING (𝕏, C, n)
2 /* 𝕏 : set of data items
3     C : number of clusters
4     n : number of tuning iterations */
5
6     group data into C clusters with FCM;
7     create rules form clusters;
8     iter := 0;
9     while iter < n do
10         for each data item elaborate derivatives (5), (7), and (8) for each descriptor
               in all rules;
11         modify the values of s and c for each descriptor in each rule;
12         calculate the parameters of conclusions with least square regression method;
13     end while;
14     return set of rules;
15 end procedure;
```

**Fig. 1.** The pseudocode for neuro-fuzzy system with increase of firing strength.

1. minimizing of the output error (as in ANNBFIS system)
2. maximizing of the firing strength for each data item.

We take into consideration the maximal firing strength of all rules

$$F_{\max}(\mathbf{x}) = \max_{l \in \mathbb{L}} \left\{ F^{(l)}(\mathbf{x}) \right\} \tag{1}$$

The main idea is not only modification of system's parameters basing in the error of the system, but also to modify the parameters to increase the firing strengths of the rules. The tuning procedure is based on gradient descent technique. Let's construct a punishment function defined for arguments from interval $(0, 1]$ whose the derivative has low values for arguments close to 1 and high values for argument close to 0. Let's define function $K$ as

$$K(F) = \ln F - F. \tag{2}$$

$F \in (0, 1]$, so $K \in (-\infty, -1]$. Each parameter $p$ is modified according to the rule

$$p \leftarrow p - \eta \frac{\partial K}{\partial p} \tag{3}$$

The differentials are:

$$\frac{\partial K}{\partial p} = \frac{\partial K}{\partial F} \cdot \frac{\partial F}{\partial p} \tag{4}$$

$$\frac{\partial K}{\partial F} = \frac{1}{F} - 1 = \frac{1 - F}{F} \tag{5}$$

The derivative (5) of function (2) has high values for low $F$ and zero for $F = 1$. So the parameters of rules fired weakly will be modified strongly. The function $K$ and its derivative are plotted in Fig. 2. The derivative is very easy to compute.

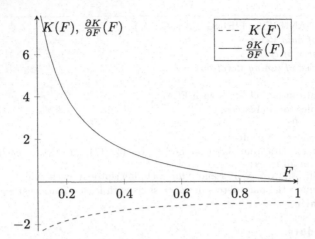

**Fig. 2.** The plots of punishment function $K$ and its derivative $\frac{\partial K}{\partial F}$ for argument $F \in [0.1, 1]$.

$$\frac{\partial}{\partial p} \max_{l \in \mathbb{L}} \left\{ F^{(l)}(\mathbf{x}) \right\} = \begin{cases} \frac{\partial}{\partial p} F_{\max}(\mathbf{x}), & F_{\max}(\mathbf{x}) = \max_{l \in \mathbb{L}} \left\{ F^{(l)}(\mathbf{x}) \right\} \\ 0, & F_{\max}(\mathbf{x}) \neq \max_{l \in \mathbb{L}} \left\{ F^{(l)}(\mathbf{x}) \right\} \end{cases} \tag{6}$$

$$\frac{\partial F}{\partial v_{ld}} = \left( 1 - F^{(l)} \right) \left[ \frac{(x_d - v_{ld})}{s_{ld}^2} \right]. \tag{7}$$

$$\frac{\partial F}{\partial s_{ld}} = \left( 1 - F^{(l)} \right) \left[ \frac{(x_n - v_{ld})^2}{s_{ld}^3} \right]. \tag{8}$$

The punishment function (Eq. 2) ensures higher modification ratios of parameters for data items with a low recognition.

The experiments show that the values of differentials (7) and (8) are sometimes quite big, we assume that the maximal change of $v$ and $s$ cannot exceed $0.001v$ and $0.001s$. Higher changes may destroy the tuning process. The algorithm is presented in Fig. 1.

## 2.2   Two Phase Domain Partition and Tuning of Firing Strengths

The second proposal (labeled 2-PHASE) uses a tuning modification (as in FIR-ING paradigm) and changes the partition of the input domain (Fig. 3). It is a hybrid partition of the input domain (scatter and hierarchical [14] partition).

1. The partition of the input domain is conducted in two steps. First the data items are clustered with the FCM [5] clustering algorithm. The number of clusters is the floor of the half of desired cluster numbers: $\lfloor \frac{C}{2} \rfloor$, where $C$ (input parameter) is a desired number of clusters.

```
 1  procedure 2PHASE (X, C, n, α)
 2  /* X : set of data items
 3     C : number of clusters
 4     n : number of tuning iterations
 5     α : weight parameter */
 6
 7     group data into ⌊C/2⌋ clusters with FCM;
 8     create rules form clusters;
 9     tune rules as in FIRING algorithm;
10
11     calculate the maximum firing of rules for each data item with (1);
12     select data items with maximum firing greater than α;
13     group the selected data items into ⌈C/2⌉ clusters with FCM;
14     create rules form clusters;
15     add new rules to previous rules;
16     tune all rules as in FIRING algorithm;
17
18     return set of rules;
19  end procedure;
```

**Fig. 3.** The pseudocode of the 2-PHASE algorithm.

2. Then the system is trained as an ANNBFIS system and the maximal firing strength of each rules is elaborated.
3. The data items are clustered once more, but now we take into consideration the data item that have firing strength (calculated with Eq. (1)) lower than parameter $\alpha$. The number of clusters in the second clustering is $C - \lfloor \frac{C}{2} \rfloor = \lceil \frac{C}{2} \rceil$.
4. New rules are added to the existing rule base.

## 3  Experiments

For experiments we used the real life data sets described below. All data sets are standardized to mean equal 0 and standard deviation equal 1.

The data set 'methane' contains the real life measurements of 8 air parameters in a coal mine in Upper Silesia (Poland) [13]. The task is to predict the concentration of the methane in 10 min. The data is divided into train set (499 tuples) and test set (523 tuples).

The 'concrete' set is a real life data set describing the parameters of the concrete sample and its strength [20]. The original data set can be downloaded from the UCI repository [6].

The data set 'chemical' represents five input and one output values of the chemical plant. The data set has 70 tuples with 6 attributes each. The date set is provided in paper [19] in two tables. We use the first table as the train set, and second table as test set.

The data set 'treasury' holds 1049 instances with 15 features [1]. The dataset 'laser' is a univariate time record of a single observed quantity, recorded from a far-infrared-laser in a chaotic state. The dataset has 993 items, each of them – 4 attributes and one decision attribute. The datasets 'treasury' and 'laser' have been downloaded from the KEEL repository [1].

## 3.1 Results

We used train-test paradigm. Each data set is divided into disjunctive train and test data sets. The train data sets were used for creating of models. We created ANNBFIS models as reference and two models for our modification. We denote the models with labels:

**ANNBFIS** a reference model created with ANNBFIS,
**FIRING** a model elaborated with firing strength tuning,
**2-PHASE** a model created with firing strength tuning and two phase domain partition.

All models have 5 rules. The premilimary experiments show that in a 2-PHASE the parameter $\alpha = 0.1$ results in the lowest error rate (Sect. 2.2). It means that in the second phase of creation of fuzzy rule base only poorly fired data items (with maximal firing strength of any fuzzy rule lower than $\alpha$) are considered. In the experiments we use the root mean square error (RMSE) defined as $RMSE = \sqrt{\frac{1}{X} \sum_{x \in \mathbb{X}}^{X} (y(x) - y_0(x))^2}$, where $\mathbb{X}$ stands for the data set with $X$ data items, $y(x)$ is a decision attribute of the data item $x$ and finally $y_0(x)$ is the answer elaborated by the system.

Table 1 gathers the error rate for ANNBFIS, FIRING, and 2-PHASE methods for all datasets. The premises of rules are presented in Fig. 6. The Figs. 4 and 5 present the histograms of the firing strength for data items of the 'methane' and 'concrete' data sets. For each data sets the train subset was used to elaborate the model. The histograms for the train data are in the left columns, and for test data sets in right columns.

**Table 1.** Comparison of root mean square error rates (RMSE) elaborated by ANNBIFS, FIRING, and 2-PHASE approaches.

| Data set | RMSE | | | | | |
|---|---|---|---|---|---|---|
| | ANNBFIS | | FIRING | | 2-PHASE | |
| | Train | Test | Train | Test | Train | Test |
| 'concrete' | 0.61860 | 1.12650 | 0.57670 | 0.70029 | 0.37229 | 0.79316 |
| 'methane' | 0.45281 | 0.40053 | 0.41456 | 0.32680 | 0.40466 | 0.34451 |
| 'chemical' | 0.08220 | 0.37502 | 0.02791 | 0.35014 | 0.03792 | 0.26499 |
| 'laser' | 0.36495 | 0.56115 | 0.17848 | 0.23339 | 0.20161 | 0.28060 |
| 'treasury' | 0.24675 | 0.33631 | 0.16799 | 0.07395 | 0.07134 | 0.11906 |

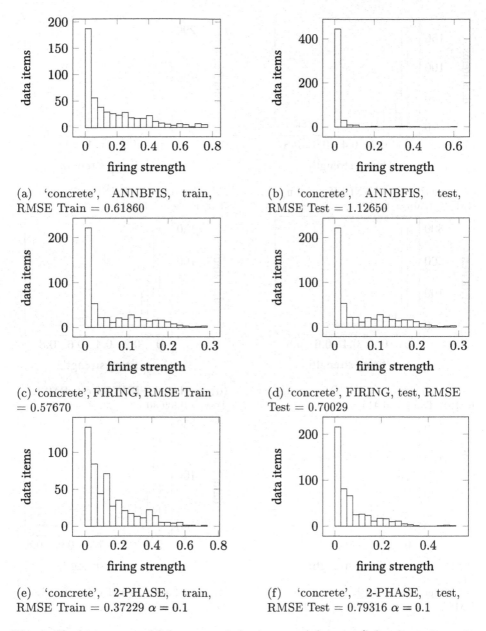

(a) 'concrete', ANNBFIS, train, RMSE Train = 0.61860

(b) 'concrete', ANNBFIS, test, RMSE Test = 1.12650

(c) 'concrete', FIRING, RMSE Train = 0.57670

(d) 'concrete', FIRING, test, RMSE Test = 0.70029

(e) 'concrete', 2-PHASE, train, RMSE Train = 0.37229 $\alpha = 0.1$

(f) 'concrete', 2-PHASE, test, RMSE Test = 0.79316 $\alpha = 0.1$

**Fig. 4.** The histograms of firing strength for 'concrete' data set (left column for train set, right – test set).

The model created for the 'concrete' data set with the ANNBFIS paradigm shows that almost 200 train data items have very low firing strength. In the test data set more than 400 data items have very low firing strength. The tuning of the firing strengths (Fig. 4c and d) deteriorates the firing strength for the

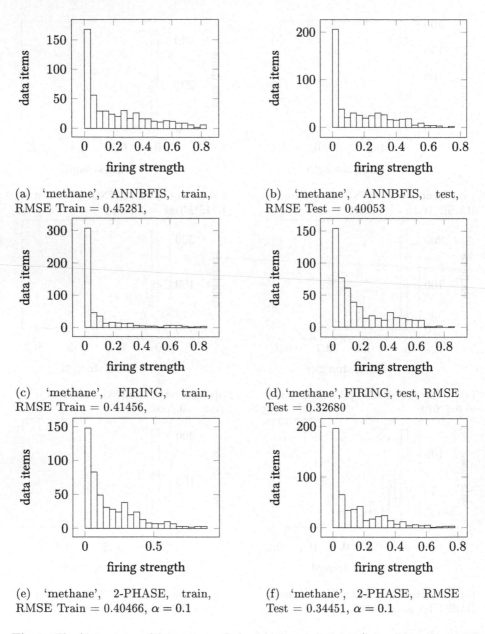

(a) 'methane', ANNBFIS, train, RMSE Train = 0.45281,

(b) 'methane', ANNBFIS, test, RMSE Test = 0.40053

(c) 'methane', FIRING, train, RMSE Train = 0.41456,

(d) 'methane', FIRING, test, RMSE Test = 0.32680

(e) 'methane', 2-PHASE, train, RMSE Train = 0.40466, $\alpha = 0.1$

(f) 'methane', 2-PHASE, RMSE Test = 0.34451, $\alpha = 0.1$

**Fig. 5.** The histograms of firing strength for 'methane' data set (left column for train set, right – test set).

train data set, but the number of test data item with very low firing strength is slightly above 200 compared to more than 400 in system without firing strength tuning. The data items are better recognized by the model and the total error is lower (0.70029 for the FIRING model and 1.12650 for the ANNBFIS model).

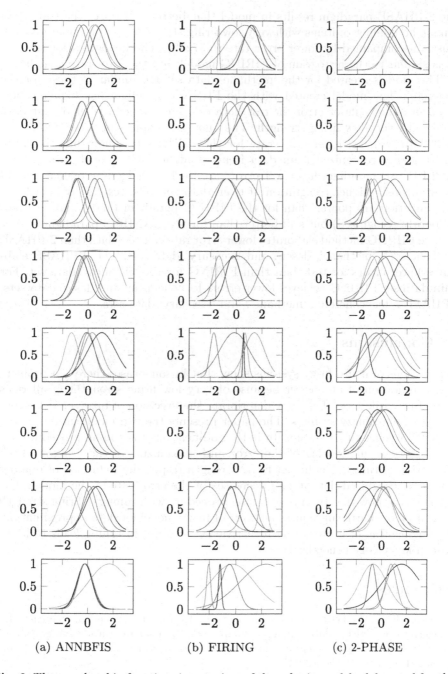

**Fig. 6.** The membership functions in premises of the rules in models elaborated for the 'concrete' data set. The left column represents the function of the ANNBFIS model, the middle one – TUNING model, and the right one – 2-PHASE model. Each row represents one attribute.

The 2-PHASE paradigm results in model that better recognizes the train data items (the number of items with very low firing strength is approximately 150). This is accompanied by lower error rate (0.37229). The histogram for the test set is similar to the histogram of FIRING model, but the error rate is higher.

The models created for the 'methane' data set are similar to those for the 'concrete'. The model created with the FIRING paradigm has higher generalization ability (lower error for test data set) and lower number of test data items with very low firing strength. The knowledge approximation (results for train data set) has lower error rate for the FIRING paradigm but the number of poorly recognized train data items is higher. The model created with the 2-PHASE paradigm has lower error rate for knowledge approximation and simultaneously higher recognition of train data set. The results for knowledge generalization are poorer than for the FIRING paradigm (and the data items are poorly recognized) but still better than for the ANNBFIS paradigm.

The FIRING method elaborates lower error rates for test data than 2-PHASE for 'concrete', 'methane', 'laser', and 'treasury' data sets. The 2-PHASE elaborates lower errors for test data than FIRING for the 'chemical' data set. For train data set FIRING achieves lower errors for 'chemical' and 'laser' data sets, 2-PHASE – for 'concrete', 'methane', and 'treasury' data sets.

## 4    Conclusions

Applications of neuro-fuzzy systems show that in some cases there is a problem with elaboration of an answer because of very low firing strengths of all rules in the fuzzy rule base. This may deteriorate the precision of fuzzy model elaborated by neuro-fuzzy systems. The paper presents the two modifications of the ANNBFIS neuro-fuzzy system. In both modifications the tuning procedure has two objectives: minimizing of the error and maximizing of the firing strength (activation of rules). The higher firing strength (especially reduction of number of poorly recognized data items) decreases of the error rate of the model. The experiments show that two phase tuning procedure with tuning of firing strength (activation of rules) can reduce the error of fuzzy model for data approximation. The augmenting of firing strength (activation of rules) in tuning leads to lower error in knowledge generalization.

## References

1. Alcalá-Fdez, J., Fernandez, A., Luengo, J., Derrac, J., García, S., Sánchez, L., Herrera, F.: KEEL data-mining software tool: data set repository, integration of algorithms and experimental analysis framework. J. Mult.-Valued Log. Soft Comput. **17**(2–3), 255–287 (2011)
2. Bezerra, R.A., Vellasco, M.M., Tanscheit, R.: Hierarchical neuro-fuzzy BSP Mamdani system. In: Neural Networks, Genetic Algorithms and Soft Computing, pp. 1321–1326 (2005)
3. Czabański, R.: Extraction of fuzzy rules using deterministic annealing integrated with $\epsilon$-insensitive learning. Int. J. Appl. Math. Comput. Sci. **16**(3), 357–372 (2006)

4. Czogała, E., Łęski, J.: Fuzzy and Neuro-Fuzzy Intelligent Systems. Studies in Fuzziness and Soft Computing. Physica-Verlag, Heidelberg (2000)
5. Dunn, J.C.: A fuzzy relative of the ISODATA process and its use in detecting compact, well separated clusters. J. Cybern. **3**(3), 32–57 (1973)
6. Frank, A., Asuncion, A.: UCI Machine Learning Repository (2010)
7. Jakubek, S., Keuth, N.: A local neuro-fuzzy network for high-dimensional models and optimalization. Eng. Appl. Artif. Intell. **19**(6), 705–717 (2006)
8. Łęski, J.: $\varepsilon$-insensitive learing techniques for approximate reasoning systems. Int. J. Comput. Cogn. **1**(1), 21–77 (2003)
9. Łęski, J., Czogała, E.: A new artificial neural network based fuzzy inference system with moving consequents in if-then rules and selected applications. Fuzzy Sets Syst. **108**(3), 289–297 (1999)
10. Mamdani, E.H.: Application of fuzzy logic to approximate reasoning using linguistic synthesis. IEEE Trans. Comput. **C–26**(12), 1182–1191 (1977)
11. Senhadji, R., Sanchez-Solano, S., Barriga, A., Baturone, I., Moreno-Velo, F.: Norfrea: an algorithm for non redundant fuzzy rule extraction. IEEE Int. Conf. Syst. Man Cybern. **1**, 604–608 (2002)
12. Setnes, M., Babuška, R.: Rule base reduction: some comments on the use of orthogonal transforms. IEEE Trans. Syst. Man Cybern. Part C: Appl. Rev. **31**(2), 199–206 (2001)
13. Sikora, M., Krzystanek, Z., Bojko, B., Śpiechowicz, K.: Application of a hybrid method of machine learning for description and on-line estimation of methane hazard in mine workings. J. Min. Sci. **47**(4), 493–505 (2011)
14. Simiński, K.: Patchwork neuro-fuzzy system with hierarchical domain partition. In: Kurzynski, M., Wozniak, M. (eds.) Computer Recognition Systems 3. AISC, vol. 57, pp. 11–18. Springer, Heidelberg (2009)
15. Simiński, K.: Neuro-fuzzy system based kernel for classification with support vector machines. In: Gruca, A., Czachórski, T., Kozielski, S. (eds.) Man-Machine Interactions 3. AISC, vol. 242, pp. 415–422. Springer, Heidelberg (2014)
16. Simiński, K.: Rough subspace neuro-fuzzy system. Fuzzy Sets Syst. **269**, 30–46 (2015). http://www.sciencedirect.com/science/article/pii/S0165011414003108
17. Siminski, K.: Ridders algorithm in approximate inversion of fuzzy model with parameterized consequences. Expert Syst. Appl. **51**, 276–285 (2016)
18. Sugeno, M., Tanaka, K.: Successive identification of a fuzzy model and its applications to prediction of a complex system. Fuzzy Sets Syst. **42**(3), 315–334 (1991)
19. Sugeno, M., Yasukawa, T.: A fuzzy-logic-based approach to qualitative modeling. IEEE Trans. Fuzzy Syst. **1**(1), 7–31 (1993)
20. Yeh, I.C.: Modeling of strength of high-performance concrete using artificial neural networks. Cem. Concr. Res. **28**(12), 1797–1808 (1998)
21. Zhou, Z.H., Chen, S.F.: Rule extraction from neural networks. J. Comput. Res. Dev. **39**(4), 398–405 (2002)

# Rough Sets in Multicriteria Classification of National Heritage Monuments

Krzysztof Czajkowski[✉]

Department of Computing Science, Faculty of Electrical and Computer Engineering,
Cracow University of Technology, Cracow, Poland
kc@pk.edu.pl

**Abstract.** The motivation of this paper are problems how to improve assessment of historic buildings, in terms of the significance of conservation activities. The protection of national heritage, so important nowadays, requires a multicriteria assessment. Seeing that different factors in varying degrees affect the comprehensive assessment of the object, it becomes necessary to use computational intelligence methods. This paper presents a rough sets approach to multicriteria rating of objects on the example of historical buildings.

**Keywords:** Rough sets · Multicriteria assessment · Heritage preservation

## 1 Introduction

Problems of the cultural heritage preservation are among the most important challenges of present days. This is due to the fact that the unprotected parts of our history are lost forever. The reconstruction, which can restore the visual aspects of objects, is not able to reproduce their real value. The restored heritage is just an imitation, which can only substitute a former value.

The size of the national heritage measured in the number of monuments illustrates the challenges that the conservators, architects and historians are facing with. Table 1 shows the number of monuments in National Register of Monuments in Poland with the division on the type of monument [10].

There is a high rate of the change in the area of the registration of monuments. In the first half of the year, approximately two hundred new objects are added, fifty of them are removed as well as three hundred and fifty decisions are changed.

Nowadays, the heritage is exposed to a large number of very different dangers. Therefore, we have to use different approaches to various parts of this heritage. One of the heritage objects category that is particularly endangered there are historic buildings. There are many factors that have a negative impact on monuments. The most significant factors can be divided into several groups, depending on the origin: caused by the human activity, caused by animals, caused by the nature.

© Springer International Publishing Switzerland 2016
S. Kozielski et al. (Eds.): BDAS 2016, CCIS 613, pp. 168–178, 2016.
DOI: 10.1007/978-3-319-34099-9_12

**Table 1.** Monuments in Poland with the division on their type

| Type of monuments | Number |
|---|---|
| sacred | 10503 |
| defensive | 840 |
| public | 3474 |
| castles | 372 |
| palaces | 1994 |
| manors | 2610 |
| parks | 6139 |
| residential | 16100 |
| economic | 4098 |
| industrial | 1562 |
| cemeteries | 3138 |
| other | 2615 |
| **total** | **53445** |

Furthermore, in the near future, many factors will affect increasingly on objects that should be protected. Thus, it is particularly important to take a fast and comprehensive action to protect endangered objects.

The result of so many threats, as well as the effect of the passage of the time and warfare is the general poor state of preservation of the historic substance. Table 2 shows how many monuments require different types of activities [10].

**Table 2.** Monuments division of work required

| Work required | Percent |
|---|---|
| not require | 10 % |
| minor repairs | 42 % |
| protection repairs | 21 % |
| general renovation | 26 % |

As we can see, nearly the half of monuments in Poland is in the state which needs a large renovation. Due to the large number of objects that require maintenance, as well as many threats and progressive degradation, there is an irreversible annihilation of the heritage and, therefore, actions for their protection are extremely important and urgent.

The complexity of aspects of the restoration and maintenance requires a structured approach. That is why considerable attention is given to applications that use the databases. Numerous studies are conducted on the use of such solutions. However, these works are usually related to a single issue, such as specialized and detailed description of the materials that are made of elements of historic buildings or encompass the wide range of topics that describe the multimedia and the GIS data [1,3].

An analysis of the actual state regarding the protection of heritage shows big problems in this regard. "In the current situation, it is not an exaggeration to say that the protection and survival of the historic resource are at risk. This is the result of the less and less effective system of the protection of monuments and the increasing pressure of modernization and investment that works in conditions of strong property rights and a full market economy. The current threats have got an unprecedented nature and scale." [16]

In addition, the national heritage protection planning must take into account the existence of many restrictions. Among the most important restriction there are:

- a limited amount of funds,
- spending time restrictions (e.g. European Union projects),
- limited human resources (specialists in history, architecture, conservation),
- ownership (not every object can be renewed with each type of funding).

In such a situation, in order to manage human and financial resources properly, it seems necessary to make the classification of historic buildings and scheduling for the order works.

For the proper assessment several main aspects should be taken into account. The first one is the value of given monuments. The issue of valuation of the monument is complex itself, because we can distinguish many types of evaluation, based on the criteria:

- historical
- uniqueness
- authenticity
- identity of place
- the overall preservation state

As a result, we have the set of values which creates the first group of monument evaluation criteria. Moreover, in the case in which more than one expert estimates the value of a given criterion, it is very probably that the results will be different.

The second group of criteria provides values for the preservation states of individual elements of the monument. For any kind of monumental object there should be considered a different structure (separately for houses, churches).

The third group of criteria constitutes values of preservation states of individual elements of the monument. The elements (structure), in the case of this group of criteria, are the same as for the second group of criteria (preservation states).

For experts evaluating the value of historic buildings, it is clear that the above factors have uneven impact on the assessment comprehensive. Additionally, between individual criteria exist some connections which are not such clearly and have complex nature. To support a multi-criteria assessment of the value of monuments and the proper comparisons, applying rough set theory can be advantageous. It is possible, by this theory, to generate rules for the classification, as well as comparing monuments with each other.

These works about evaluation of historic buildings are carried out in collaboration with the Professor B. M. Pawlicki, leader of the group of experts in the area of monuments conservation and protection of national heritage. Current works are a continuation of research in the field of monitoring the state of preservation of monuments in Cracow [14]. Optimization of human and material resource in order to properly protect the monuments, is possible only if the resource, that is the whole set of monuments, is properly valued and it is a subject to multi-criteria assessment [13]. "The potential of awareness" - an awareness of the scale and the complexity of material assets, which should be included in the multicriteria assessment as the individual elements of the monument, as well as the whole object.

The area of valuation of monuments is very complex because of the large variety of factors, and the high complexity of the monuments themselves. Looking from the broadest perspective at the scholarly discipline devoted to monuments and at conservation [one sees that] the ongoing changes encompass the key elements that are determining the system of monument protection [18]. It should be added that the ongoing transformations have a dynamic character they happen at all times, reciprocally influencing each other.

In decision problems related to the reuse of historical assets conflicts can arise and the availability of analytical frameworks able to support the process is getting more and more important. It has been generally agreed that Multicriteria Decision Analysis (MCDA) can offer a formal methodology to deal with such decision problems, taking into account the available technical information and stakeholders values [4]. Researches in this area are taken in many countries [5,9,19,20].

In Poland the creation of the new heritage preservation system is based on privatization of ownership, responsibility and financing of monuments. Low quality of any of the system elements or lack of cohesion between them results in dysfunction of a given heritage preservation system. In Poland (and other post-communist countries) the transformation continues a new heritage protection system has not yet been fully developed [17]. An important element of the system should be a tool to multicriteria evaluation of monuments.

## 2 Application the Rough Sets Theory to Classification of Monuments

The rough set theory is founded on the assumption that we associate some information (data, knowledge) with every object of the universe of discourse. Objects characterized by the same information are indiscernible (similar) in view of the available information about them. The indiscernibility relation generated in this way is the mathematical basis of rough set theory [11,12].

The approach proposed in [8] seems to be especially interesting. Basing on database operations performed on sets, such as cardinality and projection, it increases rapidity of finding the core and reducts. It is possible to take advantages of such effective solutions as indexing and sorting. Efficient implementations of SQL language give the possibility to reduce the cost of the disc access and what is more, they cope well with the large amount of data. In [2] the implementation of such an approach was presented.

Let's assume Table 3 of monuments (buildings) with the set of condition attributes $C = \{A_1, A_2, A_3, A_4\}$, which values determine quality due to the specific criterion: historical, authenticity, uniqueness and artistic. $D = \{A_5\}$ it is a set of decision attributes with one attribute, which means the priority (importance) of the maintenance works.

**Table 3.** Decision table

| Monument | $A_1$ historical value | $A_2$ authenticity value | $A_3$ uniqueness value | $A_4$ artistic value | $A_5$ decision (priority) |
|---|---|---|---|---|---|
| 1 | medium | low | medium | low | low |
| 2 | medium | low | medium | low | high |
| 3 | low | low | low | low | low |
| 4 | medium | high | medium | high | low |
| 5 | high | low | high | high | high |
| 3 | low | low | low | low | low |
| 7 | high | high | high | low | high |

In this case we have the set $X$ of monuments having a priority "high" $X = \{2, 5, 7\}$ and lower and upper approximation as well as boundary region which are:

$$\underline{C}(X) = \{5, 7\}, \quad \overline{C}(X) = \{1, 2, 5, 7\}, \quad Bn_C(X) = \{1, 2\}$$

In this way, we don't have univocal decision about priority of the maintenance works, because the boundary region is not empty. The objects 1 and 2 have the same values of all conditional attributes (the same values of assessments made by experts) but the decision about priority of works is different. In other words, the monuments 5 and 7 certainly belong to the set of high priority, but monuments 1, 2, 5, 7 can belong to the set of "high" priority. It means, that we have insufficient number of information to determine properly the elements of the set.

Analogously, in the case in which we have the set Y of monuments having a priority "low" $Y = \{2, 5, 7\}$, the approximations are accordingly:

$$\underline{C}(Y) = \{3, 4, 6\}, \quad \overline{C}(Y) = \{1, 2, 3, 4, 6\}, \quad Bn_C(Y) = \{1, 2\}$$

And then the objects 1 and 2 are a boundary region have the same values of all conditional attributes but the decision about priority of works is different.

The Table 3 we can note as a set of rules and by reduce to more concise representation:

- If $f(x, A_1)$ = low, then $f(x, A_4)$ = low
- If $f(x, A_1)$ = high, then $f(x, A_4)$ = high
- If $f(x, A_1)$ = medium and $f(x, A_2)$ = high, then $f(x, A_4)$ = low
- If $f(x, A_1)$ = medium and $f(x, A_2)$ = low, then $f(x, A_4)$ = high or low

The rules 1, 2 as well as 3 are exact rules, because they have a univocal consequence, but the rule 4 is an approximate rule, because it does not have a univocal consequence.

## 3  Multicriteria Sorting

From the multicriteria sorting point of view, the original rough set approach proved to be insufficient. The original rough set approach cannot extract all the essential knowledge contained in the decision table of multicriteria sorting problem, i.e. problems of assigning a set of objects described by a set of criteria to one of pre-defined and preference-ordered categories [7]. The case of monuments assessment with a number of criteria is a situation in which we must take into account ordinal properties of such criteria. In this case, the indiscernibility or similarity relations have specific nature and the rough set approach is not able to handle correctly such a kind of characteristic. If there is at least one criterion in the decision table, nontraditional solution is needed. The new rough set approach was proposed by Greco in [6] to evaluation of bankruptcy risk. The same solution could be applied for monuments evaluation in heritage preservation, because there are many attributes with preference-ordered categories.

In the case of Table 3 we have the following situation:

- with respect to $A_1$, "high" is better than "medium" and "medium" is better than "low",
- with respect to $A_2$, "high" is better than "low",
- with respect to $A_3$, "high" is better than "medium" and "medium" is better than "low",
- with respect to $A_4$, "high" is better than "medium"
- with respect to $A_5$, "high" is better than "medium"

According to [7] within this approach we approximate the class $Cl_1^{\leq}$ of "(at most) low" priority of works as well as the class $Cl_2^{\geq}$ of "(at least) high" priority of works. The C-lower approximations, the C-upper approximations and C-boundaries of classes $Cl_1^{\leq}$ are:

$$\underline{C}(Cl_1^{\leq}) = \{3, 6\}, \quad \overline{C}(Cl_1^{\leq}) = \{1, 2, 3, 4, 6\}, \quad Bn_C(Cl_1^{\leq}) = \{1, 2, 4\}$$

For classes $Cl_2^{\geq}$ they are respectively:

$$\underline{C}(Cl_2^{\geq}) = \{5,7\}, \quad (C) = \{1,2,4,5,7\}, \quad Bn_C(Cl_2^{\geq}) = \{1,2,4\}$$

There is only one reduct: $RED_{Cl}(C) = A_1$ and the core: $CORE_C l = RED_C l(C)$. We can determine the minimal set of decision rules as:

- If $f(x, A_1) \geq$ high, then $x \in Cl_2^{\geq}$ (objects: 5,7)
- If $f(x, A_1) \leq$ low, then $x \in Cl_1^{\leq}$ (objects: 3,6)
- If $f(x, A_1) \geq$ medium and $f(x, A_1) \leq$ medium ($f(x, A_1)$ is medium), then $x \in Cl_1^{\leq} \cup Cl_2^{\geq}$ (objects: 1,2,4)

In Table 3 we can observe that monument 4 dominates monument 2 in the sense that monument 4 is at least as valuable as monument 2 in respect to all the four criteria. Let's notice that monument 4 has a comprehensive evaluation worse than monument 3. In this way, we can observe an inconsistency, it means though better ratings in each evaluation method, the overall evaluation is different from what the logic indicates. This inconsistency is revealed by the approximation based only on dominance and cannot be revealed by the approximation based on discernibility. The approach proposed by Greco has an advantage over the original approach.

The second advantage we can notice in the case of rules which have problems with interpretation. For example, monument 2 has generally lower values of attributes, but high value of decision attribute it's logically unjustified.

The last but not least, the dominance relation gives us a more concise representation of knowledge represented by decision table, because it generate a minimal set of decision rules with smaller number of conditions.

## 4    Experiments

In the first experiment, 140 historical monuments are taken into account. The set consists of 114 tenement houses and residences, 15 churches and monasteries as well as 11 palaces and mansions [15]. We take into consideration four conditional attributes, as in the Chapter 3, but with revision of their values. For the artistic value we consider three values (not two, as in the case of the example in Chapter 3): 1–3, and the rest of attributes has the same values as in the Chapter 3: historical value: 1–2, uniqueness value: 1–3 and authenticity value: 1–3.

First, in this experiment, traditional approach is applied, in which we treat attributes as simple numbers not as a criteria. By implementation listed in Chapter 2, and presented in [2], the set of rules is generated and we carry out the classification of sample data. The whole set of data is divided into two parts in proportion 50 % for teaching data and 50 % for the validation. The content of sets is selected randomly. For greater credibility of the whole process, it is repeated 10 times and the results are averaged. The average correctness of classification is 71,4 %.

The example classifier consists of thirteen rules:

1. if hist=1 and uniq=1 then class=1
2. if hist=2 and auth=1 and uniq =1 and arti=1 then class=2
3. if hist=2 and uniq=2 then class=2
4. if hist=1 and auth=1 then class=1
5. if hist=3 then class=2
6. if hist=1 and auth=2 and uniq=2 and arti=2 then class=2
7. if hist=2 and auth=1 and uniq=1 and arti=2 then class=1
8. if uniq=3 then class=2
9. if hist=1 and auth=2 and uniq=2 and arti=2 then class=1
10. if hist=2 and auth=1 and uniq=1 and arti=2 then class=2
11. if hist=1 and arti=1 then class=1
12. if auth=2 and uniq=1 then class=1
13. if hist=2 and auth=1 and uniq=1 and arti=1 then class=1

As we can see, the rules number 2 and 13 are inconsistent; on the base of the same values of conditional attributes we have different decision. We can observe the same situation in the case of rules number 6 and 9 as well as rules number 7 and 10. As mentioned in Chapter 3, such objects are a boundary region have the same values of all conditional attributes but the decision about priority of works is different. These rules are approximate rules, because they do not have a univocal consequence.

According to an approach, presented in Chapter 4, based on the multicriteria sorting proposed by Greco, Matarazzo and Slowinski in [6], we can treat conditional attributes as criteria with preference-ordered categories. In such a situation, values 0, 1 and 3 are treated as low, medium and high (for criteria with three values, and values 0 and 1 are treated as low and high (for criteria with two values). The criterion value: medium (from the point of view of such an analysis) is more important than low (raises the importance of conservation activities) and high is more important than medium. In this way, we can approximate the class $Cl_1^{\leq}$ of "(at most) low" priority of works as well as the class $Cl_2^{\geq}$ of "(at least) high" priority of works. After correction (by an expert) of the input data set and determining univocal cases, the average correctness of classification is 74,8 %. The example of set of rules is shown below:

1. if hist=high then class=high
2. if uniq=low then class=low
3. if hist=low and uniq=medium then class=low
4. if uniq=high then class=high
5. if hist=medium then class=high

As we can observe, from the point of view of preference-ordered categories, history value (hist) = 3 (rule 1) is better than 2 (rule 5). In this way, we can reduce the number of rules to four rules:

1. if uniq<=low then class= low
2. if hist<=low and uniq<=medium then class=low
3. if uniq>=high then class=high
4. if hist>=medium then class=high

Such a rationalization of classifier not only simplifies it and accelerates classification process, but also gives possibility to classify cases which are not of the same values of criteria as values in the rule, but are similar to them (better or worse). It is particularly important in evaluation with more possible values than two or three (for example in situation, in which each criterion is marked by individual specialist in a given area: civil engineering, art etc. and the criterion can have 5–10 values).

In the second experiment, the inconsistent records are eliminated by data correction which was done by an expert. The number of objects are increased to 244: 184 tenement houses and residences, 35 churches and monasteries as well as 25 palaces and mansions. The most important difference in comparison with the first experiment is that two new attributes were added. The first one is value of identity - how strongly a specific historic building is connected with the identity of its location with two values: 1–2, and the second one is the degradation level with three values: 1–3. The example set of rules consists of 7 elements:

1. if auth=low and iden=low then class=low
2. if hist=high then class=high
3. if uniq=low and degr=low then class=low
4. if uniq=low and iden=low and degr=medium then class=low
5. if arti=medium then class=high
6. if uniq=medium and degr=low then class=low
7. if auth=high then class=high

In the above classifier the rule 3 has lower value of uniqueness criterion than the rule 6, and the same value of degradation criterion as the rule 6. In this way, we can convert these two rules into one: if uniq<=medium and degr<=low then class=low.

## 5    Conclusions and Further Works

In the paper, the approach to sorting monuments with the rough sets theory was presented. The advantages of the solution modified by Greco, with the approximation based on dominance relation instead of approximation based on inconsistency relation has been shown. Better representation of knowledge, as well as taking into account an interpretation of overall rating are particularly needed in multicriteria evaluation of such complex structures as historical building, which are characterized by many different values.

Moreover, it is undoubtedly important to take into consideration the relationship between particular evaluations. For example, if an evaluation, form the point of view of state of preservation, is getting worse, the actual artistic value, even if it is high at the moment, will decrease. It is mandatory to take into consideration time dependencies as well as relationship between evaluations and the time in heritage preservation. The above issues will be discussed in forthcoming paper by the present author. The works are still conducted in the cooperation with professor of architecture B. M. Pawlicki as well as with monuments restorers.

# References

1. Czajkowski, K., Bobowski, K.: Data anlysis using rough sets in the protection of architectural monuments. Stud. Informatica **32**, 299–313 (2011)
2. Czajkowski, K., Drabowski, M.: Semantic data selections and mining in decision tables. In: Czachórski, T., Kozielski, S., Stańczyk, U. (eds.) Man-Machine Interactions 2. AISC, vol. 103, pp. 279–286. Springer, Heidelberg (2011)
3. Czajkowski, K., Olczyk, P.: Fuzzy interface for historical monuments databases. In: Kozielski, S., Mrozek, D., Kasprowski, P., Małysiak-Mrozek, B. (eds.) BDAS 2014. CCIS, vol. 424, pp. 271–279. Springer, Heidelberg (2014)
4. Ferretti, V., Bottero, M., Mondini, G.: Decision making and cultural heritage: An application of the multi-attribute value theory for the reuse of historical buildings. J. Cult. Heritage **15**, 644–655 (2014)
5. Giove, S., Rosato, P., Breil, M.: An application of multicriteria decision making to built heritage. The redevelopment of venice arsenale. J. Multi-Criteria Decis. Anal. **17**, 85–99 (2011)
6. Greco, S., Matarazzo, B., Słowinski, R.: A new rough set approach to evaluation of bankruptcy risk. In: Zopounidis, C. (ed.) Operational Tools in the Management of Financial Risks, pp. 121–136. Kluwer, Dordrecht (1998)
7. Greco, S., Matarazzo, B., Slowinski, R.: Rough sets theory for multicriteria decision analysis. Eur. J. Oper. Res. **129**(1), 1–47 (2001)
8. Hu, X., Lin, T., Han, J.: A new rough sets model based on database systems. Fundamenta Informaticae **59**(2–3), 135–152 (2004)
9. Mazzanti, M.: Cultural heritage as multi-dimensional, multi-value and multi-attribute economic good: toward a new framework for economic analysis and valuation. J. Socio-Econ. **31**, 529–558 (2002)
10. National Heritage Board of Poland: List of immovable monuments entered into the monuments register http://www.nid.pl/en/ (2015). Accessed 30 Oct 2015
11. Pawlak, Z.: Rough set approach to knowledge-based decision support. Eur. J. Oper. Res. **99**, 48–57 (1997)
12. Pawlak, Z.: Rough sets. Int. J. Comput. Inf. Sci. **11**(5), 341–356 (1982)
13. Pawlicki, B.M.: Differences in the theory and practice of the preservation of monuments and the protection of cultural heritage sites. Tech. Trans. **112**(6–A), 141–154 (2015)
14. Pawlicki, B.M., Drabowski, M., Czajkowski, K.: Monitoring stanu zachowania obiektow zabytkowych przy wykorzystaniu wspolczesnych systemow informatycznych. Wydawnictwo Politechniki Krakowskiej, Krakow (2004)
15. Provincial Heritage Monuments Protection Office: The register of immovable monuments of Cracow http://www.wuoz.malopolska.pl/ (2015). Accessed 16 Nov 2015
16. Szmygin, B.: System ochrony zabytkw w Polsce analiza, diagnoza, propozycje. Politechnika Lubelska, Lublin Warszawa (2011)

17. Szmygin, B.: Transformation of the heritage protection system in Poland after 1989. In: Purchla, J. (ed.) Protecting and Safeguarding Cultural Heritage. Systems of Management of Cultural Heritage in the Visegrad Countries, pp. 31–38. International Cultural Centre, Krakow (2011)
18. Szmygin, B.: Theory and criteria of heritage evaluation as the basis for its protection. Wiadomosci Konserwatorskie **44**, 44–52 (2015)
19. Wang, H., Zeng, Z.: A multi-objective decision-making process for reuse selection of historic buildings. Expert Syst. Appl. **37**, 1241–1249 (2010)
20. Yildirim, M.: Assessment of the decision-making process for reuse of a historical asset: the example of diyarbakir hasan pasha khan. J. Cult. Heritage **13**, 379–388 (2012)

# Architectures, Structures and Algorithms for Efficient Data Processing

# Inference Rules for Fuzzy Functional Dependencies in Possibilistic Databases

Krzysztof Myszkorowski[(✉)]

Institute of Information Technology, Lodz University of Technology, Lodz, Poland
kamysz@ics.p.lodz.pl
http://edu.icp.lodz.pl

**Abstract.** We consider fuzzy functional dependencies (FFDs) which can exist between attributes in possibilistic databases. The degree of FFD is evaluated by two numbers from the unit interval which correspond to possibility and necessity measures. The notion of FFD is defined with the use of the extended Gödel implication operator. For such dependencies we present inference rules as a fuzzy extension of Armstrong's axioms. We show that they form a sound and complete system.

**Keywords:** Possibilistic databases · Possibility distribution · Possibility measure · Necessity measure · Fuzzy implicator · Fuzzy functional dependencies · Inference rules

## 1 Introduction

Conventional database systems are designed with the assumption of precision of information collected in them. The problem becomes more complex if our knowledge of the fragment of reality to be modeled is imperfect. In such cases one has to apply tools for describing uncertain or imprecise information [7,8]. One of them is the theory of possibility [1,3]. In the possibilistic database framework attribute values are represented by means of possibility distributions. Each value $x$ of an attribute $X$ is assigned with a number $\pi_X(x)$ from the unit interval which expresses the possibility degree of its occurrence. Different ways of determination of the possibility degree have been described in [4].

One of the most important notions of the database theory is the concept of functional dependency (FD). The classical definition of functional dependency $X \rightarrow Y$ between attributes $X$ and $Y$ of a relation scheme $R$ is based on the assumption that the equality of attribute values may be evaluated formally with the use of two-valued logic. The existence of $X \rightarrow Y$ means that $X$-values uniquely determine $Y$-values. If attribute values are imprecise one can say about a certain degree of the dependency $X \rightarrow Y$. It contains the information to what extent $X$ determines $Y$. In possibilistic databases closeness of compared values can be evaluated by means of possibility and necessity measures.

Since the notion of FD plays an important role in the design process [5], its fuzzy extension has attracted a lot of attention. Hence, different approaches

© Springer International Publishing Switzerland 2016
S. Kozielski et al. (Eds.): BDAS 2016, CCIS 613, pp. 181–191, 2016.
DOI: 10.1007/978-3-319-34099-9_13

concerning fuzzy functional dependencies (FFDs) have been described in professional literature. A number of different definitions emerged [2,6,9,10]. In the paper we extend the definition given by Chen [2]. According to [2] the degree of the fuzzy functional dependency is evaluated by means of the possibility measure. The necessity measure is not used. The equality degree equals the maximum value 1 when the two possibility distributions have the maximum degree 1 at the same element. Thus applying of the possibility measure for evaluation of the equality of two imprecise values expressed by possibility distributions is not sufficient. In the paper evaluation of closeness of imprecise values is made by means of both possibility and necessity measures. The notion of FFD is defined with the use of the extended Gödel implication operator [9]. For such dependencies we will present inference system based on the well known set of Armstrong's axioms which is an important property of FDs in classical databases.

The paper is organized as follows. In the next section we discuss the basic notions dealing with fuzzy functional dependencies in possibilistic databases and formulate the extended inference rules. Section 3 discusses properties of the extended Gödel implication operator. In Sect. 4 we proved the soundness and completeness of the inference rules.

## 2    Fuzzy Functional Dependencies in Possiblistic Databases

Let $r$ be a relation of the scheme $R(U)$ where $U$ denotes a set of attributes, $U = \{X_1, X_2, ..., X_n\}$. Let $DOM(X_i)$ denotes a domain of $X_i$. Let us assume that attribute values are given by means of normal possibility distributions:

$$t(X) = \{\pi_{t(X_i)}(x)/x : x \in DOM(X_i)\} , \quad \sup_{x \in DOM(X_i)} \pi_{t(X_i)}(x) = 1, \quad (1)$$

where $t$ is a tuple of $r$ and $\pi_{t(X_i)}(x)$ is a possibility degree of $t(X_i) = x$. The possibility distribution takes the form: $\{\pi_X(x_1)/x_1, \pi_X(x_2)/x_2, ..., \pi_X(x_n)/x_n\}$, where $x_i \in DOM(X)$. At least one value must be completely possible i.e. its possibility degree equals 1. This requirement is referred to as the normalization condition. Let $t_1$ and $t_2$ be tuples of $r$. The degrees of possibility and necessity that $t_1(X_i) = t_2(X_i)$, denoted by $Pos$ and $Nec$, respectively, are as follows:

$$Pos(\Pi_{t_1(X_i)} = \Pi_{t_2(X_i)}) = \sup_x \min(\pi_{t_1(X_i)}(x), \pi_{t_2(X_i)}(x)),$$
$$Nec(\Pi_{t_1(X_i)} = \Pi_{t_2(X_i)}) = 1 - \sup_{x \neq y} \min(\pi_{t_1(X_i)}(x), \pi_{t_2(X_i)}(y)). \quad (2)$$

The closeness degree of $t_1(X_i)$ and $t_2(X_i)$, denoted by $\approx(t_1(X_i), t_2(X_i))$, is expressed by two numbers $\approx(t_1(X_i), t_2(X_i))_N$ and $\approx(t_1(X_i), t_2(X_i))_\Pi$ from the unit interval which correspond to necessity and possibility measures. Thus $\approx (t_1(X_i), t_2(X_i)) = (\approx (t_1(X_i), t_2(X_i))_N, \approx (t_1(X_i), t_2(X_i))_\Pi)$. For identical values of $t_1(X_i)$ and $t_2(X_i)$ we have $\approx(t_1(X_i), t_2(X_i)) = (1,1)$. Otherwise,

$$\approx (t_1(X_i), t_2(X_i))_N = Nec(\Pi_{t_1(X_i)} = \Pi_{t_2(X_i)}),$$
$$\approx (t_1(X_i), t_2(X_i))_\Pi = Pos(\Pi_{t_1(X_i)} = \Pi_{t_2(X_i)}). \quad (3)$$

For estimation of tuple closeness, denoted by $=_c(t_1(X), t_2(X)) = (=_c(t_1(X),$ $t_2(X))_N, =_c(t_1(X), t_2(X))_\Pi)$, one must consider all the components $X_i$ of $X$ $(X_i \in X)$ and apply the operation $min$:

$$=_c (t_1(X), t_2(X))_N = min_i \approx ((t_1(X_i), t_2(X_i))_N,$$
$$=_c (t_1(X), t_2(X))_\Pi = min_i \approx ((t_1(X_i), t_2(X_i))_\Pi, \qquad (4)$$

In order to evaluate the degree of a fuzzy functional dependency by means of both possibility and necessity measures we will apply the following extension of the Gödel implication operator $I_G(a,b) = (I_G(a,b)_N, I_G(a,b)_\Pi)$, $a = (a_N, a_\Pi)$, $b = (b_N, b_\Pi)$, $a_N$, $a_\Pi$, $b_N$, $b_\Pi \in [0,1]$ where

$$I_G(a,b)_\Pi = \begin{cases} 1 & \text{if } a_\Pi \leq b_\Pi \\ b_\Pi & \text{otherwise,} \end{cases} \qquad (5)$$

$$I_G(a,b)_N = \begin{cases} 1 & \text{if } a_N \leq b_N \text{ and } a_\Pi \leq b_\Pi \\ b_\Pi & \text{if } a_N \leq b_N \text{ and } a_\Pi > b_\Pi \\ b_N & \text{otherwise.} \end{cases} \qquad (6)$$

**Definition 1.** *Let $R(U)$ be a relation scheme where $U = \{X_1, X_2, \ldots, X_n\}$. Let $X$ and $Y$ be subsets of $U$: $X$, $Y \subseteq U$. $Y$ is functionally dependent on $X$ in $\theta = (\theta_N, \theta_\Pi)$ degree, $\theta_N, \theta_\Pi \in [0,1]$, denoted by $X \to_\theta Y$, if and only if for every relation $r$ of $R$ the following conditions are met:*

$$min_{t_1, t_2 \in r} I(t_1(X) =_c t_2(X), t_1(Y) =_c t_2(Y))_N \geq \theta_N,$$
$$min_{t_1, t_2 \in r} I(t_1(X) =_c t_2(X), t_1(Y) =_c t_2(Y))_\Pi \geq \theta_\Pi, \qquad (7)$$

*where $=_c$ is the closeness measure (4) and $I$ is the following implicator:*

$$I(a,b) = \begin{cases} I_c & \text{if } t_1(X) \text{ and } t_2(X) \text{ are identical} \\ I_G & \text{otherwise,} \end{cases} \qquad (8)$$

*where $I_c$ is the classical implication operator and $I_G$ is the extended Gödel implicator.*

Like in classical relational databases one can formulate the following inference rules known as extended Armstrong's axioms:

A1: $Y \subseteq X \Rightarrow X \to_\theta Y$ for all $\theta$
A2: $X \to_\theta Y \Rightarrow XZ \to_\theta YZ$
A3: $X \to_\alpha Y \wedge Y \to_\beta Z \Rightarrow X \to_\gamma Z$, $\quad \gamma = (min(\alpha_N, \beta_N), min(\alpha_\Pi, \beta_\Pi))$

where $\theta = (\theta_N, \theta_\Pi)$, $\alpha = (\alpha_N, \alpha_\Pi)$, $\beta = (\beta_N, \beta_\Pi)$ and $\gamma = (\gamma_N, \gamma_\Pi)$ are pairs of numbers belonging to the unit interval $[0, 1]$.

## 3  Properties of the Extended Gödel Implicator

In order to prove the correctness of the inference rules for fuzzy functional dependencies (7) we will first show certain properties of the implicator $I_G$.

**Theorem 1.** *Let* $a = (a_N, a_\Pi)$, $a' = (a'_N, a'_\Pi)$, $b = (b_N, b_\Pi)$, $b' = (b'_N, b'_\Pi)$, $c = (c_N, c_\Pi)$, $\alpha = (\alpha_N, \alpha_\Pi)$, $\beta = (\beta_N, \beta_\Pi)$, $\gamma = (\gamma_N, \gamma_\Pi)$ *and* $\theta = (\theta_N, \theta_\Pi)$ *be pairs of numbers belonging to the unit interval* $[0, 1]$. *The implicator* $I_G$ *satisfies the following conditions:*

P1: $a_N \leq b_N \wedge a_\Pi \leq b_\Pi \Rightarrow I_G(a, b) = (1,1)$,

P2: $I_G(a,b)_N \geq \theta_N \wedge I_G(a,b)_\Pi \geq \theta_\Pi \Rightarrow I_G(a',b')_N \geq \theta_N \wedge I_G(a',b')_\Pi \geq \theta_\Pi$ *for*
$a'_N = \min(a_N,c_N)$, $a'_\Pi = \min(a_\Pi,c_\Pi)$, $b'_N = \min(b_N,c_N)$, $b'_\Pi = \min(b_\Pi,c_\Pi)$,

P3: $I_G(a,b)_N \geq \alpha_N \wedge I_G(a,b)_\Pi \geq \alpha_\Pi \wedge I_G(b,c)_N \geq \beta_N \wedge I_G(b,c)_\Pi \geq \beta_\Pi \Rightarrow$
$I_G(a,c)_N \geq \gamma_N \wedge I_G(a,c)_\Pi \geq \gamma_\Pi$ *for* $\gamma_N = \min(\alpha_N,\beta_N)$, $\gamma_\Pi = \min(\alpha_\Pi,\beta_\Pi)$.

*Proof.*
P1: This condition directly follows from the definition of $I_G$.
P2:  Let $I_G(a,b)_N \geq \theta_N$ and $I_G(a,b)_\Pi \geq \theta_\Pi$. If $I_G(a,b) = (1,1)$ then $a_N \leq b_N$ and $a_\Pi \leq b_\Pi$. It follows that $a'_N \leq b'_N$ and $a'_\Pi \leq b'_\Pi$ and so $I_G(a',b') = (1,1)$. If $I_G(a,b) \neq (1,1)$ we must prove P2 for different cases of $a$, $b$ and $c$. If $c_N < a_N$, $c_\Pi < a_\Pi$, $c_N < b_N$ and $c_\Pi < b_\Pi$ then $a' = b' = c \Rightarrow I_G(a',b') = (1,1)$. If $c_N \geq a_N$, $c_\Pi \geq a_\Pi$, $c_N \geq b_N$ and $c_\Pi \geq b_\Pi$ then ($a' = a$ and $b' = b$) $\Rightarrow I_G(a',b') = I_G(a,b)$.

I. Let $a_N > b_N$ and $a_\Pi > b_\Pi$. Thus $I_G(a,b) = b$.

1.  $a_N > b_N \geq c_N$ and ($a_\Pi > c_\Pi \geq b_\Pi$ or $a_\Pi \geq c_\Pi > b_\Pi$). $a'_N = c_N$, $a'_\Pi = c_\Pi$, $b'_N = c_N$, $b'_\Pi = b_\Pi$. If $b_\Pi = c_\Pi$ then $I_G(a',b') = (1,1)$, otherwise $I_G(a',b') = (b_\Pi, b_\Pi)$.
2.  $a_N > b_N \geq c_N$ and $c_\Pi \geq a_\Pi > b_\Pi$
$a'_N = c_N$, $a'_\Pi = a_\Pi$, $b'_N = c_N$, $b'_\Pi = b_\Pi \Rightarrow I_G(a',b') = (b_\Pi, b_\Pi)$.
3.  ($a_N > c_N \geq b_N$ or $a_N \geq c_N > b_N$) and $a_\Pi > b_\Pi \geq c_\Pi$. $a'_N = c_N$, $a'_\Pi = c_\Pi$, $b'_N = b_N$, $b'_\Pi = c_\Pi$. If $c_N = b_N$ then $I_G(a',b') = (1,1)$, otherwise $I_G(a',b') = (b_N, 1)$.
4.  ($a_N > c_N \geq b_N$ or $a_N \geq c_N > b_N$) and ($a_\Pi > c_\Pi \geq b_\Pi$ or $a_\Pi \geq c_\Pi > b_\Pi$)
$a'_N = c_N$, $a'_\Pi = c_\Pi$, $b'_N = b_N$, $b'_\Pi = b_\Pi$. If ($c_N > b_N$ and $c_\Pi > b_\Pi$) then $I_G(a',b') = b$. If ($c_N > b_N$ and $c_\Pi = b_\Pi$) then $I_G(a',b') = (b_N, 1)$. If ($c_N = b_N$ and $c_\Pi > b_\Pi$) then $I_G(a',b') = (b_\Pi, b_\Pi)$. If ($c_N = b_N$ and $c_\Pi = b_\Pi$) then $I_G(a',b') = (1,1)$.
5.  ($a_N > c_N \geq b_N$ or $a_N \geq c_N > b_N$) and $c_\Pi \geq a_\Pi > b_\Pi$. $a'_N = c_N$, $a'_\Pi = a_\Pi$, $b'_N = b_N$, $b'_\Pi = b_\Pi$. If $c_N > b_N$ then $I_G(a',b') = b$, otherwise $I_G(a',b') = (b_\Pi, b_\Pi)$.
6.  $c_N \geq a_N > b_N$ and $a_\Pi > b_\Pi \geq c_\Pi$
$a'_N = a_N$, $a'_\Pi = c_\Pi$, $b'_N = b_N$, $b'_\Pi = c_\Pi \Rightarrow I_G(a',b') = (b_N, 1)$.
7.  $c_N \geq a_N > b_N$ and ($a_\Pi > c_\Pi \geq b_\Pi$ or $a_\Pi \geq c_\Pi > b_\Pi$). $a'_N = a_N$, $a'_\Pi = c_\Pi$, $b'_N = b_N$, $b'_\Pi = b_\Pi$. If $c_\Pi > b_\Pi$ then $I_G(a',b') = b$, otherwise $I_G(a',b') = (b_N, 1)$.

II. Let $a_N > b_N$ and $a_\Pi \leq b_\Pi$ . Thus $I_G(a,b) = (b_N, 1)$.

1.  $a_N > b_N \geq c_N$ and $b_\Pi \geq c_\Pi \geq a_\Pi$
$a'_N = c_N$, $a'_\Pi = a_\Pi$, $b'_N = c_N$, $b'_\Pi = c_\Pi \Rightarrow I_G(a',b') = (1,1)$.

2. $a_N > b_N \geq c_N$ and $c_\Pi \geq b_\Pi \geq a_\Pi$
   $a'_N = c_N, a'_\Pi = a_\Pi, b'_N = c_N, b'_\Pi = b_\Pi \Rightarrow I_G(a',b') = (1,1)$.

3. $(a_N > c_N \geq b_N$ or $a_N \geq c_N > b_N)$ and $b_\Pi \geq a_\Pi \geq c_\Pi$. $a'_N = c_N, a'_\Pi = c_\Pi, b'_N = b_N, b'_\Pi = c_\Pi$. If $c_N = b_N$ then $I_G(a',b') = (1,1)$, otherwise $I_G(a',b') = (b_N,1)$.

4. $(a_N > c_N \geq b_N$ or $a_N \geq c_N > b_N)$ and $b_\Pi \geq c_\Pi \geq a_\Pi$. $a'_N = c_N, a'_\Pi - a_\Pi, b'_N = b_N, b'_\Pi = c_\Pi$. If $c_N = b_N$ then $I_G(a',b') = (1,1)$, otherwise $I_G(a',b') = (b_N,1)$.

5. $(a_N > c_N \geq b_N$ or $a_N \geq c_N > b_N)$ and $c_\Pi \geq b_\Pi \geq a_\Pi$. $a'_N = c_N, a'_\Pi = a_\Pi, b'_N = b_N, b'_\Pi = b_\Pi$. If $c_N = b_N$ then $I_G(a',b') = (1,1)$, otherwise $I_G(a',b') = (b_N,1)$.

6. $c_N \geq a_N > b_N$ and $b_\Pi \geq a_\Pi \geq c_\Pi$
   $a'_N = a_N, a'_\Pi = c_\Pi, b'_N = b_N, b'_\Pi = c_\Pi \Rightarrow I_G(a',b') = (b_N,1)$.

7. $c_N \geq a_N > b_N$ and $b_\Pi \geq c_\Pi \geq a_\Pi$
   $a'_N = a_N, a'_\Pi = a_\Pi, b'_N = b_N, b'_\Pi = c_\Pi \Rightarrow I_G(a',b') = (b_N,1)$.

III. Let $a_N \leq b_N$ and $a_\Pi > b_\Pi$ . Thus $I_G(a,b) = (b_\Pi,b_\Pi)$.

1. $b_N \geq a_N \geq c_N$ and $(a_\Pi > c_\Pi \geq b_\Pi$ or $a_\Pi \geq c_\Pi > b_\Pi)$. $a'_N = c_N, a'_\Pi = c_\Pi, b'_N = c_N, b'_\Pi = b_\Pi$. If $c_\Pi = b_\Pi$ then $I_G(a',b') = (1,1)$, otherwise $I_G(a',b') = (b_\Pi,b_\Pi)$.

2. $b_N \geq a_N \geq c_N$ and $c_\Pi \geq a_\Pi > b_\Pi$
   $a'_N = c_N, a'_\Pi = a_\Pi, b'_N = c_N, b'_\Pi = b_\Pi \Rightarrow I_G(a',b') = (b_\Pi,b_\Pi)$.

3. $b_N \geq c_N \geq a_N$ and $a_\Pi > b_\Pi \geq c_\Pi$
   $a'_N = a_N, a'_\Pi = c_\Pi, b'_N = c_N, b'_\Pi = c_\Pi \Rightarrow I_G(a',b') = (1,1)$.

4. $b_N \geq c_N \geq a_N$ and $(a_\Pi > c_\Pi \geq b_\Pi$ or $a_\Pi \geq c_\Pi > b_\Pi)$. $a'_N = a_N, a'_\Pi = c_\Pi, b'_N = c_N, b'_\Pi = b_\Pi$. If $c_\Pi = b_\Pi$ then $I_G(a',b') = (1,1)$, otherwise $I_G(a',b') = (b_\Pi,b_\Pi)$.

5. $b_N \geq c_N \geq a_N$ and $c_\Pi \geq a_\Pi > b_\Pi$
   $a'_N = a_N, a'_\Pi = a_\Pi, b'_N = c_N, b'_\Pi = b_\Pi \Rightarrow I_G(a',b') = (b_\Pi,b_\Pi)$.

6. $c_N \geq b_N \geq a_N$ and $a_\Pi > b_\Pi \geq c_\Pi$
   $a'_N = a_N, a'_\Pi = c_\Pi, b'_N = b_N, b'_\Pi = c_\Pi \Rightarrow I_G(a',b') = (1,1)$.

7. $c_N \geq b_N \geq a_N$ and $(a_\Pi > c_\Pi \geq b_\Pi$ or $a_\Pi \geq c_\Pi > b_\Pi)$. $a'_N = a_N, a'_\Pi = c_\Pi, b'_N = b_N, b'_\Pi = b_\Pi$. If $c_\Pi = b_\Pi$ then $I_G(a',b') = (1,1)$, otherwise $I_G(a',b') = (b_\Pi,b_\Pi)$.

P3: Let $\theta = (\min(I_G(a,b)_N, I_G(b,c)_N), \min(I_G(a,b)_\Pi, I_G(b,c)_\Pi))$.
If $a_N \leq c_N$ and $a_\Pi \leq c_\Pi$ then $I_G(a,c) = (1,1)$. If $a_N \leq b_N$ and $a_\Pi \leq b_\Pi$ then $I_G(a,b) = (1,1) \Rightarrow \theta = I_G(b,c)$. Since the components of $I_G$ are decreasing in the first argument [9], we obtain $I_G(a,c)_N \geq \theta_N$ and $I_G(a,c)_\Pi \geq \theta_\Pi$. If $b_N \leq c_N$ and $b_\Pi \leq c_\Pi$ then $I_G(b,c) = (1,1) \Rightarrow \theta = I_G(a,b)$. Since the components of $I_G$ are increasing in the second argument [9], we obtain $I_G(a,c)_N \geq \theta_N$ and $I_G(a,c)_\Pi \geq \theta_\Pi$. Otherwise, we must prove P3 for different cases of $a$, $b$ and $c$.

I. Let $a_N > b_N$ and $a_\Pi > b_\Pi$. Thus $I_G(a,b) = b$.

1. $a_N > b_N \geq c_N$ and $a_\Pi > b_\Pi \geq c_\Pi$. $I_G(a,c) = c$.
   If $(b_N > c_N$ and $b_\Pi > c_\Pi)$ then $I_G(b,c) = c$. If $(b_N > c_N$ and $b_\Pi = c_\Pi)$ then $I_G(b,c) = (c_N,1)$. If $(b_N = c_N$ and $b_\Pi > c_\Pi)$ then $I_G(b,c) = (c_\Pi,c_\Pi)$. If $(b_N = c_N$ and $b_\Pi = c_\Pi)$ then $I_G(b,c) = (1,1)$. Thus in all cases $\theta = c \Rightarrow I_G(a,c) = \theta$.

2. $a_N > b_N \geq c_N$ and $(a_\Pi > c_\Pi \geq b_\Pi$ or $a_\Pi \geq c_\Pi > b_\Pi)$.
   $I_G(a,c)_N = c_N$ and $I_G(a,c)_\Pi \geq c_\Pi$. $I_G(b,c)_N \geq c_N$ and $I_G(b,c)_\Pi = 1$.
   Thus $\theta = (c_N, b_\Pi) \Rightarrow (I_G(a,c)_N = \theta_N$ and $I_G(a,c)_\Pi \geq \theta_\Pi)$.
3. $a_N > b_N \geq c_N$ and $c_\Pi \geq a_\Pi > b_\Pi$. $I_G(a,c)=(c_N,1)$. $I_G(b,c)_N \geq c_N$.
   $I_G(b,c)_\Pi = 1$. Thus $\theta=(c_N,b_\Pi) \Rightarrow (I_G(a,c)_N = \theta_N$ and $I_G(a,c)_\Pi > \theta_\Pi)$.
4. $(a_N > c_N \geq b_N$ or $a_N \geq c_N > b_N)$ and $a_\Pi > b_\Pi \geq c_\Pi$.
   $I_G(a,c)_N \geq c_N$. $I_G(a,c)_\Pi = c_\Pi$. If $b_\Pi > c_\Pi$ then $I_G(b,c) = (c_\Pi,c_\Pi)$. If $b_\Pi = c_\Pi$
   then $I_G(b,c) = (1,1)$. Thus $\theta = (b_N, c_\Pi) \Rightarrow (I_G(a,c)_N \geq \theta_N$ and $I_G(a,c)_\Pi =$
   $\theta_\Pi)$.
5. $c_N \geq a_N > b_N$ and $a_\Pi > b_\Pi \geq c_\Pi$. $I_G(a,c) = (c_\Pi,c_\Pi)$.
   If $b_\Pi > c_\Pi$ then $I_G(b,c) = (c_\Pi,c_\Pi)$ and if $b_\Pi = c_\Pi$ then $I_G(b,c) = (1,1)$. Thus
   $\theta = (b_N, c_\Pi) \Rightarrow (I_G(a,c)_N > \theta_N$ and $I_G(a,c)_\Pi = \theta_\Pi)$.

II. Let $a_N > b_N$ and $a_\Pi \leq b_\Pi$ . Thus $I_G(a,b) = (b_N,1)$.

1. $a_N > b_N \geq c_N$ and $b_\Pi \geq a_\Pi \geq c_\Pi$. $I_G(a,c)_N = c_N$ and $I_G(a,c)_\Pi \geq c_\Pi$.
   If $(b_N > c_N$ and $b_\Pi > c_\Pi)$ then $I_G(b,c) = c \Rightarrow \theta = c \Rightarrow (I_G(a,c)_N = \theta_N$ and
   $I_G(a,c)_\Pi \geq \theta_\Pi)$. If $(b_N > c_N$ and $b_\Pi = c_\Pi)$ then $I_G(b,c) = (c_N,1) \Rightarrow \theta =$
   $(c_N,1)$. If $b_\Pi = c_\Pi$ then $a_\Pi = c_\Pi \Rightarrow I_G(a,c) = (c_N,1) = \theta$. If $(b_N = c_N$ and
   $b_\Pi > c_\Pi)$ then $I_G(b,c) = (c_\Pi,c_\Pi) \Rightarrow \theta = c \Rightarrow (I_G(a,c)_N = \theta_N$ and $I_G(a,c)_\Pi$
   $\geq \theta_\Pi)$. If $b = c$ then $(I_G(b,c) = (1,1)$ and $I_G(a,c) = (c_N,1)) \Rightarrow \theta = (c_N,1)$
   $\Rightarrow I_G(a,c) = \theta$.
2. $a_N > b_N \geq c_N$ and $b_\Pi \geq c_\Pi \geq a_\Pi$. $I_G(a,c) = (c_N,1)$.
   $I_G(b,c)_N \geq c_N \Rightarrow \theta_N = c_N \Rightarrow I_G(a,c)_N = \theta_N$.
3. $a_N > b_N \geq c_N$ and $c_\Pi \geq b_\Pi \geq a_\Pi$. $I_G(a,c) = (c_N,1)$.
   If $b_N > c_N$ then $I_G(b,c) = (c_N,1) \Rightarrow \theta = (c_N,1) \Rightarrow I_G(a,c) = \theta$.
   If $b_N = c_N$ then $I_G(b,c) = (1,1) \Rightarrow \theta = (c_N,1) \Rightarrow I_G(a,c) = \theta$.
4. $(a_N > c_N \geq b_N$ or $a_N \geq c_N > b_N)$ and $b_\Pi \geq a_\Pi \geq c_\Pi$
   $I_G(a,c)_N \geq c_N$ and $I_G(a,c)_\Pi \geq c_\Pi$. $I_G(b,c)_N = 1 \Rightarrow \theta_N = b_N \Rightarrow I_G(a,c)_N \geq$
   $\theta_N$. If $b_\Pi > c_\Pi$ then $I_G(b,c)_\Pi = c_\Pi \Rightarrow \theta_\Pi = c_\Pi \Rightarrow I_G(a,c)_\Pi \geq \theta_\Pi$. If $b_\Pi = c_\Pi$
   then $I_G(b,c)_\Pi = 1 \Rightarrow \theta_\Pi = 1$. If $b_\Pi = c_\Pi$ then $a_\Pi = c_\Pi \Rightarrow I_G(a,c)_\Pi = 1 =$
   $\theta_\Pi$ .
5. $(a_N > c_N \geq b_N$ or $a_N \geq c_N > b_N)$ and $b_\Pi \geq c_\Pi \geq a_\Pi$
   $I_G(a,c)_N \geq c_N$ and $I_G(a,c)_\Pi = 1$. $I_G(b,c)_N = 1 \Rightarrow \theta_N = b_N \Rightarrow I_G(a,c)_N \geq$
   $\theta_N$ .
6. $c_N \geq a_N > b_N$ and $b_\Pi \geq a_\Pi \geq c_\Pi$. $I(a,c)_N \geq c_\Pi$ and $I(a,c)_\Pi \geq c_\Pi$.
   $I(b,c)_N \geq c_\Pi \Rightarrow \theta_N = b_N \Rightarrow I(a,c)_N \geq \theta_N$. If $b_\Pi > c_\Pi$ then $I_G(b,c)_\Pi = c_\Pi$
   $\Rightarrow \theta_\Pi = c_\Pi \Rightarrow I_G(a,c)_\Pi \geq \theta_\Pi$. If $b_\Pi = c_\Pi$ then $I_G(b,c)_\Pi = 1 \Rightarrow \theta_\Pi = 1$. If
   $b_\Pi = c_\Pi$ then $a_\Pi = c_\Pi \Rightarrow I_G(a,c)_\Pi = 1 = \theta_\Pi$ .

III. Let $a_N \leq b_N$ and $a_\Pi > b_\Pi$ . Thus $I_G(a,b) = (b_\Pi, b_\Pi)$.

1. $b_N \geq a_N \geq c_N$ and $a_\Pi > b_\Pi \geq c_\Pi$
   If $a_N > c_N$ then $I_G(a,c) = c$, otherwise $I_G(a,c) = (c_\Pi,c_\Pi)$.
   If $(b_N > c_N$ and $b_\Pi > c_\Pi)$ then $I_G(b,c) = c \Rightarrow \theta = c \Rightarrow (I_G(a,c)_N = \theta_N$ and
   $I_G(a,c)_\Pi = \theta_\Pi)$. If $(b_N > c_N$ and $b_\Pi = c_\Pi)$ then $I_G(b,c) = (c_N,1) \Rightarrow \theta =$
   $c \Rightarrow (I_G(a,c)_N \geq \theta_N$ and $I_G(a,c)_\Pi = \theta_\Pi)$. If $(b_N = c_N$ and $b_\Pi > c_\Pi)$ then

$I_G(b,c) = (c_\Pi, c_\Pi) \Rightarrow \theta = (c_\Pi, c_\Pi)$. If $b_N = c_N$ then $a_N = c_N \Rightarrow I_G(a,c) = (c_\Pi, c_\Pi) = \theta$. If $b = c$ then $I_G(b,c) = (1,1) \Rightarrow \theta = (c_\Pi, c_\Pi) \Rightarrow I_G(a,c) = \theta$.

2. $b_N \geq a_N \geq c_N$ and $(a_\Pi > c_\Pi \geq b_\Pi$ or $a_\Pi \geq c_\Pi > b_\Pi)$
$I_G(a,c)_N \geq c_N$ and $I_G(a,c)_\Pi \geq c_\Pi$. $I_G(b,c)_\Pi = 1$.
If $b_N > c_N$ then $I_G(b,c) = (c_N, 1) \Rightarrow \theta = (c_N, b_\Pi) \Rightarrow (I_G(a,c)_N \geq \theta_N$ and $I_G(a,c)_\Pi \geq \theta_\Pi)$. If $b_N = c_N$ then $I_G(b,c) = (1,1) \Rightarrow \theta = (b_\Pi, b_\Pi)$. If $b_N = c_N$ then $a_N = c_N \Rightarrow I_G(a,c)_N \geq c_\Pi \Rightarrow (I_G(a,c)_N \geq \theta_N$ and $I_G(a,c)_\Pi \geq \theta_\Pi)$.

3. $b_N \geq a_N \geq c_N$ and $c_\Pi \geq a_\Pi > b_\Pi$. $I_G(a,c)_N \geq c_N$ and $I_G(a,c)_\Pi = 1$.
$I_G(b,c)_\Pi = 1$. If $b_N > c_N$ then $I_G(b,c) = (c_N, 1) \Rightarrow \theta = (c_N, b_\Pi) \Rightarrow (I_G(a,c)_N \geq \theta_N$ and $I_G(a,c)_\Pi > \theta_\Pi)$. If $b_N = c_N$ then $I_G(b,c) = (1,1) \Rightarrow \theta = (b_\Pi, b_\Pi)$. If $b_N = c_N$ then $a_N = c_N \Rightarrow I_G(a,c)_N = 1 \Rightarrow (I_G(a,c)_N > \theta_N$ and $I_G(a,c)_\Pi > \theta_\Pi)$.

4. $b_N \geq c_N \geq a_N$ and $a_\Pi > b_\Pi \geq c_\Pi$. $I_G(a,c) = (c_\Pi, c_\Pi)$.
If $(b_N > c_N$ and $b_\Pi > c_\Pi)$ then $I_G(b,c) = c \Rightarrow \theta = c \Rightarrow (I_G(a,c)_N \geq \theta_N$ and $I_G(a,c)_\Pi = \theta_\Pi)$. If $(b_N > c_N$ and $b_\Pi = c_\Pi)$ then $I_G(b,c) = (c_N, 1) \Rightarrow \theta = c \Rightarrow (I_G(a,c)_N \geq \theta_N$ and $I_G(a,c)_\Pi = \theta_\Pi)$. If $(b_N = c_N$ and $b_\Pi > c_\Pi)$ then $I_G(b,c) = (c_\Pi, c_\Pi) \Rightarrow \theta = (c_\Pi, c_\Pi) \Rightarrow I_G(a,c) = \theta$. If $b = c$ then $I_G(b,c) = (1,1) \Rightarrow \theta = (c_\Pi, c_\Pi) \Rightarrow I_G(a,c) = \theta$.

5. $b_N \geq c_N \geq a_N$ and $(a_\Pi > c_\Pi \geq b_\Pi$ or $a_\Pi \geq c_\Pi > b_\Pi)$ If $a_\Pi > c_\Pi$ then $I_G(a,c) = (c_\Pi, c_\Pi)$. If $a_\Pi = c_\Pi$ then $I_G(a,c) = (1,1)$. If $b_N > c_N$ then $I_G(b,c) = (c_N, 1) \Rightarrow \theta = (c_N, b_\Pi) \Rightarrow (I_G(a,c)_N \geq \theta_N$ and $I_G(a,c)_\Pi \geq \theta_\Pi)$. If $b_N = c_N$ then $I_G(b,c) = (1,1) \Rightarrow \theta = (b_\Pi, b_\Pi) \Rightarrow (I_G(a,c)_N \geq \theta_N$ and $I_G(a,c)_\Pi \geq \theta_\Pi)$.

6. $c_N \geq b_N \geq a_N$ and $a_\Pi > b_\Pi \geq c_\Pi$. $I_G(a,c) = (c_\Pi, c_\Pi)$.

If $b_\Pi > c_\Pi$ then $I_G(b,c) = (c_\Pi, c_\Pi) \Rightarrow \theta = (c_\Pi, c_\Pi) \Rightarrow I_G(a,c) = \theta$.
If $b_\Pi = c_\Pi$ then $I_G(b,c) = (1,1) \Rightarrow \theta = (c_\Pi, c_\Pi) \Rightarrow I_G(a,c) = \theta$.     $\square$

Moreover, the extended Gödel implicator has the following properties [9]:
P4: $a_N \leq a'_N \wedge a_\Pi \leq a'_\Pi \Rightarrow I_G(a,b)_N \geq I_G(a',b)_N \wedge I_G(a,b)_\Pi \geq I_G(a',b)_\Pi$,
P5: $b_N \geq b'_N \wedge b_\Pi \geq b'_\Pi \Rightarrow I_G(a,b)_N \geq I_G(a,b')_N \wedge I_G(a,b)_\Pi \geq I_G(a,b')_\Pi$,
P6: $I_G(1,b)_N = b_N$ and $I_G(1,b)_\Pi = b_\Pi$,
P7: $I_G(a,b)_N \geq b_N$ and $I_G(a,b)_\Pi \geq b_\Pi$,
P8: $I_G(a, I_G(b,c))_N = I_G(b, I_G(a,c))_N$ and $I_G(a, I_G(b,c))_\Pi = I_G(b, I_G(a,c))_\Pi$.

## 4   Soundness and Completeness of the Inference Rules

The set of extended Armstrong's axioms can be used to derive new fuzzy functional dependencies implied by a given set of FFDs. Let $F$ be a set of FFDs (7) with respect to the relation scheme $R$ $(U)$. Let us denote by $F^+$ the set of all FFDs which can be derived from $F$ by means of the extended Armstrong's axioms:

$$F^+ = \{X \to_\theta Y, \theta = (\theta_N, \theta_\Pi) : F \vDash X \to_\theta Y\}. \tag{9}$$

**Theorem 2.** *The extended Armstrong's axioms are sound.*

*Proof.* Let $F$ be a set of FFDs for the relation scheme $R(U)$. Let $t_1$ and $t_2$ be tuples of $r$, where $r$ is relation of $R(U)$. Let $X, Y, Z \subseteq U$.

Let $a = =_c (t_1(X), t_2(X))$, $b = =_c (t_1(Y), t_2(Y))$, $c = =_c (t_1(Z), t_2(Z))$.

A1: If $Y \subseteq X$ then by (4) $a_N \leq b_N$ and $a_\Pi \leq b_\Pi$. Since $I_G$ satisfies P1, we get $I_G(a,b) = (1,1)$ and so

$$I_G(=_c (t_1(X), t_2(X)), =_c (t_1(Y), t_2(Y)))_N = 1 \geq \theta_N,$$
$$I_G(=_c (t_1(X), t_2(X)), =_c (t_1(Y), t_2(Y)))_\Pi = 1 \geq \theta_\Pi.$$

A2: Since $X \to_\theta Y$ holds, we have $I_G(a,b)_N \geq \theta_N$ and $I_G(a,b)_\Pi \geq \theta_\Pi$. Let $a' = =_c (t_1(XZ), t_2(XZ))$ and $b' = =_c (t_1(YZ), t_2(YZ))$. From (4) we get $a'_N = \min(a_N, c_N)$, $a'_\Pi = \min(a_\Pi, c_\Pi)$ and $b'_N = \min(b_N, c_N)$, $b'_\Pi = \min(b_\Pi, c_\Pi)$. Since $I_G$ satisfies P2, we obtain $I_G(a', b')_N \geq \theta_N$ and $I_G(a', b')_\Pi \geq \theta_\Pi$ and so

$$I_G(=_c (t_1(XZ), t_2(XZ)), =_c (t_1(YZ), t_2(YZ)))_N \geq \theta_N,$$
$$I_G(=_c (t_1(XZ), t_2(XZ)), =_c (t_1(YZ), t_2(YZ)))_\Pi \geq \theta_\Pi.$$

Thus, if $X \to_\theta Y \in F^+$ then $XZ \to_\theta YZ \in F^+$.

A3: Since $X \to_\alpha Y$ and $Y \to_\beta Z$ hold, we have: $I_G(a,b)_N \geq \alpha_N$, $I_G(a,b)_\Pi \geq \alpha_\Pi$ and $I_G(b,c)_N \geq \beta_N$, $I_G(b,c)_\Pi \geq \beta_\Pi$. By P3 we obtain $I_G(a,c)_N \geq \gamma_N$ and $I_G(a,c)_\Pi \geq \gamma_\Pi$, where $\gamma_N = \min(\alpha_N, \beta_N)$ and $\gamma_\Pi = \min(\alpha_\Pi, \beta_\Pi)$ and so

$$I_G(=_c (t_1(X), t_2(X)), =_c (t_1(Z), t_2(Z)))_N \geq \gamma_N,$$
$$I_G(=_c (t_1(X), t_2(X)), =_c (t_1(Z), t_2(Z)))_\Pi \geq \gamma_\Pi.$$

Thus, if $X \to_\alpha Y \in F^+$ and $Y \to_\beta Z \in F^+$ then $X \to_\gamma Z \in F^+$ for $\gamma = (\min(\alpha_N, \beta_N), \min(\alpha_\Pi, \beta_\Pi))$. $\square$

The following rules result from Armstrong's axioms:

D1: $X \to_\alpha Y \wedge X \to_\beta Z \Rightarrow X \to_\lambda YZ$ for $\gamma = (\min(\alpha_N, \beta_N), \min(\alpha_\Pi, \beta_\Pi))$

*Proof.* By A2 we have $X \to_\alpha Y \Rightarrow X \to_\alpha XY$ and $X \to_\beta Z \Rightarrow XY \to_\beta ZY$. Then by A3 we obtain $X \to_\alpha XY \wedge XY \to_\beta YZ \Rightarrow X \to_\gamma YZ$ for $\gamma = (\min(\alpha_N, \beta_N), \min(\alpha_\Pi, \beta_\Pi))$. $\square$

D2: $X \to_\alpha Y \wedge WY \to_\beta Z \Rightarrow XW \to_\lambda Z$ for $\gamma = (\min(\alpha_N, \beta_N), \min(\alpha_\Pi, \beta_\Pi))$

*Proof.* By A2 we have $X \to_\alpha Y \Rightarrow XW \to_\alpha YW$. Then by A3 we obtain $XW \to_\alpha YW \wedge WY \to_\beta Z \Rightarrow XW \to_\gamma Z$ for $\gamma = (\min(\alpha_N, \beta_N), \min(\alpha_\Pi, \beta_\Pi))$. $\square$

D3: $X \to_\alpha Y \wedge Z \subseteq Y \Rightarrow X \to_\alpha Z$

*Proof.* By A1 we have $Z \subseteq Y \Rightarrow Y \to_\alpha Z$ for every $\alpha = (\alpha_N, \alpha_\Pi)$, $\alpha_N, \alpha_\Pi \in [0,1]$. Then by A3 we obtain $X \to_\alpha Y \wedge Y \to_\alpha Z \Rightarrow X \to_\alpha Z$. $\square$

D4: $X \to_\alpha Y \Rightarrow X \to_\beta Y$ for $\beta_N \leq \alpha_N$ and $\beta_\Pi \leq \alpha_\Pi$

*Proof.* By A1 we have $Y \to_\theta Y$ for every $\theta = (\theta_N, \theta_\Pi)$, $\theta_N, \theta_\Pi \in [0,1]$. Then by A3 we obtain $X \to_\alpha Y \wedge Y \to_\theta Y \Rightarrow X \to_\beta Y$ for $\beta = (\min(\alpha_N, \theta_N), \min(\alpha_\Pi, \theta_\Pi))$. $\square$

The closure of a set of attributes $X \subseteq U$ with respect to the set $F$ of FFDs, denoted by $X_F^+$, is defined as a set of triples $(A, \theta_N, \theta_\Pi)$, where $A \in U$, $\theta_N = \sup\{\alpha : X \rightarrow_{(\alpha,\beta)} A \in F^+\}$ and $\theta_\Pi = \sup\{\beta : X \rightarrow_{(\alpha,\beta)} A \in F^+\}$. The set of attributes occurring in $X_F^+$ will be denoted by $\mathrm{DOM}(X_F^+)$:

$$DOM(X_F^+) = \{A : (A, \theta_N, \theta_\Pi) \in X_F^+\}.$$

**Lemma 1.** *Let $F$ be a set of FFDs defined over a relation scheme $R(U)$. Let $X, Y \subseteq U$ and $Y = \{A_1, A_2, ..., A_k\}$, $A_i \in U$. Then $X \rightarrow_{(\theta_N, \theta_\Pi)} Y$ is deduced by means of the extended Armstrong's axioms if and only if $\forall i (A_i, \theta_{i,N}, \theta_{i,\Pi}) \in X_F^+$, where $\theta_{i,N} \geq \theta_N$ and $\theta_{i,\Pi} \geq \theta_\Pi$.*

*Proof.*
Necessity: If $X \rightarrow_{(\theta_N, \theta_\Pi)} Y$ is deduced by means of the extended Armstrong's axioms then by D3 we obtain $X \rightarrow_{(\theta_N, \theta_\Pi)} A_i$ for $i = 1, 2, ..., k$. Thus there exist $\theta_{i,N} \geq \theta_N$ and $\theta_{i,\Pi} \geq \theta_\Pi$ such that $(A_i, \theta_{i,N}, \theta_{i,\Pi}) \in X_F^+$ (definition of $X_F^+$). Sufficiency: If $(A_i, \theta_{i,N}, \theta_{i,\Pi}) \in X_F^+$ where $\theta_{i,N} \geq \theta_N$ and $\theta_{i,\Pi} \geq \theta_\Pi$ for $i = 1, 2, ..., k$, then $X \rightarrow_{(\theta_{i,N}, \theta_{i,\Pi})} A_i \in F^+$. By D1 we obtain $X \rightarrow_{(\theta_N, \theta_\Pi)} Y$, where $\theta_N = \min_i(\theta_{i,N})$ and $\theta_\Pi = \min_i(\theta_{i,\Pi})$. $\qquad\square$

**Theorem 3.** *The extended Armstrong's axioms are complete.*

*Proof.* In order to prove the theorem we will show that if $X \rightarrow_{(\theta_N, \theta_\Pi)} Y \notin F^+$, then it is possible to construct a relation where all FFDs in $F$ are satisfied and $X \rightarrow_{(\theta_N, \theta_\Pi)} Y$ does not hold, which means that $X \rightarrow_{(\theta_N, \theta_\Pi)} Y$ cannot be derived from $F$. Let $F$ be a set of FFDs for relation scheme $R(U)$. Suppose that $X \rightarrow_{(\theta_N, \theta_\Pi)} Y \notin F^+$. Let $X = \{X_1, X_2, ..., X_k\}$ and $\mathrm{DOM}(X_F^+) = \{X_1, X_2, ..., X_k, A_1, A_2, ..., A_l\}$. Let us construct a relation $r$ of the scheme $R(U)$, $U = \{\mathrm{DOM}(X_F^+), B_1, B_2, ..., B_m\}$, consisting of two tuples $t_1$ and $t_2$ such that:

$$t_1(X) = t_2(X) = 1,$$
$$t_1(A_i) = c_i, \quad t_2(A_i) = d_i \quad \text{for} \quad A_i \in DOM(X_F^+) - X, \quad i = 1, 2, ..., l$$
$$t_1(B_i) = 0, \quad t_2(B_i) = 1 \quad \text{for} \quad B_i \in U - DOM(X_F^+), \quad i = 1, 2, ..., m$$

where $c_i$ and $d_i$ are possibility distributions with degrees of closeness $\phi_{i,N}$ and $\phi_{i,\Pi}$. Let $\phi_{0,N} = \min_i \phi_{i,N}$ and $\phi_{0,\Pi} = \min_i \phi_{i,\Pi}$.

We will show that each FFD $V \rightarrow_{(\gamma_N, \gamma_\Pi)} W \in F$ holds in $r$. One should consider only the case when $V \subseteq \mathrm{DOM}(X_F^+)$ and $W \subseteq \mathrm{DOM}(X_F^+) - X$. (If $V \nsubseteq \mathrm{DOM}(X_F^+)$ then $t_1(V) \neq t_2(V)$ and so $V \rightarrow_{(1,1)} W$. Similarly, if $W \subseteq X$ then $t_1(W) = t_2(W)$ and so $V \rightarrow_{(1,1)} W$. Suppose that $V \subseteq \mathrm{DOM}(X_F^+)$ and $W \subseteq U - \mathrm{DOM}(X_F^+)$. Degrees of closeness of $t_1(W)$ and $t_2(W)$ are equal to 0. By Lemma 1 we obtain $X \rightarrow_{(\phi_{0,N}, \phi_{0,\Pi})} V \in F$. Since $V \rightarrow_{(\gamma_N, \gamma_\Pi)} W \in F$ we have $X \rightarrow_{(\psi_N, \psi_\Pi)} W \in F$, where $\psi_N = \min(\phi_{0,N}, \gamma_N)$ and $\psi_\Pi = \min(\phi_{0,\Pi}, \gamma_\Pi)$. Thus $W \subseteq \mathrm{DOM}(X_F^+)$: a contradiction.)

Suppose that there exists $V \rightarrow_{(\gamma_N, \gamma_\Pi)} W \in F$, which does not hold in $r$. Let $V \subseteq \text{DOM}(X_F^+)$, $V - X = \{A_{p_1}, A_{p_2}, ..., A_{p_r}\}$, $W \subseteq \text{DOM}(X_F^+) - X$ and $W = \{A_{s_1}, A_{s_2}, ..., A_{s_t}\}$, where $p_1, p_2, ..., p_r, s_1, s_2, ..., s_t \in \{1, 2, ..., l\}$. Let $\phi_{V,N} = \min_i \phi_{p_i,N}$, $\phi_{V,\Pi} = \min_i \phi_{p_i,\Pi}$ and $\phi_{W,N} = \min_i \phi_{s_i,N}$, $\phi_{W,\Pi} = \min_i \phi_{s_i,\Pi}$. Thus, $=_c (t_1(V), t_2(V)) = (\phi_{V,N}, \phi_{V,\Pi})$ and $=_c (t_1(W), t_2(W)) = (\phi_{W,N}, \phi_{W,\Pi})$. If $V \rightarrow_{(\gamma_N, \gamma_\Pi)} W$ does not hold in $r$, then $\phi_{W,N} < \min(\gamma_N, \phi_{V,N})$ or $\phi_{W,\Pi} < \min(\gamma_\Pi, \phi_{V,\Pi})$. Thus, $\phi_{p_i,N} > \phi_{W,N}$ or $\phi_{p_i,\Pi} > \phi_{W,\Pi}$ for every $p_i$. By Lemma 1 we obtain $X \rightarrow_{(\phi_{V,N}, \phi_{V,\Pi})} V \in F^+$. Since $V \rightarrow_{(\gamma_N, \gamma_\Pi)} W \in F$ we have $X \rightarrow_{(\psi_N, \psi_\Pi)} W \in F^+$, where $\psi_N = \min(\phi_{V,N}, \gamma_N)$ and $\psi_\Pi = \min(\phi_{V,\Pi}, \gamma_\Pi)$. Since $W \subseteq \text{DOM}(X_F^+) - X$ then $(A_{s_i}, \phi_{s_i,N}, \phi_{s_i,\Pi}) \in X_F^+$ for every $s_i$. According to the definition of $X_F^+$, $\phi_{s_i,N}$ and $\phi_{s_i,\Pi}$ are upper bounds. Thus, conditions $\phi_{p_i,N} > \phi_{W,N}$ or $\phi_{p_i,\Pi} > \phi_{W,\Pi}$ for every $p_i$ are not satisfied. We obtained a contradiction. Thus, $V \rightarrow_{(\gamma_N, \gamma_\Pi)} W$ holds in $r$.

Now we prove that $X \rightarrow_{(\theta_N, \theta_\Pi)} Y \notin F^+$ does not hold in $r$. We should consider only the case when $Y \subseteq \text{DOM}(X_F^+)$. (If $Y \nsubseteq \text{DOM}(X_F^+)$ then $=_c (t_1(Y), t_2(Y))_N = {}=_c (t_1(Y), t_2(Y))_\Pi = 0$ and so $X \rightarrow_{(\theta_N, \theta_\Pi)} Y$ does not hold). Let $Y \subseteq \{A_1, A_2, ..., A_l\}$. Let $\phi_{Y,N} = \min_i \phi_{i,N}$ and $\phi_{Y,\Pi} = \min_i \phi_{i,\Pi}$. By Lemma 1, it follows that $X \rightarrow_{(\theta_N, \theta_\Pi)} Y$ holds in $r$ for $\phi_{Y,N} \geq \theta_N$ and $\phi_{Y,\Pi} \geq \theta_\Pi$ and $X \rightarrow_{(\theta_N, \theta_\Pi)} Y \in F^+$ which is a contradiction to the assumption. $\qquad \square$

*Example 1.* Let us consider relation scheme $R(A,B,C,D)$ with the following set of FFDs: $F = \{ ABC \rightarrow_{(0,0.8)} D, BCD \rightarrow_{(0.5,1)} A, ACD \rightarrow_{(1,1)} B, ABD \rightarrow_{(0.8,1)} C, A \rightarrow_{(0,0.7)} C, A \rightarrow_{(1,1)} D, B \rightarrow_{(0,0.6)} AC \}$. By D2 we obtain: $AB \rightarrow_{(0,0.7)} D$, $AD \rightarrow_{(0,0.7)} B$, $AC \rightarrow_{(1,1)} B$, $AB \rightarrow_{(0.8,1)} C$ and $B \rightarrow_{(0,0.6)} D$. Since $A \rightarrow_{(1,1)} D \Rightarrow AB \rightarrow_{(1,1)} D$ (by A2 and D3) and $AB \rightarrow_{(1,1)} D \Rightarrow AB \rightarrow_{(0,0.7)} D$ (by D4) we conclude that $AB \rightarrow_{(0,0.7)} D$ is redundant. Similarly $ABC \rightarrow_{(0,0.8)} D$, $ACD \rightarrow_{(1,1)} B$, $ABD \rightarrow_{(0.8,1)} C$ are also redundant. By D3 we have $B \rightarrow_{(0,0.6)} A$ and $B \rightarrow_{(0,0.6)} C$. Thus $F_m = \{BCD \rightarrow_{(0.5,1)} A, AD \rightarrow_{(0,0.7)} B, AC \rightarrow_{(1,1)} B, AB \rightarrow_{(0.8,1)} C, A \rightarrow_{(0,0.7)} C, A \rightarrow_{(1,1)} D, B \rightarrow_{(0,0.6)} A, B \rightarrow_{(0,0.6)} C, B \rightarrow_{(0,0.6)} D\}$ is a minimal set of FFDs for the scheme $R$.

# 5    Conclusions

The paper deals with data dependencies in possibilistic databases. We applied and extended the definition of fuzzy functional dependency which was formulated by Chen [2]. Its level is evaluated by measures of possibility and necessity. For FFDs we have established inference rules which are an extension of Armstrong's axioms for conventional databases and showed that they form a sound and complete system. Similar results may be expected for other approaches. The obtained results could be generalized when using t-norms. Another line of future work is an extension of the presented considerations by taking into account unknown and inapplicable (missing) values.

# References

1. Bosc, P., Pivert, O.: Querying possibilistic databases: three interpretations. In: Yager, R.R., Abbasov, A.M., Reformat, M., Shahbazova, S.N. (eds.) Soft Computing: State of the Art Theory and Novel Applications. STUDFUZZ, vol. 291, pp. 161–176. Springer, Heidelberg (2013)
2. Chen, G.: Fuzzy Logic in Data Modeling - Semantics, Constraints and Database Design. Kluwer Academic Publishers, Boston (1998)
3. Dubois, D., Lorini, E., Prade, H.: A possibility theory viewpoint. In: Andreasen, T., et al. (eds.) FQAS 2015. Advances in Intelligent Systems and Computing, vol. 400, pp. 3–13. Springer International Publishing, Switzerland (2016)
4. Galindo, J., Urrutia, A., Piattini, M.: Fuzzy Databases: Modeling, Design and Implementation. Idea Group Publishing, London (2005)
5. Link, S., Prade, H.: Relational database schema design for uncertain data. CDMTCS-469 Research report, Centre for Disctere Mathematics and Theoretical Computer Science (2014)
6. Link, S., Prade, H.: Possibilistic functional dependencies and their relationship to possibility theory. IEEE Trans. Fuzzy Syst. (2015). doi:10.1109/TFUZZ.2015. 2466074
7. Małysiak-Mrozek, B., Mrozek, D., Kozielski, S.: Processing of crisp and fuzzy measures in the fuzzy data warehouse for global natural resources. In: García-Pedrajas, N., Herrera, F., Fyfe, C., Benítez, J.M., Ali, M. (eds.) IEA/AIE 2010, Part III. LNCS, vol. 6098, pp. 616–625. Springer, Heidelberg (2010)
8. Miłek, M., Małysiak-Mrozek, B., Mrozek, D.: A fuzzy data warehouse: theoretical foundations and practical aspects of usage. Studia Informatica **31**(2A–89), 489–504 (2010)
9. Nakata, M.: On inference rules of dependencies in fuzzy relational data models: functional dependencies. In: Pons, O., Vila, M., Kacprzyk, J. (eds.) Knowledge Management in Fuzzy Databases. Studies in Fuzziness and Soft Computing, vol. 39, pp. 36–66. Physica-Verlag, Heidelberg (2000)
10. Tyagi, B., Sharfuddin, A., Dutta, R., Devendra, K.: A complete axiomatization of fuzzy functional dependencies using fuzzy function. Fuzzy Sets Syst. **151**(2), 363–379 (2005)

# The Evaluation of Map-Reduce Join Algorithms

Maciej Penar[✉] and Artur Wilczek[✉]

Faculty of Computer Science and Management, Wrocław University of Technology,
Wybrzeże Wyspiańskiego 27, 50-370 Wrocław, Poland
penar.maciej@gmail.com, artur.wilczek@pwr.edu.pl

**Abstract.** In recent years, Map-Reduce systems have grown into leading solution for processing large volumes of data. Often, in order to minimize the execution time, the developers express their programs using
procedural language instead of high-level query language. In such cases
one has full control over the program execution, what can lead to several
problems, especially when join operation is concerned. In the literature
the wide range of join techniques has been proposed, although many
of them cannot be easily classified using old Map-Side/Reduce-Side distinction. The main goal of this paper is to propose the taxonomy of the
existing join algorithms and provide their evaluation.

**Keywords:** Map-Reduce · Join algorithms · Hadoop · Taxonomy ·
Bigdata · Use cases

## 1 Introduction

Increasing popularity of the Map-Reduce programming model encourages usage
of this model to analyse huge sets of data. Unfortunately, unless the frameworks
like Pig [2] or Hive [1] are used, the lack of declarative query languages makes the
programmer responsible for the Map-Reduce program performance. Therefore,
optimization of certain operations, which are analogous to the relational algebra
operations, is a top priority in order to achieve the minimal execution time.
One of the most expensive operations is the join operation. Due to the fact
that often the technologies based on the Map-Reduce model are cloud-based
solutions, the optimization task has even more impact as the execution time
directly corresponds to the cloud utilization cost. In this paper, the overview of
the existing join algorithms is presented, as well as design patterns for the join
jobs to help the reader with choosing the best solution for own usages.

Due to the fact that among recent years plethora of Map Reduce join algorithms were designed and many of them seem to be similar and do not fall easily
into existing taxonomy, the authors wished to contribute as follows:

- we extend current taxonomy of the Map Reduce joins, search for them among
  various sources and categorize them using proposed taxonomy
- we discuss commonly used techniques, pointing their advantages and disadvantages and highlighting the most frequent pitfalls concerning the
  implementation as well as use cases

© Springer International Publishing Switzerland 2016
S. Kozielski et al. (Eds.): BDAS 2016, CCIS 613, pp. 192–203, 2016.
DOI: 10.1007/978-3-319-34099-9_14

## 2    Map-Reduce Model

The main concept behind Map-Reduce system was to provide an easy way to express queries to large volumes of unconstrained data, which could be effortlessly paralleled and easily partially recomputed [8,10]. From the developer's point of view, Map-Reduce programs are evaluated in two steps called Map and Reduce, which have following responsibilities:

**Map** - the step executed by system agents called Mappers during which input is processed step by step (which often means line by line), as partial results are decorated with reduce key and are emitted for further processing (preferably in Reduce step)

**Reduce** - the step executed after Map phase by system agents called Reducers during which all partial results with common reduce key are being aggregated to one final result

In addition to this, developer can define quasi-step named Combine. However, one should be cautious when implementing it as the decision about its invocation is made internally by cluster, thus Combine may not be performed at all.

## 3    Taxonomy

The complexity of joining several data files in Map-Reduce model is determined by two separate factors: how many data sources are joined together and the condition itself for joining data structure. Therefore, in terms of the number of files being joined together and the type of join condition we can distinguish four kinds of the problem:

**Two-way theta Join** - denoted $R \bowtie_\theta S$. Given two datasets $R$ and $S$ and the predicate $\theta$, the theta join is defined as a subset of cartesian product of these data sets: $R \bowtie S \subseteq R \times S$, such that for every tuple $t$ in this subset the predicate over the tuple $t$ is true: $\forall_{x \in R \bowtie S} \theta(x) = true$

**Two-way equi-join** - is a special case of two-way theta join, where the predicate $\theta$ is equality comparison between values of one or more attributes from input data sources.

**Multi-way theta-join** - when two or more data sets are joined using any condition

**Multi-way equi-join** - is a special case of multi-way theta join, when the join predicate is an equality comparison between values of one or more attributes from input data sources.

These definitions describe the class of joins called Inner joins. That means that if a record originating from one of the joined table do not fit any record in other table, it will not be visible in the final result. Such records are often referred to as *dangling tuples* and they are bound to exist if they originate from a data source which has little or no referential constraints. Special kind of join named (Full) Outer join expresses the result where dangling tuples are included from both data sources. Specialized versions of the Outer join operator exist named Left/Right Outer Join which add dangling tuples from only one data set.

## 3.1   Joins in Map-Reduce Model

In contemporary sources concerning Map-Reduce join algorithms one may come across something we may call *classical* taxonomy - the categorization based on where the actual join takes place [14,18]:

**Map side joins** - these group of algorithms emits the joined record as a Mappers' output. These often leads to preprocessing or forcing the locality of one of the file used for join in Mappers'.

**Reduce side joins** - these group of algorithms emits the record as a Reducers' output. The general concept is to partition and process the groups of records with a common join key, with a way of distinguishing the source of record, to avoid joining the two records originating from the same file.

This simple taxonomy poses two major drawbacks: (1) it does not capture other algorithms which perform the join before the Map/Reduce phase, (2) it suggests that Map-Side join should always be performed during the Map phase while in some cases it is advisable to defer the tuple reconstruction to the Reduce phase - if the grouping can be performed without the join.

As the names of the algorithms seem to have enrooted and little can be done about that - we propose the extended version of the taxonomy (Fig. 1) which amortize the first reason.

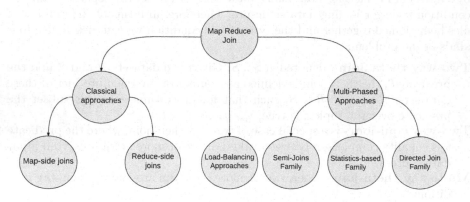

**Fig. 1.** The proposed taxonomy of Map Reduce join algorithms

The first level of our taxonomy contains a set of *Approaches*:

**Classical** which contains the Map/Reduce side taxonomy. These algorithms are mainly deterministic.

**Load-Balancing** which captures the family of multi-way theta-joins. Algorithms in this class are indeterministic.

**Multi-Phased** the techniques in this class require more than one Job to produce final output.

**Table 1.** Taxonomy of Joins in Map Reduce

| Name | Type | Also known as / Similar to | Reference |
|---|---|---|---|
| Repartition Join | Reduce Side Join | General Reduce Side Join, Standard Repartition Join | [5,9,14,16,18] |
| Improved Repartition Join | Reduce Side Join | Optimized Reduce Side Join, Optimized Repartition Join | [9] |
| Broadcast Join | Map Side Join | Map Side Join, Fragment-Duplicate Join, Fragment-Replicate Join, JDBM Join, Reverse Map Join | [5,7,9,13,16] |
| Reverse Map Join | Map Side Join | N/A | [13] |
| Map Reduce Cross Join | Map Side Join | N/A | [14] |
| MR Theta Join | Load-Balancing | 1-Bucket-Theta, Strict-Even Join | [12,15,19] |
| Semi-Join | Semi-Join Family | Per-Split Semi-Join, MR Join using Bloom Filter | [9,11,16] |
| SAND Join | Statistics-based Family | N/A | [4] |
| Directed Join | Directed Join Family | N/A | [7,9,14] |

In the table below (Table 1), we list the algorithms found in the literature and categorizing them into our taxonomy.

The following sections provides short overview of chosen join algorithms describing their common use cases.

## 4   Repartition Join

Repartition Join is concerned to be the most general approach for solving the joining problem, as it can be easily adapted to work with any number of different input files [7]. The idea of the algorithm is to tag the records with the information about their source, group them by the common joining key and finally compute the actual join. In the Fig. 2 one can observe the idea of the Repartition Join while the following sections cover the phases in detail.

### 4.1   Map Phase

Mappers responsibility is to identify the source of the record and extract the key. Therefore, if the job is configured so that every input is processed by the same

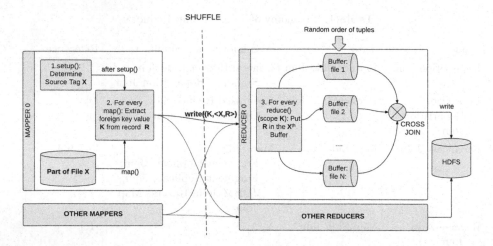

**Fig. 2.** The anatomy of N-Way repartition join

Mapper, some logic may be required to distinguish the record source. If it is not the case, and every distinct input has its own Mapper implementation, class name can be used as a source tag. Nevertheless, this approach may impose some overhead as the class names tend to be quite lengthy. As an emission key, the join attribute from the input record is used. The whole record with additional field containing the source tag is emitted as a value.

## 4.2   Reduce Phase

Reducers need to distinguish the records originating from different sources and produce the final output using any classic joining technique, preferably Nested-Loop Join. As in the current state the records from different files are most likely to come mixed up, we are forced to create as many buffers as the number of input files and defer the production of the final result till every record is buffered.

The main drawback of this solution is that if many records for the given key exist, buffering may lead to *OutOfMemoryException* (OOM), thus one may need to use the non-volatile memory to avoid raising such exception.

## 4.3   One-Shot Multi-Way Join

Although Repartition Join is not the fastest method to calculate join, its simplicity makes it a popular choice among the developers. In addition, its Reduce step can be easily generalized to handle Multi-Way join in one Job, by dynamically creating the buffers using Java *HashMap* collection. The precondition to do this is the fact that join has to be performed on the common key. Such situation is presented on Fig. 3 where one can notice the source *A* with a join attribute *id* is associated with sources *B1* and *B2* using one-to-many relation, which enables the developer to write one job that joins all sources at once.

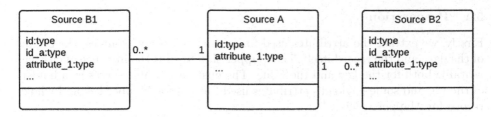

**Fig. 3.** UML diagram presenting the conditions to perform the join on multiple sources at once using repartition join

If the Standard Repartition Join suffered from OOM exception, the Multi-Way version of it does even more, as the greater number of buffers are required to be maintained.

## 5    Improved Repartition Join

As we previously stated, Repartition Join is recognized as a general solution to join problem in Map-Reduce environment. However, it is vulnerable to potential OutOfMemoryExceptions as the Reducers are forced to buffer all the incoming tuples.

The general idea behind Improved Repartition Join is to decrease the memory impact of buffering and enable producing the output as the records from the last data set are read. To minimize the impact on the memory, we wish not to buffer the largest data source. To achieve this, one is required to utilize many Hadoop mechanism, namely: *custom writables*, *Paritioners* and *secondary sorting*.

The execution of the Improved Repartition Join can be seen in the Fig. 4.

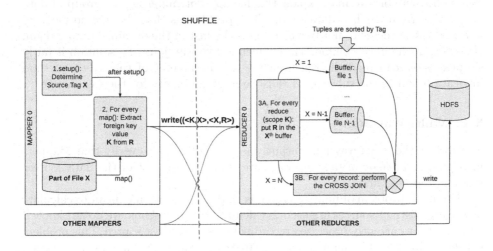

**Fig. 4.** The anatomy of N-Way improved repartition join

## 5.1  Preparation

Firstly, we extract the attributes used for joining the record sources, the origin of the data and the record itself. The Mapper should emit the instance of custom writable both for the key and the value. The custom writable serves as a holder of the tag and some payload - attributes used for join for Mapper key and whole record for Mapper value.

Secondly, the default behaviour of *Partitioner* should be overridden so Reducer is determined based on the hash value of source tag and not the hash value of grouping key.

Finally, we provide the custom Grouping Comparator which ensures that on the side of the Reduce records with same payload in the custom writable key are processed in single *reduce()* (the tag is ignored).

As we provide the prerequisites for Improved Repartition Join, we ensure that not only Reducers will posses every record for given attribute used for join, but also that the records will be processed in a predictable way.

## 5.2  Map Phase

Map phase for Improved Repartition Join only consists of retrieving the information about the attributes used for join and emitting the instance of custom writable which serves as a holder of the tag and some value - attributes used for join for Mapper key and whole record for Mapper value.

## 5.3  Reduce Phase

In the Reduce phase the data sorted by the join attribute and the source tag is obtained, therefore we may exploit this fact by not buffering the group of data originating from the last data source. The problem is that there is no mean to automate the process to notify which tag is the last in the obtained data stream. However, in the most common case, which is two-way join, the solution is fairly simple, as we can read the tag from the key parameter of the *reduce* method invocation and in the subsequent calls we may anticipate for the tag change.

## 5.4  Multi-Way Join

To produce the multi-way join output, the developer has to verify that Mappers incorporate the logic of tagging to ensure that the smallest of the data sources will be processed first.

The main problem of multi-way join is that there is no safe mean to calculate how many distinct sources are used to calculate the join. Thus, we do not know if the tag of the currently processed record belongs to the last source. This problem can be easily solved if the number of sources is counted in the job configuration process. Then, we are able to write the auxiliary value to the context and in the reduce phase use it to efficiently process the join.

# 6    Map-Side Joins

The techniques of Mapper side joins are exploiting the *DistributedCache* mechanism of Hadoop API to replicate data sources to the Mappers. This enables them to retrieve the sources as local files, load them into the memory and build the auxiliary hashtable. Finally, the join output is produced as the Mapper reads the input file record by record, which are joined using the built hashtable.

In the Fig. 5 one can observe the data flow in the Map Side Join.

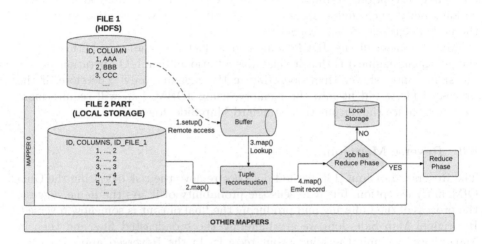

**Fig. 5.** The anatomy of 2-Way map side join

## 6.1    Broadcast Join

The most crucial part of the Broadcast join is the proper Job initialization. As we desire to distribute one of the joined sources to all Mappers, we must call proper method of the *DistributedCache* mechanism: *DistributedCache.addLocalFiles(Configuration, String)*. Doing so forces the Hadoop to replicate specified file to all Mappers. Therefore, the two-way Broadcast Join can be considered a single input file Job.

The next step of this approach is processing the file from the local storage to in-memory hash-based structure. This can be effectively implemented during Mapper's *setup(Context)* method, which serves as a constructor, as it is invoked once per Mapper's lifecycle.

After the local file is loaded and hashed in the main memory, one is required to deliver the implementation of the map function which will produce the join output.

Mapper side joins are perfect choice for performing the Star Join given the one source is served as input and rest of the sources are broadcasted via *DistributedCache* mechanism. Therefore, there is no need to perform the job chaining to produce the join output of more than two data sources. As we are able to locally read many different files, one can easily predict that the incorporation of the theta-join conditions is straightforward. What is more, this approach allows the developer to easily produce the cross product of few sources at once.

## 6.2 JDBM Join

The execution of standard Map Side Join is threatened by the raise of the Out-OfMemory exception. To remove this threat factor, one may use the Java transactional persistence engine library to enable building Java collections limited by the node persistent storage capacity.

Main problem of the JDBM join (and in fact every join exploiting any of the persistent engines) is that it generates additional disk I/O operations which are significantly slower than operating in the memory only. Therefore, if the processed file would fit into the memory, using JDBM Join will require more time to produce the output than standard Map Side Join.

## 6.3 Reverse Map Join

The Reverse Map Join [13] method does not remove the risk of raising the Out-OfMemory exception, but it lessens the probability of it. As the name suggest, the process of the buffering and producing the final output is somehow reversed. In original Map Side Join we are buffering the broadcasted files and then in Map phase, we emit the join output records. In the Reversed approach however, during the map phase the input data source is buffered and hashed and the *cleanup()* method serves as the phase of emitting the records, as the file obtained from Distributed cache is read line by line and compared.

This approach enables us to predict the amount of the memory required to buffer the records as the raw data should take up the size of the Mapper split, which in most cases is the same as the block in our Hadoop cluster. Another great advantage of this solution is the fact that the emission of the records are somehow deferred, thus the programmer is able to collect some statistics and control the shuffle process. What is even more beneficial is the fact that in the standard implementation of the Reverse Map Join during the buffering phase no disk write operations will occur. Moreover, it is guaranteed that during the write phase, no read operations will be invoked. In the standard Map Side Join, if the output buffer reaches the certain threshold it is spilled to the disk, therefore invocation of the map() may result not only in disk read operation but write also.

# 7    Directed Join

Multi-phase join algorithms are designed by chaining more than one Map-Reduce Job. Given the fact that algorithms producing the result in one Job exists, the question rises why one may want to use multi-phase approach. Often the developer posses some knowledge about the processed data which either makes it possible to achieve a shorter execution time or forces the preprocessing as the performance of the 'naive' Map/Reduce side joins may suffer [5].

The most known algorithms of this class are joins using the **bloom filter** which filters the dangling tuples. Similar approach, although using semi-join operation can be found in Map-Reduce Semi Join algorithm and its variation Per-Split Semi Join (both [9]). Also, one may consider the statistic collection as the phase of the algorithm - few such algorithms were proposed, for instance SAND [4].

In next section, we will dwell into the Directed Join - one of the first Multiphase solution. The approach itself was popular in the early versions of Hadoop not only as its implementation was available in the *contrib.utils.join* package but also as it was considered optimal solution for equi-join problem.

## 7.1   Idea

Not necessarily the algorithm itself, the Directed Join can be considered a set of conditions fulfilling which gives the developer the opportunity to minimize the amount of the data transferred over the network.

The idea behind the Directed Join is simple, as every distinct input file is split into a number of parts. The Job itself works on one file only, as the Mappers determine on which part of the file they are working and are remotely obtaining the adequate part of the second file. Therefore, one may notice that very strict conditions should be met to be able to use this method, as every file should have equal number of parts and same join attributes values sets should be found in the corresponding parts. These constraints are not so strict when one consider that in fact nearly every Hadoop Job fulfils them. That being said, the Directed Join is often preceded by specialized job which groups the data. This preprocessing is the reason why the Directed Join is considered multi-phase join. To partition the data correctly, the preprocessing job should:

– Use the same *setNumReduceTasks()* for every processed file, so that the number of the output files should be equal. In addition, the Reducer should be explicitly set, otherwise Mappers will write the output locally
– The data should be partitioned by the join attributes using the same *Partitioner* class

Additionally, one may use the comparator to sort the files by the values of the join attribute. If the data is sorted, the join can be processed in stream based fashion. If the files are not sorted, the Directed Join still can be implemented, although it will involve buffering.

## 7.2   Performance

As only the files which names contain the same part number are joined it can be easily observed that the number of the input file parts directly influence the performance of the Directed Join. Therefore, one should be aware of the following facts:

- The number of the partitions is limited by the cardinality of the values of the join attributes.
- If the parts are small enough (less than Hadoop *dfs. block. size* parameter), only one Mapper per part will be created, which guarantees that the remote parts are transferred over the network to one node only.

If the data are initially partitioned on the values of the foreign key, the Directed Join algorithm can be perceived as the optimal solution. In case of two-way join when every part is written on one HDFS block, every Mapper is required to perform the number of the remote reads equal to the number of the partitions.

The Directed Join resembles the Broadcast Join, with a notable difference that in the first algorithm only the specific part of the input file is remotely read, while in the latter one, the whole file is obtained.

## 8    Conclusion

Lack of knowledge of Map-Reduce join algorithms often forces the developers to perform the bound-to-be ineffective queries in frameworks like Pig [2] or Hive [1]. Awareness of the limitations of the existing algorithms can contribute to better decisions concerning method to be used. As at current state of art no mechanisms of collecting statistics in Map-Reduce were introduced - so the usefulness of cost-model are limited - the ability to arbitrary choose the suboptimal solution is invaluable.

This paper purpose is to group, categorize and provide the short overview of the plentiful of the Map-Reduce join algorithms which can be found in literature. In recent times, much effort was put to build cost models and optimize the existing join algorithms. The traces of these optimizations can be found in [3,16,17]. Also some progress was made with skew handling in Map-Reduce in general [6] as well as in join algorithms [4,5].

Many recent contributions were made to processing the theta-joins by subsetting the cross-join. Algorithms of this class, although break Map-Reduce programming model, are particullary attractive as they are resistant to any abnormal skews. The authors of this article are currently involved in developing optimization leveraging this process. The most known works on the field of theta-join in Map-Reduce are [12,15,19].

## References

1. Apache hive reference. https://hive.apache.org/
2. Apache pig reference. http://pig.apache.org/
3. Afrati, F.N., Ullman, J.D.: Optimizing joins in a map-reduce environment. In: Proceedings of the 13th International Conference on Extending Database Technology, pp. 99–110 (2010)

4. Atta, F., Viglas, S., Niazi, S.: SAND join - a skew handling join algorithm for google's mapreduce framework. In: 2011 IEEE 14th International Multitopic Conference (INMIC), pp. 170–175, December 2011
5. Atta, F.: Implementation and Analysis of Join Algorithms to handle skew for the Hadoop Map/Reduce Framework. Master's thesis, University of Edinburgh (2010)
6. Balazinska, M., Howe, B., Kwon, Y., Ren, K.: Managing skew in hadoop. IEEE Data Eng. Bull. **36**(1), 24–33 (2013)
7. Chandar, J.: Join Algorithms using Map/Reduce. Master's thesis, University of Edinburgh (2010)
8. Dean, J., Ghemawat, S.: MapReduce: simplified data processing on large clusters. Mag. Commun. ACM - 50th anniversary issue: 1958–2008 **51**(1), 107–113 (2008)
9. Ercegovac, V., Blanas, S.: A comparison of join algorithms for log processing in mapreduce. In: Proceedings of the 2010 ACM SIGMOD International Conference on Management of data, pp. 975–986 (2010)
10. Karloff, H., Suri, S., Vassilvitskii, S.: A model of computation for MapReduce, pp. 938–948 (2010)
11. Lee, T., Kim, K., Kim, H.J.: Join processing using bloom filter in mapreduce. In: Proceedings of the 2012 ACM Research in Applied Computation Symposium, RACS 2012, pp. 100–105. ACM, New York (2012). http://doi.acm.org/10.1145/2401603.2401626
12. Li, J., Wu, L., Zhang, C.: Optimizing theta-joins in a mapreduce environment. Int. J. Database Theory Appl. **6**, 91–108 (2013)
13. Luo, G., Dong, L.: Adaptive join plan generation in hadoop
14. Miner, D., Shook, A.: MapReduce Design Patterns. O'Reilly, Beijing (2013). http://opac.inria.fr/record=b1134500, dEBSZ
15. Okcan, A., Riedewald, M.: Processing theta-joins using mapreduce. In: Proceedings of the 2011 ACM SIGMOD International Conference on Management of Data, pp. 949–960 (2011)
16. Palla, K.: A Comparative Analysis of Join Algorithms Using the Hadoop Map/Reduce Framework. Master's thesis, University of Edinburgh (2009)
17. Pigul, A.: Generalized Parallel Join Algorithms and Designing Cost Models (2012)
18. White, T.: Hadoop: The Definitive Guide, chap. 8, 3rd edn. O'reilly, Sebastopol (2012)
19. Zhang, X., Chen, L., Wang, M.: Efficient multiway theta-join processing using mapreduce. In: Proceedings of the VLDB Endowment (PVLDB), vol. 5(11), pp. 1184–1195 (2012)

# The Design of the Efficient Theta-Join in Map-Reduce Environment

Maciej Penar[✉] and Artur Wilczek[✉]

Faculty of Computer Science and Management, Wroclaw University of Technology,
Wybrzeże Wyspiańskiego 27, 50-370 Wrocław, Poland
penar.maciej@gmail.com, artur.wilczek@pwr.edu.pl

**Abstract.** When analysing the data, the user often may want to perform the join between the input data sources. At first glance, in Map-Reduce programming model, the developer is limited only to equi-joins as they can be easily implemented using the grouping operation. However, some techniques have been developed to leverage the joins using non-equality conditions. In this paper, we propose the enhancement to cross-join based algorithms, like Strict-Even Join, by handling the equality and non-equality conditions separately.

**Keywords:** Map-Reduce · Hadoop · Join algorithms · Theta join · Inequality join · Big data

## 1 Introduction

In recent years Map-Reduce systems have proven their effectiveness to process large volumes of the data. The developers all over the world praise them for the simplicity of expressing *Jobs* - parallel programs run on the Map-Reduce cluster. Therefore, many companies have incorporated open-source version of Google's original Map-Reduce system called Hadoop [1] to leverage their analytic requirements. In such systems the data is often loaded from transactional databases without any pre-processing being done. Thus, the data originating from a single table can span across multiple files. This forces the Map-Reduce developer to express not only some computations leading to the desired final result, but also some logic to perform the join operation which ties together many separate files.

This becomes the good starting point for the Job optimization, as the join operation can be perceived as time-consuming. Whether or not the companies utilize their own cluster, it is always beneficial for Map-Reduce programs to achieve the lowest possible *execution time*. It is even more important when the cloud-based Map-Reduce clusters are concerned.

Some of the existing join algorithms could be in our opinion enhanced to ensure better performance. Our main contributions are as follow:

© Springer International Publishing Switzerland 2016
S. Kozielski et al. (Eds.): BDAS 2016, CCIS 613, pp. 204–215, 2016.
DOI: 10.1007/978-3-319-34099-9_15

- We introduce the concept of *pruning* - the enhancement to the existing Map Reduce theta join algorithms.
- We conduct the experiment which confirms the potential drop of the execution time of the join routine in Map Reduce.

The paper is organized as follows: Sect. 2 contains a short overview of Map-Reduce programming model. In Sect. 3 the idea of *Load Balancing Approach* is presented. Chapter 4 contains our contribution to the *Load Balancing Approach*. Experimental evaluation is presented in the Sect. 5. Last section is a short conclusion of the topics covered in this paper.

## 2   Joins in Map-Reduce

Map-Reduce programming model was designed to leverage and simplify the computations in the distributed environment consisting of the commodity hardware [8,11]. In fact, the philosophy of simplification in the Map-Reduce systems is often the subject of the controversy and is often criticized [9].

The idea of Map-Reduce programming model is as follows: the developer implements two (or in some cases only one) methods/functions used for the data processing - obligatory *map* function and the optional *reduce* function. In so-called *Map* phase, the system calls the *map* function for every record which has been read, so that key/value pair is emitted. If the *reduce* function was delivered, after the *Map* phase has finished, the intermediate *Shuffle* phase starts - which copies over the network the emitted key-value output to the nodes responsible for *Reduce* phase - this happens in such fashion that every record with a common key is placed on a single node. After that, the *Reduce* phase begins and the *reduce* function is applied for every set of the values with a common key. The detailed information about the Hadoop architecture and its anatomy can be found in [18].

Although many join algorithms for Map-Reduce environment were proposed, they always had limited application which varies for every algorithm. Their fundamental problem (which also is a problem in other Map-Reduce Jobs) is that the programming model can group data based only on the equality of the key.

Formally speaking, this makes only two types of join viable in Map-Reduce programming model:

**Equi-joins** - as they fit into equality based groupings.
**Band-joins** - as pointed in [15], they can be "naively" expressed if we assume that values of the foreign key are always non-negative integers. Basically, this means transforming the Inequality-join into the Equi-join.

Many methods for solving the equi-joins have been developed with notable examples of: Repartition Join, Improved Repartition Join, Bloom Filtering, Broadcast Join and etc. In general, join algorithms in Map-Reduce are divided into two categories Map-Side Joins, where tuple is reconstructed in Map phase, and Reduce-Side Joins, where tuple is reconstructed in Reduce phase. These algorithms can be found under numerous sources, for instance: [7,10,13,18].

The effectiveness of these algorithms have been deeply analysed which resulted in many cost models [3,16,17] backing up the Map-Reduce job execution.

More unorthodox approaches aim to amortize existing drawbacks of the Map-Reduce Jobs, especially data skews [5,6]. Among these, the SAND Join is worth mentioning [4] as well as joins using the technique called anti-combining [14].

## 3   Theta Join in Map-Reduce

Calculating the $\theta$-join in the plain Map-Reduce programming model can be considered a difficult task, as no explicit groups can be found at first sight. This is the obstacle which *Load Balancing* approach helps to overcome. The name refers to the fact that each Reducer in the cluster receives similar number of the records for every data source, thus same number of iterations are required to produce the subset of the final cross join. The output of every Reducer is disjunctive to each other, so their union is equivalent to cross product.

This approach can be found in [12,15,19]. The Reader should be aware that this algorithm can be found under many names like: *Strict-Even Theta*, *1-N Bucket Theta*, etc., but for the convenience, we will refer to it as the *Load-Balancing Approach*.

As the algorithm calculate the cross product, it is very convenient to perceive the $\theta$-join as the operation of subsetting on the cross-product given the input condition. That way, the two-way $\theta$-join can be expressed as $S_1 \bowtie_\theta S_2 \equiv \sigma_\theta(S_1 \times S_2)$ where $S$ means input source (i.e. a relation) and $\theta$ is a join condition.

### Idea

To demonstrate the idea of the *LoadBalancingApproach* we perfom the $N$-way theta-join on the data sources $S_1, ...S_N$ using the condition $\theta$. One of the parameters of the Map-Reduce Job which can be specified by user is the number of used Reducers $R$. Before the Job execution, one must establish the number of so-called *partitions* (or fragments [3]) for every input data source - this can be done arbitrarily or be incorporating the cost-based optimization (which will be discussed further). The number of the partitions $p_x$ for $x^{th}$ data source means that data set $S_x$ is divided into $p_x$ disjunctive parts, for example if $p_1 = 3$ then $S_1 = S_{1,1} \cup S_{1,2} \cup S_{1,3}$, so that $S_{1,1} \cap S_{1,2} = \emptyset$, $S_{1,2} \cap S_{1,3} = \emptyset$ and $S_{1,1} \cap S_{1,3} = \emptyset$.

One should be aware that the number of the partitions is constrained by the number of the Reducers, so that $R = \prod_{i=1}^{N} p_i$. For example, if two data source are joined, $R = 8$ and $p_1 = 4$, then $p_2 = 2$.

Every Reducer is assigned to process only a single part of every data source in such way that no other Reducer processes the exact same combination of parts. Then, Reducer uses the received parts to produce the Cross Product of them. Thus, if some Reducer was set up to receive the parts $S_{1,3}, S_{2,2}, ..., S_{N,1}$, the $S_{1,3} \times S_{2,2} \times ... \times S_{N,1}$ will be calculated and no other Reducer will process the same combination. This can be presented graphically as in the Fig. 1.

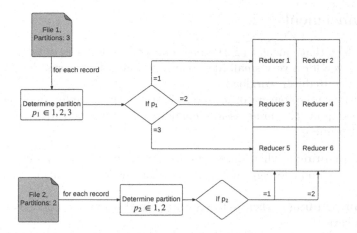

**Fig. 1.** Idea of load-balancing join algorithms for two files with 3 and 2 partitions

Given the facts that: (1) single combination of parts enables to produce subset of the cross-join, (2) every Reducer is assigned the unique partition combination and (3) every combination of parts has been assigned to Reducer, thus globally all Reducers perform the Cartesian product.

Two main disadvantages of such approach exist. Firstly, the shuffle phase is more expensive compared to classical Map-Side or Reduce-Side Joins as each record should be replicated to a number of the Reducers which increases the complexity of Shuffle phase. This can be easily measured as the records from the $i^{th}$ source are required to be sent to exactly $\frac{R}{p_i}$ Reducers. Secondly, the algorithm is problem-independent as the developer actually solves the even data distribution which, from some point of view, can be perceived as rewriting the functionality of *Partitioner* and *Mapper* classes.

## Cost-Based Optimization

The Load Balancing Approach is often considered in terms of cost-based optimization as one can easily predict the impact of different combinations of partitioning on the Shuffle phase. The most common cost model ties the size of the input file, which can be denoted by $F_i$ - for the $i^{th}$ input file, with the partitioning. The expression is presented below.

$$f(p) = R \sum_{i=1}^{N} \frac{F_i}{p_i} \tag{1}$$

The idea of this cost model is based on the assumption that every single Reducer will receive same amount of data originating from the same file. The problem is that if big input files are concerned, then Shuffle phase may be dominated by Reduce phase cross-join calculation which this model does not take into the account [15].

## 4    Enhancement

Given the fact that calculating the all-to-all comparisons is difficult and time consuming even for a very small files, our contribution concerns the methods of decreasing the Reducer workload.

We introduce the concept of *pruning* - lowering the complexity of the Reduce phase by skipping the unnecessary comparisons. One may distinguish three methods of pruning:

**Equi-join pruning** - when data sets are joined using the equality condition between common foreign keys (the foreign key referencing the same data source)

**Theta-join pruning** - when data sets are joined using any condition between foreign keys

**Hybrid pruning** - method combining the Equi-join pruning and Theta-join pruning - when data sets are joined using many conditions, one of which is equality between the common foreign keys and second can be any condition

Pruning introduces more invocations of the *reduce* method in the Load Balancing Approach program, as originally every Reducer executes *reduce* only once. More invocations often lower the requirement for the memory, as the scope of some of the buffered records is limited to a single *reduce* which is shorter than in the non-pruned version of the algorithm.

To describe our solution, let us introduce the following notation:

$R$ - index of the Reducer where the record should be sent
$F$ - index of the input file
*Equi* - foreign key values used for equi-join
*Theta* - foreign key values used for theta-join

### Equi-join Pruning

This method requires appending the information about the foreign key during the Map phase. Thus, the record is emitted using the $(R, Equi)$ key. The buffering of the last data set can be avoided by changing the key to $(R, Equi, F)$ and using the additional Comparator forcing the *reduce* on each of the $(R, Equi)$(as in the Improved Repartition Join).

This method does not differ much from the Repartition Join algorithm, apart from the fact that it is resistant to abnormal foreign key distributions in expense of more complex shuffle phase.

### Theta-join Pruning

Pruning based on theta-join condition is more complicated as no explicit grouping augmenting the join itself exists. As no assumptions can be made which records should be buffered in the memory, in the end one is forced to buffer every record. Therefore, the philosophy of the theta-join pruning differs from the previous method as the *reduce* invocation optimize the process of buffering.

The motivation is simple: many records with same foreign key values may exist, thus one may check if the condition is satisfied only on the distinct foreign key values.

To implement this solution, one is required to emit the records with a following key $(R, Theta, F)$. Then in Java implementation one can use array of Multimaps to handle the buffer: $Multimap < String, String > []\ buffer$, such that $buffer[i]$ is the record buffer for $i^{th}$ data source and the records lists in the buffer are associated with the foreign keys.

In this manner, the join can be computed either in the end of the *reduce* invocation or in the *cleanup* method - in both cases every record has to be buffered. It is worth mentioning though, that during *cleanup* Hadoop will not update the Job progress, which may be inconvenient for the user.

### Hybrid Pruning

The last optimization incorporates both previously presented so that the scope of buffering is limited only to the invocation of *reduce* - the equality conditions force the groupings - and conditions are checked only on distinct values of foreign keys used for theta join.

The method is complex, as it requires emitting $(R, Equi, Theta)$ and implementing the Grouping Comparator on $(R, Equi)$. Additionally, it may be advisable to attach redundant information about the theta-foreign key to the record - if one does not want to dynamically retrieve this information on the side of the Reducer.

Also, hybrid-pruning can be used to exploit the equi-join conditions when not all of the data sources have common foreign key. In such case, some of the records shall be emitted under the $(R, null, Theta)$ key which should be processed at the first *reduce* invocation. These records should be buffered throughout the whole Reduce phase.

Unfortunately, it is difficult to build the cost model for this method as theta join conditions are evaluated inside the groupings - such selectivity cannot be easily measured without any statistics.

## 5    Experimental Evaluation

The data which was used in the research was generated using the TPC-H data generation tool DBGen. Default behaviour of this tool forces it to generate the data which values are drawn from the uniform distribution. In all experiments, the two-way join is calculated between the two biggest files generated by TPC-H benchmark: *orders.tbl* and *lineitem.tbl*. Originally these tables are tied by one-to-many relationship. As for some of the queries we require compound keys, and the TPC-H model does not have them, we create artificial ones by generating the extra columns based on dates (for query no.3) and by rewriting the values of primary key. Additionally, as the theta-pruning approach efficiency depends on (1) the skewed data and (2) many-to-many relationship between data sources, the original primary key columns have been transformed to present the optimistic and pessimistic use cases.

As a result, we may distinguish four data sets in our experiment: 1-N (one-to-many relationship between data sets), 1-1 (one-to-one relationship between data sets), Zipf-0.5 (foreign keys drawn from the Zipf distribution with exponent equal to 0.5) and Zipf-1(foreign keys drawn from the Zipf distribution with exponent equal to 1).

The test was performed using the Amazon Elastic MapReduce cluster. The master node was running on m1.large instance while the four worker nodes were running on m1.medium. The detailed specification of the Amazon Elastic Compute Cloud instances can be found on the Amazon site [2]. All nodes were running on Hadoop 2.4.0 with the default configuration (The configuration for the worker node can be seen in the Table 1).

**Table 1.** Configuration of amazon m1.medium node

| Type | General purpose |
|------|-----------------|
| Storage space | 410 GB |
| Cores | 1 |
| Memory | 3.75 GB |
| Network performance | Moderate |

## 5.1 Assumptions

The experiment measuring the performance of the Load Balancing approach was designed with the following assumptions:

1. The following approaches were compared: joins using the Hive framework, joins using the Pig framework, Standard Cross Join, Load Balancing Approach with the equi-pruning and the hybrid-pruning
2. In case of the queries (1–3) the Scale Factor was set to 1 ($6 * 10^6$ records versus $1,5 * 10^6$ records)
3. In case of the query 4 the Scale Factor was set to 0.01 due to the compute intensive nature of the all-to-all comparisons
4. Every test case was repeated three times and the average of the measured execution time was calculated
5. The number of the Reducers and the partitioning were established using the optimizer, with the maximum number of the Reducers was 4
6. The Standard Cross Join, Pig queries and Improved Repartition Join had the number of the Reducers set to 4
7. The single test case was interrupted after 15 min

Additionally, the case of the equi-join was compared with the Improved Repartition Join and Standard Cross Join implementation was based on *MapReduce Design Patterns* [13].

## 5.2   Evaluation

In the chart for the query no. 1 (Fig. 2) we may observe that Improved Repartition Join algorithm topped in almost every case except from the data set $1 - 1$. This may be contrary to ones expectations as the data skew in data sets $Zipf - 0.5$ and $Zipf - 1$ should force some Reducers to finish up later. The problem of achieving such observation is that the data skew may be amortized by the good hashing function. In every case of equi-join, our solution, which is Load Balancing approach with hybrid-prune, finished as a runner-up. This is interesting phenomenon as both in equi-prune and hybrid-prune approaches no inequality condition was specified, thus one may think that the execution time of these approaches would be the same - or in favour of equi-prune.

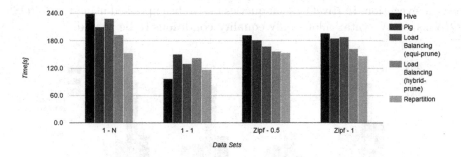

**Fig. 2.** The execution time for the query no. 1: equality condition

The case of query no. 2 (Fig. 3) proved that the hybrid-pruning imposes some overhead which may slow the execution time. This can be observed for the data set $Zipf - 1$ where evaluating the inequality conditions on the attributes with skewed values distribution does not provide any boost in the execution time. What is more, the standard Load Balancing approach with equi-prune achieves better results, although the differences in execution time between hybrid and equi pruning are marginal. Nonetheless, due to the more complex implementation of hybrid-pruning technique, the standard approach can be considered better.

Next case does not vary much from the second query (Fig. 4). The main difference is that the inequality condition is expressed using the User Defined Function (UDF) - comparing if two dates are within the same month. This implies that cost of checking it is much higher than in the previous cases, therefore we expect that our method which evaluates the condition only for unique combinations of foreign keys will achieve the best results.

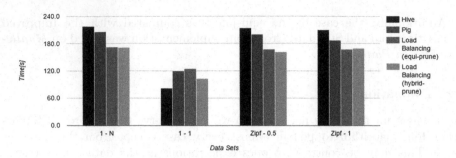

**Fig. 3.** The execution time for the query no. 2: equality + inequality condition

The results showed the poor performance of the Pig as opposed to other solutions. Again our method was the best in cases when data sets $1 - N$, $Zipf - 0.5$ and $Zipf - 1$ were processed.

The last query is expressed using inequality conditions only and does not fit well in the Map-Reduce programming model as the data cannot be grouped to ease the execution. In HiveQL such queries has to be expressed using the *CROSS JOIN* as *JOIN* syntax allows only equality conditions to be passed.

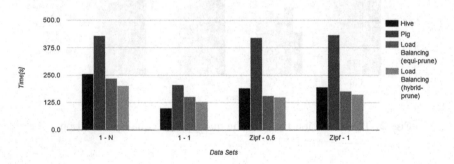

**Fig. 4.** The execution time for the query no. 3: equality + inequality condition based on User Defined Function

It should be noted that queries with only inequality conditions are the main reason behind the Load Balancing algorithms, thus this is the case where these approaches are expected to shine. Firstly, as the complexity of such queries tend to grow exponentially we run the test on a small volume with Scale Factor equal to 0.01. Therefore, our data sets contains $6 * 10^4$ records and $1,5 * 10^4$ records and requires performing $9 * 10^8$ iterations in total. The results for the aforementioned volume is presented in the Fig. 5.

Firstly, one should note that in any case the Standard Cross Join had not finished the execution. The same can be said about the Pig framework, as only in case of processing the $1 - 1$ data set it managed to finish the execution. Although the Hive finished in every scenario, its performance is poor comparing

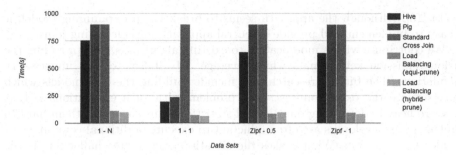

**Fig. 5.** The execution time for the query no. 4: inequality join (data volume 0.01)

to Load Balancing approaches and comparing these is difficult due to the scale imposed by the other algorithms. Nonetheless, one can see that query run on the $Zipf - 0.5$ data set was calculated faster when using the equi-pruning technique. In order to examine the Load Balancing approaches more deeply, we increased the data volume to 0.025, thus the problem on the side of the Reducer scaled by the factor of 6.25, The evaluation of this case for the Load Balancing algorithms is presented in the Fig. 6.

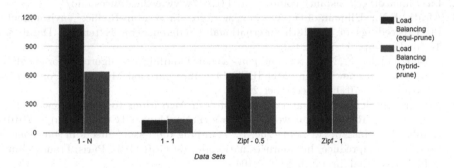

**Fig. 6.** The execution time for the query no. 4: inequality join (data volume 0.025)

The results of this experiment show that Load Balancing algorithm with hybrid pruning achieves better results than its counterpart without any pruning (as in this case equi-pruning had no effect) when processing joins on the data sources in which the values of the foreign keys reappear.

## 6   Conclusion

Many research was conducted concerning join in the Map-Reduce programming model. Recently, much effort was put into solving the inequality joins. This resulted in finding out the algorithms like Strict-Even Join [12] or 1-Bucket

Theta [15]. Although the approach seems to break the programming model, it is easy to implement and provide a general approach for calculating joins.

We contribute with a modification to existing algorithms, also providing the various emission key formats which leverage Job ability to skip unnecessary calculations. The further research may include building the cost models which will estimate the execution cost. The problem with such estimations is that they are heavily depending on collected statistics. This can make them hard to implement as developers are often reluctant to execute additional computations beside the main Job. What makes the situation even more challenging is the fact that data processed by Map-Reduce systems are often sorted, thus simple remote streaming of top $N$ records can result in the far-fetched estimations.

As our research was conducted on a relatively small cluster and data volume, additional test could be run using the real-world queries to back up obtained results. Other issue which can be tackled is extending the cost models and algorithms for other relational algebra operators - as operators may influence each other, combining them may require new cost models.

# References

1. Apache hadoop reference. http://hadoop.apache.org/
2. Easy amazon ec2 instance comparison. http://www.ec2instances.info/
3. Afrati, F.N., Ullman, J.D.: Optimizing joins in a Map-Reduce environment. In: Proceedings of the 13th International Conference on Extending Database Technology, pp. 99–110 (2010)
4. Atta, F., Viglas, S., Niazi, S.: Sand join - a skew handling join algorithm for google's MapReduce framework. In: 2011 IEEE 14th International Multitopic Conference (INMIC), pp. 170–175, December 2011
5. Atta, F.: Implementation and Analysis of Join Algorithms to handle skew for the Hadoop Map/Reduce Framework. Master's thesis, University of Edinburgh (2010)
6. Bamha, M., Hassan, A., Loulergue, F.: Handling data-skew effects in join operations using mapreduce. In: Journées nationales du GdR GPL, Paris, France, June 2014. https://hal.inria.fr/hal-00979104
7. Chandar, J.: Join Algorithms using Map/Reduce. Master's thesis, University of Edinburgh (2010)
8. Dean, J., Ghemawat, S.: MapReduce: Simplified data processing on large clusters. Mag. Commun. ACM - 50th anniversary issue: 1958–2008 **51**, 107–113 (2008)
9. Dewitt, D., Stonebraker, M.: Map-Reduce: A major step backwards. http://homes. cs.washington.edu/~billhowe/mapreduce_a_major_step_backwards.html
10. Ercegovac, V., Blanas, S.: A Comparison of join algorithms for log processing in MapReduce. In: Proceedings of the 2010 ACM SIGMOD International Conference on Management of data, pp. 975–986 (2010)
11. Karloff, H., Suri, S., Vassilvitskii, S.: A model of computation for MapReduce. pp. 938–948 (2010)
12. Li, J., Wu, L., Zhang, C.: Optimizing theta-joins in a MapReduce environment. Int. J. Database Theory Appl. **6**, 91–108 (2013)
13. Miner, D., Shook, A.: MapReduce Design Patterns. Building Effective Algorithms and Analytics for Hadoop and Other Systems. O'Reilly, Beijing (2013). http://opac.inria.fr/record=b1134500, dEBSZ

14. Okcan, A., Riedewald, M.: Anti-combining for mapreduce. In: Proceedings of the 2014 ACM SIGMOD International Conference on Management of Data, SIGMOD 2014, pp. 839–850. ACM, New York (2014). http://doi.acm.org/10.1145/2588555. 2610499
15. Okcan, A., Riedewald, M.: Processing theta-joins using MapReduce. In: Proceedings of the 2011 ACM SIGMOD International Conference on Management of data, pp. 949–960 (2011)
16. Palla, K.: A Comparative Analysis of Join Algorithms Using the Hadoop Map/Reduce Framework. Master's thesis, University of Edinburgh (2009)
17. Pigul, A.: Generalized Parallel Join Algorithms and Designing Cost Models (2012)
18. White, T.: Hadoop: The Definitive Guide, chap. 8, 3rd edn. O'Reilly, Sebastopol (2012)
19. Zhang, X., Chen, L., Wang, M.: Efficient multiway theta-join processing using MapReduce. In: Proceedings of the VLDB Endowment (PVLDB). 11, vol. 5, pp. 1184–1195 (2012)

# Non-recursive Approach for Sort-Merge Join Operation

Norah Asiri[(⊠)] and Rasha Alsulim

Computer and Information Sciences College,
King Saud University, Riyadh, Saudi Arabia
asiri.norah.m@gmail.com, r.m.z1433@hotmail.com

**Abstract.** Several algorithms have been developed over the years to perform join operation which is executed frequently and affects the efficiency of the database system. Some of these efforts prove that join performance mainly depends on the sequences of execution of relations in addition to the hardware architecture. In this paper, we present a method that processes a many-to-many multi join operation by using a non-recursive reverse polish notation tree for sort-merge join. Precisely, this paper sheds more light on main memory join operation of two types of sort-merge join sequences: sequential join sequences (linear tree) and general join sequences (wide bushy tree, also known as composite inner) and also tests their performance and functionality. We will also provide the algorithm of the proposed system that shows the implementation steps.

**Keywords:** Join operation · Bushy tree · Sequential tree · Multi-join query · RPN · Concurrent operations

## 1 Introduction

Database query operations facilitate the ease of information retrieval from one or more relations. However, binary join operation is one of the most challenging operations to be implemented efficiently. It is the only relational algebra operation that combines the related tuples from different relations among different attributes schemes [10].

The improvement of the performance of any database system necessitates improvement of the frequently executed operations such as join because it depends on transferring and moving data to/from the main memory [10]. Thus, the optimization for this process should be offered to reduce its expenses and to improve its functionality. The join operation which is denoted by ⋈ is used to combine related tuples from two relations into a single longer tuple [11].

The join operation over two datasets R and S with binary predicate t and attributes a and b is given as:

$$R \bowtie S = \{t | t = rs \wedge r \in R \wedge s \in S \wedge t(a) = t(b)\} \tag{1}$$

© Springer International Publishing Switzerland 2016
S. Kozielski et al. (Eds.): BDAS 2016, CCIS 613, pp. 216–224, 2016.
DOI: 10.1007/978-3-319-34099-9_16

There are many types of relationships among relations, such as one-to-one, one-to-many and many-to-many. A many-to-many relationship refers to a relationship between tables in a database when a parent row in one table contains several child rows in the second table and vice versa. Currently, many-to-many relationship is usually a mirror of the real-life relationship between the objects that the two tables represent. In this paper, we investigate sort-merge with multi-join queries and try to optimize the join process and to achieve efficient execution [5,13].

The rest of this paper is organized as follows: in Sect. 2, background of join strategies. In Sect. 3 methodology that are used in our approach. Our experiment and results are given in Sect. 4. Finally, conclusions from our findings are presented in Sect. 5.

## 2   Background

### 2.1   Join Strategies

The strategies used to perform join operations are discussed in this section, were some are implemented while others are just proposed. The debate over which are the comparative performance join algorithms of these approaches has been going on for decades. The main reason behind discussing these methods is to identify which one is the best and most applicable with our suggested system.

1. **Simple Nested-Loops:** is considered to be the simplest form of join. It starts from the inner loop which is designated to the inner relation and the other loop which is tied to the outer relation. For each tuple in the outer relation, all tuples in the inner relation are scanned and compared with the current tuple in the outer relation. In case of matching the condition of join, the two tuples are joined and positioned in the outer buffer [10]. In practical view, nested-loop join is performed as a nested-block join, because the tuples are retrieved in form of blocks rather than individual tuples [11].
   In this algorithm, the total amount of reduction in I/O activity depends on the size of the available main memory; it is noticed that each tuple of the inner relation is compared with every tuple of the outer relation. In this way, the execution of this algorithm requires $O(n \times m)$ time for joins execution.
2. **Nested-Loops Join with Rocking:** to optimize the simple nested-loops join method, we can use an extra step to ensure it works with more efficiency. This step is called: rocking the inner relation [8]. Rocking the inner relation means to read the inner relation from top-most to bottom for one tuple only of the outer relation and from bottom to top for the next tuple. Consequently, the I/O overhead is reduced since the last returned page of the inner relation in the current loop is also used in the next loop.
   Granting all this, the exhaustive matching strategies and the poor efficiency in both simple and rocking nested-loops join makes it inappropriate for joining large relations.

3. **Hash Join Methods:** a large number of join processes uses the hash methods. With the simple hash join methods, the join attribute(s) values at the inner relation are hashed using hash function to create a hash table with keys of inner relation and then sort them. After that the partitioning phase carried out by looping over the tuples of outer relation. For each key in the outer relation, search for matching keys in the hash table and attach the matching tuple(s) to the output. Its search execution is $O(1)$ which outperforms other join methods in some cases and makes it an accepted solution to researchers. However, it has some disadvantages such as the duplicate keys in inner relation which causes conflict in the hash table. Also, there are extra irrelevant comparisons resulting from the size of the hash table. Moreover, it has a limitation due to current CPU architecture [7].

4. **Sort-Merge Join:** relies on sorting the rows in both input tables by the join key and merges these relations. Sort-merge performance mainly depends on choosing an efficient sorting implementation where more than 98 % of sort-merge method costs lies [7]. The investigative studies show that hash join needs at a least 1.5X more memory bandwidth than sort merge join [7].
   Furthermore, authors in [3] proves that in multi-core databases and modern multi-core servers, sort-based join beats hash-based parallel join algorithms and mostly has a linear relationship with the number of cores [3]. For the future computer system architectures, if the growth gap between compute and bandwidth continues to expand, the sort-merge join will be more effective than hash join [7].

## 2.2   Multi-join Queries (Linear and Bushy Trees for Many-to-Many Multi-join Queries)

Investigators have argued and worked on the implementation and the performance of parallel and concurrent DBMS. To optimize a multi join query response time, the exploitation of concurrency by using trees is used. Hence, any differ-

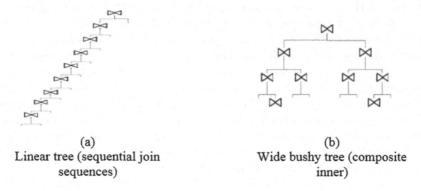

(a)
Linear tree (sequential join
sequences)

(b)
Wide bushy tree (composite
inner)

**Fig. 1.** Linear and wide bushy trees multi-join sequences shapes

ences in response time result from the differences in the shape of the tree [14]. This paper presents analytical experiments of two types of join sequences [12].

The first type is the linear tree or sequential join sequence, in which the resulting relation of an intermediate join can only be used in the next join. For instance, in Fig. 1(a) where every non-leaf node (internal node) denotes the resulting relation from joining its child nodes. The second type is known as general join sequence (wide bushy tree, also known as composite inner) [6] in which the current resulting relation of a join is not mandatory to be only used in the next join as shown in Fig. 1(b).

## 3 Methodology

Most references present tree traversal using recursion only. In [2], the literature survey shows that most references only indicate the implementations of the recursive algorithms, and only few references address the issue of non-recursive algorithms. In our investigation, we use a non-recursive algorithm that is simple, efficient which depends on a stack and a post-order binary tree traversal. We dynamically allocate the binary trees elements in a way that each element (node) has at most two potential successors.

We cover two kinds of binary trees: the wide bushy tree and the linear tree. Particular multi-join expressions are applied to the tree because all of the join operations are binary. It is also possible for a node to have only one child; as in the case with the linear tree. An expression tree can be evaluated by applying the join operators at the root and the values obtained by evaluating the left and right sub trees that contain the relations keys. We evaluate each sub tree individually with postfix traversal by using reverse polish notation (RPN). The reverse polish notation (RPN) is a well-known method for the expression notification in a postfix way compared with the typical infix notation [9].

When comparing the reverse polish notation with algebraic notation, RPN has been found to achieve faster calculations [1]. Based on that, we will convert the multi-join queries into RPN expression in a postfix bottom-up manner instead of using the normal recursive multi-queries form. We construct the binary tree from the multi-join queries expression. Then, we use RPN to traverse it depending on the operand of the tree. Table 1 presents an example of applying the RPN to SQL expression [4].

Figures 2 and 3 show an example of a wide bushy tree and a linear tree of multi-join queries that uses RPN stacks. They present multi-join queries of

**Table 1.** Example of infix and RPN expressions [4].

| Infix expression | ((Department = 'Dep1')or(Department like 'Dep2 %')) and ((Title = 't1') or (Title like 't2 %')) |
|---|---|
| RPN expression | (Department = 'Dep1') (Department,like 'Dep2 %') or (Title = 't1') (Title like 't2 %') or and |

six relations. At the beginning, we construct the binary trees where the leaves contain the operands of multi-join queries which are the relations keys. Join operation by default should contain two operands. Hence, the tree should at least contain two leaves and dynamically expands and shrinks depending on the number of join operations.

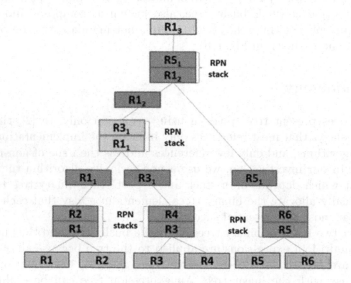

**Fig. 2.** General (bushy) join sequence.

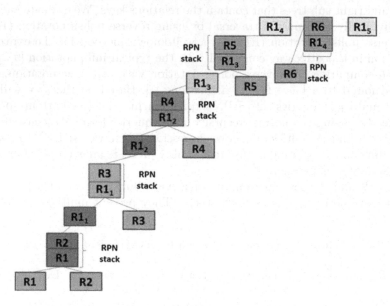

**Fig. 3.** Sequential join sequence.

This solution has a minimum space complexity of $O(n)$, where n is the number of nodes of the tree. Figure 2 presents an example of wide bushy tree of six relations R1, R2, R3, R4, R5 and R6. The final join operation between $R1_2$ and $R5_1$ would be as following:

$$R1_3 \leftarrow (R1_2, R5_1) \bowtie \equiv ((((R1, R2) \bowtie)((R3, R4)) \bowtie)((R5, R6) \bowtie) \bowtie) \quad (2)$$

In linear tree as shown in Fig. 3, we can notice that there is an extra joined relation $R1_5$ that needs one more join operation, which means more processing. The final joined relation can be obtained by:

$$R1_5 \leftarrow (R1_4, R6) \bowtie \equiv ((((((R1, R2) \bowtie)(R3) \bowtie)(R4) \bowtie)(R5) \bowtie)(R6) \bowtie) \quad (3)$$

Algorithm 1 presents the proposed technique by using subtree that adopts RPN.

---

**Algorithm 1.** Multi-join subtree that depends on RPN expression.

---

While nodes not empty

- **Read** $lc$ **from** *subtree*
- **Push** $lc \rightarrow st$
- **Read** $rc$ of $lc$ **from** *subtree*
- **Push** $rc \rightarrow st$
- **If** root is join operator $\bowtie$
  - **If** $st.lenght < 2$ and $\neg$ IsNull(*subtree*.rightChild) **Then**
    **Report** an error

  - **Else pop** $lc$, **pop** $rc$ **from** $st$
    jbt.leftChild $=(R_{lc}, R_{rc}) \bowtie {}_{R_{lc.id}=R_{rc.id}}$

---

## 4    Performance Evaluation

We conducted experiments on the proposed algorithm for concurrent join and compared it with linear sequence. We applied equi-joins which depends on equality (matching column values) where the primary keys of the tables are generated randomly from a predefined integer range. A two-phase strategy for multi-join queries is proposed in this paper.

The first phase studies a simple and cheap join algorithm, which is quicksort algorithm. Quicksort is a sorting-in place technique that applies divide and conquer algorithm. The advantage of this technique is that it is remarkably efficient on the average and it outperforms other sort-merge algorithms.

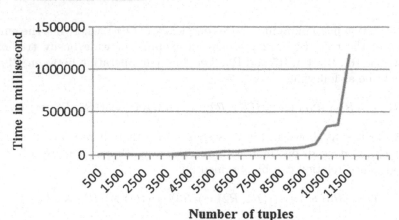

**Fig. 4.** Execution time of join operation using RPN tree regarding number of tuples.

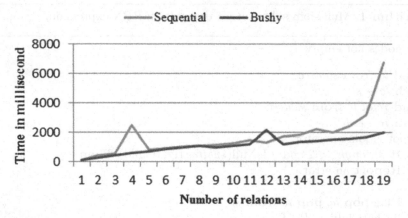

**Fig. 5.** Execution time of join operation using sequential and bushy trees regarding number of relations.

The second phase applies reverse polish notation for reading the tree in bushy or linear manner. A relation, $R_i$, with $\|R_i\|$ records, that is populated with integer values have been used in the system. The system used for our evaluation was equipped with concurrent multi-threaded processor with 3 MG cache and 6 GB of system memory.

Figures 4 and 5 present the performance of executing the sequential and general trees regarding the number of tuples and tables. Figure 4 shows the performance of quicksort with RPN tree with 2 relations. Figure 5 shows the execution time of 100 tuples when the number of relations is increased. The bushy trees would be a better choice with larger number of tuples and relations. We notice that bushy trees still outperform the sequential trees.

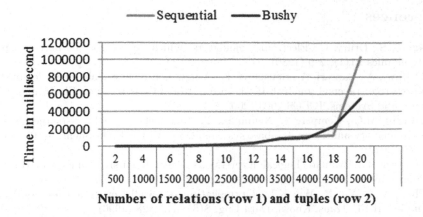

**Fig. 6.** Execution time of join operation using sequential and bushy trees regarding number of relations and tuples.

**Table 2.** Other experiments on performance of sequential and bushy trees.

| # Tuples | # Relations | Time in millisecond | | # Tuples | # Relations | Time in millisecond | |
|---|---|---|---|---|---|---|---|
| | | Sequential | Bushy | | | Sequential | Bushy |
| 300 | 4 | 916 | 2071 | 900 | 10 | 5513 | 10808 |
| 500 | 6 | 1299 | 3520 | 1000 | 11 | 7108 | 13981 |
| 600 | 7 | 2101 | 3975 | 1100 | 12 | 5712 | 14488 |
| 700 | 8 | 2347 | 4221 | 1200 | 13 | 19916 | 27311 |
| 800 | 9 | 3372 | 9801 | 1400 | 15 | 19992 | 24937 |

Figure 6 also presents the execution time when increasing both tuples and relations. However, sequential could beat the bushy trees in case of having a small number of tuples and relations as shown in Table 2.

## 5 Conclusions

Join operation is still a vital step in most DBMS and the cost of queries is highly affected by this operation, hence, optimizing join operations leads to enhancing the DBMS. This paper presents a new methodology for sort-merge join by using RPN tree to perform multi join. We consider two factors in evaluating the performance: the number of tuples and the number of relations. Our approach was tested on both sequential and general trees. We concluded that the general tree outperforms the sequential tree in case of having a huge number of relations and tuples. In our future work, we plan to run the system on multi-core processors environment and also plan to expand this methodology to other queries operations.

**Acknowledgments.** We would like to thank Nada Alzahrani for her efforts during the progress of this research.

# References

1. Agate, S., Drury, C.: Electronic calculators: which notation is the better? Appl. ergonomics **11**(1), 2–6 (1980)
2. Al-Rawi, A., Lansari, A., Bouslama, F.: A new non-recursive algorithm for binary search tree traversal. In: 10th IEEE International Conference on Electronics, Circuits and Systems, ICECS 2003 (2003)
3. Albutiu, M.C., Kemper, A., Neumann, T.: Massively parallel sort-merge joins in main memory multi-core database systems. Proc. VLDB Endowment. **5**(10), 1064–1075 (2012)
4. Capasso, T.: Evaluate Logical Expressions Using Recursive CTEs and Reverse Polish Notation. Penton Business Media, Inc (2014)
5. Chen, M.S., Yu, P., Wu, K.L.: Optimization of parallel execution for multi-join queries. IEEE Trans. Knowl. Data Eng. **8**(3), 416–428 (1996)
6. Graefe, G.: Rule-based query optimization in extensible database systems (1987)
7. Kim, C., et al.: Sort vs. hash revisited: fast join implementation on modern multi-core cpus. Proc. VLDB Endowment **2**(2), 1378–1389 (2009)
8. Kim, W.: A new way to compute the product and join of relations. In: Proceedings of the 1980 ACM SIGMOD International Conference on Management of Data (1980)
9. Krtolica, P., Stanimirovi, P.: Reverse polish notation method. Int. J. Comput. Math. **81**(3), 273–284 (2004)
10. Mishra, P., Eich, M.: Join processing in relational databases. ACM Comput. Surv. (CSUR) **24**(1), 63–113 (1992)
11. Navathe, S., Elmasri, R.: Fundamentals of Database Systems, pp. 652–660. Pearson Education, Upper Saddle River (2010)
12. Ono, K., Lohman, G.: Measuring the complexity of join enumeration in query optimization. In: VLDB (1990)
13. Taniar, D., Tan, R.B.N.: Parallel processing of multi-join expansion-aggregate data cube query in high performance database systems. In: International Symposium on Parallel Architectures, Algorithms and Networks, I-SPAN 2002 (2002)
14. Wilschut, A., Flokstra, J., Apers, P.: Parallel evaluation of multi-join queries. In: ACM SIGMOD Record (1995)

# Estimating Costs of Materialization Methods for SQL:1999 Recursive Queries

Aleksandra Boniewicz, Piotr Wiśniewski[✉], and Krzysztof Stencel

Faculty of Mathematics and Computer Science,
Nicolaus Copernicus University, Toruń, Poland
{grusia,pikonrad,stencel}@mat.umk.pl

**Abstract.** Although querying hierarchies and networks is one of common tasks in numerous business application, the SQL standard has not acquired appropriate features until its 1999 edition. Furthermore, neither relational algebra nor calculus offer them. Since the announcement of the abovementioned standard, various database vendors introduced SQL:1999 recursive queries into their products. Yet, there are popular database management systems that do not support such recursion. MySQL is probably the most profound example. If the DBMS used is contacted via an object-relational mapper (ORM), there is a possibility to offer recursive queries provided by this middleware layer. Moreover, data structures materialized in the DBMS can be used to accelerate such queries. In prequel papers, we have presented a product line of features that eventually allow MySQL users to run SQL:1999 recursive queries via ORM. They were: (1) appropriate ORM programmer interfaces, (2) optimization methods of recursive queries, and (3) methods to build materialized data structures that accelerate recursive queries. We have indicated four such methods, i.e.: full paths, logarithmic paths, materialized paths and logarithmic paths. In this paper we aim to assist a database/system architect in the choice of the optimal solutions for the expected workload. We have performed exhaustive experiments to build a cost model for each of the solutions. Their results have been analyzed to build empirical formulae of the cost model. Using this formulae and estimated properties of the expected workload, the database architect or administrator can choose the best materialization method for his/her application.

## 1 Introduction

Numerous applications require processing recursive data structures, e.g. railway networks, corporate hierarchies, bill of material and product categorization. We have a number of options to store such data in a relational database [6]. We have also a number of possible solutions to query such data. One can write a client code or use server facilities. The drawback of a dedicated client code is the need to transfer data from a database server onto the client side. On the other hand, dedicated client code is completely flexible and can encompass any form of processing. The server-side solutions have opposite properties: its flexibility is

© Springer International Publishing Switzerland 2016
S. Kozielski et al. (Eds.): BDAS 2016, CCIS 613, pp. 225–235, 2016.
DOI: 10.1007/978-3-319-34099-9_17

severely limited by the features implemented by the DBMS at hand. Nonetheless, its efficiency is not diminished by the need to transfer bulk data.

However, in our research we identified a third possibility, i.e. the usage of middleware. Various applications used object-relational mappers (ORM) to access the database [1, 12–14]. It amounted to be the middleware we can extend to facilitate numerous database features beyond simple CRUD (create, retrieve, update, delete). In our line of inventions presented in a number papers [3–5, 7, 17, 18] we showed how the ORM layer could be enriched to facilitate SQL:1999 recursive queries. We also considered other features that could be introduced into the ORM layer like partial aggregation [10] and functional indices [2].

SQL:1999 queries have already been implemented in a number of database systems [16]. A number of optimization techniques have been proposed for such queries [8, 11, 15]. However, some database management systems still do not support SQL:1999 recursion, with MySQL as the most significant example. This is why, we proposed a number of execution methods of recursive queries in such databases. Those methods were based on materialization of auxiliary data [4, 5]. In those two papers we presented initial efficiency experiments on these methods. We also analyzed the results and formulated several guidelines for database and system architects when to use each of the proposed methods.

In this paper we elaborate on those previous studies. We performed exhaustive experiments with our proof-of-concept implementation of materialized structures for SQL:1999 recursive queries. Then, we statistically analyze the results and prepare empirical formulae on the efficiency of each method. We deliberately use PostgreSQL and not MySQL, since we want to relate the efficiency of solutions with materializations to the direct usage of SQL:1999 features. Eventually, we also produce cost formulae that quantify the efficiency of our prototype ORM interface that uses plain SQL:1999 facilities.

The contributions of this paper are as follows:

- an exhaustive experimentation on previously catalogued materialization method for faster recursive querying,
- an analysis of the results that lead to empirical formulae of a cost model of each method,
- a reference empirical formula of the cost of direct recursive SQL:1999 queries.

The paper is organized as follows. In Sect. 2 we summarize the materialization methods used to accelerate SQL:1999 recursive queries. Section 3 reports the setup of the performed experiments. Sections 4 and 5 present the results of an analysis of the experimental results and also list the empirical formulae obtained by regression. Section 6 concludes.

## 2    Recursion and Materialization

Entities may have recursive relations, i.e. relations that bind an entity type more than once. In a database such relationships are implemented as a foreign key self-referencing a table or as an additional many-to-many table. The former option

```
> SELECT * FROM emp;

eid |    fname   |   sname   | bid
----+------------+-----------+-----
  1 | John       | Travolta  |
  2 | Bruce      | Willis    |  1
  3 | Marilyn    | Monroe    |
  4 | Angelina   | Jolie     |  3
  5 | Brad       | Pitt      |  4
  6 | Hugh       | Grant     |  4
  7 | Colin      | Firth     |  3
  8 | Keira      | Knightley |  6
  9 | Sean       | Connery   |  1
 10 | Pierce     | Brosnan   |  3
  ...
```

**Fig. 1.** Sample database table on corporate hierarchy.

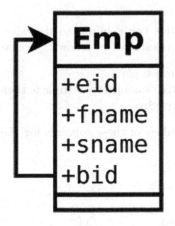

**Fig. 2.** The schema emp.

is the most straightforward and probably most efficient way to store hierarchical data. We use a corporate hierarchy of employees as an example. The number of levels of the hierarchy is not limited, thus there exists no number $n$ such that all leaves of the hierarchy are no deeper than $n$. Figure 1 shows sample data of such hierarchy stored in the table emp, while Fig. 2 depicts its schema. Figure 3 shows an example of a standard SQL:1999 query that retrieves whole subtrees.

As mentioned in Sect. 1 a number of such methods has been proposed [5,9]. We describe them briefly below. The first two methods called *nested sets* and *materialized paths* change the schema of base tables by adding new columns. The remaining two methods namely *full paths* and *logarithmic paths* leave the base table unchanged. They require creating an additional table to store materialized data.

```
WITH RECURSIVE rcte (
    SELECT eid, fname, sname, bid,
            0 as level
       FROM Emp  WHERE sname = 'Travolta'
  UNION
    SELECT e.eid, e.fname, e.sname, bid,
            level +1  as level
       FROM Emp e JOIN rcte r
         ON (e.bid = r.eid)
       WHERE r.level<4
)
   SELECT e.eid, e.fname, e.sname, bid
      FROM rcte
```

**Fig. 3.** A query that returns all (direct and indirect) subordinates of `Travolta`.

## 2.1  Nested Sets

The *nested sets* method adds two columns to the base table: `left` and `right`. The values of these columns satisfy two following constraints:

– `e.left` < `e.right` for every tuple $e$,
– If a tuple $e$ is in the subtree spanned by a tuple $b$, then it is true that `e.left` > `b.left` and `e.right` < `b.right`.

Figure 4 explains the values of these columns for the data shows Fig. 1.

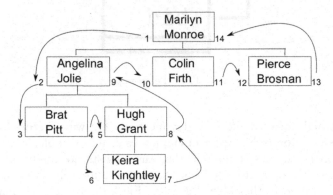

**Fig. 4.** Sample values of the redundant columns `left` and `right` for *nested set*.

The advantage of this method is that querying for descendants is nothing more than a plain inequality join. The following query returns all subordinates of `Travolta`:

```
SELECT e.eid, e.name
  FROM emp e, emp b
  WHERE b.name = 'Travolta'
    AND e.left BETWEEN b.left AND b.right
    AND e.right BETWEEN b.left AND b.right
```

## 2.2   Materialized Paths

*Materialized paths* store the path from the node to its root in an additional column **paths** as in the following sample. Unfortunately, such materialization of paths breaks the first normal form.

```
eid | fname |   sname   | bid | paths
-----+-------+-----------+-----+------
  6 | Hugh  | Grant     |   3 | [4,3]
  7 | Colin | Firth     |   3 | [3]
  8 | Keira | Knightley |   6 | [6,4,3]
```

The following query returns all subordinates of **Travolta** for this method.

```
SELECT *
  FROM emp
 WHERE path_string LIKE (
     SELECT concat(path_string,'%')
       FROM emp
      WHERE sname = 'Travolta'
```

## 2.3   Full Paths

*Full paths* have been studied in [4]. Full paths are similar to *materialized paths*, but the former stores redundant data in an extra table **fullpaths**. The table **fullpaths** contains a distinct row for every step in any path towards the root. The column **pl** (path length) shows the number of steps in each path. For **Keira Knightley** and **Colin Firth** the table *fullpaths* contains the following rows.

```
eid | bid | pl
-----+-----+----
  7 |   3 | 1
  8 |   6 | 1
  8 |   4 | 2
  8 |   3 | 3
```

If the structure is deep, the size of the table *fullpath* can even be a square of the size of the base table. If full paths are applied, the query for **Travolta** subordinates will have the following form.

```
SELECT e.eid, e.name
  FROM emp e JOIN fullpaths fp USING (eid)
    JOIN emp b ON (fp.bid = b.eid)
  WHERE b.name = 'Travolta'
```

## 2.4  Logarithmic Paths

*Logarithmic paths* or shortly *log paths* is a kind of a compromise between full paths and vertical unrolling [3]. The idea is to store only those paths whose length is a power of 2. The data on such paths is stored in the redundant table `logpaths`. For `Keira Knightley` and `Colin Firth` it contains the following rows.

```
eid | bid | pl
-----+-----+----
  7 |   3 | 1
  8 |   6 | 1
  8 |   4 | 2
```

At the query time, we have to reconstruct the information on all paths (as in *full paths*). In order to do this we can use the following auxiliary query:

```
SELECT lp1.eid, lpk.bid,
  lp1.pl + lp2.pl + ... + lpk.pl AS pl
  FROM logpaths lp1
    JOIN logpaths lp2
      ON (lp1.bid = lp2.eid)
    ...
    JOIN logpaths lpk
      ON (lp(k-1).bid = lpk.eid)
  WHERE lp1.pl < lp2.pl
    AND ...
    AND lp(k-1).pl < lpk.pl
```

This query reconstructs data on paths whose length (`p1`) has exactly $k$ ones in its binary representation. We have to run this query for $k = 1, 2, \ldots, \log m$ and combine their results using set union. Since the decomposition of each number into different powers of 2 is unique, this procedure will generate all paths and no path will be repeated. The resulting union query will be a part of the eventual user query.

## 3  Experimental Setup

We used the following experimental setup. We used a computer with the processor Intel Core i5 3570 (3.4/3.8 GHz) and 32 GiB RAM (DDR3, 1600 MHz). The system was stored on an SSD Kingston 120 GiB disk. The data was persisted on a 4x Caviar Black 1 TB in the form of a Logic Volume. The operating system used was Debian 7.3. And last but not least the database management system was DBMS PostgreSQL 9.1. We used a native PostgreSQL version shipped together with the operating system.

Altogether we used 12 database instances. We considered four sizes of tables, namely $10^4, 10^5, 10^6, 10^7$ and three depths of the hierarchies, i.e. $10, 15, 20$. Since we considered all possible combinations of these two parameters, eventually we arrived at these twelve database variants. We tested queries for spanning subtrees with various selectivities. The selectivity of a query has two aspects: the number of the returned rows and the number of levels of the resulting subtree. The number of returned rows varied between 3 and $2 \cdot 10^5$, while the height of the resulting hierarchies fell into the range $\langle 2, 13 \rangle$. For each setup we performed 1500 queries. Since we have observed that initial runs for a given setting produce unstable results we ignored first 300 executions (20 % of the workload). Those initial runs populate the database buffer. Further query runs showed to be stable enough to use them in the analysis.

Each query instance concerned a possibly different seeding row (*seed*), i.e. a row whose subtree is to be returned. For each query run we recorded the elapsed time, the number of returned rows and the depth of the returned subtree (subhierarchy). Then, we used the statistical tool SPSS to build regression models of the cost for each materialization method. We also tested and analyzed the results obtained when direct SQL:1999 queries are used.

## 4    Analysis of Queries

The experimental data collected as discussed in Sect. 3 has been analyzed using the statistical package SPSS. We used the following predictor variables:

- *dbSize* – the size of the data, i.e. the number of rows in the base table.
- *dbDepth* – the depth of the data, i.e. the depth of the deepest row in the base table.
- *qSel* – the selectivity of the query, i.e. the number of returned rows.
- *qDepth* – the depth of the query result, i.e. the depth of the deepest row in the result with respect to the seed.

Two former of these variables are properties of the database instance used. They have 12 possible combination of values. The latter two variables vary significantly between runs of the same query for different seeds.

The predicted variable was the elapsed time in seconds. We used regression to build empirical formulae for each materialization method and for direct SQL:1999 queries.

### 4.1    Full Paths

It turned out that the efficiency of the full paths method was considerably influenced by the selectivity of the query and the database size. Furthermore, it also significantly depends on the product of the number of rows returned and the depth of the fetched subtree. This is in line with the nature of this method, since the number of all redundant tuples is equal to sum of the depths of all nodes. The regression yielded the following formula with the very good value of the coefficient of determination $R^2$, namely 0.998.

$$Q_{full} = 0.001 + 1.298 \cdot 10^{-6} \cdot qSel + 4.262 \cdot 10^{-11} \cdot dbSize + 3.346 \cdot 10^{-8} \cdot qSel \cdot qDepth$$

## 4.2 Logarithmic Paths

The empirical formula for logarithmic paths depends on the logarithm of the depth of the resulting tree, the selectivity and the size of the database. This is somewhat expected since the number of records stored for each node is linear with respect to the logarithm of its depth. The value of the coefficient of determination $R^2$ is 0.882. Unfortunately, the formula has a large negative constant component that excludes using the formula for small data sets (the formula returns negative results). Regrettably, all our efforts to find a better formula failed.

$$Q_{log} = -1.863 + 6.877 \cdot 10^{-7} \cdot \ln^2(qDepth) \cdot dbSize$$
$$+ 1.980 \cdot 10^{-5} \cdot qSel \cdot \ln(qDepth) \cdot \ln(dbSize)$$

## 4.3 Materialized Paths

The empirical formula for materialized paths depends on the database size and the depth of the returned tree. The value of the coefficient of determination $R^2$ is very good, namely 1.000.

$$Q_{mat} = 0.005 + 1.365 \cdot 10^{-7} \cdot dbSize + \cdot 10^{-9} \cdot dbSize \cdot qDepth$$

## 4.4 Nested Sets

The query part for the nested set method is just a single range query. We generated a number of regression formulae for this method and none of them seemed reasonable enough. Here is one of them that has the coefficient of determination $R^2$ equal to 0.700. Unfortunately, it has a significant negative constant component.

$$Q_{nest} = -0.543 + 0.289 \cdot \frac{sel}{dbSize} + 0.058 \cdot \ln(dbSize) + 4.622 \cdot 10^{-7} \cdot \ln(dbSize) \cdot qSel$$

Theoretically, this query should depend only on the number of returned rows. Since the selection column is B-tree indexed. However, the following obvious and reasonable formula has very small $R^2$, namely 0.099. It seems that optimisation mechanisms implemented in PostgreSQL are complex enough to preclude any straightforward empirical formula for a single query.

$$Q'_{nest} = 0.053 + 4.835 \cdot 10^{-6} \cdot qSel$$

## 4.5 Direct SQL:1999

We also built the empirical formula for direct application of SQL:1999 queries. It amounted that the cost of running native SQL:1999 queries depends on the selectivity of the query and the size of the database.

$$Q_{direct} = qSel \cdot 2.993 \cdot 10^{-6} + dbSize \cdot 1.195 \cdot 10^{-9}$$

# 5    Analysis of Updates

When analyzing updates, we took a different approach. We established the number of statements and the number of rows affected by these statements for each materialization method and each operation. Furthermore, we estimated the cost of a single statement in our experimental setup. The cost of an INSERT statement showed to be $0.020 + 1.5 \cdot 10^{-5} \cdot r$ where $r$ is the number of inserted rows. On the other hand the cost of a DELETE and an UPDATE is $0.020 + 4.0 \cdot 10^{-4} \cdot r$ where $r$ is the number of removed rows.

## 5.1    INSERT and DELETE

These two operation can only concern leaves of the hierarchy due to referential integrity constraints. Therefore, for the materialized paths method, we have only one additional operation for each INSERT or DELETE (only the leave must be changed). For, nested sets unfortunately the whole structure must be rebuilt after any change of the data. Thus, we have to perform a deletion of all records and the insertion of them once again. They can be done as single operations.

For full paths and logarithmic paths we have to insert/delete the path to the root from the altered leave. It can be done as a single operation. Assume that the depth of the added/removed leave is $d$. Then we have to add/remove $d$ rows in case of full paths and respectively $\log d$ rows in case of logarithmic paths.

## 5.2    UPDATE

An interesting updating statement is one that changes the hierarchy. Therefore, it moves a node from one place to another. Let us denote its old depth in the hierarchy as $d$ and its new depth as $d'$. Furthermore, assume that its subtree has size $s$. The paths for this subtree will be affected by such an UPDATE.

For nested sets (as above) we have to rebuild the whole structure. For materialized paths we have to update all rows in the subtree of the affected node by changing the prefixes of the materialized paths. It can be done in one UPDATE statement that changes $s$ subordinate rows.

In case of full paths, we have to drop all paths from the subtree of the affected node to its ancestors. This means deleting $sd$ rows and inserting $sd'$ new rows.

For logarithmic paths, unfortunately the whole structure of paths in the subtree must be rebuilt since depths of nodes can change. Therefore, one has to delete $s \log d$ rows and to insert $s \log d'$ new rows.

# 6    Conclusions

In this paper we considered four materialized data structures that accelerate SQL:1999 recursive queries to hierarchical data. These methods have been presented in detail in our prequel paper. In this paper we focused on the assessment of the applicability of these methods for practical workloads. We set up exhaustive experiments of the proposed methods and analyzed the collected results by a statistical package. As the result we obtained a collection of empirical formulae that quantify the cost of individual queries. We also measured the efficiency of direct SQL:1999 queries to relate the materialization methods to the native SQL:1999 support. Using the presented formulae, a database/system architect can choose an optimal solution for his/her application.

# References

1. Bauer, C., King, G.: Java Persistence with Hibernate. Manning Publications Co., Greenwich (2006)
2. Boniewicz, A., Gawarkiewicz, M., Wiśniewski, P.: Automatic selection of functional indexes for object relational mappings system. IJSEA **7**(4), 189–196 (2013)
3. Boniewicz, A., Stencel, K., Wiśniewski, P.: Unrolling SQL: 1999 recursive queries. In: Kim, T., Ma, J., Fang, W., Zhang, Y., Cuzzocrea, A. (eds.) EL, DTA and UNESST 2012. CCIS, vol. 352, pp. 345–354. Springer, Heidelberg (2012)
4. Boniewicz, A., Wisniewski, P., Stencel, K.: On materializing paths for faster recursive querying. In: Catania, B. (ed.) New Trends in Databases and Information Systems. AISC, vol. 241, pp. 105–112. Springer, Switzerland (2013)
5. Boniewicz, A., Wisniewski, P., Stencel, K.: On redundant data for faster recursive querying via ORM systems. In: FedCSIS, pp. 1439–1446 (2013)
6. Brandon, D.: Recursive database structures. J. Comput. Sci. Coll. **21**(2), 295–304 (2005)
7. Burzańska, M., Stencel, K., Suchomska, P., Szumowska, A., Wiśniewski, P.: Recursive queries using object relational mapping. In: Kim, T., Lee, Y., Kang, B.-H., Ślęzak, D. (eds.) FGIT 2010. LNCS, vol. 6485, pp. 42–50. Springer, Heidelberg (2010)
8. Burzańska, M., Stencel, K., Wiśniewski, P.: Pushing predicates into recursive SQL common table expressions. In: Grundspenkis, J., Morzy, T., Vossen, G. (eds.) ADBIS 2009. LNCS, vol. 5739, pp. 194–205. Springer, Heidelberg (2009)
9. Celko, J.: Joe Celko's Trees and Hierarchies in SQL for Smarties, 2nd edn. Elsevier/Morgan Kaufmann, Boston (2012)
10. Gawarkiewicz, M., Wiśniewski, P.: Partial aggregation using Hibernate. In: Kim, T., Adeli, H., Slezak, D., Sandnes, F.E., Song, X., Chung, K., Arnett, K.P. (eds.) FGIT 2011. LNCS, vol. 7105, pp. 90–99. Springer, Heidelberg (2011)
11. Ghazal, A., Crolotte, A., Seid, D.Y.: Recursive SQL query optimization with k-iteration lookahead. In: Bressan, S., Küng, J., Wagner, R. (eds.) DEXA 2006. LNCS, vol. 4080, pp. 348–357. Springer, Heidelberg (2006)
12. Keller, W.: Mapping objects to tables. A pattern language. In: EuroPLoP, pp. 1–26 (1997)
13. Melnik, S., Adya, A., Bernstein, P.A.: Compiling mappings to bridge applications and databases. ACM Trans. Database Syst. **33**(4), 1–50 (2008)
14. O'Neil, E.J.: Object/relational mapping 2008: Hibernate and the Entity Data Model (EDM). In: SIGMOD, pp. 1351–1356 (2008)
15. Ordonez, C.: Optimization of linear recursive queries in SQL. IEEE Trans. Knowl. Data Eng. **22**(2), 264–277 (2010)

16. Przymus, P., Boniewicz, A., Burzańska, M., Stencel, K.: Recursive query facilities in relational databases: a survey. In: Zhang, Y., Cuzzocrea, A., Ma, J., Chung, K., Arslan, T., Song, X. (eds.) DTA and BSBT 2010. CCIS, vol. 118, pp. 89–99. Springer, Heidelberg (2010)
17. Szumowska, A., Burzańska, M., Wiśniewski, P., Stencel, K.: Efficient implementation of recursive queries in major object relational mapping systems. In: Kim, T., Adeli, H., Slezak, D., Sandnes, F.E., Song, X., Chung, K., Arnett, K.P. (eds.) FGIT 2011. LNCS, vol. 7105, pp. 78–89. Springer, Heidelberg (2011)
18. Wiśniewski, P., Szumowska, A., Burzańska, M., Boniewicz, A.: Hibernate the recursive queries - defining the recursive queries using Hibernate ORM. In: Eder, J., Bieliková, M., Min Tjoa, A. (eds.) ADBIS 2011. CEUR Workshop Proceedings, vol. 789, pp. 190–199 (2011)

# Performance Aspect of the In-Memory Databases Accessed via JDBC

Daniel Kostrzewa[✉], Małgorzata Bach, Robert Brzeski,
and Aleksandra Werner

Institute of Informatics, Silesian University of Technology, Gliwice, Poland
{daniel.kostrzewa,malgorzata.bach,robert.brzeski,
aleksandra.werner}@polsl.pl

**Abstract.** The conception of storing and managing data directly in RAM appeared some time ago but in spite of very good efficiency, it was impossible to massive implementation because of hardware limitations. Currently, it is possible to store whole databases in memory as well as there are some mechanisms to organize pieces of data as in-memory databases. It has been the interesting issue how this type of databases behaves when accessing via JDBC. Hence we decided to test their performance in terms/sense of the time of SQL query execution. For this purpose TPC Benchmark™ H (TPC-H) was applied. In our research we focused on the open source systems such as Altibase, H2, HyperSQL, MariaDB, MySQL Memory.

**Keywords:** Altibase · Benchmark · H2 · HyperSQL · In-memory databases · MariaDB · MySQL memory · TPC-H

## 1 Introduction

Most IT systems explicitly distinguish between two storage spaces. Primary is the RAM, in which the processor operates to store actively running programs as well as the operating system on the computer. It is so-called volatile memory because it holds an information only temporarily. It is characterised by high speed of reads and writes operations, but also the high price and limited capacity. The second type of storage is non-volatile, permanent memory, which is currently predominantly made as the mechanical-magnetic devices performing the records on the spinning magnetic disks, but is also made in the form of optic or solid (SSD) media [23].

Here, the items are retained even when power is interrupted or switched off. Compared to the fleeting memory, the capacity and price of disk storage can be taken as more favourable, however the basic speed of access time is much lower. The magnetic media are used as a data warehouse, which stores the currently unneeded data, but also as a virtual memory that emulates the missing main

© Springer International Publishing Switzerland 2016
S. Kozielski et al. (Eds.): BDAS 2016, CCIS 613, pp. 236–252, 2016.
DOI: 10.1007/978-3-319-34099-9_18

memory[1]. Since only small amounts of data from hard disk need to be loaded into the system memory at one time, computers typically have much more hard disk space than memory. The price issue of RAM is also considered as a limiting factor which causes the amount of operating memory mounted in the computing system to be lower than typical capacities of non-volatile disk memory [23]. For example a 2015 computer may come with a 2 TB hard drive and only 32 GB of RAM.

Memory distinction mentioned above is closely reflected in database systems, but it is worth noting that on-disk, traditional, database does not perform any actions directly on physical media. Memory buffers are used for read and write operations to minimize delays associated with disk operations. Thus, tasks which modify or add data to a database are first performed on buffers stored in memory until data is stored on disk. On the contrary to this type of systems, in the in-memory databases data resides permanently in the main, i.e. operating, memory. It must be underlined that this kind of storage engine is extremely vulnerable to any kind of failure: for instance power outages or hardware failure. Furthermore, it is unsuitable for the permanent data storage. In return, it features a high efficiency.

At the beginning of databases there appeared the idea of database management system (DBMS) based on memory as the main and almost the only space data collection. However, the intense increase in the size of databases caused their size exceeded quickly the small size of the RAM in those days. Because of this the idea of in-memory DBMS organization was practically completely abandoned. Nowadays, because the price of memory has decreased significantly and its capacity is quite high, the concept of in-memory databases is revived. It can be observed that currently almost every important provider of database system offers the possibility of in-memory database management [11,16,29]. Moreover, some database engines resort to a dual storage solution implementing the second storage based on disk, because such a solution can result in striking a balance between performance, cost, persistence and storage density. It would be useful to know the performance of such in-memory database, especially accessed by an application through JDBC, which is one of the most popular standards. From a practical point of view, checking a database performance in terms of speed of SQL queries' execution, seemed to be the most desirable. Therefore, this type of testing was carried out in presented study.

The content of the paper is as follow. Section 2 gives the overview of past studies comparative or relating to the in-memory database concepts and in the next section there is a short description of the particular database systems used in our research. In Sect. 4 the problem of the database system performance assessment is touched and the outline of the TPC-H benchmark for the database testing is reminded. Next part of the article contains the results of the tests performed on the actual set of data, which are shown in a tabular and graphic way, and the discussion of the outcomes is provided. In a summary, Sect. 6, the conclusions and research findings are given.

---

[1] This mechanism is an unpleasant necessity and is burdened with a large drop in system performance, but allows to maintain the system's ability to continue operating in the critical moments.

## 2  Related Work

The first study on in-memory databases appeared in the 80 s. In 1984 in the article [10] the authors considered the changes necessary to permit a relational database system to take advantage of large amounts of main memory. They evaluated the new access methods for main memory databases - hash-based query processing strategies vs sort-merge, and also studied recovery issues when most or all of the databases fit in main memory. Just one year later Ammann et al. in paper [6] defined for the first time the term of the Main-Memory Data Base Management System (MMDBMS). The idea of MMDBMS was they treated main memory as a primary storage for eliminating the cost of disk accesses.

Since then the database market has continued to grow and several database management systems for memory-resident data have been proposed or implemented.

In [13] Garcia-Molina and Salem surveyed the major memory residence optimizations and briefly discussed some of the memory-resident systems that were designed or implemented that time. These were such systems, as: MARS, HALO, OBE, TPK, System M and other, which were mainly prototypes except only one commercial - Fast Path, designed by the IBM Company. The conclusion of this article was that memory-resident database systems will be more common in the future, and hence, the mechanisms and optimizations discussed in the paper would become commonplace. Nowadays it turned out to be true. Over the time there began to appear studies in which the researchers analysed performance of in-memory databases, trying to determine the validity of their use in analytical/transaction processing in the real-time applications. For example in [5] the authors implemented the benchmark based on TPC-B on two kinds of databases: a disk-resident database SQL Server and a main memory-resident database FastDB, and studied the performance and availability of it. The performance of FastDB and SQL Server was studied, but the percentage of time that the processor was executing a non-idle thread was presented only for FastDB. There were not comparisons FastDB versus SQL Server, so it is impossible to get information about which of these two systems turned out to be faster. Similar research was performed for IBM SolidDB in [21], where there were explored the structural differences between in-memory and disk-based databases and how SolidDB works to deliver extreme speed. Also in chapter [7] included in the 6th BDAS Conference monograph the strengths and weaknesses of another two in-memory data management systems: TimesTen and SolidDB were presented. Besides, in [27] one can find the comprehensive tests performed with TimesTen used as a cache between applications and their databases. As it can be seen, almost all found publications or reports typically relate to only one, maximum two systems in the vast majority commercial ones. Observed trends clearly indicated a predominance of in-memory databases over the disk-managed ones, although the key to success was the correct configuration of the database. Therefore, in the literature one can find a lot of articles in which the authors tried to improve in-memory database performance. For example in [20] the research focused on two index implementations: a B+Tree Index and the Delta Storage

Index used in the SAP HANA database system. Scientists tried to answer the question of whether modern database implementations could take advantage of hardware support for lock elision. In conclusion they evaluated the effectiveness of the hardware support using the Intel TSX extensions available in the Intel 4th Generation Core$^{TM}$ Processors.

The problem of mixed in-memory database workloads found in enterprise applications is the primary issue in many other publications concerning in-memory databases. Especially the resource-intensive data aggregations, as the heaviest queries, are the object of research of many scholars. Skimming the available studies one can find a lot of interesting proposals to cope with above mentioned challenging queries. For instance in [24] the authors contributed a cache management system for the aggregate cache, which bases on the cache admission and replacement decisions on the novel prot metrics, while in [26] the partial decomposition of the computationally intensive queries combining with their Just-in-Time (JiT) compilation was suggested. Another solution recommended the use of the dynamic query prioritization [28].

Summarizing, the conducted research proved that memory-resident database systems could achieve tremendous speedups, but they had to be optimized for the selected classes of queries because of the limited memory bandwidth. Unfortunately there is still a lack of the comprehensive performance tests for the currently-in-use in-memory databases and there is no exact survey which queries are the heaviest for the particular systems. Moreover, as we know there are no publication with comparison of time performance for free in-memory database, accessed by an application through JDBC - one of the most popular API.

## 3    Overview of the Chosen In-Memory Databases

Several database management systems for memory-resident data have been proposed and/or implemented. In the current point some of these systems are described, but their characteristics are necessarily short due to limited space. 'Light' and free software solutions were deliberately chosen for the study, because often in-memory databases serve as auxiliary systems for the 'heavy' commercial ones. Therefore, the commercial system e.g. SolidDB, VoltDB or TimesTen were omitted for the reasons of their non-free or closed-source license. In [12] some of the systems described below may be found and their strengths as well as cautions relating to a specific vendors offering are described, so in order to supplement the knowledge we recommend to look through this study.

**MySQL.** MySQL is one of the most widely used open-source relational database management systems. It is available under the terms of the GNU General Public License as well as under a variety of proprietary agreements. There are also several paid editions that offer additional functionality. The current stable version is 5.7.9 (released on 21st October, 2015).

It is written in C and C++ and works on many platforms, including AIX, BSDi, FreeBSD, HP-UX, eComStation, i5/OS, IRIX, Linux, OS X, Microsoft Windows, NetBSD, Novell NetWare and many others.

MySQL provides support for several different storage engines, inter alia, the solution based on memory as the main space data collection. On the hard disk only one file which stores the table definition (not the data) is recorded for each table.

The main features of MySQL MEMORY tables are:

- Space for this type of tables is allocated in small blocks. Tables use 100 % dynamic hashing for inserts. No overflow area, extra key space or extra space for free lists are needed.
- MEMORY tables support hash and B-tree indexes. There can be defined up to 64 indexes per table, 16 columns per index and a maximum key length of 3072 bytes. Moreover, these tables can have nonunique keys. Columns that are indexed can contain NULL values.
- MEMORY tables use a fixed-length row-storage format so variable-length types such as e.g. *VARCHAR* are stored using the fixed length. They cannot contain *BLOB* or *TEXT* columns but include support for *AUTO_INCREMENT* columns.

MySQL offers also Cluster technology, that allows the same features as the MEMORY engine with higher performance levels, and provides additional features not available with MEMORY, such as:

- Row-level locking and multiple-thread operation for low contention between clients.
- Scalability even with statements that include writes.
- Optional disk-backed operation for data durability [22,25].

**MariaDB.** MariaDB is based on the open-source MySQL code, but many features are developed independently of MySQL. The server is available under the terms of the GNU General Public License version 2 i.e. open-source [22]. Its current stable release is 10.1.9 with the features from MySQL 5.6 and 5.7. Similarly as it is in the case of MySQL, the system is written in C and C++ and works on many system platforms. The MariaDB system architecture overview is presented in the Fig. 1 and it is noticeable that it supports a lot of engines, including the memory storage, equipped with different features. According to the Gartners research organization, [2] that placed the MariaDB system in a Magic Quadrant for Operational Database Management Systems in 2015, its strengths are: rich functionality, value and reliability and strong community and partner network. Therefore MariaDB is presented in Quadrant as one of two new entrants which are challenging the established leaders.

As it is in other in-memory systems also this one has its own limitations [8,9]. For example *BLOB* or *TEXT* columns and foreign keys are not supported for MEMORY tables. Also, like in the MySQL, the tables can have up to 64 indexes, 16 columns for each index and a maximum key length of 3072 bytes. In addition

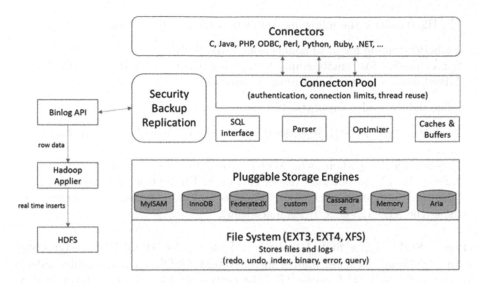

**Fig. 1.** MariaDB architecture

the maximum total size of MEMORY tables cannot exceed the $max_heap_table_size$ system server variable. Besides, when rows are deleted, space is not automatically freed. The only way to free the space is $TRUNCATE$ TABLE command. After a server restart MEMORY tables are recreated, as the definition file is stored on disk, but they are empty. It is possible to re-populate them with a query using the $init - file$ server startup option, where the name of a file containing SQL statements that should be executed by the server on startup can be specified.

**H2.** H2 database engine [14] is a relational database management system. It is available as open-source software under the Eclipse Public License or modified versions of the Mozilla Public License. The current stable version is 1.3.176 (released on 5th April, 2014). Software is written in Java and can be embedded in Java applications or run in the client-server mode on many operating systems.

This database can work as a disc or in-memory database and also as read-only one. It is possible to create 'pure' in-memory tables, but also they can be stored on disk. All typical data types are supported and can be used.

To improve the database performance the B-tree, tree and hash indexes can be used. Some of them are created automatically (for the primary key and unique constraints) while others can be created manually.

The main features of H2 database are [15]:

- Built-in clustering/replication.
- Multiple connections, table level locking and multiversion concurrency control.
- Referential integrity - foreign key constraints with cascade, check constraints.
- Transaction that supports isolation levels: read committed which is the default level, serializable and read uncommitted which means the isolation is disabled.

In H2 Database the following connection modes are supported [15]:

- Embedded mode for local connections using JDBC.
- Server mode - for remote connections using JDBC or ODBC over TCP/IP.
- Mixed mode - for local and remote connections at the same time.

The database also supports [15]:

- Protection against SQL injection by enforcing the use of parameterized statements.
- Row level locking when using MVCC - i.e. multiversion concurrency.
- Encryption of the database using Advanced Encryption Standard, SHA-256 password encryption, encryption functions, SSL (all database can be encrypted using the AES-128 encryption algorithm).

**Hyper SQL.** Hyper SQL Database [18] also called HSQLDB or HyperSQL is the relational database management system. HSQLDB is available under a license based on BSD License [17]. The current stable version is HyperSQL v. 2.3.3 (released on 30th June 2015). The software is written in Java.

The database offers both in-memory and disk-based tables and supports embedded and server modes. All typical data types are supported and can be used. It supports the primary keys, unique and foreign keys, constraints and allows defining indexes on single or multiple columns [18]. Besides, it offers:

- Transaction with $COMMIT$, $ROLLBACK$ and $SAVEPOINT$.
- SQL-92 - almost fully standard except deferrable constraint enforcement, assertion and check constraints that contain subqueries.
- Encrypted databases. Encryption services use the Java Cryptography Extensions, JCE, and the ciphers installed with the JRE. HyperSQL itself does not contain any cryptography code.

Hyper SQL has the mechanisms for three concurrency control models: two-phase-locking (2PL) - the default one, multiversion concurrency control, MVCC, and combination of 2PL plus multiversion rows MVLOCKS. Within each model it offers some of the 4 standard levels of transaction isolation: read uncommitted, read committed, repeatable read, and serializable.

The main features of HyperSQL are [19]:

- Three client server protocols: HSQL, HTTP and HSQL-BER can run as an HTTP web server - all with the SSL option.
- Default storage space of 64 GB for ordinary data. Storage space can be extended to 8 TB.

The database is fully multithreaded in all transaction models.

**Altibase.** Altibase HDB [4] in accordance with the HDB abbreviation is a Hybrid relational DataBase management system, which supports both: the data stored in memory and on disk. The current version is Altibase HDB 6.5.1. The database is available under the commercial license, but is also available as a free trial for 90 days. According to the Gartners Magic Quadrant for Operational Database Management Systems [2], which include Altibase in 2014 in Niche Players quadrant, it is one of the innovative products which delivers interesting and novel features for customers. This system is said to be the pioneer of in-memory databases with true hybrid architecture [1].

Altibase supports [4]:

- Data replication.
- ODBC, JDBC.
- SQL92 and SQL99 standard.
- Concurrency using MVCC.
- Transaction with isolation levels: read committed, repeatable read, and no phantom read.
- Row level or table level locking.
- Database backup.

Besides there is the possibility to define the primary keys, unique, not null and foreign key constraints and the B-tree or R-tree indexes on single or multiple tables columns.

Table 1 summarizes selected characteristics of the analysed systems. Blank entry in the table means that the feature is not described in the documentation. Value "Yes*" means ACID is supported except providing full Durability (in case of power failure there is no guarantee that all committed transactions will survive). Therefore, durability is not always fully supported although it is implemented.

**Table 1.** Features comparison of the MEMORY storage engines

| Feature | MySQL memory | MariaDB memory | H2 | Hyper SQL | Altibase |
|---|---|---|---|---|---|
| ACID transaction | No | No | Yes* | Yes* | Yes |
| Foreign key support | No | No | Yes | Yes | Yes |
| Multiversion concurrency control (MVCC) | No | | Yes | Yes | Yes |
| Supported indexes | Hash indexes, B-tree indexes | Hash indexes, Red-black B-tree | B-tree, tree, Hash indexes | Yes | B-tree, R-tree |
| Locking granularity | Table | Table | Row (with MVCC) | Row (with MVCC) | Table, Row |

## 4  Benchmarks

Transaction Processing Performance Council (TPC) is a non-profit organization founded in 1988 to define transaction processing and database benchmarks and to spread objective, verifiable TPC performance data to the industry. TPC benchmarks are used in evaluating the performance of computer systems.

The TPC Benchmark[TM] H (TPC-H) is a decision support benchmark. It consists of the suite of business oriented ad-hoc queries and concurrent data modifications. This benchmark illustrates decision support systems that:

– examine large volumes of data,
– execute queries with a high degree of complexity,
– give answers to critical business questions [3].

The database used in TPC-H contains eight tables (Fig. 2). The schema represents a simple data warehouse dealing with sales, customers and suppliers.

The TPC-H is a classic benchmark for analytical processing of relational database, based on 'SELECT' queries. The nature of in-memory databases is suitable exactly for this type of processing rather than the transactional one, so we decided to use this benchmark in relation to tested systems.

The benchmark includes twenty-two decision support queries that differ in the selectivity and the complexity. They are with/without tables joining, with/without aggregate functions, with GROUP BY and HAVING clauses, with LIKE/NOT LIKE, EXISTS/NOT EXISTS operators, etc. Most of them are correlated or/and nested queries. Moreover, it is difficult to group or categorize them as belonging to the specified type of query, because of their complex nature.

Due to the limitation of the paper size, the contents of all queries could not be placed in the paragraph, but they are available at [3]. Below, only the exemplary benchmark question is shown. This query identifies certain suppliers who were not able to ship required parts in a timely manner.

```
SELECT s_name, count(*) AS numwait
FROM supplier, lineitem l1, orders, nation
WHERE
  s_suppkey = l1.l_suppkey
  AND o_orderkey = l1.l_orderkey
  AND o_orderstatus = 'F'
  AND l1.l_receiptdate > l1.l_commitdate
  AND EXISTS (
    SELECT *
    FROM lineitem l2
    WHERE l2.l_orderkey = l1.l_orderkey
    AND l2.l_suppkey <> l1.l_suppkey)
  AND NOT EXISTS (
    SELECT *
    FROM lineitem l3
    WHERE l3.l_orderkey = l1.l_orderkey
    AND l3.l_suppkey <> l1.l_suppkey
    AND l3.l_receiptdate > l3.l_commitdate)
  AND s_nationkey = n_nationkey
  AND n_name = 'SAUDI ARABIA'
GROUP BY s_name
ORDER BY numwait desc, s_name;
```

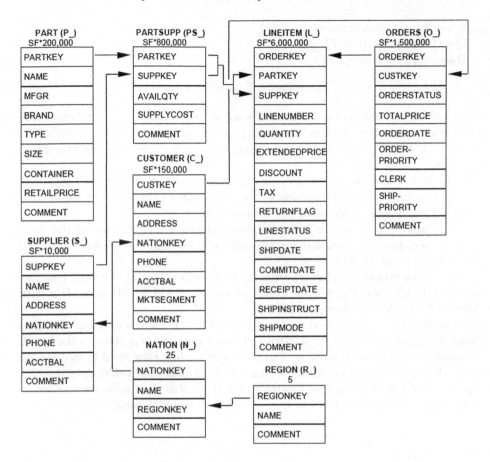

**Fig. 2.** The database schema

## 5  Tests

The main idea of all performed experiments was to prepare test environment in such a way as to ensure the greatest similarity to applications in real IT systems. The test application was written in Java language. The connection with all database management systems was implemented using JDBC libraries provided by the particular software vendors.

In case of disk-resident databases reads and writes from HDD significantly affect the overall performance of the system. In the case of in-memory databases a factor of disk operations is significantly eliminated, which should improve the performance of computing. On the other hand, not only computational but also memory performance are crucial for the entire solution of in-memory databases.

There are a few elements which can influence the overall time of access desired data ($T_D$):

- Establishing connection time ($T_{EC}$),
- Query execution time ($T_{QE}$),
- Data retrieval time ($T_{TR}$).

$$T_D = T_{EC} + T_{QE}(n) + T_{TR}(n)$$

Establishing connection time is always constant and, unlike the query execution time and data retrieval time, does not depend on processed data volume. As a result in all performed tests establishing connection time was not taken into account.

The test application measured the execution time (i.e. $T_{QE}(n) + T_{TR}(n)$) of all test queries. Retrieved data was neither buffered nor displayed in application to avoid its impact to the results of measurements.

This approach allows to compare the performance of the tested database management systems from application point of view. All queries were executed ten times on each DBMS.

The workstation used for experiments is described by the following parameters: AMD FX-8350 Eight Core 4 GHz processor, RAM 32 GB 2400 MHz with Microsoft Windows 7 operating system.

The database was created using typical 'INSERT' SQL statement executed from Java application via JDBC driver. The suitable graph is given in the Fig. 3. Typical time was about dozen of minutes. Only loading time for Altibase turned out to be extremely larger. After 5.5 h only 10 % of data was loaded. Additionally, the speed of data loading was slowing down. That was the reason we decided to cancel this process and not take into consideration this database.

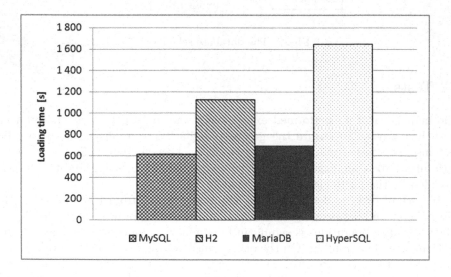

**Fig. 3.** Loading time of databases

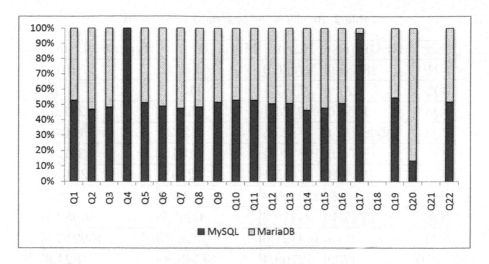

**Fig. 4.** Relative execution time compared MySQL with MariaDB

Table 2 shows the averaged results of the performed experiments. Column *No* presents the name of the test query Columns: *MySQL*, *MariaDB*, *H*2 and *HyperSQL* shows execution time averages for particular DBMS and test queries.

For the queries 7, 18, 19 and 21 time of query execution exceeded 4 h in at least one of the tested systems. After this time the query execution was stopped even if not obtained answers. It is obvious that in the in-memory database, which in their nature should be fast, no one would accept, if this time was too long. Typically, the query execution takes approximately several seconds. Sometimes several minutes. But the threshold of time length must be assumed. In our research it was the period of 4 h, although from practical point of view specified time is definitely too long. Thus, queries for which the execution time exceeded 4 h were not taken into account in our report. Among mentioned queries definitely the most problematic turned out the query 21 (its content was presented in Sect. 4). As many as 3 of the tested systems were not able to execute this query in less than 4 hs. This was achieved only in DBMS H2.

The shortest average times of ten runs obtained for individual queries are marked by grey color, the longest were underlined. As can be seen MariaDB in the case of 10 queries proved to be the fastest. The second place in terms of the number of queries completed in the shortest time has MySQL.

Results obtained for H2 were below our expectations. This system turn out to be the slowest in the case of 14 queries. However, it should not be forgotten that the H2 was the only system, which coped with the question 21, as above mentioned.

In further analysis the average times obtained by MySQL were adopted as a reference point and the times achieved in other systems were compared to them. For example Fig. 4 presents the proportions of time obtained for MySQL and MariaDB. It should be noted that for most queries included in this report,

**Table 2.** Results of the performed experiments

| No | MySQL [ms] | H2 [ms] | MariaDB [ms] | Hyper SQL [ms] |
|----|-----------:|--------:|-------------:|---------------:|
| Q1 | 18058.9 | 27046.9 | 16118.6 | 17016.6 |
| Q2 | 123.9 | 27187.4 | 139.8 | 53.5 |
| Q3 | 1458.6 | 29324.1 | 1556.1 | 1231.3 |
| Q4 | 13068809.4 | 5825.5 | 2229 | 1150.4 |
| Q5 | 5453.6 | 108239.4 | 5184.5 | 3347.3 |
| Q6 | 1214.1 | 17435.1 | 1270.5 | 2510.7 |
| Q7 | 2734.2 | 285409.9 | 3020 | max |
| Q8 | 5725.4 | 94419.8 | 6097.2 | 14980.9 |
| Q9 | 1666 | 114721.6 | 1578.7 | 109213.5 |
| Q10 | 1739 | 72768.4 | 1548.8 | 1824.8 |
| Q11 | 63.6 | 1908.2 | 57.1 | 2976.1 |
| Q12 | 1052.6 | 36768.1 | 1032.4 | 6768.8 |
| Q13 | 1314 | 49940.1 | 1265.6 | 4379.6 |
| Q14 | 942.6 | 240700.5 | 1082.6 | 2269.4 |
| Q15 | 2118.6 | 32841.4 | 2297.6 | 4801.6 |
| Q16 | 915.4 | 1264.5 | 887.1 | 2467 |
| Q17 | 2365235.7 | 1186.5 | 78021.7 | 14227.6 |
| Q18 | max | 28508.9 | 3912.2 | 473056.9 |
| Q19 | 2834.7 | max | 2367.5 | 28140.3 |
| Q20 | 25.4 | 3465.2 | 160.2 | 205.5 |
| Q21 | max | 52411.9 | max | max |
| Q22 | 104.6 | 4640.7 | 97.2 | 499.2 |

the times recorded for both systems do not differ significantly. This does not apply to queries 4 and 17 for which MySQL managed far worse. The average execution time for these queries is also worse in comparison with the H2 and Hyper SQL systems (Fig. 5). However, in the case of other queries, they are executed much faster in MySQL database than in H2.

The comparison of MySQL and Hyper SQL (Fig. 6) shows that 12 of 19 included queries were executed faster in MySQL.

The complexity of analysed queries was so large that it was impossible to determine one component of query, which was responsible for a long time of its execution by particular system. We suppose that the additional structures which are created in memory by the database systems, are responsible for the

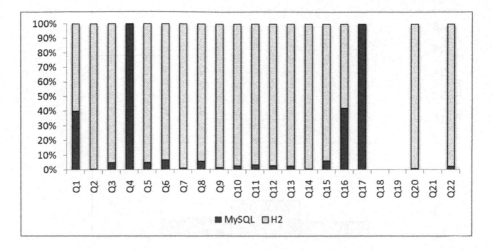

**Fig. 5.** Relative execution time compared MySQL with H2

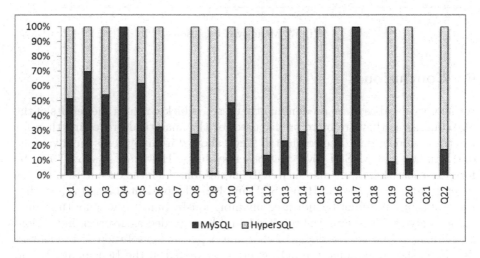

**Fig. 6.** Relative execution time compared MySQL with Hyper SQL

observed ineffectiveness. Unfortunately the information contained in the documentation of these systems in relation to their memory storage engines is not sufficient to unambiguously determine these structures. Another reason that can be responsible for obtaining such unsatisfactory results can be not optimal action of query optimizer of the system.

In the presented research the usage of memory was also studied (Fig. 7) and this comparison is favorable for MySQL and MariaDB systems. The worst result, i.e. the largest memory usage, was obtained for Hyper SQL.

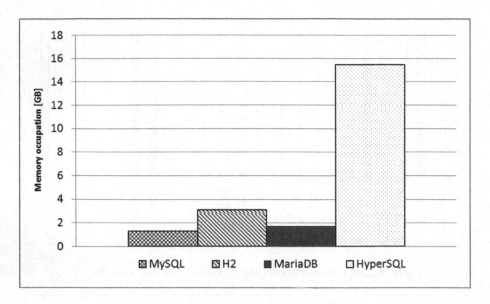

**Fig. 7.** Memory occupation

## 6  Conclusions

In-memory databases are predominantly used to quickly return the results of the SQL queries as they are more predisposed for the analytical processing than for the transactional one. The conducted study showed in many cases an advantage of MariaDB and MySQL over the other systems. However, it should not be forgotten that there were also such SQL queries for which the execution time was much shorter in the case of Hyper SQL and H2. The presented results show that the best solution cannot be indicated unambiguously. In fact it was not the aim of made analyses. Therefore, the results should be regarded as a sort of indication. Individuals facing the necessity of choosing a particular solution of in-memory database, should examine the SQL queries contained in the Benchmark H and answer the question what the created application will be used for, what queries will be the most frequently used and which of the 22 queries from Benchmark H have the most similar character to the performing one. After conducting such analysis, the results presented in this paper can help to take the final decision.

The nature of queries is so cross-cutting, that it is difficult to precisely determine, to which specific class of query they belong.

Available tracking mechanisms did not provide a clear answer to the question why the query execution lasted so long, even though the query execution plans were generated and analysed, paying particular attention to the 'problematic' ones in the individual systems.

To summarize this issue, because of the fact that execution plans did not explain the obtained results and additionally the complexity of the queries was so large, it was impossible to indicate one particular component of query, which was responsible for a long time of its execution. The same fact was observed by Simon in [27], although his research concerned TimesTen and Oracle. Taking into account the performance, the results were surprising, because Oracle on local disk proved to be faster than TimesTen, the in-memory Oracle solution. The author also claimed that the main reason of this disparity was not obvious. Therefore, in the future we plan to develop the research and define our own queries, which will belong to the specific types of queries. This might let us to identify those types of questions that are performed inefficiently in a given system.

# References

1. Altibase hybrid database included in Gartner's who's who for in-memoryDBMSs - 2014. http://www.prweb.com/releases/2014/02/prweb11541753.htm. Accessed Mar 2016
2. Magic quadrant for operational database management systems. https://www.gartner.com/doc/reprints?id=1-2PMFPEN&ct=151013. Accessed Dec 2015
3. Active TPC Benchmarks. http://www.tpc.org/information/benchmarks.asp. Accessed Dec 2015
4. Altibase. http://altibase.com/. Accessed Dec 2015
5. Amiri, H., AleAhmad, A., Rahgozar, M.: Disk resident databases versus main-memory databases. https://www.researchgate.net/publication/228363088_Main_Memory_Databases_vs._Disk-Resident_Databases. Accessed Dec 2015
6. Ammann, A.C., Hanrahan, M., Krishnamurthy, R.: Design of a memory resident DBMS. In: COMPCON, pp. 54–58. IEEE Computer Society (1985)
7. Bach, M., Duszenko, A., Werner, A.: Koncepcja pamiciowych baz danych oraz weryfikacja podstawowych zaoe tych struktur. Studia Informatica **31**(2B(90)), 63–76 (2010)
8. Bartholomew, D.: MariaDB vs MySQL. https://mariadb.com/kb/en/mariadb/mariadb-vs-mysql-features/. Accessed Dec 2015
9. Charles, C., Zoratti, I.: MariaDB 10 the complete tutorial. http://www.slideshare.net/bytebot/mariadb-10-the-complete-tutorial. Accessed Nov 2015
10. DeWitt, D., Katz, R., Olken, F., Shapiro, L., Stonebreaker, M., Wood, D.: Implementation techniques for main memory database systems. In: Proceedings of the 1984 ACM SIGMOD International Conference on Management of Data, SIGMOD 1984, vol. 14, no. 2, pp. 1–8 (1984)
11. Erickson, J.: Oracle database 12c: introducing Oracle database in-memory. http://www.oracle.com/us/corporate/features/database-in-memory-option/index.html. Accessed Dec 2015
12. Feinberg, D., Adrian, M., Heudecker, N., Ronthal, A.M., Palanca, T.: Magic quadrant for operational database management systems. http://www.gartner.com/technology/reprints.do?id=1-2PO8Z2O&ct=151013&st=sb. Accessed Dec 2015
13. Garcia-Molina, H., Salem, K.: Main memory database systems: an overview. IEEE Trans. Knowl. Data Eng. **4**(6), 509–516 (1992)
14. H2: H2 database. http://www.h2database.com/. Accessed Dec 2015
15. H2: H2 database engine documentation. http://www.h2database.com/html/features.html. Accessed Dec 2015

16. Henschen, D.: In-memory databases: do you need the speed? http://www. informationweek.com/big-data/big-data-analytics/in-memory-databases-do-you-need-the-speed/d/d-id/1114076. Accessed Dec 2015
17. Hyper SQL: Hyper SQL copyrights and licenses. http://hsqldb.org/web/hsql License.html. Accessed Dec 2015
18. Hyper SQL: Hyper SQL database. http://hsqldb.org/. Accessed Dec 2015
19. Hyper SQL: Hyper SQL features summary. http://hsqldb.org/web/hsqlFeatures. html. Accessed Dec 2015
20. Karnagel, T., Dementiev, R., Rajwar, R., Lai, K., Legler, T., Schlegel, B., Lehner, W.: Improving in-memory database index performance with Intel transactional synchronization extensions. In: 20th International Symposium on High-Performance Computer Architecture (2014)
21. Lindstrom, J., Raatikka, V., Ruuth, J., Soini, P., Vakkila, K.: IBM solidDB: in-memory database optimized for extreme speed and availability. Bull. Tech. Committee Data Eng. **36**(2), 14–20 (2013)
22. MariaDB: MariaDB server license. https://mariadb.com/kb/en/mariadb/ mariadb-license/. Accessed Nov 2015
23. Meena, J.S., Sze, S.M., Chand, U., Tseng, T.Y.: Overview of emerging non-volatile memory technologies. Nanoscale Res. Lett. **9**, 526 (2014). http://www. nanoscalereslett.com/content/pdf/1556-276X-9-526.pdf. Accessed Dec 2015
24. Muller, S., Plattner, H.: Aggregates caching in columnar in-memory databases. http://db.disi.unitn.eu/pages/VLDBProgram/pdf/IMDM/paper8.pdf. Accessed Dec 2015
25. MySQL: the memory storage engine. http://dev.mysql.com/doc/refman/5.7/en/ memory-storage-engine.html. Accessed Dec 2015
26. Pirk, H., Funke, F., Grund, M., Neumann, T., Leser, U., Manegold, S., Kemper, A., Kerste, M.: CPU and cache effcient management of memory-resident databases. http://oai.cwi.nl/oai/asset/20680/20680D.pdf. Accessed Dec 2015
27. Simon, E.A.: Evaluation of in-memory database TimesTen. CERN openlab Summer student report 2013 (2013). https://zenodo.org/record/7566/files/CERN_ openlab_report_Endre_Andras_Simon.pdf. Accessed Mar 2016
28. Wust, J., Grund, M., Plattner, H.: Dynamic query prioritization for in-memory databases. http://db.disi.unitn.eu/pages/VLDBProgram/pdf/IMDM/paper7.pdf. Accessed Dec 2015
29. Yegulalp, S.: SQL server 2014 supercharged with in-memory tables. In: Azure Connectivity, InfoWorld (2014). http://www.infoworld.com/article/2610878/data base/sql-server-2014-supercharged-with-in-memory-tables-azure-connectivity.html. Accessed Dec 2015

# Comparison of the Behaviour of Local Databases and Databases Located in the Cloud

Marcin Szczyrbowski and Dariusz Myszor[✉]

Institute of Informatics, Silesian University of Technology,
Akademicka Street 16, 44-100 Gliwice, Poland
dariusz.myszor@polsl.pl

**Abstract.** This article is dedicated to analysis and comparison of the behaviour of databases located in the cloud with databases located in the local infrastructure. Analysis presents stability of results delivery speed. Article's summary includes suggestions for utilization of particular solutions in implementation of specific types of database based applications. There will also be established border of the profitability for database migration from the local environment to the cloud. This article is aimed at specialists working on the design of IT projects as well as scientists who want to consider to utilize cloud based solutions for storage of information e.g. RNA polynucleotides sequences.

## 1 Introduction

Recently we observe tendency to migrate information storage from local environments to various kinds of clouds. Utilization of cloud based solutions becomes common practice in various areas, e.g. collecting of personal information, as it allows to quickly and easily access the data stored on multiple devices. There are various examples of companies which rely on cloud based solutions e.g. Apple uses cloud as a primary source of storage space supplied for individual customers. Such giants as Goggle, Microsoft and Amazon offer storage of information in the cloud to business customers. Cloud computing is applied in order to speed up execution of algorithms that require particularly high computing power. More than three years ago (2012) Oracle Database cloud service was launched. Currently Oracle offers three models: Database as a Service (DaaS), Exadata Service and Database Schema Service. Database Schema Service solution is easy to set up, simple to use and scalable. According to the documentation it is a perfect solution for running new projects, especially ones with weakly defend specification in the area of database performance requirements. In theory performance of cloud based database should accommodate to our temporal needs, and as a result, obtained response times should be more stable [5]. Utilization of cloud based databases might be applied in various industry applications as well as scientific ones. Among them biological problems e.g. for RNA world hypothesis in order to store RNA sequences. In such applications especially important is an ability of obtainment of high speed of search results delivery [6, 7].

© Springer International Publishing Switzerland 2016
S. Kozielski et al. (Eds.): BDAS 2016, CCIS 613, pp. 253–261, 2016.
DOI: 10.1007/978-3-319-34099-9_19

## 2   Cloud Description

In Database Schema Service three major components can be distinguished: a platform that allows to quickly develop and deploy applications, a set of easily installed built-in packaged applications, as well as the ability to conveniently access data using REST–based web services [9,10,14]. Database Schema Service is built on Oracle Database technology, running on the Oracle Exadata Database Machine [12]. In this solution access to only one schema, working on Oracle Database 11g, and maximum 50 GB of storage, is available [8,11]. Database is totally separated from external world, access to data is possible only through REST services, as a result all performed activities can be securely controlled and registered by the API.

## 3   Study

### 3.1   Description

In this study we compared behaviour of database located in the cloud with database located in local environment. For the test we used Database Schema Service provided by Oracle Corporation, the same version of database was working in local environment. Tests were carried on Oracle Database 11g Enterprise Edition without any performance increasing extensions. Local database was working on cluster composed of two computational nodes (2 CPUs / 16 GB RAM per node). Data were provided by RAID5 based on SAS hard drive. During tests conduction phase, local machine was not exposed to any additional tasks. Three types of queries were performer: select, insert and update. The purpose of conducted tests was determination of stability of data delivery. Shapiro-Wilk test was applied in order to check if collected data belongs to a Gaussian distribution. The test was carried out with a statistical significance p = 0.05. T-student test (of data belonged to Gaussian distribution) and Manna-Whitney test (when data did not belong to Gaussian distribution) were performer in order to determine whether results obtained for cloud and local based solutions are statistically different.

### 3.2   Test Results

**Warm up period.** Warm up period analysis was performed in order to find out the number of repetitions of experiments that will guarantee obtainment of statistically meaningful data [3,4]. To this end insert command was performed 2000 times (data were embedded directly in the insert query). Due to the structure of the data, and existing references between tables, during insertion of new rows integrity of references must been checked. In addition insert command triggers select query by unique indexes and foreign keys. It is one of the most complex query that is performed for the purpose of this article, therefore it is a good candidate for warm up period designation. The figure below shows changes in the average time of query conduction and standard deviation of time in subsequent attempts (Fig. 1).

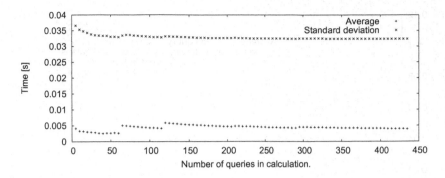

**Fig. 1.** Changes in the average time of insert query execution and the standard deviations in subsequent queries

Presented results point out that conduction of 120 experiments should be sufficient.

**Simple select operation.** In this test, data from a single table were selected. The execution plan includes accessing, by not unique index, selected part of the table and, in order to find final results set, application of range scan on the preselected part of table. The following code shows a sample query used in the test. The value `RELATIVE_HUMIDITY_INDEX` has been changed in all queries (Fig. 2 and Table 1).

```
SELECT RELATIVE_HUMIDITY_ID FROM RELATIVE_HUMIDITY WHERE
DATE_ID = TO_DATE('15/01/07','YY/MM/DD')
AND RELATIVE_HUMIDITY_INDEX = 8860;
```

Collected data belong to Gaussian distribution and therefore Manna-Whitney test was utilized. Calculated value $2.5184e^{-041}$ allows to reject the null hypothesis. Obtained results point out that, in case of a local database, single query execution times are scattered over a large areas above and below the value

**Table 1.** Results for simple select test

|  | Value for database in | |
| --- | --- | --- |
|  | Local environment | Cloud |
| Shapiro-Wilk statistic $W$ | $9.2281e^{-23}$ | $7.88e^{-8}$ |
| Standard deviation | 0.0277 | 0.0013 |
| Average execution time [s] | 0.0457 | 0.0083 |
| Median | 0.0420 | 0.0090 |
| Maximum execution time [s] | 0.3400 | 0.0100 |
| Minimum execution time [s] | 0.0370 | 0.0050 |

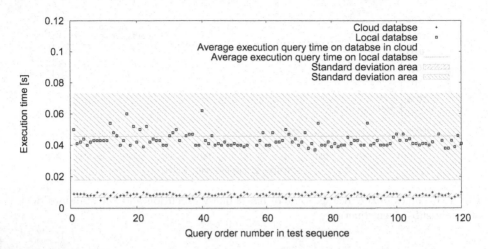

**Fig. 2.** Query execution time for simple select test

of average query execution time. The standard deviation is also much higher for local solutions. Local database seems to respond significantly slower than cloud based versions. In addition local databases access times are characterised by higher variance.

**Multi table select.** In this test, tables were joined using unique index, therefore cost of this operation was low. During test temporary table, that consumes more memory than temporary table from the previous simple select test, is created (as a result of range scan application). However, beside of temporary table creation, rest of query execution plan is similar to the simple select solution. The value of DATE_ID has been changed in all queries (Fig. 3 and Table 2).

```
SELECT COUNT(*) FROM RELATIVE_HUMIDITY RH
    INNER JOIN WIND        W ON RH.RELATIVE_HUMIDITY_ID = W.WIND_ID
    INNER JOIN TEMPERATURE T ON RH.RELATIVE_HUMIDITY_ID
    = T.TEMPERATURE_ID
    INNER JOIN SOLAR       S ON RH.RELATIVE_HUMIDITY_ID = S.SOLAR_ID
WHERE RH.DATE_ID = TO_DATE('2015/01/01', 'YYYY/MM/DD');
```

For local version of database prolongation of query execution time is visible. Although it is characterized by lower variance than for single select query. On the other hand, minimal improvements were visible at the level of cloud based database performance in the area of response time (in comparison with single select query scenario). This results from the fact that greater load in the cloud database automatically increases assigned resources. Noteworthy, this process is not linear. We observed that hysteresis based algorithm (that makes the decision about computational resources allocation) is applied [2].

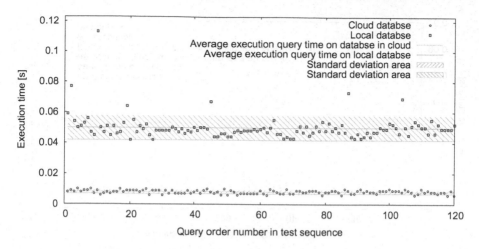

**Fig. 3.** Query execution time for multi table select

**Table 2.** Results for multi table select

|  | Value for database in | |
|---|---|---|
|  | Local environment | Cloud |
| Shapiro-Wilk statistic $W$ | $2.73e^{-17}$ | $4.56e^{-7}$ |
| Standard deviation | 0.0079 | 0.0012 |
| Average execution time [s] | 0.0498 | 0.0081 |
| Median | 0.0480 | 0.0080 |
| Maximum execution time [s] | 0.1130 | 0.0100 |
| Minimum execution time [s] | 0.0420 | 0.0060 |
| Manna-Whitney | $2.5183e^{-041}$ | |

**Update operation.** In order to find rows which should be updated, unique indexed is applied. The following example demonstrates one of the query used in the paper. Each subsequent queries contains different value in the condition (Fig. 4 and Table 3).

```
WHERE RELATIVE_HUMIDITY_ID = 9690.
   UPDATE RELATIVE_HUMIDITY SET BEST = 1
   WHERE RELATIVE_HUMIDITY_ID = 9690;
```

In this case Manna-Whitney test returns value $8.7273e^{-42}$ therefore, conclusion can be made that data obtained from two types of database differ. One more time, comparison of results points out that cloud based solution can deliver data in more stable time frame than local environment.

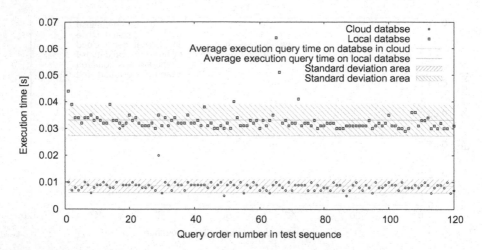

**Fig. 4.** Query execution time for updating data test

**Table 3.** Results for update test.

|  | Value for database in | |
| --- | --- | --- |
|  | Local environment | Cloud |
| Shapiro-Wilk statistic $W$ | $1.95e^{-20}$ | $1.22e^{-13}$ |
| Standard deviation | 0.0083 | 0.0016 |
| Average execution time [s] | 0.0338 | 0.0084 |
| Median | 0.0320 | 0.0080 |
| Maximum execution time [s] | 0.0930 | 0.0200 |
| Minimum execution time [s] | 0.0290 | 0.0060 |

**Insert operation.** During this operation database engine checks if some unique constraint is not violated, this is low cost operation because reference between table SOLAR_PER_DAY and RELATIVE_HUMIDITY is based on indexed data (Fig. 5 and Table 4).

```
INSERT INTO SOLAR_PER_DAY (MAX, AVERAGE, SOLAR_PER_DAY_ID)
    VALUES(3,2,TO_DATE('2015/03/01', 'YYYY/MM/DD'));
```

Mann-Whitney test was applied again. It returned value equal to $1.3346e^{-41}$, therefore conclusion can be made that the collected data differ significantly from each other. Interestingly the maximum execution time for the cloud based database is at the level of the previous test. It happens because full information must be saved on all physical nodes engaged in the storage of the table content [1,13].

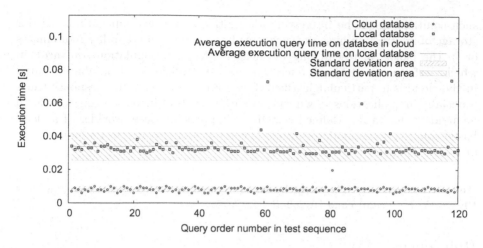

**Fig. 5.** Query execution time for inserting data test.

**Table 4.** Results for insert test.

|  | Value for database in | |
|---|---|---|
|  | Local environment | Cloud |
| Shapiro-Wilk statistic $W$ | $1.53e^{-19}$ | $3.14e^{-18}$ |
| Standard deviation | 0.0058 | 0.0025 |
| Average execution time [s] | 0.0331 | 0.0087 |
| Median | 0.0320 | 0.0090 |
| Maximum execution time [s] | 0.0780 | 0.0300 |
| Minimum execution time [s] | 0.0290 | 0.0050 |

### 3.3 Discussion

Purposefully all queries utilized in the described tests were simple, as a result database engine did not require application of modules that improve efficiency. Obtained results point out that there is significant difference between response time, as well as response time stability, obtained for local based and cloud based solutions. It was visible for all types of executed queries. In addition differences between average and median execution time are significant. Cloud database is up to ten times faster than local version. Analysis of individual charts can lead to the observation that local database delivers results with significantly higher time deviations than database located in the cloud. A significant difference in average times, obtained for local and cloud based solutions, can be justified by hardware efficiency, however there is a question regarding stability of data delivery. It should be the result of application of a more stable physical data source in cloud based solution. Clouds are characterized by high level of data dispersion, it allows for simultaneous reading from many sources. Therefore, such solutions might be interesting alternative in various applications e.g. for storage of information in sensory systems because of high reliability reqirements or in the biological researching (such as RNA world simulation software), in which performance of search queries is crucial. Utilization of database located in the cloud is in particular justified if we want to obtain stable response times especially in applications in which there are periods of increased computational demands or for weakly defined system. It happens because provider of a cloud based solution accommodates available computational power to temporal needs of DaaS customer.

**Acknowledgements.** The research leading to these results has received funding from BKM-515/ RAU2/2015 and BK263/RAU2/2015.

# References

1. Abadi, D.J.: Data management in the cloud: limitations and opportunities. IEEE Data Eng. Bull. **32**(1), 3 (2009)
2. Armbrust, M., Fox, A., Griffith, R., Joseph, A., Katz, R., Konwinski, A., Lee, G., Patterson, D., Rabkin, A., Stoica, I., et al.: Above the clouds: A berkeley view of cloud computing. uc berkeley reliable adaptive distributed systems laboratory. University of California, Berkeley (2009)
3. Grassmann, W.: Rethinking the initialization bias problem in steady-state discrete event simulation. In: Proceedings of the Winter Simulation Conference, pp. 593–599 (2011)
4. Grassmann, W.K.: Warm-up periods in simulation can be detrimental. Probab. Eng. Inf. Sci. **22**(03), 415–429 (2008)
5. Iosup, A., Ostermann, S., Yigitbasi, M.N., Prodan, R., Fahringer, T., Epema, D.H.: Performance analysis of cloud computing services for many-tasks scientific computing. IEEE Trans. Parallel Distrib. Syst. **22**(6), 931–945 (2011)
6. Myszor, D.: Influence of introduction of mitosis-like processes into mathematical-simulation model of protocells in RNA world. In: Man-Machine Interactions 4, pp. 259–268. Springer (2016)
7. Myszor, D., Cyran, K.A.: Mathematical modelling of molecule evolution in proto-cells. Int. J. Appl. Math. Comput. Sci. **23**(1), 213–229 (2013)
8. Oficial Oracle Corporation web page: your oracle database in the cloud (2015). https://cloud.oracle.com/en_US/database
9. Oracle Corporation: building oracle database applications in the cloud (2015). https://cloud.oracle.com/en_US/_downloads/eBook_Platform_DBSchema_File/Oracle_Database_Cloud_Service_Schema.pdf
10. Oracle Corporation: move to your private data center in the cloud. zerocapex. predictable opex. full control (2015). https://cloud.oracle.com/_downloads/eBook_Platform_Compute_File/Oracle_Compute_Cloud_Service.pdf
11. Oracle Corporation: oracle database cloud - database as a service your oracledatabase in the cloud (2015). https://cloud.oracle.com/en_US/database?lmResID=1410650314486&tabID=1383678929020

12. Oracle Help Center: using oracle database cloud - database schema service (2015). https://docs.oracle.com/cloud/latest/dbcs_schema/CSDBU/GUID-07F01633-0E34-40AC-A550-A549E8A62B6E.htm\#CSDBU113
13. Oracle Help Center: using oracle database cloud - database as a service, Beta Draft: 13 October 2015. https://docs.oracle.com/cloud/latest/dbcs_dbaas/CSDBI/GUID-660363B8-0E2F-4A4F-A9BD-70A43F332A16.htm#CSDBI3321
14. Pringle, T.: Data-as-a-service: the next step in the as-a-service journey 18 July 2014. http://www.oracle.com/us/solutions/cloud/analyst-report-ovum-daas-2245256.pdf

# Scalable Distributed Two-Layer Datastore Providing Data Anonymity

Adam Krechowicz[⊠]

Kielce University of Technology, Kielce, Poland
a.krechowicz@tu.kielce.pl

**Abstract.** Storing data in a public data systems (mostly in the Cloud) can lead to many considerations about the data privacy. Are our data completely safe? Inspired by these considerations the author started to develop efficient framework which can be used to improve data privacy while storing them in public data storages. The Scalable Distributed Two-Layer Datastore was used as a base for the framework because it proved to be very efficient solution to store huge data sets.

## 1 Introduction

More and more personal data like photos, videos or other documents are stored in the Cloud. Storing data in such services offers many advantages. Cloud providers may support many functions that are not available for the local data storages. Greater disk space, higher availability, data protection against failures, possibility to share data with friends cause that more and more people start using cloud provided data storage. The vast majority of people still can not afford to buy enough hardware to satisfy their needs of capacity or to build RAID systems necessary for protecting against disk failures. Many cloud providers can satisfy these needs even free of charge.

The main disadvantages of the cloud storages are connected with data privacy. Users are often bothered if their data are safe from the wrong hands. In theory, many factors can lead to the situation in which our data are exploited. User's private files may fall pray to hackers or even dishonest cloud providers may violate the privacy rules. If the data are seriously confidential the usage of cloud services is strongly discouraged. Not only dishonest activities could be dangerous to our data. Still, the vast majority of users is not aware of the methods that are used by the cloud providers. This is due to the fact that cloud providers do not like to share all their data gathering as well as processing methods and only few users search for that kind of information for example in the privacy policy manifest.

Thanks to using that kind of Cloud storages an user needs to rely on the level of privacy that the particular storage provides. The problem of privacy of the storages was widely described in [8,12,15,22]. As the NoSQL systems are focused mostly on scalability and high availability the concepts of privacy are often treated with low priority. Many NoSQL systems do not even provide

© Springer International Publishing Switzerland 2016
S. Kozielski et al. (Eds.): BDAS 2016, CCIS 613, pp. 262–271, 2016.
DOI: 10.1007/978-3-319-34099-9_20

user authentication and authorization by default. On the other hand security concepts on the service side do not need to satisfy the majority of the users while the providers can still have access to their confidential data. Because user do not need to know anything about the concepts of security they would definitely like to introduce some of the security concepts on their side.

Still, the peer-to-peer networks are serious competitors to the Cloud and typical data storages. Many users prefer to rely on the trust of other individuals rather than soulless corporation. They can think that a single user has not the required computing power and even intention to exploit their data. In fact such networks do not provide any protection at all. A single malicious or eavesdropping peer can do a lot of harm and violates privacy of other network users.

In the minds of many people the need of data security may be considered slightly exaggerated. They might think that honest and law-abiding person does not need to hide their data and actions. In reality each person needs certain intimacy. Data connected with ensuring confidentiality of the work or connected with health [13] or financial aspects are great example of things that should be highly protected.

The recent, so called, Snowden Affair put a new light into the consideration of the data privacy. Many people start to concern about the ways of the gathering people's private data [19]. The another well known scandal connected with the Cloud privacy emerged when nude photos of celebrities were stolen from the private accounts of one of the most well known mobile phone manufacturer [5]. Those are just two well known examples and other can be easily found [14]. The author believes that the privacy of the users data can be violated more and more frequent in the future.

The custom solutions for gathering data in the data storages may be a good solution to provide data privacy at an appropriate level. The most basic solutions, like data encryption, are always good ideas [18] but they can still leave room for maneuver for one that wants to exploit our data. It can still very easily trace the source of the data and also can use sophisticated data analysis that may give them some ideas about the content of the data. The author believe that the user-centric solutions are the best methods of securing privacy of our data.

This paper covers two important aspects of protecting the privacy of the data inside Scalable Distributed Two-Layered Data Storage. The most important aspect is to protect the data against unauthorized access both by other individuals and the most importantly by cloud providers that store them. The second aspect is connected with providing the anonymous access to the data so there is higher probability that the source of the data and also the data owners remain unknown.

## 2   Related Work

The vast majority of the data storage do not take great consideration to data privacy. They are developed in such a way that the storage is used in trusted environment. Some of them do not even consider security issues [4]. If the application requires they can use traditional access control techniques to manage permissions [2]. The vast majority of them store the data in a raw form so it can be easily spied on the server-side [3].

However, there are many works that takes privacy and anonymity into consideration. They are mostly focused on providing anonymous access to the structure [6, 24, 26]. Still the author believes that the content of the data can exploit users identity (e.g. faces on the photos, surnames on documents). Apart from that, the origin of the transmission can leave some clues.

The peer-to-peer network still offers the great privacy and anonymity factors. The Tor Project [23] is the most recognizable example of such network. It uses Onion Routing (OR) [7, 17] to provide anonymity and privacy of the data. The Tribler project [16] also uses OR to apply anonymous access to the network based on BitTorrent protocol [1].

## 3   Scalable Distributed Two-Layer Data Structures

Storing data in the cloud in the distributed system requires the usage of specific software solution. Almost every cloud provider developed their custom system so it is hard not only to choose the preferred solution but also to classify all of them. All of them can be classified as NoSQL systems. The BigTable [3] developed by Google or Dynamo [4] created by Amazon are the most known examples.

Recently developed Scalable Distributed Two-Layer Data Store [9, 10] proved to be very efficient solution to store huge amount of data. It allows to store the data on multicomputer nodes in so called buckets. Its most important feature is the data scalability which allows the store to grow theoretically into infinite capacity which makes it extremely useful in Cloud environment.

The main feature of SD2DS is that the structure consists of two separate layers which can be managed independently. Because of that the structure creates a division of the portion of the data (so called component) into two parts: the component header and the component body. The component header contains only basic information that helps to localize and interpret the data (meta-data) while the component body contains the actual data. The headers are located in the first layer while the bodies are stored in the second. The most important part of the header is a locator which indicates where the actual data are stored. Both the headers and the bodies are stored in the buckets while the body and the corresponding header do not need to share the same bucket.

To retrieve the data the client needs to find the appropriate header first and then the locator will point to the actual data. This indirect access proved to be very flexible and provides the way to equip the store with different functionality [11, 21]. It is done mainly by introducing other meta-data to the headers.

Due to the fact that the conception of the SD2DS does not dictate the way of managing both the first and the second layer, the structures are good solution to store data both in Cloud services and as peer-to-peer networks. Because of the division of the components into headers and bodies the usage of the structures as the Cloud storage is extremely convenient. The second layer can be located in the Cloud provided data store of practically any kind. There is also no obligation to use only single storage so it can use the benefit of many providers at once. The first layer of the structure will bind them together and can help to organize the data into different locations.

# 4   Data Privacy in SD2DS

## 4.1   Local First Layer

To ensure that the data located in the structure are safe from unauthorized access the first layer of the SD2DS should be restricted only to appropriate users. The best way to achieve that is to make the first layer local. Because of that each client should maintain its own first layer. This concept can seriously decrease the chance of the unauthorized access, because only the owner will know the exact location of all of its data.

While other users do not know where the data are stored, the dishonest store provider can still access and exploit users data. The most simple and resistant to simple threats solution is to encrypt the data. The author does not specify the method of encryption because it must be chosen depending on the nature of the stored data. The first layer of the structure is the excellent place to store the encryption key so only the owner can have access to it. So in order to get the actual content client must first access the first layer and get the locator and the key to perform decryption. To ensure that the data inside store was not changed without the user permission the hashing function is used. The SHA-1 digest algorithm is a very good solution to perform this task.

## 4.2   Data Stripping

Partitioning of the data (stripping) into many pieces also increases the level of data protection. If a user wants to hide the content of their data from storage provider it can divide the whole set of data into smaller parts. Each part of those data is encrypted and can additionally be stored in different places, possibly under jurisdiction of different storage providers. All the necessary data, like the order of pieces, are stored in the component header.

Data partitioning offers many advantages. First of all, it prevents the storage provider from knowing the contents of the data because it can store just selected parts of the data that are additionally encrypted. Even if all the pieces of data are located in the same place, which might be convenient if there is only one location available to the user, it is hard to find out the exact order of those pieces. On the other hand, if the pieces are distributed on different machines the time of uploading and downloading can be faster in comparison to storing the files uniformly.

## 4.3   Proxy Layer

In classical SD2DS there are always two layers: one for storing headers and one for bodies. To protect the privacy of the users the next layer (or layers) can be introduced. Instead of accessing the data directly from the location stored in the header, the client can use the Proxy nodes (third layer) to cover their intentions. In that scenario the cloud provider can not be sure who is the owner of the data and who actually has the access to that data. By using different proxy each time the risk of tracing of the data owner is seriously decreased.

When a client wants to insert or retrieve a piece of information from the cloud they need to randomly choose one of the possible proxy and send the request through it. It can also specify the number of proxies (hops) they want to use before requesting the actual storage and put it inside the request. If the proxy gets the request it decrements the number of hops and if the value reaches zero it connects to the storage. If the hops value is grater than zero the node randomly chooses another proxy to send the request to it.

Because the data that are transferred are encrypted, the proxy does not have possibility to determine the content of the transferred packets. The statistical analysis is also very difficult because each part of the whole data can be transferred using different proxy.

By using just one proxy the trace of the data owner and cognition of its content is seriously hampered. Still the proxy itself knows exactly the source of the data. Multiple proxy (two or even more) are introduced to protect against that threat. In that scenario each proxy does not have the idea if the user that is connecting to them is the real data owner or just another proxy.

The Author proposes P2P network for organizing proxy layer. The self organizing P2P networks are the best solutions but they are always hard to implement. In that matter the author relays on existing, well known solutions [20, 25]. The system needs additional mechanisms for protection against malicious proxies. This issue should be considered very carefully and is planned to be investigated it in the future work.

## 5   Implementation

The basic architecture of the proposed solution is presented in the Fig. 1. It consists of three main parts which can be considered as separate layers of the system. The first layer, the original first layer of the SD2DS, is located locally on the clients machine so only they have access to all its confidential content. The optional second layer can help the clients to stay anonymous in the system. This layer may consists of variable number of proxies that route the desired data from/to the Cloud provider. The third layer (the second layer in the original SD2DS) the actual storage is located in the cloud under the jurisdiction of third party company.

### 5.1   The Header

The standard header known from original SD2DS conception needs to be extended with some additional information that helps to protect the data. The header consists of the following parts that allows to identify and check all data parts.

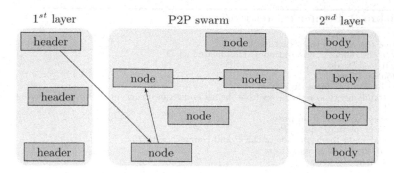

**Fig. 1.** The framework architecture

Those information consists of the sequence number of part $(P_N)$, size of that part $(P_S)$, part locator $(P_L)$, encryption key $(P_K)$ and a hash code $(P_H)$.

### 5.2  Key Algorithms

The Algorithm 1 presents the behaviour of the client in case of inserting the data to the storage with the number data checks as N is presented in Algorithm 1. The Algorithm 2 presents the behaviour of the client in case of retrieving the data from the storage. The Algorithm 3 presents the behaviour of the proxy.

## 6  Evaluation

The evaluation was carried out with a prototype implementation of the framework. The second layer consisted of 46 servers. The results was gathered during

---

**Algorithm 1.** Component insertion
_____

**Require:** $N$, $D$, $H_n$
1: Done:= **false**
2: Divide the data into set of n parts indicated as $D = \{D0, D1, ..., Dn - 1\}$
3: Choose the random permutation of $D$ indicated as $D_p$
4: **for all** $P \in Dp$ **do**
5:     Calculate the hash code $P_H$ of the $P$ and store it in the header $H$
6:     Create the encryption key $P_K$ for $P$ and store it in the $H$
7:     Encrypt the $P$ with $P_K$
8:     Choose the location of the $P$
9:     **while** Done = **false do**
10:         Randomly choose the proxy from the list and send encrypted $P$ to him with a predefined hops number $H_n$
11:         **for** $n := 0$ **to** $N - 1$ **do**
12:             Randomly choose the proxy from the list and request encrypted $P$ from him with predefined hops number $H_n$
13:             Receive the $P$ from proxy
14:             Calculate the hash code of the $P$ and compare with $P_H$
15:             **if** the $P_H$ is correct **then**
16:                 Done:= **true**
17:                 $n := N - 1$
18:             **end if**
19:         **end for**
20:     **end while**
21: **end for**
_____

**Algorithm 2.** Component retrieve

**Require:** $K$, $H_n$
1: Receive the $H$ from the structure using $K$
2: Choose the random permutation of $D$ indicated as $D_p$
3: **for all** $P \in D_p$ **do**
4:     Done:= **false**
5:     **while** Done = **false do**
6:         Randomly choose the proxy from the list and request encrypted $P$ from him with predefined hops number $H_n$
7:         Receive the $P$ from proxy
8:         Encrypt the $P$ using $P_K$
9:         Calculate the hash code of the $P$ and compare with $P_H$
10:         **if** the PH is correct **then**
11:             Done:= **true**
12:         **end if**
13:     **end while**
14: **end for**

**Algorithm 3.** Proxy service

1: **while true do**
2:     Wait for the client to connect
3:     Receive request from client
4:     Decrement $H_n$
5:     **if** $H_n > 0$ **then**
6:         Randomly choose the proxy from the list and forward the request with new $H_n$ to it
7:     **else**
8:         Send the request to the place indicated by $P_L$
9:     **end if**
10:     Receive the response and send it back to the originator
11: **end while**

randomly retrieving of 4096 components from the structure. Initially 512 components were inserted in each tests. The P2P swarm consisted of 128 proxies. For the purposes of the tests each client and proxy knows exactly the location of every other proxies. The outcomes are presented in the following figures.

There is an obvious relation between the number of proxy hops and the structure performance. The evaluation was started for that factor. This test gave the basic knowledge about the expected performance while using different sizes of the components. The obtained results are presented in the Fig. 2. In all cases the excess time was strongly correlated with the component size.

Apart from increasing privacy of the data stripping can have positive influence on the performance. Because of the data partition the whole component

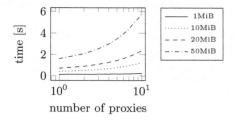

**Fig. 2.** The performance in relation to the number of proxy hops

can be simultaneously downloaded from different servers using different sets of proxies. The author presented the results of this issue in Fig. 3. Depending on the size of the component the optimal number of the strips varies.

The interesting fact is that in many cases the bigger number of proxy hops gave the better performance results. It was caused by the fact that the proxies were buffered the data so the buckets' load was reduced. Similar results were gathered in the environment where many clients operated simultaneously on the structure. In Fig. 4 there where respectively 32 and 128 additional clients tried to obtain randomly chosen components of 1MiB from the structure. In many cases the higher number of strips and higher number of proxies hops allows to achieve better results.

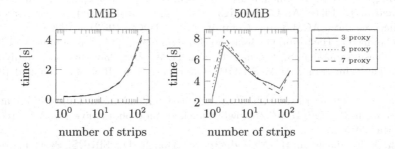

**Fig. 3.** The performance in relation to the number of strips and proxies

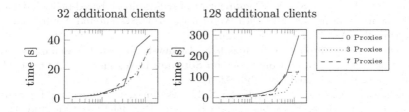

**Fig. 4.** The performance while additional clients operate on structures

## 7  Conclusions

The main goal of this work was to increase the privacy and anonymity in the Cloud storages by utilizing Scalable Distributed Two-Layer Data Structures. During the evaluation the author discovered that apart from this additional feature it was also possible to achieve very good performance. The first impression was that it might be reduction of the performance due to the usage of additional proxies. Despite from that strips parallel processing gave the opportunity to minimize the impact of the proxies.

There may be still very serious security issues in the framework that need to be improved. The author believes that the proper method of data encryption should be used so that the content of the data will not be compromised even if some eavesdropper knows all of them. In the future work the author is also planning to ensure that the whole communication between participants will be secured by utilizing Onion Routing mechanisms.

# References

1. BitTorrent: BitTorrent. http://www.bittorrent.org/. Accessed 03 Mar 2015
2. Burrows, M.: The chubby lock service for loosely-coupled distributed systems. In: Proceedings of the 7th Symposium on Operating Systems Design and Implementation, pp. 335–350. USENIX Association (2006)
3. Chang, F., Dean, J., Ghemawat, S., Hsieh, W.C., Wallach, D.A., Burrows, M., Chandra, T., Fikes, A., Gruber, R.E.: Bigtable: a distributed storage system for structured data. ACM Trans. Comput. Syst. (TOCS) **26**(2), 4 (2008)
4. DeCandia, G., Hastorun, D., Jampani, M., Kakulapati, G., Lakshman, A., Pilchin, A., Sivasubramanian, S., Vosshall, P., Vogels, W.: Dynamo: amazon's highly available key-value store. ACM SIGOPS Operating Syst. Rev. **41**, 205–220 (2007). ACM
5. Duke, A.: 5 Things to know about the celebrity nude photo hacking scandal. http://edition.cnn.com/2014/09/02/showbiz/hacked-nude-photos-five-things/. Accessed 03 Mar 2015
6. Goh, E.J., Shacham, H., Modadugu, N., Boneh, D.: SiRiUS: Securing remote untrusted storage. In: NDSS, vol. 3, pp. 131–145 (2003)
7. Goldschlag, D., Reed, M., Syverson, P.: Onion routing. Commun. ACM **42**(2), 39–41 (1999)
8. Jiang, Z., Zheng, Y., Shi, Y.: Document-oriented database-based privacy data protection architecture. In: 2013 10th Web Information System and Application Conference (WISA), pp. 19–22. IEEE (2013)
9. Krechowicz, A., Chrobot, A., Deniziak, S., Łukawski, G.: Sd2ds-based datastore for large files. In: Federated Conference on Software Development and Object Technologies (2015, accepted for publication)
10. Krechowicz, A., Deniziak, S., Bedla, M., Arkadiusz, C., Grzegorz, L.: Scalable distributed two-layer block based datastore. In: Parallel Processing and Applied Mathematics (2015, accepted for publication)
11. Krechowicz, A., Deniziak, S., Łukawski, G., Bedla, M.: Preserving data consistency in scalable distributed two layer data structures. In: Kozielski, S., Mrozek, D., Kasprowski, P., Kostrzewa, D., Małysiak-Mrozek, B. (eds.) BDAS 2015. CCIS, vol. 521, pp. 126–135. Springer International Publishing, Switzerland (2015). http://dx.doi.org/10.1007/978-3-319-18422-7_11
12. Lenard, T.M., Rubin, P.H.: The big data revolution: privacy considerations (2013)
13. Mazurek, M.: Applying NoSQL databases for operationalizing clinical data mining models. In: Kozielski, S., Mrozek, D., Kasprowski, P., Małysiak-Mrozek, B., Kostrzewa, D. (eds.) BDAS 2014. CCIS, vol. 424, pp. 527–536. Springer, Heidelberg (2014). http://dx.doi.org/10.1007/978-3-319-06932-6_51
14. Minkus, T., Ross, K.W.: I know what you're buying: privacy breaches on eBay. In: Cristofaro, E., Murdoch, S.J. (eds.) PETS 2014. LNCS, vol. 8555, pp. 164–183. Springer, Heidelberg (2014)

15. Okman, L., Gal-Oz, N., Gonen, Y., Gudes, E., Abramov, J.: Security issues in NoSQL databases. In: 2011 IEEE 10th International Conference on Trust, Security and Privacy in Computing and Communications (TrustCom), pp. 541–547. IEEE (2011)
16. Pouwelse, J.A., Garbacki, P., Wang, J., Bakker, A., Yang, J., Iosup, A., Epema, D.H., Reinders, M., Van Steen, M.R., Sips, H.J.: TRIBLER: a social-based peer-to-peer system. Concurrency Comput. Pract. Experience **20**(2), 127–138 (2008)
17. Reed, M.G., Syverson, P.F., Goldschlag, D.M.: Anonymous connections and onion routing. IEEE J. Sel. Areas Commun. **16**(4), 482–494 (1998)
18. Renaud, K., Volkamer, M., Renkema-Padmos, A.: Why doesn't jane protect her privacy? In: Murdoch, S.J., Cristofaro, E. (eds.) PETS 2014. LNCS, vol. 8555, pp. 244–262. Springer, Heidelberg (2014)
19. Richelson, J.T.: The Snowden Affair. Web Resource Documents the Latest Firestorm over the National Security Agency. http://www2.gwu.edu/~nsarchiv/NSAEBB/NSAEBB436/. Accessed 03 Mar 2015
20. Rowstron, A., Druschel, P.: Pastry: scalable, decentralized object location, and routing for large-scale peer-to-peer systems. In: Guerraoui, R. (ed.) Middleware 2001. LNCS, vol. 2218, pp. 329–350. Springer, Heidelberg (2001)
21. Sapiecha, K., Lukawski, G., Krechowicz, A.: Enhancing throughput of scalable distributed two-layer data structures. In: 2014 IEEE 13th International Symposium on Parallel and Distributed Computing (ISPDC), pp. 103–110. IEEE (2014)
22. Takabi, H., Joshi, J.B., Ahn, G.J.: Security and privacy challenges in cloud computing environments. IEEE Secur. Priv. **8**(6), 24–31 (2010)
23. Tor: Tor project. https://www.torproject.org/. Accessed 03 Mar 2015
24. Zarandioon, S., Yao, D.D., Ganapathy, V.: K2C: cryptographic cloud storage with lazy revocation and anonymous access. In: Rajarajan, M., Piper, F., Wang, H., Kesidis, G. (eds.) SecureComm 2011. LNICST, vol. 96, pp. 59–76. Springer, Heidelberg (2012)
25. Zhao, B.Y., Kubiatowicz, J., Joseph, A.D., et al.: Tapestry: an infrastructure for fault-tolerant wide-area location and routing (2001)
26. Zhi-hua, Z., Jian-jun, L., Wei, J., Yong, Z., Bei, G.: An new anonymous authentication scheme for cloud computing. In: 2012 7th International Conference on Computer Science and Education (ICCSE), pp. 896–898. IEEE (2012)

# Coordination of Parallel Tasks in Access to Resource Groups by Adaptive Conflictless Scheduling

Mateusz Smolinski[✉]

Institute of Information Technology,
Lodz University of Technology,
Wolczanska 215, 90-924 Lodz, Poland
mateusz.smolinski@p.lodz.pl
http://it.p.lodz.pl

**Abstract.** Conflictless task scheduling is dedicated for environment of parallel task processing with high contention of limited amount of resources. For tasks that each one requires group of resources presented solution can prepare schedule of tasks execution without occurrence of any resource conflict. As a task it can be used any selected sequence of operation that for execution requires access for resource group, to which access is controlled by conflictless scheduling. Any resource group required by task has own FIFO queue, where tasks are waiting for access of those resources. Queues are emptying according to prepared conflictless schedule in such a way that there is no starvation of waiting tasks. Presented scheduling concept for tasks and resource group bases on resource representation model which allows to efficient detect a resource conflict using dedicated data structures like task classes and conflict matrix and algorithms which allows to prepare adaptive conflictless schedule. Prepared conflictless schedule adapts to current environment state like number of resource groups and tasks in their queues and also waiting times of tasks. Prepared schedule ensures task execution without resource conflicts and therefore there is no tasks deadlock. As example of environments where conflictless scheduling can be applied is transaction processing in databases or OLTP systems, processes or threads competing for resources. In transaction processing environment deadlock elimination by using proposed conflictless scheduling reduces the number of transaction rollbacks.

**Keywords:** Conflict free task execution · Transaction processing without resource conflict · Deadlock avoidance · Concurrency control · Mutual exclusion

## 1 Introduction

Many IT systems allow parallel tasks processing, which in general provides better performance in modern computer architectures. Processing environments differ

© Springer International Publishing Switzerland 2016
S. Kozielski et al. (Eds.): BDAS 2016, CCIS 613, pp. 272–282, 2016.
DOI: 10.1007/978-3-319-34099-9_21

in task specifics and relationships between them. If tasks are completely independent then they can be executed in parallel without any order. Otherwise relationships between tasks have impacts of their execution. Popular relationship between tasks are use of global resources which each of them requires to finish own execution. When two parallel executed tasks require to use the same resource instance and at least one of them perform write operation then resource conflict occurs. The occurrence of many resource conflicts can cause a task deadlock. To avoid resource conflict between parallel tasks its execution should be synchronized in competitive concurrent processing or by planned global resource allocation when task are coordinated [7].

The proposed solution bases on resources groups allocation to tasks that no resource conflict occurs between executed tasks. Using mutual exclusion in temporary allocation each of resource groups to selected task eliminates resource conflicts, however requires prior conflictless schedule preparation. Proposed scheduling concept uses dedicated model of task representation to fast resource conflict detection and data structures like task classes or conflict matrix to efficiently prepare conflictless schedule. All specified elements are presented later in this paper. The next chapter includes assumptions for task processing environment, which fulfillment allows to apply proposed conflictless scheduling. The conflictless schedule is prepared using proposed algorithm, each time it is adapted to current environment state which includes active tasks that are executed and suspended tasks that waiting for execution. Elimination of task starvation problem in proposed algorithm of adaptive conflictless scheduling is also discussed in this paper.

## 2    Assumptions of Task Processing Environment

In IT systems the task is defined by sequence of operation and its input data set i.e. tasklet, job, transaction, thread or process [2,7]. In conflictless task scheduling the task is defined as the shortest sequence of operations with all private resources and group of global resources necessary to perform those operations. Software developers in task processing environment with high contention of resources need to prepare dedicated algorithms for proper resource allocation to fixed group of tasks. Conflictless schedule is universal and can be used in various task processing environment with high contention of limited global resources, which meets all criteria:

- execution of any requested task can be delayed,
- execution time of each task is finite, but it is not known,
- task executions order is not fixed,
- all instances of global resources are reusable,
- set of all global resources required by task is known when task is requested for execution,
- no task priorities, that affect execution order,
- each execution finish of active task has to be reported, even if it is terminated as a result of an error.

It should be noted that task execution time is not known and is not limited by deadline. Many tasks that require the same resource group can be requested and each one instance of those tasks is discriminated. In conflictless scheduling concept a task designation is performed by programmer, which for any separated task has to select all its global resources. Mutual exclusion between tasks in access to global resources is controlled by conflictless schedule, which prepares order of conflicted task execution. Conflictless schedule has to be prepared for all situation like a execution finish one of active tasks or new task is requested. Conflictless schedule preparation needs an efficient method of resource conflict detection. Proposed resource conflict detection between tasks bases on special model of task global resources representation.

## 3    Efficient Resource Conflict Detection Between Tasks

Proposed representation of task global resources includes two binary resource identifiers $IRW$ and $IR$, which have to be assigned individually to any task instance. Each instance of reusable global resource has uniquely assigned the bit position, which is selected in $IRW$ when task includes at least one operation that requires those global resource. Second identifier $IR$ represents by bits those global resources, that are only read by task. According to task processing environment characteristics all bit positions in binary identifiers can be assigned to global resources statically by programmer or dynamically by central resource controller. Length of task binary identifiers that represents its required global resources is determined by number of all used global resources. Additionally each requested task $t_k$ has assigned a logical time $T_k$ number, which higher value represents later granting of binary resource identifiers $IRW$ and $IR$ to task. Logical time is used in conflictless scheduling to prevent task starvation.

Verification that two tasks are free of resource conflict requires to check condition 1, and if this condition is satisfied there is no resource conflict between task $t_i$ and $t_j$.

$$IRW_i \text{ and } IRW_j = 0 \qquad (1)$$

Detection of resource conflict between two tasks require to check condition (2), and if this condition is satisfied at least one resource conflict exists between task $t_i$ and $t_j$ [6].

$$(IRW_i \text{ and } IRW_j)x \text{ or } (IR_i \text{ and } IR_j) \neq 0 \qquad (2)$$

Using condition 2 detection of resource conflict between tasks is efficient, because it requires to perform maximally three simple operations if all binary resource identifiers are stored in single words. Once detected resource conflict between tasks can be remembered and used multiple times due to proposed data structures like task classes and conflict matrix.

# 4    Task Classes and Conflict Matrix

A efficient resource conflict verification between tasks requires additional data structures. In conflictless scheduling are used task classes and conflict matrix. The task class $C_k$ represents group of requested and not executed tasks, that have the same values of binary resource identifiers:

$$C_k = \{t_i : IRW_i = IRW_k \wedge IR_i = IR_k\} \tag{3}$$

Each task class is associated with a resource group and the way how they are used by task, taking into account the read and write operations on global resources required by task. Task class $C_k$ includes all active tasks that are executed and suspended tasks that are waiting for execution, that meets the condition 3. The number of task in class $C_k$ changes over time, therefore in $n$-th point in time $C_k^n$ represents set of waiting task, that is subset of $C_k$. Each task class $C_k^n$ has its own FIFO queue for waiting tasks, which determines the order in resuming task execution. There is no limitation of size for class FIFO queue and number of waiting tasks in class queue is represented by $count(C_k)$. Task class has to have FIFO queue only if $count(C_k) > 0$. Emptying task class queues is determined by adaptive conflictless schedule, which is individually prepared for any case when active task from set $R^n$ execution finish. Many FIFO queues for task classes presents Fig. 1.

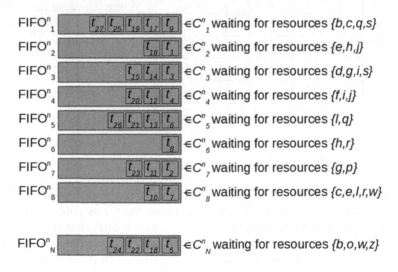

**Fig. 1.** FIFO queues for task classes.

For task class $C_k^n$ if condition 4 is satisfied, then class queue can be flushed (all its waiting tasks are resumed and begin parallel execution without resource conflicts), otherwise only longest waiting task leaves class queue and begin its execution.

$$IRW_k = IR_k \qquad (4)$$

The $R^n$ represents set of active tasks in $n$-th point in time, that are executed without resource conflicts. If set $R^n$ includes at least one active task from class $C_k$ then class $C_k^n$ is known as active class.

When new task is requested its class is determined and if this class has not exist yet it is created. If new requested task from class has conflict with any task from active task set it is suspended in its class queue, otherwise it execution begins and task belong to $R^n$.

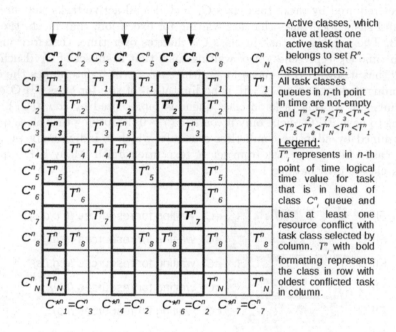

Fig. 2. A conflict matrix example.

The resource conflict existence between task classes are determined using conditions 2 and stored in two dimensional array known as conflict matrix $M$. The resource conflict between task classes represents a fact, that a common part exists between their resource groups. Conflict matrix $M$ is determined for the number of existed task classes $N^n$, therefore in other $n$-th points in time matrix $M^n$ dimension $N^n \times N^n$ can be different. Using conflict matrix $M = [a_{ij}]$ where $i, j = 1....N^n$ there is no need to repeat resource conflict detection between the same classes. Resource conflict between tasks can be simply detected by reading single value from conflict matrix $M$ i.e. at least one resource conflict between tasks $t_x \in C_i$ and $t_y \in C_j$ exists only if $m_{ij} \neq 0$. If $m_{ij} = 0$ then no resource conflict exists between tasks $t_x \in C_i$ and $t_y \in C_j$. The conflict matrix

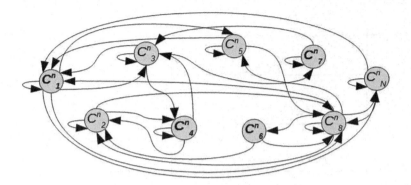

**Fig. 3.** Wait for graph for tasks.

is symmetric because $M^T = M$, so only half of its values has to be stored in memory. The Fig. 2 show example conflict matrix.

The conflict matrix is representation of WFG (Wait For Graph) for task classes, where WFG can be used to deadlock detection and avoidance [7,8]. The Fig. 3 show WFG for sample conflict matrix presented on Fig. 2. Verification of resource conflicts between task classes are necessary in preparing each adaptive conflictless schedule. Conflict matrix is prepared for all classes but some of class queues can be empty, therefore the conflict matrix reduction was proposed through the elimination from conflict matrix $M^n$ rows related with classes that have empty queues. The conflict matrix $A^n = [a_{ij}]$ after reduction process has not to be symmetric because its dimensions can change in relation to conflict matrix $M^n$. After conflict matrix reduction for all active task the class with a longest waiting conflicted task has to be determined, this task in $n$-th point in time for $C_k$ active class will be indicated as $t_k^{*n}$ and its class will be denoted $C_k^{*n}$. Determination of $C_k^{*n}, k = 1..count(R^n)$ for all active classes is required to prevent task starvation in conflictless task scheduling concept.

## 5    Conflictless Schedule

### 5.1    Adaptive Conflictless Schedule

Preparation of conflictless schedule requires to designate a collection $S_k^n$ of task classes with non-empty FIFO queues from which task (or tasks if condition 4 is satisfied) can start its execution without resource conflicts after active task $t$ from set $C_k \cap R^{n-1}$ finish its execution. Every finish of active task execution determines next $n$-th point in time, when prepared adaptive conflictless schedule need to be immediately applied, as shown in Fig. 4.

If for $n$-th point in time the adaptive conflictless schedule is prepared earlier then there is no additional delay. Each conflictless schedule is adaptive, because it is prepared to specific environment state, that includes: finish of fixed active task, active task set $R^n$, state of queues of classes that are in conflict with

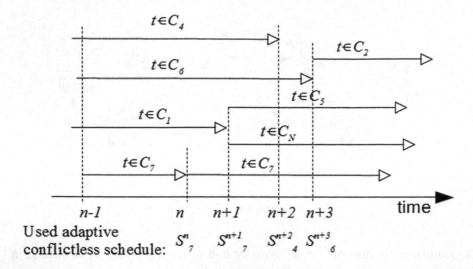

Fig. 4. Example conflictless schedule.

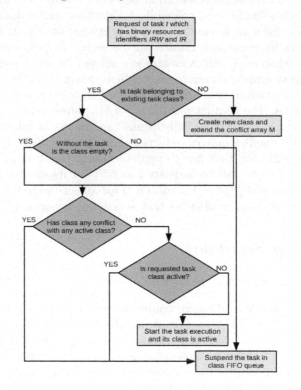

Fig. 5. Actions for new requested task.

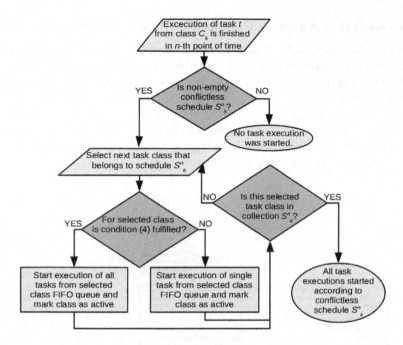

**Fig. 6.** Actions for execution finish of active task.

that finished task and even logical time values of task that are waiting in class queues. This means that prepared adaptive schedule $S_k^n$ is correct in $n$-th point in time when task $t$ from set $C_k \bigcap R^{n-1}$ finish its execution and is useless in other situations. Even in two other $n$ and $m$ points in time when task $t \in C_k$ finish its execution prepared adaptive conflictless schedules do not have to be the same $S_k^n \neq S_k^m$.

Number of adaptive conflictless schedules for $n$-th point in time is determined by number of active tasks that belongs to $R^{n-1}$. Each change in the active task set or in reduced conflict matrix requires preparation of a new adaptive conflictless schedules. This can be aa result of a new task request or finish one of active tasks, as shown on Figs. 5 and 6. Created earlier schedules can be invalid or not optimal in new state of task processing environment. Frequent preparation of many conflictless schedules needs isolated computing environment, that is separated from task processing and its resources. For example modern GPGPU (General Purpose computing on Graphic Processing Units) can be used to store resource identifiers for task classes and conflict matrix and efficient prepare many adaptive conflictless schedules. With GPGPU there is no need to transfer conflict matrix between host and dedicated GPU memory. Many work introduce concept of using GPGPU to efficient transaction processing or graph algorithms [1,4,5].

## 5.2    Preparation Adaptive Conflictless Task Scheduling Algorithm

The proposed algorithm of task scheduling determines subset of classes that tasks can be executed in parallel without resource conflict. Preparation of the conflictless schedule is step process because for all active tasks their execution time is unknown. In every step many adaptive conflictless schedules has to be prepared for fixed $n$-th point in time. Only if one of active tasks finishes its execution, then appropriate adaptive conflictless schedule can be chosen. The simplified block diagram on Fig. 7 presents algorithm for adaptive conflictless schedule preparation. This algorithm can be improved by using association rules [3].

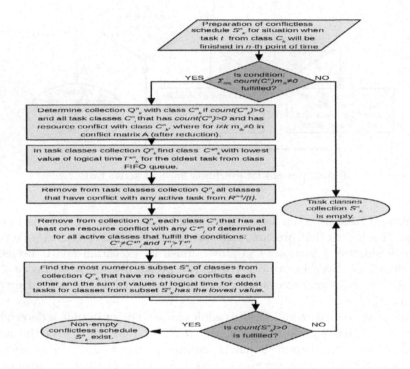

**Fig. 7.** Algorithm of adaptive conflictless schedule preparation.

## 5.3    Task Starvation Problem in Conflictless Scheduling

To prevent task starvation in class FIFO queue, additional condition has been proposed for each class $C_j^n$ belonging to conflictless schedule $S_i^n$: if resource conflict exists between different classes $C_j^n$ and $C_k^{*n}$ determined for each remaining active class $C_k^n$ then its logical time values has to satisfy condition $T_j^n > T_k^{*n}$ for oldest task waiting in those class queues. This additional condition prevents situation when conflictless schedule $S_i^n$ includes task from class that blocks execution of older task from class queue $C_k^{*n}$ in conflictless schedule determined in next $(n+1)$ point in time where active task from class $C_k^n$ finish its execution.

If many conflictless schedules exist for situation when active task finishes its execution, then arbitration rule has to be used. As arbitration rule was proposed most numerous collection of classes $S_i^n$ with the lowest sum of logical time values of its oldest waiting tasks. Proposed arbitration rule prevents task starvation in class FIFO queue. Choosing from classes the group of oldest waiting tasks that each other have none resource conflict and are resource conflict free with active tasks guarantees that all next prepared conflictless schedules will not include task always from the same classes if other conflicted class queues are not empty. This assures liveness requirement of for each waiting task, that will be executed in task processing environment according to the conflictless schedule.

## 6    Conclusions

This paper presents alternative approach to task concurrency control in high contention environment with limited number of global resources. It bases on preparation of separate FIFO queues for any global resource groups and using them to control executions of tasks without occurrences of resource conflicts. The proposed algorithm for preparation of conflictless schedules assures there is no task starvation, according to this schedule task FIFO queues can be emptying and tasks execution is started without interruptions in access to global resources. Efficient preparation of conflictless schedule needs dedicated structures like proposed task classes and conflict matrix. These structures enable preparation of many adaptive conflictless schedules, which are use for determination order of task executions without resources conflicts.

Elimination of resource conflict between tasks executed in parallel causes, that there will be no deadlock due to access to global resources. Software developer can prepare many tasks for parallel execution, which access various global resources from different groups and simply process them according to conflictless schedule without any resource conflict. Using conflictless scheduling no other synchronization mechanism is required to synchronization of task access to global resources. Additionally in transaction processing environments like OLTP or DBMS using conflictless scheduling assures no transaction rollback due to resource conflict occurrence.

## References

1. Bakkum, P., Skadron, K.: Accelerating SQL database operations on a GPU with CUDA. In: Proceedings of the 3rd Workshop on General-Purpose Computation on Graphics Processing Units, GPGPU 2010, pp. 94–103. ACM, New York (2010)
2. Bernstein, P.A., Newcomer, E.: Morgan Kaufman (2009)
3. Duraj, A.: Confictless task scheduling using association rules. In: Kozielski, S., Mrozek, D., Kasprowski, P., Małysiak-Mrozek, B., Kostrzewa, D. (eds.) BDAS 2016. CCIS, vol. 613, pp. 283–292. Springer, Heidelberg (2016)
4. Harish, P., Narayanan, P.J.: Accelerating large graph algorithms on the GPU using CUDA. In: Aluru, S., Parashar, M., Badrinath, R., Prasanna, V.K. (eds.) HiPC 2007. LNCS, vol. 4873, pp. 197–208. Springer, Heidelberg (2007)

5. He, B., Yu, J.X.: High-throughput transaction executions on graphics processors. Proc. VLDB Endowment **4**(5), 314–325 (2011)
6. Smolinski, M.: Rigorous history of distributed transaction execution with systolic array support. In: Information Systems Architecture and Technology New Developments in Web-Age Information Systems, pp. 235–245. Oficyna Wydawnicza Politechniki Wroclawskiej, Wroclaw (2010)
7. Tanenbaum, A., Bos, H.: Modern Operating Systems, 4th edn. Prentice Hall, Upper Saddle River (2014)
8. Wu, H., Chin, W.N., Jaffar, J.: An efficient distributed deadlock avoidance algorithm for the and model. IEEE Trans. Softw. Eng. **28**(1), 18–29 (2002)

# Conflictless Task Scheduling Using Association Rules

Agnieszka Duraj[✉]

Institute of Information Technology, Technical University of Lodz,
Wolczaska 215, 90-924 Lodz, Poland
agnieszka.duraj@p.lodz.pl
http://it.p.lodz.pl

**Abstract.** The proposed rules of conflictless task scheduling is based on binary representation of tasks. Binary identifiers promote the process of rapid detection of conflicts between tasks. The article presents the concept of conflictless tasks scheduling using one of the data mining methods, namely association rules.

**Keywords:** Conflictless tasks · Association rules · The conflict of parallel tasks · Mutual exclusion · Task synchronization

## 1  Introduction

In view of rapid development of technology, modern information systems, and in particular decision support systems and expert systems have to solve two common problems. The first problem relates to a large amount of information that originates mostly from many sources. The problem associated with the integration of input from multiple sources is due to the strong distribution of data. Currently, data fusion is increasingly used in technical, medical and economic solutions, see for example in [4,6,7]. The second one concerns the parallel execution of tasks in the system. Tasks can be divided into conflictless and conflicting. For the conflicting task the so-called cooperative concurrency or competitive concurrency are used. In cooperative concurrency the tasks work together using communications and synchronization mechanism. In competitive concurrency tasks do not know each other and are rivals in obtaining resources necessary for further processing [17,18,23]. The selection of appropriate methods of communication and synchronization is very important in the case of simultaneous processing. Synchronization mechanism used as a lock or semaphore could also lead to a queue of waiting tasks. Improper management of queues synchronization mechanism may lead to the risk of stagnation of tasks. Eliminating the resulting conflicts can be realized by preparing schedules. However, performance of tasks is planned in the schedule in advance and therefore any newly emerging task must be initially suspended and may be resumed in a finite time. Therefore, there is a kind of deadlock of tasks, as discussed in [5,22]. See also [19–21].

© Springer International Publishing Switzerland 2016
S. Kozielski et al. (Eds.): BDAS 2016, CCIS 613, pp. 283–292, 2016.
DOI: 10.1007/978-3-319-34099-9_22

Algorithms for discovering association rules are very popular tool for detecting dependencies. The association rules were explained in [8,13] or [10]. See for instance [9,12,14,15]. These rules are in the form of conditional sentences. Their structure is based on the logical implication "if sentence $Z_1$, then sentence $Z_2$" $Z_1 \Rightarrow Z_2$. In the mathematical sense, it should be described as the distribution of events both in time and space and, therefore, structural relationships and cause-effect relationships. Association rules are an example of learning without a teacher, which can be divided according to their application or utility. This is discussed more widely in the works e.g. [11,16,24].

This study attempts to find conflictless tasks using association rules and scheduling tasks using association rules. The work consists of the following sections. After a brief introduction in Sect. 2, the basic principles of conflictless tasks are discussed. The definition of conflictless task is given. Section 3 describes the procedure of performing conflictless tasks. The next section regards the proposed new procedure for the detection and scheduling of conflictless tasks using association rules. Section 5 presents the simulations (examples) of the implementation of procedures for different cases.

## 2    Definitions and Assumptions of Conflictless Tasks

Let us define the task as a series of operations processed without the users intervention. The task be defined as the shortest sequence of operations with all the private resources to perform those operations. The collection of all the resources of the task is known. In addition, it is known that:

(a) the duration of the task is not known and is not limited,
(b) the order of execution of the required task is not important,
(c) the task can be repetitive and the set of all global resources of the task is known before it is executed,
(d) the choice of operation for one job defines the selection of necessary resources and affects the time of duration of the task,
(e) the task can be divided into subtasks.

Each task is assigned a binary identifier IRW or IR. IRW stands for resources that are read-write and IR represents resources that are only read. The conflict between two tasks expressed in the form of formula (1) was defined and proven in [19].

$$[(IRW_i \text{ and } IRW_j)x \text{ or } (IR_i \text{ and } IR_j)] \neq 0 \tag{1}$$

Let there be a group of R running tasks. It is then possible to create an array of conflicts of the tasks as follows. Each task is sorted by the specified class definition as in (2).

$$C_k = \{t_i : IRW_i = IRW_j \wedge IR_i = IR_j\} \tag{2}$$

The array of conflicts created on the basis of formula (2) allows checking whether there is a conflict between the tasks of two classes. If there is a conflict between

**Table 1.** Sample array of conflicts. The value 1 means there is a conflict between the classes of tasks. The remaining cells of the array with a value of 0 mean the absence of conflict.

|       | $C_1$ | $C_2$ | $C_3$ | $C_4$ | $C_5$ | $C_6$ | ... | $C_N$ |
|-------|-------|-------|-------|-------|-------|-------|-----|-------|
| $C_1$ | 1     | 0     | 0     | 0     | 1     | 0     | ... | 0     |
| $C_2$ | 0     | 1     | 0     | 1     | 0     | 0     | ... | 0     |
| $C_3$ | 0     | 1     | 1     | 0     | 0     | 0     | ... | 0     |
| $C_4$ | 0     | 0     | 0     | 1     | 0     | 0     | ... | 1     |
| ...   | ...   | ...   | ...   | ...   | ...   | ...   | ... | ...   |
| $C_N$ | 0     | 0     | 0     | 0     | 0     | 0     | ... | 1     |

the tasks of the classes $C_i$ and $C_j$, the value in the array of conflicts is different from 0. In other words, if the value in the array of conflicts $m_{ij} \neq 0$ is different from zero, then every task of class $C_i$ is in conflict with each task of $C_j$ class. If a new class $C_n$ is added to the array of conflicts, the conflict between the class $C_n$ and the already existing classes is verified. All the tasks of the new class $C_n$ are suspended until the conflicts are determined. Graphic interpretation of the conflict matrix is shown in Table 1. Specific aspects regarding task classes and conflict matrix are given in [20,21].

## 3   The Procedure for Execution of Tasks

The rules of executing conflictless tasks can be reduced to a few steps. Each new task must be verified whether its execution will cause conflict or not. The procedure is triggered by the occurrence of a new task $Z_i$ with binary identifiers $IRW$ or $IR$. Class $C_k$ is called active if one of the tasks which it contains is being executed at the given moment. The call up the procedure is followed If there is a request of task with binary identifiers IRW or IR. This procedure, therefore, can be written as follows:

(1) Check whether there is a class to which the requested task belongs.
  (1a) If the class exists, check if it is empty.
    (2a) If it is empty, check whether there is a conflict between other active classes.
      If there is a conflict, suspend the tasks.
      Otherwise, set the class of the new task as active.
    (2b) If the class to which the task belongs is not empty, the task should be suspended.
  (1b) If the class does not exist, create a new class and enter it to the matrix of conflicts and go to step 2a)(See e.g. [20]).

# 4  The Concept of Execution (Scheduling) of Conflictless Jobs

The concept of scheduling conflictless tasks presented in this paper is based on association rules.

Association rules, sometimes also referred to as associative, take the form of conditional sentences. Their structure is based on logical implications. For example, "if the task $Z_1$, then task $Z_2$". The mathematical implication is interpreted as a causal relationship, in the result of which it is possible to arrange events in time and space. Association rules describe the relationship between attributes, not having a predetermined correct answer. It could be said that they are an example of learning without a teacher. The level of support and confidence must be determined for each such rule. In traditional systems that use association rules, the concept of a frequent set is introduced, and its frequency is determined. The most common application of association rules is analysis of a shopping cart. We analyze individual products and we look for rules that may determine whether there are any dependencies in the purchases, and if so, what they are. In the case of conflict detection transactions are appropriate task classes $C_1, C_2, ..., C_N$ shown in Table 1. Let us transform the conflict matrix presented in Table 1 into the transaction matrix used for association rules, for example, when analyzing a shopping cart. The transactional associative array of conflicts is shown in Fig. 1 in Table A. Each record of Table A with the value 1 denotes a class that is in conflict with another class. Because we are interested in conflictless tasks, we transform Table A in Fig. 1 into an associative array of conflictless tasks by changing the designation. This will enhance the understanding of further steps of the procedure of detection of conflictless tasks classes. So now let 1 denote the class of conflictless tasks, and table A is translated into Table B in Fig. 1. In this case we detect conflictless classes and therefore the set of events will be conflictless tasks classes. According to the definition of a frequent set, as for example in [11, 24], we assume that a frequent set will be a set of events that occur minimum a certain number of times, eg. 4.

Table A

|   | $C_1$ | $C_2$ | $C_3$ | $C_4$ | $C_5$ | $C_6$ | ... | $C_N$ |
|---|---|---|---|---|---|---|---|---|
| 1 | 1 | 0 | 1 | 1 | 0 | 1 | ... | 0 |
| 2 | 0 | 1 | 0 | 0 | 1 | 0 | ... | 1 |
| 3 | 1 | 0 | 1 | 1 | 0 | 1 | ... | 0 |
| 4 | 1 | 0 | 0 | 1 | 0 | 0 | ... | 0 |
| ... | ... | ... | ... | ... | ... | ... | ... | ... |
| N | 1 | 0 | 0 | 0 | 1 | 0 | ... | 1 |

Table B

|   | $C_1$ | $C_2$ | $C_3$ | $C_4$ | $C_5$ | $C_6$ | ... | $C_N$ |
|---|---|---|---|---|---|---|---|---|
| 1 | 0 | 1 | 0 | 0 | 1 | 0 | ... | 1 |
| 2 | 1 | 0 | 1 | 1 | 0 | 1 | ... | 0 |
| 3 | 0 | 1 | 0 | 0 | 1 | 0 | ... | 1 |
| 4 | 0 | 1 | 1 | 0 | 1 | 1 | ... | 1 |
| ... | ... | ... | ... | ... | ... | ... | ... | ... |
| N | 0 | 1 | 1 | 1 | 0 | 1 | ... | 0 |

**Fig. 1.** Table A Sample transactional associative array of conflicts. Symbols: 1 conflicting class, 0 conflictless class. Table B An example of transactional associative array of conflictless tasks, where the value 1 means the conflictless class, and 0 class with conflict.

In the presented case we detect conflictless classes; therefore the set of events will be a conflictless task class. According to the definition of a frequent set given for instance in [1], we assume that the frequency of a set is the number of instances of the active class. It can be said that the set of active classes is a frequent set if it occurs at least the minimum number of times (which is defined by the expert). For each association rule, two very important statistical measures, namely confidence and support, are determined.

Support is regarded as a measure evaluating the specific association rule. It specifies the number of active classes in the analyzed sets confirming a given rule. This measure is symmetric for the rules of the type $Z_i \longrightarrow Z_j$. Confidence is a measure defining the certainty of a given rule and it is an asymmetric measure. The support of an association rule $Z_i \longrightarrow Z_j$ is the ratio of the number of observations that satisfy the condition $Z_i \cap Z_j$ to the total number of observations, while the support of a rule is equal to the probability of occurrence of the event $Z_i \cap Z_j$.

Confidence of an association rule $Z_i \longrightarrow Z_j$ is the ratio of the number of observations which satisfy the condition $Z_i \cap Z_j$ to the number of observations that satisfy the condition $Z_j$. The confidence of a rule can also be regarded as the conditional probability of the occurrence of the even $Z_i$ assuming that the even $Z_j$ also took place. It is a measure of the accuracy of the rule. In our example, we will describe the confidence by (3) and the support by (4).

$$confidence = \frac{the\ number\ of\ conflictless\ tasks Z_i\ and Z_j}{the\ number\ of\ conflictless\ tasks\ containing Z_i} \tag{3}$$

$$support = \frac{the\ number\ of\ conflictless\ tasks\ containing Z_i\ and Z_j}{the\ total\ number\ of\ conflictless\ tasks} \tag{4}$$

The threshold, which is assumed by the designer or analyst, and which defines the rule as strong, is very important. The support and confidence must be greater than or equal to the minimum value assumed before the research. Depending on the nature of the data set and the character of the events detected by the analyst, these values may be different. In general, the threshold is very high, but there are association rules whose support threshold is set at 1 %, eg. in the investigation of tax evasion.

## 5    The Procedure for Scheduling

The basic approach [1–3] to the detection of association rules presupposes the use of a minimum support threshold at each level of abstraction. The procedure involves the use of different thresholds (points 6a and 6b). Determining one threshold (regardless of the value of minsup) would lead, in the case considered in this paper, to performing the tasks that were previously performed and about which we have some knowledge. The new task of the new class would not have a chance to activate because it would not be any rule for this task. It is therefore justified for such a task to be performed as first.

**Table 2.** Examples of active tasks classes in the subsequent intervals.

| | Classes active in a given moment | | Classes active in a given moment |
|---|---|---|---|
| 1 | $C_1, C_2, C_6$ | 8 | $C_1, C_3, C_2$ |
| 2 | $C_2, C_5, C_7$ | 9 | $C_7, C_5, C_4$ |
| 3 | $C_3, C_4, C_6, C_7$ | 10 | $C_4, C_6$ |
| 4 | $C_2, C_3, C_4, C_6$ | 11 | $C_2, C_1, C_4, C_7$ |
| 5 | $C_1, C_4, C_5$ | 12 | $C_5, C_4, C_7$ |
| 6 | $C_7, C_5, C_4, C_3$ | 13 | $C_7, C_6, C_5, C_4$ |
| 7 | $C_3, C_6$ | 14 | $C_6, C_2, C_3, C_4, C_1$ |

Similar procedure applies if there are simple rules $Z_i \longrightarrow Z_j$. The algorithm prefers performing the task that is not on the list of rules. Then it will perform the tasks which belong to the classes of simple association rules (two-class) with the lowest level of confidence.

1. Enter a new task t of the class $C_k$.
2. Check the currently running tasks.
3. Detect the association rules for existing conflictless tasks.
4. Verify if the class $C_k$, to which the task $t$ belongs, is in the list of the discovered association rules.
   4a. If not, then do the task t belonging to the class $C_k$ in the first place.
   4b. otherwise go to step 5.
5. Start the execution of the tasks on the list of association rules.
6. Check how many rules class $C_k$ belongs to
   6a. If the association rules containing the class $C_k$ are simple (two-class), start carrying out the tasks starting from the rule with the lowest level of confidence.
   6b. If the association rules containing the class $C_k$ are complex rules, the rule of the highest level of confidence is executed.
   6c. If the association rules have the same value of confidence level, the choice is usually random.

The procedure of searching rules should be automatically updated after a user-defined time or when the next $z$ new tasks appear.

## 6 Illustration of the Proposed Method with an Example

Let us assume that the records of Table 2 contain classes containing the tasks which are to be executed. They are conflictless. The matrix format of Table 2 is shown in Table 3.

The frequency of a set is the number of active instances of the class. Let us assume, in this case, that the frequent set $F_k$ is a frequent set if it appears at least 4 times. Thus, the set of active classes will be a frequent set if they

**Table 3.** The matrix data format

| Transaction | $C_1$ | $C_2$ | $C_3$ | $C_4$ | $C_5$ | $C_6$ | $C_7$ |
|---|---|---|---|---|---|---|---|
| 1 | 1 | 1 | 0 | 0 | 0 | 1 | 0 |
| 2 | 0 | 1 | 0 | 0 | 1 | 0 | 1 |
| 3 | 0 | 0 | 1 | 1 | 0 | 1 | 1 |
| 4 | 0 | 1 | 1 | 1 | 0 | 1 | 0 |
| 5 | 1 | 0 | 0 | 1 | 1 | 0 | 0 |
| 6 | 0 | 0 | 1 | 1 | 1 | 0 | 1 |
| 7 | 0 | 0 | 1 | 0 | 0 | 1 | 0 |
| 8 | 1 | 1 | 1 | 0 | 0 | 0 | 0 |
| 9 | 0 | 0 | 0 | 1 | 1 | 0 | 1 |
| 10 | 0 | 0 | 0 | 1 | 0 | 1 | 0 |
| 11 | 1 | 1 | 0 | 1 | 0 | 0 | 1 |
| 12 | 0 | 0 | 0 | 1 | 1 | 0 | 1 |
| 13 | 0 | 0 | 0 | 1 | 1 | 1 | 1 |
| 14 | 1 | 1 | 1 | 1 | 0 | 1 | 0 |
| Frequent set | 5 | 6 | 6 | 10 | 6 | 7 | 7 |

**Table 4.** Two-element sets

| $C_1, C_2$ | 4 | $C_2, C_3$ | 3 | $C_3, C_5$ | 1 | $C_1, C_3$ | 2 | $C_2, C_4$ | 2 | $C_3, C_6$ | 4 | $C_5, C_6$ | 1 |
|---|---|---|---|---|---|---|---|---|---|---|---|---|---|
| $C_1, C_4$ | 3 | $C_2, C_5$ | 1 | $C_3, C_7$ | 2 | $C_1, C_5$ | 1 | $C_2, C_6$ | 3 | $C_4, C_5$ | 5 | $C_5, C_7$ | 5 |
| $C_1, C_6$ | 2 | $C_2, C_7$ | 2 | $C_4, C_6$ | 5 | $C_1, C_7$ | 1 | $C_3, C_4$ | 4 | $C_4, C_7$ | 6 | $C_6, C_7$ | 2 |

appear at least 4 times. The activity of a single class is large. Each one-element set, containing only one class, is a frequent set. It turns out that the frequency of each one-element set is $\geq 4$. Thus, each one-element set is a frequent set $F_1 = \{C_1, C_2, C_3, ..., C_7\}$.

We create the possible two-element sets (in general, in order to find $F_k$, the A priori algorithm first constructs $k$-element candidates by merging $F_{k-1}$ sets, then cuts them according to property). Possible two-element sets $F_2$. See Table 4. We select those two-element sets for which the frequency is greater or equal than 4. Therefore, we obtain the set
$F_2 = \{\{C_1, C_2\}; \{C_3, C_4\}; \{C_3, C_6\}; \{C_4, C_5\}, \{C_4, C_6\}; \{C_4, C_7\}; \{C_5, C_7\}\}$. We create three-element sets based on $F_2$. We must remember that the sets of events are merged if their first $k-1$ elements are the same. We did not find 3-class frequent sets, therefore we proceed to determine the association rules. (See Table 5). Table 6 is presents the determined values of the confidence and support for the rules.

The detection of the above rules allows the execution of the tasks from these classes in the first place. If there are jobs from classes $C_1, C_2, C_4, C_5, C_7$, thanks to the detected rules it is known which tasks from another class can be executed.

**Table 5.** Three class frequent sets

| $C_3, C_4, C_6$ | 2 | $C_4, C_5, C_7$ | 3 |
|---|---|---|---|
| $C_4, C_5, C_6$ | 1 | $C_4, C_6, C_7$ | 1 |

**Table 6.** The determined values of the confidence and support for the rules.

| Rule | Support | Confidence | Rule | Support | Confidence |
|---|---|---|---|---|---|
| IF C1 THEN C2 | $4/14 = 0.28$ | $4/5 = 0.8$ | IF C6 THEN C3 | $4/14 = 0.28$ | $4/7 = 0.57$ |
| IF C1 THEN C3 | $2/14 = 0.14$ | $2/5 = 0.4$ | IF C4 THEN C5 | $5/14 = 0.35$ | $4/10 = 0.4$ |
| IF C2 THEN C1 | $4/14 = 0.28$ | $4/6 = 0.66$ | IF C5 THEN C4 | $5/14 = 0.35$ | $4/6 = 0.66$ |
| IF C3 THEN C4 | $4/14 = 0.28$ | $4/6 = 0.66$ | IF C4 THEN C6 | $5/14 = 0.35$ | $5/10 = 0.5$ |
| IF C4 THEN C3 | $4/14 = 0.28$ | $4/10 = 0.4$ | IF C6 THEN C4 | $5/14 = 0.35$ | $5/7 = 0.71$ |
| IF C3 THEN C6 | $4/14 = 0.28$ | $4/6 = 0.66$ | IF C4 THEN C7 | $6/14 = 0.42$ | $6/10 = 0.6$ |
| IF C5 THEN C7 | $5/14 = 0.35$ | $5/6 = 0.83$ | IF C7 THEN C4 | $6/14 = 0.42$ | $6/7 = 0.85$ |
| IF C7 THEN C5 | $5/14 = 0.35$ | $5/7 = 0.71$ | | | |

Starting the procedure of detection the rules of conflictless tasks should be performed once in a while, so that there is no so called starvation of tasks. Let us consider the following cases.

CASE 1. Let $C_1, C_2, ..., C_{10}$ denote the current classes. The new task $t \in C_9$. We assume that the association rules given in Table 6 have been detected. In Table 6 we will not find a rule that contains class $C_9$, to which the new task $t$ belongs. Therefore, this task will be executed first. If we enter more tasks to be executed and no class to which the newly introduced tasks belong is on the list of the discovered association rules, the tasks are executed in the order of entry.

CASE 2. We assume that a new task $t \in C_3$ was introduced to be performed. We also assume that the association rules given in Table 6 were detected. Therefore, the rules in the following form were found for class $C_3$:
IF $C_1$ THEN $C_3$; confidence=0.4; IF $C_3$ THEN $C_4$; confidence=0.66;
IF $C_4$ THEN $C_3$; confidence=0.4; IF $C_3$ THEN $C_6$; confidence=0.66;
IF $C_6$ THEN $C_3$; confidence=0.57;
The tasks from the classes $C_1, C_4$ or $C_6$ can be performed with the task $t \in C_3$. According to the calculated level of confidence we start scheduling tasks from the lowest level of confidence. We will get:
IF $C_1$ THEN $C_3$; confidence=0.4; IF $C_4$ THEN $C_3$; confidence=0.4;
IF $C_6$ THEN $C_3$; confidence=0.57;
Therefore, the task $t \in C_3$ will be performed after the task from classes $C_1$ and $C_4$ is completed, and, subsequently, the task from class $C_6$ will be taken into account.

CASE 3. We assume that a new task $t \in C_3$ was introduced to be performed.
For the class $C_3$ the following rules were found:
IF $C_1, C_2, C_5$ THEN $C_3$; confidence=0.4;
IF $C_4, C_5, C_7$ THEN $C_3$; confidence=0.61;
IF $C_1, C_4, C_6$ THEN $C_3$; confidence=0.35;
IF $C_2, C_5, C_7$ THEN $C_3$; confidence=0.57;
Assuming that tasks from classes $C_1, ..., C_7$ are executed, then, according
to the highest level of confidence, if one task from classes $C_4, C_5, C_7$ is per-
formed, then the task from class $C_3$ can be executed.

# 7    Conclusion

Determining statistical measure to optimize task scheduling is known to be difficult. A large number of tasks increases, among others, processing cycle time,
response time and delay time. Besides minimizing the time another important
element is predictability. This paper attempts to schedule conflictless tasks. An
alternative approach to cooperative and competitive concurrency solutions was
presented. An attempt was made to determine the measure of predictability of
detection and scheduling of tasks using association rules.

Automatic detection of association rules will determine the order of their
execution. At the same time, the procedure provides for situations when a new
task, which is not shown in the list of found rules, appears. In this case, the
priority privilege of executing such a task applies. The new request-new class is
executed first. Such functionality of the algorithm seems justified. Since there is
no class in the list of rules to which the task belongs, the algorithm must learn
what kind of limitations and requirements this class has. It should also be noted
that the proposed procedure for scheduling tasks using association rules does
not require using any additional task synchronization mechanisms. There is no
deadlock or starvation of tasks.

In subsequent works an attempt will be made to replace the Apriori association rules detection algorithm with the FP-Growth algorithm or other modifications. An implementation of the proposed procedure for the GPU and its
practical applications will also be presented.

# References

1. Agrawal, R., Imieliński, T., Swami, A.: Mining association rules between sets of
   items in large databases. ACM SIGMOD Rec. **22**(2), 207–216 (1993)
2. Agrawal, R., Mannila, H., Srikant, R., Toivonen, H., Verkamo, A.I., et al.: Fast discovery of association rules. Adv. Knowl. Discov. Data Min. **12**(1), 307–328 (1996)
3. Agrawal, R., Srikant, R., et al.: Fast algorithms for mining association rules.
   In: Proceedings of the 20th International Conference on Very Large Data Bases,
   VLDB. vol. 1215, pp. 487–499 (1994)
4. Bedworth, M., O'Brien, J.: The omnibus model: a new model of data fusion? IEEE
   Aerosp. Electron. Syst. Mag. **15**(4), 30–36 (2000)
5. Belik, F.: An efficient deadlock avoidance technique. IEEE Trans. Comput. **39**(7),
   882–888 (1990)

6. Duraj, A.: Application of fussyclassify of data classification. J. Appl. Comput. Sci. **21**, 39–52 (2013)
7. Duraj, A., Krawczyk, A.: Outliers detection of signals in biomedical information systems fusion. Electr. Rev. **12b**, 56–60 (2012)
8. Guillet, F., Hamilton, H.J.: Quality Measures in Data Mining, vol. 43. Springer, Heidelberg (2007)
9. Han, E.H., Karypis, G., Kumar, V., Mobasher, B.: Clustering based on association rule hypergraphs. University of Minnesota, Department of Computer Science (1997)
10. Hilderman, R., Hamilton, H.J.: Knowledge Discovery and Measures of Interest, vol. 638. Springer, Heidelberg (2013)
11. Hipp, J., Güntzer, U., Nakhaeizadeh, G.: Algorithms for association rule mining – a general survey and comparison. SIGKDD Explor. Newsl. **2**(1), 58–64 (2000)
12. Lin, W., Alvarez, S.A., Ruiz, C.: Efficient adaptive-support association rule mining for recommender systems. Data Min. Knowl. Discovery **6**(1), 83–105 (2002)
13. McGarry, K.: A survey of interestingness measures for knowledge discovery. Knowl. Eng. Rev. **20**(01), 39–61 (2005)
14. Mobasher, B., Dai, H., Luo, T., Nakagawa, M.: Effective personalization based on association rule discovery from web usage data. In: Proceedings of the 3rd International Workshop on Web Information and Data Management. WIDM 2001, pp. 9–15 (2001)
15. Rajendran, P., Madheswaran, M.: Hybrid medical image classification using association rule mining with decision tree algorithm. CoRR abs/1001.3503 (2010)
16. Sarawagi, S., Thomas, S., Agrawal, R.: Integrating association rule mining with relational database systems: alternatives and implications. SIGMOD Rec. **27**(2), 343–354 (1998)
17. Silberschatz, A., Galvin, P., Gagne, G.: Applied Operating System Concepts. Wiley, Hoboken (2001)
18. Silberschatz, A., Galvin, P., Gagne, G.: Operating System Concepts. Wiley, Hoboken (2012)
19. Smolinski, M.: Rigorous history of distributed transaction execution with systolic array support. XXXI ISAT Conf. Inf. Syst. Archit. Technol. New Dev. Web-Age Inf. Syst. **28**(1), 235–254 (2010)
20. Smolinski, M.: Conflictless task scheduling concept. In: Borzemski, L., Grzech, A., Świątek, J., Wilimowska, Z. (eds.) Information Systems Architecture and Technology: Proceedings of 36th International Conference on Information Systems Architecture and Technology - ISAT 2015 - Part I. Advances in Intelligent Systems and Computing, vol. 429. Springer, Heidelberg (2016)
21. Smolinski, M.: Coordination of parallel tasks in access to resource groups by adaptive conflictless scheduling. In: Kozielski, S., Mrozek, D., Kasprowski, P., Malysiak-Mrozek, B., Kostrzewa, D. (eds.) Beyong Databases, Architectures and Structures, BDAS 2016, Poland. Communications in Computer and Information Science, pp. 272–282. Springer, Heidelberg (2016)
22. Stetsyura, G.G.: Fast decentralized algorithms for resolving conflicts and deadlocks in resource allocation in data processing and control systems. Autom. Remote Control **71**(4), 708–717 (2010)
23. Tanenbaum, A.S., Bos, H.: Modern Operating Systems. Prentice Hall Press, Upper Saddle River (2014)
24. Zhang, C., Zhang, S.: Association Rule Mining: Models and Algorithms. Springer, Heidelberg (2002)

# Distributed Computing in Monotone Topological Spaces

Susmit Bagchi[✉]

Department of Aerospace and Software Engineering (Informatics),
Gyeongsang National University, Jinju 660-701, South Korea
susmitbagchi@yahoo.co.uk

**Abstract.** In recent times, an alternate approach to model and analyze distributed computing systems has gained research attention. The alternate approach considers higher-dimensional topological spaces and homotopy as well as homology while modeling and analyzing asynchronous distributed computing. This paper proposes that the monotone spaces having ending property can be effectively employed to model and analyze consistency and convergence of distributed computing. A set of definitions and analytical properties are constructed considering monotone spaces. The inter-space relationship between simplexes and monotone in topological spaces is formulated.

**Keywords:** Distributed computing · Topology · Monotone spaces · Shared object · Consistency

## 1 Introduction

The applications of distributed computing systems are pervasive in nature. The modern distributed database systems are prime examples of applications of distributed computing platforms for data storage and processing. The distributed computing systems contain concurrently executing distributed processes mapped into a set of nodes. The nodes are connected by network and, the distributed processes communicate through synchronous as well as asynchronous protocols. The network-topology of the interconnection graph in the systems involving distributed nodes can be static or dynamic in nature depending on the architecture. In general, the interactive distributed processes compute following the iterated shared memory model [3]. According to this model, distributed processes access and update a sequence of shared objects asynchronously following a deterministic sequence. The asynchronous and concurrent accesses to shared memory locations by multiple distribute processes invite the issues related to race condition and concurrency control. The control of concurrency is highly complex in distributed computing systems due to the availability of partial information about global states of computations. However, the algebraic topology and higher dimensional automata can be applied to model complexity of concurrent computation [9,10]. The employment of algebraic topological theory in computation architecture

© Springer International Publishing Switzerland 2016
S. Kozielski et al. (Eds.): BDAS 2016, CCIS 613, pp. 293–300, 2016.
DOI: 10.1007/978-3-319-34099-9_23

enables the handling of concurrency control in concurrently executing programs [6]. Researchers have indicated that, the distributed computing systems have a closer interplay with algebraic topology [5]. However, the monotone spaces are the generalized form of topological spaces [7]. This paper argues that, the structural dynamics of monotone spaces have direct applications in modeling and analyzing distributed computing systems. The monotone property of topological spaces may be used to determine consistent cuts and to analyze the convergence of distributed computing in monotone topological spaces. This paper proposes the modeling and analysis of distributed computation in monotone topological spaces. It is illustrated that, the global consistency of a distributed computation can be maintained in the presence of closure in the monotone spaces. The concept of boundary elements within the state-space of distributed processes is introduced. The topological spaces of distributed computation are partitioned into two sub-spaces and, connectedness of the monotone topology is analyzed. A set of analytical properties is presented following the model of distributed computing in monotone topological spaces. The closure property of monotone spaces to maintain consistent cuts in a distributed computation is constructed. Rest of the paper is organized as follows. Section 2 describes related work. Section 3 presents definitions and constructions of the model. Section 4 explains a set of analytical properties. Section 5 presents discussions on the topics along with applications and, Sect. 6 concludes the paper.

## 2    Related Work

The applications of concepts of algebraic and combinatorial topology into the domain of distributed computing have gained appreciations offering promising results. The algebraic topology is about higher dimensional geometrical shapes of objects [1]. The topological structure of higher dimensional automata can model the properties of concurrent processes and distributed computation [9]. The combinatorial and algebraic topological methods are applicable to model and analyze distributed computing systems following asynchronous iterated shared memory model [3,5]. However, the issues related to concurrency control in distributed asynchronous systems are highly complex. Interestingly, the homotopy theory along with topology can be applied to solve the mutual exclusion issues related to concurrency of locally executing processes [6,10]. It is shown that stability properties of distributed concurrent processes can be proved by considering homotopy [8]. The semaphore objects can be formed by using partially ordered topological spaces and, can be extended to analyze deadlock as well as serializability properties of concurrent processes [4]. However, the general homotopy does not prevent time-reversal and thus, the directed homotopy is required in model construction.

The structural forms of algebraic topology can be equally applied to the synchronous as well as asynchronous distributed computing systems [11]. The topological objects are generalizations of graphs with higher dimensions and, the connectivity property of topological objects can be employed to analyze the computability problems of processes in distributed computing [2,11]. The simulation methods utilizing topological objects are proposed by researchers to prove impossibility results in distributed computing systems [2]. Furthermore, the algebraic topological concepts are employed in deriving the time complexity bounds to compute approximate agreement for the iterated immediate snapshot model [13]. In this approach, the time complexities of the distributed processes are mapped into the degree of sub-division of input complex in order to generate output complex. In general, these algebraic topological objects form simplical complexes to model distributed processes and protocols, which are having very complex and rigid structural geometries [12]. The combinatorial topological model of asynchronous processes and the wait-free computation involve defining combinatorial relations in topological spaces [12,14]. However, these relations are static and do not consider operational mode of interleaving computations. In case of immediate snapshot model of distributed computation, the simplical complexes can be reduced to manifolds [2]. This method reduces the complexity of structures of simplical complexes. These algebraic topological approaches do not consider monotone property of topological spaces, where monotone spaces are the more general form of topological spaces [7]. The extension of monotone topological spaces into distributed computing systems can determine the convergence and closure of consistent distributed computation.

## 3   Definitions and Model

In this section, the basic concepts are explained and a set of definitions is constructed in order to formulate the model. Let a distributed computation be represented as $D$ comprised of a set of distributed processes given by $P = \{pj : 1 \leq j \leq N, j \in \mathbf{Z}^+\}$. All the distributed processes are considered as finite state machines and a pj has a set of deterministic execution states denoted by $Sj(\phi \notin Sj)$. Hence, $D$ can be identified by process-states,

$$S(D) = \bigcup_{j=1}^{N} Sj \qquad (1)$$

The distributed processes execute as state machines and as a result state transitions occur in a process, which are governed by a function given by,

$$f : S(D) \rightarrow S(D)$$
$$f(sj \in Sj) \in Sj \backslash \{sj\} \qquad (2)$$

The distributed computation is considered to be deterministic if it converges to a space and such converging space of $D$ is defined as,

$$\overline{S(D)} = \bigcap_{j=1}^{N} Sj \qquad (3)$$

The set of all possible consistent cuts in $D$ is denoted by $C(D) \subset \Omega(S(D))$ where, $\Omega(.)$ is a power set. A process $pj \in P$ contains a set of channels connecting it to other nodes. The states of channels are given by, $Qj \subset P^2$ where, $Qj = \{(pj, pk) : pk \in P\}$. The process pj maintains the number of messages in transit within its channels by following a channel function given by, $\chi : Sj \times \mathbf{Z}^+ \rightarrow \mathbf{Z}^+$. The consistent cut of a distributed computation is highly dependent on the values of $\chi(.)$ at every node.

## 3.1  Definition: Boundary Elements

The set of boundary elements $Bj$ of a process pj is defined as, $\forall pj \in P$,

$$f(Bj \subset Sj) \in \overline{S(D)} \tag{4}$$

## 3.2  Definition: Computation in $D$

If $Bj \subset Sj$ then a computation in $D$ is defined as,

$$\overrightarrow{D} = S(D) \backslash f(\bigcup_{j=1}^{N} Bj) \tag{5}$$

## 3.3  Definition: Monotone Space in Distributed Computing

A monotone topological space over $\overrightarrow{D}$ is $(\overrightarrow{D}, g)$ having following properties:

$$g : \Omega(\overrightarrow{D}) \rightarrow \Omega(\overrightarrow{D})$$
$$g(\phi) = \phi$$
$$\forall A \in \Omega(\overrightarrow{D}), A \subset g(A) \tag{6}$$
$$\forall A, B \in \Omega(\overrightarrow{D}), A \subset B \Rightarrow g(A) \subset g(B)$$

Thus, if $\sum_{x \in a} \chi(x, z_x) = 0$ where, $a \in A$, $z_x \in \mathbf{Z}^+$ and, $A \subset \Omega(\overrightarrow{D}) \backslash C(\overrightarrow{D})$ then, $A$ is a set of consistent subcuts iff $|a| < N$.

## 3.4  Definition: Connected Monotone

A monotone topological space $(\overrightarrow{D}, g)$ of distributed computation $D$ is connected if following properties hold:

$$\forall sj \in \overrightarrow{D}, n \geq 1$$
$$A = \Omega(f^n(sj)) \backslash \{\phi\}$$
$$g(A) \backslash A \subset (\Omega(\overrightarrow{D}) \backslash \bigcup_{j=1}^{N} \Omega(Sj)) \tag{7}$$

## 3.5    Definition: Convergent Monotone

In a connected monotone topological space $(\overrightarrow{D}, g)$, $\forall \mathrm{pj} \in P$, $\forall sj \in \overrightarrow{D}$, if $\exists H \in g$ such that, $Bj \subset H$ then $(\overrightarrow{D}, g)$ is a convergent monotone of distributed computation $D$.

## 3.6    Definition: Consistent Monotone

A monotone topological space $(\overrightarrow{D}, g)$ of a distributed computation $D$ is called consistent when following property holds:

$$\forall A \in \Omega(\overrightarrow{D}), \qquad (8)$$
$$A \in C(\overrightarrow{D}) \Rightarrow g(A) \subset C(\overrightarrow{D})$$

# 4    Analytical Properties

A distributed computing system in monotone topological spaces has dynamic structures. The convergence and consistency properties of distributed computing vary in topological spaces over time depending on the executions of distributed processes. Moreover, the termination of a distributed computation can be validated in monotone topological spaces if a certain set of properties are maintained. In this section, a set of analytical properties of the distributed computing in monotone topological spaces is formulated.

## 4.1    Theorem 1: In a Convergent Distributed Computation $D$, $\overline{S(D)} \neq \phi$

**Proof:** The proof is by contradiction. Let $\overline{S(D)} = \phi$ in $D$. This indicates that, $\bigcup_{j=1}^{N} Bj = \phi$ in the distributed computation $D$. However, $(\overrightarrow{D}, g)$ is a convergent monotone. Let $\exists sj \in \overrightarrow{D}$ such that, $A = \{\{sj*\} : sj* \in Sj, \, sj* = f^n(sj), n \geq 1\}$ and, $E \in g(A)$. Thus, $\exists x \in E$ where, $f(x) \in S(D)$. However, $f(x) \neq \phi$, which is a contradiction. Hence, $\overline{S(D)} \neq \phi$.

## 4.2    Theorem 2: If $D$ is in convergent monotone topological spaces then $|Bj| \geq 1$ for all j

**Proof:** The proof is by contradiction. Let $\exists \mathrm{pj} \in P$ such that, $\forall sj \in Sj$, $f(sj) \notin \overline{S(D)}$. Now, $(\overrightarrow{D}, g)$ is a connected monotone if $D$ is convergent. Thus, $\exists A \in \Omega(\overrightarrow{D}) \backslash \bigcup_{j=1}^{N} \Omega(Sj)$, where $\exists x \in A$, $f(x) \notin \overline{S(D)}$. This further indicates, $\exists E \in g(A)$, $\exists x \in E$, $f(x) \notin \overline{S(D)}$. It concludes that $(\overrightarrow{D}, g)$ is not convergent, which is a contradiction. Hence, $\forall \mathrm{pj} \in P$, $Bj \neq \phi$.

## 4.3    Theorem 3: A Consistent Monotone Topological Space is Closed

**Proof:** The proof is by induction. Let $(\overrightarrow{D}, g)$ be a consistent monotone over distributed computation $D$ and, $C(\overrightarrow{D}) \subset \Omega(\overrightarrow{D})$. Let $A \in C(\overrightarrow{D}) \backslash X$ where, a consistent cut in distributed computation is $X \in C(\overrightarrow{D})$. Thus, if $E \subset A$ then, $g(E) \subset g(A)$ following monotone topological property. However, $g(A) \subset C(\overrightarrow{D})$. Hence, $(\overrightarrow{D}, g)$ is closed under consistency.

## 4.4    Theorem 4: A Distributed Computation Will Terminate if the Monotone Topological Spaces Are Consistent and Convergent

**Proof:** The proof is by induction. Let $\exists pj \in P$ such that, $|Bj| \geq 1$ and, $(\overrightarrow{D}, g)$ is a consistent as well as convergent monotone. Let $A \subset \Omega(\overrightarrow{D})$ such that, $A = C(\overrightarrow{D}) \backslash X$ where, $X \subset C(\overrightarrow{D})$ and, $\exists E \in X$, $Bj \cap E = \phi$. However, $A \subset g(A)$ and, $g(A) \subset C(\overrightarrow{D})$. Thus, $\exists Hj \in g(A)$ where, $Bj \cap Hj \neq \phi$ and, $f(Bj \cap Hj) \in \overline{S(D)}$. As $(\overrightarrow{D}, g)$ is consistent and convergent, hence, $\forall pj \in P$, $\exists Hj \in g(A)$ such that, $f(Bj \cap Hj) \in \overline{S(D)}$ iff $|Bj| \geq 1$. This completes the proof.

# 5    Applications and Discussions

The traditional modeling approaches of distributed computing systems consider set-theoretic discrete structures. The basic form of distributed computations in discrete structural domain is based on the concepts of monotonically increasing integers representing logical clocks. The analysis of consistency and determination of consistent cuts in a distributed computation employ the notion of logical clocks. Distributed computing and concurrency are traditionally considered as operational phenomena having discrete structures. However, specific structural forms and the associated changes in the corresponding structures can not be determined in traditional models of distributed computation. However, topology is about objects having higher dimensional geometrical structures, whereas the homotopy is about objects with continuous structures. Thus, algebraic topology of simplexes and the homotopy theory can be applied to model and analyze distributed asynchronous computations. The specific applications of monotone topological approaches to model distributed computation can be listed as, (1) *Determination of set of consistent structural forms of a distributed computation,* (2) *Easy detection of non-conformal states,* (3) *Accurate modeling of concurrent distributed computation* and, (4) *Simplification of fault detection and debugging of a distributed computation by determining non-conformal states and associated structures.*

In general, the conformal topological mapping between the input structures and output structures of simplical complexes of distributed computing is static in nature. The closure property of the dynamics of distributed computation can be evaluated in the presence of monotone spaces. The convergence of the distributed computation requires the presence of boundary elements within the state-space of the distributed processes. Moreover, the connectedness of monotone spaces maintains the structure of distributed computation within topological spaces.

The interrelation and mapping between the monotone spaces and simplexes can be formed. Let $X$ be a basis set and, $X_S = \Omega(F \in X)$ be represented in simplex structure in topological spaces. Thus, $K \subseteq F \Rightarrow K \in X_S$ and, $K \neq \phi$. However, if $X_m = g(F \in X)$ in monotone spaces, then $F \subset X_m$. This indicates, $X_m \subseteq X_S$. Again, by following monotone property it can be said that, $K \subset F \Rightarrow g(K) \subset g(F)$. On the other hand, $K \subset F \Rightarrow X_{mK} \subset X_m$ where, $X_{mK} = g(K)$. Hence, considering the base case of $K \subset F$, the mapping can be established between simplex and monotone space such that, $K \in X_S \Leftrightarrow X_{mK} \subset X_m$. Thus, there exists a bi-connective mapping between the monotone spaces and simplexes. The entire monotone topological spaces of distributed computation can be partitioned into two sub-spaces, where one sub-space contains the execution of distributed processes and another one contains the boundary of convergence. The consistency of the distributed computation is maintained in the connected and consistent monotone topological spaces. The convergent property of monotone space enforces the stability of distributed computation within the space of convergence. The consistency property of distributed computation effectively partitions the monotone spaces. The closure property of monotone spaces within the partition enables the determination of consistent cuts of distributed computation.

# 6 Conclusions

The concepts of monotone spaces can be employed to model and analyze the properties of distributed computation. The monotone along with ending property of topological spaces enables the evaluations of consistency and convergence of distributed computation. The static and logic based program analysis facilitates the determination of presence of boundary elements, which indicates the eventual convergence as well as stability of distributed computation. The consistency property of computation in monotone spaces enforces the determination of consistent cuts in the distributed computations.

# References

1. Armstrong, M.A.: Basic Topology. Springer, Berlin (1983)
2. Borowsky, E., Gafni, E.: Generalized FLP impossibility result for t-resilient asynchronous computations. In: The 25th Annual ACM Symposium on Theory of Computing. ACM (1993)
3. Borowsky, E., Gafni, E.: A simple algorithmically reasoned characterization of wait-free computation. In: The Sixteenth Annual ACM Symposium on Principles of Distributed Computing, pp. 189–198 (1997)
4. Carson, S.D., Reynolds, J.P.F.: The geometry of semaphore programs. ACM Trans. Program. Lang. Syst. 9(1), 25–53 (1987)

5. Conde, R., Rajsbaum, S.: An introduction to topological theory of distributed computing with safe-consensus. Electron. Notes Theoret. Comput. Sci. **283**, 29–51 (2012)
6. Fajstrup, L., Rauben, M., Goubault, E.: Algebraic topology and concurrency. Theoret. Comput. Sci. **357**, 241–278 (2006)
7. Ghosh, S.R., Dasgupta, H.: Connectedness in monotone spaces. Bull. Malays. Math. Sci. Soc. **27**(2), 129–148 (2004)
8. Goubault, E.: Some geometric perspectives in concurrency theory. Homology Homotopy Appl. **5**(2), 95–136 (2003)
9. Goubault, E., Jensen, T.P.: Homology of higher dimensional automata. In: Cleaveland, W.R. (ed.) CONCUR 1992. LNCS, vol. 630, pp. 254–268. Springer, Heidelberg (1992)
10. Gunawardena, J.: Homotopy and concurrency. Bull. EATCS **54**, 184–193 (1994)
11. Herlihy, M., Rajsbaum, S.: New perspectives in distributed computing. In: Kutyłowski, M., Wierzbicki, T.M., Pacholski, L. (eds.) MFCS 1999. LNCS, vol. 1672, pp. 170–186. Springer, Heidelberg (1999)
12. Herlihy, M., Shavit, N.: The topological structure of asynchronous computability. J. ACM **46**, 858–923 (1999)
13. Hoest, G., Shavit, N.: Toward a topological characterization of asynchronous complexity. SIAM J. Comput. **36**(2), 457–497 (2006)
14. Saks, M., Zaharoglou, F.: Wait-free k-set agreement is impossible: the topology of public knowledge. SIAM J. Comput. **29**(5), 1449–1483 (2000)

# Data Warehousing and OLAP

# AScale: Auto-Scale in and out ETL+Q Framework

Pedro Martins[✉], Maryam Abbasi, and Pedro Furtado

Department of Computer Sciences, University of Coimbra, Coimbra, Portugal
{pmom,maryam,pnf}@dei.uc.pt

**Abstract.** The purpose of this study is to investigate the problem of providing automatic scalability and data freshness to data warehouses, while simultaneously dealing with high-rate data efficiently. In general, data freshness is not guaranteed in these contexts, since data loading, transformation and integration are heavy tasks that are performed only periodically.

Desirably, users developing data warehouses need to concentrate solely on the conceptual and logic design such as business driven requirements, logical warehouse schemas, workload and ETL process, while physical details, including mechanisms for scalability, freshness and integration of high-rate data, should be left for automated tools.

In this regard, we propose a universal data warehouse parallelization system, that is, an approach to enable the automatic scalability and freshness of warehouses and ETL processes. A general framework for testing and implementing the proposed system was developed. The results show that the proposed system is capable of handling scalability to provide the desired processing speed and data freshness.

**Keywords:** Algorithms · Architecture · Performance · Distributed · Elastic · Parallel processing · Distributed systems · Database · Scalability · Load-balance

## 1 Introduction

Today's data warehouses demand fast, reliable tools that help user to acquire and manage data. We also need the flexibility to load large volumes of data from any source at any time to meet the business demands. Challenges come from everywhere: more data sources, growing data volumes, dynamically changing business requirements, and user demands for fresher data. Extract - Transform - Load (ETL) tools are special purpose software artifacts used to populate a data warehouse with up-to-date, clean records from one or more sources.

When defining the ETL+Q the user must keep in mind the existence of data sources, where and how the data is extracted to be transformed (e.g. completed, cleaned, validated), loading into the data warehouse and finally the data warehouse schema; each of these steps requires different processing capacity, resources and data treatment.

© Springer International Publishing Switzerland 2016
S. Kozielski et al. (Eds.): BDAS 2016, CCIS 613, pp. 303–314, 2016.
DOI: 10.1007/978-3-319-34099-9_24

After extraction, data must be re-directed and distributed across the available transformation nodes, again since transformation involves heavy duty tasks (heavier than extraction), more than one node should be present to assure acceptable execution/transformation times. Consequently, once more new data distribution policies must be added.

After the data transformed and ready to be load, the load period time (e.g. every night, every hour) and a load time control (e.g. maximum load time = 5 h) must be scheduled. Which means the data have to be held between the transformation and loading process in some buffer.

Eventually, regarding the data warehouse schema, the entire data will not fit into a single node, and if it fits, it will not be possible to execute queries within acceptable time ranges. Thus, more than one data warehouse node is necessary with a specific schema which allows to distribute, replicate, and finally query the data within an acceptable time frame.

In this work we studied automatic ETL+Q scalability with ingress high-data-rate in big-data warehouses. We propose a software tool, AScale, to parallelize and scale the entire ETL+Q process. The system does not require intensive human effort from the designers and the user must only focus on the conceptual design of the ETL+Q processes as like for a single server.

## 2    Related Work

Work in the area of ETL scheduling includes efforts towards the optimization of the entire ETL workflows [11] and of individual operators in terms of algebraic optimization; (e.g., joins or data sort operations). The work presented in [6] deals with the problem of scheduling ETL workflows at the data level and in particular scheduling protocols and software architecture for an ETL engine in order to minimize the execution time and allocated memory needed for a given ETL workflow. A second aspect in ETL execution that the authors addressed is how to schedule flow execution at the operational level when there is a blocking or non-parallelizable operations and how to improve it with pipeline parallelization [5].

The paper [1] studies how to manage large ETL processes by implementing a set of basic management operators, such as "MATCH", "MERGE", "INVERT", "SEARCH", "DEPLOY". The framework is web-based. The user creates the ETL flow using drag-and-drop with the available filters, then the framework determines the best execution order for the ETL using a set of optimization algorithms.

In [8] the authors describe an Extract-Transform-Load programming framework using Map-Reduce to achieve scalability. Data sources and target dimensions need to be configured and deployed. The framework has built-in support for star schemas and snowflakes. Users have to implement the parallel ETL programs using the framework constructors. The authors use pygrametl [12], a Python based framework for easy ETL programming. The flow consists of two phases, dimension processing and fact processing. Data is read from sources (files) on a Distributed File System (DFS), transformed and processed into dimension values and facts by the framework instances, which materialize the data into the DW. The framework requires users to declare (code) target tables and transformation functions. Then, it uses a master/worker architecture (one master, many workers), each worker running jobs in parallel. The master distributes data, schedules tasks, and monitors the workers.

ETLMR [7] is an academic tool which builds the ETL processes on top of Map-Reduce to parallelize the ETL operation on commodity computers. ETLMR contains a number of novel contributions. It supports high-level ETL-specific dimensional constructs for processing both star-schemas and snowflake-schemas, and data-intensive dimensions. Due to its use of Map-Reduce, it can automatically scale to more nodes (without modifications to the ETL flow) while at the same time also providing automatic data synchronization across nodes (even for complex dimension structures like snowflakes). Apart from scalability, Map-Reduce also gives ETLMR a high degree of fault-tolerance. ETLMR does not have its own data storage (note that the offline dimension store is only for speedup purpose), but is an ETL tool suitable for processing large scale data in parallel. ETLMR provides a configuration file to declare dimensions, facts, User Defined Functions (UDFs), and other run-time parameters. Despite a good approach to ETL scalability it does not support complex transformation processes and the entire ETL process and its coordination and scheduling becomes very complex.

The work [9] focuses on finding approaches for the automatic code generation of ETL processes which is aligning the modeling of ETL processes in data warehouse with MDA (Model Driven Architecture) by formally defining a set of QVT (Query, View, Transformation) transformations.

Related problems studied in the past include the scheduling of concurrent updates and queries in real-time warehousing and the scheduling of operators in data streams management systems. However, we argue that a fresher look is needed in the context of ETL technology. The issue is no longer the scalability cost/price, but rather the complexity it adds to the system.

The point that previews presented recent works in the field do not address is the automatic scalability to make ETL scalability easy and automatic. The authors focus on mechanisms to improve scheduling algorithms and optimizing work-flows and memory usage. In our work we assume that scalability in number of machines and quantity of memory is not the issue, we focus in offering automatic ETL scalability, without the nightmare of operators relocation and complex execution plans. Thus in our work we focus on automatic scalability, given a single server ETL process, our proposed system (AScale) will scale it automatically. AScale based on generic ETL process thought for a single server (no concern with performance or scalability) will scale it to provide the users configured performance with minimum complexity and implementations.

## 3  Architecture

In this section we describe the main components of the propose architecture, AScale, for automatic ETL+Q scalability. Figure 1 shows the main components of the framework.

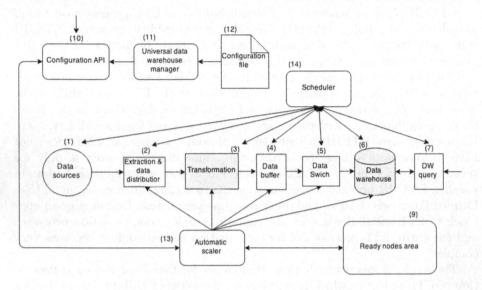

**Fig. 1.** Automatic ETL+Q scalability

- Components (1) to (7), except (5) are part of the traditional Extract, Transform, Load and Query (ETL+Q) process;
- The "Automatic Scaler" (13), is the component responsible for performance monitoring and scaling the system when necessary;
- The "Configuration file" (12) represents the location where all user configurations are saved;
- The "Universal Data Warehouse Manager" (11), uses the configurations provided by the user and the available "Configurations API" (10) to set the system to perform according with the desired parameters and selected algorithms. The "Universal Data Warehouse Manager" (11), also sets the configuration parameters for automatic scalability at (13) and the policies to be applied by the "Scheduler" (14);
- The "Configuration API" (10), is an access interface which allows to configure each part of the proposed Universal Data Warehouse architecture, automatically by (11) or manually by the user;
- The "Scheduler" (14), is responsible for applying the data transfer policies between components (e.g. control the on-demand data transfers).
- The "Ready nodes area" (9) represent nodes that are not being used. This nodes can be added to any part (2) to (6) of the system to scale-out, improving performance where needed, or removed to scale-in, saving resources that can be used in other places.

All these components when set to interact together provide automatic scalability to the ETL+Q and to the data warehouse processes without the need for the user to concern about its scalability or management. The user can use it with any data warehouse project and without having to deal with how to scale each part fo the ETL and query processing stage. Instead of having to program the entire ETL project, when using the proposed framework the user can focus only in programming the transformations and data warehouse schema (Fig. 1, highlighted in grey color), leaving the other scalability detail to be handle automatically by the proposed framework.

Additionally he can choose any data warehouse engine to store data (e.g. Relational data warehouses, column oriented, noSQL, Map-Reduce architectures).

## 4    Scalability Mechanisms

In this section we introduce how each part of the (ETL+Q) auto-scale framework scales individually to obtain the necessary performance. Figure 2 depicts each part of the ETL scaling including:

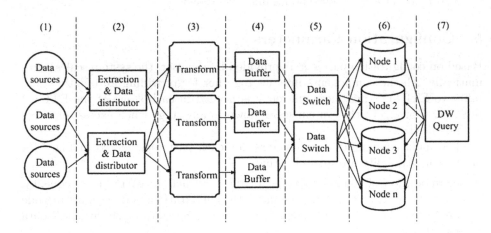

**Fig. 2.** Total ETL scalability

(1) The increase of Data Sources (1), implies the increase of data, leading to the need of scale other parts of the proposed framework. Each (1) has associated a extraction frequency (e.g. every minute);

(2) The "Extraction & Data Distributor" nodes forwarding and/or replicating the extracted (raw) into the transformer nodes. Scaling (2) is detected by monitoring the extraction time performance. If the extraction time is larger than a maximum configured extraction time, or if data is not extracted until the next extraction frequency period (e.g. every minute), more data distributor nodes (2) are required;

(3) Transformation nodes include the data transformations programmed by the user. This nodes include a buffer queue to monitor ingress data. If that queue increases size above a certain limit the transformation node is scaled, by replicating the transformation scrip into other node;

(4) The data buffer holds transformed data, it can be in memory or disk. These nodes are scaled based on memory monitoring parameters. If at any time the memory reaches the 90 % the maximum configured memory size a new node must be added (we choose 90 % of the memory so that hardware resources are not so much wasted);

(5) The data switches are responsible for the data distribution (pop/extract) from the "Data Buffers" and set it for loading into the data warehouse. Each data switch is configured to support a maximum data-rate (e.g. 100 MB/sec) if that limit is passed, more data switch nodes are needed;

(6) The data warehouse can be in a single node, or parallelized by many nodes. Scalability is based in two parameters. The loading time and query response time. If the data warehouse nodes take more time to load data than the maximum configured time, more nodes are added and data is re-distributed. If queries average execution time is superior to the desired response time, also more data warehouse nodes must be added.

## 5    Configuration Parameters

Based on configuration parameters provided by the user, the system scales automatically. All the components interact together for providing automatic scalability to ETL+Q when more efficiency is needed in each stage of the ETL+Q pipeline. Conversely, the system scales down in any stage when excess resources are not needed.

The main configuration parameters, for each part represented in Fig. 2 for automatic scalability are related with:

(1) configuration of sources location for data extraction; extraction frequency, data size to extract and maximum extraction duration; Sources must provide an API for data extraction including information regarding the available data to extract;

(2) distribution algorithm (when none configured default it is used, such as on-demand and round-robin);

(3) transformations to be applied. User can program the transformations directly to the auto-scale framework or program in any other language connecting to the provide API or connect an external application to do the data transformation by using the available web service API;

(4) data buffers size (memory and disk);

(5) maximum supported distribution rate;

(6) load frequency, maximum load duration, load online (while performing queries at the same time), load offline (without queries executing and during low data rate moments, e.g. during the night), data warehouse schema (made by the user), definition of fact tables and dimension tables (i.e. replication parameters);

(7) maximum desired queries execution time parameters.

# 6   Data Scalability and Query Processing

Automatic query processing scalability can be achieved from a set of guiding parameters.

When data arrives from "Transformation" nodes, to the "Data buffer" nodes the "Data Switches" (5) automatically partition the data into the data warehouse log files to be loaded in the configured time periods. Configuration parameters must define which tables are dimensions and facts, since this influences whether the data must be partitioned or replicated. The data model assumes that the data warehouse is organized as a set of star schemas. Horizontal partitioning of data is used for scalability [2].

When scaling all dimension tables are replicated into all nodes and the fact table is partitioned across the nodes. Data inside the nodes is re-balanced automatically when nodes are added or removed. Queries must be decomposed to run in the partitioned model, and the results from each partition are merged into the final result.

Other alternatives can be considered to partition data. All tables can be fully replicated by the data warehouse nodes, or partitioned using round-robin, hash by some field, random or manually [4,13].

# 7   Decision Algorithms for Scalability Parameters

In this section we define the scalability decision methods as well as the algorithms which allow the framework to automatically scale-out and scale-in.

**Extraction and Data Distributors - Scale-out.** Depending on the number of existing sources and data generation rate and size, the nodes that process the extraction of the data from the sources might need to scale.

The addition of more "extraction and data distributors" (2) depends if the current number of nodes is being able to extract and process the data with the correct frequency (e.g. every 5 min) and inside the limit maximum extraction time (without delays). For instance, if the extraction frequency is specified as every 5 min and extraction duration 10 s, every 5 min then the "Extraction and Data distributor" nodes cannot spend more than 10 s extracting data. Otherwise a scale-out is needed, so the extraction size can be reduced and the extraction time improved.

If the maximum extraction duration is not configured, then the extraction process must finish before the next extraction instant. If not processed until the next extraction instant, as defined by the extraction frequency, a scale-out is also required, to add more extraction power.

**Extraction and Data Distributors - Scale-in.** To save resources when possible, nodes that perform the data extraction from the sources can be set in standby or removed. This decision is made based on the last execution times. If previous execution times of at least two or more nodes are less than half of the maximum extraction time (or if the maximum extraction time is not configured, the frequency), minus a configured variation parameter (X), one of the nodes is set on standby or removed, and the other one takes over.

**Transform - Scale-out.** The transformation process is critical, if the transformation is running slow, data extraction at the referred rate may not be possible, and information will not be available for loading and querying when necessary. The transformation step has an important queue used to determine when to scale the transformation phase. If this queue reaches a limit size (by default 50 %), then it is necessary to scale, because the actual transformer node(s) is not being able to process all data that is arriving.

**Transform - Scale-in.** The size of all queues is analyzed periodically. If this size at a specific moment is less than half of the limit size for at least two nodes, then one of those nodes is set on standby or removed.

**Data buffer - Scale.** The data buffer nodes scale-out based on the incoming memory queue size and the storage space available to hold data. When the available memory queue becomes full above 50 % the configured maximum size, data starts being swapped into disk, until the memory is empty. If even so the data buffer memory reaches the limit size the data buffer must be scaled-out. This means that the incoming data-rate (going into memory storage) is not being swapped to the disk storage fast enough and more nodes are necessary.

If the disk space becomes full above a certain configured size, the data buffers are also set to scale-out.

Flowchart describes the algorithm used to scale-out the data buffer nodes.

By user request the data buffers can also scale-in, in this case the system will do so it if the data from any data buffer can be fitted inside another data buffer.

**Data Switch - Scale.** The Data Switch nodes scale based on a configured data-rate limit. If the data-rate rizes above the configured limit the data switch nodes are set to scale-out.

The data switches can also scale-in, in this case the system will allow it if the data-rate is less than the configured maximum by at least 2 nodes, minus a Z configured variation, for a specific time.

**Data Warehouse - Scale.** Data warehouse scalability needs are detected after each load process or by query execution time.

The data warehouse load process has a configured limit time to be executed every time it starts. If that limit time is exceeded, then the data warehouse must scale-out.

The data warehouse scalability is not only based on the load & integration speed requirements, but also on the queries desired maximum execution time. After each query execution, if the query time to the data warehouse is more than the configured maximum desired query execution time, then the data warehouse is set to scale-out.

Scale-in process is more simple, data just needs to be extracted and loaded across the available nodes as if it is new data.

# 8   Experimental Setup and Results

In this section we prove the ability of the proposed system, AScale, to automatic scale-out the ETL process when more performance is necessary and to scale-in when resources are not necessary to provide the configured performance. Then by applying AScale system we observed how it scales automatically to provide the configured performance.

We defined the following scenario:

- Our sources are based on the TPC-H benchmark [3] generator, which generate data at the highest possible rate (on each node) and through a web service make it available for extraction.
- The transformation process consisted on transforming the TPC-H relational model into a star-schema, which resulted in the recreation of the SSB benchmark [10]. This process involves heavy computational memory and temporary storage systems, look-ups and data transformation to assure consistency.
- The data warehouse tables schema consists on the same schema from SSB benchmark. Replication and partitioning is assured by the proposed system, whereas, dimension tables are replicated and fact tables are partitioned across the data warehouse nodes.
- The E (extraction), T (transformation) and L (load) were set to perform every 2 s, and can not last more than 1 s. Thus the ETL process will last at the worst case 3 s total.
- The load process was made in batches of 10 MB maximum.

The data-rate was gradually increased at the sources (adding more data sources when necessary). The next base configuration were used:

- Each time we increment the data rate, we let the system run for 60 min, to allow it to stabilize and scale.
- The data-rate was increased from 1.000 rows per second, up to, 500.000 rows per second (50 tests, each test 60 min).
- Each part of the ETL process was not allowed to exceed more than 1 s. Thus the entire ETL process should take 3 s.
- Processing nodes were added, by the system request, gradually when the queues reached a limit size, or the processing time was exceeded.
- For the load method we considered a data warehouse without indexes or views and the data load was made in batch files with 10 MB size.

Figures 3 and 4 shows the obtained results during our automatic scalability, in and out, tests. In the experimental results test for scale-out, Fig. 3, is possible to observe the system scaling-out automatically each part of the ETL in order to assure the configured parameters. In some stages of the tests the 3 s limit execution time was exceeded in 0.1, 0.2 s, this happened due to the high usage of the network system connecting all nodes. We conclude that, when gradually increasing the processing load, the proposed architecture is capable to scale the most overloaded parts that are affecting the global ETL performance. Figure 4,

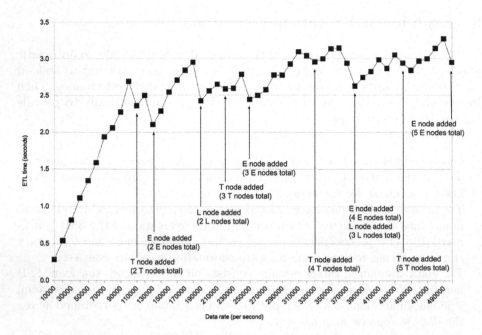

**Fig. 3.** Full ETL system **scale-out**. Limit configured time for the full ETL process 3 s. Each run for each data rate 60 min. 50 tests performed.

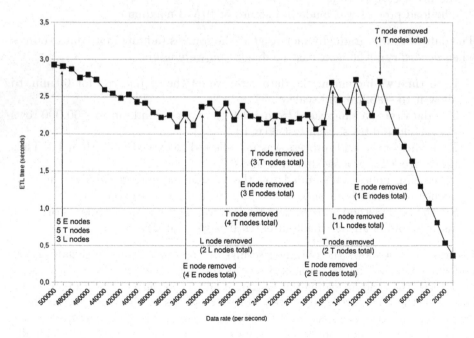

**Fig. 4.** Full ETL system **scale-in**. Limit configured time for the full ETL process 3 s. Each run for each data rate 60 min. 50 tests performed.

shows how the system scales-in when resources are not necessary to guarantee the configured performance. Important to note that scaling-in the system is slower (requires less data rate) than the scale-out, also note that when removing heavier processing nodes the ETL processing time increases faster (e.g. when removing T (transform) nodes). At the end because there is only one node for each ETL task, thus no more can be removed, as the data rate decreases, so the processing time decreases/improves.

For both results, the scaling out and in requires time to become stable since some parts of the system are triggered based on average times as scheduled/defined in configurations.

## 9   Conclusions and Future Work

In this work we propose mechanisms and algorithms to achieve automatic scalability for complex ETL+Q, offering the possibility to the users to think solely in the conceptual ETL+Q models and implementations for a single server (only the data transformations, database schema and queries). Then our proposed framework, AScale, will automatically scale each ETL+Q part when required, and assure performance according with the user configuration.

Future work will involve setting the load process to be performed during low activity periods (e.g. during the night) and this way improving the tests on the dataBuffers capacity to hold data for longer periods of time before distributing the data across the data warehouse for load. At the same time it will allow to test the data warehouse scalability to assure execution times, when running heavy queries (e.g. if the query takes too long to execute the data warehouse must scale-out). Other stage of future steps include dealing with real-time ETL for querying and at the same time providing data freshness. Finally as future work we intend to compare the approach with an ETL process fully implemented using Map-Reduce application.

**Acknowledgement.** This project is part of a larger software prototype, partially financed by, Portugal, CISUC research group from the University of Coimbra and by the Foundation for Science and Technology.

## References

1. Albrecht, A., Naumann, F.: Metl: managing and integrating ETL processes. In: VLDB PhD Workshop (2009)
2. Ceri, S., Negri, M., Pelagatti, G.: Horizontal data partitioning in database design. In: Proceedings of the 1982 ACM SIGMOD International Conference on Management of Data, pp. 128–136. ACM (1982)
3. Council, T.P.P.: Tpc-h benchmark specification (2008). http://www.tcp.org/hspec.html
4. Furtado, P.: Efficient and robust node-partitioned data warehouses. In: Data Warehouses and OLAP: Concepts, Architectures, and Solutions, p. 203 (2007)

5. Halasipuram, R., Deshpande, P.M., Padmanabhan, S.: Determining essential statistics for cost based optimization of an ETL workflow. In: EDBT, pp. 307–318 (2014)
6. Karagiannis, A., Vassiliadis, P., Simitsis, A.: Scheduling strategies for efficient ETL execution. Inf. Syst. **38**(6), 927–945 (2013)
7. Liu, X.: Data warehousing technologies for large-scale and right-time data. Ph.D. thesis, dissertation, Faculty of Engineering and Science at Aalborg University, Denmark (2012)
8. Liu, X., Thomsen, C., Pedersen, T.B.: Mapreduce-based dimensional ETL made easy. Proc. VLDB Endowment **5**(12), 1882–1885 (2012)
9. Muñoz, L., Mazón, J.N., Trujillo, J.: Automatic generation of ETL processes from conceptual models. In: Proceedings of the ACM Twelfth International Workshop on Data Warehousing and OLAP, pp. 33–40. ACM (2009)
10. O'Neil, P.E., O'Neil, E.J., Chen, X.: The star schema benchmark (ssb). Pat (2007)
11. Simitsis, A., Wilkinson, K., Dayal, U., Castellanos, M.: Optimizing ETL workflows for fault-tolerance. In: 2010 IEEE 26th International Conference on Data Engineering (ICDE), pp. 385–396. IEEE (2010)
12. Thomsen, C., Bach Pedersen, T.: pygrametl: a powerful programming framework for extract-transform-load programmers. In: Proceedings of the ACM Twelfth International Workshop on Data Warehousing and OLAP, pp. 49–56. ACM (2009)
13. Vassiliadis, P., Simitsis, A.: Near real time ETL. In: Vassiliadis, P., Wrembel, R. (eds.) New Trends in Data Warehousing and Data Analysis. Annals of Information Systems, vol. 3, pp. 1–31. Springer, New York (2009)

# AScale: Big/Small Data ETL and Real-Time Data Freshness

Pedro Martins[✉], Maryam Abbasi, and Pedro Furtado

Department of Computer Sciences, University of Coimbra, Coimbra, Portugal
{pmom,maryam,pnf}@dei.uc.pt

**Abstract.** In this paper we investigate the problem of providing timely results for the Extraction, Transformation and Load (ETL) process and automatic scalability to the entire pipeline including the data warehouse. In general, data loading, transformation and integration are heavy tasks that are performed only periodically during specific offline time windows. Parallel architectures and mechanisms are able to optimize the ETL process by speeding-up each part of the pipeline process as more performance is needed. However, none of them allow the user to specify the ETL time and the framework scales automatically to assure it.

We propose an approach to enable the automatic scalability and freshness of any data warehouse and ETL process in time, suitable for small-Data and bigData scenarios. A general framework for testing and implementing the system was developed to provide solutions for each part of the ETL automatic scalability in time. The results show that the proposed system is capable of handling scalability to provide the desired processing speed for both near-real-time results ETL processing.

**Keywords:** Scalability · ETL · Freshness · High-rate · Performance · Parallel processing · Distributed systems · Database · bigData · small-Data · Business management

## 1 Introduction

ETL tools are special purpose software used to populate a data warehouse with up-to-date, clean records from one or more sources. The majority of current ETL tools organize such operations as a workflow. At the logical level, the E (Extract) can be considered as a capture of data-flow from the sources with more than one high-rate throughput. T (Transform) represents transforming and cleansing data in order to be distributed and replicated across many processes and ultimately, L (Load) convey by loading the data into data warehouses to be stored and queried. For implementing these type of systems besides knowing all of these steps, the acknowledge of user regarding the scalability issues is essential, which the ETL might be introduced.

When defining the ETL the user must consider the existence of data sources, where and how the data is extracted to be transformed, loading into the data warehouse and finally the data warehouse schema; each of these steps requires

© Springer International Publishing Switzerland 2016
S. Kozielski et al. (Eds.): BDAS 2016, CCIS 613, pp. 315–327, 2016.
DOI: 10.1007/978-3-319-34099-9_25

different processing capacity, resources and data treatment. Moreover, the ETL is never so linear and it is more complex than it seems. Most often the data volume is too large and one single extraction node is not sufficient. Thus, more nodes must be added to extract the data within the configured time bound.

After extraction, data must be re-directed and distributed across the available transformation nodes, again since transformation involves heavy duty tasks (heavier than extraction), more than one node should be present to assure acceptable execution/transformation times.

After the data transformed and ready to be load, the load period time and a load time control must be scheduled. Which means that the data have to be held between the transformation and loading process in some buffer. Eventually, regarding the data warehouse schema, the entire data will not fit into a single node, and if it fits, it will not be possible to execute queries within acceptable time ranges. Thus, more than one data warehouse node is necessary with a specific schema which allows to distribute, replicate, and query the data within an acceptable time frame.

In this paper we study how to provide parallel ETL automatic scalability with ingress high-data-rate in big data and small data warehouses. We propose a set of mechanisms and algorithms, to parallelize and scale each part of the entire ETL process, within configured time ranges for each part of the ETL pipeline and for the global ETL process.

## 2    Related Work

Works in the area of ETL scheduling includes efforts towards the optimization of the entire ETL workflows and of individual operators in terms of algebraic optimization; e.g., joins or data sort operations). However many works focus on complex optimization details that only apply to very specific cases. The work [3] focuses on finding approaches for the automatic code generation of ETL processes which is aligning the modeling of ETL processes in data warehouse with MDA (Model Driven Architecture) by formally defining a set of QVT (Query, View, Transformation) transformations. ETLMR [2] is an academic tool which builds the ETL processes on top of Map-Reduce to parallelize the ETL operation on commodity computers. ETLMR does not have its own data storage (note that the offline dimension store is only for speedup purpose), but is an ETL tool suitable for processing large scale data in parallel. ETLMR provides a configuration file to declare dimensions, facts, User Defined Functions (UDFs), and other run-time parameters. ETLMR toll has the same problem as the MapReduce architectures, too much hardware resources are required to guaranty basic performance.

In [5] the authors consider the problem of data flow partitioning for achieving real-time ETL. The approach makes choices based on a variety of trade-offs, such as freshness, recoverability and fault-tolerance, by considering various techniques. In this approach partitioning can be based on round-robin (RR), hash (HS), range (RG), random, modulus, copy, and others [6].

In [1] the authors describe Liquid, a data integration stack that provides low latency data access to support near real-time in addition to batch applications.

There is a vast related work in ETL field. Although main related problems studied in the past include the scheduling of concurrent updates and queries in real-time warehousing and the scheduling of operators in data streams management systems. However, we argue that a fresher look is needed in the context of ETL technology. The issue is no longer the scalability cost/price, but rather the complexity it adds to the system. Previews presented recent works in the field do not address in detail how to scale each part of the ETL automatically in order to assure an configured execution time bound. We focus on offering scalability for each part of the ETL pipeline process, without the nightmare of operators relocation and complex execution plans. Our main focus is automatic scalability to provide the users desired performance, based on configured time bounds, with minimum complexity and implementations. In addition, we also support queries execution.

## 3   Architecture

In this section we describe the main components of the proposed architecture for ETL scalability. Figure 1 shows the main components to achieve automatic scalability.

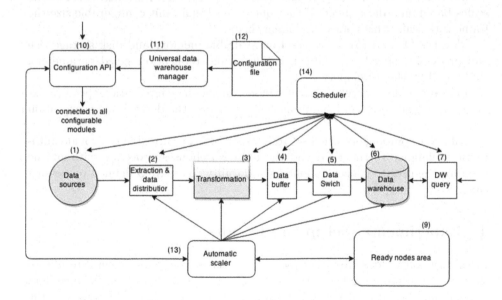

**Fig. 1.** Framework for automatic ETL scalability

- All components from (1) to (7) are part of the Extract, Transform, Load and query (ETL) process. All can auto-scale automatically when more performance is necessary.

- The "Automatic Scaler" (13), is the node responsible for performance monitoring and scaling the system when is necessary.
- The "Configuration file" (12) represents the location where all user configurations are defined by the user.
- The "Universal Data Warehouse Manager" (11), based on the configurations provided by the user and using the available "Configurations API" (10), sets the system to perform according with the desired parameters and algorithms. The "Universal Data Warehouse Manager" (11), also sets the configuration parameters for automatic scalability at (13) and the policies to be applied by the "Scheduler" (14).
- The "Configuration API" (10), is an access interface which allows to configure each part of the proposed Universal Data Warehouse architecture, automatically or manually by the user.
- The "ready nodes area" (9) represent a pool of nodes, from where the proposed framework can request more resources.
- Finally the "Scheduler" (14), is responsible for applying the data transfer policies between components (e.g. control the on-demand data transfers).

For module (2) the developer defines the extraction frequency (e.g. every one hour) and the maximum extraction time (e.g. 30 min).

Module (3) does not require any time configuration for scaling, this module scales based on a data queue. If that queue reaches a limit configurable size the framework scales this nodes by cloning them.

Modules (4) and (5) scale based on available memory and disk storage size and processing speed (i.e. data distribution and replication speed parameters, configured by the developer).

Finally the data warehouse scales based on the data integration speed (maximum data load time) and simultaneously based on the desired query execution time.

All these components when set to interact together are able to provide automatic scalability to the ETL and the data warehouses processes without the need for the user to concern about its scalability to assure a defined processing time.

## 4    Experimental Setup and Results

In this section we describe the experimental setup, and experimental results to show that the proposed system, AScale, is able to scale and load balance data in small and big data scenarios for near real-time and offline ETL, assuring the desired time bounds.

The experimental tests were performed using 30 computers, denominated as nodes, with the following characteristics: Processor Intel Core i5-5300U Processor (3 M Cache, up to 3.40 GHz); Memory 16 GB DDR3; Disk: western digital 1 TB 7500 rpm; Ethernet connection 1 Gbit/sec; Connection switch: SMC SMCOST16, 16 Ethernet ports, 1 Gbit/sec; Windows 7 enterprise edition 64 bits; Java JDK 8; Netbeans 8.0.2; Oracle Database 11g Release 1 for Microsoft Windows (X64) - used in each data warehouse nodes; PostgreSQL 9.4 - used for look ups during the transformation process; TPC-H benchmark - representing the operational log data used at the extraction nodes. This is possible since TPC-H data is still normalized; SSB benchmark - representing the data warehouse. The SSB is the star-schema representation of TPC-H data. Data transformations consist loading from the data sources the "lineitem" and "order" TPC-H data logs and besides the transformation applied to achieve the SSB benchmark star-schema [4] configuration we added some data quality verification and cleansing.

## 4.1 Performance Limitations Without Automatic Scalability

In this section we test both ETL and data warehouse scalability needs when the entire ETL process is deployed without automatic scalability options. The system is stressed with increasing data-rates until it is unable to handle the ETL and query processing in reasonable time. Automatic scalability which we evaluate in following sections, is designed to handle this problem.

The following deployment is considered: One machine to extract, transform data and store the data warehouse; extraction frequency is set to perform every 30 s; desired maximum allowed extraction time, 20 s; data load is performed in offline periods.

Based on this scenario we show the limit situation in which performance degrades significantly, justifying the need to scale the ETL (i.e. parts of it) or data warehouse.

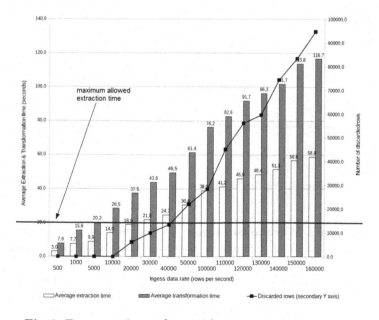

**Fig. 2.** Extract and transform without automatic scalability

**Extraction & Transformation:** Considering only extraction and transformation, using a single node, Fig. 2 shows: in the left Y axis is represented the average extraction and transformation time in seconds; in the right Y axis is represented the number of discarded rows (data that was not extracted and not transformed); in the X axis we show the data-rate in rows per-second; white bars represent extraction time; gray bars represent the transformation time; lines represent the average number of discarded rows (corresponding values in the right axes).

For this experiment we generated log data (data to be extracted) at a rate $\lambda$ per second. Increasing values of $\lambda$ were tested and the results are shown in Fig. 2.

Extraction is performed every 30 s. This means that in 30 s there is 30x more data to extract. Extraction must be done in 20 s maximum. As the data-rate increases, a single node is unable to handle so much data. At a data-rate of 20.000 rows per second, buffer queues become full and data starts being discarded at sources because the extraction time is too slow. The transformation process is slower than transformation, requiring more resources to perform at the same speed as the extraction.

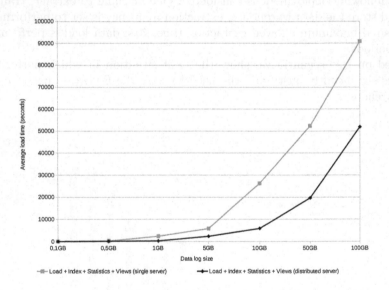

**Fig. 3.** Loading data, one server vs two servers

**Loading the Data Warehouse:** Figure 3 shows the load time as the size of the logs is increased. It also compares the time taken with single node versus two nodes.

Figure 3 shows: in the Y axis is represented the average load time in seconds; in the X axis is represented the loaded data size in GB; the black line represents two servers; the grey line represents one server; all times were obtained with the following load method: destroy all indexes and views, load data, create indexes,

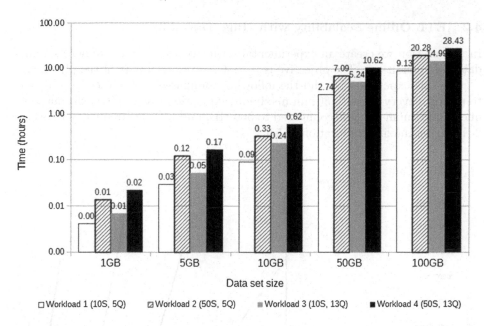

**Fig. 4.** Average query time for different data sizes and number of sessions

update statistics and update views; data was distributed by replicating and partition the tables.

Differences are noticeable when loading more than 10 GB. When adding two data warehouse nodes, performance improves and the load time becomes almost less than half.

**Query Execution:** Figure 4, shows the average query execution time for a set of tested workload (using the SSB benchmark queries): workload 1, 10 sessions, 5 Queries (Q1.1, Q1.2, Q2.1, Q3.1, Q4.1); workload 2, 50 sessions, 5 Queries (Q1.1, Q1.2, Q2.1, Q3.1, Q4.1); workload 3, 10 sessions, 13 Queries (All); workload 4, 50 sessions, 13 Queries (All); for all workloads, queries were executed in a random order; the desired maximum query execution time was set to 60 s.

The Y axis shows the average execution time in seconds. The X axis shows the data size in GB. Each bar represents the average execution time per query fro each workload. Note that, Y axis scale is logarithmic for better results representation.

Depending on the data size, number of queries and number of simultaneous sessions (e.g. number of simultaneous users), execution time can vary from a few seconds to a very significant number of hours or days, especially when considering large data sizes and simultaneous sessions or both. In these results, and referring to 10 GB and 50 GB, we see that an increase of 5x of the data size resulted in an increase of approximately 20x in response time. An increase in the number of number of queries of 5x resulted in an increase of approximately 2x in query response time.

## 4.2    ETL Offline Scalability with Huge Data Sizes

In this section we create an experimental setup to stress AScale under extreme
data rate conditions. The objective is to test scaling each part of the pipeline.

For this experiment we did the following configured: E (extraction) was set
to perform every 1 h with 30 min maximum extraction time, T (transformation)
queue maximum size was configured to 500 MB, and L (load) frequency to every
24 h with a maximum duration of 5 h.

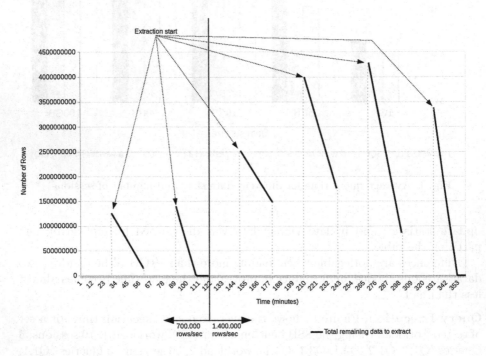

**Fig. 5.** Extraction (60 min frequency and 30 min maximum extraction time)

**Extraction:** Figure 5 shows the AScale extraction process when using an extraction frequency of 60 s and 30 s for the maximum extraction time. The figure
is divided into two sections: first we use a data rate of 700.000 rows/sec and
scale the extraction until all rows are extracted successfully; second we increase
the data rate to 1.400.000 rows/sec and automatically scale until all rows are
extracted within the configured time bound.

In Fig. 5 we show: in the left Y axis, the number of rows to extract; in the X
axis, the time in minutes; black line represents the total number or rows left to
be extracted at each extraction period.

By analyzing results from Fig. 5 we conclude that the extraction process
is able to scale efficiently until all data can be extracted within the desired
maximum extraction time. However, if the data-rate increases very fast

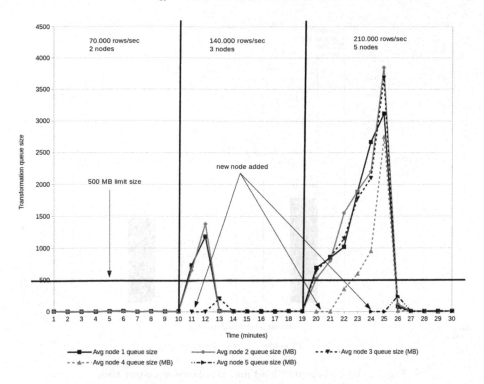

**Fig. 6.** Transformation scale-out

(e.g. into the double) in a small time window, AScale requires additional extraction cycles to restore the normal extraction frequency.

**Transformation:** In Fig. 6 we test the transformation scale-out. The scale-out decision is based on monitoring the data queue size in each node. Every time a queue exceeds the maximum configured size AScale scales that part automatically. The monitoring process allows to scale-out very fast, even if the data rate increases suddenly.

Figure 6 is divided in three parts, each one with a different data-rate. The data-rate is increased in each part in order to show AScale scaling-out the transformation nodes every time a queue reaches above the limit size, by adding one and two extra nodes.

For each scale-out it was necessary less than 1 min until the node is processing data, the time delay refers to the copy and replication of the staging area.

**Load:** AScale load needs to be scaled at the end of a load cycle if the last load cycle did not respect the maximum load time. The number of nodes to add is calculated linearly based on previews load times. For instance, if the load time using 10 nodes was 9 h, in order to be ablo to load in 5 h, we need X nodes, Eq. 1.

$$\frac{loadTime}{targetTime} \times n \tag{1}$$

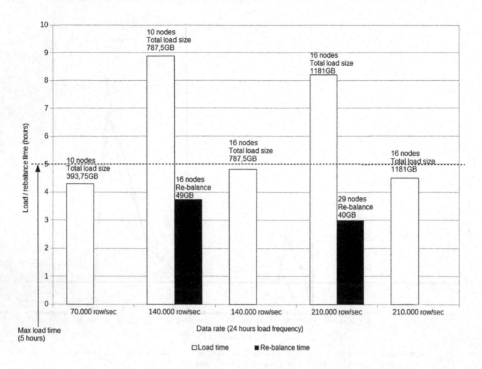

**Fig. 7.** Data warehouse load and rebalance scale-out time

where "loadTime" represents the last load time, "targetTime", represents the desired load time and "n" represents the current number of nodes.

Figure 7 shows the data warehouse nodes load time for different data-rates over 24 h generation, on top of each bar is represented the total loaded data size and the number of nodes used. Every time the maximum load time was exceeded more data warehouse nodes were added and the data warehouse rebalanced.

We conclude that the data warehouse nodes can be scaled efficiently in a relatively short period of time given the large amounts of data being considered.

### 4.3  Near-Real-time DW Scalability and Freshness

In this section we assess the scale-out and scale-in abilities of the proposed framework in near-real-time scenarios requiring data to be always updated and available to be queried (i.e. data freshness).

The near-real-time scenario was set-up with: E (extraction) and L (load) were set to perform every 2 s; T (transformation) was configured with a maximum queue size of 500 MB; the load process was made in batches of 100 MB maximum size. The ETL process is allowed to take 3 s.

Figures 8 and 9 show AScale, scaling-out and scaling-in automatically, respectively, to deliver the configured near-real-time ETL time bounds, while the data rate increases/decreases. The charts show: the X axis represents the data-rate,

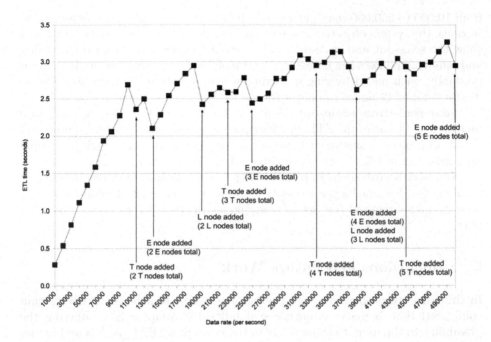

**Fig. 8.** Near-real-time, full ETL system scale-out

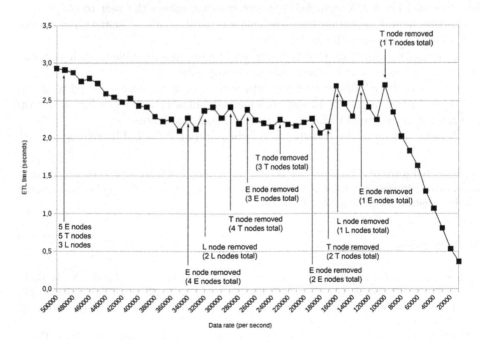

**Fig. 9.** Near-real-time, full ETL system scale-in

from 10.000 to 500.000 rows per second; the Y axis is the ETL time expressed in seconds; the system objective was set to deliver the ETL process in 3 s; the charts show the scale-out and scale-in of each part of the AScale, obtained by adding and removing nodes when necessary; A total of 7 data sources were used/removed gradually, each one delivering a maximum average of 70.000 rows/sec; AScale used a total of 12 nodes to deliver the configured time bounds.

**Near-real-time scale-out** results in Fig. 8 show that, as the data-rate increases and parts of the ETL pipeline become overloaded, by using all proposed monitoring mechanisms in each part of the AScale framework, each individual module scales to offer more performance where and when necessary.

**Near-real-time scale-in** results in Fig. 9 show the instants when the current number of nodes is no longer necessary to ensure the desired performance, leading to some nodes removal (i.e. set as ready nodes in stand-by, to be used in other parts).

## 5    Conclusions and Future Work

In this work we propose mechanisms and algorithms to achieve in time, within configured time bounds, automatic scalability for complex ETL, offering the possibility to the users to think solely in the conceptual ETL models and implementations for a single server.

We provide tests showing, with and without our proposal. Unfortunately because we were not able to find any system that allows the user to choose the ETL time bound execution time and automatically scale to achieve it we were not able to compare with any other tool.

Our tests demonstrate that the proposed techniques are able to scale-out and scale-in when necessary to assure the necessary efficiency.

Future work includes real-time event processing integration oriented to alarm and fraud detection. Other future work included making an visual drag and drop interface, improve monitoring and scale decision algorithms.

A beta version of the framework is being prepared for public release.

**Acknowledgement.** This project is part of a larger software prototype, partially financed by, Portugal, CISUC research group from the University of Coimbra and by the Foundation for Science and Technology.

## References

1. Fernandez, R.C., Pietzuch, P., Koshy, J., Kreps, J., Lin, D., Narkhede, N., Rao, J., Riccomini, C., Wang, G.: Liquid: unifying nearline and offline big data integration. In: Biennial Conference on Innovative Data Systems Research (CIDR), Asilomar, CA, USA. ACM, January 2015
2. Liu, X.: Data warehousing technologies for large-scale and right-time data. Ph.D. thesis, dissertation, Faculty of Engineering and Science at Aalborg University, Denmark (2012)

3. Muñoz, L., Mazón, J.N., Trujillo, J.: Automatic generation of ETL processes from conceptual models. In: Proceedings of the ACM Twelfth International Workshop on Data Warehousing and OLAP, pp. 33–40. ACM (2009)
4. O'Neil, P.E., O'Neil, E.J., Chen, X.: The star schema benchmark (ssb). Pat (2007)
5. Simitsis, A., Gupta, C., Wang, S., Dayal, U.: Partitioning real-time ETL workflows (2010)
6. Vassiliadis, P., Simitsis, A.: Near real time ETL. In: Kozielski, S., Wrembel, R. (eds.) New Trends in Data Warehousing and Data Analysis. Annals of Information Systems, vol. 3, pp. 1–31. Springer, New York (2009)

# New Similarity Measure for Spatio-Temporal OLAP Queries

Olfa Layouni$^{(\boxtimes)}$ and Jalel Akaichi

Computer Science Department, BESTMOD Laboratory,
Institut Supérieur de Gestion de Tunis, Tunis, Tunisia
layouni.olfa89@gmail.com, j.akaichi@gmail.com

**Abstract.** Storing, querying, and analyzing spatio-temporal data are becoming increasingly important, as the availability of volumes of spatio-temporal data increases. One important class of spatio-temporal analysis is computing spatio-temporal queries similarity. In this paper, we focus on assessing the similarity between Spatio-Temporal OLAP queries in term of their GeoMDX queries. However, the problem of measuring Spatio-Temporal OLAP queries similarity has not been studied so far. Therefore, we aim at filling this gap by proposing a novel similarity measure. The proposed measure can be used either in developing query recommendation, personalization systems or speeding-up query evolution. It takes into account the temporal similarity and the basic components of spatial similarity assessment relationships.

**Keywords:** Spatio-temporal OLAP queries · Spatio-temporal data warehouse · Data · Cube · Similarity

## 1  Introduction

ST-OLAP users interactively navigate a spatio-temporal data cube by launching a sequence of ST-OLAP queries, which is often tedious since the user may have no idea of what the forthcoming query should be. Adding to that, spatio-temporal data cubes store a big amount of data that's become increasingly complex to be explored and analyzed [13,14]. The notion of similarity has been considered as an important component for the development of recommendation systems. In our context, similarity measures are used to identify the degree of similarity between two ST-OLAP queries. To the best of our knowledge, there is no proposed similarity measure between ST-OLAP queries (GeoMDX queries [27]). So, in this paper we aim at filling this gap.

The paper is organized as follows: Sect. 2 briefly reviews related work, this section presents the different similarity assessment models proposed in the literature for comparing between queries. Section 3 presents the basic definitions in the context of spatio-temporal data warehouses and ST-OLAP systems. Section 4 presents our proposal of the new spatio-temporal similarity measure, Sect. 5 presents the performance evaluation. Finally, Sect. 6 concludes this paper.

© Springer International Publishing Switzerland 2016
S. Kozielski et al. (Eds.): BDAS 2016, CCIS 613, pp. 328–337, 2016.
DOI: 10.1007/978-3-319-34099-9_26

## 2   Related Works

Comparing queries has attracted a lot of attention in different areas like information retrievals [3,15,18,22,25], bioinformatics [9,19], etc. We note that the most proposed approach focused on assessing the similarity between queries. This section reviews the literature for similarity functions that could possibly be used to compare ST-OLAP queries, in order to be a support in the development of Spatio-Temporal personalization and recommendation approaches. Adding to that, the best of our knowledge, this is the first work dealing with the problem of ST-OLAP queries similarity measures expressed by GeoMDX manipulation language. So, in this section we begin by presenting some methods for comparing queries.

In the literature, we found two different motivations that could be used for comparing OLAP queries. The first one is query optimization [22]. This motivation is based on comparing a query $q$ to another $q'$ in order to find a better way to evaluate the query $q$. The second motivation is the most interesting for us because it used to compare a query $q$ to another $q'$ in order to help the user explore and analyze data, by recommending him queries, without focusing on the query evaluation [3,11,15,20,24,25].

We define a taxonomy for comparing OLAP queries as follows (see Fig. 1). We start by considering three basic classes for comparing OLAP queries according to the query models, the query expression and the distance functions used for computing similarity.

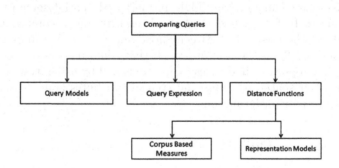

**Fig. 1.** A taxonomy for comparing OLAP queries

We find that the most proposed approaches for recommending OLAP queries, in order to help the user in his exploration, used at least one of the three classes mentioned in the taxonomy. The works proposed by [5,6,11,15,22,24,25] used the query models for comparing between OLAP queries. The works proposed by [11, 18,21–26] compare between OLAP queries according to the query expression. And finally, the works proposed by [3,11,16,24,25] compare between OLAP queries according to the distance functions. We notice that the most popular functions used for measuring similarity, can be classified in two different categories, the first one is based on the query corpus measures and the second is based on the query representation model. We find that the most popular functions used for measuring similarity based on the query corpus measures are: the Cosine similarity which was used in the works proposed by [24,25], the Dice coefficient which was used in the work proposed by [3] and the Hausdorff distance which was used in the work proposed by [11]. Adding to that, we find the most popular functions used for comparing between OLAP queries based on queries representation models, are the Levenshtein distance and the TF-IDF representation which were used in the work proposed by [16], and the Smith-Waterman algorithm which was used in the work proposed by [3]. In fact the work proposed by [3] demonstrates that the Smith-Waterman algorithm is the best method to be used for comparing between OLAP queries and sessions. Furthermore, we remark that those methods do not take into account both spatial relations and temporal data types in the same time, essentially Spatio-temporal OLAP queries (GeoMDX queries).

In order to measure the similarity between ST-OLAP queries, we need to compare between spatial objects and scenes invoked in ST-OLAP queries, by taking into account spatial relations: topological relation, distance relation and orientation relation. So, in order to measure the spatial relations distance between queries, we need to measure the topological distance, metric distance and direction distance between spatial objects and scenes invoked in queries [1,2,7,12,17]. In the literature, we distinguish five main approaches which are adopted to compare spatial relations. Those approaches are: the conceptual neighborhood approach, the projection-based approach, the combination of the conceptual neighborhood approach and the projection-based approach, the spatial relations-oriented model (the TDD model) and spatial semantic-oriented models/measures. In fact, a review of the literature and a comparative study was described in the book [28] between those approaches. We distinguish that the TDD model [17,28] is the best approach that could be possible to use in our case because it's applicable in queries expressions and by using it, we could compare between their spatial relations and spatial attributes launched in different queries without any transformations.

## 3　Basic Formal Definitions

In order to introduce and itemize the ST-OLAP queries similarity measure, we give, in this section the formal definitions of the basic concepts used in our proposal.

**Definition 1.** *Spatio-Temporal Dimensions and Hierarchies*

　*A dimension, in our case can be: a spatial dimension (SDim), a temporal dimension (TDim) and a classical dimension (Dim).*

$$STDim = \{SDim_1, ...., SDim_n, Dim_{n+1}, ...., Dim_{m-1}, TDim_m\}$$

*STDim represents a set of dimensions.*

*A dimension has different levels of members, so these members are arranged into a spatio-temporal hierarchy $ST\_H$.*

$$ST\_H = \{ST\_h_1, ST\_h_2, ...., ST\_h_n\}$$

*$ST\_h_i \in ST\_H$ is the level Lev of a hierarchy for a dimension $SDim_i$, $l \in Lev(ST\_h_i) \, \forall \, i \in [1, n]$, $l$ represents a level that belongs to the sets of levels for a dimension hierarchy.*

$$Allmeb\{STDim_i\} = \{meb_0, ...., meb_n\}$$

*For each $j \in [0, n]$, $j$ represents the level of a member $meb_j$.*

**Definition 2.** *Spatio-Temporal Cube*
   *A spatio-temporal cube is:*

$$ST\_C = \langle STDim, ST\_F \rangle$$
$$ST\_C = \langle SDim_1, ...., SDim_n, Dim_{n+1}, ...., Dim_{m-1}, TDim_m, ST\_F \rangle$$

*$STDim$ represents the set of dimensions and $ST\_F$ represents the fact table.*

$$ST\_F = \{SPK_1, ...., SPK_n, SM_1, ....., SM_n\}$$

*For each $i \in [1, n]$, $SPK_i$ represents the primary key of the dimension $STDim_i$ and $SM_i$ represents the different values and measures.*

The ST-OLAP query model based on the intentional level represents the query expression written in a particular query language, especially the GeoMDX language.

## 4  Similarity Measure for Comparing Spatio-Temporal OLAP Queries

In the literature, we distinguish the gaps of the similarity measures between ST-OLAP queries. So, we propose a similarity measure between ST-OLAP queries by taking into account not only the spatial data with specific characteristics such as topological, directional and metric distances, but also temporal data.

   We define the similarity function used in our approach to compare ST-OLAP queries. In fact, this function must consider the peculiarities of the multidimensional spatio-temporal data model and be calculable based on query expression only, GeoMDX query expression. In order to compare similarity of ST-OLAP queries, we propose three new similarity measures: spatial similarity measure to compute spatial distance between ST-OLAP queries, temporal similarity measure to compute temporal distance between ST-OLAP queries, and spatio-temporal similarity measure to compute spatio-temporal aspects of the ST-OLAP queries. Furthermore, the spatio-temporal similarity measure is a combination of three components: one related to measure sets, one to the set selection and one to where set.

## 4.1    Similarity Between Spatial Relations

In a ST-OLAP queries, we found the use of the three main categories of spatial relations, those relations are defined in the literature as follows: topological relation, orientation relation and metric distance relation [1,2,7,12,17].

So, to compare between two ST-OLAP queries, we need to measure the similarity between the topological relation, orientation relation and metric distance relation, invoked in each query.

### Spatial Distance:

Given two spatial scenes $SpatialR$ invoked by two ST-OLAP queries $q$ and $q'$. The distance between two spatial relations $Dist_{SpatialR}(q, q')$ is as follow:

$$Dist_{SpatialR}(q, q') = Dist_{Top}(q, q') + Dist_{Dir}(q, q') + Dist_{Met}(q, q') \qquad (1)$$

With:

- $Dist_{Top}(q, q')$ represents the topological distance between two ST-OLAP queries $q$ and $q'$ based on the distance proposed in [17].
- $Dist_{Dir}(q, q')$ represents the orientation distance between two ST-OLAP queries $q$ and $q'$ based on the distance proposed in [17].
- $Dist_{Met}(q, q')$ represents the distance in term of the metric distance between two ST-OLAP queries $q$ and $q'$ based on the distance proposed in [17].

## 4.2    Similarity Between Measure Sets

Our definition of measure sets similarity takes into account both the spatial calculated measures and measures that are represented in the schema of the spatio-temporal data warehouse. So, to define the measure similarity that is represented in the schema, we use the Jaccard index [4] and for the spatial calculated measures similarity we propose a $Dist_{SCMeas}$ distance that takes into account spatial relations.

### Distance Between Spatial Calculated Measures:

Given two ST-OLAP queries $q$ and $q'$ with spatial calculated measure $CMeas$ and $CMeas'$, respectively. The spatial calculated measure distance between $q$ and $q'$ is:

$$Dist_{SCMeas}(q, q') = Dist_{SpatialR}(CMeas, CMeas') \qquad (2)$$

### Distance Between Measure Sets:

Given two ST-OLAP queries $q$ and $q'$ with measure sets $MeasSM$ and $MeasSM'$, respectively. The measure sets similarity between $q$ and $q'$ is:

$$Dist_{Meas}(q, q') = \frac{|MeasSM \cap MeasSM'|}{|MeasSM \cup MeasSM'|} + Dist_{SCMeas}(q, q') \qquad (3)$$

## 4.3   Similarity Between Selection Sets

Our definition of measure similarity between selection sets is based on comparing between two sets of members and spatial scenes. We first introduce the notion of distance between members. Given a dimension $STDim_i \in STDim$ with its hierarchy $ST\_H$, the distance between two members $meb_1$ and $meb_2$ in this dimension is the shortest path [10] from $meb_1$ and $meb_2$ in $ST\_H$ is noted as $Dist_{meb}(meb_1, meb_2)$.

### Distance Between Selection Sets:
Given two ST-OLAP queries $q$ and $q'$ with selection sets $Sel$ and $Sel'$, respectively. The selection sets similarity between $q$ and $q'$ is:

$$Dist_{Sel}(q, q') = Dist_{meb}(meb_1, meb_2) + Dist_{SpatialR}(q, q') \tag{4}$$

With:

- $Dist_{meb}(meb_1, meb_2)$ represents the distance between members $meb_1$ invoked in $q$ and $meb_2$ invoked in $q'$.
- $Dist_{SpatialR}(q, q')$ represents the distance between spatial relations invoked in the selection sets in each query $q$ and $q'$.

## 4.4   Similarity Between Where Sets

Our definition of measure similarity between where sets is based on the temporal similarity distance between ST-OLAP queries. In our contribution, we adopt the similarity temporal distance defined in [8]. So, we need to compute the time range distance, the day distance and the week distance between two queries $q$ and $q'$ by comparing between the time values $t$ invoked in $q$ and $t'$ invoked in $q'$.

### Temporal Distance:
The temporal similarity distance $Dist_{Temp}$ between two queries $q$ and $q'$ is:

$$Dist_{Temp}(q, q') = \alpha.Dist_{TRng}(t, t') + \beta.Dist_{TDay}(t, t') + \gamma.Dist_{TWeek}(t, t') \tag{5}$$

With:

- $Dist_{TRng}$, $Dist_{TDay}$ and $Dist_{TWeek}$ stand for time range distance, day distance and week distance, respectively.
- $\alpha$, $\beta$ and $\gamma$ are temporal weights (The near optimal values of $\alpha$, $\beta$ and $\gamma$ are: $\alpha = 60$, $\beta = 20$ and $\gamma = 1$ [8]).

## 4.5   Similarity Measures Between ST-OLAP Queries

For measuring the similarity between two ST-OLAP queries $q$ and $q'$ we need to calculate the distance between them. Thus, we define the similarity function between two ST-OLAP queries $q$ and $q'$ as follows:

*Distance between two ST-OLAP queries:*
Given two ST-OLAP queries $q$ and $q'$. The similarity distance between $q$ and $q'$ is:

$$Dist_{STq}(q, q') = \alpha . Dist_{Meas}(q, q') + \beta . Dist_{Sel}(q, q') + \gamma . Dist_{Temp}(q, q') \quad (6)$$

With $\alpha$, $\beta$ and $\gamma$ are normalized to 1.

## 5    Performance Evaluations

In the experimental evaluation, we developed the *ST-OLAPSIM* system using Java language. *ST-OLAPSIM* system implements our proposal of the spatio-temporal similarity measure. It identifies the references of two given ST-OLAP queries (GeoMDX queries) and computes the spatio-temporal distance between them. For more explanation, we illustrate an example of a spatio-temporal similarity measure computed between two ST-OLAP queries using *ST-OLAPSIM*.In the example, we have two ST-OLAP queries $q$ and $q'$. The GeoMDX formulation of each query is as follows:

- $q$: with member [Measures].[Geom_Union] as 'ST_UnionAgg([Store].[All Stores]. CurrentMember.Children, "geom")' select [Measures].[Unit Sales], [Measures]. [Geom_Union] on columns [Store].[All Stores].[USA] on rows from [Sales] where [Time].[1997]
- $q'$:with member [Measures].[$Geom\_area\_in_k m2$] as 'ST_Area(ST_Transform (ST_UnionAgg ([Store].[All Stores].CurrentMember.Children, "geom"), 4326, 2991)) /1E6' select [Measures].[Unit Sales], [Measures].[$Geom\_area\_in\_km2$] on columns, [Store].[All Stores].[USA] on rows from [Sales] where [Time].[1997]

So, after launching those two GeoMDX queries in our system *ST-OLAPSIM*, we find out that the similarity measure between ST-OLAP queries $Dist_{STq}(q, q')$ is equal to the distance between measure set $Dist_{Meas}(q, q')$; and the distance between the selection sets and where sets are equal to zero because it is the same between $q$ and $q'$.

In order to evaluate the performance of our proposed similarity measure, we submitted a questionnaire to 20 persons with different ST-OLAP skills. The results have been used to understand how ST-OLAP queries similarity is perceived by the users, and they will be used to verify if the proposed methods capture the users' perception of similarity. To enable a better interpretation of the results, for each questionnaire test, we show the consensus $\delta$ which represents the percentage of users who gave the degree of agreement.

The tests of the questionnaire (T1, T2 and T3) were focusing on ST-OLAP queries comparison. In each test, the user was asked to rate the similarity between a given query $q$ and two other queries $q'$ and $q$" in both absolute using four scores: $\{low, fair, good \ and \ high\}$ and ranking queries in order of similarity.

We used the results obtained in two ways:

1. To compare $Dist_{STq}$ with the Smith-Waterman algorithm($\sigma_{SM}$) mentioned in Sect. 2 in terms of compliance with the users' judgments.
2. To set the weights of the three components of our query similarity function $Dist_{STq}$.

As to the first way, we defined two matching factors as follows:

- The score matching factor SMF for $Dist_{STq}$ is the percentage of times the score given by a user is the same returned by $Dist_{STq}$.
- The rank matching factor RMF for $Dist_{STq}$ is the percentage of cases in which the rankings $Dist_{STq}$ provides match with those given by users.

The comparison results are reported in Table 1. For all the tests $Dist_{STq}$ matches the users' judgment at least like $\sigma_{SM}$.

**Table 1.** Consensus and matching factors for ST-OLAP query comparison user tests.

| Test number | $\delta$ | | $\sigma_{SM}$ | | $Dist_{STq}$ | |
|---|---|---|---|---|---|---|
| | Score | Rank | SMF | RMF | SMF | RMF |
| T1 | 80% | 93% | 75% | 90% | 80% | 93% |
| T2 | 64% | 70% | 57% | 62% | 64% | 71% |
| T3 | 76% | 86% | 45% | 58% | 60% | 84% |

Overall, these results confirm a strong correlation between ST-OLAP queries similarities computed through $Dist_{STq}$ and the one perceived by users prove the efficiency of our proposed measure. Also $Dist_{STq}$ is more sensitive than $\sigma_{SM}$ and it shows better results.

As to second way, we tuned the weights through an optimization process whose goal function was the maximization of the correspondence with the questionnaire results. In this case, we suggest that the measure set is the most relevant

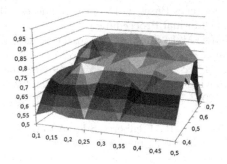

**Fig. 2.** Questionnaire matching for $Dist_{STq}$ as a function of weights $\alpha$ and $\beta$.

component, followed by the selection component and the less important component is the temporal set. The ranges for the weights were chosen consistently as follows: $\alpha \in [0.1, 0.5]$, $\beta \in [0, 4, 0.75]$ and $\gamma \in [0.05, 0.5]$. The function to be optimized was the average value of RMF for $Dist_{STq}$, that measures the percentage of cases in which the rankings provided by $Dist_{STq}$ match with those given by users. Figure 2 shows the average RMF as a function of $\alpha$ and $\beta$ (with $\gamma$ is equal to 1).

## 6    Conclusion

The contribution of the paper can be summarized as follows. First, we propose a new spatio-temporal similarity measure between ST-OLAP queries in the context of ST-OLAP manipulations. To the best of our knowledge, our proposal is the first work proposing a spatio-temporal similarity measure between ST-OLAP queries. Second, the present experimental evaluations validate the efficiency of our work. In particular, the next step will be to use our proposal to set up a method for recommending queries.

## References

1. Aissi, S., Gouider, M.S.: Spatial and spatio-temporal multidimensional data modelling: survey. CoRR abs/1208.0163 (2012)
2. Aissi, S., Gouider, M., Sboui, T., Bensaid, L.: Enhancing spatial datacube exploitation. In: Dregvaite, G., Damasevicius, R. (eds.) ICIST 2014. CCIS, vol. 465, pp. 121–133. Springer, Switzerland (2014)
3. Aligon, J., Golfarelli, M., Marcel, P., Rizzi, S., Turricchia, E.: Similarity measures for OLAP sessions. Knowl. Inf. Syst. **39**(2), 463–489 (2014)
4. Bank, J., Cole, B.: Calculating the jaccard similarity coefficient with map reduce for entity pairs in wikipedia, December 2008
5. Bellatreche, L., Giacometti, A., Marcel, P., Mouloudi, H., Laurent, D.: A personalization framework for OLAP queries. In: ACM 8th International Workshop on Data Warehousing and OLAP, DOLAP 2005, Bremen, Germany, 4–5 November 2005, pp. 9–18 (2005)
6. Bellatreche, L., Mouloudi, H., Giacometti, A., Marcel, P.: Personalization of MDX queries. In: 22èmes Journées Bases de Données Avancées, BDA 2006, Lille, 17–20 Octobre 2006, Actes (Informal Proceedings) (2006)
7. Bruns, H.T., Egenhofer, M.J.: Similarity of spatial scenes. In: 7th Symposium on Spatial Data Handling, pp. 31–42 (1996)
8. Chang, J.-W., Bista, R., Kim, Y.-C., Kim, Y.-K.: Spatio-temporal similarity measure algorithm for moving objects on spatial networks. In: Gervasi, O., Gavrilova, M.L. (eds.) ICCSA 2007, Part III. LNCS, vol. 4707, pp. 1165–1178. Springer, Heidelberg (2007)
9. Cohen, W.W., Ravikumar, P., Fienberg, S.E.: A comparison of string distance metrics for name-matching tasks, pp. 73–78 (2003)
10. Dijkstra, E.W.: A note on two problems in connexion with graphs. Numerische Mathematik **1**(1), 269–271 (1959)

11. Giacometti, A., Marcel, P., Negre, E., Soulet, A.: Query recommendations for OLAP discovery-driven analysis. IJDWM **7**(2), 1–25 (2011)
12. Glorio, O., Mazón, J.N., Garrigós, I., Trujillo, J.: A personalization process for spatial data warehouse development. Decis. Support Syst. **52**(4), 884–898 (2012)
13. Gorawski, M.: Extended cascaded star schema and ECOLAP operations for spatial data warehouse. In: Corchado, E., Yin, H. (eds.) IDEAL 2009. LNCS, vol. 5788, pp. 251–259. Springer, Heidelberg (2009)
14. Gorawski, M.: Multiversion spatio-temporal telemetric data warehouse. In: Grundspenkis, J., Kirikova, M., Manolopoulos, Y., Novickis, L. (eds.) ADBIS 2009. LNCS, vol. 5968, pp. 63–70. Springer, Heidelberg (2010)
15. Jerbi, H.: Personnalisation danalyses dcisionnelles sur des donnes multidimensionnelles. Ph.D. thesis, Institut de Recherche en Informatique de Toulouse UMR 5505, France (2012)
16. Layouni, O., Akaichi, J.: A novel approach for a collaborative exploration of a spatial data cube. IJCCE: Int. J. Comput. Commun. Eng. **3**(1), 63–68 (2014)
17. Li, B., Fonseca, F.: TDD: a comprehensive model for qualitative spatial similarity assessment. Spat. Cogn. Comput. **6**(1), 31–62 (2006)
18. Marcel, P., Missaoui, R., Rizzi, S.: Towards intensional answers to OLAP queries for analytical sessions. In: ACM 15th International Workshop on Data Warehousing and OLAP, DOLAP 2012, Maui, HI, USA, 2 November 2012, pp. 49–56 (2012)
19. Moreau, E., Yvon, F., Cappé, O.: Robust similarity measures for named entities matching. In: Proceedings of the 22nd International Conference on Computational Linguistics, COLING 2008, vol. 1. pp. 593–600. Association for Computational Linguistics, Stroudsburg (2008)
20. Negre, E.: Exploration collaborative de cubes de donnes. Ph.D. thesis, Universit Franois Rabelais of Tours, France (2009)
21. Sapia, C.: On modeling and predicting query behavior in OLAP systems. In: Proceedings of the International Workshop on Design and Management of Data Warehouses (DMDW 1999), pp. 1–10. Swiss Life (1999)
22. Sapia, C.: PROMISE: predicting query behavior to enable predictive caching strategies for OLAP systems. In: Kambayashi, Y., Mohania, M., Tjoa, A.M. (eds.) DaWaK 2000. LNCS, vol. 1874, pp. 224–233. Springer, Heidelberg (2000)
23. Sapia, C., Alexander, F.: Promise: modeling and predicting user behavior for online analytical processing applications. Ph.D. Thesis submitted, Technische Universitt Mnchen (2001)
24. Sarawagi, S.: Explaining differences in multidimensional aggregates. In: Proceedings of the 25th International Conference on Very Large Data Bases, VLDB 1999, pp. 42–53. Morgan Kaufmann Publishers Inc., San Francisco (1999)
25. Sarawagi, S.: User-adaptive exploration of multidimensional data. In: VLDB, pp. 307–316. Morgan Kaufmann (2000)
26. Sathe, G., Sarawagi, S.: Intelligent rollups in multidimensional OLAP data. In: Proceedings of the 27th International Conference on Very Large Data Bases, VLDB 2001, pp. 531–540. Morgan Kaufmann Publishers Inc., San Francisco (2001)
27. Tranchant, M.: Capacits des outils solap en termes de requêtes spatiales, temporelles et spatio-temporelles. Technical report, Conservatoire National des Arts et Metiers Centre Regional Rhône- Alpes Centre Denseignement de Grenoble (2011)
28. Yan, H., Li, J.: Spatial Similarity Relations in Multi-scale Map Spaces. Springer, Switzerland (2015)

# Natural Language Processing, Ontologies and Semantic Web

# Enhancing Concept Extraction from Polish Texts with Rule Management

Piotr Szwed[✉]

AGH University of Science and Technology,
Mickiewicza Av. 30, 30-059 Kraków, Poland
pszwed@agh.edu.pl

**Abstract.** This paper presents a system for extraction of concepts from unstructured Polish texts. Here concepts are understood as n-grams, whose words satisfy specific grammatical constraints. Detection and transformation of concepts to their normalized form are performed with rules defined in a language, which combines elements of colored and fuzzy Petri nets. We apply a user friendly method for specification of samples of transformation patterns that are further compiled to rules. To improve accuracy and performance, we recently introduced rule management mechanisms, which are based on two relations between rules: partial refinement and covering. The implemented methods include filtering with metarules and removal of redundant rules (i.e. these covered by other rules). We report results of experiments, which aimed at extracting specific concepts (actions) using a ruleset refactored with the developed rule management techniques.

**Keywords:** NLP · Text mining · Concepts extraction · Unstructured text · Inflection · Rules · Petri nets

## 1  Introduction

Although the term *concept* is widely used in various disciplines, it lacks of precise definition and is often interpreted in various, conflicting ways [22]. Concepts can be defined as abstract ideas, generalizations describing objects sharing common features, names in various languages, synonyms having common meaning or ontology classes being named representations of sets of instances.

In this paper we define a concept as a meaningful sequence of words (an n-gram), which can be used to represent objects, ideas, events or activities. Hence, the n-gram (in fact a concept name) should form a structure of words, following certain language conventions. For example, its main element should be a noun or a verbal noun, possibly extended with some complements.

The goal of our work is to develop a software tool allowing to extract concept names from unstructured Polish texts based on descriptions provided by users specifying, how such names may look like. Moreover, we assumed that a key requirements are that n-grams representing concepts should have correct

© Springer International Publishing Switzerland 2016
S. Kozielski et al. (Eds.): BDAS 2016, CCIS 613, pp. 341–356, 2016.
DOI: 10.1007/978-3-319-34099-9_27

morphological form according to Polish grammar rules and that they should be normalized, typically, by transforming a central noun to the Nominative case.

We were motivated by two potential applications of the extracted multi-word concept names. The first is related to expected improved categorization of text document. A good example can be a compound term "table wine". A document referencing it is more likely to discuss wines, than pieces of furniture. It should be noted that a compound term built around a noun may appear in Polish texts inflected for seven declination cases. Hence, normalization to the Nominative case, apart from aesthetic issues, may positively influence accuracy of classification algorithms.

The second potential application consists in building structured representation of document content. It may have a form of thesaurus of most referenced terms, a taxonomy or even a shallow ontology. Figure 1 gives an example of a structure that can be derived from an n-gram *Analizowanie dokumentów projektowych* (analyzing design documents). *Analizowanie* is an action, which is subclassed by the term *Analizowanie dokumentów* and further *Analizowanie dokumentów projektowych*. Each descendant term can be defined using an object property linked to subsequently refined object of the action: *Dokument* (document) and *Dokument projektowy* (design document).

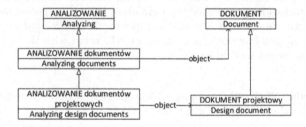

**Fig. 1.** Example ontology structure for the n-gram Analizowanie dokumentów projektowych (Analyzing design documents)

In our previous work [26] we described the first prototype implementation of the concept extraction system. It uses a rule based approach to perform extraction and transformation of concept names. The rule language, inspired by Petri nets, allows to define patterns of input tokens and required transformations. The rules are not coded directly, instead, they result from compilation of text translation patterns specified in a user friendly form. However, formal rules derived from prototype translation schemes, can suffer from certain drawbacks including ambiguities of their output and redundancy.

In this paper we discuss recently developed features related to management of rules. They include refinement of rules with metarules and checking redundancy. The new functionality is based on two formally defined relation between rules: partial refinement and covering. We report results of focused tests, restricted to concepts describing actions, performed on two documents. The first, already processed in [26], was a Polish translation of a medical guideline for asthma treatment. The second was a collection of RSS feeds retrieved from Polish portals in September 2015.

The paper is organized as follows: Sect. 2 discusses the problem of concept extraction, as well as several tools dedicated to Polish language processing. It is followed by Sect. 3, which presents key features of the concept extraction system. Next Sect. 4 presents mentioned earlier solutions related to rule management. Results of initial experiments are given in Sect. 5. Last Sect. 6 provides concluding remarks.

## 2    Related Works

Ontology learning is a process of building ontologies from various data representations. It includes such efforts as identification of concepts, their relations and attributes and arranging them into hierarchies. The mentioned tasks are apparently easier in the case of structured or semi-structured data. A challenging problem, however, is to build ontologies (usually taxonomies) comprising concepts extracted from unstructured text documents [14]. For this purpose various text mining techniques can be used: syntactical analysis, Formal Concept Analysis [4] and clustering [7].

Concepts (sometimes also referred as compound terms or phrases) are important features used in Text Mining [23]. Compound terms processing is a technique aiming at improving accuracy of search engines by indexing documents according to compound terms, i.e. combinations of 2 or more single words. During query execution searched compound terms are also extracted from queries (which can be phrases in natural language) and then matched with compound terms attributed to documents. Such approach outperforms solutions that use in queries keywords combined with boolean operators. The idea of statistical compound terms processing was proposed and commercialized by Concept Search company [5]. On the other hand, a syntactical technique based on rules defining patterns for various European languages was developed within CLAMOUR project (see reports at http:// webarchive.nationalarchives.gov.uk/20040117000117/, http://statistics.gov.uk/ methods_quality/clamour/default.asp).

In [9] Dalvi, Kumar et al. from Yahoo Labs. coined an idea of a *web of concepts*. They claimed that the current model of the web in the form of hyperlinked pages represented by bags of words should be augmented by extracting concepts and creating a new rich view of all available resources for each concept instance. Authors discussed several use cases enabled by the new approach, including: more accurate web search, browsing optimization, session optimization and advertising. The paper also indicated several challenges, related mainly to information extraction, potential uncertainty and changing data. An algorithm for concept extraction following this idea was proposed in [20].

Blake and Pratt [2] used automatically retrieved concepts to build searchable representations of medical texts. In [3] authors extracted ontologies from medical documents and showed that the use of concepts improves search results.

Osinski and Weiss described Lingo, a concept driven document clustering algorithm [19]. Concepts (frequent sequences of terms), were used to label clusters in a document-term matrix.

An important task in NLP language classification is *tagging*, i.e. assigning part of speech (POS) information to inflected forms of words. This is a challenging task for a highly inflected languages like Polish. According to [1] words in English language can be described with about 200 tags whereas for Polish their number ranges at 1000.

In our work we used Morfologik stemming library [17]. The software and the dictionary found application in several projects including Language tool [18], carrot search [19] and PSI toolkit [10].

Morfologik dictionary was integrated as a part of PoliMorf [28].

Language tool [16,18] is a proofreading program supporting a number of languages including Polish. It allows to define text correction rules referring to explicit token values, lemmatized words and part of speech tags. Rules can be specified either in XML format or coded directly in Java. The tool provides also a visual rule editor, which allows to construct rules in an interactive manner. As the rules yield suggestions, their core functionality is somehow similar to the concepts extraction rules described in this paper.

An exhaustive survey of relations between rules and properties of rulesets can be found in [13]. Although the definitions are formulated for tabular systems and First Order Logic, they can be applied in other settings. The discussed relations include various types of redundancy, e.g. equivalence and subsumption (a rule subsumes other, if it has weaker precondition and stronger conclusion). Another properties that can be collectively characterized as inconsistency of rulesets are: nondeterminism (at least two rules can fire for the same state) and conflict (their results are contradictory). It should be mentioned that such types of inconsistency does not apply to fuzzy rules and fuzzy inference systems [15,21], as conflict resolution is handled at the aggregation (defuzzification) stage.

## 3    Concepts Extraction

Figure 2 shows the logical model of the concept extraction system indicating main tasks and data flows. The basic text processing chain (shown in the lower part of the diagram) includes three activities: splitting an input document into sentences, then applying transformation rules from a *RuleSet* and, finally, aggregating and sorting concept candidates. The solution uses also Morfologik dictionary, in particular it adheres to its tagging conventions.

The *RuleSet* set is stored in a file. The XML based language used to define it is relatively complex, moreover, to specify word transformations, the rules reference rather sophisticated POS tagging symbols used in the Morfologik dictionary. Although the rules can be coded manually, this would be a tedious and potentially erroneous task. Instead, samples of transformation patterns in a user friendly, textual form can be prepared and translated into rules. We refer to those samples as *Annotations*. In Fig. 2 they are input to the *Compilation* process that produces an initial ruleset, which can by further enhanced by applying *Filtering*.

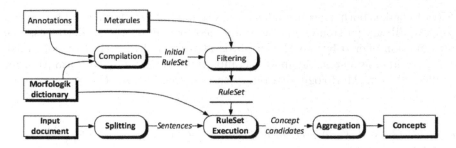

**Fig. 2.** Process of concept extraction

## 3.1 Annotations

Annotations are specified using a simple syntax @(*input pattern* = *output pattern*). One of output tokens can be indicated with the at sign @ as a *key*, i.e. a term that may be used alone as a superclass of the referenced concept. Additionally, input tokens marked with dollar sign $ should be matched exactly with words in an input document. Examples of annotations are:

- @(zielonym samochodom = zielony @samochód) – the annotation (referencing the term *green car*) specifies a transformation of a string, which is a combination of an adjective and a plural masculine noun in Datum case into the Nominative, singular form. Hence, the resulting rule can be applied to such n-grams, as "ostrym ołówkom" (*sharp pencils*) or "wysokim wzgórzom" (*high hills*).
- @(prowadzenia pojazdów = @prowadzenie pojazdów) – defines a transformation of a verbal noun with a noun complement from the Genitive to Nominative case using *car driving* as a sample.
- @ (koń $, $który lata = latający @ koń) – this annotation defines a translation pattern that can be expressed in English as: *horse, which flies* → *flying horse*. The dollar sign is used to mark exact token value expected at input. The transformation involves changing a verb "lata" (*flies*) into a corresponding gerund form "latający" (*flying*).

The original idea behind the annotations was that they would be used to mark expected translation patterns inside an input document. Such usage is still possible, however, it turned out during experiments that more practical is to gather an exhaustive set of systematically prepared annotations in a separate file and compile them to a rule set, which can be reused in several concept extraction tasks.

## 3.2 Morfologik

During annotations processing and rules execution input words are checked for specific part of speech properties and undergo morphological transformations. This function is provided by the Morfologik.

Morfologik is both a a comprehensive dictionary of Polish inflected forms and a software library written in Java accompanied by a number of utility tools. The main function offered by the Morfologik is *stemming* (lemmatization) of Polish words, i.e. finding a stem (lemma) accompanied by a grammar information for an inflected form. Morfologik dictionary can be seen as a relation

$$D \subset IF \times S \times \mathcal{P}, \tag{1}$$

where $IF$ is a set of inflected forms, $S$ is a set of stems, $S \subset IF$, and $\mathcal{P}$ is a set of POS tags defining properties of inflected forms (part of speech, gender, singular vs. plural, declination case, etc.)

It should be observed, that a given inflected form may appear in multiple dictionary entries. An example can be the word *czytaniu* (Eng. reading) for which the corresponding lemma and POS combinations are:

- ("czytaniu", "czytać", "ger:sg:dat.loc:n2:imperf:aff:refl.nonrefl") – verbal noun derived from the verb
- ("czytaniu", "czytanie", "subst:sg:dat:n2") - a noun in Datum case
- ("czytaniu", "czytanie", "subst:sg:loc:n2") - a noun in Locative case

Based on the Morfologik dictionary scheme two functions can be defined: *stem* (2) and *synth* (3). The first takes as input an inflected form and returns a set of stems with accompanying tags, the second synthesizes inflected forms from an input stem and tag.

$$stem\colon IF \to 2^{S \times \mathcal{P}} \tag{2}$$

$$synth\colon S \times \mathcal{P} \to 2^{IF} \tag{3}$$

Morfologik fully supports stemming. The library uses internally an efficient dictionary representation based on Finite State Machine (FSM) model, which is characterized by compact data size and short access times [8].

Unfortunately, the synthesizing function is not supported directly by the library. We developed a synthesizing function, which uses the dictionary data stored in a local PostgreSQL database. It was populated with 288657 stems, 1173 tags and 7410145 triples built from inflected forms, stems and tags.

## 3.3   Rules

A rule defines transformations of input n-grams (sequences of words) by specifying a class of patterns, to which it can be applied, as well as transformations for individual words.

The language used to define rules is based on Petri nets. Each rule comprises ordered sets of input and output places, which are linked by transitions. Figure 3 gives an example of the rule suitable to process 2-grams. It comprises two input places, two output places and three transitions. Multiple transitions, here t1 and t2, may link a pair of places.

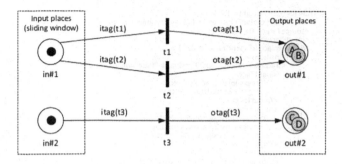

**Fig. 3.** Rule as a Petri net

Transitions are assigned with two sets of tags: input *itag* and output *otag*. Input tags are used as guards, they allow to check, if the transition applies to a word tagged with part of speech information. Output tags are used to synthesize target inflected form from a word stem (lemma). Additionally, each transition has assigned weight, that can be used to differentiate less and more likely translation schema. Below we give a formal definition of the language used to define rules.

**Definition 1.** *Concept extraction rule is a tuple*
$R = (\mathcal{P}, I, O, T, F, itag, otag, itoken, \mu)$, *where:*

- $\mathcal{P}$ *is a set of tags*
- $I = \{i_1, i_2, \ldots, i_n\}$ *is an ordered set of input places,*
- $O = \{o_1, o_2, \ldots, o_m\}$ *is an ordered set of output places,*
- $T$ *is a set of transitions* $T \cap (I \cup O) = \emptyset$,
- $F \subset I \times T \cup T \times O$ *is a set of arcs. For each input place* $i \in I$, *output places* $o, o' \in O$ *and transitions* $t_1, t_2 \in T$ *the following condition holds:* $((i, t_1), (t_1, o), (i, t_2), (t_2, o') \in F \Rightarrow o = o')$
- $itag: T \rightarrow 2^{\mathcal{P}}$ *is a function assigning to a transition a set of* input *tags,*
- $otag: T \rightarrow 2^{\mathcal{P}}$ *is a function assigning to a transition a set of* output *tags,*
- $itoken: I \cup O \rightarrow \mathcal{A}$ *is a function assigning to an input place an exact (possible empty) string form* $\mathcal{A}$
- $\mu: T \rightarrow [0, 1]$ *is a transition weight function*

Figure 4 shows XML code for a rule obtained from the annotation @(czytaniu książek = @czytanie książek) (*book reading*). The algorithm used in the compilation process [26] creates input and output places corresponding to words in both patterns and adds transitions based on stem matching. Here, the words "czytaniu" and "czytanie" have two common stems: "czytać" and "czytanie" (see examples in Sect. 3.2), thus two transitions were created. Moreover, POS tags in input and output words are used as sets *itag* and *otags* assigned to the transitions.

```
<rule id="r:15" weight="1.0">
  <source>@( czytaniu książek ) = ( @key(czytanie) książek ) @line:31</source>
  <sourceTarget>0</sourceTarget>
  <inputPlaces>
    <inputplace isExact="false" ord="0">
      <transitions>
        <transition target="op:29" isGuard="false" weight="1.0">
          <inputTags>
            <intag>subst:sg:loc:n2</intag>
            <intag>subst:sg:dat:n2</intag>
          </inputTags>
          <outputTags>
            <outtag>subst:sg:nom:n2</outtag>
            <outtag>subst:sg:acc:n2</outtag>
            <outtag>subst:sg:voc:n2</outtag>
          </outputTags>
        </transition>
        <transition target="op:29" isGuard="false" weight="1.0">
          <inputTags>
            <intag>ger:sg:dat.loc:n2:imperf:aff:refl.nonrefl</intag>
          </inputTags>
          <outputTags>
            <outtag>ger:sg:nom.acc:n2:imperf:aff:refl.nonrefl</outtag>
          </outputTags>
        </transition>
      </transitions>
    </inputplace>
    <inputplace isExact="false" ord="1">
      <transitions>
        <transition target="op:30" isGuard="false" weight="1.0">
          <inputTags>
            <intag>subst:pl:gen:f</intag>
          </inputTags>
          <outputTags>
            <outtag>subst:pl:gen:f</outtag>
          </outputTags>
        </transition>
      </transitions>
    </inputplace>
  </inputPlaces>
  <outputPlaces>
    <outputplace id="op:29" ord="0" isKey="true" isExact="false"/>
    <outputplace id="op:30" ord="1" isKey="false" isExact="false"/>
  </outputPlaces>
</rule>
```

**Fig. 4.** A rule in XML format

## 3.4   Rule Execution

During rule execution a window having the length equal to the number of the rule's input places slides through a sequence of words forming a sentence. For each window position, input places are filled with tokens, which are obtained from words in the sentence by applying to them the *stem* function. Tokens are tuples from the set $TK = IF \times S \times \mathcal{P} \times \mathbb{R}^{0+}$ (see Sect. 3.2). Components of a token tuple $(if, s, p, w) \in TK$ are the following: $if$ denotes an inflected form, $s$ is a stem, $p$ is a part of speech tag and $w$ is a non-negative weight.

A transition $t \in T$ is enabled (can fire) for a token $\tau = (if, s, p, w)$ located in its input place, if $p \cap itag(t) \neq \emptyset$, i.e. the POS tagging for a token matches the set of transition's input tags $itag(t)$. As the transition fires, a number of output tokens can be created. For an output token $\tau' = (if', s', p', w')$ the following relationships hold: $s' = s$, $p' \in otag(t)$, $if' = synth(s', p')$ and $w' = \frac{p \cap itag(t)}{p \cup itag(t)} \cdot \mu(t)$.

The application of the Jaccard index in the formula for weight $w'$ calculation allows to assess grammatical similarity between an input term and a prototype word appearing in the annotation, which was the rule origin.

## 3.5   Aggregation

Aggregation of results constitutes the final text processing stage. Actually, it is rather a chain of aggregations performed at subsequent levels: for places, rules, sentences and documents.

In a general case, after a rule is executed, its output places may contain a number of tokens. This is due to multiple transitions reaching the same output place or several output tags assigned to transitions. During aggregation at the *place level* identical tokens in output places are combined.

In the next step, referred as the *rule level*, strings of symbols are formed and attributed with weights. For the situation presented in Fig. 3, assuming distinct tokens A, B, C and D, four strings: AC, AD, BC, and BD would be formed.

Various rules may return identical concepts. Aggregation at the *sentence level* combines identical strings extracted from a sentence. Finally, the aggregation at the *document level* keeps track of occurrences of concepts in sentences, counts them and aggregates the weights.

## 4   Rule Management

At this stage of the software development we stated that the quality of a rule set is a key factor influencing accuracy and performance of concept extraction process. Due to high inflection level of the Polish language the number of translation patterns defined by annotations and corresponding rules is quite high, at present it ranges at about four hundreds. In many cases rules obtained from the compilation stage may be redundant or define translation patterns, which are undesired for certain words.

This section discusses introduced recently mechanisms related to rule management aiming at at removing ambiguities, which compromise accuracy, and redundant rules, which hinder performance. Rule management is achieved within the *Filtering* task shown in Fig. 2. It includes two activities executed sequentially: filtering with metarules followed by removal of redundant rules.

### 4.1   Relations Between Rules

A rule transforming an input string of length $k$ into a set of output strings of length $m$ can be considered a partial function $R\colon IF^k \to 2^{IF^m}$ mapping a Cartesian product of $k$ sets of inflected forms $IF$ into a powerset of $IF^m$. We will refer to the numbers $(k, m)$ as the rule *arity*.

**Definition 2.** *Rule $R_1$ covers rule $R_2$ of the same arity, if $dom(R_1) \supseteq dom(R_2)$ and $\forall \alpha \in dom(R_2)\colon R_1(\alpha) \supseteq R_2(\alpha)$.*

**Definition 3.** *Rule $R_1$ partially refines rule $R_2$ of the same arity, if $dom(R_1) \subseteq dom(R_2)$ and $\forall \alpha \in dom(R_1): R_1(\alpha) \subseteq R_2(\alpha)$.*

The introduced relations are explained by the diagram in Fig. 5. If a rule $R_1$ covers $R_2$, then it is applicable to a wider set of input strings and it may produce more output strings (concept candidates), hence, $R_1$ is more general and may replace $R_2$. Equality of rules, which is quite a common situation, is a particular case of covering: two rules are equal, if they cover each other. On the other hand, if a rule $R_1$ partially refines $R_2$, then it is more specific. Although it is applicable to a narrower input, it may produce better results. Moreover, we may consider a set of rules $\rho = \{R_{1i}\}$, each partially refining $R_2$. If their domains cover the the domain of $R_2$, i.e. $dom(R_2) \subseteq \bigcup_{R \in \rho} dom(R)$, then the rule $R_2$ can be replaced by the set of rules $\rho$. Such relation (for single rules) is called in [13] subsumption.

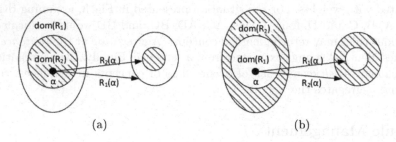

(a)                              (b)

**Fig. 5.** Two relations between rules: (a) covers (b) partially refines

## 4.2 Metarules

Metarules allow to transform a rule set by replacing selected rules by one or more rules that partially refine them. They are used to restrict the output to particular sequences of words satisfying grammatical constraints. For example, they may remove undesired translations to Accusative cases, where Nominative is expected or ignore translations producing particular part of speech elements.

Technically, metarules are defined as sequences of regular expressions, which are applied to filter output tags (*otag* sets) assigned to transitions. Additionally, they may specify weights, which will be used for resulting rules. If an application of the metarule empties the *otag* set of a transition, the transition will be removed. If all original transitions reaching an output place are removed, the resulting rule will not be produced.

For example, the application of a metarule ``ger.* subst.* 0.22'' to the rule presented in Fig. 4 will remove the transition referencing subst part of speech (noun) and set the weight to 0.22. Thus, the resulting rule will become more specific (focused on actions described by verbal nouns) and not applicable to such strings as, e.g. *zarobkom piosenkarek* (*earnigs of* [female] *singers*).

The filter based on metarules actually does not act on existing rules, but produces new ones. Applying appropriatelly designed multiple metarules to a given rule $R$ may produce a set of resulting rules $R'_1, R'_2, \ldots, R'_n$, which can subsume it.

## 4.3  Removing Redundant Rules

Other method of rules filtering consists in detecting redundant rules based on the covering relation. To check presence of the covering relation for both rules we analyze the structure formed by transitions and places, as well as the sets of tags assigned to transitions.

Let us assume that two rules $R_1 = (\mathcal{P}, I_1, O_1, T_1, itag_1, otag_1, itoken_1, \mu_1)$ and $R_2 = (\mathcal{P}, I_2, O_2, T_2, itag_2, otag_2, itoken_2, \mu_2)$ have equal numbers of input and output places. Let $T_1(s, e)$ be a set of transitions linking places $i_{1s} \in I_1$ and $o_{1e} \in O_1$ belonging to the rule $R_1$. The set $T_2(s, e)$ is defined for the rule $R_2$ analogously. For a set of transitions $T$ let us define: $itag(T) = \bigcup_{t \in T} itag(t)$ and $otag(T) = \bigcup_{t \in T} otag(t)$.

Now we consider two sets of transition $T_1(s, e)$ and $T_2(s, e)$ linking places with numbers $s$ and $e$ in both rules. If $itag(T_1(s, e)) \supseteq itag(T_2(s, e))$, then $T_1(s, e)$ is applicable (enabled) for a larger set of input tokens than $T_2(s, e)$. Due to the fact that during rule execution all enabled transition fire, relations between $otag(T_1(s, e))$ and $otag(T_2(s, e))$ also hold for the sets of tokens produced in output places. This can be further extended for the set of resulting strings.

**Corollary 1.** *Rule $R_1$ covers $R_2$ if for each input place number $s$ and output place number $e$ holds:*
$itag(T_1(s, e)) \supseteq itag(T_2(s, e)) \wedge otag(T_1(s, e)) \supseteq otag(T_2(s, e)).$

Corollary 1 provides a formal basis for the procedure that checks covering (or equality) of rules and removes rules being covered by another. In the current implementation rules left are assigned with higher weight from $\{R_1, R_2\}$. The filter is applied after refinement with metarules, hence, first the rules are made more specific as regards output and then more general as regards their domain.

## 5  Experiments and Results

In this section we present results of concepts extraction with use of rules selected specifically to recognize 2-grams and 3-grams comprising verbal nouns. Hence, resulting concepts may express *actions* with complementary description specifying their objects.

We used a gradually developed ruleset, which counted 355 rules. Then 8 metarules having the form similar to ``ger.* subst:sg:gen.* .7'' were applied. This reduced the active ruleset to 124 rules, their transitions were also truncated. Removal of redundant rules reduced further the ruleset to 89 elements.

During reported here experiments two input texts were processed. The first, used in earlier works aiming at building ontologies of medical guidelines [24], was a Polish translation of a guidelines for asthma treatment issued by Global Initiative for Asthma (GINA) in 2011. The file size is 308 KB, it contains about 40000 words and 2000 sentences. In the experiments described in [26] the same document was processed to detect noun based concepts giving about 5000 items.

The second document was a collection of 28740 RSS feeds in Polish (titles and short content) registered between 2015.09.11 and 2015.09.18. The size of cleansed text file was about 6 MB.

Experiments were conducted on Intel Core i7-2675QM laptop at 2.20 GHz, 8 GB memory under Windows 7 with local PostgreSQL 9.1 database supporting *synth* function. Processing of the GINA text took about 5.91 s and returned 240 concepts. For the RSS extract the processing time was about 96.30 s and 400 concepts were recognized.

Table 1 shows top 25 results for both documents ordered by their scores. Words in capital letter indicate keys, i.e. the parts of concept names referring to prospective superclasses in the concepts hierarchy. Analyzing the results, it can be stated, that in the GINA document, which is generally a scientific text, verbal nouns appeared quite often. Moreover, the obtained concepts (asthma treatment, application of glucocorticosteroids, treatment of exacerbation, bronchodilation, etc.) rather accurately characterize the document content. On the other hand, references to actions expressed as verbal nouns in RSS feeds are relatively rare and, what is not surprising, a diversity of topics can be observed.

Examination of resulting concepts, also those obtained in other experiments, which, due to the limited space, are not reported here, indicated that filtering with metarules improved significantly accuracy. However, at this stage, which is still a development and a prototyping phase, we have not applied any technique to quantify quality of results. Problems that can be observed are often related to occurrences of not meaningful words, e.g. *jaki, taki* (which, such as), *swój* (my, his, her), *pewien* (certain), etc., which are classified in the dictionary as adjectives. To handle them an additional dictionary of *stop words* is used and gradually extended.

Another problem is related to multiple POS tagging of words in the dictionary. For example, processing of RSS feeds returned at position 235 the concept *MIESZKANIE dziewczyny*. The word *mieszkanie* can be both interpreted as a noun *apartment* (in most cases) and a verbal noun indicating an action *living* (rare). As we used rules focusing on verbal nouns, an occurrence of this term can be considered a kind of error. Without statistical information embedded in the dictionary, such misinterpretations are hard to avoid.

Removal of redundant rules based on detection of covering relation resulted in performance increasing. For the GINA document reduction of processing time was about 26 % (from 8.04 to 5.91 s). For the RSS feeds it was about 6.6 % (from 103.13 to 96.3 s).

**Table 1.** Extracted action concepts from the asthma treatment medical guideline (GINA) and RSS feeds

| | GINA | | | RSS | | |
|---|---|---|---|---|---|---|
| | Concept | Cnt | Score | Concept | Cnt | Score |
| 1 | LECZENIE astmy | 87 | 60.9 | OBCINANIE funduszy strukturalnych | 4 | 4.0 |
| 2 | STOSOWANIE glikokortykosteroidów wziewnych | 24 | 24.0 | PROWADZENIE kampanii | 7 | 3.8 |
| 3 | LECZENIE zaostrzeń | 21 | 14.7 | PRZEKRACZANIE granicy | 21 | 3.6 |
| 4 | ROZSZERZANIE oskrzeli | 11 | 7.7 | BUDOWANIE dobrego wizerunku | 10 | 3.3 |
| 5 | WDRAŻANIE wytycznych | 20 | 7.0 | ROZPOZNAWANIE mowy | 4 | 2.8 |
| 6 | STOSOWANIE glikokortykosteroidów doustnych | 5 | 5.0 | TWORZENIE gier | 3 | 2.1 |
| 7 | STOSOWANIE teofiliny | 7 | 4.9 | ODRABIANIE strat | 3 | 2.1 |
| 8 | MONITOROWANIE astmy | 6 | 4.2 | MODYFIKOWANIE gry | 3 | 2.1 |
| 9 | PRZESTRZEGANIE zaleceń | 6 | 4.2 | UNIKANIE opodatkowania | 3 | 2.1 |
| 10 | LECZENIE rozszerzające oskrzela | 4 | 4.0 | PILNOWANIE granic | 3 | 2.1 |
| 11 | WYSTĘPOWANIE astmy | 10 | 3.5 | POSZUKIWANIE trasy | 3 | 2.1 |
| 12 | KONTROLOWANIE objawów podmiotowych | 3 | 3.0 | DOSKONALENIE gry obronnej | 2 | 2.0 |
| 13 | WYSTĘPOWANIE objawów podmiotowych | 6 | 3.0 | ZMNIEJSZANIE unijnych funduszy | 2 | 2.0 |
| 14 | STOSOWANIE dużych dawek | 3 | 3.0 | PODEJMOWANIE decyzji | 14 | 2.0 |
| 15 | ZMNIEJSZANIE dawki | 4 | 2.8 | WYPRZEDAWANIE polskich przedsiębiorstw | 5 | 1.7 |
| 16 | STOSOWANIE leków doraźnych | 5 | 2.5 | PODNOSZENIE bramek | 2 | 1.4 |
| 17 | WYKONYWANIE ćwiczeń | 3 | 2.1 | BLOKOWANIE reklam | 2 | 1.4 |
| 18 | KOŁATANIE serca | 3 | 2.1 | BADANIE zapotrzebowania | 2 | 1.4 |
| 19 | LECZENIE ciężkich zaostrzeń | 2 | 2.0 | SZACOWANIE strat | 2 | 1.4 |
| 20 | PŁUKANIE jamy ustnej | 2 | 2.0 | PRZYJMOWANIE list | 2 | 1.4 |
| 21 | LECZENIE astmy zawodowej | 2 | 2.0 | WSPIERANIE rodzin | 2 | 1.4 |
| 22 | DRŻENIE mięśni szkieletowych | 4 | 2.0 | TWORZENIE prawa | 2 | 1.4 |
| 23 | LECZENIE chorych | 5 | 1.7 | TRANSMITOWANIE rozgrywek | 2 | 1.4 |
| 24 | WYSTĘPOWANIE świszczącego oddechu | 5 | 1.6 | USTALANIE walk | 2 | 1.4 |
| 25 | UNIKANIE palenia | 2 | 1.4 | POBIERANIE daniny | 2 | 1.4 |

# 6 Conclusions

In this paper we discuss a system for concepts extraction from unstructured Polish texts. Concepts are defined as n-grams having specific structure formed by tokens belonging to particular grammatical categories and having appropriate morphological forms. As Polish is a highly inflected language, detected concept names may appear in a few declination variants. Hence, they are expected to be normalized.

Detection and transformation of concepts is done with rules, whose design combines elements of colored [12] and fuzzy [6] Petri nets. During a rule execution words in an input n-gram undergo morphological transformation producing in a general case a set of output strings. Hence, rules are by design nondeterministic [13] and their results are aggregated using techniques widely applied for fuzzy inference systems. The similar approach was applied in our previous works related to semantic event recognition and it turned out to be very efficient [25,27]. It should be noted, that idea of specifying rules as colored Petri nets and use them for detection is quite attractive. It was successfully used in various fields, e.g. computer security [11].

Extraction rules are never coded manually. They are derived from a user friendly specifications giving samples of transformation patterns called annotation. Using annotations to specify rules indirectly, simplifies their definition, but sometimes introduces additional unwanted nondeterminism.

The main contribution of this paper is related to recently introduced rule management mechanisms. We defined two relations between rules: partial refinement (a rule can be applied to narrower input, but it produces more specific results) and covering (a rule has both larger input and set of results than the other). The described filtering technique with metarules consists in applying partial refinement. For an exhaustive set of metarules, a sum of refined rules may be equivalent to subsumption [13]. Detection of covering relation is a basis for removing redundant rules.

Performed experiments show that the application of rule management resulted in improved accuracy of extracted concepts. Moreover, the rules became more deterministic. It should be observed that the redundancy of rules practically has no influence on the accuracy (scores assigned to concepts), at least for the *max* norm, which is the default choice for aggregation at the sentence level, [26]. In the tests focusing on a particular group of concepts (actions) removal of redundant rules slightly reduced the processing time. Probably, the effect can be more visible for larger rulesets. Nevertheless, the new functionality for cleansing a ruleset facilitates its definition.

# References

1. Acedański, S.: A morphosyntactic brill tagger for inflectional languages. In: Loftsson, H., Rögnvaldsson, E., Helgadóttir, S. (eds.) IceTAL 2010. LNCS, vol. 6233, pp. 3–14. Springer, Heidelberg (2010)
2. Blake, C., Pratt, W.: Better rules, fewer features: a semantic approach to selecting features from text. In: Proceedings IEEE International Conference on Data Mining, ICDM 2001, pp. 59–66. IEEE (2001)
3. Bloehdorn, S., Cimiano, P., Hotho, A.: Learning ontologies to improve text clustering and classification. In: Spiliopoulou, M., Kruse, R., Borgelt, C., Nürnberger, A., Gaul, W. (eds.) From Data and Information Analysis to Knowledge Engineering. Studies in Classification, Data Analysis, and Knowledge Organization, pp. 334–341. Springer, Heidelberg (2006). http://dx.doi.org/10.1007/3-540-31314-1_40

4. Carpineto, C., Romano, G.: Concept Data Analysis: Theory and Applications. John Wiley & Sons, New York (2004)
5. Challis, J.: Lateral thinking in information retrieval white paper. Technical report, Concept Searching (2003)
6. Chen, S.M., Ke, J.S., Chang, J.F.: Knowledge representation using fuzzy Petri nets. IEEE Trans. Knowl. Data Eng. **2**(3), 311–319 (1990)
7. Cimiano, P., Hotho, A., Staab, S.: Learning concept hierarchies from text corpora using formal concept analysis. J. Artif. Intell. Res. (JAIR) **24**, 305–339 (2005)
8. Daciuk, J.: Incremental construction of finite-state automata and transducers, and their use in the natural language processing. Ph.D. thesis, Gdansk University of Technology, ETI faculty, Gabriela Narutowicza 11(12), pp. 80–233 Gdansk Poland (1998)
9. Dalvi, N., Kumar, R., Pang, B., Ramakrishnan, R., Tomkins, A., Bohannon, P., Keerthi, S., Merugu, S.: A web of concepts. In: Proceedings of the Twenty-Eighth ACM SIGMOD-SIGACT-SIGART Symposium on Principles of Database Systems, pp. 1–12. ACM (2009)
10. Graliński, F., Jassem, K., Junczys-Dowmunt, M.: PSI-Toolkit: A natural language processing pipeline. In: Przepiórkowski, A., Piasecki, M., Jassem, K., Fuglewicz, P. (eds.) Computational Linguistics. SCI, vol. 458, pp. 27–40. Springer, Heidelberg (2013). http://dx.doi.org/10.1007/978-3-642-34399-5_2
11. Jasiul, B., Szpyrka, M., Sliwa, J.: Detection and modeling of cyber attacks with Petri nets. Entropy **16**(12), 6602–6623 (2014). http://dx.doi.org/10.3390/e16126602
12. Jensen, K.: Coloured Petri Nets: Basic Concepts, Analysis Methods and Practical Use, vol. 1. Springer, Berlin Heidelberg (1996)
13. Ligeza, A.: Logical Foundations for Rule-Based Systems. Studies in Computational Intelligence, vol. 11, 2nd edn. Springer, Heidelberg (2006)
14. Maedche, A., Staab, S.: Ontology learning for the semantic web. Intell. Syst. IEEE **16**(2), 72–79 (2001)
15. Mamdani, E.H., Assilian, S.: An experiment in linguistic synthesis with a fuzzy logic controller. Int. J. ManMach. Stud. **7**(1), 1–13 (1975). http://linkinghub.elsevier.com/retrieve/pii/S0020737375800022
16. Miłkowski, M.: Developing an open-source, rule-based proofreading tool. Softw.: Pract. Exp. **40**(7), 543–566 (2010)
17. Miłkowski, M.: Morfologik (2015). http://morfologik.blogspot.com/. Accessed May 2015
18. Naber, D.: Language tool style and grammar check (2015). https://www.languagetool.org/. Accessed May 2015
19. Osinski, S., Weiss, D.: A concept-driven algorithm for clustering search results. Intell. Syst. IEEE **20**(3), 48–54 (2005)
20. Parameswaran, A., Garcia-Molina, H., Rajaraman, A.: Towards the web of concepts: Extracting concepts from large datasets. Proc. VLDB Endow. **3**(1–2), 566–577 (2010)
21. Ross, T.: Fuzzy Logic with Engineering Applications. Wiley, New York (2009)
22. Smith, B.: Beyond concepts: ontology as reality representation. In: Proceedings of the Third International Conference on Formal Ontology in Information Systems (FOIS 2004), pp. 73–84 (2004)
23. Stavrianou, A., Andritsos, P., Nicoloyannis, N.: Overview and semantic issues of text mining. ACM Sigmod Rec. **36**(3), 23–34 (2007)

24. Szwed, P.: Application of fuzzy ontological reasoning in an implementation of medical guidelines. In: 2013 The 6th International Conference on Human System Interaction (HSI), pp. 342–349, June 2013

25. Szwed, P.: Video event recognition with fuzzy semantic petri nets. In: Gruca, A., Czachórski, T., Kozielski, S. (eds.) Man-Machine Interactions 3. AISC, vol. 242, pp. 431–439. Springer, Heidelberg (2014). http://dx.doi.org/10.1007/978-3-319-02309-0_47

26. Szwed, P.: Concepts extraction from unstructured Polish texts: A rule based approach. In: Federated Conference on Computer Science and Information Systems (FedCSIS), pp. 355–364, September 2015

27. Szwed, P., Komorkiewicz, M.: Object tracking and video event recognition with fuzzy semantic petri nets. In: Proceedings of the 2013 Federated Conference on Computer Science and Information Systems, Kraków, Poland, 8–11 September 2013, pp. 167–174 (2013)

28. Wolinski, M., Milkowski, M., Ogrodniczuk, M., Przepiórkowski, A.: Polimorf: a (not so) new open morphological dictionary for polish. In: LREC, pp. 860–864 (2012)

# Mapping of Selected Synsets to Semantic Features

Tomasz Jastrząb[✉], Grzegorz Kwiatkowski, and Paweł Sadowski

Institute of Informatics, Silesian University of Technology, Gliwice, Poland
{Tomasz.Jastrzab,Grzegorz.Wojciech.Kwiatkowski,Pawel.Sadowski}@polsl.pl

**Abstract.** In the paper we devise a novel algorithm related to the area
of natural language processing. The algorithm is capable of building a
mapping between the sets of semantic features and the words available
in semantic dictionaries called wordnets. In our research we consider
wordnets as ontologies, paying particular attention to hypernymy relation. The correctness of the proposal is verified experimentally based
on a selected set of semantic features. plWordNet semantic dictionary
is considered as a reference source, providing required information for
the mapping. The algorithm is evaluated on an instance of a decision
problem related to data classification. The quality measures of the classification include: false positive rate, false negative rate and accuracy.
A measure of a strength of membership (SOM) in a semantic feature
class is proposed and its impact on the aforementioned quality measures
is evaluated.

**Keywords:** Natural language processing · Ontologies · Wordnets ·
Semantic features

## 1 Introduction

Natural language processing (NLP in short) is a well-studied domain of computer science. The main area of interest is the design of solutions providing the
ability to explore information described in natural language. There is a wide
variety of applications of NLP algorithms and methods, including but not limited to: computer-generated translation, network content analysis, searching and
extraction of information and many others [9,18]. The task of natural language
processing usually involves the following steps, namely morphological, syntactic
and semantic analysis. The first two elements can be handled by already existing tools dealing with morphological and syntactic analysis, e.g. the Linguistic
Analysis Server (LAS) [13,23]. The last problem, i.e. semantic analysis, which is
discussed in this paper, is one of the key elements of natural language processing
systems. Its goal is the extraction of the information contained in text.

In the paper, the following problem is considered. Given a set of words we
seek for a mapping that assigns to each word a semantic feature. In this context, a
semantic feature may be viewed as a descriptive term (label) associating semantic

© Springer International Publishing Switzerland 2016
S. Kozielski et al. (Eds.): BDAS 2016, CCIS 613, pp. 357–367, 2016.
DOI: 10.1007/978-3-319-34099-9_28

information with words. The ability to provide such a mapping is an important aspect of semantic analysis since it may become a valuable source of data in the form of a semantic dictionary.

Our research is motivated by the belief that having a well-customized semantic feature set could aid various natural language processing tasks, e.g. valence schema matching. A valence schema describes the syntax of a sentence by defining the arguments of its predicates. A predicate in turn is usually a verb. Therefore the valence schema is a source of information on the phrases linked with the predicate [1,24]. Predicates and valence schemas are combined together in valence dictionaries, which provide the definition of associations between them [8,21,22].

The problem of valence schema matching can be interpreted in two correlated ways. On the one hand it is the process of choosing the best matched valence schema for the sentence being analyzed. On the other hand it can be viewed as fitting individual parts of the sentence to a particular valence position. The valence position defines a requirement or possibility of occurrence of a particular argument with given predicate. A complete definition of the valence position involves many attributes such as case, person, number and selectional restrictions [21]. Selectional restrictions allow to specify the semantic requirements for the valence position and are represented by semantic features. Consequently, semantic features can be viewed as necessary elements to solve the problem of valence schema matching.

Taking into account the importance of semantic information for the task of valence schema matching we seek a method of automatic or semi-automatic semantic data generation. Several different approaches are known from the literature, including corpora-based data extraction [12,15], query-based approaches [2] or translation-based approaches using existing valence dictionaries [7]. The method closest to ours is the one discussed in [7]. However, our approach differs in the following respects. Firstly, it is applied to Polish language while [7] uses Japanese and English. Secondly, we propose to use wordnets instead of existing valence dictionaries to gather the information on word semantics, which may be useful in valence schema matching.

A wordnet is a network of words modelling semantic knowledge in the form of a graph. In ontological terms wordnets are perceived as terminological fundamental ontologies [4,6]. A node of wordnet graph is equivalent to a synset which denotes a set of synonyms (**syn**onyms **set**). Thus the synset is the basic piece of knowledge representation in wordnet [16,17]. The key relations influencing wordnet structure are:

1. synonymy, which joins all lexical units with the same meaning into synsets,
2. hypernymy, which can be treated as a generalization, providing links to less specific synsets, and
3. hyponymy, which is the opposite of hypernymy, i.e. a specialization linking to more specific synsets.

Out of the two Polish wordnets, namely plWordNet [20] and Polnet [25] the former one was selected for the purposes of this paper.

The main contribution of the paper is the proposal of a mapping algorithm discovering semi-automatically the associations between selected semantic features and wordnet's data. We believe that once the mapping function is appropriately defined it will be able to assign semantic features to previously unseen words available in the wordnet. The main aim of the paper is to present a new method, which, given limited number of semantic feature representatives, will be able to automatically identify other members of the semantic feature class. The goal of the method is to support the tasks of natural language processing and artificial intelligence, such as the aforementioned valence schema matching as well as database schema matching [3] or querying database using natural language [5,11]. The latter problem is of particular interest and can benefit from the proposed method. As discussed in [5], semantic features and related elements such as semantic categories or roles are crucial in database semantics identification.

The devised algorithm is trained on a set of examples, i.e. words described by the feature (determined manually). Algorithm's correctness is experimentally evaluated based on selected features and sets of words classified as belonging or not belonging to the feature set. The quality of the classification is measured and the results are reported.

The paper is divided into 4 sections. Section 2 introduces the mapping algorithm. We discuss the conducted experiments and obtained results in Sect. 3. The final Section contains conclusions and future research perspectives.

## 2    Algorithm by Example

The discovery of semantic features for a given word can be viewed as a convenient and compact way of providing its definition. It may become particularly useful in case of words described by many hypernym sets, resulting in relatively complex structures of mutual relations. The proposed algorithm deals with this problem providing a much more concise, yet comparably meaningful description.

The algorithm has several key advantages. First of all, it is **semi-automatic** in the meaning that, except for the initial configuration phase, it proceeds without the need for external user interaction. Secondly, to avoid the need for continuous data retrieval from the plWordNet system, we provide a simple local caching mechanism making the algorithm **optimized** in this matter. The cache may be updated periodically in case some changes are introduced in the aforementioned knowledge database. One of the next important characteristics is that the algorithm is **universal** in two major respects. Firstly, although we use the resources from plWordNet, the algorithm allows for using various other dictionaries with little additional effort. Secondly it makes it possible to map any semantic feature sets, provided that corresponding data is available in the wordnet. Furthermore, the algorithm is **robust**. By robust we mean here that the words used during the initialization phase are contributing with equal strength. Consequently the mapping between words and features is free from any bias possibly imposed by words with multiple meanings (senses) and/or multiple hypernym sets. Finally, the algorithm is **widely applicable**, for instance to the problems of data classification, automatic semantic features discovery, automatic valence dictionaries generation and others, to name just a few. Also it can aid the task of database schema matching, following the already existing approaches that use wordnets discussed in [10, 14].

Before we proceed with the description of the algorithm let us comment on our assumptions related to dealing with synonymy and ambiguity of words. In terms of synonymy we observe that exact synonyms[1] are included in some common synset, thus their hypernym sets are exactly the same. As an example one can consider words samochód, auto, wóz and pojazd samochodowy with the English meaning of *a car*, which are all exact synonyms included in one synset. On the other hand for non-exact synonyms our expectation is that at least some major part of the hypernym sets should be the same. Again, as an example let us consider words most (*bridge*) and wiadukt (*overpass*), which have exactly the same hypernym sets, but are not exact synonyms. Consequently, our algorithm will be able to assign these words to the same semantic feature class.

The ambiguity is tackled by considering all synsets for given word. However, taking several words to be the feature's representatives, we effectively select only these hypernyms (and consequently meanings) which are dominant, i.e. most frequently appearing among the representatives. Therefore we can say that even if the particular words have some meanings synonymous to some other words not related to the feature, these unrelated meanings and their hypernyms will be rejected, as not contributing with enough strength to be the dominant part.

For the purpose of algorithm introduction let us consider the following example. Let $F = $ zawód (*profession*) be a semantic feature. Let also $W = \{$stolarz (*woodworker*), informatyk (*computer expert*)$\}$ be the set of representatives of the feature. Finally let $W_+ = \{$urzędnik (*official*)$\}$ and $W_- = \{$pacjent (*patient*)$\}$ be singleton sets of words expected to belong and not to belong to the feature class, respectively. Then the algorithm proceeds as follows:

1. Collect the hypernyms appearing on the paths between words in $W$ and their highest hypernym for each meaning of the word.

    (a) Stolarz (*woodworker*) has a single meaning with the following hypernyms:

    $H_w = \{$rzemieślnik (*craftsman*), człowiek ze względu na swoje zajęcie (*person with respect to its occupation*), człowiek, który coś robi (*person who does something*), człowiek (*person*), osoba (*intelligent being*), istota (*being*)$\}$.

    (b) Informatyk (*computer expert*) has two meanings, with the second one having two separate paths to the highest hypernym. Together they produce the following set of unique hypernyms:

---

[1] We use here the term *exact* synonyms to denote words with identical meanings. On the other hand by *non-exact* synonyms we mean words with similar yet not identical meanings.

$H_{ce}$ = {specjalista (*expert*), człowiek charakteryzowany ze względu na swoje kwalifikacje (*person with respect to its qualifications*), człowiek (*person*), osoba (*intelligent being*), istota (*being*), nauczyciel przedmiotowy (*subject teacher*), nauczyciel szkolny (*school teacher*), nauczyciel (*teacher*), pracownik oświaty (*employee of education system*), pracownik umysłowy (*brainworker*), pracownik ze względu na rodzaj pracy (*employee with respect to job type*), pracownik (*employee*), człowiek ze względu na swoje zajęcie (*person with respect to its occupation*), człowiek, który coś robi (*person who does something*), popularyzator (*popularizer*)}.

2. Count the number of occurrences of each hypernym in sets $H_w$ and $H_{ce}$. Sort hypernyms in descending order according to their number of occurrences. This gives a ranking of the form:

$R$ = {(człowiek ze względu na swoje zajęcie (*person with respect to its occupation*); 2), (człowiek, który coś robi (*person who does something*); 2), (człowiek (*person*); 2), (osoba (*intelligent being*); 2), (istota (*being*)); 2), (rzemieślnik (*craftsman*); 1), (specjalista (*expert*); 1), ..., (popularyzator (*popularizer*); 1)}.

3. Select the topmost part of ranking $R$, containing hypernyms with the highest number of occurrences and consider it as a representative set of hypernyms for the semantic feature:
$R_F$ = {człowiek ze względu na swoje zajęcie (*person with respect to its occupation*), człowiek, który coś robi (*person who does something*), człowiek (*person*), osoba (*intelligent being*), istota (*being*)}.

To classify a word as belonging or not belonging to the feature class, the following steps are performed:

1. Collect the hypernyms appearing on the paths between words in $W_+$, $W_-$ and their highest hypernym for each meaning of the word.
   (a) Urzędnik (*official*) has a single meaning with the following hypernyms:

   $H_o$ = {pracownik instytucji publicznych (*public servant*), pracownik ze względu na rodzaj pracy (*employee with respect to job type*), pracownik (*employee*), człowiek ze względu na swoje zajęcie (*person with respect to its occupation*), człowiek, który coś robi (*person who does something*), człowiek (*person*), osoba (*intelligent being*), istota (*being*)}.

   (b) Pacjent (*patient*) has a single meaning with two separate paths to the highest hypernym producing the following set of unique hypernyms:

   $H_p$ = {chory (*sick person*), człowiek znajdujący sie w określonym stanie (*person being in a certain state*), nosiciel cechy (*feature carrier*), człowiek (*person*), osoba (*intelligent being*), istota (*being*)}, klient (*customer*), usługobiorca (*service recipient*), podmiot (*subject – a person or institution*)}.

2. Compare sets of hypernyms $H_o$ and $H_p$ with feature ranking $R_F$ and if all ranking elements are found, classify word as belonging to the feature. Since $R_F \subset H_o$ and $R_F \not\subset H_p$, then indeed urzędnik (*official*) belongs to the feature class and pacjent (*patient*) does not.

In the experiments we relax the requirement of finding *all* ranking elements. By successively lowering the number of elements that have to be matched we evaluate the strength of class membership and observe how it impacts the quality of classification.

# 3 Experiments

## 3.1 Experimental Setup

The correctness of the algorithm was verified based on the problem of classification of data. The classification algorithm takes two inputs: the word to be classified and the feature against which the classification occurs. The outcome is a decision variable stating if the word is described by the feature. In other words the result defines the type of the word, being *positive* if the word is described by the feature or *negative* otherwise.

The experiments were conducted for a set of semantic features using a Java-based implementation[2] of the algorithm introduced in Sect. 2. Three different features were selected for the purpose of the paper, namely: zawód (*profession*), ssak (*mammal*) and zwierzę domowe (*pet*). Four representatives were also chosen for each feature to build the ranking discussed in Sect. 2. Table 1 summarizes the information on used features and their representatives.

The test sets used for the classification were generated according to the following scenario. At first, all words assigned by plWordNet to categories: ludzie (*people*) and zwierzęta (*animals*) were extracted. Next, three sets of words were produced at random, using the words contained in the aforementioned categories. In particular, the category ludzie (*people*) was used for the zawód (*profession*) and ssak (*mammal*) semantic features while the category zwierzęta (*animals*) was used for the ssak (*mammal*) and zwierzę domowe (*pet*) features. Finally, each word was manually given the information on whether it is described by the

Table 1. Semantic features and their representatives

| Semantic feature | Representatives |
|---|---|
| zawód (*profession*) | stolarz (*carpenter*), lekarz (*doctor*), informatyk (*computer expert*), dyrygent (*conductor*) |
| ssak (*mammal*) | człowiek (*human*), dziecko (*child*), kot (*cat*), owca (*sheep*) |
| zwierzę domowe (*pet*) | pies (*dog*), kot (*cat*), kanarek (*canary*) chomik (*hamster*) |

---

[2] The executable file, short manual and input sets are available at https://github.com/tjastrzab/semantics.

feature (marked with '+' in the input files) or not (marked with '−' in the input files). This information was later used to compare the expected classification results with the results produced by the algorithm.

Table 2 summarizes the characteristics of generated test sets, including the total number of test words $|W_{total}|$, the number of positive words $|W_+|$ and the number of negative words $|W_-|$. There are two different versions of the test set used for the ssak (*mammal*) semantic feature, differing in the expected classification results. The difference results from the fact that in case of the first set (denoted *ssak-v1*) all professions (or occupations) were considered negative, due to indirect nature of the relation profession → human → mammal. Set *ssak-v2* assumed that such a transitive relation is acceptable and thus included these elements in the set $W_+$.

**Table 2.** Test sets characteristics

| Semantic feature | $|W_{total}|$ | $|W_+|$ | $|W_-|$ |
|---|---|---|---|
| zawód (*profession*) | 478 | 131 | 347 |
| ssak-v1 (*mammal-v1*) | 488 | 56 | 432 |
| ssak-v2 (*mammal-v2*) | 488 | 390 | 98 |
| zwierzę domowe (*pet*) | 440 | 27 | 413 |
| Totals | 1894 | 604 | 1290 |

### 3.2 Results

Based on the comparison of expected and actual results of classification the following statistical measures were computed, namely false positive rate (FPR), false negative rate (FNR) and accuracy (ACC) [19]:

$$FPR = \frac{FP}{FP + TN} \tag{1}$$

$$FNR = \frac{FN}{FN + TP} \tag{2}$$

$$ACC = \frac{TP + TN}{FP + TP + FN + TN} \tag{3}$$

where FP (resp. FN) denotes the number of false positives (resp. false negatives) i.e. negative examples classified as belonging to the feature (resp. positive examples classified as not belonging to the feature) and TP, TN stand for true positives and true negatives, i.e. correctly classified words.

Additionally, we introduce a measure of the strength of membership (SOM) in a semantic feature class specified by the following formula:

$$SOM = \frac{|H_w \cap R_F|}{|R_F|} \times 100\% \tag{4}$$

where $H_w$ denotes the set of hypernyms found for given word $w$ and $R_F$ is the topmost part of the ranking built according to the algorithm described in Sect. 2. It can be said that the measure describes how well the semantic feature ranking is covered by the word's hypernyms. We evaluate the proposed measure for each test word and observe its impact on the statistics given by (1)–(3).

Figures 1, 2, 3 and 4 present the statistical measures given by (1)–(3) obtained for the test sets described in Table 2. The X axis of each figure is the SOM parameter given by (4). It can be observed that the number of analyzed SOM values differ for particular features, which results from various cardinalities of the topmost part of the ranking. The impact of the SOM value follows a well defined trend, namely it makes the FPR parameter value smaller with increasing SOM. Such a behaviour is well justified by the fact that with stronger requirement of ranking coverage (represented by larger SOM value) the negative examples are no longer described by the feature. On the other hand FNR is directly proportional to SOM, which signals possible inconsistencies in the wordnet structure.

In terms of accuracy, it can be stated that for zawód (*profession*) and ssak-v1 (*mammal-v1*) features (see Figs. 1 and 2), ACC is satisfactory (for SOM values close to 100 %) while for the other two test sets it should be considered poor (see Figs. 3 and 4). In case of ssak-v2 (*mammal-v2*) feature (Fig. 3) the best accuracy is obtained for SOM equal to 6.25 %, since $|W_+|$ is much larger than $|W_-|$, which makes the number of true positives a dominant factor in the accuracy measure. Comparing the results for ssak-v1 and ssak-v2, let us note that FPR and FNR values increased significantly for ssak-v2 group, which may suggest that the transitive relation is not fully reflected in plWordNet.

The general conclusion is that although with stronger requirement for ranking coverage the negative examples can be properly excluded from the semantic feature class, for larger SOM values the positive examples tend to be excluded as well. It can be treated as a signal of wordnet inconsistencies and consequently may become a suggestion for wordnet contents modification.

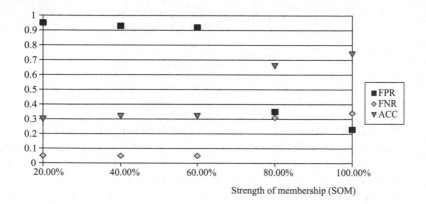

**Fig. 1.** Classification quality measures for zawód (*profession*) feature

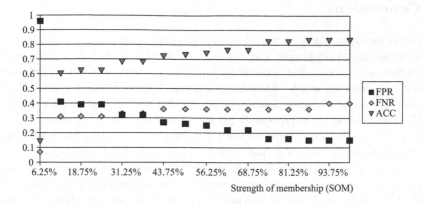

**Fig. 2.** Classification quality measures for ssak-v1 (*mammal-v1*) feature

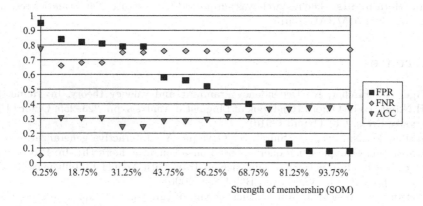

**Fig. 3.** Classification quality measures for ssak-v2 (*mammal-v2*) feature

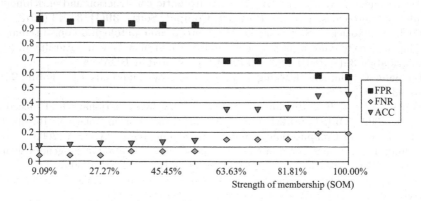

**Fig. 4.** Classification quality measures for zwierzę domowe (*pet*) feature

## 4   Conclusions

The paper introduces a novel algorithm for creating a mapping between selected semantic features and words. The resources available in Polish wordnet called plWordNet have been used, providing for the automatic expert knowledge of associations between words. The algorithm has been evaluated experimentally on selected sets of semantic features and words. The strength of membership (SOM) measure has been proposed and its importance has been discussed.

In the future we plan to use negative representatives during the initial phase of the algorithm. We presume that such an approach may lead to devising even more precise mapping function, assigning the most characteristic word(s) to the feature, which may bring some improvements to algorithm's performance. We would also like to apply the algorithm to the task of valence schema matching or to a semi-automatic wordnets' inconsistencies detection.

**Acknowledgments.** The research was supported by Institute of Informatics research grant no. BKM-515/RAU2/2015.

## References

1. Ágel, V., Fischer, K.: Dependency grammar and valency theory. In: Heine, B., H.N.(eds.) The Oxford Handbook of Linguistic Analysis, pp. 223–255. Oxford University Press, UK, Oxford (2010)
2. Akiba, Y., Nakaiwa, H., Shirai, S., Ooyama, Y.: Interactive generalization of a translation example using queries based on a semantic hierarchy. In: Proceedings of 12th IEEE International Conference on Tools with Artificial Intelligence (ICTAI 2000), pp. 326–332. Vancouver, BC, USA (2000)
3. Arfaoui, N., Akaichi, J.: Automating schema integration technique case study: generating data warehouse schema from data mart schemas. In: Kozielski, S., Mrozek, D., Kasprowski, P., Małysiak-Mrozek, B., Kostrzewa, D. (eds.) BDAS 2015. CCIS, vol. 521, pp. 200–209. Springer, Heidelberg (2015)
4. Astrakhantsev, N.A., Turdakov, D.Y.: Automatic construction and enrichment of informal ontologies: a survey. Program. Comput. Softw. **39**(1), 34–42 (2013)
5. Bach, M., Kozielski, S., Świderski, M.: Zastosowanie ontologii do opisu semantyki relacyjnej bazy danych na potrzeby analizy zapytań w języku naturalnym. Stud. Informatica **30**(2A(83)), 187–199 (2009). presented at BDAS'09
6. Biemann, C.: Ontology learning from text: a survey of methods. LDV Forum **20**(2), 75–93 (2005)
7. Fujita, S., Bond, F.: Extending the coverage of a valency dictionary. In: Proceedings of the 2002 COLING workshop on Machine translation in Asia, vol. 16, pp. 1–7. Association for Computational Linguistics, Stroudsburg, PA, USA (2002)
8. Grund, D.: Komputerowa implementacja słownika syntaktyczno-generatywnego czasowników polskich. Stud. Informatica **21**(3(41)), 243–256 (2000)
9. Hajnicz, E.: Automatyczne tworzenie semantycznych słowników walencyjnych. Akademicka Oficyna Wydawnicza EXIT, Warszawa (2011)
10. Hossain, J., Sani, F., Affendey, L.S., Ishak, I., Kasmiran, K.A.: Semantic schema matching approaches: A review. J. Theor. Appl. Inf. Technol. **62**(1), 139–147 (2014)

11. Jagielski, J.: Język naturalny w systemach baz danych. Stud. Informatica **31**(2B(90)), 281–290 (2010). presented at BDAS'10
12. Kawahara, D., Kurohashi, S.: Japanese case frame construction by coupling the verb and its closest case component. In: Proceedings of First International Conference on Human Language Technology Research (HLT 2001), pp. 204–210. Association for Computational Linguistics, Stroudsburg, PA, USA (2001)
13. Kulików, K.: Implementacja serwera analizy lingwistycznej dla systemu THETOS - translatora tekstu na język migowy. Stud. Informatica **24**(3(55)), 171–178 (2003)
14. Mahdi, A.M., Tiun, S.: Utilizing wordnet for instance-based schema matching. In: Proceedings of the International Conference on Advances in Computer Science and Electronics Engineering (CSEE 2014), pp. 59–63. Institute of Research Engineers and Doctors (2014)
15. Manning, C.D.: Automatic acquisition of a large subcategorization dictionary from corpora. In: Proceedings of 31st Annual Meeting of the Association for Computational Linguistics (ACL-1993), pp. 235–242. Ohio State University, Columbus, Ohio, USA (1993)
16. Miller, G.A.: Nouns in wordnet: a lexical inheritance system. Int. J. Lexicogr. **3**(4), 245–264 (1990)
17. Miller, G.A., Beckwith, R., Fellbaum, C., Gross, D., Miller, K.: Introduction to wordnet: an on-line lexical database. Int. J. Lexicogr. **3**(4), 235–244 (1990)
18. Mykowiecka, A.: Inżynieria lingwistyczna: komputerowe przetwarzanie tekstów w języku naturalnym. Wydawnictwo PJWSTK, Warszawa (2007)
19. Olson, D., Delen, D.: Performance evaluation for predictive modeling. In: Olson, D., Delen, D. (eds.) Advanced Data Mining Techniques, pp. 137–147. Springer, Berlin Heidelberg (2008)
20. Piasecki, M., Szpakowicz, S., Broda, B.: Toward plWordNet 2.0. In: Bhattacharyya, P., Fellbaum, C., Vossen, P. (eds.) Proceedings of the 5th Global Wordnet Conference Principles, Construction and Application of Multilingual Wordnetsm, pp. 263–270. Narosa Publishing House (2010)
21. Polański, K.: Słownik syntaktyczno-generatywny czasowników polskich. Zakład Narodowy im. Ossolińskich, Wrocław (1980)
22. Przepiórkowski, A., Hajnicz, E., Patejuk, A., Woliński, M., Skwarski, F., Świdziński, M.: Walenty: Towards a comprehensive valence dictionary of Polish. In: Proceedings of the Ninth International Conference on Language Resources and Evaluation (LREC), pp. 2785–2792. Reykjavik, Iceland (2014)
23. Suszczańska, N., Szmal, P., Simiński, K.: The deep parser for polish. In: Vetulani, Z., Uszkoreit, H. (eds.) LTC 2007. LNCS, vol. 5603, pp. 205–217. Springer, Heidelberg (2009)
24. Tesnière, L.: Elements of Structural Syntax. John Benjamins Publishing Company, New York (2015)
25. Vetulani, Z.: Komunikacja człowieka z maszyną. Akademicka Oficyna Wydawnicza EXIT, Warszawa (2014)

# A Diversified Classification Committee for Recognition of Innovative Internet Domains

Marcin Mirończuk$^{(\boxtimes)}$ and Jarosław Protasiewicz

Laboratory of Intelligent Information Systems,
National Information Processing Institute, al. Niepodległości 188b,
00-608 Warsaw, Poland
marcin.mironczuk@opi.org.pl
http://lis.opi.org.pl

**Abstract.** The objective of this paper was to propose a classification method of innovative domains on the Internet. The proposed approach helped to estimate whether companies are innovative or not through analyzing their web pages. A Naïve Bayes classification committee was used as the classification system of the domains. The classifiers in the committee were based concurrently on Bernoulli and Multinomial feature distribution models, which were selected depending on the diversity of input data. Moreover, the information retrieval procedures were applied to find such documents in domains that most likely indicate innovativeness. The proposed methods have been verified experimentally. The results have shown that the diversified classification committee combined with the information retrieval approach in the preprocessing phase boosts the classification quality of domains that may represent innovative companies. This approach may be applied to other classification tasks.

**Keywords:** Text mining · Classification · Text classification · Information retrieval · Committee classification

## 1 Introduction

It is believed that innovativeness plays an important role in the development of modern economies. Since it depends on cooperation between companies and researchers, we have decided to create an information platform called Inventorum, which is aimed to boost information flow between these two sides. The platform is a recommender system that proposes innovations, projects, partners and experts that suit both, companies and research teams best. Among many processes that this system contains, there is one especially important, i.e. finding potentially innovative companies on the Internet, which should help to find more participants of the Inventorum. The system collects data from the Internet. Then each acquired domain (a group of web pages) is analyzed in order to find out whether it is related to an innovative company. This study deals with the classification issue of such domains into innovative companies and the others.

© Springer International Publishing Switzerland 2016
S. Kozielski et al. (Eds.): BDAS 2016, CCIS 613, pp. 368–383, 2016.
DOI: 10.1007/978-3-319-34099-9_29

The term "an innovative domain" is an abstract idea similar to such concepts like spam, pornography, or sport, etc. These domains may be automatically recognized on the Internet by supervised machine learning methods. Unfortunately, the innovativeness is hard to define; however, the research conducted by Leon Kozminski Academy provided some definitions and depicted attributes that may characterize an innovative company. The research was based on questionnaires that covered several useful indicators of innovativeness; thus, they helped to classify companies. However, it is almost impossible to retrieve these indicators from company's domains automatically, and then create the profiles of considered businesses. Beside directly built models of innovativeness, there are annual rankings of leading companies on the market, e.g. Forbes rankings, awards like Business Gazelle, etc. We assumed that these rankings are highly reliable and decided to utilize these data to verify our hypothesis as follows. We assume that it is possible to create a classification model that can decide whether a company is innovative or not based on automatic analysis of its Internet domain.

The review of the recent literature showed that there are some works [9–11,15] concerning the detection of innovative themes on the Internet; however, the analyzed approaches are inappropriate to solve sufficiently the problem posed in our article. Of course, there are a plenty of works concerning the problem of documents classification in various areas like spam detection, porn sites recognition, security issues, or medical documentation analysis. Nonetheless, these solutions are designed for limited purposes and are unsuitable for the problem of innovative sites detection. Thus, we experimentally constructed a classification model that can recognize innovative domains on the Internet with sufficient quality.

In this study, we use some text mining techniques. The text mining is related to data mining. David Hand et al. [5] defined data mining as an analysis of observable and often large data sets to find unsuspected relationships and to summarize data in novel ways that will provide understandable and useful information to a data owner. Text mining relates to data mining methodologies applied to textual sources. It covers various approaches to text analysis, which mainly are [16,19] classification, clustering, translation, information retrieval, and summarization or information extraction.

The main aim of this work is to classify domains collected from the Internet to *innovative* or *no innovative* groups based on the documents that they include. To resolve this task, we propose an experimental Naïve Bayes (NB) classification committee. The committee is based on a diversified feature space, feature weights and two models of feature distribution, i.e. Bernoulli and Multinomial models. In the experiments, we verify whether this committee is enough diversified to resolve the classification task mentioned above and if the committee improve the classification quality in comparison to a single NB classifier. We provide the results of the whole classification committee as well as the performance of the individual NB classifiers composing the committee. Thus, it is easy to notice advances of using the committee in comparison to single classification models.

Furthermore, the presented study is theoretically well grounded to give a deep understanding of the proposed methods as well as to be easy to reproduce.

It is worth to note that as a result of several experiments, we elaborated the classification model consisting a classifiers committee. This non-trivial model covers advanced methods of features construction. We have to underline that the various single classifiers (k-NN, decision trees, support vector machines, etc.) have been verified experimentally prior to construction of the final model. Moreover, the experiments involved dimensionality reduction of the feature space by using typical methods like principal component analysis, singular value decomposition, and filters. Unfortunately, the examined classifiers produced such poor decisions that we decided to exclude their results from the article. Since the use of a single classifier turned out to be inappropriate to the task under examination, we worked out the more complicated and sophisticated solution that finally gave the sufficient results. The model was tested on a new and unique test set constructed by us only for the purpose of this study.

The paper is structured as follows. Section 2 contains the definitions and mathematical background of text mining techniques and components, which are used to resolve the defined classification problem. Section 3 presents an overview of the proposed innovative domains classification system. Next, Sect. 4 describes the evaluation process of the proposed system and the results obtained during experiments. Finally, Sect. 5 concludes the findings.

## 2    Text Mining and Mathematical Background

This section presents all necessary definitions and mathematical background that are used in the proposed classification system. Definition 1 presented below explains the term *text mining* [16].

**Definition 1.** *The term "text mining" is used analogously to data mining when data are text. As there are some data specificities when handling text compared to handling data from databases, text mining has some specific methods and approaches. Some of these are extensions of data mining and machine learning methods while other are rather text-specific. Text mining approaches combine methods from several related fields, including machine learning, data mining, information retrieval, natural language processing, statistical learning, and the Semantic Web.*

Usually, we build a processing pipeline to process a text. Figure 1 presents the basic pipeline of text processing.

Text corpora → Feature construction → Text representation →
Feature selection → Text mining tasks

**Fig. 1.** The basic pipeline of text processing.

A *text corpora* (*a corpus of texts* or shortly *a corpus*) $D$ is a collection of documents $d \in D$. Solka [19] defined a *document* as a sequence of words and punctuations following the grammatical rules of language. A document is any relevant segment of a text and can be of any length. Examples of documents include sentences, paragraphs, sections, chapters, books, web pages, emails, etc. A *term* (feature) is usually a word (uni-gram), but it can also be a word-pair (bi-gram) or a phrase (n-gram). Terms are constructed by a *Feature construction* component, and they build an appropriate *Text representation* (Subsect. 2.1). Usually, when the *Text representation* is created we try to reduce the number of terms using a *Feature selection* component. Subsect. 2.2 presents several feature selection methods. Generally, after the *Feature selection* phase follows the realization of an appropriate *Text mining task*. In our solution, we applied two *Text mining tasks*. The first task is related to *Information retrieval*, which also supports a realization of the second task, i.e. classification. Subsect. 2.3 briefly describes the information retrieval approach and presents the Okapi BM25 ranking function used by search engines. We use this ranking function to find the best innovative web documents per domain and build a classification model. The proposed classification process is based on the Naïve Bayes classifier, which is described in Subsect. 2.4.

## 2.1 Feature Construction and Text Representation

Definition 2 presented below explains the term *feature construction* in the context of text mining [16].

**Definition 2.** *Feature construction in text mining consists of various techniques and approaches, which convert textual data into a feature-based representation. Since traditional machine learning and data mining techniques are generally not designed to deal directly with textual data, feature construction is an important preliminary step in text mining converting source documents into a representation that a data mining algorithm can then work with it. Various kinds of feature construction approaches are used in text mining depending on a task that is being addressed or data mining algorithms and the nature of the dataset in question.*

Usually, the document preprocessing methods are used in the *feature construction* task. These methods use *natural language processing* (NLP) techniques. Figure 2 presents the basic pipeline of document pre-processing methods.

Document → Tokenization → End sentence recognition →
Morphological analysis → Morphological disambiguation → Labeling

**Fig. 2.** The basic pipeline of document pre-processing methods.

As shown in Fig. 2, each *document* from a *Text corpora* is tokenized. The *tokenization* process splits all documents into words. A *word* is a sequence of letters in a text written in a natural language and usually separated by either spaces or punctuation marks [21]. Usually, after the *tokenization* we need to know where is the end of a sentence. It is recognized by the *End sentence recognition* process. After that, each sentence is processed by the *Morphological analysis* component. The *morphological analysis* relies on the determination of all morphological forms of all particular words. For each word, we try to find all *lemmas* (base forms) and all tags. A tag contains values of grammatical categories specifying the form. The *morphological disambiguation* determine the form realized by a particular occurrence of a word in its context [21]. The sequence of a morphological analysis and disambiguation is in jargon referred as *tagging* [21]. The *labeling* component is used to create the final features set. In the simplest way, we may take to the analysis (to resolve *Text mining tasks*) all *words* after *tokenization* process. We can reduce this set by using a *stop list*. The *stop list* contains words/features that we remove from document, i.e. we remove from *document* all occurrences of each word/feature from the stop list. Also, we can use some heuristics methods (manual methods) to create the set of features. Moreover, we can create own ontology or use a more sophisticated ontology like *Słowosieć* (a Polish *Word-Net*) [12], for example, to unify words (to find one basic synonym of words). Also, we may use a more sophisticated feature construction based on a statistical word co-occurrence analysis [4] or NLP shallow parsing technique [13]. In the first case, we create the set of features (*n*-grams) by finding a *frequent occurrence* of $2, 3, ..., n$ words like a *business intelligence* or *commercial enterprise*. In the second case, we create the set of features by finding, for example, noun or verb phrases like an *innovative technology* or a *fast car*, etc.

Table 1 presents the results, i.e. an example of the *tagging* process. We considered the document example *d* – *"Mam próbkę analizy morfologicznej."* (*"I have a morphological analysis sample."*)

**Table 1.** The example of the tagging process. Source [1].

| 0 | Mam | mama [mother] | fin:sg:ter:subst:pl:gen:f |
|---|---|---|---|
| | | mamić [to beguile] | impt:sg:sec:imperf |
| | | mieć [to have] | fin:sg:pri:imperf |
| 1 | próbkę | próbka [sample] | subst:sg:acc:f |
| 2 | analizy | analiza [analysis] | subst:sg:gen:f |
| | | | subst:pl:nom.acc.voc:f |
| 3 | morfologicznej | morfologiczny [morphological] | adj:sg:gen.dat.loc:f:pos |
| 4 | . | . | interp |

According to [1], we can consider each line of Table 1 as one morphological interpretation, the horizontal lines separate the groups of analysis of particular words. The input document was segmented into words (particularly the full stop was separated from the word "morfologicznej"). On the right, corresponding lemmas (entries) are provided. The next column contains tags describing values of grammatical categories (IPI PAN morphological tagset [20]) of particular forms.

After the *Morphological disambiguation* phase, and when we take only *lemmas*, is received the following document $d' = \{\{ mie\acute{c}, [fin:sg:pri:imperf]\}, \{pr\acute{o}bka,$ $[subst:sg:acc:f]\}, \{analiza, [subst:sg:gen:f]\}, \{morfologiczny [adj:sg:gen.dat.loc:$ $f:pos]\}, \{., [interp]\}\}$. Based on this document and the *Labeling* techniques mentioned above, we can construct the following example set of features: $s =$ $\{mie\acute{c}, pr\acute{o}bka, analiza, morfologiczny\}$ - this set will be created when we will use *stop words set* like a $S_w = \{.\}$; $s = \{pr\acute{o}ba, analiza, morfologiczny\}$ - this set will be created when we will use $S_w = \{.,mie\acute{c}\}$ and when we will replace feature *próbka* by its synonym, i.e. *próba*; $s = \{pr\acute{o}bka\text{-}analiza\text{-}morfologiczny,$ $analiza\text{-}morfologiczny\}$ - this set will be created when we will use NLP shallow parsing technique to recognition *noun phrase NP*; $s = \{pr\acute{o}bka\text{-}analiza, analiza\text{-}$ $morfologiczny\}$ - this set will be created when we will use the co-occurrence recognition techniques; $s = \{fin, subst, adj, interp, NP, NP\}$ - this set will be created when we will use for example only the *part of speech* and labels of the recognized $NP$; $s$ as the combination of the mentioned above sets.

After the *feature construction* phase, we may construct an appropriate text representation. There are two main representations of document [5,8,17]: a graph or a vector space model VSM (a document-term matrix or a term-document matrix). Schenker and all [17] modeled the document as a graph, where features are vertexes, and edges model connections between features. The VSM approach represents the collection of documents as a matrix.

The document-term matrix has $n$ ($1 \leq i \leq n$) rows and $m$ ($1 \leq j \leq m$) columns, where $m$ represents the number of features in the corpus $D$, and $n$ is the number of documents. The $w_{ij}$ element of the matrix $D$ relates to the weight of the term $f_j$ in the document $d_i$. There are a few typical weighting functions such as [8] binary, term-frequency (TF), invert document frequency (IDF) or mixed $TF \times IDF$, etc. In this study, we use binary and TF weighting functions. The binary weighting function sets the weight $w_{ij}$ equal to 1 ($w_{ij} = 1$) if and only if the feature $f_j$ occurs in the document $d_i$. If the feature $f_j$ does not occur in the document $d_i$, the weight $w_{ij}$ is set to 0 ($w_{ij} = 0$). The TF weighting function counts the number of times the $j$-th feature appears in the $i$-th document. Encoding the corpus as the matrix allows to utilize the power of linear algebra and quickly analyze the documents collection. All presented solutions like the feature selection, the ranking functions used by search engines or classification methods are based on the VSM model discussed above.

## 2.2   Feature Selection Methods and Filter Methods

Definition 3 presented below explains the term *feature selection* in the context of text mining [16].

**Definition 3.** *The term "feature selection" is used in machine learning for the process of selecting a subset of features (dimensions) used to represent the data. Feature selection can be seen as a part of data pre-processing potentially followed or coupled with feature construction, but can also be coupled with the learning phase if embedded in the learning algorithm. An assumption of feature selection* *is that we have defined an original feature space that can be used to represent the data, and our goal is to reduce its dimensionality by selecting a subset of original features.*

We can divide the feature selection into three main groups [2], namely filters, wrappers, and embedded methods. In this study, we use the filter methods. These methods utilize a feature ranking function to choose the best features. The ranking function gives a relevance score based on a sequence of examples. Intuitively, more relevant features will be higher in the rank. Thus, we may keep $n$-top features or remove $n$-worst ranked features from the dataset. These methods are often univariate and consider each feature independently or with regard to a dependent variable. For example, some filter methods include the $\chi^2$ squared test, information gain, and correlation coefficient scores. In the presented research the following filter methods are used [3,6,8,18]: fisher ranking (Fisher), correlation coefficients (Cor), mutual information (MI), normalized punctual mutual information (NPMI), $\chi^2$ (Chi), Kolmogorov–Smirnov test (KS), and Mann–Whitney U test (WC).

## 2.3    Information Retrieval and The Okapi BM25 Search Engine

Definition 4 presented below explains the term *information retrieval* in the context of text mining [8,16].

**Definition 4.** *Information retrieval (IR) is a set of techniques that extract from a collection of documents those that are relevant to a given query. Initially addressing the needs of librarians and specialists, the field has evolved dramatically with the advent of the World Wide Web. It is more general than data retrieval, which purpose is to determine which documents contain occurrences of the keywords that make up a query. Whereas the syntax and semantics of data retrieval frameworks are strictly defined, with queries expressed in a totally formalized language, words from a natural language give no or limited structure are the medium of communication for information retrieval frameworks. A crucial task for an IR system is to index the collection of documents to make their contents efficiently accessible. The documents retrieved by the system are usually ranked by expected relevance, and the user who examines some of them might be able to provide feedback so that the query can be reformulated and the results improved.*

The Okapi BM25 retrieval function is state-of-the-art of information retrieval systems [8,14]. It was first implemented in London's City University in the 1980s and 1990s by Stephen E. Robertson, Karen Spärck Jones, and others. The Okapi BM25 is a bag-of-words retrieval function that is used by search engines to rank matching documents $d$ according to their relevance to a given search query $q$. It is not a single function, but the whole family of scoring functions with slightly different components and parameters. One of the most popular instantiations of this function is the BM25 scoring function that is defined by Eq. 1.

$$scoring(q,d) = \sum_{i=1}^{|q|} idf(q_i) \cdot \frac{tf(q_i, d) \cdot (k_1 + 1)}{tf(q_i, d) + k_1 \cdot (1 - b + b \cdot \frac{|d|}{avg_{docs}})} \tag{1}$$

where: $q$ is the query that consists of features/terms; $d$ is the document (the bag of features/terms/words); $f(q_i, d)$ correlates to the term's frequency defined as the number of times that query term $q_i$ appears in the document $d$; $d$ is the length of the document $d$ in words; $avg_{docs}$ is the average document length over all documents in the collection; $k_1$ and $b$ are free parameters usually equal to $k_1 = 2.0$ and $b = 0.75$; $idf(q_i)$ is the inverse document frequency weight of the query term $q_i$.

The Eq. 2 describes how $idf(q_i)$ is computed.

$$idf(q_i) = log(\frac{N - df(q_i) + 0.5}{df(q_i) + 0.5}) \tag{2}$$

where: $N$ is the total number of documents in the collection; $df(q_i)$ is the number of documents containing the query term $q_i$.

## 2.4   Text Classification and the Naïve Bayes classifier

Definition 5 presented below explains the term *text classification* in the context of text mining [7,8].

**Definition 5.** *Text classification or categorization is the problem of learning classification models from training documents labeled by pre-defined classes. Then, such models are used to classify new documents. For example, we have a set of web page documents that belongs to two classes or topics, e.g. companies and no-companies. We want to learn a classifier that is able to classify new documents into these classes.*

We can compute the probability $P$ of the document $d$ belonging to a class $c$ thanks to Naïve Bayes Theorem. Equation 3 shows how we can compute this probability.

$$P(c|d) = \frac{P(d|c)P(c)}{P(d)} \propto P(c)P(d|c) = P(c) \prod_{1 \leq k \leq n_d} P(f_k|c) \tag{3}$$

where: $P(f_k|c)$ is the conditional probability of the feature $f_k$ occurring in the document of the class $c \in C$; $P(c)$ is the prior probability of the document occurring in class $c$; $n_d$ is the number of features $f$ in the document $d$.

In the text classification task, our goal is to find the *best* class $c$ for the document $d$. The best class in NB classification is the most likely or *maximum a posteriori* (MAP) class $c_{map}$. Equation 4 presents $c_{map}$ for the Bernoulli model feature distribution, and Eq. 7 presents $c_{map}$ for the Multinomial model feature distribution.

$$c_{map,Bernoulli} = \arg\max_{c \in C}[\log \hat{P}(c)$$
$$+ \sum_{1 \leq k \leq n_d} (\log(b_{f_k}\hat{P}(f_k|c)) + \log((1 - b_{f_k})(1 - \hat{P}(f_k|c)))] \quad (4)$$

where: $b_{f_k}$ - $b_{f_k} = 1$ if the feature $f_k$ is present in the document $d$, otherwise $b_{f_k} = 0$.

We can estimate $\hat{P}(c)$ and $\hat{P}(f_k|c)$ by using Eqs. 5 and 6 respectively.

$$\hat{P}(c) = \frac{N_c}{N} \quad (5)$$

where: $N_c$ is the number of documents in the class $c$; $N$ is the total number of documents.

$$\hat{P}(f_k|c) = \frac{N_{c,f} + 1}{N_c + 2} \quad (6)$$

where: $N_{c,f}$ is the number of document in the class $c$ that contain the feature $f$; $N_c$ see Eq. 5.

$$c_{map,Multinomial} = \arg\max_{c \in C}[\log \hat{P}(c) + \sum_{1 \leq k \leq n_d} \log(\hat{P}(f_k|c))] \quad (7)$$

where: $\hat{P}(f_k|c)$ we can be estimated by Eq. 8.

$$\hat{P}(f_k|c) = \frac{T_{c,f} + 1}{\sum_{f' \in F}(T_{c,f'} + 1)} \quad (8)$$

where: $F$ is the set of features (vocabulary) of the text corpora; $T_{c,f}$ is the number of occurrences of the feature $f$ in training documents belonging to the class $c$ including multiple occurrences of this feature in the document.

## 3 Innovative Domains Classification System - Overview

This section describes the proposed classification system of innovative domains. Figure 3 presents a basic flow and components of the system.

A component for *Crawling of the potentially innovative domains* (see Fig. 3), based on the initial list of companies (domains), is used for browsing the World Wide Web (Internet) and download its content for each domain, i.e. all domains of web pages. All downloaded pages are stored in *The potential innovative domains of companies and their documents* data-store $D_{all}$. After that, a *Recognition of the innovative logo for each domain* component recognizes if a domain contains any innovative logo. The innovative logo is a logotype such as "gazela biznesu", "diamenty forbesa", Europe Union logotypes, etc. All results of this recognition are stored in *The innovative logo of companies* data-store $D_{logo}$.

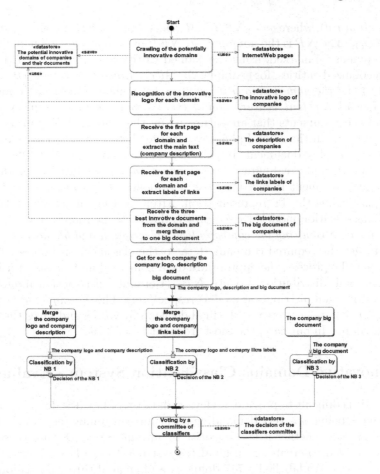

**Fig. 3.** The proposed classification system of innovative domains.

Next, the *Receive the first page for each domain and extract the main text (company description)* component processes the main text, i.e. tokenizes it to a simple feature set (a set of features of a type $s_2$) and saves it in *The description of companies* data-store $D_{description}$ using the binary representation. The *Receive the first page for each domain and extract labels of links* component gets all labels of links from the first page, i.e. it extracts the phrases between $< a > (.+?) < /a >$ HTML tags and joins the extracted phrase to the single feature. We can extract the feature $f = our - company$ from $< a > our\ company < /a >$ and transform it to $f' = our - company$. The results are saved in *The links labels of companies* data-store $D_{labels\ of\ links}$ using the binary representation. The next component named *Receive the three best innovative documents from the domain and merge them to one big document* utilizes the Okapi BM25 to search the best innovative documents in the domain. This process is divided into two phases. In the first phase, we find three best documents, which are on the top of a rank $scoring(q, d)$, where $q = \{NP, NER\,labels, the\,named\,innovative\,phrases\}$ and $d \in D_{all}$. The $NP$ is the *noun phrase*, $NER$ are the recognized name entities (in a document we marked, using the Named Entity Recognition $NER$ labeller, all the recognized entities, for example, the *Albert Einstein* was labelled as PERSON, etc.) and *the named innovative phrases* are manually marked "innovative" phrases like a *b-r*, *patent*, *venture-capital*, etc. We can interpret the query $q$ as follows: find the documents that are saturated by the *noun*, $NER$, and *manually* created phrases. In the second phase, we merge three best documents from the domain into one big document. In this case we transform all $NP$, which occur in the document, into real phrases, e.g. *production-line*, *innovative-solution*, etc. The merged document is saved in *The big document of companies* data-store $D_{big\,document}$ using the TF representation. After creating all necessary data-sets and NB classification models (based on the data-sets mentioned above), we use the *Get for each company the company logo, description and big document* component to get the required company data to the classification process. Finally, in three parallel process, the appropriate data is merged and classified using the $NB_1$ (Bernoulli distribution model), $NB_2$ (Bernoulli distribution model), and $NB_3$ (Multinomial distribution model) respectively. The *Voting by a committee of classifiers* component creates the final decision by voting and saves the results into *data stores*. In our case, the most frequent label wins.

## 4    Innovative Domains Classification System - Evaluation

*Data set.* It is difficult to provide a relevant data set of labelled domains because the manual labeling is a very demanding and time-consuming task. Each domain must be assessed whether it is *innovative* or *no innovative*. Despite these problems, we decided to create an original test set due to the lack of such data in the other works. We labelled 2,747 domains and created three sets as follows:

- $D_1[2,747 \times 140,699]$ - each example contains a company description $D_{description}$ and its logo $D_{logo}$, which gives 140,699 features;
- $D_2[2,747 \times 140,271]$ - each example contains link labels originating from the first page of a company domain $D_{labels\,links}$ and its logo $D_{logo}$, which gives 140,271 features;
- $D_3[2,747 \times 663,015]$ - each example contains a big document created form three the most relevant documents selected form a company domain $D_{big\,document}$, which gives 663,015 features.

The classification quality is evaluated according to the 10-fold cross-validation procedure and measured by precision, recall, and F measure (in fact, it is F1 measure, which assumes an equal balance of precision and recall) [8].

*Naïve Bayes classifiers and feature selection methods.* The first set of experiments covered comparisons of various feature selection methods (see Subsection 2.2) by using Naïve Bayes classifiers based on the Bernoulli distribution model ($NB_1$, $NB_2$) and the Multinomial distribution model ($NB_3$).

Figure 4 contains the F-score values achieved by above classifiers. Namely, the results depicted in Fig. 4A are produced by the $NB_1$ on the set $D_1$, Fig. 4B shows the outcomes of the $NB_2$ on the set $D_2$, and Fig. 4C includes the results of the $NB_3$ on the set $D_3$. Both $NB_1$ and $NB_2$ classifiers use the Bernoulli distribution model, whereas the $NB_3$ uses the Multinomial distribution model.

The set $D_1$, which contains descriptions and logos of companies, gives better results in comparison to the set $D_2$ regardless of almost all methods used for feature selection. The set $NB_3$ causes higher differences in performance among methods of feature selection in comparison to two previous sets. The $NB_1$ classifier works the best in the case of using the $\chi^2$ feature selection method (Fig. 4A). On the contrary, the classifier $NB_2$ produces the best results when using *Fisher* feature selection method. However, the differences among tested methods are rather not very significant in this case (Fig. 4B). Finally, the classifier $NB_3$ performs the best when using the *Fisher* feature selection method (Fig. 4C).

*The diversified classification committee.* The second series of experiments was intended to compare the best Naïve Bayes classifiers from the previous experiments with the proposed classification committee.

Figure 5 compares the F measure values of the committee and three the best $NB$ classifiers according to the previous experiments. The committee outperforms every single classifier when the number of features varies between 600 and 4,000. The further increase in the number of features leads to the decrease of F measure of the committee. Moreover, there are observed higher F measure values of the $NB_1$ classifier with the $\chi^2$ feature selection method than the committee. However, the analysis of precision and recall (Fig. 6) shows that the increase in the number of features causes overfitting of almost all classifiers and the committee. This is observed as the precision increase and the recall decrease.

The precision increase and the recall decrease are equivalent to the increase of correct decisions that the domains really represent innovative or no innovative companies and concurrently the increase of the number of really innovative domains that are not marked as innovative companies. This is improper behavior because many innovative companies are not selected, whereas it is better for our purposes when some non-innovative companies are treated as innovative (lower precision) but the high number of truly innovative businesses is found on the Internet (higher recall).

Worth to note are the results of the $NB_3$ with the set $D_3$ created by the proposed information retrieval system. This classifier outperforms the committee when the number of features is very high (Fig. 5) and, at the same time, it is robust to overfitting (Fig. 6). However, the classifier requires two times higher number of features than the committee.

Although each classifier in the committee uses the different feature space, the number of features is the same for all of them. It is the sub-optimal solution because we can easily notice that the classifiers produce the best results in the distinct ranges of feature numbers (Fig. 6). The optimal system may involve an adaptive selection of the features number for each classifier in the committee. However, our approach is simple and produces the acceptable results.

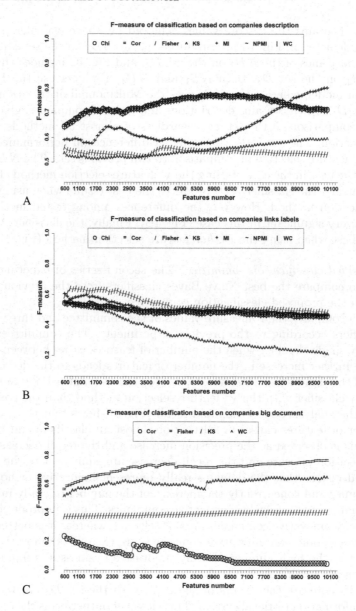

**Fig. 4.** F-measure values achieved by (A) the Naïve Bayes classifier $NB_1$ on the set $D_1$, (B) the Naïve Bayes classifier $NB_2$ on the set $D_2$, and (C) the Naïve Bayes classifier $NB_3$ on the set $D_3$ when using various methods of feature selection.

We can conclude that the proposed committee produces the most stable decisions, even if some its classifiers are unstable is some range of features, e.g. $NB_2$ (see Fig. 5). Moreover, the proposed information retrieval system improves

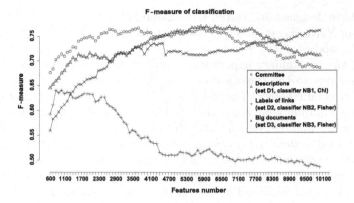

Fig. 5. F measure of the best Naïve Bayes classifiers and the committee.

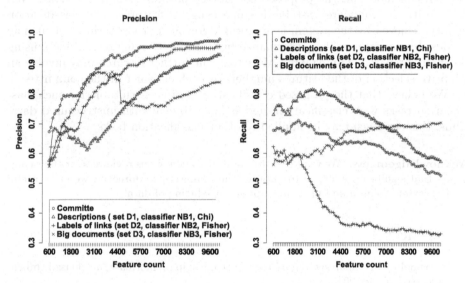

Fig. 6. Precision and recall of the best Naïve Bayes classifiers and the committee.

the classification quality, and, what is more important, it prevents both classifiers and the whole committee from overfitting.

## 5   Conclusions

The objective of this work was to propose a classification method of potentially innovative domains acquired from the Internet to estimate whether they represent companies that are innovative or not. Since it is not clear which attributes indicate innovativeness of a company, the classification model had to utilize the numerous features without exact information regarding their importance. The preliminary tests showed that a single classifier is not suitable for this task.

Thus, we have designed and tested a diversified classification committee that is composed of Naïve Bayes classifiers based on Bernoulli and Multinomial feature distribution models.

During the preliminary experiments we found out that the use of a whole domain as input to the classification system is inappropriate. Thus, we have applied retrieval methods at the preprocessing phase to extract the most innovative documents from analyzed domains. These methods were the Okapi BM25 ranking function and the Named Entity Recognition labeler. The proposed information retrieval system improved the classification quality. Moreover, it prevented both the classifiers and the whole committee from overfitting.

In the experiments, the performance of simple Naïve Bayes classifiers and the proposed system was analyzed in respect of feature numbers. It has to be noted that the number of features was selected according to Fisher and $\chi^2$ distributions. We can conclude that the proposed committee produces stable decisions, even if some its classifiers are unstable is some range of features. The classification quality achieved by the system is acceptable for our purposes. However, it may be possible to improve the performance by iterative training of classifiers using manually prepared training data. Moreover, the further research may involve an adaptive selection of the features number for each classifier in the committee.

We believe that the proposed classification system is suitable for such classification tasks were classification models have to deal with unstructured data. Thus, this approach may be applied to other classification tasks.

**Acknowledgements.** We would like to thank the anonymous reviewers for their comments that significantly helped to improve the manuscript. Moreover, we are grateful for Krzysztof Wolinski and his team support in labeling of data.

# References

1. Morphological analyser morfeusz. http://sgjp.pl/morfeusz/morfeusz.html.en. Accessed 28 Oct 2015
2. Bellotti, T., Nouretdinov, I., Yang, M., Gammerman, A.: Feature selection, pp. 115–130. Elsevier (2014)
3. Bouma, G.: Normalized (pointwise) mutual information in collocation extraction. In: Biennial GSCL Conference 2009, Tübingen, pp. 31–40 (2009)
4. Figueiredo, F., Rocha, L., Couto, T., Salles, T., Gonçalves, M., Meira, W.: Word co-occurrence features for text classification. Inf. Syst. **36**(5), 843–858 (2011)
5. Hand, D., Smyth, P., Mannila, H.: Principles of Data Mining. MIT Press, Cambridge (2001)
6. Li, S., Xia, R., Zong, C., Huang, C.: A framework of feature selection methods for text categorization. AFNLP **2**, 692–700 (2009)
7. Liu, B.: Web Data Mining: Exploring Hyperlinks, Contents, and Usage Data. Data-Centric Systems and Applications. Springer, Heidelberg (2006)
8. Manning, C., Raghavan, P., Schutze, H.: Introduction to Information Retrieval. Cambridge University Press, Cambridge (2008)

9. Nakatsuji, M.: Identifying novel topics based on user interests. In: Elçi, A., Koné, M.T., Orgun, M.A. (eds.) Semantic Agent Systems. SCI, vol. 344, pp. 273–292. Springer, Heidelberg (2011)

10. Nakatsuji, M., Miyoshi, Y., Otsuka, Y.: Innovation detection based on user-interest ontology of blog community. In: Cruz, I., Decker, S., Allemang, D., Preist, C., Schwabe, D., Mika, P., Uschold, M., Aroyo, L.M. (eds.) ISWC 2006. LNCS, vol. 4273, pp. 515–528. Springer, Heidelberg (2006)

11. Nakatsuji, M., Yoshida, M., Ishida, T.: Detecting innovative topics based on user-interest ontology. Web Semant. Sci. Serv. Agents World Wide Web 7(2), 107–120 (2009)

12. Piasecki, M., Szpakowicz, S., Broda, B.: Toward plWordNet 2.0. Principles, Construction and Application of Multilingual Wordnets, pp. 263–270 (2010)

13. Przepiórkowski, A., Buczyński, A.: Shallow parsing and disambiguation engine. In: Proceedings of the 3rd Language and Technology Conference, Poznań (2007)

14. Robertson, S., Walker, S., Jones, S., Hancock-Beaulieu, M., Gatford, M.: Okapi at TREC-3. In: Proceedings of The Third Text REtrieval Conference, TREC 1994, Gaithersburg, Maryland, USA, pp. 109–126 (1994)

15. Romanov, D., Ponfilenok, M., Kazantsev, N.: Potential innovations (new ideas/trends) detection in information network. Int. J. Future Comput. Commun. 2(1), 63–66 (2013)

16. Sammut, C., Webb, G.: Encyclopedia of Machine Learning. Springer, New York (2011)

17. Schenker, A., Bunke, H., Last, M., Kandel, A.: Graph-Theoretic Techniques for Web Content Mining. World Scientific Publishing, Singapore (2005)

18. Schurmann, J.: Pattern Classification - A Unified View of Statistical and Neural Approaches. Wiley, New York (1996)

19. Solka, J.: Text data mining: theory and methods. Statist. Surv. 2, 94–112 (2008)

20. Woliński, M.: Morphological tagset in the ipi pan corpus, Polonika, pp. 39–54 (2004)

21. Wolinski, M.: Morfeusz - a practical tool for the morphological analysis of polish. Intell. Inf. Process. Web Min. Adv. Soft Comput. 35, 511–520 (2006)

# The Onto-CropBase – A Semantic Web Application for Querying Crops Linked-Data

Abba Lawan[1], Abdur Rakib[1(✉)], Natasha Alechina[2], and Asha Karunaratne[3]

[1] School of Computer Science, The University of Nottingham Malaysia Campus,
Semenyih, Malaysia
{khyx3alw,Abdur.Rakib}@nottingham.edu.my
[2] School of Computer Science, The University of Nottingham, Nottingham, UK
Natasha.Alechina@nottingham.ac.uk
[3] Crops For the Future Research Centre (CFFRC), Semenyih, Malaysia
Asha.Karunaratne@cffresearch.org

**Abstract.** The lack of formal technical knowledge has been identified as one of the constraints to research and development on a group of crops collectively referred to as underutilized crops. Some information about these crops is available in informal sources on the web, for example on Wikipedia. However, this knowledge is not entirely authoritative, it may be incomplete and/or inconsistent and for these reasons it is not suitable as a basis of decision making by crop growers. Alternatively, we present an ontology-driven tool as a web-based access point for underutilized-crops ontology model. Major discussion points in this paper highlight our design choices, the tool implementation and a preliminary validation of the tool with brief discussion of related developments.

**Keywords:** Semantic web · Ontology · Semantic search · Knowledge base · Underutilized crops

## 1 Introduction

Utilizing the rapidly growing collection of semantic data is one of the contemporary challenges in the field of life sciences. In the last decade, with the advancement of semantic technologies in the design and development of domain-specific decision support systems, various RDF datasets, SPARQL endpoints, OWL ontologies as well as generic taxonomies have been developed. This success is more pronounced in the field of life-sciences, where various biomedical as well as agricultural ontologies have been developed and with existing thesaurus converted from simple vocabs showing taxonomic relations to well-structured linked data-sets [7], describing complex relationships between concepts. Ontologies being explicit specifications of domain knowledge, as widely quoted in [13], are known to enhance the semantics of data so that it can be easily interpreted by machines and often enable consistent, precise and human-understandable queries over such data. While these developments strengthen semantic knowledge modeling and representation, of equal importance is the retrieval and dissemination

© Springer International Publishing Switzerland 2016
S. Kozielski et al. (Eds.): BDAS 2016, CCIS 613, pp. 384–399, 2016.
DOI: 10.1007/978-3-319-34099-9_30

of the knowledge to non-technical stakeholders for informed decision-making. We developed the underutilized crops ontology (UC-ONTO), a preliminary version of the ontology presented in [21], which is an OWL 2 RL and Semantic Web Rule Language (SWRL)-enabled domain ontology conceptualizing the relations between concepts in the field of crops with emphasis on those crops categorized as underutilized. As usual, the size of the ontology continues to grow as more underutilized-crops data becomes available. Underutilized crops are those that though previously grown and consumed with considerable nutritional advantages, are currently being neglected [24]. Developed using Protégé[1] desktop application, the ontology has been designed to serve as a quick guide for the end users (researchers and farmers) seeking first-hand information on underutilized crops – a research mandate currently overseen by the "Crops For the Future Research Centre (CFFRC)"[2]. As such, a user-friendly tool with communal access and customized presentation views needs to be considered beyond the usability offered by the desktop applications, often considered suitable as development tools for knowledge engineers and domain experts. In essence, the Onto-CropBase tool presented in this paper is a web-app developed to serve as an interactive gateway to the UC-ONTO enhanced with additional linked-datasets in RDF. The tool allows for federated search on the resulting knowledge pool through shared concepts in the ontologies.

The remainder of the paper is structured as follows: we discuss the concept hierarchy of our ontology and briefly discuss ontology-based information retrieval with relevant works in Sect. 2. In Sect. 3 we present related work. Notable design choices and methodologies involved in developing the Onto-CropBase are discussed in Sect. 4. We present an overview of the basic functionalities of the Onto-CropBase tool in Sect. 5, and discuss the preliminary validation of the tool to ensure conformity with its initial requirements in Sect. 6. We conclude in Sect. 7 highlighting some upcoming works.

## 2  Preliminaries

### 2.1  The Concept Hierarchy

This section describes the meta-ontology of our UC-ONTO, highlighting the domain of discourse and scope of our knowledge base, among others.

Being the knowledgebase component of our Onto-CropBase tool, the development of our UC-ONTO follows the conceptualization of relevant terms in the underutilized crops domain using the OWL 2 ontology language and the addition of SWRL rules for added expressiveness [21]. Though special domain concepts were introduced in the development process, however, basic crops terminologies are reused from existing agricultural ontologies, such as the Plant Ontology (PO) [9] and Crops Ontology (CO) [23], among others. Basic ontologies added as

---

[1] http://protege.stanford.edu/.
[2] http://www.cropsforthefuture.org/.

direct imports include the OWL-temporal ontology to utilize the time concepts, FAOs geopolitical ontology, and the SWRL built-in functions.

While existing crops ontologies are mainly in the Open Biomedical Ontology (OBO) format, our UC-ONTO is developed in OWL 2 to utilize the expressive powers of the language. And with OWL being the standard ontology language of the Semantic Web [4], efforts to provide crops ontologies in OWL format will undoubtedly enhance knowledge-sharing and standardization in the domain of discourse. Moreover, semantic web applications, such as the proposed Onto-CropBase, can easily be developed to utilize such ontologies in conjunction with other OWL ontologies.

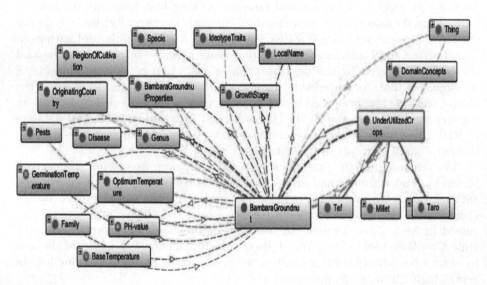

**Fig. 1.** A conceptual hierarchy showing fragment of the UC-ONTO developed using Protégé

The ontology, depicted in Fig. 1, is modularized into specialized ontologies such as, *Naming Ontology, Agronomy Ontology, Nutrition Ontology, Farming Ontology*, and so on. These are aligned together to form a larger ontology model for the underutilized crops domain. Individual instances are added to the UC-ONTO through the local ontologies in RDFS, including *Bambara groundnut, Taro, Moringa, Tef,* and *Wing Bean* [21].

## 2.2   Ontology-Based Search

Information Retrieval (IR), which involves searching and/or browsing of stored data, is the process of discovering specific portion of a large collection of stored information that satisfies certain user requirements [26]. Various forms of information searching for logically stored data do exist and are described in various terms. Commonest term is the *keyword search*, a strategy frequently employed by traditional search engines, where indexed documents can be retrieved based on their lexical matches to the search keyword. Another term is the *ontology-based semantic search*, which involves defining a possible metadata for the stored contents using ontologies or other semantic annotations to retrieve matching concepts, their relationships and instances via unique identifiers (URIs). Thirdly, a combination of these two search approaches was proposed in [5], and termed as *hybrid search*. The hybrid search simply combines the functionalities by employing the semantic search where search terms are explicitly defined in the ontology and where the ontology falls short of describing the search keyword, the system works as a traditional search engine.

Another form of search approach, often employed to compliment keyword-based searches, is the *faceted search* [2]. In this approach, to searching or browsing, to be precise, information discovery is achieved through filtering of facet values. A faceted search combines the techniques of direct keyword search and a navigational search. Navigational search is a form of directory-based search approach and as argued by [2], the use of faceted searches for querying RDF data is known to have solid background and various theoretical frameworks that work fairly well in the life-sciences domain. As our knowledge collection basically involves RDF datasets that are semantically enhanced and linked together using OWL ontology, we simply employ the semantic search approach to query our knowledge bases, as detailed in Sect. 4.

## 3  Related Works

In the literature, various tools for semantic data exploration and browsing have been developed to allow utilization of available RDF datasets and ontological knowledge bases. In the field of crops and life sciences domain, tools such as CO Curation tool [23], SemFacet [2], do exist among many others. The crop ontology curation tool is a curation and annotation website that provides a search engine for exploring the CO including various URL lists for browsing the ontology. In addition, the tool also allows uploading and creating ontologies in the OBO format with available RDF dumps and a web service API for download. While the CO curation tool may have been motivated by the need to disseminate CO knowledge, a stand-alone tool, SemFacet was developed by Zhou and others, for querying arbitrary life sciences ontologies. According to the authors, the tool allows the use of keyword search as well as faceted navigation for querying ontology-enhanced RDF datasets. The tool's interface shows a navigation map, which enables search refocusing from one object to another and search results can be filtered based on relevant facet names and values.

Another domain-specific, ontology-based semantic search engine is presented in [11], which is developed for querying agricultural information. The search engine relies on a domain ontology developed in RDF and a web-based interface consisting of a keyword based search engine. With little or no structure, text-based query results are also returned in another text box on the interface. Other more generic semantic search engines include SemSearch [22], SHOE [15], SWSE [16] and Swoogle [12]. The Swoogle search engine, which is one of the early search and meta data engine for the Semantic Web, is able to discover matching keywords from thousands of ontologies, documents, and terms. It allows end users to select preferred data sources before submitting a query and construct relations between resulting documents.

In line with the basic features provided by these systems, we employ the keyword based search engine and complement it with a faceted navigation for browsing the initial search results from the keyword search. Initial search results from the knowledge base are returned as lists of subjects with navigational links categorized for each data source. In doing so, the user is given a choice to navigate the search results based on available datasets.

## 4    Onto-CropBase Development Methodology

In this section, we discuss the design strategy and methodology employed in the development of the Onto-CropBase tool. As explained earlier, the Onto-CropBase tool was designed to comprise three major components: (i) the *domain ontologies* component—consisting of a global ontology developed in OWL 2 RL & SWRL and local ontologies in RDFS, (ii) the *mediator* component— provided by the Jena API, and (iii) the *ontology-based semantic search engine*— developed using J2EE[3]. We introduced the domain ontology component in the previous section and here, for brevity, we intend to discuss only the ontology-based integration strategy employed in developing the Onto-CropBase tool.

Details of the development of the domain ontology are discussed elsewhere [21], and the process of converting available crops data/metadata from XML to OWL format to generate relevant domain concepts is discussed in [25]. The overall framework of the project was based on an open-source solution and is shown in Fig. 2. Methodology and detailed description of these components are presented in the following subsections.

### 4.1    Integration of Ontology Components and Data Sources

To allow interoperability between the UC-ONTO and other crops data involved in our knowledge base and due to the modularized nature of the ontologies involved, a connection needs to be made between the various components data sources to allow writing the federated queries. This is particularly useful when answering user queries that require pulling information from two or more data sources. Common procedure of integration involves the use of *mappings* between terms in the relevant ontologies, i.e., where two or more ontologies are involved, termed as, inter-ontology mapping. Another approach involves the *linking* of the ontologies with the actual information within the data sources, i.e., where data sources other than ontologies are involved in the integration process. By definition, *ontology mapping* is a directed matching of two or more ontologies in which entities of one ontology are mapped to at-most one entity of another ontology [6].

---

[3] http://www.oracle.com/technetwork/Java/javaee/overview/index.html.

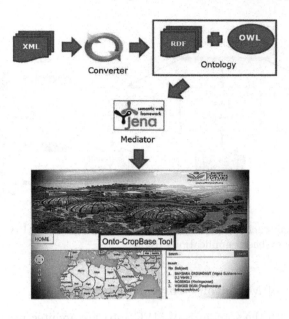

**Fig. 2.** Onto-CropBase tool components

Ontology-based integration of data sources can be generally categorized into three approaches, namely, *the single ontology approach, multiple ontologies approach* and *the hybrid approach* [28]. The *single ontology* method follows the centralization of all data sources and shared vocabulary into a global ontology. This method is suitable when all data sources used for the integration share the same view of the domain. While in the *multiple ontologies* approach, each data source is separately described by its own local ontology with inter-ontology mapping between relevant terms. This method is suitable for decentralized data sources that do not need a common vocabulary. The third approach, *hybrid ontology integration*, involves a combination of the single ontology and multiple ontology approaches. In this approach, the ontology of each data source is developed separately and mapped to a shared vocabulary (global ontology) to allow interoperability.

We chose the *hybrid* approach to integrate our UC-ONTO with the RDF data sources or ontologies due to the particular advantage that new sources can easily be added without modifying existing mappings. This suits comfortably with our ever-evolving underutilized crops knowledge base, the UC-ONTO, as it ensures that the knowledge base can be easily extended in the future to accommodate data from other heterogeneous sources like RDFs, relational databases, excel and web documents. The architecture of the hybrid ontology approach is shown in Fig. 3 (left) and a more detailed discussion on the three approaches can be found in [28]. It should be noted that apart from achieving interoperability between data sources, ontology-based integration has the following added advantages: *global conceptualization, mapping support, metadata representation*, as well as *support for high-level queries*, among others [10].

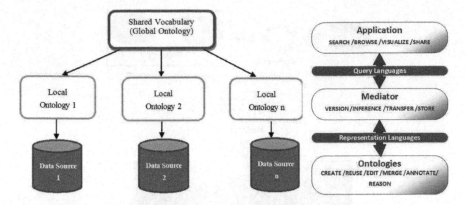

**Fig. 3.** Hybrid approach to ontology integration (left) and an MVC adapted design concept for ontology-based applications (right).

## 4.2   Selection of Mediator Component

In compliance with the standards of MVC software architecture [27], the Onto-CropBase was also designed to have a robust control module handling the interaction between the thin web interface and the ontology-based semantic data model. Considering the design paradigm for ontology based applications in Fig. 3 (right), it is possible to make the following assertion: "In order to develop a semantic web application (SWAP), a mediator component—*the controller*, is needed as the programmatic environment that can interact with the semantic data—*the model*, to allow visualization on the states of the model—*the views*". Common open-source choices for a mediator component in developing SWAPs include the OWL API [18], Protégé-OWL API [17], and the Apache Jena framework [1], among others. Detailed comparison of these common tools and applications can be found in [29].

**The Jena API Distribution.** Apache Jena [1], is a Java-based, free, as well as open-source programming tool for developing semantic web and linked-data applications. Firmly rooted in RDF, the framework consists of inter-connected APIs that can be invoked into an application code to manipulate an RDF data or knowledge model. Basic Jena APIs include Ontology, RDF, SPARQL, Inference and Storage APIs. The generic nature of the Jena's Ontology API, *OntModel*, makes it capable of handling other ontology languages that can be serialized in RDF format. A scalable Triple Store Database (TDB), which implements SPARQL specifications, is also provided in the Jena distribution for storing and retrieving RDF data. With regards to inference, a set of predefined Reasoners such as: the RDFS rule reasoner, Transitive reasoner, OWL-Lite Reasoners, and a Generic rule reasoner are available in the Jena distribution [20]. Moreover, Reasoners need not always be explicitly specified while working with the Jena Ontology API, as appropriate reasoner can be accessed to generate inferences by the query engine based on existing ontology configuration.

**The OWL API.** This is also an open source and Java-based reference tool for creating as well as manipulation of OWL ontologies [3]. Centered on the web ontology language (OWL), it allows working directly with OWL axioms without having to be serialized in to RDF, as required by the Jena API. Moreover, the OWL API provides interfaces for OWL 2 with *parsers* and *writers* available for: RDF/XML, OWL/XML, OWL Functional Syntax, Turtle, KRSS and the OBO format. Various Reasoners such as Fact++, HermiT, Pellet, etc. are also embedded for inferencing [18].

**The Protégé-OWL API.** Considered as an extension to the OWL API, the Protégé-OWL API [17], is another open source Java-based tool designed for manipulating both OWL and RDF models. The Protégé-OWL API also uses Jena for *parsing*, thereby providing some of the available Jena services. The API is provided in a standard Protégé installation. Notable improvements in the OWL API over its predecessors, is its ease of implementing graphical user interfaces (GUIs) for users based on their working ontology or data models and also the ability to control internal representation of the ontologies using the GUIs. Various DL-based Reasoners are also available in the Protégé-OWL API.

On the other hand, due to its rich documentation, stability in handling RDF data and the availability of rule-based and non-rule-based reasoners, we chose the Apache-Jena to serve as the connecting point between our ontology-based model and the user interfaces or views. Moreover, our primary choice of a Java-based development for the Onto-CropBase tool, is another deciding factor for choosing Apache Jena as the mediator. Nevertheless, there are other Java-based tools exist that provide a complete development environment for SWAPs, by shielding the user from the need to bother with the above connections. Commonly known examples include Stardog RDF database [8].

## 4.3   Interface Design and Search Engine Approach

The home page of the Onto-CropBase tool is designed with a form-based search engine and as location data is critical to crop-based knowledge systems especially that of the underutilized crops, a map interface is provided to display the crops location data. The map interface is embedded using the Google Maps JavaScript API version 3, which enables map features in a web application, including styled maps, place data, 3D buildings, and geocoding, among other features. The map information is extracted from the named locations *east-west bound longitude* and *north-south bound latitude* elements.

**Search Approach.** In our Onto-CropBase search engine, a search begins with a keyword entered in the search area and the query results are returned as a set of navigational links based on the subjects of the data sources— using their title annotation provided in the corresponding local ontology, see Fig. 4. Users can then browse the list of subject titles to explore the remaining information. Clicking on a particular subject will present the RDF assertions as subjects and objects pairs. For example, in Fig. 4, subject number 3 is a caption for *Carbohydrate* of Bambara groundnut and the corresponding object asserted as its value is *High (65 %)*.

**Query Language.** SPARQL is used as the query language, which is a recursive acronym for SPARQL Protocol And RDF Query Language (SPARQL), and a standard query language for retrieving information stored in RDF graph or triples [14]. A basic structure of a SPARQL query includes the SELECT, CONSTRUCT, ASK, and DESCRIBE statements followed by the WHERE clauses and GROUPBY clauses where applicable, see example in Sect. 5. As mentioned earlier, we employ the Jena-ARQ [20], a query engine for Jena that supports SPARQL RDF Query language, to allow federated user queries across the local ontologies – using the corresponding URIs of concepts found in our global ontology as search phrases.

```
1."PREFIX rdfs: <http://www.w3.org/2000/01/rdf-schema#>"+
2. "PREFIX rdf: <http://.../1999/02/22-rdf-syntax-ns#>"+
3. "PREFIX xsd: <http://www.w3.org/2001/XMLSchema#>"+
4. "PREFIX ucnames: <http://.../ontologies/2015/Naming#>"+
5. "PREFIX agronomy: <http://.../ontologies/2015/agrono#>"+
6. SELECT distinct ?subject ?object "+
7. WHERE {" + ""+"
8. OPTIONAL" + "{?subject rdf:value ?object ."+"
9. ?subject rdf:type ucnames:"+ className +".}"+""+""+"
10. OPTIONAL" + "{?subject rdf:value ?object ."+"
11. ?subject rdf:type agronomy:"+ className + ".}"}
```

**Listing 1.1.** Example use of SELECT and OPTIONAL constructors in a SPARQL query

**Query Design.** Search keywords from the Onto-CropBase interface are embedded in a Java code with a SPARQL query, parsed as a text string to the *QueryFactory* method. The SELECT query is first executed on the base or global ontology, which in this case, is the UC-ONTO. This helps to retrieve the concepts matching the search keywords in the global ontology and an OPTIONAL query pattern, see Listing 1.1, is provided to recursively explore all interlinked RDF datasets. It shows an example use of SELECT and OPTIONAL constructors in a SPARQL query to retrieve distinct subject-object pair from two local ontologies, ucnames and agronomy. Note that the prefix declarations, lines 1 through 6, allow for abbreviating URIs, so that the short names can be used instead of the URIs in the query body.

It allows the Jena-ARQ query engine to search for all the relevant triples with inferences from the local ontologies and without failing in the event that the optional data does not exist. The use of *optional* query also helps to ensure that the non-optional information is returned from at least, the global ontology. For the map query, in the event that one of the RDF subjects returned is a location data, a *setMap* object is then used to obtain the set of *longitudes* and *latitudes* of the location, for display on the map interface. The size of the resulting subject-predicate-object (S, P, O) triples, is also calculated for each dataset to determine the number of pages to be returned – we set the capacity to 10 triple set per page to allow the map view stay in focus.

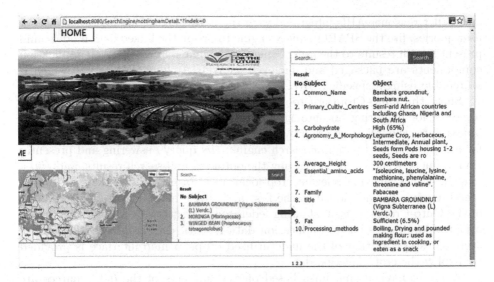

**Fig. 4.** The Onto-CropBase tool showing search result titles *(inserted left)* and detailed assertions *(background right)*.

**Query Processing.** Jena-ARQ is employed as a query engine for executing our SPARQL queries. This is achieved through a sequence of five iterative steps as follows: (i) String to Query parsing, where the text string parsed to the *QueryFactory* method is structured from a query string to a query object, (ii) Next is the Algebra Generation, which involves translation of the query object to a SPARQL algebra expression using the SPARQL specification algorithm, (iii) Third is the Optimization of the algebra expression generated and is called high-level optimization and transformation. Where a *Transformer* class applies a transform code to convert or replace the algebra expression tree with more efficient expressions. Example of transform code function is replacing the equality filter with more efficient graph pattern, (iv) In the next step, the query plan is determined and is called the Low-level optimization, which involves deciding the order in which to evaluate the basic graph patters transformed earlier. However, this stage can be carried out concurrently with the last, which is (v) Evaluation of the query plan, which involves executing the algebra expressions to generate the solution graph patterns, returned as sets of (S, P, O) information. These steps can be extended and modified to allow searching different graph-pattern implementations and the evaluation strategies can also be enhanced to suit specific requirements of application developers [20].

## 5   The Onto-CropBase: Architecture and Functionalities

The Onto-CropBase tool consists of a web interface, depicted in Fig. 4, which in the background utilizes the Java Server Pages (JSP) and *servlets* components of the web application. These components contain Java codes that invoked the Jena and Pellet reasoner APIs integrated with the data model (ontologies), to accept user queries, fire the SPARQL query engine to probe the linked data models, and present search results to the user. Using the Apache Server, the Web application components are packaged and deployed as Web Archive (WAR) file, which is searchable using a specified URL.

Since the Onto-CropBase tool aims at providing an information retrieval interface for exploring an ontological knowledge base integrated with relevant linked-data, the following are provided as its major functionalities: *(i)* provision for a keyword-based semantic search engine, *(ii)* query answering and presentation of query results, *(iii)* navigating the ontology and search results, and *(iv)* a map interface, showing relevant crop location information. The keyword-based search engine was developed as the first and single most important component of the Onto-CropBase tool, allowing federated searches with a single user query over the knowledge bases. As location data is critical to crop-based knowledge systems especially that of the underutilized crops, a map interface is provided to display the crops location data.

A set of navigational links based on the subjects of the data sources are returned as a search result based on the keyword entered by a user. Then he can browse the list of subject titles to explore the detailed subject-object pair assertions matching the initial query. Information provided by the tool is designed to be presented in a simple and straight forward fashion with minimum ambiguity. However, more functionalities will be added as the Onto-CropBase tool evolves and we leave the discussion of added features to the future works section. We briefly highlight here, some of the existing functionalities:

**Keyword Search.** Exploring the knowledge base in the Onto-CropBase tool typically starts with a keyword search, which is matched against the concepts in the global ontology. Where a match is found, the RDF triples in the corresponding local ontologies are probed through the linked URIs and the associated triples are returned with a map view of the origin or farming area of the specific underutilized crop(s). The map interface is searched separately according to the asserted location data through Google Maps JavaScript API[4] version 3.

**Query Answering.** User queries containing keywords entered in the search engine are compiled into a SPARQL query, as shown in the Listing 1.2. The queries are designed to provide search results based on *classes, entities,* and textitmap area. The Jena Ontology API first loads the ontologies and the ARQ query engine uses the search keywords to generate a query plan. The results of executing the query plan generate an RDF data satisfying the query pattern, called the *Result Set.* The result sets are then passed to the *filtering object,* which returns only matching triple objects from each dataset as the final answer to the query. Ordering of the answer triples is achieved using a *binding hierarchy* generated by the query engine. This binding hierarchy reflects the query patterns and the final query output are presented based on the initial values to be resolved.

---

[4] https://developers.google.com/maps/documentation/javascript/.

```
...
"PREFIX uconto: <http://www.nottingham.edu.my/ontologies
        /2014/Ontology-uco#>" +
"PREFIX ucnutrition: <http://.../ontologies/2015/nut#>" +
"PREFIX ucnames: <http://.../ontologies/2015/Naming#>" +
"PREFIX agronomy: <http://.../ontologies/2015/agrono#>" +
  SELECT distinct ?subject ?object "+
    WHERE {"+"?subClass rdfs:subClassOf uconto:" +
        className + "."+""+"}" ;
```

**Listing 1.2.** A SPARQL example query

An example SPARQL query to request all classes in the global OWL ontology is shown in Listing 1.2, and to query all the instances assertions matching the above subject from any of the local ontologies (e.g., "agronomy"), the example SPARQL query is written as:

```
...
   SELECT DISTINCT ?subject ?property ?object"+"
     WHERE {"+" ?subject a agronomy:"+"queryString
       "+"."+" ?subject ?property ?object . "+"}
       ORDER BY ?subject" ;
```

The results of these user queries are presented as RDF triples and grouped under the titles of their source ontologies— shown as list of subjects in Fig. 4 (left). Clicking on a particular dataset title shows the detailed *subject* and *object* information pairs derived from the RDF statements or assertions— shown in Fig. 4 (right).

**Paging.** As shown in Fig. 4, search results are designed to be presented in a series of numbered rows showing the RDF result sets in the form of subject-object pair and with a maximum of 10 rows per page. Where the resulting information exceeds 10 rows, a new page is automatically created with *page counts* similar to those found in the commercial search engines. Similarly, user can navigate these pages by clicking on the desired page number at the bottom of the search results. This allows the map interface to stay in focus throughout the navigation process giving users continuous access to the location data.

**Map Interface.** In line with location data imported from FAO's geopolitical ontology [19], the Onto-CropBase tool is also designed to show the underutilized crops location data where asserted concepts include *Crop Origin* and/or *Cultivation Region*, among others. As explained earlier, Google Map has been used as a base map for the location-based data and the information is extracted from the named location's east/west bound longitude and north/south bound latitude elements. The map location data is important in the Onto-CropBase tool as it helps users to get a clear picture of the crops origin, cultivation regions and/or locations where similar crops can be cultivated, among other information.

## 6    Preliminary Evaluation and Discussion

The preliminary evaluation follows an unstructured evaluation approach, where the evaluation success is entirely based on the end user's perception and ability to effectively utilize the tool – something that is difficult to be accurately measured. We also evaluate the Onto-CropBase tool conventionally, by considering its pre-designed requirements and comparing it with other existing tools that are considered successful in the related field, as discussed in the following scenarios.

As earlier assumed, the Onto-CropBase tool has been primarily designed to be used by the users who are seeking first-hand information on underutilized crops, rather than software agents. As such, for evaluating the suitability of our approach and the information provided by our knowledge system, we employ the services of five domain experts including those actively involved in the domain knowledge modelling. In their response, all users agreed on the fact that the information provided by the tool is relevant and that the tool is simple to use even for non-technical experts. Similarly, almost all the end users exposed to the Onto-CropBase tool are able to freely navigate its functionalities without much intervention.

Moreover, in order to verify that desired functionalities are provided by the tool, we compare features of the Onto-CropBase tool with other ontology-based knowledge systems in the field of life-sciences, such as the Crop Ontology curation tool [23], and SemFacet [2], see Table 1. The evaluation shows that our Onto-CropBase tool's basic functionalities, which involve the keyword-based search and navigational presentation of search results is a common approach employed in those systems. However, we observed the need to provide a visualization of hierarchical relations between concepts to enhance the information presentation— a need already mentioned by some of our end users.

Another useful functionality not currently covered by our tool is the implementation of an ontology-based faceted search mechanism. This is considered important in ontology-based searches as it helps to portray the inner structure of the ontology, which is useful for efficient probe of the knowledge base. However, these functionalities are considered to be of lower priority as compared to the integration of the ontological knowledge models and the current features of the search engine, and therefore, will be included in the future releases of the tool.

**Table 1.** Comparison of features for Onto-CropBase, crop ontology and semfacet.

| Features | Onto-CropBase | Crops Ontology [23] | SemFacet [2] |
|---|---|---|---|
| Knowledge Domain | Crops (Specific) | Crops (Generalized) | Life-Sciences |
| Search Type | Keyword | Keyword + hierarchy navigation | Keyword + Faceted navigation |
| Ontology Format | OWL2 + RDF | OBO | OWL2 + RDF |
| Result Visualization | Text + Map | Text + Image | Text + Image |
| Data Curation | None | Keyword + hierarchy | None |

## 7  Conclusion and Future Works

In this paper, we have presented the Onto-CropBase tool, a flexible semantic web application for browsing our previously developed underutilized-crops ontology

(UC-ONTO) [21], extended with relevant linked data in RDFS. The tool provides a web-based user interface with a search engine for querying an ontology-based knowledge model. At the center of the knowledge base, is an OWL 2 ontology, serving as a *global ontology* containing shared concepts from integrated local ontologies or RDF datasets. From the preliminary evaluation of the tool, it can be concluded that the Onto-CropBase can without doubts serve as a first-hand information portal for information on underutilized crops – a basic requirement of the project. It was also found to be usable even to non-technical or domain experts. Though, there are still open issues to be considered in the future.

Our future works are focused on enriching the back end model with more crops-related RDF datasets, SPARQL endpoints and other text documents, to allow for more comprehensive search results. Similarly, the user interface will be enhanced with navigation hierarchy or *facets* to guide users on the structure of the knowledge base for enhanced queries, *lookup and comparison tables* to compare data on the popular underutilized crops, and a more specialized *location-based information* on crops, among other features.

**Acknowledgement.** This work is partially supported by a CFFRCPLUS CropBase PhD scholarship scheme.

# References

1. Apache Jena: Apache jena - a free and open source java framework for building semantic web and linked data applications (2015). https://jena.apache.org/
2. Arenas, M., Cuenca Grau, B., Kharlamov, E., Marciuska, S., Zheleznyakov, D., Jimenez-Ruiz, E.: Semfacet: Semantic faceted search over yago. In: Proceedings of the 23rd International Conference on World Wide Web WWW 2014 Companion, pp. 123–126 (2014)
3. Bechhofer, S., Volz, R., Lord, P.: Cooking the semantic web with the OWL API. In: Fensel, D., Sycara, K., Mylopoulos, J. (eds.) ISWC 2003. LNCS, vol. 2870, pp. 659–675. Springer, Heidelberg (2003)
4. Berners-Lee, T., Hendler, J., Lassila, O.: The semantic web. Sci. Am. **284**(5), 29–37 (2001)
5. Bhagdev, R., Chapman, S., Ciravegna, F., Lanfranchi, V., Petrelli, D.: Hybrid search: effectively combining keywords and semantic searches. In: Bechhofer, S., Hauswirth, M., Hoffmann, J., Koubarakis, M. (eds.) ESWC 2008. LNCS, vol. 5021, pp. 554–568. Springer, Heidelberg (2008)
6. Bohring, H., Auer, S.: Mapping XML to OWL ontologies. In: Proceedings of 13. Leipziger Informatik-Tage (LIT 2005). Lecture Notes in Informatics (LNI), 21–23 September 2005
7. Caraccioloa, C., Stellatob, A., Morsheda, A., Johannsena, G., Rajbhandaria, S., Jaquesa, Y., Keizera, J.: The agrovoc linked dataset. Semant. Web **4**(3), 341–348 (2013)
8. Cerans, K., Barzdins, G., Liepins, R., Ovcinnikova, J., Rikacovs, S., Sprogis, A.: Graphical schema editing for stardog OWL/RDF databases using OWLGrEd/S. In: CEUR Workshop Proceedings OWLED, vol. 849 (2012)

9. Cooper, L., Walls, R.L., Elser, J., Gandolfo, M.A., Stevenson, D.W., Smith, B., Preece, J., Athreya, B., Mungall, C.J., Rensing, S., Hiss, M., Lang, D., Reski, R., Berardini, T.Z., Li, D., Huala, E., Schaeffer, M., Menda, N., Arnaud, E., Shrestha, R., Yamazaki, Y., Jaiswal, P.: The plant ontology as a tool for comparative plant anatomy and genomic analyses. Plant Cell Physiol. **54**(2), e1 (2013)

10. Cruz, I.F., Xiao, H.: The role of ontologies in data integration. J. Eng. Intell. Syst. **13**, 245–252 (2005)

11. Debajyoti, M., Aritra, B., Sreemoyee, M., Jhilik, B., Kim, Y.: A domain specific ontology based semantic web search engine. CoRR abs/1102.0695 (2011)

12. Ding, L., Finin, T., Joshi, A., Pan, R., Cost, R.S., Peng, Y., Reddivari, P., Doshi, V., Sachs, J.: Swoogle: A search and metadata engine for the semantic web. In: Proceedings of the Thirteenth ACM International Conference on Information and Knowledge Management CIKM 2004, pp. 652–659. ACM (2004)

13. Gruber, T.R.: A translation approach to portable ontology specifications. Knowl. Acquis. **5**(2), 199–220 (1993)

14. Harris, S., Seaborne, A.: SPARQL 1.1 query language: W3c recommendation (2013). https://www.w3.org/TR/sparql11-query/

15. Heflin, J., Hendler, J.: In artificial intelligence for web search. In: Papers from the AAAI Workshop, WS-00-01, pp. 35–40. AAAI Press (2000)

16. Hogan, A., Harth, A., Umbrich, J., Kinsella, S., Polleres, A., Decker, S.: Searching and browsing linked data with SWSE: The semantic web search engine. Web Semant. Sci. Serv. Agents World Wide Web **9**(4), 365–401 (2011)

17. Holger, K.: Protege-OWL API programmer's guide (2010). http://protegewiki. stanford.edu/wiki/ProtegeOWL_API_Programmers_Guide

18. Horridge, M., Bechhofer, S.: The OWL API: A java API for OWL ontologies. Semant. Web **2**(1), 11–21 (2011)

19. Iglesias-Sucasas, M., Kim, S., Viollier, V.: The FAO geopolitical ontology: a reference for country-based information (2015). http://www.fao.org/countryprofiles/geoinfo/en/

20. Jena-ARQ: ARQ - extending query execution (2015). https://jena.apache.org/documentation/query/arq-query-eval.html

21. Lawan, A., Rakib, A., Alechina, N., Karunaratne, A.: Advancing underutilized crops knowledge using SWRL-enabled ontologies-a survey and early experiment. In: Proceedings of the 2nd International Workshop on Linked Data and Ontology in Practice (LDOP 2014), vol. 1312, pp. 69–84 (2014)

22. Lei, Y., Uren, V.S., Motta, E.: Semsearch: a search engine for the semantic web. In: Staab, S., Svátek, V. (eds.) EKAW 2006. LNCS (LNAI), vol. 4248, pp. 238–245. Springer, Heidelberg (2006)

23. Matteis, L., Chibon, P., Espinosa, H., Skofic, M., Finkers, R., Bruskiewich, R., Hyman, G., Arnaud, E.: Crop ontology: vocabulary for crop-related concepts. In: Larmande, P., Arnaud, E., Mougenot, I., Jonquet, C., Libourel, T., Ruiz, M. (eds.) Proceedings of the First International Workshop on Semantics for Biodiversity, vol. 979 (2013)

24. Padulosi, S., Hodgkin, T., Williams, J., Haq, N.: Underutilised crops: trends, challenges and opportunities in the 21st century. In: Engels, J., Rao, V., Jackson, M. (eds.) Managing Plant Genetic Diversity, pp. 323–338. CAB International (2002)

25. Rakib, A., Lawan, A., Walker, S.: An ontological approach for knowledge modeling and reasoning over heterogeneous crop data sources. In: Abraham, A., Muda, A.K., Choo, Y.-H. (eds.) Pattern Analysis, Intelligent Security and the Internet of Things. AISC, vol. 355, pp. 35–47. Springer, Heidelberg (2015)

26. Salton, G., McGill, M.J.: Introduction to Modern Information Retrieval. McGraw-Hill Inc., New York (1986)
27. Stoughton, A.: A functional model-view-controller software architecture for command-oriented programs. In: Proceedings of the ACM SIGPLAN Workshop on Generic Programming, pp. 1–12. ACM (2008)
28. Wache, H., Voegele, T., Visser, T., Stuckenschmidt, H., Schuster, H., Neumann, G., Huebner, S.: Ontology-based integration of information - a survey of existing approaches. In: Stuckenschmidt, H. (ed.) IJCAI-01 Workshop: Ontologies and Information, pp. 108–117 (2001)
29. Waterfeld, W., Weiten, M., Haase, P.: Ontology management infrastructures. In: Hepp, M., Leenheer, P.D., de Moor, A., Sure, Y. (eds.) Ontology Management, vol. 7, pp. 59–87. Springer, Heidelberg (2008)

# TripleID: A Low-Overhead Representation and Querying Using GPU for Large RDFs

Chantana Chantrapornchai[1]([✉]), Chidchanok Choksuchat[2],
Michael Haidl[3], and Sergei Gorlatch[3]

[1] Department of Computer Engineering, Kasetsart University, Bangkok, Thailand
`fengcnc@ku.ac.th`
[2] Department of Computing, Silpakorn University, Bangkok, Thailand
`cchoksuchat@hotmail.com`
[3] University of Münster, Münster, Germany
`{michael.haidl,gorlatch}@uni-muenster.de`

**Abstract.** Resource Description Framework (RDF) is a commonly used format for semantic web processing. It basically contains strings representing terms and their relationships which can be queried or inferred. RDF is usually a large text file which contains many million relationships. In this work, we propose a framework, *TripleID*, for processing queries of large RDF data. The framework utilises Graphics Processing Units (GPUs) to search RDF relations. The RDF data is first transformed to the encoded form suitable for storing in the GPU memory. Then parallel threads on the GPU search the required data. We show in the experiments that one GPU on a personal desktop can handle 100 million triple relations, while a traditional RDF processing tool can process up to 10 million triples. Furthermore, we can query sample relations within 0.18 s with the GPU in 7 million triples, while the traditional tool takes at least 6 s for 1.8 million triples.

**Keywords:** Query processing · Parallel processing · GPU · RDF · CUDA

## 1 Introduction

Semantic web provides a standard and common framework to share and reuse data across the Internet [6,7]. These data contain millions of triples in a form: (subject, predicate, object). The predicates describe the relations between subjects and objects. Each of these terms is usually a Unified Resource Identifier (URI), which are commonly very long strings. Thus, RDF files have a significant size and it is time consuming to load and query such triples.

Traditional methods of querying RDF data are based on graph libraries such as Redland [5], RDFlib [23], and RDFsh [2]. These consume lots of memory for constructing models and it is very time consuming to load them and create a data model as well as to query the model. With the current parallel technology, it is possible to utilize multithreading to perform such tasks to speed up the processing time.

© Springer International Publishing Switzerland 2016
S. Kozielski et al. (Eds.): BDAS 2016, CCIS 613, pp. 400–415, 2016.
DOI: 10.1007/978-3-319-34099-9_31

Current parallel hardware technology has been advanced which allows to process applications using multi-core CPUs or many-core architectures, Graphics Processing Units (GPUs) that contain hundreds to thousands of cores. Due to the low cost of GPUs, it becomes affordable for a desktop user to gain high-speed processing using them. Nowadays, GPUs can be used for general-purpose computing in many application areas [21]. However, to use GPUs, applications must be designed properly to support the GPU architecture.

The above mentioned RDF libraries like Redland and RDFlib have complicated data structures, such as graphs, and they use heap storages for RDF triples. These storages are usually not thread-safe and have deep pointers which are not easily mapped to the GPU memory layout.

In this paper, we provide a simplified approach, *TripleID*, to convert the representation to store RDF data in the GPU memory directly and utilise GPU threads for querying them. For reasons of simplicity, the framework takes the RDFs in the triple form (N-Triples and/or N3) and converts them into the triples of identifiers (TripleID). All IDs are loaded to the GPU memory for the parallel search. The converted tripleID files are about 2–4 times smaller than the original NT files and the conversion time to TripleID is 1.5–2.4 times shorter than the loading time using the original NT file. Furthermore, the querying using GPUs with TripleID is much faster than using the traditional NT file.

The outline of the remainder of this paper is as follows. First, the background about RDF and GPU as well as related work is presented. After that, our approach is presented in Sect. 3. Then, the GPU representation as well as the querying algorithm are presented in Sect. 4. The experiments comparing the data size and conversion time, and the query time are also presented. Section 5 concludes the paper and describes the future work.

## 2    Background

Resource Description Framework (RDF) is a common format used to describe the relations of items in a triple form: (*subject, predicate, object*). Each term is usually a Uniform Resource Identifier (URI) which can be linked to another web resource [24]. Let us consider an example of the RDF triple.

```
<http://dbpedia.org/resource/Plasmodium_hegneri>
<http://www.w3.org/1999/02/22-rdf-syntax-ns#type>
<http://dbpedia.org/ontology/Species>
```

Here, three terms represent the relationship: `Plasmodium_hegneri` is a Species, where <http://dbpedia.org/resource/Plasmodium_hegneri> is the link data source of `Plasmodium_hegneri`, <http://dbpedia.org/ontology/Species> is the link data of Species. Both of these are from *dbpedia* [7]. The predicate <http://www.w3.org/1999/02/22-rdf-syntax-ns#type> describes the relationship as `type` which is a type definition by W3C [1].

Searching for the RDF data is usually done by a semantic web query using SPARQL [10]. Typical queries can ask for subjects, predicates, or objects of the triples. Thus, it is possible that the data can be indexed based on subjects, predicates, or objects, or any combination of them to speed up the query processing. However, it takes a lot of time to index millions of triples for subjects, predicates, or objects, or any combinations of them.

**Table 1.** Example of indexing usages

| No | Query | CumulusRDF index |
|----|-------|------------------|
| 1  | (spo) | SPO, POS, OSP    |
| 2  | (sp?) | SPO              |
| 3  | (?po) | POS              |
| 4  | (s?o) | OSP              |
| 5  | (?p?) | POS              |
| 6  | (s??) | SPO              |
| 7  | (??o) | OSP              |
| 8  | (???) | SPO, POS, OSP    |

Table 1 shows examples of queries in triples (one per row). Column "Query" is the possible query. Column "CumulusRDF index" shows the index for each query. The first letter is the primary index. For example, the query (sp?) means "list all objects of the given subject $s$ with predicate $p$". For this type of query, it is better to use the data indexing in the order SPO. The data should be stored ordered by subject $S$ first, and then by predicate $P$ accordingly. The query (?po) is better to be applied on the data indexed in the order POS; the data should be stored according to the index $P$ and then $O$. Generating all possible indices requires powerful computers which can take hours, days or weeks because of the size of RDF data.

SPARQL is a query language for RDF. A SPARQL's SELECTing query of the same dataset looks like SQL in the same table. WHERE clause constrains the query result. A query such as "find all issued years of The Journal of Supercomputing" can be written in SPARQL as:

```
SELECT ?yr
WHERE {
''The Journal of Supercomputing''^^xsd:string dcterms:issued ?yr}
```

Such a query can be executed fast when data are indexed by subjects as row 2 of Table 1. If we consider another query, adapted from [22], with details in Sect. 5.2, "find all years and authors of The Journal of Supercomputing":

```
SELECT ?yr ?authors
WHERE {
?journal dc:title ''The Journal of Supercomputing''^^ xsd:string.   (1)
?journal dcterms:issued ?yr.                                        (2)
?journal dc:creator ?authors}                                       (3)
```

then, if data is indexed with POS as in row 3 in Table 1, the subquery in (1) can be executed fast. For subqueries in lines (2) and (3), the better index corresponds to row 5 of Table 1.

Graphics Processing Units (GPUs) were originally used to process graphics objects. With the advance of hardware technology, they now contain many thousands of cores which can be used to do any kind of general-purpose computations. Though they have a lower clock speed than the general-purpose CPU, thousands of cores can compute faster if they are utilized properly. Previous work addressed using GPUs to compute in the semantic web area e.g. [8,9,13,15,16,18].

To utilize GPUs, proper programming frameworks are needed. Compute Unified Device Architecture (CUDA) is one of them to program NVIDIA GPU cards [21]. In the CUDA architecture, threads are organized in so-called grids and each grid is divided into blocks. Threads in a block are executed concurrently.

CUDA cores are grouped into Streaming Multiprocessor (SM) units. One GPU card usually contains 4–26 SMs. Memories of the GPU are categorized into levels: *shared, global, texture, constant, register*, and *local*. Global memory (off-chip) can be accessed by all threads in all blocks while the shared memory (on-chip) can be accessed only by threads in the same block. Texture memory and constant memory are read-only memory. Global memory has the largest size, varying from 2G to 24 GB depending on the card model. Though the access time of shared memory is much lower, the shared memory usually has a limited size of up to 112 KB. While most computations utilize the global memory since it is large, for small frequently accessed data, the shared memory is used. For that, the data from the global memory must be copied to shared memory before accessing it.

The GPU memory transfer latency can be a serious obstacle for improving the execution time: the data from the main memory must be copied to the GPU global memory, and/or from the global memory to the shared memory before the GPU can start computing. Efficient algorithms must be designed so that the data can be retained inside the GPU memory as long as possible to reduce transfer time, thus reducing execution time.

In our case, the TripleID array is transferred to the GPU global memory. Then, the search is performed on the array and the related positions are returned.

## 2.1 Related Work

Several approaches aimed at the relational database query on GPUs [8,13,16,18]. He et al. perform *join* operations on the relational database on GPUs [12]. They provide data-parallel primitives such as *split, merge, map, gather-scatter, sort* and *join*, for memory optimization. A useful data structure is the *lock-free* scheme for storing result outputs. Two phases are used: the first phase is to scan the total size of the results for GPU memory allocation, the second phase is to perform the operation on the results in GPU. Groppe et al. focused on the *join* operations for semantic web query on the GPU [11].

Utilizing indices can improve query throughput. Wei and Jaja used the single dimension array index for the tree representation on the GPU for faster preparation of web page documents [27]. Kim et al. focused on parallel indexing [14]. They adapted the braided parallelism to improve the multi-dimensional query processing throughput, and compared it with the pure data parallel partitioned indexing scheme. The same single query was processed by all streaming multiprocessors concurrently with different divided indices. The data structure used was a tree that has the advantage of indexing more data; thus, the data-parallel partitioned tree outperformed the braided parallel brute-force when the number of indices of data is larger than 3 millions.

Centralized indexing and hierarchical two-level indexing are examined by Nam and Sussman [19]. If more data are added, the size of the centralized index and the global index for two-level indexing can increase linearly. Data-intensive unstructured data can also be represented in an inverted file with a trie structure (an ordered tree data structure) on heterogeneous platforms [26]. As discussed earlier, constructing indices give advantages on fast search but it is still time consuming to construct them.

Some efforts focused on transforming and compressing the RDF representation. One of the well-known structures is BitMat. Atre et al. stored the relations in the bit matrix: one matrix is created for one predicate [3,4]. They demonstrated the queries with *join* operations. Madduri and Wu presented BitMat compression [17]. BitMat dictionaries are created first. Then, the indices are created for each subject, predicate, object and composite indices such as (predicate, object) are created. Kim et al. considered the binary Header-Dictionary-Triple (HDT) form and processed RDF query and *join* within the GPU. The BitMaps as well as IDs in HDT are directly loaded to the GPU memory. The prefix sum is applied to compute predicate and object positions in BitMaps [15]. Though HDT is a compressed format, converting to this form takes a lot of time and memory. A common computer desktop with 16 G RAM needs to convert about 9 million triples many hours. For larger numbers of triples over 9 million, HDT with Java language is required to increase Java heap memory for handling more elements in the set and it will take even more conversion time.

Again, format conversion takes a lot of time and memory. They also did not address the issue of speeding up the conversion.

One may come up with a file splitting solution to handle many searches at the same time. File splitting script takes a lot of time to run. Using MapReduce to process large files is also another possibility of conversion on a cluster. However, this is a choice of large batch processing, which we may consider in the future for handling external search.

## 3    Processing Framework

The challenges of this research are to process a big data set using the limited memory of the GPU and to simplify the representation properly for GPU computation. The design goals are the following: (1) The format should be simple so that the overhead of conversion time is not large. (2) It should not occupy too large space since GPU has limited memory. (3) Since GPU has a lot of threads for parallelizing search, we do not focus on index construction, rather we encourage the use of a large number of worker threads for searching. Our proposed framework to manage input mapping, look for query answers, and return the results is based on such a representation as presented in the following.

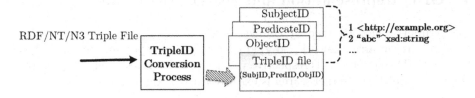

**Fig. 1.** First step of conversion to TripleID

Figure 1 presents the overall process. The RDF file (N3/N-Triple type) is transformed into ID files. Subject ID, Predicate ID, and Object ID files are separated. These files are in the format (keyID, value), where the key is an integer and the value is a string. Our TripleID file now contains only triples in the triple form (SubjID, PredID, ObjID). The TripleID file is a binary file assuming that each ID is unsigned (32-bit). When loading these ID files, *zlib* [29] can be used to encode the name to save space in hash memory.

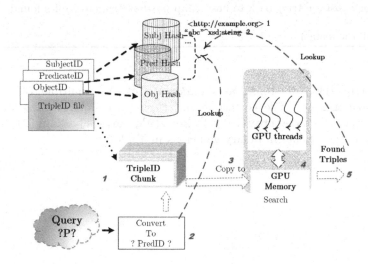

**Fig. 2.** Overall process

Figure 2 shows the process of querying by using the GPU. First, all the ID files (Subject ID, Predicate ID, and Object ID files) are loaded to hash tables and then loaded into the host memory (1). We use the hash memory implementation to look up the ID in constant time. We load the (key, value) pair where key is an ID and value is a string. The given query is transformed into triple form (? P ?) (2). For example, we look for the two equivalent classes, where P=owl:sameAs. The string owl:sameAs is lookuped against Pred hash to obtain the relevant ID. To perform the search for the query, the TripleIDs are copied to the GPU memory (3) and the query ID is searched by GPU threads (4). The found triples are returned from GPU to the host for reverse conversion (5).

## 4   GPU Representation and Search Algorithm

The TripleIDs files are read by chunks in Algorithm 1. Assume that the keys to search are in array key[0], key[1], key[2], corresponding to Subject ID, Predicate ID, and Object ID respectively, where value 0 is reserved to represent ?. For each thread, the kernel search is executed using Algorithm 2.

---

**Algorithm 1.** RDF Parallel Search for TripleID

---
   **Input:** *dataArray, keyArray*
   **Output:** *positionArray*
1 Allocate device memory for *dataArray, keyArray, positionArray*.
2 **while** *not EOF* **do**
3     Read the RDF data file chunk in *dataArray*.
4     Copy *dataArray, positionArray* (initialized to false) and copy *keyArray* to GPU memory
5     Call *GPUSearch* (Algorithm 2) with *dataArray, keyArray*, and *positionArray*
6     Copy *positionArray* back to host. Map *positionArray* to triples found.
7 **end**
8 Free all the memory.

---

The TripleIDs are stored as dataArray, of size multiple of three, in GPU global memory. The corresponding thread $i$ compares dataArray[i], dataArray[i+1], dataArray[i+2] to keys[0], keys[1], keys[2]. We represent the output values in binary values as in Table 2.

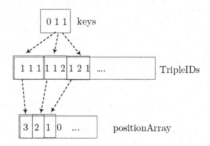

**Fig. 3.** Triple Mapping

Figure 3 presents the matching for each thread between keys array and dataArray. Only thread IDs that are multiple of 3 perform the matching in Line

---

**Algorithm 2.** GPU Search for TripleID (see description from Table 2)

---

**Input**: *dataArray, keys*
**Output**: *positionArray*
1  **for** *i = threadIdx*; *i < n*; *i+ = blockDim * gridDim* **do**
2      **if** *i % 3 == 0* **then**
3          posidx = (int) i/3;
4          **case** *keys[0]==data[i] and keys[1]==data[i+1] and keys[2]==data[i+2]*
5           |  positionArray[ posidx ] =7;
6          **case** *keys[1]==data[i+1] and keys[2]==data[i+2]*
7           |  positionArray[ posidx ] =3;
8          **case** *keys[0]==data[i] and keys[2]==data[i+2]*
9           |  positionArray[ posidx ] =5;
10         **case** *keys[0]==data[i] and keys[1]==data[i+1]*
11          |  positionArray[ posidx ] =6;
12         **case** *keys[0]==data[i]*
13          |  positionArray[ posidx ] =4;
14         **case** *keys[1]==data[i+1]*
15          |  positionArray[ posidx ] =2;
16         **case** *keys[2]==data[i+2]*
17          |  positionArray[ posidx ] =2;
18         **otherwise**
19          |  positionArray[ posidx ] =0;
20         **end**
21     **end**
22 **end**

---

2 in Algorithm 2. This arrangement may be modified so that every thread performs the same checking. However, this would require the data rearrangement or maybe the splitting of **dataArray**, enabling memory coalescing, which can cost more preprocessing overhead. The tradeoff between such preprocessing and the speedup due to memory coalescing needs to be further investigated. The output **positionArray** is shortened by 1/3 because we only mark the triples that match, in order to save the GPU memory usage. By the representation in Table 2, the checking can be done as nested conditions. For instance, the value 011 implies both matched predicates and objects which can imply only the matched predicate, or the matched object. As in Fig. 3, the first triple has both matches, the second triple has a predicate match and the third triple has an object match.

Currently our implementation requires GPU memory for **dataArray** of chunk size $N$ and the **positionArray** of size $N/3$. Each element of **dataArray** is **unsigned** and the element of **dataArray** is **unsigned char**. The **keyArray** contains three elements of **unsigned**.

Based on Algorithm 1, it is possible to extend memory space to work on multi-GPUs and a cluster of GPUs. When the host memory is available, in Line 3 of Algorithm 1 we can read each chunk and send each one to each GPU in Line 6. Then, the search is called for each GPU. CUDAaware [20] can be setup to combine **MPI_Send** and **cudaMemcpy** together in one command and chunks are sent to each node. Also, it is easy to extend our approach to query multiple **keys** in GPU kernels to handle many subqueries in one query simultaneously. For very large RDF data, we will attempt to compare to tools such as Virtuoso [25].

**Table 2.** Answer bits

| Binary value | Decimal | Meaning |
|---|---|---|
| 000 | 0 | All matches |
| 001 | 1 | object match (??O) |
| 010 | 2 | predicate match (?P?) |
| 100 | 4 | subject match (S??) |
| 011 | 3 | predicate and object match (?PO) |
| 110 | 6 | subject and predicate match (SP?) |
| 101 | 5 | subject and object match (S?O) |
| 111 | 7 | subject, predicate, object match (SPO) |

# 5   Experiments

In the experiments, we demonstrate how the framework can be applied to query the RDF triple. First, the conversion time is presented as well as the size of TripleID files. Then, the RDF search time is presented. We compare the query time with a traditional tool, in our case with Redland.

The tested machine has the following specification: Intel($R$) Core($TM$) $i7 - -5820K$ CPU @ 3.30 GHz, 6 cores, and 16 GB RAM. Two NVIDIA Tesla K40 cards are used. Each Tesla card contains 15 Multiprocessors, 192 CUDA cores per MP (totally 2,880 CUDA cores) with maximum clock rate 745 MHz (0.75 GHz). Memory bus width is 384-bit. Total amount of global memory is 12,288 MB. We perform the test using a thread block size of 1024, a grid size of 480, two streams are used. We perform the experiments on two data sets: Billion Triples Challenge and SP$^2$Bench.

## 5.1   Billion Triples Challenge Data Sets

The first data set is obtained from Billion Triples Challenge 2009. The major part of the dataset was crawled during February/March 2009 based on datasets provided by Falcon-S, Sindice, Swoogle, SWSE, and Watson using the MultiCrawler/SWSE framework. To ensure wide coverage, we included for this set also a (bounded) breadth-first crawl of depth 50 starting from http://www.w3.org/People/Berners-Lee/card.

The downloaded content was parsed using the Redland toolkit with rdfxml, rss-tag-soup, rdfa parsers. The blank nodes were re-written identifiers to include the data source in order to provide unique blank nodes for each data source, and appended the data source to the output file. The data is encoded in NQuads (three-column n-triples file with one graph-name column) format and split into chunks of 10 million ($10^7$) statements each. The size of the combined original dataset (gzipped) is around 17 GB.

A smaller crawled data sample for testing is available at btc-2009-small.nq.gz, when unzipping, the size is equal to 2.172 GB which is the largest in this set. From the file, btc-2009-small.nq with size 2,172 MB, it is converted into btc-2009-small.nt with size 1,611 MB and into RDF with size 1,556 MB.

We separated this data set into 7 parts, 01,02,..07, each of which has 350 MB. We generate the combinations of these files by grouping them to obtain different triple sizes. For example, (1) 01.nq with size 350 MB becomes 01.nt with size 270 MB, (2) 01–03.nq with size 1,050 MB becomes 01–03.nt with size 796 MB, (3) 012347.nq with size 1,472 MB becomes 012347.nt with size 1,103 MB. However, after the generating process, some part of btc-2009-small.rdf has invalid RDF syntax type from the original transformation, so we eliminate these parts (72 MB), leading to a size of 1,408 MB.

The conversion program (command-line tool), `rdf-convert-0.4` (http://sourceforge.net/projects/rdfconvert/), is used. It is based on the Open-RDF Rio parser toolkit, and currently supports RDF/XML, Trig, Trix, Turtle, N3, N-Triples, RDF/JSON, JSON-LD, Sesame Binary RDF and N-Quads. Hence, we selected N3, N-Triples, and RDF/XML for our work.

**Table 3.** Data set characteristics

| Data set | #subj | #pred | #obj | #triples |
|---|---|---|---|---|
| 01 | 314,285 | 3,458 | 583,555 | 1,868,651 |
| 0103 | 778,772 | 5,849 | 1,383,943 | 5,160,648 |
| 0203 | 504,082 | 4,477 | 990,414 | 3,291,997 |
| 0207 | 366,654 | 3,563 | 688,019 | 2,017,469 |
| 012347 | 1,113,824 | 7,542 | 1,674,407 | 7,083,790 |
| btc-2009-small | 1,383,542 | 8,205 | 2,260,819 | 9,627,877 |

Table 3 displays the characteristics of the tested data sets. The maximum one contains 9 million triples. Table 4 displays the processing time for each query. The simple query is tested: `SELECT distinct ?subject ?object WHERE ?subject owl:sameAs ?object`. When using Redland library, we reached the memory heap limit for allocation a graph model storage of size larger than 01 case (1.8 M triples), due to the growth of the hash table. In our machine, it is impossible to allocate new hash larger than 2G size due to the multi-user environment where a user process is not allowed to consumes too much resources. We attempt to make a modification that splits the model into smaller submodels and holds the pointers of smaller submodels. The model is split into chunks of 500 MB equally and submodel pointers are held as list iterators. The splitting is done during the reading and parsing of the input RDF file by Rasqal parser. However, the submodel is a list of iterators which is not thread-safe. When running the test using the Redland library, we found out that the time consuming portion is due to the time to load the RDF file and to construct the graph model (shown in Table 4). The query time of the Redland model lies between 1/2 or 1/3 of the model loading time.

Column "Hexastore" shows the load time and query time using Hexastore with MPI [28] (https://github.com/kasei/hexastore/). We assume using 4 nodes of MPI clusters of Intel(R) Xeon(R) CPU X3470 @ 2.93 GHz. Hexastore loading time is much longer than that of Redland while the query processing can obtain benefits from MPI.

**Table 4.** Comparison between Redland, Hexastore, and TripleID Query

| Data set | Redland (s) | | Hexastore(s) | | TripleID time (s) | | | | Speedup | |
| (triples) | load | query | load | query | convert | load | data | query | Redland | Hexastore |
| 01 | 14.89 | 6.29 | 111.87 | 0.59 | 3.25 | 0.52 | 0.12 | 0.13 | 32.58 | 173.02 |
| 0103 | 46.84 | 28.22 | 279.57 | 1.82 | 8.57 | 1.33 | 0.15 | 0.20 | 49.06 | 171.92 |
| 0203 | 31.66 | 21.31 | 166.84 | 1.16 | 6.15 | 0.92 | 0.15 | 0.23 | 46.06 | 146.09 |
| 0207 | 16.46 | 11.55 | 126.77 | 0.70 | 3.06 | 0.66 | 0.12 | 0.12 | 35.92 | 137.09 |
| 012347 | 68.64 | 36.98 | 414.34 | 1.90 | 10.38 | 1.95 | 0.12 | 0.18 | 49.59 | 174.56 |
| btc-2009-small | 83.77 | 35.39 | N/A | N/A | 28.86 | 2.40 | 0.06 | 0.09 | 47.86 | N/A |

From this observation, since each triple file is large, it is clear that the straight-forward program, which parses the RDF and constructs very simple models would save this construction overhead. We first convert the triple in NT format to TripleID files, as in Fig. 1. The conversion processing time is compared against the Redland model loading time. Our conversion process takes around 1/4 of the loading time (Column "convert"). Once the data is converted, the query is done by using the ID. In the above query which looks for owl:sameAs, it is in the form ?P?. owl:sameAs is converted into predicate ID to search using GPU search in Algorithms 1. Column "load" is the time to load the ID files of subjects, predicates, objects into three hash tables, for ID lookup. Column "data" is the time to transfer triple IDs in a chunk to GPU memory, assuming (250 M of unsigned for each chunk), and transfer the found positions back to host memory. Column "query" shows the overall time including the data transfer time and kernel search time. Note that the kernel search time is very small so we cannot report only this number. Using our framework utilises thousands of threads (with grid size = 480 and block size = 256) and a simple representation can gain much advantage for speedup.

Column "Speedup Redland" shows the speedup of TripleID over Redland (combining all loading time and search time). The significant speedup over the traditional approach is about 49 times. Compared to the speedup over Hexastore in overall, TripleID search using the GPU is 174 times faster. Note that we cannot perform the largest test for BTC-small for Hexastore, since it uses up the memory allowed in our cluster environment. Consider the comparison to RDFlib [23]. On the same machine, the process of 5 million-triple data (with N3 size 826,904,622 bytes) takes 778.22 s where we observe that the loading time is already 776.96 s and the query time is 1.25 s. Thus, we cannot perform the larger test using RDFlib since the process uses much more memory resource than allowed.

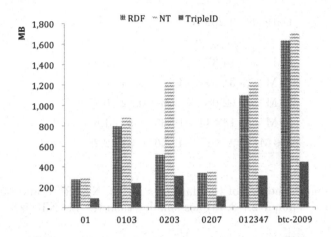

**Fig. 4.** Comparison of data size for test set I (BTC)

Figure 4 compares the file size after the conversion to TripleIDs. As in the previous section, after conversion, the four files are generated. The file size in Column "TripleID" is the summation of the triple file size plus the subject, predicate, object ID files' size. Our size compared to NT size is around 3–4 times smaller.

## 5.2  SP$^2$*Bench* Data Sets

In the second data set, SP$^2$*Bench* [22], we generate the data set with different numbers of triples. Table 3 shows the data set characteristics where the number of triples is up to 100 million. SP$^2$*Bench* generates the data sets in N3 format. The triples contain various numbers of subjects, predicates and objects as shown in Table 5.

**Table 5.** Data set characteristics (SP$^2$*Bench*)

| Data set (triples) | #subj | #pred | #obj | #triples |
|---|---|---|---|---|
| 5 M | 896,359 | 76 | 2,400,922 | 5,000,120 |
| 10 M | 1,712,642 | 77 | 4,662,411 | 10,000,091 |
| 20 M | 3,404,855 | 153 | 9,379,299 | 20,000,429 |
| 50 M | 8,639,994 | 306 | 24,058,862 | 50,000,100 |
| 100 M | 17,652,609 | 613 | 48,965,319 | 100,000,144 |

Table 6 shows the total query time for our TripleID form in Column "total time". The query is in the form "?PO", e.g., P is `rdf:type` and O is `foaf:Person`. The total time obtained for querying 100 million triples is 2.28 s, where the triple conversion time is 298.1 s.

**Table 6.** Execution time Using TripleID Query on SP$^2$Bench

| Data set | TripleID time (s) | | | Total time |
|----------|---------|-------|------|------------|
|          | convert | load  | copy |            |
| 5 M      | 14.37   | 1.86  | 0    | 0.25       |
| 10 M     | 31.04   | 4.1   | 0.04 | 0.65       |
| 20 M     | 62.08   | 8.66  | 0.12 | 0.77       |
| 50 M     | 148.44  | 19.65 | 0.2  | 1.66       |
| 100 M    | 298.1   | 42.56 | 0.31 | 2.28       |

Fig. 5 compares the size of the data sets in N3, NT, RDF and our TripleID forms. Our size is about 2.7-3.4 times smaller than the file NT sizes.

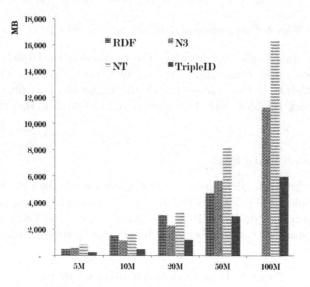

**Fig. 5.** Comparison for data size for SP$^2$Bench

# 6    Conclusion and Future Work

In this paper, we present a simplified framework based on TripleIDs to process RDF queries. The straightforward conversion from RDF triples to TripleIDs was developed. The separated ID files are generated as subject, predicate, object ID files. Then the triple file contains only rows of triple IDs. The whole storage required is much smaller than the total storage used by NT or N3 file. The TripleIDs are imported to GPU in chunks to be searched for particular subject, predicate, and/or object IDs. The found results are returned based on the given encoding. We minimise the memory usage required for GPU for storing results by only considering the mapping thread index divisible by three. Using the GPU search can speed up the query by 393 times over the traditional RDF query. The total ID file size is about 2–4 times smaller than NT files. The simple TripleID representation enables us to query up to 2–3 hundred millions of triples, which is more than traditional RDF tools can do, in a desktop PC within seconds while keeping low pre-processing overhead.

Our framework relies on the hash data structure where we construct three hash tables for subjects, predicates, and objects during the TripleID generation. This gives the upper limit of the total maximum subjects, predicates, and objects that the heap memory can allocate for. In our desktop machine, we reach this limit when generating $SP^2Bench$ triples larger than 300 million. Thus, to overcome this, we would like to convert the file of any triple sizes without relying on the available in-memory hash table. The generation process will make use of external disk and omit the use of hash. The GPU parallel external sort and merge will speed up the generation process.

Furthermore, we will consider to perform more operations such as union, intersection and perform matching multiple key arrays for the query with many simultaneous subqueries in GPU. Extension using multiple GPUs and a cluster of GPUs are also easily adopted.

**Acknowledgement.** This work was supported in part by the following institutes and research programs: The Thailand Research Fund (TRF) through the Royal Golden Jubilee Ph.D. Program under Grant PHD/0005/2554, DAAD (German Academic Exchange Service) Scholarship project id: 57084841, NVIDIA Hardware grant, and the Faculty of Engineering at Kasetsart University Research funding contract no. 57/12/MATE.

# References

1. RDF Vocabulary Description Language 1.0: RDF Schema. http://www.w3.org/TR/2004/REC-rdf-schema-20040210/#ch_type
2. ref.sh. (2015). https://github.com/seebi/rdf.sh. Retrieved Nov 2015
3. Atre, M., Chaoji, V., Zaki, M.J., Hendler, J.A.: Matrix "bit" loaded: A scalable lightweight join query processor for RDF data. In: Proceedings of the 19th International Conference on World Wide Web WWW 2010, pp. 41–50. ACM, New York (2010)
4. Atre, M., Hendler, J.A.: BitMat: A main memory bit-matrix of RDF triples. In: Proceedings of the 5th International Workshop on Scalable Semantic Web Knowledge Base Systems (2009)
5. Beckett, D.: The design and implementation of the Redland librdf RDF API Library. In: Proceedings of WWW10, Hong Kong, May 2001

6. Berners-Lee, T., Hendler, J., Lassila, O.: The Semantic Web: Scientific American (2001). citeulike-article-id:1176986 http://www.sciam.com/article.cfm?articleID=00048144-10D2-1C70-84A9809EC588EF21&pageNumber=1&catID=2
7. Bizer, C., Lehmann, J., Kobilarov, G., Auer, R., Becker, C., Cyganiak, R., Hellmann, S.: DBpedia - A crystallization point for the Web of Data. Web Semant. **7**(3), 154–165 (2009)
8. Choksuchat, C., Chantrapornchai, C.: Large RDF representation framework for GPUs case study key-value storage and binary triple pattern. In: International Computer Science and Engineering Conference (ICSEC), pp. 13–18, September 2013
9. Choksuchat, C., Chantrapornchai, C., Haidl, M., Gorlatch, S.: Accelerating keyword search for big RDF web data on many-core systems. In: Fujita, H., Guizzi, G. (eds.) SoMeT 2015. CCIS, vol. 532, pp. 190–202. Springer, Heidelberg (2015)
10. Grant, C.K., Lee, F., Torres, E.: SPARQL Protocol for RDF.W3c Recommendation (2008). http://www.w3.org/TR/rdf-sparql-protocol/
11. Groppe, J., Groppe, S.: Parallelizing join computations of SPARQL queries for large semantic web databases. In: Proceedings of the 2011 ACM Symposium on Applied Computing SAC 2011. pp. 1681–1686. ACM, New York (2011). http://doi.acm.org/10.1145/1982185.1982536
12. He, B., Yang, K., Fang, R., Lu, M., Govindaraju, N., Luo, Q., Sander, P.: Relational joins on graphics processors. In: Proceedings of the 2008 ACM SIGMOD International Conference on Management of Data SIGMOD 2008, pp. 511–524. ACM, New York (2008). http://doi.acm.org/10.1145/1376616.1376670
13. Heino, N., Pan, J.Z.: RDFs reasoning on massively parallel hardware. In: Cudré-Mauroux, P., Heflin, J., Sirin, E., Tudorache, T., Euzenat, J., Hauswirth, M., Parreira, J.X., Hendler, J., Schreiber, G., Bernstein, A., Blomqvist, E. (eds.) ISWC 2012, Part I. LNCS, vol. 7649, pp. 133–148. Springer, Heidelberg (2012)
14. Kim, J., Kim, S.G., Nam, B.: Parallel multi-dimensional range query processing with R-trees on GPU. J. Parallel Distrib. Comput. **73**(8), 1195–1207 (2013)
15. Kim, Y., Lee, Y., Lee, J.: An efficient approach to triple search and join of HDT processing using GPU. In: Proceedings of The Seventh International Conference on Advances in Databases, Knowledge, and Data Applications (DBKDA), pp. 70–74. IARIA (2015)
16. Liu, C., Urbani, J., Qi, G.: Efficient RDF stream reasoning with graphics processing units (GPUs). In: Proceedings of the Companion Publication of the 23rd International Conference on World Wide Web Companion Steering Committee, Republic and Canton of Geneva, Switzerland, pp. 343–344. WWW Companion 2014, International World Wide Web Conferences (2014)
17. Madduri, K., Wu, K.: Massive-scale RDF processing using compressed bitmap indexes. In: Bayard Cushing, J., French, J., Bowers, S. (eds.) SSDBM 2011. LNCS, vol. 6809, pp. 470–479. Springer, Heidelberg (2011). doi:10.1007/978-3-642-22351-8_30
18. Makni, B.: Optimizing RDF stores by coupling general-purpose graphics processing units and central processing units. In: Proceedings of ISWC (2013). http://ceur-ws.org/Vol-1045/paper-06.pdf
19. Nam, B., Sussman, A.: Analyzing design choices for distributed multidimensional indexing. J. Supercomputing **59**(3), 1552–1576 (2012). doi:10.1007/s11227-011-0567-7
20. NIVIDIA: An introduction to CUDA-Aware MPI. (2013). http://devblogs.nvidia.com/parallelforall/introduction-cuda-aware-mpi/. Retrieved July 2015

21. NVIDIA: NVIDIA GPU programming guide (2015). https://developer.nvidia.com/ nvidia-gpu-programming-guide. Retrieved July 2015
22. Schmidt, M., Hornung, T., Meier, M., Pinkel, C., Lausen, G.: SP2Bench: A SPARQL performance benchmark. In: de Virgilio, R., Giunchiglia, F., Tanca, L. (eds.) Semantic Web Information Management, pp. 371–393. Springer, Heidelberg (2010). doi:10.1007/978-3-642-04329-1_16
23. Teams, R.: rdflib 4.2.1. (2015). http://rdflib.readthedocs.org/. Retrieved November 2015
24. W3C.: Resource description framework (2004). http://www.w3.org/RDF/. Retrieved July 2015
25. W3C.: Virtuosouniversalserver (2009). http://www.w3.org/wiki/VirtuosoUniversal Server. Retrieved Dec 2015
26. Wei, Z., Jaja, J.: A fast algorithm for constructing inverted files on heterogeneous platforms. In: 2011 IEEE International Parallel Distributed Processing Symposium (IPDPS), pp. 1124–1134, May 2011
27. Wei, Z., JaJa, J.: A fast algorithm for constructing inverted files on heterogeneous platforms. J. Parallel Distrib. Comput. **72**(5), 728–738 (2012). doi:10.1016/j.jpdc. 2012.02.005
28. Weiss, C., Karras, P.J.D., Martínez-Prieto, M.A., Bernstein, A.: Hexastore: Sextuple indexing for semantic web data management. In: Proceedings of PVLDB, pp. 1008–1019. ACM (2008). http://www.vldb.org/pvldb/1/1453965.pdf
29. zlib.: zlib usage example (2012). http://www.zlib.net/. Retrieved Nov 2015

# Bioinformatics and Biomedical
# Data Analysis

# eQuant - A Server for Fast Protein Model Quality Assessment by Integrating High-Dimensional Data and Machine Learning

Sebastian Bittrich$^{(\boxtimes)}$, Florian Heinke, and Dirk Labudde

University of Applied Science Mittweida, 09648 Mittweida, Germany
bittrich@hs-mittweida.de

**Abstract.** In molecular biology, reliable protein structure models are essential in order to understand the functional role of proteins as well as diseases related to them. Structures are derived by complex and resource-demanding experiments, whereas *in silico* structure modeling and refinement approaches are established to cope with experimental limitations. Nevertheless, both experimental and computational methods are prone to errors. In consequence, small local regions or even the whole tertiary structure can be unreliable or erroneous, leading the researcher to formulate false hypotheses and draw false conclusions.

Here, we present eQuant, a novel and fast model quality assessment program (MQAP) and server. By utilizing a hybrid approach of established MQAPs in combination with machine learning techniques, eQuant achieves more homogeneous assessments with less uncertainty compared to other established MQAPs. For normal sized protein structures, computation requires less than ten seconds, making eQuant one of the fastest MQAPs available. The eQuant server is freely available at https://biosciences.hs-mittweida.de/equant/.

**Keywords:** Protein structure · Structure quality · Quality assessment · Quality scoring · eQuant

## 1 Introduction

Ever since the discovery of the cell made by Robert Hooke in 1665, humans try to understand the underlying microscopic and molecular processes facilitating various forms of life. In addition, deducing pathological changes at molecular level can provide the basis for treating severe and lethal diseases [15, 29, 46, 70, 74]. In the complex and dynamic molecular networks, proteins play a key role by regulating and supporting these processes. Mutations in proteins can thus lead to adverse effects and result in serious illnesses and disorders, such as various forms of cancer, Alzheimer's disease, cystic fibrosis, retinitis pigmentosa, hereditary deafness, familial hypercholesterolaemia, Lesch-Nyhan syndrome, autism, and diabetes insipidus [27, 33, 42, 79].

A crucial step toward understanding a protein's functional role is to determine its structural features [15, 24, 25, 64, 70], e.g. by deducing functional

© Springer International Publishing Switzerland 2016
S. Kozielski et al. (Eds.): BDAS 2016, CCIS 613, pp. 419–433, 2016.
DOI: 10.1007/978-3-319-34099-9_32

substructures [39,40]. In this context, experimental structure determination techniques have become of great value, of which X-ray diffraction [43,69] and nuclear magnetic resonance spectroscopy [77] are the most commonly used. Although sensitivity and capabilities of these techniques have increased, experimental structure determination remains challenging and resource demanding. Due to these limitations, those methods cannot meet the demand present in current life sciences, resulting in an enormous discrepancy between the number of available protein sequences (53,000,000, see UniProtKB/TrEMBL statistics [71]) and only 37,000 distinct experimentally solved structures [61]. In addition, obtained information often cover only a narrow excerpt of reality. More precisely, experimentally derived protein structures represent only rigid snap-shots of the conformational space. Naturally, proteins are flexible bodies that can undergo long-range spatiotemporal rearrangements induced by interactions with their environment—an aspect that is difficult to capture experimentally [5,6]. A wide range of tools exist, which aim at providing additional information addressing these issues. On the one hand, molecular dynamics [17] or normal mode analysis [5,28] can give insights into the nature of a protein's flexibility. On the other hand, automated protein modeling pipelines, such as SWISS-MODEL [14,30], MODELLER [63,73], or I-TASSER [2], provide convenient and relatively fast strategies to obtain structural insight into proteins with yet to be determined structures.

However, deciding whether an experimentally determined or *in silico* modeled structure resembles the functional native structure of a protein still remains a major issue. Non-native structures are potentially unreliable on a global level or contain regions exhibiting a non-native conformation of amino acids. Studies based on such models would likely result in false conclusions; thus, protein structure quality determines the limitations for investigations. E.g. in order to study reaction mechanisms of enzymes as well as protein-ligand interactions—the bases of knowledge-based drug design—structures of high quality are necessary. Thereby, the position of individual atoms which constitute the protein could be determined or predicted with high precision and also the overall fold of the protein is correctly captured, meaning it resembles the native conformation. When the uncertainty regarding the placement of individual atoms is rather high, only coarse guesses are possible e.g. on the proteins function or the nature of macromolecular ensembles [65]. Thus, there is a significant demand for model quality assessment programs (MQAPs) estimating the reliability of structures—both experimental and theoretical models—in an unbiased, easy interpretable, fast, and meaningful manner.

In this paper, we introduce eQuant, a novel MQAP and server for fast and reliable protein model quality assessment. We give a brief overview over established MQAPs, upon which the theoretical background of eQuant is introduced. We further demonstrate comparative results of eQuant and MQAPs considered in our studies. Finally, we report preliminary results from the CAMEO project (Continuous Automated Model Evaluation, [31]). As an independent initiative introduced by the Protein Structure Initiative Knowledgebase, CAMEO aims at benchmarking MQAPs independently and, thus, provides a valuable resource for objective performance evaluation.

## 2    Model Quality Assessment Programs

Two major motivations exist to develop MQAPs: some aim at helping crystallographers approaching the problem from the experimental point of view, while others originated from theoretical methods developed to predict protein structures *in silico*. The latter mostly follow the paradigm that a native structure exhibits a nearly minimal total free energy. Interpretation of these contact energies is intuitive, since interactions within the protein also directly govern initial folding [3]. On residue level high potentials indicate unfavorable states; which are observed in hydrophobic groups of residues being exposed to the solvent [10,34,66]. Even though calculated energies are only approximations, their usefulness is proven [38,66,67]. Knowledge-based mean force potentials (also commonly referred to as pseudo potentials) play a major role in the context of threading and fold recognition [38,66], homology modeling [67], molecular docking [72], and last but not least protein structure evaluation [22]. Structure prediction pipelines, such as SWISS-MODEL [4,14], produce structure models and iteratively perform refinement cycles in order to improve model quality. Thereby, various MQAPs such as QMEAN [7–10], ANOLEA [52–54], and GROMOS [56] are consulted. Their predications indicate whether structural modifications made in refinement iterations have resulted to a desirable change in overall model quality. This implies that a superior MQAP can enhance current modeling strategies; not by directly performing predictions but by serving as meta-predictor selecting native confirmations. Nevertheless, there is significant variability in the exact implementation of each MQAP regardless of the general motivation.

Generally speaking, MQAPs define residue features or descriptors which are suitable to quantify structure quality. In our study, feature values specific for investigated MQAPs were computed for high-quality structures and implications with respect to per-residue modeling error were evaluated. Eventually, features and strategies for capturing the characteristics of reliable structures were identified and incorporated in the eQuant assessment pipeline. In this section, a brief summary on the theoretical background of investigated MQAPs is given, followed by introducing the basic concepts considered by eQuant. Finally, the eQuant web server is presented.

### 2.1    Overview on Considered MQAPs

**Verify3D** [16,49] analyzes residue surface area, presence of water and polar groups located near the side chain and local secondary structure propensity, whereas predefined environment classes are determined for each residue in the structure. Eventually, unfavored packings and environments are detected. In contrast, **PROCHECK**, one of the most widely applied MQAPs, operates at atomic level, assessing bond lengths and angles as well as atom contacts and non-bonded interactions [47, 48]. Given reference statistics of bond lengths, angles and non-bonded interactions [23], unfavored side chain conformations are identified. In addition, unfavored dihedral side chain angles are reported and identified by the classic Ramachandran projection [59]. The combination of both methods has been introduced as **VADAR** (volume, area, dihedral angle reporter) [76]. Further, hydrogen bond and solvation energies as well as analyzes of cavities are considered here. **PROSESS** (protein structure evaluation suite & server) uses over 100 features and criteria, some directly derived from Verify3D, PROCHECK, and VADAR [11]. It additionally allows to incorporate raw experimental data, whereas consistency with the structure model is tested. In contrast to previously mentioned MQAPs, **ProSA** (protein structure analysis) follows a completely different paradigm [19, 67, 68, 75]. Here, a minimalistic structure model is obtained from backbone and side chain atoms from which the local geometrical arrangement of residues is evaluated by means of pseudo-potentials. As an enhancement of ProSA, **ANOLEA** (atomic non-local environment assessment) considers each heavy atom in the structure individually. The focus in analyzes lies on non-local residue-residue interactions (sequence separation between residues is greater than eleven positions). Again, pseudo-potentials are used in assessment [52–54]. Finally, **QMEAN** (qualitative model energy analysis) is essentially constituted by a combination of potentials, agreement terms, relative accessible surface area, and the ratio of random coil-forming residues in the local sequence region around each residue [7–10]. By employing reference statistics as well as agreement terms of observed to predicted secondary structure and solvent accessibility, QMEAN reports an intuitively interpretable per-residue score. This score corresponds to the predicted spatial deviation of the observed residue position to the actual location in the (unknown) native structure. Thus, low scores (small deviations) are favored. Table 1 gives an overview of the features used by these MQAPs (including eQuant) in assessing quality.

## 2.2  eQuant—Energy-Based Quality Assessment

An analysis of investigated MQAPs introduced above was conducted to extract successful strategies which are of high importance to predict protein structure quality. In an iterative process of non-linear regression model generation and refinement, per-feature impact on model power was tested in combination with the random subspace method. Yielded correlation coefficients derived from the model were evaluated and optimized with respect to feature selection. Thus, a subset of discriminative features was determined and implemented, which ultimately constitutes eQuant's quality assessment routine whereas the trained random subspace model [18, 41, 57] is utilized in prediction (see Fig. 1 for a schematic overview).

eQuant employs knowledge-based pair potentials utilized in protein energy profiling as discussed in [20, 33, 34]. Protein energy profiling follows the assumption that each amino acid has specific packing preferences. Polar residues achieve energetically favorable states when they interact with the solvent. In contrast non-polar amino acids are more often found embedded in the hydrophobic protein core. The residue solvation energy is obtained by considering residue packing within an 8 Å[1] sphere surrounding the residue in question. Also, a contact energy term is employed, which characterizes the preferences of amino acids being in close proximity. Another quantity motivated by the energy profile approach are long range interactions with an interaction distance below 10 Å and a sequence separation greater than eight residues. Furthermore, it is not only possible to calculate per-residue solvation energies for a given protein structure, but also to predict them solely based on the protein's sequence [34]. The agreement between actually observed and predicted per-residue energy is considered as an additional feature in the quality assessment process. Disagreement implies problems in the spatial arrangement of the residues [10]. It was shown that QMEAN performed exceptionally well and especially the relative accessible surface area as well as the fraction loop are strongly correlated to per-residue error (spatial residue displacement). In consequence, a set of residue descriptors was determined, which fairly well represent single residue characteristics. The Weka framework [26,32,37] was utilized to train a local and global quality assessment model. All descriptors were combined by the random subspace method to derive a composite score for each residue as well as the entire structure. Similar to QMEAN, the obtained local per-residue score aims at approximating error for each residue as a prediction of spatial deviation.

To assess the global quality of a structure, the GDT (global distance test) value between the processed structure and the expected native structure is predicted. The GDT is a measure for global structure similarity [62,78]. Since the raw GDT score tends to be correlated to the protein's sequence length, a Z-score is computed by taking into account eQuant GDT scores of pre-processed structures with a length deviation of $\pm 10\,\%$. For GDT and Z-score, high values indicate structures evaluated as of good quality. An implementation of eQuant is freely available as a Web server at https://biosciences.hs-mittweida.de/equant/. Additionally, in contrast to other MQAPs, eQuant does not require additional parameterization prior to computation, which results in explicit assessment reports and avoids parameter-dependent ambiguity.

### 2.3   The eQuant Web Server

The structure quality assessment routine as well as major parts of the back-end are written in Java. Integral parts of the implementation are realized by means of BioJava [36,58], a robust Java framework widely used in bioinformatics. Since it provides robust file parsing capabilities and object-oriented handling of structure data, BioJava forms the basis for assessment routine implementations, modules performing feature computations, data-to-structure mappings, as well as data in- and output.

The front end is written in HTML5, CSS3 and jQuery respectively JavaScript. All information is propagated to the front-end by a JSON-like

---

[1] One Ångström corresponds to a length of 0.1 nm.

annotation and actually visualized by the browser. Highcharts [35] is a JavaScript library that is utilized to illustrate box plots and line charts as SVG images. The three-dimensional structure is rendered by PV (Protein Viewer [13]) and allows the user to interactively explore the structure. PV is written in JavaScript and utilizes WebGL [50]. Highcharts data is embedded in the HTML page, while PV fetches referenced structure data from the server via AJAX. A screenshot of the Web server's front-end is shown in Fig. 2.

**Table 1.** Features used by investigated MQAPs.

| feature | Verify3D | PROCHECK | VADAR | PROSESS | ProSA | ANOLEA | QMEAN | eQuant |
|---|---|---|---|---|---|---|---|---|
| bond lengths & angles | | ✓ | ✓ | ✓ | | | | |
| torsion angles | | ✓ | ✓ | ✓ | | | ✓ | |
| non-bound atom contacts | | ✓ | ✓ | ✓ | | | | |
| polar groups in environment | ✓ | | ✓ | ✓ | | | | |
| hydrogen-bond energy | | | ✓ | ✓ | | | | |
| solvation energy | | | ✓ | ✓ | ✓ | | ✓ | ✓ |
| interaction energy | | | ✓ | ✓ | ✓ | ✓ | ✓ | ✓ |
| accessible surface area | ✓ | | ✓ | ✓ | | | ✓ | ✓ |
| secondary structure | ✓ | | ✓ | ✓ | | | ✓ | ✓ |

In principle, eQuant traverses each residue of the structure and investigates its spatial neighborhood and residue packings (A). Residue-solvent interaction and residue-residue contact energies are also considered by the utilized approach. In addition, long range interaction energy terms are included in making assessments. Additional residue descriptors utilized are accessible surface area (ASA) as well as loop fraction of unordered secondary structure in the sequential neighborhood of individual residues. Furthermore, per-residue solvation energies are predicted based on the protein's sequence; these values are eventually compared to structure-derived energy values. Agreement between both values indicates reliable protein regions, whereas disagreement occurs for unfavorable conformations. The integrated features have proven to be correlated to the local as well as global protein structure quality. Final assessment is obtained by the random subspace method which combines the multi-dimensional vectors to intuitively interpretable composite scores. Thus, per-residue local errors (see Fig. B on the left) are obtained. The line chart depicted here illustrates the error score of each residue over the protein's sequence. The box plot on the right-hand side shows the overall structure quality expressed by GTD scoring (indicated by the blue horizontal line).

eQuant's quality assessment process is based on predicting the degree of similarity of the model to the optimal—though unknown—native conformation. More precisely, for each residue the assessment routine aims at predicting the

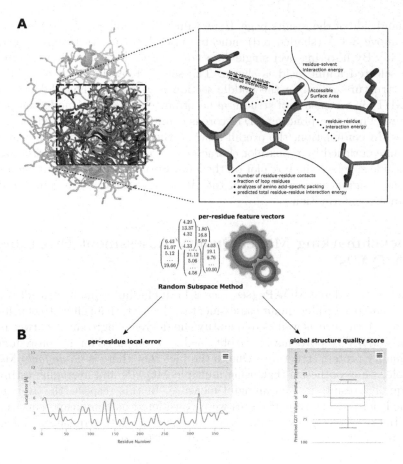

**Fig. 1.** Workflow of eQuant

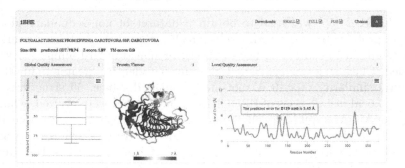

**Fig. 2.** User interface of the eQuant web server (Color figure online).

structural deviation between its location in the model and the unknown location in the native structure, and thus leading to an assessment of local modeling error. Local per-residue modeling errors are presented as a line chart on the right hand side of the results page. Here, values close to 0 are favorable, whereas scores above 3.8 Å (shaded red) indicate major local per-residue discrepancies [12,44,45]. By hovering over single data points, detailed evaluation information of the selected amino acid is shown. The selected residue is also highlighted in the structure viewer in the middle section of the page. The global distance test (GDT) is a measure of structural accordance of between model and native structure. The corresponding GDT score is computed from predicted local error features. In conjunction, the overall score of the processed model indicated by a blue line enriched by a box plot characterizing the population of structures of comparable size (see the box plot on the left). Furthermore, a Z-score indicating GDT score significance is given. For the global assessment, high values indicate a favorable conformation.

## 3    Benchmarking Model Quality Assessment Programs (MQAPs)

In general, considered MQAPs (see Sect. 2.1), including eQuant, provide a measure of local, per-residue quality assessment to the user. In an effort to benchmark MQAP performance as well as to quantify the degree of agreement between used methods, MQAP reports have been obtained and compared for 40 non-redundant high-quality protein structures. Due to the fact that there is no *de facto* MQAP available, we hypothesize that using multiple MQAPs and averaging estimated local quality results in a meaningful measure. In this respect, the averages of obtained per-residue MQAP scores were considered as consensus scores in our study. In the following text, the strategy for structure data acquisition and agreement quantification of per-residue scores are discussed.

### 3.1    Benchmarking Local Quality Assessment Scores

PDB-REPRDB [55] was used to obtain a dataset of non-redundant globular (soluble) protein structures with resolutions less than 1 Å. Proteins missing side chain coordinates, containing nucleic acids, or non-standard amino acids were excluded. Clustering was conducted at sequence identity greater than 30 % or structural similarity of <10 Å root-mean-square distance. 56 structures were selected. Local installations of PROCHECK and ProSA were utilized, while the remaining MQAPs automatically received tasks by querying their web interface. Output files were collected, parsed and their results arranged to match observed index offsets, especially occurring for terminal residues. However, 16 structures could not be processed by at least one tool, mostly if multiple chains are present in the structure, resulting in 40 structures that were finally used for benchmarking.

Assessment reports between considered MQAPs were analyzed by means of Spearman's rank correlation coefficients. For each of the 40 proteins, the pairwise correlation coefficients of all MQAP scores and consensus scores were calculated.

This data was finally used to compute the average correlation coefficients as well as standard deviations for each pair of MQAPs.

Analysis for the local evaluation scores (see Table 2) resulted to surprisingly small correlation coefficients between most regarded variables. Even though all MQAPs aim at quantifying the local and global quality of a structure, and some follow quite similar paradigms, obtained reports are rather inconclusive. As indicated, QMEAN, eQuant, and ANOLEA show the strongest, yet overall moderate correlations to the consensus scores. Even though most variables do not correlate well, the proposed consensus shares some agreement with all contributing MQAPs. Especially per-residue scores obtained by ProSA are uncorrelated to Verify3D scores or scores deduced from any MQAP focused on stereochemical features. The same holds true for all methods premised on knowledge-based potentials. QMEAN and ProSA feature the most tangible correlation. Towards the consensus however, correlation observed for ProSA is of the same magnitude as all other programs. Interestingly, the standard deviations are of the same magnitude. Assessment performance of eQuant is observed to be similar to QMEAN in average with higher standard deviation. eQuant thus could serve as an additional assessment routine for supporting and cross-evaluate reports obtained from QMEAN.

## 3.2 Automated Benchmarking by CAMEO

The CAMEO project [31] provides an independent service for benchmarking MQAPs. It further allows the researchers to register their assessment web server *via* a defined web interface which is used by the CAMEO benchmarking pipeline for automated access to the server. By that CAMEO automatically distributes query structures to all registered services, and obtained reports are evaluated

**Table 2.** Spearman's rank correlation coefficients between local quality scores obtained from 40 benchmark structures. In the upper half (highlighted in blue for visual clarity) the average of 40 correlation coefficients are listed, while the lower half shows standard deviations of the correlation coefficients.

| | Verify3D | PROCHECK | VADAR | PROSESS | ProSA | ANOLEA | QMEAN | eQuant | Consensus |
|---|---|---|---|---|---|---|---|---|---|
| Verify3D | - | 0.01 | 0.26 | -0.03 | 0.02 | -0.07 | -0.18 | -0.20 | -0.38 |
| PROCHECK | 0.12 | - | -0.14 | 0.28 | 0.00 | 0.03 | 0.07 | 0.09 | 0.40 |
| VADAR | 0.26 | 0.11 | - | -0.16 | -0.09 | -0.01 | -0.10 | -0.07 | -0.40 |
| PROSESS | 0.11 | 0.12 | 0.12 | - | 0.00 | 0.12 | 0.15 | 0.16 | 0.45 |
| ProSA | 0.12 | 0.12 | 0.12 | 0.14 | - | 0.04 | 0.17 | 0.08 | 0.35 |
| ANOLEA | 0.20 | 0.12 | 0.18 | 0.15 | 0.12 | - | 0.56 | 0.49 | 0.52 |
| QMEAN | 0.22 | 0.14 | 0.18 | 0.15 | 0.14 | 0.16 | - | 0.56 | 0.66 |
| eQuant | 0.23 | 0.13 | 0.24 | 0.17 | 0.15 | 0.23 | 0.25 | - | 0.62 |
| Consensus | 0.22 | 0.14 | 0.15 | 0.13 | 0.13 | 0.16 | 0.13 | 0.20 | - |

independently. Researchers can track the performance data during the entire process and can review their MQAP's performance with respect to other MQAP services attending the CAMEO benchmark—although names of MQAPs and researchers remain anonymous, unless respective MQAPs have already been published. CAMEO thus provides a valuable resource for testing work-in-progress MQAPs. During the last three-month benchmark between August and October 2015, eQuant processed about 1,550 query structures submitted by CAMEO.

As shown in Table 3, eQuant outperforms Verify3d [21] and Dfire v1.1 [4], but is observed to provide less accurate assessments in comparison to ModFOLD4 [51] and ProQ2 [60]. However, eQuant is observed to be significantly faster by a factor of 60–70 in comparison to ModFOLD4 and ProQ2. This discrepancy can be attributed to the extensive additional modeling pipeline utilized by the latter.

**Table 3.** Results of the three-month assessment benchmark conducted by CAMEO [31]. As stated, eQuant processed all queries successfully. AUC (area under curve) are obtained from query lists ranked according to assessment quality. Ranked lists are then compared to actual structure quality and specificity-sensitivity ROC plots are computed, from which AUC values are eventually determined. $AUC^*_{0,0.2}$ indicates a weighted AUC obtained by trimming the corresponding curve at the baseline true-positive rate of 0.2 achieved by PSI-BLAST [1].

| MQAP | Processed queries [%] | $AUC_{0,1}$ | $AUC^*_{0,0.2}$ |
|---|---|---|---|
| ModFOLD4 | 99.7 | 0.86 | 0.58 |
| ProQ2 | 93.1 | 0.78 | 0.51 |
| eQuant | 100.0 | 0.70 | 0.32 |
| Verify3d smoothed | 94.8 | 0.67 | 0.29 |
| Dfire v1.1 | 96.3 | 0.63 | 0.23 |
| Naive PSI-BLAST | 81.3 | 0.42 | 0.00 |

## 4    Conclusion

The knowledge of protein structures and theoretical models is crucial in modern molecular biology. However, structures vary in quality respectively reliability. Especially *in silico* generated structure predictions are prone to errors. Thus, there is a significant need of methods which objectively quantify the flaws and errors hidden in protein structures. Model quality assessment programs (MQAPs) are a means of addressing this task. Having reliable MQAPs at hand is also a stepping stone: beside being an essential tool in evaluating data necessary in structure based investigations, they can also aid in improving current homology modeling methodologies as well as support the development of urgently needed *ab initio* modeling techniques. Furthermore, structure quality assessment tools should allow users to actually utilize them without further necessary re-evaluations.

In this work, we have shown that there is little to no common agreement between widely used MQAPs, and also introduce a strategy for formulating a consensus based on a wide range of different quality assessment methods. This combination is proposed as being able to capture structural issues, and yet this method leaves room for improvements which requires further investigations. Furthermore, eQuant is shown to be successfully applicable to approximate the proposed MQAP consensus reasonable well and thus could support reports retrieved from other MQAPs. Future investigations should be focused on explaining varying assessment performance in this respect. Gained insights could aid in improving the per-residue local error prediction technique, but also could provide the basis for deducing measures for assessment confidence.

Further improvements in the eQuant pipeline should be made concerning the performance of global assessment scores. As shown by the CAMEO benchmark, the current implementation of eQuant is inferior compared to exhaustive assessment techniques in this respect. In summary, every MQAP has its individual advantages and drawbacks—thus, users should consult as many tools as possible. Especially the combination of MQAPs based on different paradigms should be persuaded, whereas a computational protocol is strongly needed. Here, combining MQAPs should be done carefully, as they focus on varying aspects and point out divergent problems in a structure. Here, eQuant could provide an parameter-free and easy-to-use additional source of information.

**Acknowledgments.** The authors thank the Free State of Saxony and the Saxon Ministry of Science and the Fine Arts for funding.

**Conflict of Interest.** The authors declare that there are no conflicts of interest.

# References

1. Altschul, S.: Gapped BLAST and PSI-BLAST: a new generation of protein database search programs. Nucleic Acids Res. **25**(17), 3389–3402 (1997)
2. Ambrish, R., Kucukural, A., Zhang, Y.: I-TASSER: a unified platform for automated protein structure and function prediction. Nucleic Acids Res. **5**(4), 725–738 (2010)
3. Anfinsen, C.B.: Principles that govern the folding of protein chains. Science **181**(4096), 223–230 (1973)
4. Arnold, K., Bordoli, L., Kopp, J., Schwede, T.: The SWISS-MODEL workspace: a web-based environment for protein structure homology modelling. Bioinformatics **22**(2), 195–201 (2006)
5. Bahar, I., Rader, A.J.: Coarse-grained normal mode analysis in structural biology. Bioinformatics **15**(5), 586–592 (2005)
6. Bastolla, U.: Detecting selection on protein stability through statistical mechanical models of folding and evolution. Bioinformatics **4**(1), 291–314 (2014)
7. Benkert, P., Biasini, M., Schwede, T.: Toward the estimation of the absolute quality of individual protein structure models. Bioinformatics **27**(3), 343–350 (2011)
8. Benkert, P., Kunzli, M., Schwede, T.: QMEAN server for protein model quality estimation. Nucleic Acids Res. **37**(Web Server), W510–W514 (2009)

9. Benkert, P., Schwede, T., Tosatto, S.: QMEANclust: estimation of protein model quality by combining a composite scoring function with structural density information. Bioinformatics **9**(1), 35 (2009)
10. Benkert, P., Tosatto, S.E., Schomburg, D.: QMEAN: a comprehensive scoring function for model quality assessment. Bioinformatics **71**(1), 261–277 (2008)
11. Berjanskii, M., Liang, Y., Zhou, J., Tang, P., Stothard, P., Zhou, Y., Cruz, J., MacDonell, C., Lin, G., Lu, P., et al.: PROSESS: a protein structure evaluation suite and server. Nucleic Acids Res. **38**(Web Server), W633–W640 (2010)
12. Bhattacharya, A., Tejero, R., Montelione, G.T.: Evaluating protein structures determined by structural genomics consortia. Bioinformatics **66**(4), 778–795 (2006)
13. Biasini, M.: Pv-WebGL-based protein viewer (2014)
14. Biasini, M., Bienert, S., Waterhouse, A., Arnold, K., Studer, G., Schmidt, T., Kiefer, F., Cassarino, T.G., Bertoni, M., Bordoli, L., Schwede, T.: SWISS-MODEL: modelling protein tertiary and quaternary structure using evolutionary information. Nucleic Acids Res. **42**(W1), W252–W258 (2014)
15. Blundell, T., et al.: Structural biology and bioinformatics in drug design: opportunities and challenges for target identification and lead discovery. Bioinformatics **361**(1467), 413–423 (2006)
16. Bowie, J., Luthy, R., Eisenberg, D.: A method to identify protein sequences that fold into a known three-dimensional structure. Science **253**(5016), 164–170 (1991)
17. Bradley, P., Malmström, L., Qian, B., Schonbrun, J., Chivian, D., Kim, D., Meiler, J., Misura, K., Baker, D.: Free modeling with Rosetta in CASP6. Science **61**(S7), 128–134 (2005)
18. Bryll, R., Gutierrez-Osuna, R., Quek, F.: Attribute bagging: improving accuracy of classifier ensembles by using random feature subsets. Science **36**(6), 1291–1302 (2003)
19. Domingues, F., Lackner, P., Andreeva, A., Sippl, M.J.: Structure-based evaluation of sequence comparison and fold recognition alignment accuracy. Science **297**(4), 1003–1013 (2000)
20. Dressel, F., Marsico, A., Tuukkanen, A., Schroeder, M., Labudde, D.: Understanding of SMFS barriers by means of energy profiles. In: Proceedings of German Conference on Bioinformatics, pp. 90–99 (2007)
21. Eisenberg, D., Lüthy, R., Bowie, J.U.: Verify3D: assessment of protein models with three-dimensional profiles. Science **277**, 396–404 (1997)
22. Elofsson, A., Le Grand, S.M., Eisenberg, D.: Local moves: an efficient algorithm for simulation of protein folding. Science **23**(1), 73–82 (1995)
23. Engh, R.A., Huber, R.: Accurate bond and angle parameters for x-ray protein structure refinement. Science **47**(4), 392–400 (1991)
24. Fersht, A.: Structure and Mechanism in Protein Science: A Guide to Enzyme Catalysis and Protein Folding, 3rd edn. W H Freeman & Co, New York (1995)
25. Forster, M.J.: Molecular modelling in structural biology. Science **33**(4), 365–384 (2002)
26. Frank, E., Hall, M., Trigg, L., Holmes, G., Witten, I.H.: Data mining in bioinformatics using Weka. Bioinformatics **20**(15), 2479–2481 (2004)
27. Fujiwara, T.M., Bichet, D.G.: Molecular biology of hereditary diabetes insipidus. Bioinformatics **16**(10), 2836–2846 (2005)
28. Go, N., Noguti, T., Nishikawa, T.: Dynamics of a small globular protein in terms of low-frequency vibrational modes. Bioinformatics **80**(12), 3696–3700 (1983)
29. Grabowski, M., Chruszcz, M., Zimmerman, M.D., Kirillova, O., Minor, W.: Benefits of structural genomics for drug discovery research. Bioinformatics **9**(5), 459–474 (2009)

30. Guex, N., Peitsch, M.C., Schwede, T.: Automated comparative protein structure modeling with SWISS-MODEL and Swiss-PdbViewer: a historical perspective. Bioinformatics **30**(S1), S162–S173 (2009)
31. Haas, J., Roth, S., Arnold, K., Kiefer, F., Schmidt, T., Bordoli, L., Schwede, T.: The protein model portal – a comprehensive resource for protein structure and model information. Database **2013**, bat031 (2013)
32. Hall, M., Frank, E., Holmes, G., Pfahringer, B., Reutemann, P., Witten, I.H.: The Weka data mining software. Bioinformatics **11**(1), 10 (2009)
33. Heinke, F., Labudde, D.: Membrane protein stability analyses by means of protein energy profiles in case of nephrogenic diabetes insipidus. Bioinformatics **2012**, 1–11 (2012)
34. Heinke, F., Schildbach, S., Stockmann, D., Labudde, D.: eProS-a database and toolbox for investigating protein sequence-structure-function relationships through energy profiles. Bioinformatics **41**(D1), D320–D326 (2013)
35. A Highsoft Solutions: Highcharts JS (2012)
36. Holland, R.C.G., Down, T.A., Pocock, M., Prlic, A., Huen, D., James, K., Foisy, S., Drager, A., Yates, A., Heuer, M., et al.: BioJava: an open-source framework for bioinformatics. Bioinformatics **24**(18), 2096–2097 (2008)
37. Holmes, G., Donkin, A., Witten, I.: Weka: a machine learning workbench. In: Proceedings of ANZIIS 94 - Australian New Zealand Intelligent Information Systems Conference, pp. 357–361 (1994)
38. Jones, D.T., Taylort, W.R., Thornton, J.M.: A new approach to protein fold recognition. Nature **358**(6381), 86–89 (1992)
39. Kaiser, F., Eisold, A., Bittrich, S., Labudde, D.: Fit3D - a web application for highly accurate screening of spatial residue patterns in protein structure data. Bioinformatics **32**(5), 792–794 (2015)
40. Kaiser, F., Eisold, A., Labudde, D.: A novel algorithm for enhanced structural motif matching in proteins. Nature **22**(7), 698–713 (2015)
41. Ho, T.K.: The random subspace method for constructing decision forests. Nature **20**(8), 832–844 (1998)
42. Kang, J., Lemaire, H., Unterbeck, A., Salbaum, J.M., Masters, C.L., Grzeschik, K.H., Multhaup, G., Beyreuther, K., Müller-Hill, B.: The precursor of Alzheimer's disease amyloid A4 protein resembles a cell-surface receptor. Nature **325**(6106), 733–736 (1987)
43. Kendrew, J.C., Bodo, G., Dintzis, H.M., Parrish, R.G., Wyckoff, H., Phillips, D.C.: A three-dimensional model of the myoglobin molecule obtained by x-ray analysis. Nature **181**(4610), 662–666 (1958)
44. Kryshtafovych, A., Barbato, A., Fidelis, K., Monastyrskyy, B., Schwede, T., Tramontano, A.: Assessment of the assessment: evaluation of the model quality estimates in CASP10. Nature **82**, 112–126 (2014)
45. Kryshtafovych, A., Monastyrskyy, B., Fidelis, K.: CASP prediction center infrastructure and evaluation measures in CASP10 and CASP ROLL. Nature **82**, 7–13 (2014)
46. Kuntz, I.D.: Structure-based strategies for drug design and discovery. Science **257**(5073), 1078–1082 (1992)
47. Laskowski, R., Rullmann, J., MacArthur, M., Kaptein, R., Thornton, J.M.: AQUA and PROCHECK-NMR: programs for checking the quality of protein structures solved by NMR. J. Biomol. NMR **8**(4), 477–486 (1996)
48. Laskowski, R.A., MacArthur, M.W., Moss, D.S., Thornton, J.M.: PROCHECK: a program to check the stereochemical quality of protein structures. Science **26**(2), 283–291 (1993)

49. Lüthy, R., Bowie, J.U., Eisenberg, D.: Assessment of protein models with three-dimensional profiles. Nature **356**(6364), 83–85 (1992)
50. Marrin, C.: WebGL Specification. Khronos WebGL Working Group (2011)
51. McGuffin, L.J., Buenavista, M.T., Roche, D.B.: The ModFOLD4 server for the quality assessment of 3D protein models. Nature **41**(W1), W368–W372 (2013)
52. Melo, F., Devos, D., Depiereux, E., Feytmans, E.: ANOLEA: a WWW server to assess protein structures. Nature **5**, 187–190 (1997)
53. Melo, F., Feytmans, E.: Novel knowledge-based mean force potential at atomic level. Nature **267**(1), 207–222 (1997)
54. Melo, F., Feytmans, E.: Assessing protein structures with a non-local atomic interaction energy. Nature **277**(5), 1141–1152 (1998)
55. Noguchi, T.: PDB-REPRDB: a database of representative protein chains from the Protein Data Bank (PDB). Nature **29**(1), 219–220 (2001)
56. Oostenbrink, C., Villa, A., Mark, A.E., van Gunsteren, W.F.: A biomolecular force field based on the free enthalpy of hydration and solvation: the GROMOS force-field parameter sets 53A5 and 53A6. Nature **25**(13), 1656–1676 (2004)
57. Panov, P., Dzeroski, S.: Combining bagging and random subspaces to create better ensembles. In: Berthold, M., Shawe-Taylor, J., Lavrač, N. (eds.) IDA 2007. LNCS, vol. 4723, pp. 118–129. Springer, Heidelberg (2007)
58. Prlic, A., et al.: BioJava: an open-source framework for bioinformatics in 2012. Bioinformatics **28**(20), 2693–2695 (2012)
59. Ramachandran, G., Ramakrishnan, C., Sasisekharan, V.: Stereochemistry of polypeptide chain configurations. Bioinformatics **7**(1), 95–99 (1963)
60. Ray, A., Lindahl, E., Wallner, B.: Improved model quality assessment using ProQ2. BMC Bioinform. **13**(1), 224 (2012)
61. Rose, P.W., et al.: The RCSB Protein Data Bank: new resources for research and education. Nucleic Acids Res. **41**(Database issue), D475–D482 (2013)
62. Sadowski, M.I., Jones, D.T.: Benchmarking template selection and model quality assessment for high-resolution comparative modeling. Proteins: Struct. Funct. Bioinform. **69**(3), 476–485 (2007)
63. Sali, A., Blundell, T.L.: Comparative protein modelling by satisfaction of spatial restraints. BMC Bioinform. **234**(3), 779–815 (1993)
64. Schulz, G.E., Schirmer, R.H.: Principles of Protein Structure, 5th edn. Springer, New York (1984)
65. Schwede, T., et al.: Outcome of a workshop on applications of protein models in biomedical research. BMC Bioinform. **17**(2), 151–159 (2009)
66. Sippl, M.J.: Boltzmann's principle, knowledge-based mean fields and protein folding. An approach to the computational determination of protein structures. J. Comput.-Aided Mol. Des. **7**(4), 473–501 (1993)
67. Sippl, M.J.: Recognition of errors in three-dimensional structures of proteins. BMC Bioinform. **17**(4), 355–362 (1993)
68. Sippl, M.J.: Knowledge-based potentials for proteins. BMC Bioinform. **5**(2), 229–235 (1995)
69. Strandberg, B.: Chapter 1: building the ground for the first two protein structures: myoglobin and haemoglobin. J. Mol. Biol. **392**(1), 2–10 (2009)
70. Surade, S., Blundell, T.L.: Structural biology and drug discovery of difficult targets: the limits of ligandability. BMC Bioinform. **19**(1), 42–50 (2012)
71. The UniProt Consortium: Activities at the universal protein resource (UniProt). Nucleic Acids Res. **42**(Database issue), D191–D198 (2014)

72. Verkhivker, G., Appelt, K., Freer, S., Villafranca, J.: Empirical free energy cal-culations of ligand-protein crystallographic complexes. I. Knowledge-based ligand-protein interaction potentials applied to the prediction of human immunodeficiency virus 1 protease binding affinity. Protein Eng. Des. Sel. **8**(7), 677–691 (1995)
73. Webb, B., Sali, A.: Protein structure modeling with modeller. BMC Bioinform. **1137**, 1–15 (2014)
74. Whittle, P.J., Blundell, T.L.: Protein structure-based drug design. BMC Bioinform. **23**, 349–375 (1994)
75. Wiederstein, M., Sippl, M.J.: ProSA-web: interactive web service for the recog-nition of errors in three-dimensional structures of proteins. Nucleic Acids Res. **35**(Web Server), W407–W410 (2007)
76. Willard, L.: VADAR: a web server for quantitative evaluation of protein structure quality. BMC Bioinform. **31**(13), 3316–3319 (2003)
77. Wüthrich, K.: Protein structure determination in solution by nmr spectroscopy. BMC Bioinform. **265**(36), 22059–22062 (1990)
78. Zemla, A.: LGA: a method for finding 3D similarities in protein structures. BMC Bioinform. **31**(13), 3370–3374 (2003)
79. Zhao, N., Han, J.G., Shyu, C., Korkin, D.: Determining effects of non-synonymous SNPs on protein-protein interactions using supervised and semi-supervised learn-ing. PLoS Comput. Biol. **10**(5), e1003592 (2014)

# Evaluation of Descriptor Algorithms
# of Biological Sequences and Distance Measures
# for the Intelligent Cluster Index (ICIx)

Stefan Schildbach[1,2]($\boxtimes$), Florian Heinke[1], Wolfgang Benn[2], and Dirk Labudde[1]

[1] Department of Applied Computer Sciences and Biosciences,
University of Applied Sciences Mittweida,
Technikumplatz 17, 09648 Mittweida, Germany
stefan.schildbach@hs-mittweida.de

[2] Department of Computer Science, Chemnitz University of Technology,
Straße der Nationen 62, 09111 Chemnitz, Germany

**Abstract.** In hindsight of the previous decades, a rapid growth of data in all fields of life sciences is perceptible. Most notably is the general tendency of retaining well established techniques regarding specific biological requirements and common taxonomies for data classification. Therefore a change in perspective towards advanced technological concepts for persisting, organizing and analyzing these huge amounts of data is essential. The Intelligent Cluster Index (ICIx) is a modern technology capable of indexing multidimensional data through semantic criteria, qualified for this challenge. In this paper methodical approaches for indexing biological sequences with the ICIx are discussed and evaluated. This includes the examination of established methods concentrating on vector transformation as well as outlining the efficiency of different distance measures applied to these vectors. Based on our results, it becomes apparent that position conserving methods are superior to other approaches and that the applied distance measures heavily influence performance and quality.

**Keywords:** DNA sequences · Intelligent cluster index · ICIx · Vector transformation · Feature vector · Distance measure

## 1 Introduction

During the last three decades, bioinformatics evolved to become the major cornerstone in modern life sciences. Its field of research ranges from purely theoretical methodologies of biological data processing, visualization, annotation, and classification to persisting experimentally obtained data. However, with the ongoing exponential growth of biological data caused by the rise of high-throughput methods in proteomics and genomics, the demand for inventive solutions for persistence management and organizing is thriving. For example, during the last 20 years the number of nucleotide sequences at GenBank [5] has been growing exponentially by 5 % in average between every two-monthly update, with the

© Springer International Publishing Switzerland 2016
S. Kozielski et al. (Eds.): BDAS 2016, CCIS 613, pp. 434–448, 2016.
DOI: 10.1007/978-3-319-34099-9_33

total number of sequences exceeding the 200 million mark in 2016. This illustrates non-trivial problems for traditional technologies associated with data persistence, annotation and searching. In addition to the growing amount of data, current biological questions require databases to be flexible with respect to data integration and openness.

The Intelligent Cluster Index (ICIx) [12,13,19,21–23] provides solutions for these requirements. Compared to common metrical and technical methods, ICIx employs a set of semantic criteria, thus allowing efficient handling of high-dimensional data. More specifically, the system requires a transformation of arbitrary data objects into feature vectors utilized to generate a hierarchical clustering by an artificial neural network. Similarity groups discovered through this process are used to create an artificial taxonomy dividing the data space into a multidimensional, content-based ordering of objects. The ICIx employs this hierarchy to derive an index structure providing a fast data retrieval and similarity search of related objects.

In this paper, we demonstrate the capabilities of ICIx with respect to nucleic acid (DNA / RNA) sequence searching as a characteristic example of biological sequences, although all techniques can also be applied to protein sequences. Sequence similarity searching – the identification of identical or similar sequences in sequence databases – has wide applicability in bioinformatics. Here, gaining insights on functionally uncharacterized sequences can be achieved by identifying similarities to sequences with known function from which annotations can be retrieved and distinctive functional mechanisms can be deduced. Exemplary scenarios in application include genome sequencing and annotation, sequence clustering, molecular phylogenetics, PCR primer design, and microarray probe design (for a comprehensive overview see [30]). In the majority of these scenarios researchers have to deal with large sequence datasets and the resulting methodological bottleneck. Thus, fast heuristic sequence search programs, such as FASTA [24], BLAT [17], and BLAST (including its derivatives, [1,2]), are commonly employed over exhaustive search strategies. Although more sensitive in comparison, the latter rely on computing sequence similarity from exact global and local sequence alignments [27], which proves impractical in most cases due to their high time complexity [28,30].

In our study the combination of five sequence vector transformation algorithms and 37 distance measures was evaluated with respect to performance in sequence similarity searching. Therefore an elaborate comparison of sensitive strategies supporting nucleic acids sequences on ICIx is presented.

In the first section of this paper, a brief overview is given on the ICIx technology, followed by elucidations on employed sequence vector transformation algorithms. Furthermore, distance measures utilized in sequence search assessment, which showed promising performance, are outlined. We conclude this paper by discussing results, potential sequence search strategies for ICIx, and future work.

## 2    Sequence Similarity Searching in the Context of ICIx

In this study, sequence searching in the context of ICIx concentrates on indexing biological sequences with nucleic acid sequences as objects of interest.

Deoxyribonucleic acid (DNA) serves as essential information storage of genetic instructions, as a blueprint for protein synthesis and consequently the building pattern maintaining life on the lowest, nanoscopic level. Although DNA and protein molecules can form complex three-dimensional structures, both are intrinsic one-dimensional chains comprised by a set of smaller, linearly connected molecules. In nature, there are only four nucleotides that constitute every chain of DNA. Due to intrinsic linearity of these molecules, they are commonly represented as strings comprised of four single letters. Each letter thus symbolizes a different molecular entity with its own physicochemical properties. Due to molecular variations and selective evolutionary processes, mutations are introduced to DNA sequences and passed on to the successor generation – a biological principle that has given rise to a plethora of shapes and functions. In molecular biology deciphering (functional) sequence relations is of great interest in many applications. Investigations rely essentially and initially on detecting (sub)-sequence similarity, which however remains often challenging due to sequence divergence and diversity yielding biologic correspondences 'hidden' within sets of thousands or even millions of sequences. In this study it is investigated if vector transformations of DNA sequences in combination with vector distances measures could provide the basis for fast sequence retrieval by means of the Intelligent Cluster Index.

The Intelligent Cluster Index can be characterized as an enhanced index technology for existing database structures. It is located as an independent system between the user interface and the database itself and therefore does not greatly interfere with existing systems. In general, benefit can be obtained from the ICIx as soon as large amounts of data need to be processed and especially if multidimensional and complex queries are in the system's focus, e.g. range queries and similarity or distance queries across multiple attributes without a natural order. The index translates incoming complex requests into accurate targeted and particularly simpler queries for the associated database, respectively it greatly reduces the search space and therefore the executed database accesses. Internally the ICIx extracts features from the database using an encoder and generates a taxonomy-like hierarchical index tree through utilizing artificial intelligence. This index tree is employed for clustering the data space and allows fast data processing and/or harnessing feature similarities. A schematic of its internal modules is shown in Fig. 1. This paper focuses on the encoder of the ICIx technology, primarily on vector transformation algorithms applicable to this module and distance measures utilized by the index and the data access module.

The conjunction of this modern technology with well-established sequence search algorithms exhibits challenges as well as it poses requirements. There are many available algorithms for sequence comparison, but the most restrictive demand for an algorithm integration in the context of ICIx is the capability of transforming any sequence into a vector of consistent size respectively to extract a predefined set of features. Subsequently an appropriate measure of similarity or distance measure between two extracted feature vectors needs to be evaluated. To assess suitable algorithms and pertinent measures of distance a benchmark for cross-evaluation (cf. Fig. 2) was prepared. In this benchmark, sets

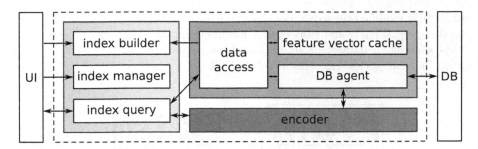

**Fig. 1.** Schematic representation of the ICIx modules surrounded by user interface (UI) and database (DB). Three interacting primary sections compose the ICIx core. The index unit administrates the tree structure and handles query processing and redirection. Transparent data access through database communications and caching is managed by the data unit. Feature extraction respectively vector transformation and vector preparation is handled by the encoder.

of similar sequences related to defined query sequences are ought to be identified in a large set of unrelated dissimilar sequences using various sequence vector transformation algorithms and distance measures. Assessment of sequence search performance was evaluated by means of the widely-applied BLAST [2] method. The benchmark does not incorporate the ICIx technology itself, it is designed to extract feature vectors and rank their usability with different distance measures to assess specific performance on similarity conservation. Theoretical and experimental aspects of the benchmark are elucidated in the following sections.

## 2.1   Data Acquisition

To evaluate the vector transformation algorithms independent and neutral datasets have been created. Therefore the Pfam (protein families) database [25] was used to extract the twenty largest families, respectively their seed sequences. These families are grouped by common functions and thus by functionally identical or similar sequences. The biological prototypes of corresponding protein families (seed sequences) were retrieved from the database. In extension, random sequences from the NCBI BLAST nucleotide collection *nt* database [7] were included to produce additional noise and to enlarge the search space. Consequently, the dataset consists of similar intra-family sequences combined with dissimilar inter-family sequences and random unrelated sequences.

To cope with different search scenarios, four representative sequences from the Pfam families were chosen and prepared to estimate four different difficulty categories in sequence search. The first and simplest scenario is the recovering of complete sequences (Complete), followed by a slightly more difficult search scenario, where marginally modified sequences were supposed to be retrieved (Complete$^{m}$). These modifications were realized by editing the representative sequences with removing, inserting and substituting operations of 5 % of all their letters. The last two scenarios contain the identification of complete (Sub) and slightly modified subsequences (Sub$^m$), where the representatives were sliced into three equal parts and respectively modified analog to the second scenario.

As a result, a dataset of 10,000 database sequences and 32 test sequences within four categories was created (cf. Fig. 2A).

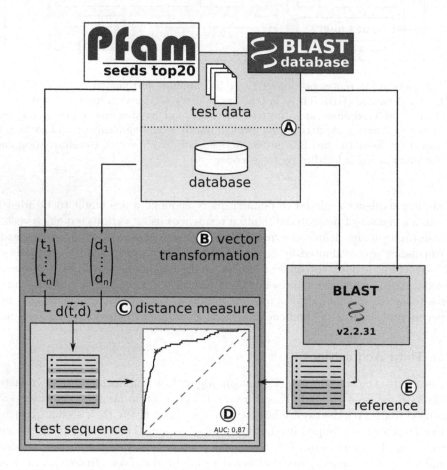

**Fig. 2.** The general workflow of the benchmarking process conducted in this study. The initial stage of the benchmark is the data acquisition (**A**) of test sequences and database sequences. Therefore, Pfam's [25] top twenty family seed sequences enriched by sequences from the NCBI BLAST nucleotide collection *nt* database [7] were retrieved. Afterwards, all sequences are transformed by the first vector transformation algorithm (**B**) and their distance is calculated by utilizing the first distance measure (**C**). A reference list is created once for every test sequence by BLAST (**E**) and consistently compared to the current distance matrix generated from (**B**) and (**C**). The consequent generated ROC curve and its AUC value are stored for further evaluation (**D**). The process continues iteratively for every test sequence, every distance measure and every vector transformation algorithm.

## 2.2   Vector Transformation Algorithms

The vector transformation algorithm collection (Fig. 2B) was realized according to the ICIx requirements (extraction of a defined feature vector) and the resulting algorithms are outlined in the following paragraphs. A summary of investigated algorithms in combination with the employed parameters and the resulting feature vector dimensions are given in Table 1. The vector dimension can be considered as an indicator of required memory and processing time for utilizing the appropriate approach.

**Table 1.** Vector transformation algorithms and their parameters utilized for the nucleic acid sequences. Abbreviations are used throughout the paper for conclusive identification of the algorithms. The specified parameter ranges characterize the used boundaries in this study, not the limits of the algorithms. A coarse measure of complexity for processing and storing the result vectors can be derived from the vector dimension.

| Algorithm | | Abbreviation | Parameters | Dimension |
|---|---|---|---|---|
| Nucleic Acids | $n$-gram word count | $n$-gram$^{wc}$ | $1 \leq n \leq 9$ | $4^n$ |
| | $n$-gram frequency | $n$-gram$^f$ | $1 \leq n \leq 9$ | $4^n$ |
| | Return time distribution | RTD | $1 \leq n \leq 9$ | $2 \cdot 4^n$ |
| | Category-Position-Frequency | CPF | $1 \leq n \leq 9$ | $3 \cdot 2^n$ |
| | Moments of inertia | Inertia | n/a | 6 |

**$n$-gram Word Counts and Frequencies**[1]. Biological sequences can be analyzed relatively similar to human language. They cover a really large data space and they can be structured into subunits. Therefore, many methods from language processing can be applied to the biological sequence domain [11, 26]. One widely used key feature in language technologies is the use of word $n$-grams and related $n$-gram frequencies [6]. In general, they describe short overlapping sequences of building blocks or fragments of constant length $n$ over a fixed alphabet. The probably most widely known application of $n$-grams in bioinformatics is its use within the BLAST [2] algorithm. This clearly shows, that $n$-grams – respectively their distribution and frequencies – can be utilized to characterize biological sequences. Beside straightforward $n$-gram counting (e.g. word existence, word count, word frequency), the length $n$ has a major influence on the incorporated information and is therefore a significant parameter in the transformation.

**Return Time Distribution (RTD).** The return time distribution of $n$-grams accounts for the sequence composition as well as their relative order [18]. It can be described as the number of symbols between successive appearances of $n$-grams. The frequency distribution of these so called return times for a particular $n$-gram is referred as its return time distribution (RTD). Statistical parameters, more specific the arithmetic mean and standard deviation, are calculated for each return time distribution, resulting in two features for every $n$-gram.

---

[1] Other commonly used notions for $n$-grams are $k$-, $t$- or $n$-tuples and $k$-, $t$- or $n$-mers.

**Category-Position-Frequency (CPF) Model.** The Category-Position-Frequency (CPF) model has been proposed by Bao and colleagues [4] and is also based on $n$-grams. It utilizes the word frequency, position and classification information of the underlying biological building blocks. The classification is based on physicochemical properties of the respective biological components. The algorithm converts the original sequence into three sequences based upon three distinct classifications, consequently reducing the alphabet size. These intermediate sequences are processed with a sliding window marking the occurrence of $n$-grams while creating a binary sequence. This occurrence sequence of every $n$-gram of every classification is converted into a single value by employing the Shannon Entropy. Thus a feature vector is computed by an entropy based model combining local word frequencies and position information.

**Principle Moments of Inertia on 3D Property Graphs.** A dynamic 3D representation of sequences based on their physicochemical properties representing the innate structure of the sequences has been introduced by Yao et al. [29]. This approach classifies the biological components based on their physicochemical properties into three categories. Each category is used as an axis in a three dimensional coordinate system, where the occurrence of a property is mapped as 1 and the absence is mapped as -1. Consequently, this projects a property cube into the three-dimensional space with biological components at its corners. The sequence transformation starts with creating a path by following the relative directions of this property cube. Subsequently the principal moments of inertia and the range of axis coordinate values are extracted from this path, resulting in six feature values. To avoid the influence of different sequence lengths the descriptor vectors are normalized and can therefore be seen as a quantitative measurement of similarity. This model was extended to harness its transformation for nucleic acid sequences utilizing the physicochemical properties of the nucleotides (purine/pyrimidine, amino/keto group, weak/strong hydrogen bond).

## 2.3   Distance Measures

In mathematical sense, distance functions have to satisfy strict conditions in order to be considered as 'true' measures of distance. Although often incorrectly termed as distance measures in the literature, some of these functions are rather measures of dissimilarity. Furthermore, some functions provide measures of similarity, whereas conversions can readily be computed. Equations (1) and (2) show two common conversions. Note that Eq. (1) is only valid if $s \in [0, 1]$. In this paper, although formally incorrect, all measures are termed as distance measures for textual clarity.

$$d = 1 - s \tag{1}$$
$$s = (1 + d)^{-1} \tag{2}$$

The authors of the considered vector transformation algorithms all suggest to use the Euclidean distance (15) as a sufficient distance measure. Nevertheless, there is a wide variety of additional distance measures available [3,8,10,16], which however are rarely discussed with respect to considered vector transformations. Based on the restrictive characteristic of a single measure for the algorithm evaluation, a set of currently 37 measures was acquired (Fig. 2C). An excerpt of these distance measures is listed in Table 2. Although some of these equations are sensitive in regards to arguments $\leq 0$, they can still be utilized by omitting the respective indices, the potential distortion of the result should be considered by choosing such distance measures.

## 2.4 Cross-Evaluation

A cross-evaluation approach was employed to score the acquired algorithms in combination with different distance measures (Fig. 2D). Every sequence in the dataset was transformed into its vector representation (Fig. 2B) and the distance of every test sequence to every database sequence was calculated (Fig. 2C). Subsequently the determined distance matrix was converted to an ordered list and processed in hindsight of the reference list. Therefore a hide-and-seek benchmark was utilized, where a subset of reference sequences ordered according to sequence similarity determined by BLAST [2] (with an e-value cutoff of 10, cf. Fig. 2E) is compared to the ordered distance list of each algorithm. By comparing the obtained lists, rank-based true positive rates (sensitivity) as well as false positive rates (1 - specificity) are determined for each transformation.

Based on these computed rate statistics, Receiver Operating Characteristic (ROC) curves [14] were computed to extract the area under the curve (AUC) values as an estimation of overall performance for every test sequence. The ROC curve is a common method to evaluate analysis strategies whereby true-positive rates are projected as a function over corresponding false-positive rates. AUC values are subsequently obtained from these projections. An AUC value close to 0.5 corresponds to the baseline performance achieved by random guessing. A value of 1.0 indicates exact correspondence to the reference list, where in contrast an AUC of 0.0 indicates inverse correspondence. The latter can be a result of applying similarity measures without appropriate conversions as exemplarily shown in Eqs. (1) and (2).

Finally, the AUC values were assigned to their respective categories to acquire a more detailed result while a category-spanning result for overall performance was considered additionally in the analysis. The median AUC value of each category was selected as a representative for the respective category-algorithm-measure combination.

**Table 2.** Excerpt of the utilized distance measures for the cross-evaluation. The first column classifies the presented measures into families as suggested by Cha [8], unclassified measures are grouped as *Other\**. The $i$th elements of the input vectors are denoted as $x_i$ and $y_i$, where $\bar{x}$ and $\bar{y}$ represent the corresponding arithmetic mean.

| Distance Measure | | Equation | |
|---|---|---|---|
| Absolute difference ($L_1$) | Canberra [9] | $\sum \left( \dfrac{\lvert x_i - y_i \rvert}{\lvert x_i \rvert + \lvert y_i \rvert} \right)$ | (3) |
| | Lorentzian [8] | $\sum \log\left(1 + \lvert x_i - y_i \rvert\right)$ | (4) |
| | Wave Hedges [8] | $\sum \left( 1 - \dfrac{\min(x_i, y_i)}{\max(x_i, y_i)} \right)$ | (5) |
| | Hassanat [15] | $\sum \left( 1 - \dfrac{1 + \min(x_i, y_i)}{1 + \max(x_i, y_i)} \right)$ | (6) |
| Fidelity | Bhattacharyya [3] | $-\log\left(\sum \sqrt{x_i \cdot y_i}\right)$ | (7) |
| | Matusita [8] | $\sqrt{\sum \left(\sqrt{x_i} - \sqrt{y_i}\right)^2}$ | (8) |
| Inner Product | Carbo [9] | $\dfrac{\sum (x_i \cdot y_i)}{\sqrt{\sum x_i^2 + \sum y_i^2}}$ | (9) |
| | Cosine [8] | $1 - \dfrac{\sum (x_i \cdot y_i)}{\sqrt{\sum x_i^2} \cdot \sqrt{\sum y_i^2}}$ | (10) |
| | Dice [3] | $2 \cdot \dfrac{\sum (x_i - y_i)^2}{\sum x_i^2 + \sum y_i^2}$ | (11) |
| | Inner Product [8] | $\sum (x_i \cdot y_i)$ | (12) |
| Intersection | Intersection [8] | $\sum \min(x_i, y_i)$ | (13) |
| | Ruzicka [8] | $\dfrac{\sum \min(x_i, y_i)}{\sum \max(x_i, y_i)}$ | (14) |
| Minkowski | Euclidean [8] | $\sqrt{\sum (x_i - y_i)^2}$ | (15) |
| | Manhattan [8] | $\sum \lvert x_i - y_i \rvert$ | (16) |
| Other\* | Exponential [10] | $\sum \dfrac{\lvert x_i - y_i \rvert}{1 + e^{-\lvert x_i - y_i \rvert}}$ | (17) |
| | Geometric Average Maximum [10] | $\dfrac{\sum \max(x_i, y_i)}{\sum \sqrt{x_i \cdot y_i}}$ | (18) |
| | Correlation [8] | $1 - \dfrac{\sum ((x_i - \bar{x}) \cdot (y_i - \bar{y}))}{\sqrt{\sum (x_i - \bar{x})^2 \cdot \sum (y_i - \bar{y})^2}}$ | (19) |

# 3  Results and Discussion

After benchmark processing (cf. Fig. 2), median AUC values of each category for every combination of vector transformation algorithm and distance measure were obtained. A summary of the highest attained values for every transformation is shown in Table 3. The evaluation is based on median AUC values of all underlying test sequences (instead of the arithmetic mean) to reduce the influence of possible outlying sequences and to characterize a result by a centered representative regardless of the data distribution. In general, every vector transformation algorithm can be combined with at least one distance measure in order to achieve satisfying search performance (AUC $\geq 0.80$).

**Table 3.** Highest AUC values for each search category and sequence vector transformation algorithm (see Sects. 2.1 and 2.2) observed in the benchmark. The amount of distance measures capable to achieve the AUC is superscripted, and the corresponding parameters $n$ are subscripted to each value.

| | Overall | Complete | Complete$^{\mathrm{m}}$ | Sub | Sub$^{\mathrm{m}}$ |
|---|---|---|---|---|---|
| $n$-gram$^{\mathrm{wc}}$ | $0.93^1_{7 \leq n \leq 9}$ | $0.94^3_{n \in \{3,4,8,9\}}$ | $0.94^2_{n=9}$ | $0.93^1_{n \in \{7,8\}}$ | $0.93^2_{n=8}$ |
| $n$-gram$^{\mathrm{f}}$ | $0.93^5_{7 \leq n \leq 9}$ | $0.94^{20}_{3 \leq n \leq 9}$ | $0.94^7_{n=9}$ | $0.93^9_{n \in \{7,8\}}$ | $0.93^5_{n \in \{7,8\}}$ |
| RTD | $0.90^3_{n \in \{1,2,6\}}$ | $0.92^{11}_{n \in \{2,4,6,7\}}$ | $0.92^5_{n \in \{2,4,6\}}$ | $0.91^1_{n=2}$ | $0.91^1_{n=6}$ |
| CPF | $0.89^1_{n=9}$ | $0.92^1_{n=9}$ | $0.92^1_{n=9}$ | $0.89^1_{n=9}$ | $0.88^1_{n=9}$ |
| Inertia | $0.80^1$ | $0.85^1$ | $0.85^1$ | $0.81^1$ | $0.80^1$ |

For the presented scenario the $n$-gram approach scores conspicuously good, despite the usage of word counts ($n$-gram$^{\mathrm{wc}}$) or frequencies ($n$-gram$^{\mathrm{f}}$). Nevertheless, a greater range of distance measures can be applied to the latter to achieve comparable high AUC values. For the return time distribution (RTD) it is significant that lower values of $n$ result in better performances. Inertia and CPF are both heavily dependent on the distance measure in question, additionally, in case of CPF, search performance is observed to be increased with larger values of $n$.

A summary of all results including parametrization is shown in Fig. 3. An immediate implication of the increasing difficulty for the respective search scenarios can be perceived in a minor decrease of AUC values regarding modified sequences contrary to their unmodified variants as well as by comparing complete sequences to subsequences (Fig. 3**C**). Nevertheless, a general trend of increasing AUCs can be observed by incrementing $n$ (Fig. 3**A**). Especially the CPF transformation benefits significantly from higher parameter values $n \geq 7$, whereas the $n$-gram transformers report only a minor increase in performance. On the contrary, search performance achieved by RTD transformation is not as dependent to the majority of investigated choices of $n$ as other transformation algorithms. More specifically, for $n > 6$ in general and additionally for $n = 4$ in the case of subsequences a decrease in performance is observable for RTD. However, even though there seems to be no decline in performance with increasing $n$ for the other transformations, further incrementing is not favorable due to rising memory and processing time consumption (cf. Table 1).

The selection of a vector distance measure has a major influence on the resulting AUC values as shown in Fig. 3**A**. Although, the $n$-gram frequency approach and Inertia perform almost independent regarding the utilized distance measures, the $n$-gram word count presents a bimodal distribution of rather good and worse measures. The RTD transformation on the contrary is comprised of a comparably uniform distribution throughout the complete range. With a generally low AUC for the most distance measures the CPF transformer seems to be very responsive to a single measure (19). Considering the classification proposed by Cha [8], distance measures from the *Inner Product* family are well distributed on the upper AUC values of every search scenario and transformation (except

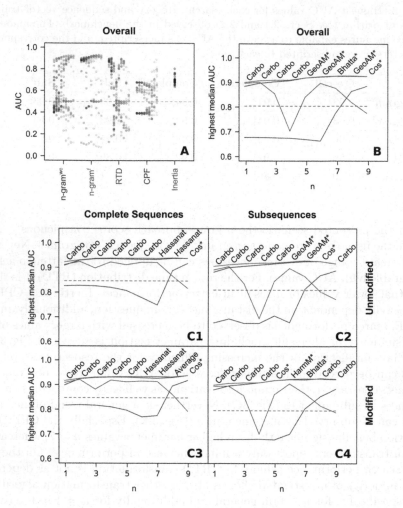

**Fig. 3.** The plots illustrate the benchmark results, where the maximum median AUCs appear on the ordinate and the vector transformation algorithms respective the investigated choices of parameter $n$ appear on the abscissa. **A** summarizes the dependencies of the transformers towards all distance measures. The colored columns of each algorithm represent $n$ in ascending order ($1 \leq n \leq 9$), where every point symbolizes a measure with its scored AUC value. Due to the independence from $n$, Inertia is depicted as a single column. Subfigures **B** and **C** show the correlation of the highest scored median AUC and the parameter $n$ for each vector transformation algorithm, regardless of the utilized distance measure. The overall result is shown in **B**, detailed results of the four search categories are presented in **C1–4**. Inertia is depicted as a dotted line to emphasize its independence to parameter $n$. The best score of the leading vector transformation algorithm for every $n$ is labeled with its corresponding distance measure. The shortened distance measures correspond to Bhattacharyya (Bhatta*), Cosine (Cos*), Geometric Average Maximum (GeoAM*) and Harmonic Mean (HarmM*).

for CPF) and are therefore a good choice as default measures for the presented study (data not shown). Complete sequences additionally respond well to distance measures from the *Absolute Difference* family and the Correlation distance (19). The *Fidelity* family and the Geometric Average Maximum distance (18) are only observed to achieve best results in combination with $n$-gram$^f$ transformations, however throughout all categories, indicating a good performance of these measures on normalized vectors. The *Intersection* family scores the highest AUC values nearly exclusively in the return time distribution transformation, similar to the Correlation distance (19) for CPF. Remarkably, the Manhattan distance (16) is the only occurring representative of the *Minkowski* family and is only observable for complete sequences (Complete, Complete$^m$) of the $n$-gram frequency approach ($n$-gram$^f$).

In summary, both $n$-gram vector transformation algorithms outperform the other transformers, whereas the word frequencies ($n$-gram$^f$) are slightly more accurate (Fig. 3**B**) throughout nearly all values of $n$. This examination can be confirmed for all search scenarios and therefore endorses the employment of a single transformation for computing a single index instead of potential multiple indices for the respective search scenarios. The overall highest AUC value of 0.9275 is observable by employing the $n$-gram$^f$ transformer with $n = 8$ and the Geometric Average Maximum distance (18) or a representative from the *Fidelity* family. In comparison, the overall highest AUC value for utilizing the Cosine distance (10) is 0.9261 for $n$-gram$^f$ with $n = 8$. Even though the improvement of the AUC value is not substantial, this exhibits that substituting commonly used distance measures from literature can enhance overall performance and sensitivity. This statement can further be supported by comparing the recommended or common distance measure to the overall highest scoring distance measure observed for all vector transformation algorithms (cf. Table 4).

**Table 4.** Comparison of recommended or commonly proposed distance measures from literature (see column proposed measures) for the vector transformation algorithms and the highest scoring distance measures from this study for every approach. The AUC difference between two measures is shown in the last column.

| Algorithm | Proposed measure | Highest scoring measure | Gain |
|---|---|---|---|
| $n$-gram$^{wc}$ | 0.9261 – Cosine (10) | 0.9261 – Cosine (10) | n/a |
| $n$-gram$^f$ | 0.9261 – Cosine (10) | 0.9275 – Geometric Avg. Max. (18) | +0.0014 |
| RTD | 0.8657 – Euclid (15) | 0.9014 – Intersection (13) | +0.0357 |
| CPF | 0.6128 – Euclid (15) | 0.8851 – Correlation (19) | +0.2723 |
| Inertia | 0.6997 – Euclid (15) | 0.8040 – Inner Product (12) | +0.1043 |

Further work applicable to this study should incorporate additional reference lists despite BLAST (e.g. from Needleman-Wunsch [20] and Smith-Waterman [27] algorithms). Further extension to verify the reproducibility can be promoted by supplemental distance measures as well as vector transformation algorithms and multiple datasets. Aside from that, an analogous setup incorporating protein sequences as well as protein structures is currently under development.

## 4    Concluding Remarks

In this study, we evaluated five methodical approaches for preparing biological sequences in four different search scenarios to be indexed by the Intelligent Cluster Index (ICIx). A transformation of these sequences into defined feature vectors is a necessity and therefore the primary selection criterion for the assessed algorithms. Furthermore we utilized a set of distance measures to investigate their impact on the transformer performance regarding a widely employed reference algorithm (BLAST).

It is shown, that the applied distance measures significantly influence performance and quality of the vector transformation algorithms with respect to sensitivity, and therefore should be well evaluated for respective systems. The different search scenarios are nearly equally treated by each transformation and consequently do not require divergent processing. The incrementation of the observation frame $n$ for $n$-gram based approaches can improve the result quality at the expense of increasing required memory and processing overhead. Although, even small values for $n$ can yield sufficient results. Moreover we utilized an $n$-gram independent approach, which shows comparable search performance to $n$-gram based transformation techniques, considering its really small and constant feature vector.

In conclusion, we suggest the use of $n$-gram word frequencies with an $n \leq 8$ for indexing purposes of DNA sequences with the ICIx technology to incorporate high sequence vector similarity with acceptable computational overhead. We recommend using either Bhattacharyya distance (7) from *Fidelity* family, Carbo (9) or Cosine (10) distance from *Inner Product* family or the Geometric Average Maximum distance (18) as methods for measuring vector distance due to their consistent high performance in combination with this transformation.

### Disclosure Policy

The authors declare that there is no conflict of interests regarding the publication of this paper.

**Acknowledgement.** The study has been supported by the Free State of Saxony, the University of Applied Sciences Mittweida and Chemnitz University of Technology.

## References

1. Altschul, S.F., Madden, T.L., Schäffer, A.A., Zhang, J., Zhang, Z., Miller, W., Lipman, D.J.: Gapped BLAST and PSI-BLAST: a new generation of protein database search programs. Nucleic Acids Res. **25**(17), 3389–3402 (1997)
2. Altschul, S.F., Gish, W., Miller, W., Myers, E.W., Lipman, D.J.: Basic local alignment search tool. J. Mol. Biol. **215**(3), 403–410 (1990)
3. Baby, J., Kannan, T., Vinod, P., Gopal, V.: Distance indices for the detection of similarity in C programs. In: International Conference on Computation of Power, Energy, Information and Communication (ICCPEIC), pp. 462–467. IEEE (2014)

4. Bao, J., Yuan, R., Bao, Z.: An improved alignment-free model for dna sequence similarity metric. BMC Bioinform. **15**(1), 321 (2014)
5. Benson, D.A., Karsch-Mizrachi, I., Lipman, D.J., Ostell, J., Sayers, E.W.: Genbank. Nucleic Acids Res. **39**(suppl 1), D32–D37 (2011)
6. Bogan-Marta, A., Hategan, A., Pitas, I.: Language engineering and information theoretic methods in protein sequence similarity studies. Computational Intelligence in Medical Informatics, pp. 151–183. Springer, Heidelberg (2008)
7. Boratyn, G.M., Camacho, C., Cooper, P.S., Coulouris, G., Fong, A., Ma, N., Madden, T.L., Matten, W.T., McGinnis, S.D., Merezhuk, Y., Raytselis, Y., Sayers, E.W., Tao, T., Ye, J., Zaretskaya, I.: BLAST: a more efficient report with usability improvements. Nucleic Acids Res. **41**(W1), W29–W33 (2013)
8. Cha, S.H.: Taxonomy of nominal type histogram distance measures. In: Proceedings of the American Conference on Applied Mathematics, pp. 325–330. World Scientific and Engineering Academy and Society (WSEAS) (2008)
9. Deza, M.M., Deza, E.: Encyclopedia of Distances. Springer, Heidelberg (2012)
10. Doreswamy, Manohar, M.G., Hemanth, K.S.: A study on similarity measure functions on engineering materials selection. AIAA **1**, 157–168 (2011)
11. Ganapathiraju, M., Manoharan, V., Klein-Seetharaman, J.: BLMT - statistical sequence analysis using N-grams. Appl. Bioinform. **3**(2–3), 193–200 (2004)
12. Gilg, S., Neubert, R.: Semantische Indexierung mittels dynamisch-hierarchischer Neuronaler Netze. Master's thesis, Chemnitz University of Technology (1999)
13. Görlitz, O., Neubert, R., Benn, W.: Access to distributed environmental databases with ICIx technology. Online Inf. Rev. J. **24**(5), 364–370 (2000)
14. Hanley, J.A., McNeil, B.J.: The meaning and use of the area under a receiver operating characteristic (ROC) curve. Radiology **143**(1), 29–36 (1982)
15. Hassanat, A.B.: Dimensionality invariant similarity measure. J. Am. Sci. **10**(8), 221–226 (2014)
16. Hatzigiorgaki, M., Skodras, A.N.: Compressed domain image retrieval: a comparative study of similarity metrics. In: Visual Communications and Image Processing 2003, pp. 439–448. International Society for Optics and Photonics (2003)
17. Kent, W.J.: BLAT - the BLAST-like alignment tool. Genome Res. **12**(4), 656–664 (2002)
18. Kolekar, P., Kale, M., Kulkarni-Kale, U.: Alignment-free distance measure based on return time distribution for sequence analysis: applications to clustering, molecular phylogeny and subtyping. Mol. Phylogenet. Evol. **65**(2), 510–522 (2012)
19. Leuoth, S., Adam, A., Benn, W.: Profit of extending standard relational database with the intelligent cluster index (ICIx). In: 11th ICARCV International Conference ond Control, Automation, Robotics and Vision, vol. 1, pp. 1198–1205 (2010)
20. Needleman, S.B., Wunsch, C.D.: A general method applicable to the search for similarities in the amino acid sequence of two proteins. J. Mol. Biol. **48**(3), 443–453 (1970)
21. Neubert, R., Görlitz, O., Benn, W.: Incorporating knowledge technology in databases. In: KnowTech 2000 Conference (2000)
22. Neubert, R., Görlitz, O., Benn, W., Teich, T.: Obstacles for application of neural networks in the ICIx database index. Int. Joint Conf. Neural Networks **1**, 2351–2356 (2002)
23. Neubert, R., Görlitz, O., Benn, W.: Towards content-related indexing in databases. Datenbanksysteme in Büro, Technik und Wissenschaft. Informatik aktuell, pp. 305–321. Springer, Heidelberg (2001)
24. Pearson, W.R., Lipman, D.J.: Improved tools for biological sequence comparison. PNAS USA **85**(8), 2444–2448 (1988)

25. Punta, M., Coggill, P.C., Eberhardt, R.Y., Mistry, J., Tate, J., Boursnell, C., Pang, N., Forslund, K., Ceric, G., Clements, J., Heger, A., Holm, L., Sonnhammer, E.L.L., Eddy, S.R., Bateman, A., Finn, R.D.: The pfam protein families database. Nucleic Acids Res. **40**(D1), D290–D301 (2012)
26. Searls, D.B.: The language of genes. Nature **420**(6912), 211–217 (2002)
27. Smith, T.F., Waterman, M.S.: Identification of common molecular subsequences. J. Mol. Biol. **147**(1), 195–197 (1981)
28. Sun, W.K.: Algorithms in Bioinformatics - A practical Introduction. CRC Press, Boca Raton (2010)
29. Yao, Y., Han, J., Dai, Q., He, P.: A novel descriptor of protein sequences and its application. J. Theor. Biol. **347**, 109–117 (2014)
30. Zvelebil, M., Baum, J.O.: Understanding Bioinformatics. Garland Science (2008)

# A Holistic Approach to Testing Biomedical Hypotheses and Analysis of Biomedical Data

Krzysztof Psiuk-Maksymowicz[1]([✉]), Aleksander Płaczek[2], Roman Jaksik[1],
Sebastian Student[1], Damian Borys[1], Dariusz Mrozek[3], Krzysztof Fujarewicz[1],
and Andrzej Świerniak[1]

[1] Institute of Automatic Control,
Silesian University of Technology, ul. Akademicka 16, 44-100 Gliwice, Poland
krzysztof.psiuk-maksymowicz@polsl.pl
[2] Research and Development Department,
WASKO S.A., ul. Berbeckiego 6, 44-100 Gliwice, Poland
a.placzek@wasko.pl
[3] Institute of Informatics, Silesian University of Technology,
ul. Akademicka 16, 44-100 Gliwice, Poland

**Abstract.** Testing biomedical hypotheses is performed based on advanced and usually many-step analysis of biomedical data. This requires sophisticated analytical methods and data structures that allow to store intermediate results, which are needed in the subsequent steps. However, biomedical data, especially reference data, often change in time and new analytical methods are created every year. This causes the necessity to repeat the iterative analyses with new methods and new reference data sets, which in turn causes frequent changes of the underlying data structures. Such instability of data structures can be mitigated by the use of the idea of data lake, instead of traditional database systems.

The aim of this paper is to show system for researchers dealing with various types of biomedical data. Such a system provides a functionality of data analysis and testing different biomedical hypotheses. We treat a problem in a holistic way giving a researcher freedom in configuration his own multi-step analysis. This is possible by using a multiversion dynamic-schema data warehouse, performing parallel calculations on the virtualized computational environment, and delivering data in MapReduce-based ETL processes.

**Keywords:** NoSQL database · Multiversion dynamic-schema · Data warehouse · Biomedical data processing · Big Data · MapReduce · ETL

## 1 Introduction

Modern age biomedical studies often involve high throughput measurement techniques that require complex data processing to provide biologically meaningful conclusions. State of the art genomic studies aim at identifying regulatory interactions between various types of molecules that either drive the biochemical processes in response to external or internal stimuli or characterize a specific

© Springer International Publishing Switzerland 2016
S. Kozielski et al. (Eds.): BDAS 2016, CCIS 613, pp. 449–462, 2016.
DOI: 10.1007/978-3-319-34099-9_34

disease state [3,19]. However to identify possible regulatory mediators among hundreds of thousands nucleic acids and proteins a large scale screening must be conducted in order to reduce the search space for the following more precise and single molecule oriented methods. Large scale measurement techniques which among others include microarrays, next generation sequencing (NGS) or reverse-phase protein arrays (RPPA) require appropriately trained staff to both carry out the experiment and process the results. In many cases, significant difficulties with the data processing limit the advancements in the biomedical research wasting the potential which the data can carry [4,23].

While there are many publicly available algorithms and analysis workflows which aim at providing high-quality reproducible results [2,17], most of them require knowledge in bioinformatics and/or biostatistics providing difficulties for the persons trained to carry out the experiment itself. The reasons for that are:

– Large number of data, stored in various formats, some of which are not readable by human;
– The necessity to utilize statistical tests in order to assess the significance of the findings;
– The need to include additional data in the study that either allows to determine the structural features of specific DNA regions or provides information on the known function of identified molecules;
– The need to combine various analysis methods with incompatible input/outputs.

The number of possible combinations of experiment scenarios and measurement techniques is almost limitless although the experimental setups and data processing workflows usually follow a constant scheme. Single sample studies are extremely rare since the most of the analysis techniques focus solely on between sample comparisons. Figure 1 shows the most common analysis setups, which can be divided into two main types:

– A1-3 study of the effects of various chemical or physical treatments. This setup usually utilizes a small number of samples some of which represent a specific treatment. The use of biological/technical replicates is essential for this setup. Comparative analysis is performed versus a control sample (no treatment) or a different treatment, *e.g.*, cells treated with various doses of a chemical compound versus control cells (A1), cells treated with damage inducing factor and tested at various time points, versus control cells gathered at the same time (A2), cells treated with two or more regulatory molecules and a combination of both treatments, compared between each other and control (A3);
– Identification of similarities in the expression profile of mutation signature characteristic to a preselected group of cells (B) - this setup usually utilizes a high number of samples with a low number or no technical/biological replicates. In a typical experiment, two patient groups are compared which represent various tumor subtypes or samples isolated from tumor and healthy cells, second of which serves as a control.

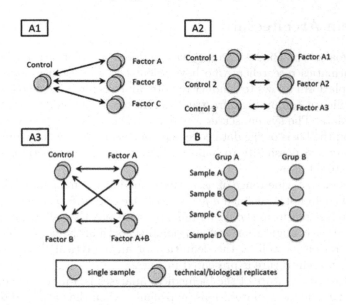

**Fig. 1.** Common experimental setups used in biomedical studies.

Data processing workflows usually also follow a similar scheme despite significant differences between data types and specific algorithms used for data analysis [15]. A typical workflow comprises the following steps:

- Quality control of the data - assessment of technical variability in the experiment, identification of outlying samples, validation of the experimental procedure;
- Data standardization - reduction of between-sample variability of technical origin which aims at separating it from the biological differences;
- Identification of distinctive features – selection of molecules or DNA regions that show a specific pattern in the data, unique to groups of analyzed samples;
- Searching for common attributes of the features identified - identification of similarities in either the structure or function based on previous studies deposited in publicly available databases;
- Data visualization - a graphical presentation of the results obtained as a result of the study.

The similarities provide the possibility to create an integrated system which could be used to test various biomedical hypotheses, under the experiment setups described above, using the concept of intelligent data analysis. This motivated the creation of the testing platform presented in the paper. The platform was developed withing the BioTest project. In the following subchapters we will describe the architecture of the computational system built for testing biomedical hypotheses, the dynamic-schema multiversion data warehouse, which is a central part of the data layer, and MapReduce-based ETL process that feeds the data warehouse with data for further analyses.

## 2     System Architecture

The main goal of the BioTest project was to develop a platform for biomedical and bioinformatics researchers who have scientific or biomedical hypothesis to proof. The platform allows to conduct remote analysis for those who do not have access to sufficient computing infrastructure. Moreover, it enables for integration with other data. The system serves a multidimensional analysis and generation of reports on the basis of the data collected over years during the clinical experiments. There is a possibility to build and run individual algorithms operating on already stored data.

The demands of wide range of users (undefined at the moment of the start of the project) could not be defined *a priori*. Because of that fact, it was considered that the BioTest platform should be built by solutions that allow an individual user to define the configuration and functionality, both in the scope of calculation flow configuration as well as the definition of an analytical data model. The overall system architecture is presented in Fig. 2.

A characteristic feature of the biomedical analyses is a multitude of calculations and a large number of iterations. Especially, when they are used to confirm or negate a hypothesis, this entails a need to perform parallel calculations. The project team has assumed that different data sets, not necessarily in the same structure, could be created within the dynamic-schema data warehouse. Additionally, confirmation of the authenticity of the medical hypothesis requires an analysis of the function of genes or proteins. It is usually the last, but one of the most important stages of data processing obtained in molecular biology experiments. At that stage it is necessary to use publicly available databases for referential data to obtain additional information about identified molecules, regarding their structure (*e.g.* allowing to identify protein domains with a specific function), regulatory dependency and role in specific biochemical processes taking place inside a cell.

### 2.1     Computation Area

It was decided to build a system architecture that combines solutions for medical researchers and solutions for the BigData segment (see Fig. 2). In the area of computing the opportunities offered by the Galaxy Computing Server (GalaxyServer) were combined with the computing capabilities of the Silesian BIO-FARMA Center for Biotechnology, Bioengineering and Bioinformatics, in a transparent manner to the final user. Such a combination provides the opportunity to prepare any set of algorithms and calculation procedures publicly available from the GalaxyServer platform - very rich list of processing modules of genomic and proteomic data, to build custom algorithms and embed them in an isolated area of the BIO-FARMA environment to carry out calculations. BIO-FARMA is an environment created as a platform to support large-scale bioinformatics systems that process large amounts of data, it provides the network infrastructure with high capacity with the interface to information systems of the Oncology Centre in Gliwice, which is the primary source of medical and

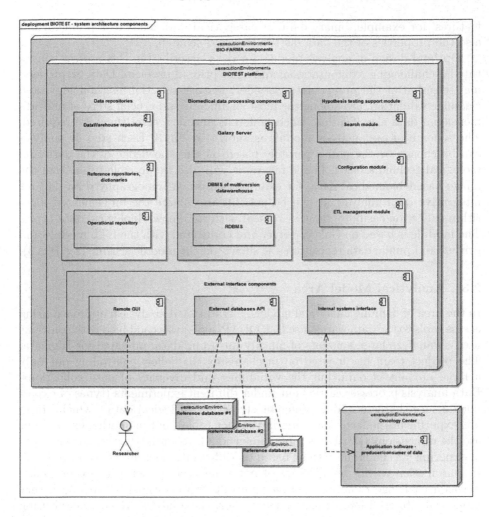

**Fig. 2.** Overall system architecture.

scientific data for our system. Data processing layer architecture has the ability to use high-performance parallel processing solutions. Additionally, the remote access environment has implemented virtualization to allow scientists to use the project solutions through Remote Desktops.

## 2.2 Intelligent Analysis Module

The crucial step in intelligent biomedical data analysis software construction is proper data preprocessing prepared separately for each biological technique used. For example, we must choose different scenarios for micro-RNA, DNA mutation, metabolomics, proteomics or histopathological markers. Our experience with methods in the integration of such data suggests that not only the type of used features, for example, binary data, or categorical data leads to a problem, but also the properties of the data distribution and scale [6,7,21].

Optimal design of analysis module for data described in this project constituted a challenging, computational and conceptional problem. Data preprocessing, selection of the proper data classifier, feature selection methodology, and classifier validation, are among the most significant and crucial steps of multidimensional data classification, where different and new algorithms are required. The scheme of our analysis module is shown in Fig. 3. Each step of analysis can be chosen by the user separately based on the knowledge of the data and biological technique used. In the future, it will also be possible to analyze the image-based biomedical data, for example, emission tomography data [20] for integrative data analysis. It will be possible to enlarge the implemented algorithm collection easily by the user to deal with not known biological data types. The module is based on Galaxy Server and connected with the ETL module and private and public data repositories as shown on the module scheme (see Fig. 3).

## 2.3   Analytical Model Area

In the area of defining analytical models, a traditional analytical approach using access tools to the data warehouse (*i.e.* OLAP tools) was combined with solutions used to analyze large amounts of quickly incoming data, prepared for BigData. The project team has focused on mechanisms that allow controlling not only a large amount of data, but the variability and diversity of data collections. Data analysis processes carried out under different experiments (types of experiments) are characterized by different structures of logical data - which are in one experiment the facts (measures in the fact table), but in another experiment are the dimensions (attributes table dimensions). Accordingly, the various types of experiments should have to be based on different models of a data warehouse. In conjunction with a broad range of potential types of scientific experiments, it was a challenge to ensure multiversion data warehouse structures on a very large scale. In addition, as mentioned earlier, the analysis are conducted using databases of information about genes that are updated directly by their users or employees of institutions responsible for their development not by researchers themselves. The main problem associated with the functional analysis of genes based on such databases are frequent updates to the information. The reason of it is straightforward - the database of genes and their functions as a knowledge database are under continuous development. This feature is very important, because of the fact that access at another time may lead to a different version of the data, and thus, may result in different conclusions about the examined biological processes. Therefore, it was decided to apply the approaches of the construction of the Data Lake using the NoSQL database and engine to handle large data sets such as Apache Spark [14]. In traditional data warehouses, data must be converted in accordance with a predetermined *Extract, Transform, and Load* (ETL) scheme [24] before loading into the data warehouse. In traditional data warehouses decision of the data usage is taken at the stage of the ETL process - unused source data are not loaded into the data warehouse.

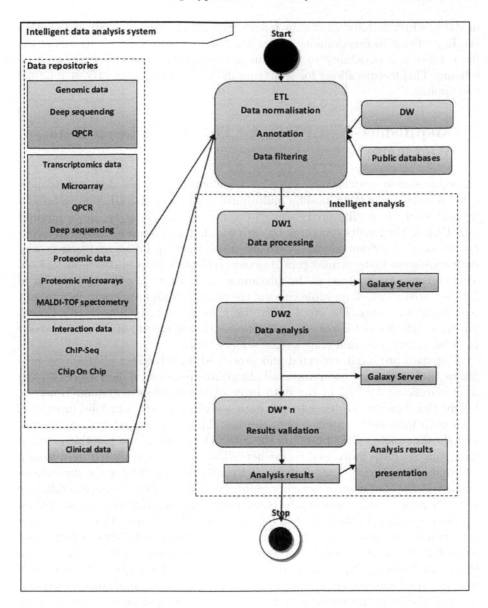

**Fig. 3.** The intelligent analysis module scheme.

Whereas data lake solutions aim at using the data in raw form, and ETL-like processes occur at the stage when applications need access to the data. Such an approach of creating data structures at the stage of reading is called *schema-on-read*, while traditional data warehouses use the approach that creates data structures at the stage of writing, called *schema-on-write*. The approach used in a traditional warehouse requires careful pre-planning and rethinking the use of data, which is hard to obtain before formulation any biomedical hypothesis. In addition to functionality that allows building alternative patterns of the data, there is a possibility to maintain alternative versions of data in a single scheme. This feature allows for analyzing data described differently at multiple time points.

## 3   MapReduce-Based ETL for Dynamic-Schema Data Warehouse

Data warehouse systems are complex information systems designed to perform advanced reporting, on-line multidimensional analysis (OLAP), trend and *what-if* analysis, results prediction and advanced data analysis (data mining) [5,8,10,16]. The results generated by data warehouse systems are often an essential element of enterprise level strategic management process, and that is why data warehouse systems must provide exact, reliable and up-to-date information. Multidimensional analysis is also becoming increasingly popular in the area of Life Sciences, in which multiple views of the same data allow to get insights about real living organisms. The proof of the this gives projects such as BioWarehouse [9], Atlas [18], BioDWH [22], and GPKB [11] that make use of data warehouses in order to integrate and analyze Life sciences data.

The data produced, collected and processed on a BioTest platform by programs that implement computational algorithms represent a data source for the data warehouse located in the data layer of the developed system. However, before the data are loaded into the data warehouse, they must be integrated with data from other external data sources. The data derived from referential biological databases distributed on the Internet are one of the possible external data sources. Each individual researcher is looking at the collected data from a different point of view, puts a different hypothesis, i.e., defines the dimensions and attributes that describe a fact or builds hierarchies of attributes in a different way. Therefore, in this project, the team focused on building the data schema on demand according to the idea of schema-on-read. This assumes that the data are kept in a source format in most cases, as they are produced. When a user wants to analyze the data, a special procedure is run to transform source data into the format and schema expected by the user. Data transformations are performed in a high-performance computing environment, which ensures quick processing of data when they are necessary for analysis. Transformations are based on the MapReduce [13] methods. Such an approach allows to handle individual analyses, which use the same stored data but require completely different data scheme or a different version of an existing scheme or data model. Versions are distinguished here by the difference in perception of the relationships between data: a hierarchy of data in a given dimension built in a different way, validity or non-validity of a range of data, dynamic data typing or even changes in metadata (*e.g.*, some values once treated as attributes of a dimension and another time as measures). The system allows users to create new ETL processes based on the collected data. These ETL processes are responsible for transforming data and recording them in the proposed scheme, according to user's point of view. Data warehouse is hosted on the NoSQL database management system [12]. NoSQL database systems allow to store data in the unstructured form while ensuring efficient mechanisms for querying and analysis for large amounts of data. Storing data without clearly defined structure allows one to place data coming from various data sources and stored in different formats in a single logical data set. The developed data warehouse accepts any type, structured and unstructured data. Relationships between data tables are dynamic, and they are not the result of the pre-planned scheme [1], but they are consequences of the need to perform analysis based on the specific data model (related to the posed hypotheses). In the data warehouse, storage is organized based on key-value relationship, where the key is an indexed pointer to the data and the value can be of any type. Using the database management system, which does not presuppose any predefined data structure (and thus the data model) and also provides a broad range of analytical queries, allowed the project team to deliver a platform for verifying various hypotheses that are unknown at the moment of designing the system. The data loaded into the data warehouse staging area are the source data that can be analyzed by a researcher during the verification of the posed hypothesis. Source data for any analysis performed in the system are derived not only from external sources, but also from results of previously performed ETL operations that generate data stored in an OLAP cube-like format.

The proposed methodology of data transformation is based on the assumption that data warehouses are optimized for reporting purposes, and therefore, their data structures should be adapted for querying operations, for which analytical requirements are defined ad hoc. The data should be redundant where it is beneficial for the performance of executed queries. When searching for answers to raised questions about the scientific or medical hypotheses (which are nor predefined), it is not possible to optimize data structures having only the source data. Therefore, while developing the ETL processes we used the approach used for building computational processes in environments resistant to hardware failures. This approach uses MapReduce programming model, which relies on dividing tasks into smaller parts processed in the following operations:

- Mapping data - this operation allows to copy data stored in one form to another and perform data classification based on the applied filter function. Data are converted to sequences of pairs in the key - value form, and then, collected and sorted by given specific key.
- Reducing data - this operation allows to aggregate data (reduce an amount of data) into a form, in which values would have the same indexing key according to applied filter function.

In order to perform data mapping operation researcher can build a Map function, which is a series of operations that convert the input arguments to key-value pairs. The purpose of the function is to classify the data given in the input, determine what key identifies them and then assign appropriate value to the key. Keys do not have to be unique. During the analysis of individual data

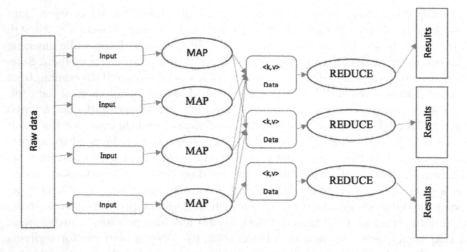

**Fig. 4.** MapReduce process scheme.

records this function may classify data as matching to a pattern of a given type of cancer (this is the key) and assign the clinical data of the patient from whom the sample was taken to the value. The new set of data obtained in this way can be reduced based on the key – all the pairs having the same key are passed to a Reduce function, which is series of operations that aggregate data to achieve the final result (Fig. 4). For example, the Reduce function can calculate the average of the reported values or number of occurrences of each sequence genes.

Map & Reduce functions may follow one another repeatedly, and the output data set from one function becomes the input data set for the other function. The data warehouse computation layer responsible for managing and running Map & Reduce operations was built by using Apache Spark framework for a rapid analysis of large amounts of data. The framework is installed on the computational cluster in the BIO-FARMA environment. At any time when a researcher needs to carry out additional computations, build ad-hoc queries, calculate aggregates taking into account different types of analytical hierarchies, he defines appropriate MapReduce processes or uses one of existing processes stored in the system. The use of distributed data processing systems allows for parallelization of computations, and thereby fast production of intermediate results in the form of key - value pairs, which are then reduced and stored in the final, dynamically generated multidimensional structure - the analogy of tables corresponding to OLAP cube. All Map & Reduce operations are responsible for defining the required in the particular research project version of data model (not necessarily only data structure), which is used to conduct statistical analyzes to verify the posed biomedical hypothesis. It is possible to design and execute operations that dynamically build a different multidimensional data model each time, starting from the same source data. The extra benefit is that new researcher can use the model developed by another researcher and introduce modifications to the model or create a new version of the data model based on the existing one.

## 4    Concluding Remarks

A holistic approach to testing biomedical hypotheses and analysis of biomedical data needs a general-purpose platform for performing associated computations. Moreover, these analyzes require intermediate data that should be stored somewhere, as they may be a source of data for further analyzes. The presented system

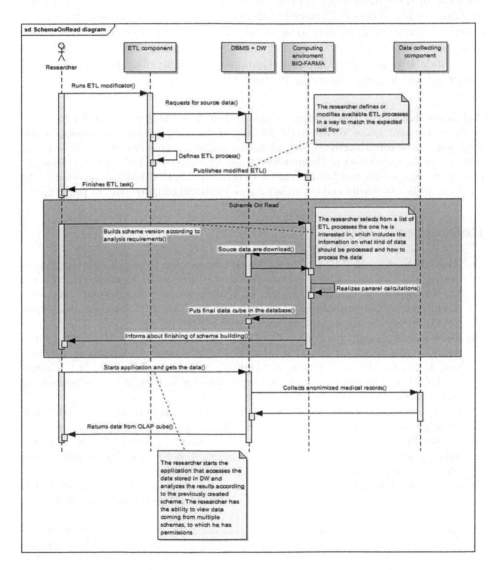

**Fig. 5.** Dynamic construction of multidimensional data structures (schema-on-read) through customizations of the ETL processes and building dynamic-schema data warehouse (DW)

with the multiversion, dynamic-schema data warehouse guarantees the storage place and capabilities of relatively fast multidimensional analysis of biomedical data, which is supported by MapReduce-based ETL processes that deliver the data to the data warehouse and build appropriate data model on-demand (Fig. 5).

Future works will cover further development of the presented system and creation of data models for typical analyzes that are performed by biologists and oncologists. These data models will allow to test already known hypotheses based on the current referential data and with the use of existing algorithms. They will also be a base for building and deriving other versions of analytical data models by incorporating algorithms of intelligent analysis and referential data that will be available in the future.

**Acknowledgments.** This work was supported by The National Centre for Research and Development grant No PBS3/B3/32/2015. Presented system was developed and installed on the infrastructure of the Ziemowit computer cluster (www.ziemowit.hpc. polsl.pl) in the Laboratory of Bioinformatics and Computational Biology, The Biotechnology, Bioengineering and Bioinformatics Centre Silesian BIO-FARMA, created in the POIG.02.01.00-00-166/08 and expanded in the POIG.02.03.01-00-040/13 projects.

# References

1. Arfaoui, N., Akaichi, J.: Automating schema integration technique case study: generating data warehouse schema from data mart schemas. In: Kozielski, S., Mrozek, D., Kasprowski, P., Malysiak-Mrozek, B., Kostrzewa, D. (eds.) Beyond Databases, Architectures and Structures. CCIS, vol. 521, pp. 200–209. Springer, Heidelberg (2015). http://dx.doi.org/10.1007/978-3-319-18422-7_18
2. DePristo, M., Banks, E., Poplin, R., Garimella, K., Maguire, J., Hartl, C., Philippakis, A., del Angel, G., Rivas, M., Hanna, M., McKenna, A., Fennell, T., Kernytsky, A., Sivachenko, A., Cibulskis, K., Gabriel, S., Altshuler, D., Daly, M.: A framework for variation discovery and genotyping using next-generation dna sequencing data. Nature Genet. **43**, 491–498 (2011)
3. Govindarajan, R., Duraiyan, J., Kaliyappan, K., Palanisamy, M.: Microarray and its applications. J. Pharm. Bioallied Sci. **4**(Suppl 2), S310–S312 (2012)
4. Gullapalli, R., Desai, K., Santana-Santos, L., Kant, J., Becich, M.: Next generation sequencing in clinical medicine: Challenges and lessons for pathology and biomedical informatics. J. Pathol. Inform. **3**, 40 (2012)
5. Inmon, W., Linstedt, D.: Data Architecture: A Primer for the Data Scientist: Big Data, Data Warehouse and Data Vault. 1st edn. Morgan Kaufmann, Waltham, MA, USA (2014)
6. Jaksik, R., Bensz, W., Smieja, J.: Nucleotide composition based measurement bias in high throughput gene expression studies. In: Gruca, A., Brachman, A., Kozielski, S., Czachórski, T. (eds.) Man–Machine Interactions 4. AISC, vol. 391, pp. 205–214. Springer, Heidelberg (2016)
7. Jaksik, R., Iwanaszko, M., Rzeszowska-Wolny, J., Kimmel, M.: Microarray experiments and factors which affect their reliability. Biology Direct **10**, 1–14 (2015). http://dx.doi.org/10.1186/s13062-015-0077-2

8. Kimball, R., Reeves, L., Margy, R., Thornthwaite, W.: The Data Warehouse. Lifecycle Toolkit. 3rd edn. John Wiley & Sons, Indianapolis, IN, USA (2013)
9. Lee, T., Pouliot, Y., Wagner, V., Gupta, P., Stringer-Calvert, D., Tenenbaum, J., Karp, P.: Biowarehouse: a bioinformatics database warehouse toolkit. BMC Bioinform. **7**(170), 1–14 (2006)
10. Małysiak-Mrozek, B., Mrozek, D., Kozielski, S.: Processing of crisp and fuzzy measures in the fuzzy data warehouse for global natural resources. In: García-Pedrajas, N., Herrera, F., Fyfe, C., Benítez, J.M., Ali, M. (eds.) IEA/AIE 2010, Part III. LNCS, vol. 6098, pp. 616–625. Springer, Heidelberg (2010)
11. Masseroli, M., Canakoglu, A., Ceri, S.: Integration and querying of genomic and proteomic semantic annotations for biomedical knowledge extraction. IEEE/ACM Trans. Comput. Biol. Bioinform. **PP**, 1–11 (2015). http://dx.doi.org/10.1109/TCBB.2015.2453944
12. Mazurek, M.: Applying NoSQL databases for operationalizing clinical data miningmodels. In: Kozielski, S., Mrozek, D., Kasprowski, P., Malysiak-Mrozek, B., Kostrzewa, D. (eds.) Beyond Databases, Architectures, and Structures: 10th InternationalConference, BDAS 2014, Ustron, Poland, May 27-30, 2014. Proceedings, Communications in Computer and Information Science, vol. 424, pp.527–536. Springer International Publishing (2014). http://dx.doi.org/10.1007/978-3-319-06932-6_51
13. Mrozek, D., Daniłowicz, P., Małysiak-Mrozek, B.: HDInsight4PSi: Boosting performance of 3D protein structure similarity searching with HDInsight clusters in Microsoft Azure cloud. Inform. Sci. (2016). http://dx.doi.org/10.1016/j.ins.2016.02.029
14. Official web page of Apache Spark: accessed on dec 10, 2015. http://spark.apache.org/
15. Pabinger, S., Dander, A., Fischer, M., Snajder, R., Sperk, M., Efremova, M., Krabichler, B., Speicher, M., Zschocke, J., Trajanoski, Z.: A survey of tools for variant analysis of next-generation genome sequencing data. Brief. Bioinform. **15**, 256–278 (2014)
16. Ponniah, P.: Data Warehousing Fundamentals. A Comprehensive Guide for IT Professionals. John Wiley & Sons, Hoboken, New Jersey, USA (2001)
17. Ritchie, M., Phipson, B., Wu, D., Hu, Y., Law, C., Shi, W., Smyth, G.: limma powers differential expression analyses for RNA-sequencing and microarray studies. Nucleic Acids Res. **43**(7), e47 (2015). http://dx.doi.org/10.1093/nar/gkv007
18. Shah, S., Huang, Y., Xu, T., Yuen, M., Ling, J., Ouellette, B.: Atlas - a data warehouse for integrative bioinformatics. BMC Bioinform. **6**(34), 1–16 (2005)
19. Shyr, D., Liu, Q.: Next generation sequencing in cancer research and clinical application. Biol. Proced. Online **15**(1), 4 (2013)
20. Student, S., Danch-Wierzchowska, M., Gorczewski, K., Borys, D.: Automatic segmentation system of emission tomography data based on classification system. In: Ortuño, F., Rojas, I. (eds.) IWBBIO 2015, Part I. LNCS, vol. 9043, pp. 274–281. Springer, Heidelberg (2015)
21. Student, S., Fujarewicz, K.: Stable feature selection and classification algorithms for multiclass microarray data. Biol. Direct **7**(33), 1–20 (2012)
22. Topel, T., Kormeier, B., Klassen, A., Hofestädt, R.: Biodwh: A data warehouse kit for life science data integration. J. Integr. Bioinform. **5**(2), 1–9 (2008)
23. Ulahannan, D., Kovac, M., Mulholland, P., Cazier, J.B., Tomlinson, I.: Technical and implementation issues in using next-generation sequencing of cancers in clinical practice. Br. J. Cancer **109**, 827–835 (2013)

24. Wycislik, L., Augustyn, D.R., Mrozek, D., Pluciennik, E., Zghidi, H., Brzeski, R.:
E–LT concept in a light of new features of Oracle Data Integrator 12c based on
data migration within a Hospital Information System. In: Kozielski, S., Mrozek,
D., Kasprowski, P., Małysiak-Mrozek, B., Kostrzewa, D. (eds.) Beyond Databases,
Architectures and Structures: 11th International Conference, BDAS2015, Ustroń,
Poland, May 26-29, 2015, Proceedings, Communications in Computer and Infor-
mation Science, vol. 521, pp. 190–199. Springer International Publishing (2015).
http://dx.doi.org/10.1007/978-3-319-18422-7_17

# Distributed Monte Carlo Feature Selection: Extracting Informative Features Out of Multidimensional Problems with Linear Speedup

Lukasz Krol[✉]

Faculty of Automatic Control, Electronics and Computer Science,
Data Mining Group, Silesian University of Technology, Gliwice, Poland
lukasz.krol@polsl.pl

**Abstract.** Selection of informative features out of ever growing results of high throughput biological experiments requires specialized feature selection algorithms. One of such methods is the Monte Carlo Feature Selection - a straightforward, yet computationally expensive one. In this technical paper we present architecture and performance of a development version of our distributed implementation of this algorithm, designed to run in multiprocessor as well as multihost computing environments, and potentially controllable through a web browser by non-IT staff. As a simple enhancement, our method is able to produce statistically interpretable output by means of permutation testing. Tested on reference *Golub et al.* leukemia data, as well as on our own dataset of almost 2 million features, it has shown nearly linear speedup when executed with an increased amount of processors. Being platform independent, as well as open for extensions, this application could become a valuable tool for researchers facing the challenge of ill-defined high dimensional feature selection problems.

**Keywords:** Feature selection · Dimensionality reduction · Parallel computing · Actor systems · Akka · Spark · Scala · Java

## 1 Introduction

Modern high throughput biological experiments like RNA microarray analysis or Genome Wide Association Studies (GWAS) provide a special class of datasets. Often having a few orders of magnitude more features than observations, these problems are ill-defined in the sense that they present a great amount of noise. Furthermore, due to high number of False Discoveries (FD), the results obtained from them are hard to reproduce. Put straightforward, the more features, the bigger the chance that one of them is correlated with the decision vector solely by chance.

The reason for this feature to observation imbalance is the nature of the collected data. Biological processes are extremely complex, and as the research designer is often not aware what to look for, all the technically available features are provided. On the other hand, observations are scarce. It is often difficult to

© Springer International Publishing Switzerland 2016
S. Kozielski et al. (Eds.): BDAS 2016, CCIS 613, pp. 463–474, 2016.
DOI: 10.1007/978-3-319-34099-9_35

find the group of interest (like patients with a certain type of disease). Price of a microarray chip is another factor limiting the number of observations. Initiatives like the *1000 Genomes Project* [12] or *HapMap Project* [5] can provide more data, but force the researcher to use a predefined, often not representative ethnical group, with the number of observations still being orders of magnitude lower than the number of features.

What is more convenient about these datasets is that their features (besides patient's additional phenotypic data like age or sex) are homogenous - they are either all categorical, all ordinal or all real variables, making automatic feature selection easier.

When working with ill-defined datasets, it is an obligation to be aware of the problem of False Discoveries (FD) - non reproducible correlations existing in the sample, but not in the population. A typical scenario of dealing with FD is adjusting individual p-values (probabilities of observing an extreme value of a feature "by chance") by one of more or less liberal procedures [8,10,11]. A nonparametric and computationally intensive approach to obtain statistically interpretable results is to perform permutation testing in order to verify if a statistic observed for certain feature is "extreme enough" when compared to analogous values obtained for randomized classes. With increasing availability of multiprocessor and multihost systems, as well as libraries facilitating parallel programming, this second approach is becoming more and more attractive.

As author's contribution in the area of feature selection, a new distributed implementation of existing Monte Carlo Feature Selection (MCFS) algorithm [1] is presented. This new implementation scales up linearly with number of allocated processors, meaning that computations can be shortened almost exactly by the factor of increased amount of resources. As a further enhancement, the feature ranking created by the algorithm is enhanced by p-values obtained by means of permutation testing. At the moment of writing, to our best knowledge, our implementation is the only one aimed to run in distributed environment. Furthermore, our architecture allows extending the algorithm to other classes of problems than those of discrete decision vector. After the development is completed, and full functionality of [2] is implemented, our version is going to be publically available for researchers facing the problem of feature selection.

Next sections describe an overview of feature selection approaches and compare them with MCFS algorithm. This is followed by description of our own implementation, and by presentation of a performance comparison between it and the latest version of *dmLab* software [2] tested on the reference *Golub et al.* [3] microarray dataset. Finally, capability of our software to perform feature selection on a dataset of nearly 2 mln features is demonstrated.

## 2    Feature Selection Approaches

### 2.1    Overview

As a general introduction to Monte Carlo Feature Selection, we present a short non-exhaustive list of feature selection approaches. All of them are supervised, meaning that the primary reason for choosing a feature is its usefulness in explaining the decision class rather than dissimilarity to other features.

**Single Feature Analysis.** A common approach to feature selection in multi-dimensional data is to evaluate each feature individually based on its usefulness in explaining the decision vector. A typical scenario when dealing with continuous features and discrete output would be to perform Analysis of Variance (ANOVA) for each feature, and then rank the features by increasing order of p-values. While simple to perform, this approach has the obvious drawback of ignoring inter-feature interactions. Outcome of a biological process is often a result of cooperation of whole sets of modified genes, none of which is able to explain the outcome alone.

**Sequential Feature Selection.** A well known example of this group are the Forward and Backward Feature Selection algorithms. These methods rely on an external model to provide the score for a certain subset of features, and modify this subset by adding or removing a single variable. A potential problem with these methods is a fixed search path - a Forward Feature Selection algorithm will begin by the best single feature, and then by the best feature complementing the first one in the current dataset (not necessarily the population second best, and not even the sample second best feature). Also, relying on an external scoring model, these methods tend to be biased towards it.

**Dimensionality Reduction.** Methods like PCA [7] and SVD allow to project the original feature space into one rotated towards the axis of highest variability, thus allowing to transform the original set of features towards a set of its combinations (Principal Components). The number of Principal Components can be controlled, so that a required portion of variability in the original space can be explained by these new features. Although powerful, this method also suffers from drawbacks. Primarily, the results obtained from the transformed feature space are hard to interpret by domain experts. The new feature space is also biased by the sample dataset. Finally, high variability does not have to mean - in all cases - high usefulness of a feature.

**Genetic Algorithms.** An interesting feature selection approach is to use Genetic Algorithms (GA). Distributed approaches to GA implementation have been practiced for years, they have been also used for microarray feature selection with superior performance it terms of classification accuracy when compared to Sequential Feature Selection [6]. A drawback of this approach is that it does not generate a global feature ranking, but a population of "species" of feature sets, with less optimal feature sets being overshadowed or brought into extinction by the winning sets.

## 2.2 Monte Carlo Feature Selection

Monte Carlo Feature Selection (MCFS) was originally proposed in 2008 by Draminski et al. [1] as a method of extracting informative features out of microarray datasets supervised by a categorical decision vector. The method was originally tested for microarray datasets with number of features below 10 000, and number of patients below a 100. As the size of datasets as well as computational power increased, nowadays that amount of features could be an output of a feature selection algorithm rather than an input. Nevertheless the method is powerful for detecting feature interactions, and its principals remain valid for larger datasets like Genome Wide Association Studies or Copy Number Variation analysis. The fundamental idea of MCFS is to create a global ranking of features, sorted by their Relative Importance (RI) in explaining the decision vector. This is achieved by allowing to check performance of the features free of influence of their competition, but at the same time giving them a chance to cooperate with each other. This is achieved by drawing many - in fact as many as possible - random subsets of the original feature set. Then, for each such a subset, another - as high as possible - number of train-test splits is created. Under each split, a decision tree is trained and evaluated on the test set. The performance of the tree measured as class weighted accuracy when predicting the class of the test observations serves as a measure of usefulness of the features included in the tree. Furthermore, the very structure of the created tree is used to measure the importance of the selected features. The higher in the tree and the greater the information gain on splitting on a specific feature, the greater its importance relative to other features. The outer loop of MCFS responsible for performing distinct feature samples iterates $s$ times, while the inner one - responsible for performing different training and testing splits of observations - $t$ times, resulting in a complete MCFS run requiring training $st$ classifiers on $st$ different training sets. The layout of data selected for an iteration is presented on Fig. 1. The information about feature's Relative Importance derived from such a single iteration for a single feature can be expressed using Eq. 1. Setting $u$ to values greater than 1 will favor features included in trees of high weighted accuracy. Setting it close to 0 would reduce the impact of the accuracy on the measure. High $v$ results in discriminating features that appear low in the decision trees, while setting it close to 0 removes the influence of coverage of the feature on its Relative Importance. For a more detailed explanation of the formula, the Reader may refer to the original paper of Draminski et al. [1].

$$RI_{f_i s_j t_k} = wAcc^u \sum_{node}^{nodes(f_i)} IG(node)(\frac{n.\ obs.\ in\ node}{n.\ obs.\ in\ tree})^v \qquad (1)$$

$RI_{f_i s_j t_k}$ - *the importance of feature i obtained from training set k of feature sample j (those parameters are omitted on lower levels of the equation)*
*wAcc - the weighted accuracy of the trained decision tree assessed on the temporary test set*
$nodes(f_i)$ - *a set of nodes of the decision tree where the split is made on feature i*
*IG(node(f)) - information gain on the node f of the tree, ex. obtained by using Gini purity measure*
*u,v - non-negative real tuning parameters*

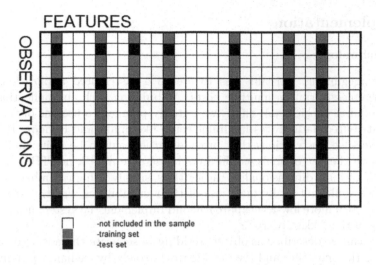

**Fig. 1.** Range of data selected in a single MCFS iteration

The RI information obtained from multiple trees trained over multiple samples can be put together simply by summing over all $st$ trees:

$$RI_{f_i} = \sum_{j}^{s} \sum_{k}^{t} RI_{f_i s_j t_k} \qquad (2)$$

This is extremely important when considering a parallel implementation, as there is no sequential relationship between the iterations - they can be all executed independently, and summed over asynchronously while the algorithm is running. After this is completed, all the features can be ranked based on their total RIs.

An important aspect of MCFS is determining what size of $s$ and $t$ parameters is enough. $s$ is especially important, since it is affected by the complex and hard to predict relationships of the features. The solution used in the original paper is to repeat the whole MCFS procedure for a growing value of $s$, thus obtaining a sequence of improving feature rankings. A distance between the feature rankings is defined, following Draminski et al., as a sum of absolute differences between positions of analogous features in two rankings, divided by the number of features taken into account - being upper p% features of the older ranking.

One can argue if the subsequent feature rankings should be completely independent, or it is acceptable to create the new ranking by combining RIs of the old one with new results. The first approach seems safer, as it proves that truly enough samples have been drawn to capture the characteristics of the dataset. On the other hand, combining subsequent rankings reduces time complexity. In the *dmLab* software by dr Draminski the latter approach is used. In our implementation, we have provided both options.

# 3  Implementation

## 3.1  Technologies

From the beginning, our aim was to create a distributed application being able to seamlessly run on multiple hosts. As building such an application from scratch requires considerable effort, we initially planned to use one of the JVM-based distributed data processing frameworks like Spark and Hadoop. However, MapReduce-like abstractions they provide turned out to be quite limiting for us, and with the size of analyzed datasets being several GB, not TB, and facing a problem of high dimensionality rather than simply a high number of observations, we could not really make use of their capabilities. For this reason, we decided to use a more low-level approach, and implement the system using actors, with help of the Akka library [9].

Actors can be described as objects residing in separate threads, and communicating with each other and the outside world solely by exchanging immutable messages. After sending a message to an actor, it is placed in his input queue. The Actor System guarantees that only one message is processed by the actor at a time. The actors are arranged in a hierarchy, with the first actor being created outside of the actor system - the rest of actors should be his offspring. This hierarchy defines error handing in case of actor failures - besides that, all the actors are equals and can communicate inside the Actor System freely.

## 3.2  Architecture

In our system - at the moment of writing - we use four major actor types, depicted on Fig. 2: *Master*, *Workers*, *Collector* and *DataActors*. Besides them, there exists the original thread which started the Actor System. After initialization, it exposes an instance of Application's API - the *Listener* - in Java RMI. The *Listener* acts as a bridge between the outside world and the Actor System by sending blocking synchronized messages to *Master* and awaiting answer. *Master* is the parent of all the other actors, and is responsible for handling their lifecycle. Furthermore, it is the only entry point for handling requests from the outside world. Communicating with the actor system is initiated from the outside - *Master* responds to these requests (like request for a status update, or for a current feature ranking) by sending back a message. *Master* can control an arbitrary number of *Workers*. *Workers* are ignorant of their task, and perform abstract "iterations" on an object interface pointing to the actual single-threaded "MCFS Engine" class created and configured in the *Listener* thread, and then copied and distributed by the *Master*. This "MCFS Engine" is operating using the Java Statistical Analysis Tool Library. (Original MCFS implementation by Draminski et al. is using Weka [4].) In a single "big iteration", the "MCFS Engine" outputs results of performing $\Delta$st real MCFS iterations. All the algorithm configuration (besides the input data) is contained within the "MCFS Engine", and created

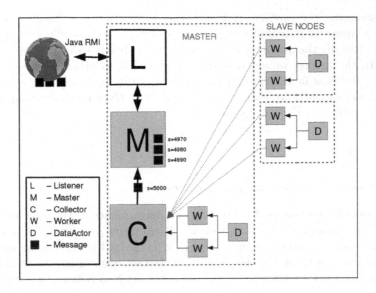

**Fig. 2.** Architecture of the Actor System, with data flow direction displayed by arrows.

in the *Listener* thread as a result of RMI calls from the outside world (being most likely a Web Server residing on the same machine). After performing an iteration, *Worker* sends a status message containing - among other things - the RI information to the *Collector*, and follows by sending a loopback message to itself. This allows it to remain responsive, in case *Master* requires it to stop or change behaviour (ex. to start permutation testing).

A very important actor is the *Collector* responsible for coalescing obtained partial RIs into full feature rankings, and calculating distances between those rankings. If the *Collector* detects reaching convergence (s being big enough), it starts to ignore incoming data, and signals the condition to the *Master*, which in turns switches the *Workers* to perform permutation testing. The results of permutation tests arrive at the *Collector* again, which coalesces them together, calculates p-values and sends the results to *Master*. As an additional crash countermeasure, the *Collector* serializes and stores itself on disk on each received update. As two most recent *Collector* images are always stored, a seamless restart of the whole system even if it is interrupted while performing the backup is possible.

While the computations are progressing, consequent statuses are being queued at the *Master*. Upon a status update request from the outside world, they are being sent outside and the queue is emptied. It is possible, that for a large number of features and *Workers*, the *Collector* will not be able to process incoming partial feature rankings on time. In this case, it is necessary to increase the step ($\Delta s$) or $t$ parameter, so that the *Workers* would send status updates less often.

For accessing the problem datasets, the application is relying on external systems. Currently, it handles text input stored on a local file system - it is assumed that all the physical nodes are able to reach the same network file system. In the future, we are planning to extend it to other data sources like external databases or distributed file systems. Still, accessing the data would happen independently at each node through the *DataActor*, creating feature samples in reply to *Workers* requests.

## 4   Results

### 4.1   Datasets and Testing Environment

Two datasets have been used for testing the performance of the new implementation. First dataset is the *Golub et al.* [3] microarray dataset. It is a small dataset for today's standards, nevertheless it is useful for running benchmarks. The other one is a Copy Number Variation dataset containing almost 2 million complete features measured on 52 radiosensitive and 83 radioresistant patients.

### 4.2   Comparison with DmLab [2]

Performance comparison, as well as basic speedup tests are presented on Fig. 3. Observed times are highly repetitive, however each presented value is a mean of three runs. Computations were performed on a small 4 core (i5 750) machine. Both algorithms were using overlapping feature rankings, sample sizes of 5 % of

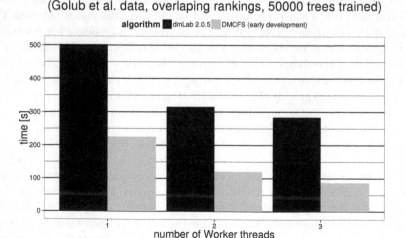

**Fig. 3.** Comparison of performance of *dmLab* and our implementation measured on *Golub et al.* datasets.

the total number of features, and were set to stop after reaching $s=5000$. Other parameters were set in the same way as in [1]. The distributed implementation was using 1–3 *Workers*, while *dmLab* was configured to use 1–3 threads.

### 4.3 Speedup Assessment

Speedup was measured by running the algorithm using independent feature rankings, that is with each new level of $s$ requiring training $st$ new classifiers. Computations were performed on one node of a computational cluster (see Acknowledgements), using consequently 6, 9 and 12 Worker threads. In addition, when running on 12 threads, 108 permutations ($ceil(100/12)*12$) were performed. A number like 10 000 permutations would be more appropriate for real analysis, however the reason for these calculations is only to provide a benchmark. Results are presented on Fig. 4. As before, nearly linear speedup can be observed.

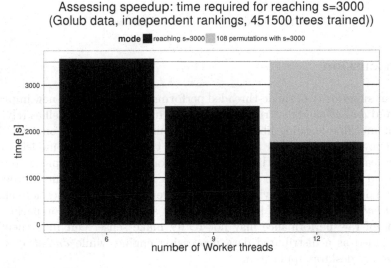

**Fig. 4.** Speedup on *Golub et al.* data observed while increasing the amount of threads available.

### 4.4 High Dimensional Dataset

At the end, the software was tested on a *Copy Number Variation* dataset containing almost 2 mln features. Initial distances of upper 1 % of consequent feature rankings obtained with this approach are presented on Fig. 5. A small sample size of 1000 features has been used.

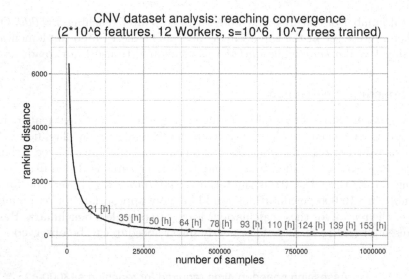

**Fig. 5.** Initial ranking distances obtained for the CNV dataset.

## 5   Discussion

Apparent superiority of single-threaded performance between our new implementation and *dmLab* can be attributed to smaller functionality, as well as relying on different machine learning libraries. These results could change once the Interdependency Discovery [2] is added, and new types of internal models able to perform feature selection with continuous response vector are implemented. Our distributed implementation's speedup when increasing the amount of processors is independent of the underlying library however, and seems to be more efficient than *dmLab*. Philosophies of both applications are different, so comparing them in terms of raw performance may not really make sense. Our implementation was designed as a distributed data processing engine, while *dmLab* is a more user friendly desktop application.

Setting the size of feature sample is critical for the algorithm's capability of finding feature interactions. In case of a dataset containing $2*10^6$ observations, setting the sample size to 1000 would make the chance of a single feature not being selected in a single sample equal 0.9995. Performing $10^6$ samples reduces this chance below $10^{-20}$, and makes the expected number of times each feature being selected around 500. The aim of MCFS is discovering interdependencies rather than individual strong features however - the latter can be obtained simply by performing individual one-dimensional tests. A $2*10^6$ feature dataset contains almost $2*10^{12}$ potential 2-feature interactions (number of 2-element combinations of a set). Similarly, a 1000 feature sample contains almost $5*10^5$ potential interactions. Performing $10^6$ samples makes the expected number of times for each 2-element combination of features being selected around 0.25 - so the population of potential interactions remains barely touched with these settings. A valuable preprocessing step would be to add a routine of testing system's performance with varying sample size - then a sample size that would satisfy a demanded quality criterion (like the expected number of hits for each feature pair or triplet) in the cheapest way would be automatically selected.

Current implementation of the *DataActor* loads the entire dataset into memory, and uses it to provide feature samples to *Workers*. However, sooner or later it may be required to process datasets that do not fit in the memory even in a single copy. This could be overcome by relying on external DBMSs storing the data in a transposed form (so that indexing by features is possible), and fetching individual samples into memory when required. Using high values of $t$ parameter would allow to limit the frequency of such requests.

Finally, for datasets combining the problems of high dimensionality and high volume of data, for which not even a feature sample fits the memory (and reducing the size of the feature sample is not acceptable), it could be necessary to rely on Big Data tools like Spark or Hadoop, possibly introducing two levels of parallelism - inner "Big Data" applications responsible for running distributed decision trees, and the outer asynchronous actor system.

# 6   Conclusion

Monte Carlo Feature Selection was originally created for handling datasets containing thousands of features. Our new implementation enhances this classical approach by allowing to run a satisfactory number of permutation tests. Furthermore it extends the size of analyzable problems by two orders of magnitude. The size of the feature sample should not be to small however - otherwise results would provide little more information than parallel single feature analysis.

**Acknowledgements.** I would like to thank dr Draminski for providing the latest version of dmLab software for evaluation, as well as Najla Al-Harbi, Sara Bin Judia, dr Salma Majid, dr Ghazi Alsbeih (Faisal Specialist Hospital & Research Centre, Riyadh 11211, Kingdom of Saudi Arabia), and furthermore Bozena Rolnik (Data Mining Group) for providing the CNV data. Calculations were carried out using the computer cluster Ziemowit (http://www.ziemowit.hpc.polsl.pl) funded by the Silesian BIO-FARMA project No. POIG.02.01.00-00-166/08 in the Computational Biology and Bioinformatics Laboratory of the Biotechnology Centre in the Silesian University of Technology. The work was financially supported by NCN grant HARMONIA UMO-2013/08/M/ST6/00924 (LK).

Finally I would like to thank Anonymous Reviewers who helped to increase quality of the paper.

# References

1. Draminski, M., Rada-Iglesias, A., Enroth, S., Wadelius, C., Koronacki, J., Komorowski, J.: Monte carlo feature selection for supervised classification. Bioinformatics **24**, 110–117 (2008)

2. Dramiński, M., Kierczak, M., Koronacki, J., Komorowski, J.: Monte carlo feature selection and interdependency discovery in supervised classification. In: Koronacki, J., Raś, Z.W., Wierzchoń, S.T., Kacprzyk, J. (eds.) Advances in Machine Learning II. SCI, vol. 263, pp. 371–385. Springer, Heidelberg (2010)
3. Golub, T., et al.: Molecular classification of cancer: Class discovery and class prediction by gene expression monitoring. Science **286**, 531–537 (1999)
4. Hall, M., Frank, E., Holmes, G., Pfahringer, B., Reutemann, P., Witten, I.H.: The weka data mining software: An update. SIGKDD Explor. **11**, 10–18 (2009)
5. International HapMap Consortium: The international hapmap project. Nature **426**, 789 (2003)
6. Luque-Baena, R.M., Urda, D., Subirats, J.L., Franco, L., Jerez, J.M.: Application of genetic algorithms and constructive neural networks for the analysis of microarray cancer data. Theor. Biol. Med. Model. **11**, 7 (2014)
7. Pearson, K.: On lines and planes of closest fit to systems of points in space. Philos. Mag. Series **6**(2), 559–572 (1901)
8. Perneger, T.: What wrong with Bonferroni adjustments. BMJ **316**, 1236–1238 (1998)
9. Quinlan, J.R.: Effective Akka. MO'Reilly Media, Inc. ISBN: 1449360076 9781449360078 (2013)
10. Sidak, Z.: Rectangular confidence regions for the means of multivariate normal distributions. J. Am. Stat. Assoc. **62**, 626–633 (1967)
11. Storey, J.D.: A direct approach to false discovery rates. J. R. Stat. Soc. Series B (Stat. Methodol.) **64**, 479–498 (2002)
12. The: An integrated map of genetic variation from 1,092 human genomes. Nature 491, 56–65 (2012)

# Architectural Challenges
# of Genotype-Phenotype Data Management

Michał Chlebiej[1], Piotr Habela[2], Andrzej Rutkowski[1], Iwona Szulc[1],
Piotr Wiśniewski[1(✉)], and Krzysztof Stencel[1]

[1] Faculty of Mathematics and Computer Science,
Nicolaus Copernicus University, Toruń, Poland
{meow,rudy,iwa,pikonrad,stencel}@mat.umk.pl
[2] Polish-Japanese Academy of Information Technology, Warsaw, Poland
habela@pja.edu.pl

**Abstract.** Medical research initiatives more and more often involve processing considerable amounts of data that may evolve during the project. These data should be preserved and aggregated for the purpose of future analyses beyond the lifetime of a given research project. This paper discussed the challenges concerned with the construction of the storage management layer for genotype-phenotype data. These data were used to research neurodegeneration disorders and their therapy. We outline the functionality of data processing services. We also present a flexible data-storage structure. Finally, we discuss the choices regarding database schema management and input sanitation and processing.

## 1 Introduction

The work presented in this paper has been as a part of the software development component of the NeuStemGen project. This project includes the research and clinical activities aimed at the development of new, more effective forms of therapy. The data being collected in the course of project involve the fields of neurology, ophthalmology and genetics. A dedicated software solution is required to assure data storage for the analysis during the project. It is also necessary for further data collection and sharing among the researches after the conclusion of the project. Cardionet [7] is an example of a successful medical data processing system.

Our current approach does not assume the direct storage in the complete genotype data sequenced. However, even despite this fact, the system needs to be prepared to handle large volumes of data, mainly due to the image data to be stored. The typical usage pattern represented by the applications of this kind [3] can be summarized as follows:

1. The shape of the repository evolves intensively in the course of the project.
2. Initially, the number of data contributing members is small. However, it is likely to increase along the subsequent product life.

© Springer International Publishing Switzerland 2016
S. Kozielski et al. (Eds.): BDAS 2016, CCIS 613, pp. 475–484, 2016.
DOI: 10.1007/978-3-319-34099-9_36

3. The number of the read-only users is large. They have varying privilege sets, not only with respect to update rights, but also in terms of establishing distinctive data sets within the system.
4. The repository after the research project conclusion is maintained, in order to collect further data and serve it for the purpose of other research initiatives, including e.g. the reading centre functionality.
5. The system is used by geographically distributed users.
6. Future extensions are likely, possibly by creating new, specialized client applications.

Hence, although there is a broad selection of mature technologies for data storage as well as for distributed system construction, a number of tough architectural decisions needs to be made. A significant amount of custom-build functionality is necessary to face the following specific challenges:

**Irregular data model**
It includes several core notions of the problem domain. However, the set of the features in newly added objects changes with time and across different data sets.
**High confidentiality and security requirements for sensitive data**
For some use cases it is unacceptable to anonymize patient data. Hence, appropriate security and data access logging mechanisms are unavoidable in the data access layer,
**Heterogeneous nature of the data content**
It requires combining large number of primitive values and voluminous binary data.
**Flexible schema mechanisms for data validity**
Although the system needs to be open for data structure variability, the data management layer should assure a degree of schema validation, like e.g. properly identifying data types of data submitted, enforcing the presence of particular obligatory attributes or files (and their naming patterns) in the sets being submitted.
**Database-managed consistency**
Given the fact the system is to be extended with another client applications, a proper amount of consistency control needs to be centralized within the data management layer.

The contributions of this article are twofold. Firstly, we present a hybrid storage architecture that offers advantages of schema-aware and schema-less techniques. Data whose schema can be firmly established at early design time are stored in a more efficient and less flexible way, i.e. in classic relational tables. Other data are stored in a more flexible, yet slightly less efficient way, namely in the EAV model. Such data are required by ever-changing client applications in our research project. Secondly, we present a novel module that extracts metadata from various medical imaging results.

The paper is organized as follows. Section 2 outlines the architecture of the storage system for the NeuStemGen project. Section 3 shows the specific database model used by this storage system. Section 4 sketches the data processing routines used in this system. Section 5 enumerates possible extensions to our system. Section 6 concludes.

## 2     System Architecture

Clearly, the abovementioned description indicates that we face the task of building a data-centric software. The primary concern is to keep the data consistent by exercising user privilege control and data validity check. Of course, some workflows dealing with data collection and approval might go beyond the simple CRUD-style transactions [5]. However, given the current set of requirements, the REST interface has been found sufficient to maintain the proper authorization and consistency policy. The REST-style remote interface introduces the necessary layer of indirection, allowing to achieve: (1) platform independence with respect to the client (messages sent as JSON over secure HTTP), and (2) persistence mechanism transparency, i.e. the ability to alter the storage technology without any impact on the client.

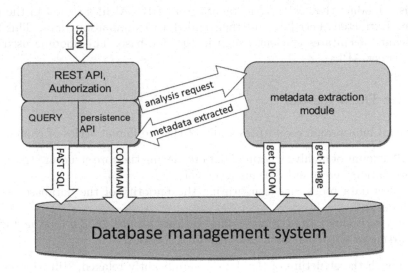

**Fig. 1.** Architectural components of the data storage layer

The resulting architecture includes the following components at the top level:

**REST interface module**
It handles the requests and performs authentication in order to determine if the requesting user is granted the respective privileges. The request is logged for audit purposes.

**Query interface**

It supports read-only operations against the data store. Thus, the query and update paths are separated at the top level. This is extremely important given the performance considerations. A separate database connection is maintained for this component.

**Persistent API**

It serves as the sole entry point for any data to be stored in the data-base. Hence, it is responsible for basic data validation. This component supports both the storage of explicitly handed named features, as well as the upload of BLOBs representing the image files. In the course of processing the latter, the metadata extraction module can be invoked in order to extract further features from the files.

**Database management system**

It is a replaceable component that may be changed accordingly to scalability requirements. Given the data structure flexibility requirements, only a part of traditional DBMS responsibilities is to be handled directly by it Others responsibilities require custom support from the service layer wrapping it.

**Metadata extraction module**

It is a module that extracts metadata from DICOM files stored in the system. Extracted metadata are then added to examination data. This significantly facilitates efficient search in the database. This module used the Persistent API.

## 3 The Data Model

The stored information constitutes a mix of three different styles of content:

1. small amount of highly structured data regarding the core structure (patients, examinations, etc.) and user management,
2. a binary data repository, constituting the majority of the database overall volume, and
3. loosely-structured data representing various groups of data or metadata occurring in particular data instances or categories.

Although the structure of the data is significantly relaxed, still, the precise knowledge of a data type may be essential for effective searching and processing of that data. The above variability may cause the dilemma of choosing the underlying database tool. On one hand, the scalability issues and the necessary flexibility may speak in favour of a some kind of NoSQL storage engine as is proposed in [6]. On the other hand, the security and data consistency concerns make an extended relational system a potentially attractive choice.

Taking the above considerations into account, we have picked the relational system, namely PostgreSQL for the initial exploration. The risk associated with that choice is mitigated by the ability to replace the data storage technology, as mentioned earlier. Within the chosen system we use its features regarding e.g. the transactional processing. However, we push much of the concerns of schema

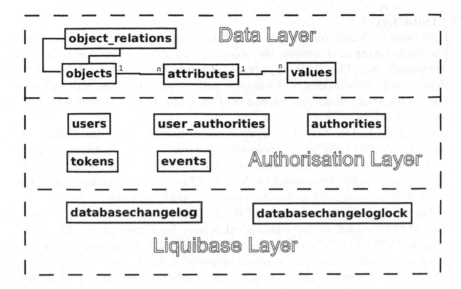

**Fig. 2.** The structure of the data model

management to the application layer. The data on users, doctors, administrators and the data keeping track of the system usage are stored in dedicated, traditional tables. On the other hand, the data describing patients and examination data are stored in an extensible EAV (Entity-Attribute-Value) structure inspired by database SONCA for the SYNAT project [4]. Data security is supported by dedicated RDBMS features. Additionally, the system follows the principle of no physical deletes. It means that is any update and removal is actually just the appropriate tagging of a given object or property.

Before, we dived deeply into the EAV storage model, we performed initial efficiency experiments. We used the following experimental setup. The test were executed on a database instance with 100 000 records in the table `objects`, 6 000 000 rows in the table `attributes` and 6 000 000 entries in the table `values`. We used a computer with the processor Intel Core i5 3570 (3.4/3.8 GHz) and 32 GiB RAM (DDR3, 1600 MHz). The system was stored on an SSD Kingston 120 GiB disk. The data was persisted on a 4x Caviar Black 1TB in the form of a Logic Volume. The operating system used was Debian 7.3. And last but not least the database management system was DBMS PostgreSQL 9.1. We used a native PostgreSQL version shipped together with the operating system.

The testing query was an equality selected based on two attribute values. This query was to return all attributes of the objects found. The running time of the testing queries were below half a second. The result of this test has been assumed to be satisfactory. The expected number of objects stored in the eventual database is roughly a dozen of thousands.

The data model of our storage system is divided into three following layers.

**The Data Layer**

This layer is based on an extended EAV model (Entity-Attribute-Value). The table `objects` represents the objects of patients, medical examinations, treatments etc. The table `object_relations` stores relationships between objects, e.g. associations of a patients and examinations or series of them. The tables `attributes` and `values` contain respectively attributes of objects and their values.

Our system does not store genetic data directly. If it did, significantly more date would have to be processed. That would notably increase the cost of the system's maintenance. However, the system collects results of generic examinations in the form useful for doctors. Outcomes of genetic examinations are stored as examination objects associated with patients.

The separation of attributes and values is necessary in order to be able to store lists of values. In particular a value may be a medical imaging scan or a whole DICOM [1]. The other functionality that need this splitting of tables is data archiving. The system never removes or alters the data in response to *update* or `delete` requests. Instead, new value records are inserted, while obsolete data is tagged so. It is also augmented with the information who performed the operation and when it happened.

**The Authorization Layer**

The tables of this layer define the schema of user management in the classic relational way. The table `authorities` contains the access rights. The table `users` stores data on users, while the table `user_authorities` contains the assignment of access rights to the users. The table `tokens` is exploited by the Spring token mechanism. The table `events` stores all events occurring in the system. It is used to generate audit reports, especially on data changes.

**The Liquibase Layer**

This is a technical layer used by the Liquibase mechanism used to manage the schema evaluation of the database.

# 4    Data Processing in the System

Various levels of transparency of particular kinds of machinery is used to perform the medical examinations. Moreover, there are different possible scenarios of acquiring that data. Therefore, there might be large amounts of data already collected that is not easily extractable from the different file formats.

If a contributor of data is required to manually input them into the database, he/she will frequently make errors. Additionally, if he/she is unaware of the file content, it may be an obstacle for effective anonymization of data served to the certain groups of users for which it is assumed. Hence, we introduce the metadata extraction module as a data extraction filter. Currently only data extraction is performed, but the same component may be potentially made responsible for anonymizing/filtering the file content. There is a broad range of the image data that might be submitted as the result of medical examinations.

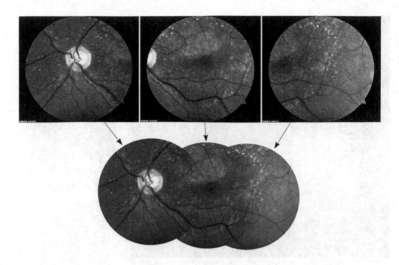

**Fig. 3.** An imaging of blood vessels.

For a typical, standardized DICOM formatted data [1] we have implemented a configurable mechanism for extracting all the information defined in this format's specification. However, numerous types of files are purely graphical. Thus they offer no straightforward way of extracting the metadata. A specific subcategory of such images are the graphical files that contain not only the image but also some textual and geometrical data.

In order to aid processing such image files, we have initiated a work on the development of advanced methods of OCR-based metadata extraction Its goal is to store it in the database in a processing friendly form. The data types to be the subject of extraction might be designated accordingly to cater the needs of the users. They may include both character and numerical data, as well as the plots and other objects consisting of graphical primitives.

Another subject of image data analysis may be the processing of image stereopairs to extract the spatial data. Additionally, for example in the eye imaging data, the vessel geometrical structure may be reconstructed. Such a preparation of data may allow specific analyses involving patient comparison, therapy progress assessment for a single patient, as well as 2D/3D/4D visualizations.

We present some examples of the visualization possibilities that can be included in our system (see Figs. 3–5).

Figure 3 presents a two dimensional Fusion as a result of partial registration procedures of three images delivered by our medical partners in this project. Similar procedures can be performed also in 3D for extension of 3D image space to improve visual inspection possibilities of the fused datasets.

Figure 4 presents the reconstruction of aerial spaces in human head. The paranasal sinuses were segmented in a semi-automatic way and extracted as separated objects for further numerical and visual analysis.

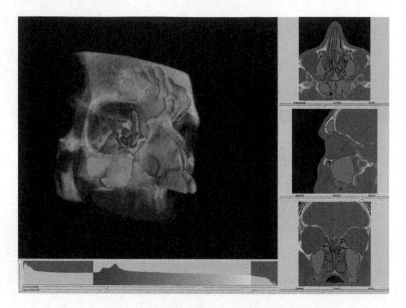

**Fig. 4.** An imaging of paranasal sinuses.

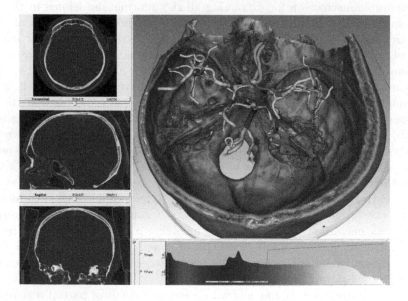

**Fig. 5.** An imaging of blood vessels in human head.

Figure 5 shows a geometrical reconstruction possibilities based on segmented vessels and arteries from 3D images of human head [8]. The objects were reconstructed as 3D tubes enabling numerical analysis of separated tubular segments and also a quantitative comparison of symmetrical objects. In case of eye 3D images such reconstruction can also be performed.

Our system can be easily extended by adding other automatic mechanisms to analyze medical imaging [2].

## 5   Possible Extensions of the System

From the above overview it is clear that the metadata extraction component constitutes an example of an extensible part of our architecture. Such a part should be treated as a generic mechanism rather than being a hard wired into the import functionality. Hence, this is one (and probably the most complex one) of the several extendibility aspects we have identified for the system under consideration. Below we enumerate and briefly comment on other possible extensions of the system.

**User groups and data groups**
They are needed to independently support different project initiatives and varying access policies within the system. It is assumed that (potentially many-to-many) assignments between users, user groups, data groups and data items may serve as the basis of the access control policy.

**Examination types**
They allow defining and subsequently exercising the validity and completeness constraints on the submitted data. For example, a type describing a set of neurological examination data might indicate the names and types of obligatory and optional attributes.

**File (re)naming, content type and completeness rules**
They facilitate validating a submitted file set for a given defined examination kind.

**Image metadata extraction and filtering profiles**
This aspect may involve (apart from possible validation) the metadata extraction and/or subtraction routines to be applied against the images being uploaded.

**Data submission, supplementation and approval workflows**
The respective activities might involve several steps to be performed by designated user roles in appropriate sequence. Hence, an additional coordination rules may need to be defined as a constraint over the sequence REST transactions applicable to a given resource.

## 6   Conclusions

In this paper we discussed challenges we have faced in the NeuStemGen project during the design and implementation of the data storage layer. We had to prepare a system that will flexibly accept a really wide range of data sorts, that furthermore significantly evolve in time. As the result of our efforts we have come out with a robust system architecture that allows satisfying all functional and non-functional requirements. The implementation of this architecture facilitates fast query processing and supports effective metadata extraction. We have also rolled out a plethora of further extensions of the system.

**Acknowledgements.** This work is sponsored by the project "Innovative strategy for diagnostics, prevention and adiuvant therapy of selected neurodegenerative disorders in population of Polish ancestry" co-funded by The National Centre for Research and Development, contract no. STRATEGMED1/234261/2/NCBR/2014.

# References

1. Bidgood, W.D., Horii, S.C.: Modular extension of the ACR-NEMA DICOM standard to support new diagnostic imaging modalities and services. J. Digit. Imaging **9**(2), 67–77 (1996)
2. Chlebiej, M., Nowinski, K., Scislo, P., Bala, P.: Development of heart motion reconstruction framework based on the 4d echocardiographic data. Ann. UMCS, Inform. **8**(2), 43–50 (2008)
3. Filipowicz, W., Habela, P., Kaczmarski, K., Kulbacki, M.: A generic approach to design and querying of multi-purpose human motion database. In: Bolc, L., Tadeusiewicz, R., Chmielewski, L.J., Wojciechowski, K. (eds.) ICCVG 2010, Part I. LNCS, vol. 6374, pp. 105–113. Springer, Heidelberg (2010)
4. Grzegorowski, M., Pardel, P.W., Stawicki, S., Stencel, K.: SONCA: Scalable semantic Processing of rapidly growing document stores. In: Pechenizkiy, M., Wojciechowski, M. (eds.) New Trends in Databases and Information Systems. AISC, vol. 185, pp. 89–98. Springer, Heidelberg (2012)
5. Kulbacki, M., Segen, J., Habela, P., Janiak, M., Knieć, W., Fojcik, M., Mielnik, P., Wojciechowski, K.: Collaborative tool for annotation of synovitis and assessment in ultrasound images. In: Chmielewski, L.J., Kozera, R., Shin, B.-S., Wojciechowski, K. (eds.) ICCVG 2014. LNCS, vol. 8671, pp. 364–373. Springer, Heidelberg (2014)
6. Mazurek, M.: Applying NoSQL Databases for Operationalizing Clinical Data Mining Models. In: Kozielski, P., Stanislaw, K., Mrozek, D., Kasprowski, P., Boz, M. (eds.) (2014)
7. Sierdzinski, J., Bala, P., Rudowski, R., Grabowski, M., Karpinski, G., Kaczynski, B.: KARDIONET: Telecardiology based on GRID technology. In: Medical Informatics in a United and Healthy Europe - Proceedings of MIE 2009, The XXIInd International Congress of the European Federation for Medical Informatics, Sarajevo, Bosnia and Herzegovina, Agust 30 - September 2, 2009, pp. 463–467 (2009)
8. Żurada, A., Gielecki, J., Shane Tubbs, R., Loukas, M., Maksymowicz, W., Chlebiej, M., Cohen-Gadol, A., Zawiliski, J., Nowak, D., Michalak, M.: Detailed 3d-morphometry of the anterior communicating artery: potential clinical and neurosurgical implications. Surg. Radiol. Anat. **33**(6), 531–538 (2011)

# Appling of Neural Networks to Classification of Brain-Computer Interface Data

Malgorzata Plechawska-Wojcik$^{(\boxtimes)}$ and Piotr Wolszczak

Lublin University of Technology, Nadbystrzycka 38D, 20-618 Lublin, Poland
{m.plechawska,p.wolszczak}@pollub.pl

**Abstract.** The paper presents application of neural networks to the construction of a brain-computer interface (BCI) based on the Motor Imagery paradigm. The BCI was constructed for ten electroencephalographic (EEG) signals collected and analysed in real time. The filtered signals were divided into three groups corresponding to the information displayed to users on the screen during the experiments. ANOVA analysis and automatic construction of a neural network (NN) classification were also performed. Results of the ANOVA analysis were confirmed by the neural networks efficiency analysis. The efficiency of NN classification of the left and right hemisphere activities reached almost 70 %.

**Keywords:** Neural networks · Brain-computer interface · EEG data · ANOVA

## 1 Introduction

Electroencephalography (EEG) is a method of measuring the synchronous activity of neurons [4]. A characteristic feature of EEG signals is the presence of rhythms (waves) defined as structures with a characteristic frequency range and repeatable shape. They are associated with the state of patient activity. There are five basic types of brain waves. In the presented study two of them were applied: alpha and beta waves. Alpha waves (8–12 Hz) are activities occurring during a state of relaxation, detectable especially in occipital lobe, responsible for visual information processing. Beta waves are low amplitude activities occurring in the 12–30 Hz range. Beta waves are reflected primarily in the frontal lobe and represent poorly synchronised work of neurons specific to everyday, typical cortical activity. A single EEG study is carried out usually in the period between several minutes to several hours. It may serve as a diagnosis tool (coma, epilepsy, sleep disorders, monitoring the patient during the operation) or might be helpful in studies of neurological and functional properties of the brain. EEG is also used in the study of schizophrenia and personality disorder tests [4,5]. Modern applications of EEG are also dedicated to the construction of brain-computer interfaces (BCI) and applying them to supporting not only disabled people, but also users of computer games.

© Springer International Publishing Switzerland 2016
S. Kozielski et al. (Eds.): BDAS 2016, CCIS 613, pp. 485–496, 2016.
DOI: 10.1007/978-3-319-34099-9_37

The term brain-computer interface appeared in the 1970s [25], but the recent development of technology has made practical realisations of the BCI idea possible. A brain-computer interface [26] is "a communication system that does not depend on the brain's normal output pathways of peripheral nerves and muscles." So, the BCI is a system of direct communication between man and machine without nervous and muscular way, allowing user to control devices without verbal or physical (muscular) interaction [16].

The primary objective of BCI systems is to enable communication for paralyzed people suffering from so-called closing syndrome [20,26]. The BCI is also successfully applied in treatment of patients with neurological diseases [23]. A popular application of BCIs is also neurofeedback therapy [9,17] supporting the process of learning, concentration, and treatment of attention deficit hyperactivity disorder [6]. It is becoming increasingly popular to use BCIs for entertainment in computer games controlled by brain waves [22,24]. The construction of BCI is a complex task, because signal needs to be preprocessed, reduced and classified accroding to particular features selected depending on the type of BCI. What is more, differences between individuals need to be taken into account. Online analysis requires well-prepared signal, adjusted classifier parameters [27], incompleteness data handling [19].

Among the BCI systems based on EEG signals one can distinguish systems based on different paradigms. The presented study is concentrated on the brain-computer interfaces based on ERS / ERD. These interfaces use the phenomenon of brain waves oscillation coming from the vicinity of the sensorimotor cortex by imagining the movement of various body parts (for example limbs, tongue, thumb). Applied phenomenon is based on Event-related synchronization (ERS) and Event-related desynchronisation (ERD) [18,26].

The aim of the paper is to compare EEG data classification with classical supervised learning algorithm with ANOVA analysis and neural networks. The analysis is perform for the needs of the BCI based on the motor imagery paradigm. The BCI used in the study was constructed based on the ERD/ERS ratio and the LDA algorithm. Data gathered during BCI sessions were then analyzed offline using ANOVA and neural networks. The classification efficiency comparative study was also performed.

The rest of the paper is structured as follows. The Sect. 2 is dedicated to the process of the EEG data analysis for the purpose of BCI based on the motor imagery paradigm. It covers all stages of the analysis including data preprocessing consisting in signal filtering and preparing, feature extraction related to the ERD/ERS ratio and supervised data classification with LDA algorithm. The Sect. 3 describes details of the BCI used in the study whereas Sect. 4 presents the performed experiment details. Section 5 presents data analysis results. It covers both ANOVA analysis and neural networks classification and analysis. Summering discussion is presented in Sect. 6.

# 2    Analysis of EEG Data

Analysis of EEG data is a complex process composed of several steps. These steps depend on the purpose of the analysis, but in the case of Brain-Computer Interfaces one can distinguish such steps as signal preprocessing, features extraction and classification leading to obtaining a control signal.

## 2.1    Data Preprocessing

Preprocessing is a necessary stage of EEG data analysis. It covers such stages as artifact elimination, frequency filtration and spatial filtration. All preprocessing steps are performed to raise the signal-to-noise ratio and strengthen the EEG signal which has relatively low amplitude.

A serious problem limiting the quality of the EEG signals are artifacts, which are unwanted signals of non-cerebral origin. Artifacts in the signal can be misleading. During the EEG signal recording one can distinguish several types of artifacts. Depending on their origin, they can be divided into two groups: technical artifacts related to the registration signal and biological artifacts originating from the person examined. They all can cause different shape artefacts.

Detection and correction of artefacts is carried out under the preliminary signal processing. Digital EEG measurement allows for frequency filtering and appropriate selection of the EEG assembly, which makes it easier to interpret low quality signal. However, the total elimination of artefacts is not possible.

Among the most commonly used for artifacts eliminating methods one can find Principal Component Analysis (PCA) [3,11] and Independent Component Analysis (ICA) [7], which are used primarily for solving the problem of muscle artifacts. The PCA method decomposes the signal into uncorrelated components [12]. However, for separarion of similar amplitude artifacts, the ICA method is applied. It decomposes the signal into mutually independent components. However, applying the ICA might cause loss of part of the EEG signal [8], especially if removed artefacts are similar to those of the EEG signal [2].

## 2.2    Feature Extraction

Feature extraction is dedicated to extracting certain characteristics of the EEG signal [10]. Feature extraction in BCIs based on the Motor Imagery Paradigm rests on the calculation of the ERD/ERS value [21]. Calculation of ERD/ERS in the time domain for a given frequency band.

## 2.3    Classification

Linear Discriminant Analysis (LDA) is the basic classifier applied to the tested BCI. It implements so called allocation of feature space [15]. LDA is a method applied to both feature reduction and classification [1]. The LDA method is based on searching of new coordinates that will be used to project and explain the data. The idea is to distinguish class membership. The new coordinates should be defined to maximise the between-class scatter and minimise the within-class scatter. In the LDA method new components may not be orthogonal.

In the classification task associated with the division of the feature space, it is assumed that each category is represented in the feature space by a certain limited set of standard features (characteristics) [28]. Another strategy is to identify the separating surface. Usually, such surfaces are N-dimensional hyperplanes, where N is the number of features used. For the purposes of classification, in addition to determining the best direction of projection a to maximise the distance between the projected averages for classes (regarding component variances), class separating hyperplanes need to be indicated.

## 3   BCI Construction

The brain-computer interface applied in the study was based on the motor imagery (MI) paradigm. This paradigm is based on the so called *pre-motor potential* or *readiness potential* which might be detected for about a second before the body movement [14]. This potential, however, is weak and multiple averaging of the set of EEG movement signals is needed to see it.

Sensorimotor rhythms applied in the MI-based BCI are $\mu$ (8–12 Hz) and $\beta$ (18–26 Hz) EEG rhythm. The maximum amplitude value of these rhythms appear in the sensorimotor cortex at the rest stage. Research shows [7] that a person does not need to perform a move, its imagery is enough.

During the study event-related EEG responses were gathered. Event-related synchronisation (ERS) and event-related desynchronisation (ERD) are analyzed in the form of the ERD/ERS ratio. Changes in this ratio result from changes in the activity of a neuron population and are characterised by a short-term change of power in certain EEG frequency bands. Changes of ERD and ERS may occur simultaneously in different areas of the cortex and might be analysed as a function of both time and frequency.

Due to a very good time resolution of electroencephalography one can observe changes occurring in the brain at different stages of the body movement: preparation to move, execution of movement and return to a resting state.

Due to the fact that single signals are too weak to detect significant changes frequently used multiple repeats are needed for averaging the results and to improve the classification [13].

Construction of the BCI applied in the study was based on the motor imagery of the left and right hand. The BCI was designed in OpenVibe environment connected with a 21 channel EEG amplifier - Mitsar EEG 201.

The signal was gathered from ten electrodes: C3, C4, FC3, FC4, C5, C1, C2, C6, CP3, CP4 placed in a special EEG cap to keep them in the right position. In addition, the ground electrode was placed in the centre of the frontal lobe and two reference electrodes (A1, A2) were placed on ears.

The samples were recorded with frequency of 360 Hz and processed in real time. The signal was gathered from the device Mitsar 201 using electrodes attached to the cap located on the head. The EEG amplifier transmitted data to a computer. The BCI scenario was prepared and realised in the OpenVibe application. Implementation of the experiment was based on four scenarios:

1. *Online test* dedicated to check the quality of the signal. For a good quality signal is one which low impedances and visible changes during blinking and clamping teeth.
2. *Online calibration* allowing to adapt the parameters of the BCI interface to the particular user. The user was asked to imagine left and right hand movements in random order imposed by the scenario.
3. *Offline training* where the applied classifier was learned. The signal was filtered with a Butterworth Band pass filter to select only alfa and beta bounds (frequency range of 8–24 Hz). Surface small Laplacian filter was also calculated to reduce the number of dimensions from ten to two, representing the left and right hemisphere. Next, the signal was epoched regarding to stimulation (4 seconds of epoch duration) and time (1 second of epoch duration). Next, the ERD/ERS ratio was calculated and the LDA classifier was trained. K-fold cross validation was applied to check the training performance.
4. *Online testing* processing of EEG signals in real-time. The signal was filtered, epoched and averaged and processed by the classifier learned in the training. The results are displayed in the form of the distance to the separation plane. One class is represented with a negative value for one class whereas the second one with a positive value.

## 4 The Experiment

The experiment was tested on five users of different sex, age, and handedness in age of 22–40 years. Each person was tested several times. Jointly 36 single studies were performed.

The aim of the BCI scenario was to move the stripe to the left or right side of the screen using motor imagery. The desired side is presented to a user in the form of an arrow. The scenario is composed of repetitions of a series: arrow and moving stripe. There are twenty repetitions of left and twenty of right hand movements. The EEG signals registered during the observation and imagination are presented at the computer screen, the arrows indicating the directions of the left and right movements. Raw EEG signal was gathered from ten electrodes distributed symmetrically in both hemispheres, representing images of the activity of the respective brain hemispheres. The signal was reduced to two channels with Laplacian Filter. Resulted signals were subjected to statistical analysis including neural networks based classification. Average and variability of data as well as results of ANOVA were also compared.

Two filtered signals were divided into groups corresponding to the motor imagery obtained on the basis of signs presented on the computer screen during the simulation. The results were divided into two basic categories related to study stages: left arrow (L) and right arrow (R), which were displayed to inform the user of the desired direction of the imagined move. There was also an additional category: black screen (B) displayed before the stimulus, defined as a rest time, no arousal state. Comparison of the results was performed separately for each test. Each person was calibrated individually, so the analysis also was processed separately. The comparison covered representative average values (average and median) and variability measures of three groups of the filtered signal.

## 5    Results

The result section is divided into two parts: ANOVA analysis and Neural Network based classification.

### 5.1    ANOVA Analysis of Representative Signals

The signal analysed was the EEG Surface Laplacian for left and right side of the brain. Figure 1 illustrates an exemplary distribution (study test no. 101) of the filtered signal values recorded during the left-hand imagery preceded by displaying the left arrow (group L). Signal values were subjected to analysis of variance (ANOVA).

The analysis of variance was performed due to the fact that the histogram indicated a normal distribution. The ANOVA analysis showed a statistical significance ($pvalue < 0.05$) of differences between the filtered signals for groups L, R and B in all 36 studies. Only in four tests (of the same person) p value achieved the appropriate level of intergroup differences for only one of the two representative signals.

The results of the ANOVA analysis was verified on the basis of box-plot diagrams presenting the localisation of the mean values. Figure 2 presents the position of the mean value of the filtered signals 0 and 1 for the groups L, R and B (Left arrow, Right arrow, and Black screen representing relaxation).

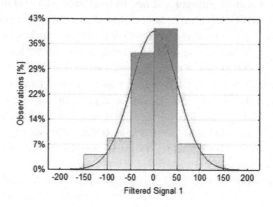

**Fig. 1.** Example of a histogram of filtered signal 1in group L during examination no 101

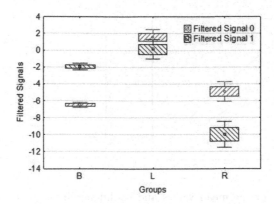

**Fig. 2.** Box-plot obtained for examplary study results

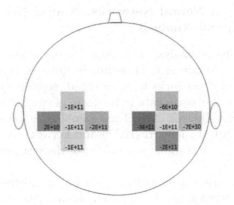

**Fig. 3.** Example of hemisphere activities: the two images represent the activity of the brain hemispheres. The colour of pixels corresponds to the values of signals from the electrodes

The hatched field define the interval mean standard error, whereas the whiskers represent intervals: $mean \pm 1.96 * standarderror$.

Moreover, signals from particular hemispheres were compared. Figure 3 presents a map of the hemisphere activities.

Analysis of the box-plot example presented in Fig. 5 shows that intervals determined by the mean value $\pm$ standard error of the mean are separable. A similar situation was observed in the vast majority of the tests. However, the position of the group means with regard to the other groups were changed. It is exemplified in Fig. 4, where interactions are presented. Figure 4 presents exemplary interaction of the mean values and confidence intervals calculated for representative filtered signals.

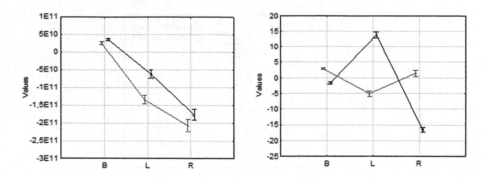

**Fig. 4.** Interaction of the mean values and confidence intervals for two examples of study tests

## 5.2  Construction of Neural Networks. Neural Network Based Classification and Analysis

The Neural Network-based analysis was dedicated to the assessment of the ability to recognise the type of brain arousal (motor imagery or rest) during the study. For this purpose, neural networks were designed, separately for each test. The network was designed to classify data in the form of two feature vectors obtained after preprocessing and filtering of the EEG signal. The role of classification was to separate three groups: L (left), R (right) and B (rest). A random sample was divided into: (70 %) of the training set, (15 %) of the test set and (15 %) a set of validation.

Neural Network construction and analysis was performed in Statistica software. The network entrance was two filtered signals representing the activation of the left and right brain hemispheres. As the network output, three neurons corresponding to the three groups L, R and B were applied. In the study both radial networks (RBF) and linear networks (MLP) were tested. In the case of MLP networks, network testing was performed with the number of neurons in the hidden layer in the range of 3 to 9, whereas for radial networks in the range of 21 to 30. To evaluate the network performance (as the error function) the sum of squared errors and cross entropy the was applied. The sum of squared errors was applied to the thirteen networks whereas cross entropy was applied to twenty three networks. Due to the application of cross-entropy as the function of the error, network exits showed direct probability of belonging to the classes. This improved the accuracy of the classification. The most commonly used functions dedicated to hidden neurons' activations were: logistics functions (applied fourteen times) and hyperbolic tangent (Tanh, 1 applied eleven times). In contrast, the most common function of three output neurons activation was gradient-log-normaliser of the categorical probability distribution (Softmax). In the group of networks representing the best input signal classification performance the most

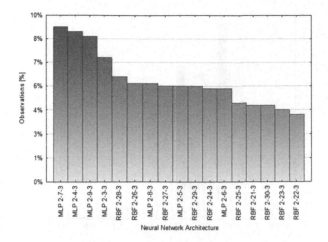

**Fig. 5.** Pareto analysis of the architecture of 720 neural networks (36 studies x the best 20 networks) distinguished during the assessment of the filtered signals classification according to the directions of motor imagery

frequently applied network architectures were MLP 2-9-3 (9-fold) and MLP 2-7-3 (9-fold) having respectively nine and seven neurons in the hidden layers. In the originally constructed collection of 720 network (36 studies x 20 networks), where for each study the best 20 neural networks were retained, the most commonly used were the architectures with the following number of neurons in the hidden layer: 7 (65), 4 (63), and 9 (61).

Table 1 and Fig. 5 present the statistics and distributions of testing and validation results of the best neural network obtained for thirty six motor imagery EEG studies of left-right arrow simulations. The mean values of the quality of learning, testing and validation of networks were in the range of 67.42 to 70.73 % of correct classification.

The modes (the highest bars) and average values (with values marked with maxima of normal distributions) were presented in Fig. 6 the average value of the validation of the network reaching 68.83 %.

**Table 1.** Descriptive statistics of the best neural network designed for individual EEG studies of left-right arrow simulations. The number of networks: 36.

|  | Mean | Median | Minimum | Maximum | Standard deviation |
|---|---|---|---|---|---|
| Quality (learning) | 68,59 | 68,44 | 67,94 | 70,11 | 0,513 |
| Quality (testing) | 68,59 | 68,51 | 67,42 | 70,30 | 0,705 |
| Quality (validation) | 68,84 | 68,84 | 67,67 | 70,73 | 0,582 |

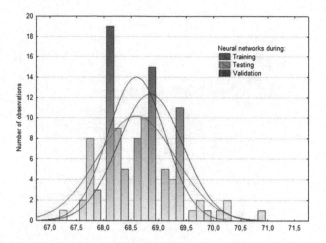

**Fig. 6.** Neural networks performance of training (blue), testing (red) and validation (green) (Color figure online)

# 6    Summering Discussion

The paper presents analysis results of the data obtained in the study of applying the brain-computer interface based on the motor imagery paradigm. In the study, brain activity was gathered in the form of EEG data. The experiment was dedicated to left and right hand movement imagery. The series of imagery tasks were preceded with the presentation of an arrow indicating the desired direction. In the study five persons of different sex, age, and handedness were tested. As a result of the study, 36 records were gathered and subjected to individual processing and analysis. Brain activity maps of the left and right hemispheres were generated on the basis of the signal recorded from ten electrodes. These signals were preprocessed: filtering, artifact removal and ERS/ERD ratio was calculated. Spatial filtering was also applied to reduce the data dimensionality to feature vectors. During the study the LDA classifier was performed. After the study postprocessing analysis was done. The filtered signals were divided into three groups corresponding to information displayed to users on the screen during the experiments. ANOVA analysis and automatic construction of a neural network classification were also performed.

The ANOVA analysis was calculated for filtered EEG signals divided into groups (L, R and B) according to user activity (Left imagery, Right imagery, Rest). The analysis allows to conclude that the applied measurement might be used to distinguish different signals generated by the human brain, indicating the direction of movement imagery. In addition, the differences in the EEG activity of the right and left hemispheres during different activities (left hand motor imagery, right hand motor imagery, resting) were confirmed. These results were confirmed by the neural network efficiency analysis. Neural networks were applied in the task of classification of left and right motor imagery in EEG data. The efficiency of neural networks classification of left and right hemisphere activities reached almost 70 %.

The present research is a stage in the construction process of a robot system controlled by processed EEG data. The results obtained do not as yet ensure adequate repeatability of the driving mechanism. However they justify further tests and improvements. Future works could also cover integrating EEG with additional techniques, such as fMRI and eye-tracking to increase the accuracy of the results and confirm their reliability.

# References

1. An, X., Kuang, D., Guo, X., Zhao, Y., He, L.: A deep learning method for classification of EEG data based on motor imagery. In: Huang, D.-S., Han, K., Gromiha, M. (eds.) ICIC 2014. LNCS, vol. 8590, pp. 203–210. Springer, Heidelberg (2014)
2. Barbati, G., Porcaro, C., Zappasodi, F., Rossini, P., Tecchio, F.: Optimization of an independent component analysis approach for artifact identification and removal in magnetoencephalographic signals. Clin. Neurophysiol. **115**, 1220–1232 (2004)
3. Blinowska, K., Kaminski, M.: Multivariate signal analysis by parametric models. In: Schelter, B., Winterhalder, M., Timmer, J. (eds.) Handbook of Time SeriesAnalysis. WILEY-VCH Verlag GmbH & Co. KGaA, Weinheim (2006)
4. Bromfield, E., Cavazos, J., Sirven, J.: Basic mechanisms underlying seizures and epilepsy. In: An Introduction to Epilepsy (2006)
5. Broniec-Wojcik, A.: Ph.D. dissertation. AGH, Krakow (2013)
6. Cho, B., Lee, J., Ku, J., Jang, D., Kim, J., Kim, I., Kim, S.: Attention enhancement system using virtual reality and EEG biofeedback. In: Virtual Reality, Proceedings, IEEE, pp. 156–163 (2002)
7. Croft, R., Barry, R.: Eog correction: a new perspective. Electroencephalogr. Clin. Neurophysiol. **107**, 387–394 (1998)
8. Croft, R., Barry, R.: Removal of ocular artifact from the EEG: a review. Neuro. Physiol. Clin. **30**, 5–19 (2000)
9. Diab, M., Ismail, G., Al-Jawha, M., Hsaiky, A., Moslem, B., Sabbah, M., Taha, M.: Biofeedback for epilepsy treatment. In: Mechatronics and its Applications (ISMA), pp. 1–4. IEEE (2012)
10. Geethanjali, P., Mohan, Y., Sen, J.: Time domain feature extraction and classification of eeg data for brain computer interface. In: 2012 9th International Conference on Fuzzy Systems and Knowledge Discovery (FSKD). IEEE (2012)
11. Joyce, C., Gorodnitsky, I., Kutas, M.: Automatic removal of eye movement and blink artifacts from EEG data using blind component separation. Psychophysiol. **41**, 313–325 (2004)
12. Jung, T., Makeig, S., Humphries, C., Lee, T.W., Mckeown, M.J., Iragui, V., Sejnowski, T.J.: Removingelectroencephalographic artifacts by blind source separation. Psychophysiol. **37**, 163–178 (2000)
13. al-Ketbi, O., Conrad, M.: Supervised ANN vs. unsupervised SOM to classify EEG data for BCI: Why can GMDH do better? Int. J. Comput. Appl. **74**(4), 37–44 (2013)
14. Kornhuber, H.H., Deecke, L.: Changes in human brain potentials before and after voluntary movement studied by recording on magnetic tape and reverse analysis (1964)

15. Koronacki, J., Cwik, J.: Statisticcal learning systems (in Polish: Statystyczne systemy uczace sie), Exit (2008)
16. Lee, S., Abibullaev, B., Kang, W., Shin, Y., An, J.: Analysis of attention deficit hyperactivity disorder in EEG using wavelet transform and self organizing maps. In: Control Automation and Systems (ICCAS), pp. 2439–2442 (2010)
17. Mingyu, L., Jue, W., Nan, Y., Qin, Y.: Development of EEG biofeedback system based on virtual reality environment. In: Engineering in Medicine and Biology Society, pp. 5362–5364 (2005)
18. Neuper, C., Miller, G., Kebler, A., Birbaumer, N., Pfurtscheller, G.: Clinical application of an eeg-based brain-computer interface: a case study in a patient with severe motor impairment. Clin. Neurophysiol. **114**(3), 399–409 (2003)
19. Nowak-Brzezińska, A., Jach, T.: The incompleteness factor method as a support of inference in decision support systems. In: Kozielski, S., Mrozek, D., Kasprowski, P., Małysiak-Mrozek, B. (eds.) BDAS 2014. CCIS, vol. 424, pp. 201–210. Springer, Heidelberg (2014)
20. Pfurtscheller, G., Neuper, C.: Motor imagery and direct brain-computer communication. Proc. IEEE **89**(7), 1123–1134 (2001)
21. Pfurtscheller, G., da Silva, L.: Functional Meaning of Event-Related Desynchronization (ERD) and Synchronization (ERS), pp. 51–65 (1999)
22. Shim, B., Lee, S.W., Shin., J.H.: Implementation of a 3-dimensional game fordeveloping balanced brainwave. In: Software Engineering Research, Management & Applications, SERA 2007 (2007)
23. Suresh, K., Heng, J.: Quantitative eeg parameters for monitoring and biofeedback during rehabilitation after stroke. In: IEEE/ASME International Conference on Advanced Intelligent Mechatronics (2009)
24. Van Vliet, M., Robben, A., Chumerin, N., Manyakov, N., Combaz, A., Van Hulle, M.: Designing a brain-computer interface controlled video-game using consumer grade EEG hardware. In: Biosignals and Biorobotics Conference (BRC 2012), pp. 1–6. ISSNIP (2012)
25. Vidal, J.: Toward direct brain-computer communication. Ann. Rev. Biophys. Bioeng. **2**(1), 157–180 (1973)
26. Wolpaw, J., Birbaumer, N., Heetderks, W., McFarland, D., Peckham, P., Schalk, G., Vaughan, T.: Brain-computer interface technology: a review of the firstinternational meeting. IEEE Trans. Rehabil. Eng. **8**(2), 164–173 (2000)
27. Żbikowski, K.: Time series forecasting with volume weighted support vector machines. In: Kozielski, S., Mrozek, D., Kasprowski, P., Małysiak-Mrozek, B. (eds.) BDAS 2014. CCIS, vol. 424, pp. 250–258. Springer, Heidelberg (2014)
28. Zielosko, B.: Optimization of inhibitory decision rules relative to coverage - comparative studys. In: Kozielski, S., Mrozek, D., Kasprowski, P., Małysiak-Mrozek, B., Kostrzewa, D. (eds.) BDAS 2014. CCIS, vol. 521, pp. 267–276. Springer, Heidelberg (2014)

# Data Processing Tools

# Content Modelling in Radiological Social Network Collaboration

Riadh Bouslimi[1,2](✉), Mouhamed Gaith Ayadi[1,2], and Jalel Akaichi[2]

[1] Higher Institute of Technological Studies, Jendouba, Tunisia
bouslimi.riadh@gmail.com, mouhamed.gaith.ayadi@gmail.com
[2] BESTMOD Lab, Computer Science Department,
ISG-University of Tunis, Le Bardo, Tunisia
j.akaichi@gmail.com

**Abstract.** We present in this paper a model representation of a report extracted from a radiological collaborative social network, which combines textual and visual descriptors. The text and the medical image, which compose a report, are each described by a vector of TF-IDF weights following an approach "bag-of-words". The model used, allows for multimodal queries to research medical information. Our model is evaluated on the basis imageCLEFMed' 2015 for which we have the ground truth. Many experiments were conducted with various descriptors and many combinations of modalities. Analysis of the results shows that the model, which is based on two modalities allows to increase the performance of a search system based on only one modality, that it be textual or visual.

**Keywords:** Multimodal modeling of medical information · Fusion multimodale · Radiological social network · Multimodal information retrieval · Bag-of-words

## 1 Introduction

In recent years, the development of social health networks such as Patients-LikeMe[1], DailyStrength[2], MedPics[3], Carenity[4], etc. has led to an explosion in the number of media collections. The wealth and diversity of these collections makes access to useful information increasingly difficult. It has become essential to develop indexing and search methods appropriate to social networks taking into account the different modalities contained (text, image, video, etc.). Standard collections as, TREC, ImagEVAL or ImageCLEFMed, for evaluating such systems. The largest number of systems handling multimedia documents exploits only the textual part of the documents. The standard approach is to represent a text as a

---

[1] https://www.patientslikeme.com/.
[2] http://www.dailystrength.org/.
[3] https://www.medpics.fr/.
[4] https://www.carenity.com/.

© Springer International Publishing Switzerland 2016
S. Kozielski et al. (Eds.): BDAS 2016, CCIS 613, pp. 499–506, 2016.
DOI: 10.1007/978-3-319-34099-9_38

"bag of words" [11] and to associate with each word a weight that characterizes his frequency by the TF-IDF method. In wineskins, terminology extraction in collaborative social networks is to seek the most frequent words in the comments. The challenge now is to extend the system to other modalities, including medical images, since they are extensively used in the social networks in healthcare. A natural approach is to use the representation based on "bag-of-word" for modeling the image. This approach has already shown its effectiveness in particular for medical image annotation applications [1,8]. We propose a model that combines both visual and textual information in collaborative social health networks. The pertinence of our model is evaluated on a search for medical information. This is to compare the results obtained with our model to those obtained with a single modality, textual or visual. In this article, we firstly present our data representation model extracted from a collaborative social health network and its use in the context of research-oriented health information. Next, we show the results of application on ImageCLEFMed' 2015 data set.

## 2 Data Representation Model in Collaborative Social Health Networks

The proposed model is to describe text and images with textual and visual terms. The two modalities are firstly processed separately for each using the "bag-of-words" approach for visual and textual description. Indeed, they are represented as a vector of TF-IDF weights characterizing frequency of each of the visual or textual words. The vector describing textual content is cleaned by using the UMLS thesaurus. To use the same mode of representation for the two modalities can combine them with a late fusion method, to make after multimodal queries to retrieve information. This general methodology is presented in Fig. 1.

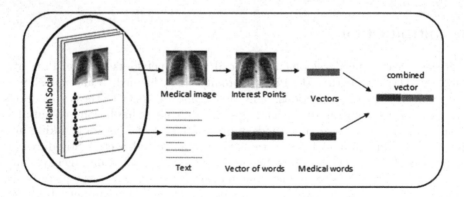

**Fig. 1.** Representation of content collaborative social health network

## 2.1   Representation of Textual Modality

To represent a text report in the form of a weight vector, it is first necessary to define an index of textual or vocabulary terms. For that, we will apply initially a stemming algorithm with Snowball and we delete the black words from all reports. The indexing will be performed by the Lemur-software[5]. In bottles, the terms selected are then filtered using the thesaurus UMLS. Each report is then represented following the model of Salton [11], is as a weight $vector r_i^T = (w_{i,1}, ..., w_{i,j}, ...w_{i,|T|})$, where $w_{i,j}$ represents the weight of the term $t_j$ in a report $r_i$. This weight is calculated as the product of two factors $tf_{i,j}$ and $idf_j$. The factor $tf_{i,j}$ is the frequency of occurrence of the term $t_j$ in the report $r_i$ and the factor $idf_j$ measures the inverse of the frequency of the word in the corpus. Thus, the weight $w_{i,j}$ is even higher than the term $t_j$, and frequent in the report $r_i$ and rare in the corpus. For the calculation of $tf$ et $idf$, we are using the formulations defined by Robertson [10]:

$$tf_{i,j} = \frac{k_1 n_{i,j}}{n_{i,j} + k_2(1 - b + b\frac{|r_i|}{r_{avg}})} \tag{1}$$

where $n_{i,j}$ is the number of occurrences of the term $t_j$ in the report $r_i$, $|r_i|$ is the report size of $r_i$ and $r_{avg}$ is the average size of all reports in the corpus. $k_1$, $k_2$ and $b$ are three constants that take the respective values 1, 1 and 0.5.

$$idf_j = \log \frac{|R| - |r_i|t_j \in r_i| + 0.5}{|r_i|t_j \in r_i| + 0.5} \tag{2}$$

where $|R|$ is the size of the corpus and $|r_i|t_j \in r_i|$ is the number of documents in corpus, which the term $t_j$ appears at least once. A textual request $q_k$ can be considered a very short text report, it can also be represented by a weight vector. This vector is noted $q_k^T$, will be calculated with formulas of Roberson but with b = 0. To calculate the relevance score of a report $r_i$ opposite an request $q_k$, we apply the formula given by Zhai in [13] and defined as:

$$score_T(q_k, r_i) = \sum_{j=1}^{|T|} r_{i,j}^T q_{k,j}^T \tag{3}$$

## 2.2   Representation of Medical Image

The representation of the visual modality is carried out in two phases: the creation of a visual vocabulary and the representation of the medical image using there of. The vocabulary V of the visual modality is obtained using the approach "bag-of-words" [2]. The process consists of three steps: the choice of regions or interest points,the description by calculating a descriptor of points or regions and grouping of descriptors into classes constituting the visual words. We use

---

[5] https://www.carenity.com/.

two different approaches for the first two steps.The first approach uses a regular cutting of the image into $n^2$ thumbnails. Then a color descriptor with 6 dimension, denoted Meanstd, is obtained for each thumbnail, by calculating the mean and standard deviation of normalized components $\frac{R}{R+G+B}$, $\frac{G}{R+G+B}$ et $\frac{R+G+B}{3\times 255}$ where R, G and B are the colors components. The second approach uses the characterization of images with regions of interest detected by the MSER [7] and presented by their bounding ellipses (according to the method proposed by [9]). These regions are then described by the descriptor SIFT [6]. For the third step, the grouping of classes is performed by applying the k-means algorithm on the set of descriptors to obtain k clusters descriptors. Each cluster center then represents a visual word. The representation of an image using the vocabulary defined previously for calculating a weight vector $r_i^V$ exactly as for the text modality. To obtain a visual words from the medical image, we first calculate the descriptors on the points or regions of the image, and then is associated,at each descriptor, the word vocabulary, the nearest in the sense of the Euclidean distance.

### 2.3   Fusion Model

From our two vocabularies T and V, we calculate a scores obtained which is a linear combination of scores for each category: $score(q_k, d_i) = \propto score_V(q_k, d_i) + (1+ \propto)score_T(q_k, d_i)$.

This score corresponds to a scalar product of two vectors representing respectively the query and the document and each having a textual part and a visual part. The $\propto$ parameter used to weight the amount of information flowing through each modality.

## 3   Experimental Evaluation

### 3.1   Test Data and Evaluation Criteria

The pertinence of our model is evaluated on the collection provided for the competition ImageCLEFMed' 2015 [3]. This collectionis composedover 45.000 biomedical research articles of the PubMed Central (R). Each document is composed of an image and a text part. The images are very heterogeneous in size and content. The text part is relatively short with an average of 33 words per document. The goal of the information search task is to return to the 75 queries supplied by ImageCLEFMed' 2015 a list of pertinent documents. All requests have a textual part, but many do not have a query image. Order to have a visual part for each query, we use the first two pertinent medical images returned by our system when we use only the textual part. This corresponds to a relevance feedback fact by the system user. The criteria of average accuracy (Mean Average Precision - MAP), which is a classic criteria in information retrieval, is then used to evaluate the pertinence of the results.

**Table 1.** Result of average precision obtained for different modalities.

| Modality type | MAP |
|---|---|
| SIFT | 0.1287 |
| Meanstd | 0.0962 |
| Text | 0.2346 |
| Fusion :Text + SIFT | 0.2762 |
| Fusion : Text + Meanstd | 0.2494 |

## 3.2   Results and Discussion

To demonstrate the contribution of the use of our model compared to only textual or visual model,we realized experiments using a single modality, textual or visual,then experiments combining two modalities, modality text with visual descriptor,this visual features for both Meanstd and SIFT previously presented. The text vocabulary is consists of approximately 200000 wordswhereas the two visual vocabularies are constituted of 10000. Table 1 summarizes the MAP values obtained for each experiment. On the one hand, it can be stated that the use of the single visual modality irrespective of the descriptor used leads to poorer results that the use of the only modality text. On the other hand, combine a visual descriptor with the text improves search performance with the only textual descriptor. These overall observations are confirmed by the precision/recall

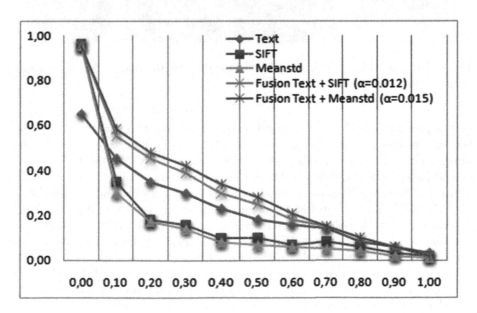

**Fig. 2.** Precision / Recall curve for different modalities (text only, visual only and fusion text / visual).

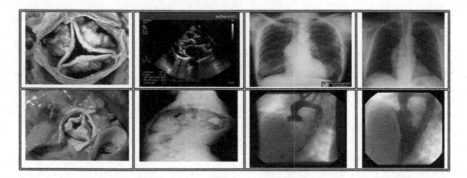

**Fig. 3.** Results obtained with the textual modality for the query "Aortic stenosis", images 3 and 4 are selected for visual query.

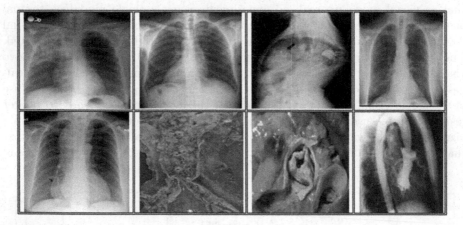

**Fig. 4.** Results obtained with the textual modality for the query "Aortic stenosis", images 3 and 4 are selected for visual query.

curves presented in Fig. 2. A detailed analysis per request show that, for some, the first results returned by the visual modality is best for text modality. For illustration, Figs. 3, 4 and 5 show the results for the query "Aortic stenosis". We can add,about the performance obtained with a visual modality, that the regular division of the associated image at Meanstd color descriptor is more robust than MSER+SIFT. We explain this behavior by clustering problems.With the color descriptor, we work with 6 characteristic parameters and 4 million thumbnails to consolidate vocabulary words. With SIFT descriptor, we have 128 features and settings 54 million thumbnails. In the second case, the thumbnails are divided-very irregularlyin the space of descriptors, because of the use of MSER, the large size and the large amount of data. This situation is very unfavorable for cluster-ing algorithms such as K-means [5]. Also, it has been shown in [4,12] that the descriptors of the most densely spaces of the parameter space are not necessarily the most informative.

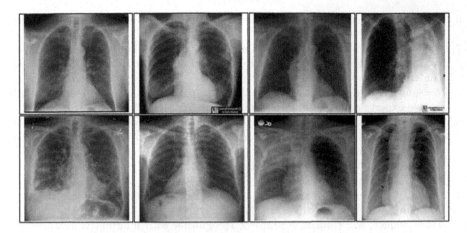

**Fig. 5.** Results obtained for the modalityfusion of the query "Aortic stenosis".

## 4 Conclusions

We have presented in this paper a representing model of multimedia data extracted from radiological social networks where they are used for annotating medical images. This model is based on a late fusion of the textual and visual information using the "bag-of-words" approach. The performance of the indexing and search system has been evaluated based on real dataset of Image-CLEFMed' 2015 and the obtained results were very promising to use a media model specialized radiology, like the one proposed to retrieve information from a collection of radiological media. Indeed, the fusion of both textual and visual methods allows each time to increase the performance of the system. In this context, Larlus [5] proposes a clustering method that allows uniform quantifications of spaces contrary to K-means which focuses on dense spaces. This method could be used to improve our system when creating the visual vocabulary.

## References

1. Bouslimi, R., Akaichi, J.: Automatic medical image annotation on social network of physician collaboration. J. Netw. Model. Anal. Health Inform. Bioinform. **4**(10), 219–228 (2015)
2. Csurka, G., Dance, C., Fan, L., Willamowski, J., Bray, C.: Visual categorization with bags of keypoints. In: ECCV 2004 Workshop on Statistical Learning in Computer Vision, pp. 59–74 (2004)
3. de Herrera, S.G.A., Muller, H., Bromuri, S.: Overview of the imageclef 2015 medical classification task. In: Working Notes of CLEF (Cross Language Evaluation Forum) (2015)
4. Jurie, F., Triggs, W.: Creating efficient codebooks for visual recognition. In: ICCV (2005)
5. Larlus, D., Dork, G., Jurie, F.: Cration de vocabulaires visuels efficaces pour la catgorisation d'images. Reconnaissance des Formes et Intelligence Artificielle (2006)

6. Lowe, D.: Distinctive image features from scale-invariant keypoints. Int. J. Comput. Vis. **60**(2), 91–110 (2004)
7. Matas, J., Chum, O., Martin, U., Pajdla, T.: Robust wide baseline stereo from maximally stable extremal regions. In: Proceedings of the British Machine Vision Conference, pp. 384–393 (2002)
8. Messoudi, A., Bouslimi, R., Akaichi, J.: Indexing medical images based on collaborative experts reports. Int. J. Comput. Appl. **70**(5), 0975–8887 (2013)
9. Mikolajczyk, K., Tuytelaars, T., Schmid, C., Zisserman, A., Matas, J., Schaffalitzky, F., Kadir, T., Van Gool, L.: A comparison of affine region detectors. Int. J. Comput. Vis. **65**, 43–72 (2005)
10. Robertson, S., Walker, S., Hancock-Beaulieu, M., Gull, A., Lau, M.: Okapi at trec-3. In: Text REtrieval Conference, pp. 21–30 (1994)
11. Salton, G., Wong, A., Yang, C.: A vector space model for automatic indexing. Commun. ACM **18**(11), 613–620 (1975)
12. Vidal-Naquet, M., Ullman, S.: Object recognition with informative features and linear classification. In: ICCV, pp. 281–288 (2003)
13. Zhai, C.: Notes on the lemur tfidf model (2001)

# Features of SQL Databases for Multi-tenant Applications Based on Oracle DBMS

Lukasz Wycislik[⊠]

Institute of Informatics, Silesian University of Technology,
16 Akademicka Street, 44-100 Gliwice, Poland
lukasz.wycislik@polsl.pl

**Abstract.** The paper presents several architectural aspects of data layer for developing of multi-tenant applications. Multi-tenancy is a term describing the application model of services delivering in which a single or many instances of a software run on a server and serves multiple tenants. This feature has an impact on several nonfunctional aspects of system such as security, availability, backup, recovery and more. This article clarifies the specificity of these aspects in approach to multi-tenant applications building and points how different data layer architecures address them. On an example of Oracle database serveral build-in features and concepts that could be helpful to increase the quality of mensioned nonfunctional aspects were discussed.

**Keywords:** Oracle · Multi-tenancy · Applications · Cloud computing

## 1 Introduction

Contemporary trends in a field of data computing move toward centralization.

A hundred years ago companies were producing electricity by their own to support their needs. But soon they realized that it is unprofitable. Entrusting the production of eletricity to a large specialized unit reduces the cost of its production and also allows to do it at a higher level of quality and reliability.

Similarly today, thanks to the rapid development of information technology, business units are more often deciding to transfer data processing from it's data centers to specialized infrastructure maintained by large companies. This specialization allows ones to focus on business goals and functional aspects of software, leaving responsibility for infrastructure requirements to specialized service providers. This business model falls within the definition of cloud computing.

The whole basic concept of cloud computing is presented for example in [3].

A multilayered software architecure is a software architecture that uses many layers for allocating the different responsibilities of an application. Data layer of software is a piece of system and infrastructure responsible for storing of persistent data. This part of system is often considered the most troublesome due to problems in maintenance, performance, scalability or availability. Common

S. Kozielski et al. (Eds.): BDAS 2016, CCIS 613, pp. 507–517, 2016.
DOI: 10.1007/978-3-319-34099-9_39

issues related to layered architecture can be found in [5] and in particular to Microsoft database layer in [2].

Multi-tenancy is a term describing the application model of services delivering in which a single or many instances of a software run on a server and serves multiple tenants.

There are several definitions of multi-tenancy computing [6] but most of them assume 'logical isolation' of data and/or business processes for each tenant - eg. [4] and this situation is the subject of the article. Situation when several entities have common data may be considered as one tenant with several departments (when there is responsibility for shared data on an owner of departments) or several tenants (when there is responsibility for shared data on an outside institution (e.g. dictionaries or registers of administration or legal institutions). The last situation is usually resolved by loose coupling approach where data replication or web services integration is used.

Taking into account that cloud computing could support the great scalability and availability it seems to be the best place to drive multi-tenant applications. This deployment scenario usually gives the best benefit resulting from the fact that software services/applications are served from one specialized place to many different entities. But building multi-tenant applications requires special approach to system architecture that must ensure an adequate level of data separation between tenants, one's data migration to another cloud and others [1, 17].

Multi-tenancy is an aspect that affects the whole system architecture and all its layers. Considerations relating to the application layer and business processes are presented e.g. in [13–15].

The main point of the article is to show possible ways of software development toward the concept of multi-tenancy to Oracle database architects. Documentation of Oracle database is complete and comprehensive but Oracle as a business entity is interested in promotion the most complete solution, which is also the most expensive. In aspect of multitenancy there are several articles written by Oracle's engineers but most of them limit considerations to PDB option (pluggable databases) - e.g. [8] where Oracle admits it has such an approach by following words:

"The Oracle view on multitenancy is that user data becomes the tenant - and as you can run multiple user data stores (or containers as Oracle calls them) in the same database - you have a multi-tenant database. Oracle complements this by separating the metadata from the user data and can point multiple user data stores to a common set of metadata, thus achieving better hardware utilization and with that better elasticity of the database. Or in other words - you can run more database on the same server with 12c."

Unfortunatelly this is the most expensive option because it requires the Enterprise Edition of database and PDB option that is payed seperately. The article describes also other approaches toward multi-tenant development based on cheaper or even free options. Here one can find tips what architecure will be suitable to one's non-functional requirements and budget limits.

## 2 Advantages and Challenges of Building Multi-tenant Applications

One of 'traditional' deployment model assumes that company owns technical infrastructure and deploy there aplications needed to support their business proceses. In such a model this company to themselves is responsible for all aspects of running system. It must then act proactively what means anticipation in demand for computing power, disk space etc. Also all security and availability issues are its concern.

Moving data processing toward specialized data centers depending on the model of operating [16] (e.g. IaaS, PaaS, SaaS, ..) allows one to transfer responsibility for some aspects to supplier of infrasturture. The company may act then reactively what means setting the parameters of the instrastructure to the current (not future) requirements for processing power or disk space, what allows for paying the costs of resources actually used but not allocated for future. The most mature model often involves payments for business indicators such as the number of system users or the number of transactions made. This is often called 'pay-as-you-go'.

Dependly of deployment model (PaaS, SaaS, IaaS) the application is being built by a supplier of infrastructure or by an independent software vendor. But no matter who is bulding muli-tenant applications the problems still remain the same and many of them also relates to the data layer.

The whole situation is further complicated by the fact that different tenants may have different expectations for SLA (Service Level Agreement) so the architecture of system should be flexible enough to ensure SLA for each tenant at the level it expects.

### 2.1 Scalability

Scalability is the capability of system to handle a growing amount of work.

In cloud infrastructure this can be achived both by vertical and horizontal way. Vertical way means adding computing resources to one application instance while vertical way means multiplying application instances.

Multi-tenant concept assumes many tenants being supported by single/ unified infrastructure but this may be done by single application instance or by many application instances - each dedicated to concrete tenant. In practice mixed models may be useful when many 'microtenants' are supported by one instance of application but many instances of application support one 'macrotenant'.

### 2.2 Availability

Availability is the probability that a system will work as required during the period of a mission.

For computer systems higher availability may be achived by elimination of single points of failure but concrete approach depends on the required level of availability and higher level means higher costs.

For multi-tenant application it must be considered that each tenant may has different requirements. System architecture should be flexible enough to meet high requirements but on the other hand fulfilling lower requirements for less demanding tenants should entail lower costs.

## 2.3  Manageability

Multi-tenant environments are supposed to operate in a competitive market thus from the perspective of tenants to change supplier of infrastructure should not be associated with a significant undertaking.

For cloud owners this means the need of elasticity not only in model of mapping tenants to application instances but primarily in the management of persisten data.

The data, often with embarrassingly large volume, should preferably be stored separately for each of tenants so as to allow independent backups or giving back the tenant its historical data when it resigns from cooperation with that cloud.

## 2.4  Security

Multi-tenancy brings new challenges in terms of data security. Providing services to multiple tenants from a single application or a single infrastructure requires a change in approaches to the architecture of the system so as to ensure that data belonging to one tenant will not be made avaible to the other.

This could be done on several data access levels. Easiest way is to stop there on embedding relevant business rules at tha application level what gives only basic security. The best level of security can only be achieved by controlling and separating access to data at the dababase management system level.

# 3  Database Models

Relational databases are used as permanent application data marts for years, and nothing indicates that in the near future this will change. Individual suppliers of DBMS compete in qualities such as performance, security, scalability etc. The result is that only well-established functions are subject to standardization organizations (such ANSI) and the remaining part of the functionality is specific to each provider. As the multi-tenant applications are a relatively new concept they have not developed uniform mechanisms ready to specifics in different variants of deployment.

The following subsections describe the use of different options of An Oracle Database in the context of different variants of multi-tenant applications implementation, taking into account earlier described non functional requirements.

## 3.1   Common Schema

Logical model of applications running as multi-tenant is quite trivial and shown in Fig. 1.

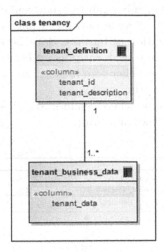

**Fig. 1.** Tenancy model

As we can see the multi-tenant system has to identify each tenant and store all business data related to it wherein the business data in practice consist of structures of complex relationships.

The simplest data model that implements the above problem is shown in Fig. 2.

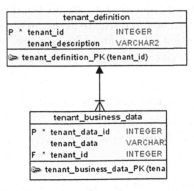

**Fig. 2.** Common schema

This implementation requires the least effort on the side of the data layer leaving security and access control aspects to be implemented at the application layer side.

With already developed application working in a sigle-tenant one has to put the least effort to adapt it to multi-tenant model.

But this solution does not address the problems of scalability, availability and manageability. Business data of all tenants are collected together and only the foreing key determines membership to a particular tenant.

By using the technique of weak entities, what is illustrated in Fig. 3, we gain complete logical separation of business data between tenans but all other problems remain untouched. A weak entity is an entity that cannot be uniquely identified by its attributes and therefore it must use a foreing key in conjuction with its attributes to create primaty key.

This can be helpful in case where tenants' data can sometimes be identical in terms of values of some rows of a table.

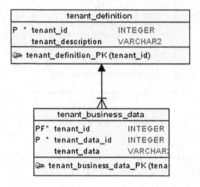

**Fig. 3.** Common schema with weak entities

Because above solutions assume data of all tenants physically are stored together, the only improvment may relate to security issues where thanks to Oracle Label Security option [7] we can enforce access control to each row (i.e. to validate whether the tenant operates on the data really belonging to him) at the database level.

**Listing 1.1.** View subsetting

```
create view 123_business_data
          as select * from tenant_business_data
                              where tenant_id = '123';

grant select, update, delete on 123_business_data to 123;
```

The more generic but also more complicated technique could be creating a subset of database perpectives selecting rows by foreign key dedicated to each tenant individually and granting permission for each subset to dedicated tenant (e.g. Listing 1.1).

## 3.2  Partitioning

Partitioning is a very powerful Oracle Database option [10] which can be used to improve the performance and manageability for handling large volumes of data. For multi-tenant applications particularly useful is partitioning by list what allows for collecting data in separate files for each tenant (e.g. Listing 1.2).

**Listing 1.2.** Partitioning by list

```
alter table tenant_business_data
        add partition 123 values (123)
        tablespace 123;
```

This technique improves scalability, because the time needed to find information for specific tenant depends almost entirely on the amount of data stored by this particular tenant.

Partitioning also makes the system more easily manageable. Storing data in the physical distribution of tenant makes it easy to start services for new tenants or stop them for tenants with expired contract.

With the ability to do backups of individual tablespace both backupping and restoring processes can be done exactly with the requirements of the SLA. The option of transportable tablespaces [9] allows tenants to move data from one cloud to another (when the target cloud is also based on Oracle Database).

## 3.3  Separated Schemas

Database schema is a logical subset of database objects (e.g. tables, indexes). This objects are usually grouped in schema to support well definded business domain. In Oracle database each schema can have one (default) tablespace and must have one owner. Although the access rights can be defined to concrete database objects usually it is easier to manage them at the whole schema level.

All of above causes separation using schemas to be a useful tool in the implementation of multi-tenant applications. In this model each tenant has its own schema what gives great manageability and quite good security. The idea is ilustrated in Fig. 4.

To easily adapt existing applications to the above model, one should use the default schema mechanism. This is based on switching to the desired schema immediately after establishing database connection (e.g. Listing 1.3).

**Listing 1.3.** Setting default schema

```
alter session set current_schema = 123;
```

Obviously all subsequent SQL commands should contain the names of objects not preceded by name of the owner.

As in the case of partitioning so here we can get physical separation of data (dedicated tablespaces) what improves manageability and availability. The disadvantage, however, is more complicated administration when adding or removing

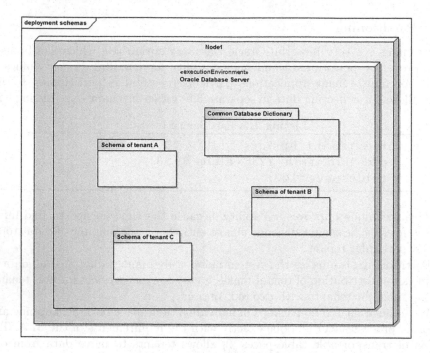

**Fig. 4.** Separated schemas

tenants to the system because it will require to create or drop database schemas. One must also remeber that identical SQL phrases are treated as different when they relate to different schemas what has bad consequences for parsing and query execution time.

## 3.4   Pluggable Databases

Pluggable databases [11] is the option introduced in the newest version of Oracle database 12c. This native muli-tenant option allows a single container database to host multiple separated pluggable databases. The basic concept is shown in Fig. 5.

Container database works inside instance processes and acts as traditional database with data dictionary, memory structures and standard users as SYSTEM or SYS. This database can be a container to many separated and independent pluggable databases. Each of pluggable database is an autonomous unit that can be unplugged and plugged to any container.

From designer and developer point of view the pluggable database is fully compatible with previous version of Oracle database what allows to port an application build as single-tenant directly into multi-tenant world without need of any modifications.

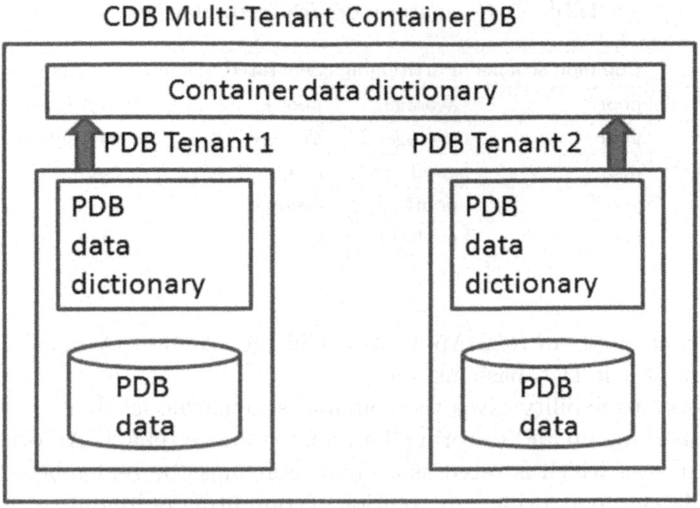

## CDB Multi-Tenant Container DB

**Fig. 5.** Pluggable databases

From administrator point of view each pluggable database can be maintained separately. It can be opened, shutted down and backed up independently of the other.

## 4   Summary

The article shows problems specific to building data layer of multi-tenant application ready to deploy in the cloud infrastructure.

A summary assessments of various architectural approaches are presented in Table 1.

These assessments actually do not come from experimental studies but for non-funtional requirements such as scalability, manageability, security it would be difficult to come up with metrics (how contrary it could be done e.g. in respone times). The assessments are determined on the basis of comparisons of option to one another. E.g. it seems obvious and also results from experience that solution based on access control at application server side is less secure then solution based on access control both at application and database sever side, and thus it was worse rated.

As regards the horizontal scalability at the storage level the best options seem to be partitioning and pluggable databases - both of them are extra payed option of enterprise edition of Oracle Database. Pluggable databases seem a bit ahead of the partitiong due to total transparency for the database administrator. Due to this transparency pluggable databases can always even coexists with all other approaches (partitioning, common schema, separated schema). But when we talk about horizontal scalability at the processing power level we must remember that each option is drived by single operating system so adding computing

**Table 1.** Comparison of architectural aspects

|               | Common schema | Partitioning | Separated schemas | Pluggable databases |
|---------------|---------------|--------------|-------------------|---------------------|
| scalability   | poor          | average      | poor+             | average+            |
| availability  | poor          | average      | average           | average             |
| manageability | poor          | good         | good              | good                |
| security      | poor*         | poor*        | average           | good                |
| cost          | low           | medium       | low               | high                |

nodes is possible only in Real Application Cluster model or by multiplication of independent Oracle Database instances.

The worst availability gives the common schema model due to inability to selectively backing up and restoring data for a contrete tenant. All other models offer similar level which is rated as average as it must be remembered that the higher availability may bring only solution of computing redundation (e.g. Oracle Data Guard).

Good manageability offer partitioning, separated schemas and pluggable databases options but it must be noticed that separated schemas will require the most concern from the database administrator.

The best security is offered by pluggable databases option which offers completly seperated local user for each of pluggable database. The option of seperated schemas can also provide sufficient protection. The worst evaluated options are common schema and partitioning where access control must be implemented at application level. However, there exist custom methods to improve safety e.g. through the use of Oracle Label Security option.

Taking into account the cost of each option [12] is worth mentioning that both common schema and separated schemas model come for free (one only have to pay for standard edition licence). But mentioned above Oracle Label Security option requires enteprise edition and is payed separately thus implementing it to common schema gives the 'high' cost while the combination Oracle Label Security with partitioning option (also payed seperately) makes this solution the most expensive ('highest') - even more expensive than PDB option that also requires Enterpise Edition and additional payment.

As it can be seen the architectural approach to database layer of multi-tenant applications is still not a trivial task and should be carried out on the basis of both non-functional requirements and economic calculations regarding cost of licenses, SLA, etc.

**Acknowledgements.** This work was supported by NCBiR of Poland (No INNOTECH-K3/IN3/46/ 229379/NCBR/14).

# References

1. Al Sheshri, W.: Cloud database: database as a service. Int. J. Database Manage. Syst. **5**(2), 1 (2013)
2. Bernstein, P.A., Cseri, I., Dani, N., Ellis, N., Kallan, A., Kakivaya, G., Lomet, D.B., Manne, R., Novik, L., Talius, T.: Adapting microsoft SQL server for cloud computing. In: ICDE. IEEE Computer Society, April 2011. http://research.microsoft.com/apps/pubs/default.aspx?id=152656
3. Durao, F., Carvalho, J., Fonseka, A., Garcia, V.: A systematic review on cloud computing. J. Supercomputing **68**(3), 1321–1346 (2014). http://dx.doi.org/10.1007/s11227-014-1089-x
4. Gartner: It glossary. (visited on November 2015). http://www.gartner.com/it-glossary/multitenancy
5. Jain, H., Reddy, B.: Layered architecture for assembling business applications from distributed components. J. Syst. Sci. Syst. Eng. **13**(1), 60–77 (2004). http://dx.doi.org/10.1007/s11518-006-0154-2
6. Kabbedijk, J., Bezemer, C., Jansen, S., Zaidman, A.: Defining multi-tenancy: a systematic mapping study on the academic and the industrial perspective. J. Syst. Softw. **100**, 139–148 (2015). http://dx.doi.org/10.1016/j.jss.2014.10.034
7. Oracle: Oracle label security with oracle database 12c. (visited on November 2015), June 2013. http://www.oracle.com/technetwork/database/options/label-security/label-security-wp-12c-1896140.pdf
8. Oracle: Oracle multitenant. (visited on November 2015), June 2013. http://www.oracle.com/technetwork/database/multitenant/overview/multitenant-wp-12c-2078248.pdf
9. Oracle: Oracle database 12c: full transportable export/import. (visited on November 2015), January 2014. http://www.oracle.com/technetwork/database/enterprise-edition/full-transportable-wp-12c-1973971.pdf
10. Oracle: oracle partitioning with oracle database 12c. (visited on November 2015), September 2014. http://www.oracle.com/technetwork/database/options/partitioning/partitioning-wp-12c-1896137.pdf
11. Oracle: Plug into the cloud with oracle database 12c. (visited on November 2015), October 2015. http://www.oracle.com/technetwork/database/plug-into-cloud-wp-12c-1896100.pdf
12. Oracle: Oracle technology global price list. (visited on January 2016), January 2016. http://www.oracle.com/us/corporate/pricing/technology-price-list-070617.pdf
13. Pathirage, M., Perera, S., Kumara, I., Weerawarana, S.: A multi-tenant architecture for business process executions. In: Proceedings of the 2011 IEEE International Conference on Web Services. ICWS 2011, pp. 121–128. IEEE Computer Society, Washington (2011). http://dx.doi.org/10.1109/ICWS.2011.99
14. Seltsikas, P., Currie, W.: Evaluating the application service provider (ASP) business model: the challenge of integration. In: Proceedings of the 35th Annual Hawaii International Conference on System Sciences. HICSS, pp. 2801–2809, January 2002
15. Tao, L.: Shifting paradigms with the application service provider model. Computer **34**(10), 32–39 (2001). http://dx.doi.org/10.1109/2.955095
16. Zarate, M.S., Mendoza, L.C.: Cloud computing: a review of PAAS, IAAS, SAAS services and providers. Lampsakos **7**(1), 47–57 (2012)
17. Zhang, Q., Cheng, L., Boutaba, R.: Cloud computing: state-of-the-art and research challenges. J. Internet Serv. Appl. **1**(1), 7–18 (2010). http://dblp.uni-trier.de/db/journals/jisa/jisa1.html#ZhangCB10

# A New Big Data Framework for Customer Opinions Polarity Extraction

Ammar Mars[(✉)], Mohamed Salah Gouider, and Lamjed Ben Saïd

Institut Suprieur de Gestion de Tunis, Le Bardo, Tunisia
ammar.mars@gmail.com, MS.GOUIDER@yahoo.fr, lamjed.bensaid@isg.rnu.tn

**Abstract.** Recently, we are talking about opinion mining: It refers to extract subjective information from text data using the natural language processing, text analysis and computational linguistics. Micro-blogging is one of the most popular Web 2.0 applications, such as Twitter which is evolved into a practical means for sharing opinions around different topics. It becomes a rich data sources for opinion mining and sentiment analysis.

In this work, we interest by to study users opinions about an object in social networks, for example studying the opinion of users about "the Samsung brand" or "the nokia brand", using text mining and NLP (Natural language processing) technologies. We propose a new *ontological approach* able to determinate the polarity of user post. This approach classify the users posts to negative, positive or neutral opinions. To validate the effectiveness of our approach, we used a dataset published by Bing Liu's group in our approach experimentation.

## 1  Introduction

Recent years have witnessed a dramatic increase in our ability to collect data from various sensors, devices, in different formats, from independent or connected applications. This expansion has made a variety of databases, including user profile database and social network data. The search of desired information for making decision then became a challenge because this data flood has outpaced our capability to process, analyze, store and understand these datasets. A report by IDC (International Data Corporation) was published in 2011 [10], shows that the overall volume of data worldwide is 1.8 ZB (equal to 1.6 trillion gigabytes). Two new terms was born: (1) "Big Data" which is used to identify sets of data that we cannot treat them with traditional methodologies and tools, and (2) "Big Data mining" which is the ability to extract relevant information from a large amount of data or from the data stream, which due to its volume and its variability, it n 'was not possible to do so before.

"What other people think" has always been an important piece of information decision-making processes. Today people frequently make their opinions available in the different social media such as Facebook and Twitter. And as a result, the Web has become an excellent source for gathering consumer opinions. There are now numerous Web resources containing such opinions, e.g., social networks,

© Springer International Publishing Switzerland 2016
S. Kozielski et al. (Eds.): BDAS 2016, CCIS 613, pp. 518–531, 2016.
DOI: 10.1007/978-3-319-34099-9_40

microblogging tools, and Blogs. It is important for the manufacturer or seller to study the opinion of customers about a product. But, due to the large amount of information collected from this data sources, it is essentially impossible for a supplier or decision maker to read all of this data collected and make a decision. For this reason, the opinion mining has become an important issue for research in Web information extraction.

We take Twitter as a data source: it is a microblogging service addressed in our work, is a communication means and a collaboration system that allow users to share short text messages, which does not exceed 140 characters with a defined group of users called followers.

In this work, we aim to design a big data framework able to extract opinions and their polarity about a product from the microblogging platform Twitter. We present the related work in Sect. 1. In Sect. 2, we discuss our proposed approach for customer opinions extraction from social network. Next, we present the experimentation and result founded in Sect. 3. Finally, we give some conclusions in Sect. 4.

## 2  Related Work

In this work, we will propose big data framework able to extract opinions of customer about a product from open data. It uses mainly two domains: "Opining Mining" and "Big Data".

### 2.1  Related Work in "Big Data"

To face the explosion in the volume of data, a new field of technology has emerged: Big Data [2]. It Invented by the web giants, these solutions are designed to provide real-time access to big databases.

It is a blanket term for any collection of data sets so large and complex, that it becomes difficult to process using on-hand data management tools or traditional data processing applications.It refers to a wide range of large data sets almost impossible to manage and process using traditional data management tools due to their size, but also their complexity [23,25].

The concept of big data has been characterized by "3V": Volume, Velocity and Variety [11].

1. Volume: With the lower cost storage, the trend today is to store everything you can store, and to treat after.
2. Variety: Nowadays,the mass of data generated arrives today in a variety of formats.The data cam from social network, mobile devices and TV do not fit naturally into existing Data architectures. There are many different types of data, as text, sensor data, audio, video, graph, and more.
3. Velocity: Data is arriving continuously as streams of data, and we are interested in obtaining useful information from it in real time. Collection and sharing of data in real time becomes an absolute prerequisite of Big Data approach.

The value chain of big data, which can be generally divided into four phases: data generation, data acquisition, data storage, and data analysis [6]. We will detail each phase following:

*Big Data Generation.* Data Generation as all Decision Support System (DSS) is s the first step of the analysis process. It is large-scale, highly diverse, and complex datasets generated through longitudinal and distributed data sources. Such data sources include sensors, videos, click streams, and/or all other available data sources [3, 19, 21].

*Big Data Acquisition.* The sets of data collected can sometimes include redundant or unnecessary data, which unnecessarily increases the storage space and affects retardation in the analysis processes. This phase of the big data system, big data acquisition includes data collection, data transmission, and data pre-processing [3, 22].

*Big Data Storage.* The exponential growth of data has requirements for storage and management, and also of treatment. This third phase is mainly concerned by the data storage keeping the reliability and availability of data access. The storage mechanisms existing data in big data can be classified into two types: file systems, databases [7].

*Big Data Analysis.* Data analysis is the final and the most important phase in the value chain of big data, with the purpose of extracting useful values, providing suggestions or decisions [7, 8, 24].

## 2.2   Related Work in "Opining Mining"

An important part of our information-gathering behavior has always been to find out what other people think. With the growing availability and popularity of opinion-rich resources such as online review sites and personal blogs, new opportunities and challenges arise as people can, and do, actively use information technologies to seek out and understand the opinions of others.

Opinion mining, also known as sentiment analysis, is the process aiming to determine whether the polarity of a textual corpus (document, sentence, paragraph etc.) tends towards positive, negative or neutral.

Kim [14] present a system that, given a topic, automatically finds the people who hold opinions about that topic and the sentiment of each opinion. This system contains a module for determining word sentiment from a sentence using various models of classifying.

Currently, the most popular on-line micro-blogging service is Twitter, which enables its users to send and receive text-based posts, known as "tweets", consisting of up to 140 characters. Kontopoulos *et al.* [15] have proposed an ontology-based technique towards sentiment analysis of Twitter posts. This approach uses machine learning classifiers.

In [9,16] the author have proposed a classifier based approach for discovering opinions to improve the firms marketing strategies. In addition, Agrawal [1] have presented a method to examine sentiment analysis on Twitter data. The idea of this paper is to propose a POS-tagging specific to analyze the role of linguistic features. Cardie *et al.* proposes an approach to multi-perspective question answering that views the task as one of opinion-oriented information extraction. The author describe an annotation schema representing an opinion [4].

[12] have presented a new method for product features opinion mining of product reviews. Ku et al. have proposed in [17] three algorithms for opinion extraction at word, sentence and document level. Another work have been proposed by [20] for an unsupervised information extraction system named OPINE witch mines in order to build a model of important product features, their evaluation by reviewers, and their relative quality across products.

## 3   Proposed Approach

We noticed that researchers tend to study the opinions and sentiments of users about a subject. Knowing the data is more and more huge, variable in space and in time, heterogeneous as they come from different sources. In this work, we aim at extracting opinions polarities of customer about a product. We present a new and effective approach for extracting opinion polarity from enormous data user using the combine of big data and Text mining technologies. Our approach composed essentially by four phases:

1. **Opinions Words Learning:** it aims to construct a tree of opinions word from training data.
2. **Tweets Collection and Integration:** it aims to collect and integrate data published by customers in the web and more precisely from social networks and microblogging platforms.
3. **Big Data Analysis and Opinions Extraction:** in this step, we will analyse data collected using Big data and text mining techniques for customers opinions extraction.
4. **Results Presentation:** Finally, we will show the positive and negative opinions founded about products.

### 3.1   Opinions Words Learning

In order to resolve automatic summarization problem of customer reviews about product, Minqing Hu and Bing Liu [12] present a list of negative and positive word used by the customer to express her opinion [12]. We will use this database in our work.

We can cite some examples in the table below:

| Words | Polarity |
|---|---|
| Afraid, aggressive, agony, angry, batty, bad, bait, boil, buggy | Negative |
| Accommodative, accomplish, achievement, admire, affable, alluring, amazed, amusing | Positive |

In this work, we propose using ontology to present the negative and positive word to simplify the search in next step. Figure 1 present two ontologies designed from words list cited previously.

**Fig. 1.** Words opinions ontology

Ontologies are considered one of the pillars of the Semantic Web. It can be defined as an explicit specification of conceptualization. Ontologies capture and describe the structure of the domain, in our case we present negative and positive words list. In our case we have classified this list in three categories: Verbs, Adjectives and adverbs.

To adapt an ontology to our case, we considered the verbs, adjectives and adverbs as **classes**. Each class divided to sub-categories using the first letter of word. Its considered as **sub-classes**. Each word is considered as an **individual**.

In fact, this ontology is used in our work is to identify word polarity (negative or positive). This ontology can simplify finding the polarity of word by knowing its grammatical class and the word searching will be easy and more fast. We have used OWL[1] (Web Ontology Language) language for building this ontology.

## 3.2   Tweets Collection and Integration

As we mentioned, one of the fundamental characteristics of the Big Data is the huge volume of data represented by heterogeneous and diverse dimensionalities such as unstructured data, flat File, article and database. This data is not adapted to the treated of analysis and extract information. We need to integrate and store this data in an environment witch support Volume, Variety and Velocity of data.

In this step, we aim to conceive a model able to store the data collected. We have a program that collect and stock Tweets from Twitter in our proposed structure. In the world of Web applications, like Facebook for example or Twitter, the data are not described in the same way. In fact it would lead to store data with zero values. Thus the use of relational databases is not always appropriate.

When the data scale has surpassed the capacities of traditional relational databases, the massive analysis is needed. At present, most massive analysis utilize HDFS (Hadoop Distributed File System) of Hadoop to store data.

We notice the work of S. Chalmers and C. Bothorel [5] that present an investigation into the current state of the art with respect to "Big Data" frameworks and tools. This work proposes several criteries to choose the adequate tool. We have proposed to use HDFS model to store tweets collected. This structure is alimented from Twitter.

Next, Data Integration is the first step of big data mining process. Specifically, it is large-scale, diverse and complex datasets integrated that coming from different data sources. Such data sources include flat File, data base, streams or other available data sources. This operation also called Data Acquisition. During this acquisition, once the raw data is collected, an efficient transformation mechanism should be used to store it in storage system for using in the analytic process.

We can define data integration as a basic step to prepare data before processing. It is an important and complex step because of the staggering growth in the amount of data comes from different sources.

**Algorithm of Tweets Collection.** We conceived an algorithm to integrate the data come from Twitter. We will illustrate this task in Algorithm 1 bellow, it able to integrate tweets in our proposed structure.

The algorithm of Data Stream Integration creates a listener that can detect the new tweet posted. These algorithm stay collect tweets while the program executing. For each new tweet, it extract different properties of tweet and it store in the NoSql database.

---

[1] http://www.w3.org/2001/sw/wiki/OWL.

---

**Algorithm 1.** Algorithm of tweets Integration from data stream

---

**Require:** Access to stream
**Ensure:** Data integration in HDFS file
 1: **Stream** ← Twitter STREAM
 2: **HF** ← HDFS File
 3: Open(**HF**)
 4: Open(**Stream**)
 5: **if** (**New** Tweet comes from Stream) **then**
 6:     **Store** (Tweet) **in** HF
 7: **end if**

---

### 3.3   Data Analysis and Opinions Polarity Extraction

Generally, Data analysis research can be classified into six fields: structured data analysis, text data analysis, website data analysis, multimedia data analysis, network data analysis, and mobile data analysis.

This phase is the most important phase of big data process. As we mentioned that in this approach we extract opinions from tweets collected from Twitter. The format of tweets stored is text. Therefore, we need use text analysis in this step. Text analysis, also called text mining, is a process to extract useful information and knowledge from unstructured data. Most text mining systems are based on text expressions and natural language processing (NLP).

NLP can enable computers to analyze, interpret, and even generate text. Some common NLP methods are: lexical acquisition, word sense disambiguation, part of-speech tagging, and probabilistic context free grammar. Some NLP-based technologies have been applied to text mining, including information extraction, topic models, text summarization, classification, clustering, question answering, and opinion mining. Information mining shall automatically extract specific structured information from texts.

We illustrate this phase in this figure below in Fig. 2.

**Using Map Reduce in Opinion Mining.** MapReduce is a programming model designed for processing large volumes of data in parallel by dividing the work into a set of independent tasks. MapReduce programs are written in a particular style influenced by functional programming constructs, specifically idioms for processing lists of data. MapReduce can take advantage of locality of data, processing it on or near the storage assets in order to reduce the distance over which it must be transmitted.

In fact MapReduce is divided to two tasks:

– **"Map"** step: Each worker node applies the "MAP" function to the local data, and writes the output to a temporary storage. A master node orchestrates that for redundant copies of input data, only one is processed.
– **"Reduce"** step: Worker nodes now process each group of output data, per key, in parallel.

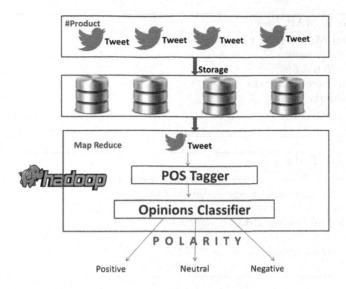

**Fig. 2.** Data analysis and opinions polarity extraction

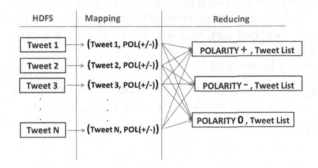

**Fig. 3.** MapReduce for opinion polarity extraction

In our case, we aims to use the MapReduce programming model to extract opinion polarity from a huge amount of data. We adapted this model in our work, we show algorithm tasks in Fig. 3.

Our system aims to find the negative and positive opinions from tweets collected from twitter about a product.

In the first task (MAP) our system iterate the set of tweets and determinate the polarity of each tweet. We illustrate this task in the algorithm bellow:

Conventionally the polarity of a Tweet is defined as follows:

$$Polarity(T) = \begin{cases} 1 & \text{if it is a positive opinion} \\ -1 & \text{if it is a negative opinion} \\ 0 & \text{if it is a neutral opinion} \end{cases}$$

The next step (REDUCE), our system will group the tweets according to their polarity.

**Algorithm 2.** MAP Task

1: **Function Map** (String name, String file):
2: // name: file name
3: // File : HDFS file
4: **for each** Tweet **in** file **do**
5:    P=**Polarity** (tweet)
6:    **emit** (tweet, P)
7: **end for**

**Algorithm 3.** Reduce Task

1: **Function Reduce** ( int Polarity, Iterator TweetList):
2: // Polarity: Opinion Polarity +,- or Neutral
3: // TweetList : List of tweets
4: **for each** t **in** TweetList **do**
5:    Sum+=1
6:    **emit** (Polarity, Sum)
7: **end for**

**Polarity Determination.** Using the ontology defined previously, the POLARITY task able to determine the polarity of a tweet. We illustrate this method in Algorithm 4. It is a method used by Map task to identify if a tweet is a negative or positive opinion.

This task is divided by two task:

**POS-tagging.** Before discussing the application of part-of-speech tagging from natural language processing, we first give some example sentences.

**Example 1**

| Input | The pictures are very clear |
|---|---|
| Output | The/**DT** pictures/**NNS** are/**VBP** very/**RB** clear/**JJ** |

**Example 2**

| Input | it will not easily fit in pockets |
|---|---|
| Output | it/**PRP** will/**MD** not/**RB** easily/**RB** fit/**VB** in/**IN** pockets/**NNS** |

In corpus linguistics, part-of-speech tagging (POS tagging or POST), also called grammatical tagging or word-category disambiguation, is the process of marking up a word in a text (corpus) as corresponding to a particular part of speech, based on both its definition, as well as its context i.e. relationship with adjacent and related words in a phrase, sentence, or paragraph.

| Algorithm 4. Polarity method |
|---|
| 1: **Function Polarity** (String Tweet): |
| 2: **for each** W **AS** [**verb/adjective/adverb**] **in Tweet Do** TweetList **do** |
| 3:     **if** W **Exist in** NegativeOntology **then** |
| 4:         **Return** -1 |
| 5:     **else if** W **Exist in** PositiveOntology **then** |
| 6:         **Return** 1 |
| 7:     **else** |
| 8:         **Return** 0 |
| 9:     **end if** |
| 10: **end for** |

**Matching Ontology.** For each word (verb, adverb or adjective) found in a tweet, our systems try to verify if its a negative/positive words from the ontology constructed previously. We present bellow the method of polarity determination.

### 3.4   Results Presentation

Result presentation is the final and most important phase of the value chain of big data. Big data opinion mining can provide useful values and statics for decision making. It is a summarisation and result of data processing. In our work, it is summary representation of the opinions collected from Twitter about a product.

## 4   Experimentation and Results

As we mentioned, we extract data from platform of microblogging Twitter. To validate our approach described above, we conceive three following modules:

1. Module 1 : It to learn opinion words and construct ontologies.
2. Module 2 : It covers the data collection and integration.
3. Module 3 : It designed to opinion polarity determination.
4. Module 4 : It is an algorithm based on Map reduce model to analyse tweets

### 4.1   Working Environment

To implement the algorithms that we have shown previously, we used a hardware configuration that has served to validate our approach. We have used a node with the following configuration:

– Operating System : Ubuntu 12.02 LTS
– RAM: 8 GB
– CPU: Core i7, 8 MB Hidden memory and 3.6 GHZ frequency.

## 4.2    Experimental Data and Result

As we mentioned above, we conceived an a tree of negative and positive lexicon from list of words published by Minqing Hu and Bing Liu [12,18].

Many data set available in the web, we uses three datasets to validate our approach in the first step. Next we are collected a set of tweets from twitter and tested their with our approach proposed.

**Opinion Mining Dataset of Joshi and Rose** [13]. This dataset is a subset of the opininon mining datasets released by Dr. Bing Liu's group from University of Illinois at Chicago. Their dataset is available in this web site[2]. This subset consists of 200 review comments each for 11 different products.

This dataset has classification labels (from the manual annotation process done by Bing Liu's group) for the "opinion" class which marks whether or not a review comment consists of any subjective evaluation of one or more features of the product or the product itself.

This dataset is stored in XML format. The comment is stored between a tag named "<text>" and the polarity (POS/NEG) of review is given in "cname" attribute. This is an example of a review in Fig. 4.

```
<instance id="1" subpop="Canon-G3">
    <class cname="POS"/>
    <text>
        overall i 'm happy with my toy .
    </text>
</instance>
```

Fig. 4. Review example of dataset

We measured the effectiveness of our system by using this Dataset. This table illustrate the experimentation results: (Figs. 5 and 6)

**Tweets Collected from Twitter.** To retrieve tweets from twitter, you must have a subject determined by a hashtag. For this case, we used some hash-tag to find tweets about some products from twitter; we chose the name of products to find and retrieve tweets about these products : Samsung and Nokia. We have collected 123 000 tweets around this. This table illustrate the data analysis results:

---

[2] http://www.cs.uic.edu/~liub/FBS/sentiment-analysis.html.

| Products | Comments | | | Cost(ms) |
|---|---|---|---|---|
| | Misclassified | Well classified | Effectiveness | |
| Canon-G3 | 59 | 141 | 70.5% | 252 |
| Canon-PS-SD500 | 43 | 157 | 78.5 % | 193 |
| Canon-S100 | 22 | 178 | 89 % | 189 |
| Hitachi-Router | 54 | 166 | 73 % | 223 |
| Linksys-Router | 41 | 159 | 79,5 % | 271 |
| Micro-MP3 | 33 | 177 | 84,28 % | 343 |

**Fig. 5.** Experimentation results

| Products | Tweets | | | Cost(s) |
|---|---|---|---|---|
| | Positive | Negative | unknown | |
| Samsung | 18,79% | 24,13 % | 57,07 % | 31 |
| Nokia | 34,32 % | 21,77 % | 43,92 % | 27 |
| Orange | 17,22 % | 33,07 % | 49,72 % | 28 |
| IPhone | 25,42 % | 14,55 % | 60,03 % | 35 |

**Fig. 6.** Experimentation results

### 4.3 Tools

**Twitter4j.** Twitter4J[3] is a Twitter API binding library for the Java language licensed under Apache License 2.0. It is an open source library. It is able to integrate the Twitter service on a program.

**Hadoop (HDFS).** Hadoop[4] is an open source Apache project that aims to provide "reliable and scalable distributed computing". The Hadoop package aims to provide all the tools one could need for large scale computing: Fault Tolerant Distributed Storage in the form of the HDFS. A job scheduling and monitoring framework, as well as the popular MapReduce Java libraries that allow for scalable distributed parallel processing.

## 5 Conclusion

In this work, we proposed a framework for extracting customer opinions using Big Data technologies combined with machine learning and text mining tools. First,we proposed an algorithm of opinion polarity determination based on MapReduce. Next, we test and validate the effectiveness of this framework with

---

[3] www.twitter4j.org.
[4] https://hadoop.apache.org/.

datasets available in the web in first time and next we use data cames from platform of microblogging Twitter. Our experimental results show that it is necessary to test this system with other benchmarks. Another line of research will be to extract opinion polarity not only about a product but the characteristics or features of product. It is also very complex operation that need to combine machine learning and text mining tools.

# References

1. Agarwal, A., Xie, B., Vovsha, I., Rambow, O., Passonneau, R.: Sentiment analysis of twitter data. In: Proceedings of the Workshop on Languages in Social Media, pp. 30–38. Association for Computational Linguistics (2011)
2. Baru, C., Bhandarkar, M., Nambiar, R., Poess, M., Rabl, T.: Big data benchmarking. In: Proceedings of the 2012 Workshop on Management of Big Data Systems, MBDS 2012, pp. 39–40. ACM, New York (2012). http://doi.acm.org.sci-hub.org/10.1145/2378356.2378368
3. Brinkmann, B.H., Bower, M.R., Stengel, K.A., Worrell, G.A., Stead, M.: Large-scale electrophysiology: acquisition, compression, encryption, and storage of big data. J. Neurosci. Meth. **180**(1), 185–192 (2009)
4. Cardie, C., Wiebe, J., Wilson, T., Litman, D.J.: Combining low-level and summary representations of opinions for multi-perspective question answering. In: New Directions in Question Answering, pp. 20–27 (2003)
5. Chalmers, S., Bothorel, C.: Big data - state of the art. Research gate (2012)
6. Chen, M., Mao, S., Liu, Y.: Big data: a survey. Mobile Netw. Appl. **19**(2), 171–209 (2014). Springer Science + Business Media
7. Chen, M., Mao, S., Zhang, Y., Leung, V.C.: Big data storage. In: Big Data, pp. 33–49. Springer, Heidelberg (2014)
8. Cohen, J., Dolan, B., Dunlap, M., Hellerstein, J.M., Welton, C.: Mad skills: new analysis practices for big data. Proc. VLDB Endow. **2**(2), 1481–1492 (2009)
9. Dave, K., Lawrence, S., Pennock, D.M.: Mining the peanut gallery: opinion extraction and semantic classification of product reviews. In: Proceedings of the 12th international conference on World Wide Web, pp. 519–528. ACM (2003)
10. Gantz, J., Reinsel, D.: Extracting value from chaos. IDC iview **1142**, 1–12 (2011)
11. Halevi, G., Moed, H.F.: Special issue on big data. Research Trends (2012)
12. Hu, M., Liu, B.: Mining opinion features in customer reviews. AAAI **4**, 755–760 (2004)
13. Joshi, M., Penstein-Rosé, C.: Generalizing dependency features for opinion mining. In: Proceedings of the ACL-IJCNLP 2009 Conference Short Papers, pp. 313–316. Association for Computational Linguistics (2009)
14. Kim, S.M., Hovy, E.: Determining the sentiment of opinions. In: Proceedings of the 20th International Conference on Computational Linguistics, pp. 1367. Association for Computational Linguistics (2004)
15. Kontopoulos, E., Berberidis, C., Dergiades, T., Bassiliades, N.: Ontology-based sentiment analysis of twitter posts. Expert Syst. Appl. **40**(10), 4065–4074 (2013)
16. Ku, L.W., Lee, L.Y., Wu, T.H., Chen, H.H.: Major topic detection and its application to opinion summarization. In: Proceedings of the 28th Annual International ACM SIGIR Conference on Research and Development in Information Retrieval, pp. 627–628. ACM (2005)

17. Ku, L.W., Liang, Y.T., Chen, H.H.: Opinion extraction, summarization and tracking in news and blog corpora. In: AAAI Spring Symposium: Computational Approaches to Analyzing Weblogs, pp. 100–107 (2006)
18. Liu, B., Hu, M., Cheng, J.: Opinion observer: analyzing and comparing opinions on the web. In: Proceedings of the 14th International Conference on World Wide Web, pp. 342–351. ACM (2005)
19. Marx, V.: Biology: the big challenges of big data. Nature **498**(7453), 255–260 (2013)
20. Popescu, A.M., Etzioni, O.: Extracting product features and opinions from reviews. In: Kao, A., Poteet, S.R. (eds.) Natural Language Processing and Text Mining, pp. 9–28. Springer, London (2007)
21. Rabl, T., Jacobsen, H.-A.: Big data generation. In: Rabl, T., Poess, M., Baru, C., Jacobsen, H.-A. (eds.) WBDB 2012. LNCS, vol. 8163, pp. 20–27. Springer, Heidelberg (2014)
22. Tanner Jr., J.F.: Big data acquisition. In: Analytics and Dynamic Customer Strategy: Big Profits from Big Data, pp. 85–101. Wiley, Hoboken (2014)
23. Fan, A.B.W.: Mining big data: current status, and forecast to thefuture. ACM SIGKDD Explor. Newsl. Arch. **14**(2), 1–5 (2012)
24. Wu, X., Zhu, X., Wu, G.Q., Ding, W.: Data mining with big data. IEEE Trans. Knowl. Data Eng. **26**(1), 97–107 (2014)
25. Zhang, L., Stoffel, A., Behrisch, M., Keim, D.: Visual analytics for the big data era comparative review of state-of-the-art commercial systems. In: VAST 2012 Proceedings of the 2012 IEEE Conference on Visual Analytics Science and Technology (VAST), pp. 173–182 (2012)

# Evidence Based Conflict Resolution
# for Independent Sources
# and Independent Attributes

Walid Cherifi$^{(\boxtimes)}$ and Bolesław Szafrański

Faculty of Cybernetics, Military University of Technology, Warsaw, Poland
walid.cherifi@wat.edu.com, b.szafranski@milstar.pl

**Abstract.** The gigantic use of digital information has changed compre-
hensively the way we live. People rely more and more on information
collected from various sources in every aspect of life. However, due to
the natural variety and autonomy of these sources, finding relevant and
accurate information is becoming increasingly difficult. Indeed, several
sources can provide different conflicting facts for the same real-world
object. Moreover, most modern-day applications often provide imperfect
information. Therefore, it is strenuous to distinguish the true facts from
the false ones. To deal with this problem, we propose in this paper a new
evidential conflict resolution method for independent sources and independent attributes. Our method exploits the power of Dempster-Shafer
theory so as to find the most trustable facts when data sources provide
imperfect information.

**Keywords:** Conflict resolution · Dempster-Shafer theory · Evidential
graph

## 1 Introduction

Since we are living in the digital age, an enormous quantity of data is generated
by humans and machines. Indeed, with the significant development of the Internet, information about the same real-world object can be gathered from various
independent data sources. For instance, in health care system, patient's medical
information can be captured from multiple patient entry points, including the
emergency room, during clinic visits and hospital stays, from workplace health
and well-beings programs, from social media, at call centres and through mobile
fitness applications. Each one of these entry points has its own autonomous database where the facts about patients are stored. Using all these data collected from
different databases will definitely improve the quality of the care. Another example can be found in the business sector, where companies, as well as investors,
have been gathering information for years about potential customers, the financial performance of other companies, economic trends and the economies of other
countries. All these pieces of information are collected from diverse sources as
so to make the best investment possible.

© Springer International Publishing Switzerland 2016
S. Kozielski et al. (Eds.): BDAS 2016, CCIS 613, pp. 532–544, 2016.
DOI: 10.1007/978-3-319-34099-9_41

Unfortunately, data sources are not often of the same quality, and some of them frequently provide wrong and contradictory information for the same real-world object. Yet, decisions based on inaccurate information usually lead to severe harm. For example, the wrong diagnosis based on incorrect measurements of a patient will absolutely lead to potentially fatal consequences, and an investment decision based on false and biased financial information will obviously cause the investors to make irrational decisions by investing in an unprofitable area in the economy, thus suffering from huge financial losses. Therefore, the GIGO principle (Garbage In, Garbage Out) is highly applicable here, where inaccurate inputs for any processing problem will certainly lead to inaccurate outputs. For this reason, resolving the conflict, and finding the true facts among the collected, duplicated and contradictory ones is a crucial step before providing any information to requesters.

In order to get around the above problem, conflict resolution [9], known also as truth discovery [14,16], data fusion [4], or information corroboration [6], has been widely investigated recently in several real-world applications, such as web data integration [16], social sensing [15], crowdsourcing aggregation [1] and knowledge fusion [3], where vital decisions have to be taken in presence of conflicting information collected from various sources. Most of these investigated approaches for conflict resolution problem only deal with perfect or uncertain information [4,10,14]. However, modern-day applications often provide imperfect (uncertain, imprecise and incomplete) information, which make the problem of finding the correct facts more difficult. In this regard, we propose in this paper a new evidential model for the conflict resolution problem. Our Evidence Based Conflict Resolution for Independent Sources and Independent Attributes (EBCR.IS.IA) is based on Dempster-Shafer Theory (DST), which is a powerful theory that provides a robust framework to represent uncertain, imprecise and incomplete information in better way than probability functions, and offers an adequate framework to combine multiple opinions given by different experts.

The rest of this paper is organized as follows: the DST is briefly introduced in Sect. 2. In Sect. 3, the problem of conflict resolution is described. The proposed EBCR-IS-IA method is presented in Sect. 4. Next, evaluation results are presented in Sect. 5. Finally, Sect. 6 concludes the paper with an outlook into ongoing research work.

## 2    Review of the Dempster-Shafer Theory

The DST, also known as the theory of belief function or the evidence theory, is based on the pioneering work of Dempster [2] and Shafer [12]. More recent advances in this theory have been introduced in the Transferable Belief Model (TBM) proposed by Smets [13]. We present in this section some basic concepts of it.

In the evidence theory, the frame of discernment (FoD) also called universe of discourse $\Theta = \{H_1, H_2, ..., H_N\}$ is a set of N mutually exclusive and exhaustive hypotheses. These hypotheses are all the possible and eventual solutions of the problem under study. The set of all subsets of $\Theta$ is its power set $2^\Theta$. A subset of those $2^\Theta$ sets may consist of a single hypothesis or a conjunction of hypotheses.

The main element of this theory is the Basic Belief Assignment ($BBA$), known also as mass function. It represents the degree of support allocated to $A \in 2^\Theta$ justified by the available evidence, the $BBA$ is defined as a mapping $m : 2^\Theta \rightarrow [0, 1]$ satisfying the following properties:

$$\begin{cases} m() = 0 \\ m(A) \geq 0 \quad , \forall A \in 2^\Theta \\ \sum_{A \in 2^\Theta} (A) = 1 \end{cases} \tag{1}$$

One or many subsets $A \in 2^\Theta$ may have a non-null mass, and are considered as focal elements. This mass is the source's degree of belief that the solution to the studied problem is in that subset. In the framework of the DST, the belief function ($bel$), the plausibility function ($pl$) and the disbelief function ($dis$) are in one to one correspondence with the $BBA$ ($m$) [12]. For a given subset $A \in 2^\Theta$, the ($bel$), ($pl$) and ($dis$) are defined as follows:

$$\begin{cases} bel(A) = \sum_{B \subseteq A, B \neq \emptyset} m(B) \\ pl(A) = \sum_{B \in 2^\Theta, A \cap B \neq \emptyset} m(B) \\ dis(A) = \sum_{B \in 2^\Theta, A \cap B = \emptyset} m(B) = 1 - pl(A) = bel(\overline{A}) \end{cases} \tag{2}$$

For the subset $A$, $bel(A)$ and $pl(A)$ represent upper and lower belief bounds, and the interval $[bel(A), pl(A)]$ represents the degree to which the evidence set is uncertain whether to support $A$ or $\overline{A}$. The relationships between $bel$ value, $dis$ value and *uncertainty* are described in Fig. 1.

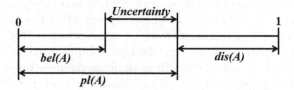

**Fig. 1.** The uncertainty interval for a hypothesis A.

It is possible to have multiple mass functions on the same domain $\Theta$ that correspond to different experts' opinions. A great number of combination rules are proposed such as the Dempster's Rule of combination [11] which can be used to combine several independent sources. Given two $BBAs$ $m_1$ and $m_2$ that are associated with two independent evidence sources, the combined mass, denoted $m_{1 \oplus 2}(A) = m_1(A) \oplus m_2(A)$, is defined as follows:

$$m_{1 \oplus 2}(A) = \begin{cases} \frac{\sum_{B \cap C = A} m_1(B) * m_2(C)}{1 - \sum_{B \cap C = \emptyset} m_1(B) * m_2(C)} & \forall A \subseteq \Theta, A \neq \emptyset \\ 0 & if A = \emptyset \end{cases} \tag{3}$$

The denominator interpreted as a measure of conflict between the pieces of evidence and evaluating the quality of combination.

In practice, sources of evidence may not be completely reliable. Concidering that, we can weaken the $BBA$ by introducing a discount rate parameter $\alpha \in [0, 1]$ [5] by which the mass function may be discounted to reflect the accuracy of the source. The discounted mass function can be represented as:

$$\begin{cases} m^{\alpha}(A) = \alpha m(A) & for\, A \neq \Theta \\ m^{\alpha}(\Theta) = (1 - \alpha) + \alpha m(\Theta) \end{cases} \qquad (4)$$

When $\alpha = 1$, the source is considered as absolutely reliable. Whereas when $\alpha = 0$, the source is viewed as completely unreliable. After discounting, the source is treated as totally reliable.

To make the best decision, it is usually preferable to use a well-defined probability function. the pignistic transformation [13] was proposed for that purpose. it maps $BBAs$ to so called pignistic probability functions $BetP$. The pignistic transformation of $m$ is given by:

$$BetP(A) = \sum_{B \subseteq \Theta, A \cap B \neq \emptyset} \frac{|A \cap B|}{|B|} m(B), \quad \forall A \in \Theta \qquad (5)$$

In the framework of DST, a distance measure computes the discord between two $BBAs$, namely, it computes how much the two $BBAs$ are disimilar. Jousselme distance [7] has been widely used in this purpose. It is defined as follows:

$$d(m_1, m_2) = \sqrt{\frac{1}{2}(m_1 - m_2)^t \underline{\underline{D}}(m_1 - m_2)}, \quad \underline{\underline{D}} = \begin{cases} 1 & if\, A = B \\ \frac{|A \cap B|}{|A \cup B|} \forall A, B \in 2^{\Theta} \end{cases} \qquad (6)$$

## 3   Conflict Resolution Problem

Conflict resolution is the predominant class of methods that must be applied whenever contradictory pieces of information (facts) about the same real-world object are collected from various sources. The main goal of this class is to define methods that are able to determine the most trustable facts among the conflicting ones.

Commonly, the input of a traditional conflict resolution method is a large number of facts about certain attributes, as well as the set of sources providing them. The relationship between the sources, facts and attributes can be described as follows: Each fact about an attribute is at least given by one source. Simultaneously, different sources can assign different facts for the same attribute. This relationship is usually represented by a 4-partite graph $G = (S \cup F \cup A \cup O, E_{SF} \cup E_{FA} \cup E_{AO})$ of sources $S$, facts $F$, attributes $A$ and objects $O$, with edges $E_{SF}$ connecting each source to the facts it asserts, $E_{FA}$ linking each attribute to its corresponding facts, and $E_{AO}$ relating the objects to their attributes. For the sake of simplicity, we define the bipartite subgraph $G' = (S \cup F, E_{SF})$ that shows the relationship between the sources and the facts they provide. $G'$ is not generally a complete bipartite graph because an edge only exists if and only if a source $S_i$ provides the fact $F_{j,k,l}$.

The objective of a conflict resolution algorithm is to distinguish between true and false facts. This could potentially be achieved by assigning a correctness score to each fact so that the true facts have the highest scores. Obviously, not all sources have the same quality. Furthermore, many of these sources often provide wrong information that may influence the aggregation results. Thus, including the sources reliability score in the calculation of the facts' correctness will certainly improve the algorithm accuracy. Unfortunately, the source reliability is usually unknown a priori and hence it has to be inferred from the provided facts. To overcome this problem, conflict resolution algorithms usually operate iteratively so as to calculate both sources reliability and facts correctness simultaneously, by following the principle that reliable sources provide correct facts, and at the same time, correct facts are given by reliable sources.

The majority of conflict resolution methods assume that the data sources can only provide precise facts with total certainty for each attribute [4,10,14]. However, most modern-day applications, such as data integration, information extraction, scientific data management, sensor deployment and social sensing often generate imperfect information. In fact, imperfection is ubiquitous and hence any method that intends to manage real-world objects is required to cope with the effect of imperfection.

Information imperfection can be caused by several reasons including:

- Imprecision: which is a characteristic of an information that cannot be expressed by a single value, but by a set of possible values where the real value is one of the elements of this set. Imprecise facts come in many forms, such as disjunctive facts, e.g., *"the patient has either a cold or the flu"*; negative facts, e.g., *"the patient does not have the flu"*; range facts, e.g., *"the patient temperature is between 38°C and 41°C"* or *"the patient temperature is over 40°C"*.
- Uncertainty: represents the state of knowledge about a particular fact. It arises when a source has only partial knowledge of the correctness of the fact, e.g., *"the patient probably has a cold"* or *"the patient has a cold with confidence 0.7"*.
- Incompleteness: it occurs when a data source does not provide some facts for different attributes, e.g. a patient can suffer from many diseases simultaneously, but the data source only provides some of them.

## 4   The Evidence-Based Conflict Resolution Model

In this section, we first start by formulating our problem. Then we introduce the evidential graph which is considered as the pillar of the EBCR-IS-IA. After that, we show how the evidential graph can be constructed given imperfect information. Next, we propose two methods that determine the facts correctness degree and the sources reliability respectively. Finally, we present the overall process of the EBCR-IS-IA.

## 4.1   Problem Formulation

We consider a set of objects $O = \{O_1, O_2, ..., O_{N_O}\}$, where each object $O_l$ is described by a set of independent attributes $A_l = \{A_{1,l}, A_{2,l}, ..., A_{N_{A_l},l}\}$. Each attribute $A_{k,l}$ pertains to a particular domain enumerating all its possible facts $F_{k,l} = \{F_{1,k,l}, F_{2,k,l}, ..., F_{N_{F_{k,l}},k,l}\}$, where $N_{F_{k,l}}$ represents the number of facts contained in that domain. A fact $F_{j,k,l}$ could be either true (meaning that it is correct) or false (meaning that it is erroneous). In addition to the domain, each attribute has a specific cardinality that indicates the necessary facts the attribute needs so as for it to be valid. For the sake of simplicity, we only consider the single truth attribute i.e., one and only one true value exists for each attribute $A_{k,l}$. We let the case of multi-truths attribute for future work.

On the other hand, we suppose a group $S$ of $N_S$ independent data sources $S = \{S_1, S_2, ..., S_{N_S}\}$, where each source $S_i$ provides for each attribute $A_{k,l}$ imperfect information describing its opinion about the correctness of each fact. In such case, information can be imprecise, uncertain, incomplete or any possible combination of each of them, depending on the application domain. Besides, we consider the close world assumption, where each source has the complete knowledge about the domain for each attribute $A_{k,l}$.

Our main goal here is to determine the correct fact among the conflicting ones in the presence of imperfect information. This, in reality, can be done by using the DST. Indeed, unlike probability theory, the DST is able to express, in a more faithful manner, the whole continuum of information availability: from complete or partial ignorance to total knowledge. In addition, it offers a mathematical way to combine evidence from different experts without the need to know a priori or conditional probabilities. Therefore, this theory seems to provide an excellent tool for our conflict resolution problem.

## 4.2   Evidential Graph Definition

The key benefits of using the evidential graph (Fig. 2) are firstly its possibility to encode elegantly both perfect and imperfect information provided by different sources. Moreover, its ability to represent various attribute cardinalities (single and multivalued attributes). Finally, its capability to consider the specific possible world semantic for each data source. All these can be obtained by replacing the bipartite subgraph $G'$ of the traditional conflict resolution with a new evidential complete bipartite subgraph $EG' = (S \cup F, E_{SF}^B)$, where $E_{SF}^B$ represents the set of edges relating every source $S_i \in S$ to every fact $F_{j,k,l} \in F$. The edge $E_{SF_{i,j,k,l}}^B \in E_{SF}^B$ describes the source $S_i$ degree of belief in the correctness of the fact $F_{j,k,l}$.

We assign for each source $S_i$, fact $F_{j,k,l}$ and edge $E_{SF_{i,j,k,l}}^B$ a BBA $m_i^\Theta$, $m_{j,k,l}^\Omega$ and $m_{i,j,k,l}^\Omega$, respectively. These BBAs take their values respectively from the power set of the FoD $\Theta_i$, $\Omega_{j,k,l}$ and $\Omega_{i,j,k,l}$.

– Let $\Theta_i = \{R_i, \overline{R_i}\}$ be a FoD for each data source $S_i$. $\Theta_i$ contains two hypothesis: $R_i$ which means that the source $S_i$ is reliable and every time gives correct facts, and $\overline{R_i}$ which expresses that the source is unreliable and always provides wrong facts. We define the BBA $m_i^\Theta$ over the FoD $\Theta_i$, where $m_i^\Theta(R_i)$ represents the support degree for the reliability of the source $S_i$, $m_i^\Theta(\overline{R_i})$ is the support degree for the unreliability and $m_i^\Theta(R_i \cup \overline{R_i})$ is the support degree for the ignorance ( we do not know whether the source is reliable or unreliable).

- $\Omega_{j,k,l} = \{C_{j,k,l}, \overline{C_{j,k,l}}\}$ be the FoD for each fact $F_{j,k,l}$. $\Omega_{j,k,l}$ holds the two hypotheses that describe the correctness of the fact. Here, $C_{j,k,l}$ expresses that the fact is correct, whereas $\overline{C_{j,k,l}}$ shows that it is incorrect and thus wrong. Over this FoD, we define the BBA $m_{j,k,l}^\Omega$ which provides the support degree for each subset of $\Omega_{j,k,l}$ for the correctness of each fact $F_{j,k,l}$. Here, $m_{j,k,l}^\Omega(C_{j,k,l})$ represents the support degree for the fact correctness, $m_{j,k,l}^\Omega(\overline{C_{j,k,l}})$ is the support degree for the fact incorrectness and $m_{j,k,l}^\Omega(C_{j,k,l} \cup \overline{C_{j,k,l}})$ is the support degree for the ignorance. We define also the BBA $m_{i,j,k,l}^\Omega$ which shows the degree of belief of the source $S_i$ in the trustworthiness of the fact $F_{j,k,l}$. Then, we have $m_{i,j,k,l}^\Omega(C_{j,k,l})$ which is the degree of belief of the source $S_i$ in the correctness of the fact $F_{j,k,l}$, $m_{i,j,k,l}^\Omega(\overline{C_{j,k,l}})$ is the belief degree of the source $S_i$ in the incorrectness of the fact and $m_{i,j,k,l}^\Omega(C_{j,k,l} \cup \overline{C_{j,k,l}})$ is the ignorance.

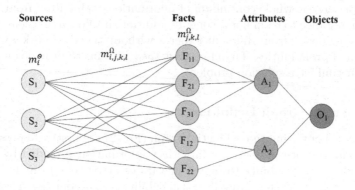

**Fig. 2.** Evidential graph for Evidence-based Conflict Resolution.

## 4.3 Evidential Graph Construction

Ideally, each source $S_i$ provides directly its belief in the correctness of each fact $F_{j,k,l}$ in the form of BBA $m_{(i,j,k,l)}^\Omega$ over the FoD $\Omega_{j,k,l}$. However, data sources oftentimes give imperfect information in the form of evidential tables, known also as evidential databases [8], where the facts about an attribute $A_{k,l}$ are expressed by means of BBA $m_{i,j,k,l}^\Psi$ over the FoD $\Psi_{k,l} = \{F_{1,k,l}, F_{2,k,l}, ..., F_{N_{F_{k,l}},k,l}\}$ containing all elements of the domain $F_{k,l}$. Note that the BBA $m_{i,j,k,l}^\Psi$ must verify the conditions presented in (1). As an illustration, let us consider the example of an evidential table that contains information about patients. In this example, we are only interested in the patients blood group, which is a single truth attribute (each patient has one and only one blood type). There are four major blood groups (we do not consider the Rh factor), thus the FoD is $\Psi_{k,l} = \{O, A, B, AB\}$. Table 1 contains five possible evidential representation of the information.

**Table 1.** The evidential table for the doctors' diagnosis.

| Patient | Blood group |
|---------|-------------|
| $P_1$ | $\{O\}$ (0.4) \| $\{A\}$ (0.6) |
| $P_2$ | $\{O\}$(0.1) \| $\{O,AB\}$(0.7) \| $\{O,AB,A\}$(0.2) |
| $P_3$ | $\{A\}$ (1) |
| $P_4$ | $\{O,A,B,AB\}$ (1) |
| $P_5$ | $\{O\}$(0.5) \| $\{AB\}$(0.3) \| $\{O,AB\}$(0.2) |

An evidential table is a powerful mean that allows the representation of different levels of uncertainty, such as *probabilistic information* ($P_1$), *possibilistic information* ($P_2$), *perfect information* ($P_3$), *incomplete information* ($P_4$), and *evidential information* ($P_5$).

In order to infer the belief degree of the source $S_i$ in the correctness of each fact $F_{j,k,l}$ given the evidential table, we define the following mapping:

$$\begin{cases} m_{i,j,k,l}^{\Omega}(C_{j,k,l}) = bel_i(F_{j,k,l}) \\ m_{i,j,k,l}^{\Omega}(\overline{C_{j,k,l}}) = des_i(F_{j,k,l}) \\ m_{i,j,k,l}^{\Omega}(C_{j,k,l} \cup \overline{C_{j,k,l}}) = 1 - (bel_i(F_{j,k,l}) + des_i(F_{j,k,l})) \end{cases} \tag{7}$$

where $bel_i(F_{j,k,l})$ and $des_i(F_{j,k,l})$ are computed from (2).

### 4.4    Fact Correctness Calculation

We suppose that both the sources reliability $m_i^\Theta$ and their belief in the correctness of each fact $m_{i,j,k,l}^{\Omega}$ are known. The goal here is to calculate the facts' degree of correctness $m_{j,k,l}^{\Omega}$. To do so, the idea that reliable sources always give accurate opinions about the correctness degree of each fact is applied. The proposed method comprises two steps:

**Step 1:** discounting.
In the DST, the discount operation aims to reduce the *BBA* on the frame with respect to the degree of reliability of the source as showed in (4). In our EBCR-IS-IA, we propose a new operation which is slightly different from the discount operation proposed in the literature, in the sense that it fits the single truth assumption. The idea is that if an unreliable source believes that a fact is correct, the new discounted *BBA* considers it as wrong. On the other hand, if the same

unreliable sources believe that the fact is wrong, the discounted *BBA* does not know whether it is correct or wrong. This can be formulated as follows:

$$\begin{cases} m_{i,j,k,l}^{\Omega*}(C_{j,k,l}) = m_i^{\Theta}(R_i) * m_{i,j,k,l}^{\Omega}(C_{j,k,l}) \\ m_{i,j,k,l}^{\Omega*}(\overline{C_{j,k,l}}) = m_i^{\Theta}(R_i) * m_{i,j,k,l}^{\Omega}(\overline{C_{j,k,l}}) + m_i^{\Theta}(\overline{R_i}) * m_{i,j,k,l}^{\Omega}(C_{j,k,l}) \\ m_{i,j,k,l}^{\Omega*}(C_{j,k,l} \cup \overline{C_{j,k,l}}) = 1 - (m_{i,j,k,l}^{\Omega*}(C_{j,k,l}) + m_{i,j,k,l}^{\Omega*}(\overline{C_{j,k,l}})) \end{cases} \tag{8}$$

**Step 2:** opinion aggregation.

By applying the Dempster's Rule of combination (3) over the same FoD $\Omega_{j,k,l}$, we build a new evidence representing the consensus of the evidence obtained from the disparate opinions of sources $S$. For $N_S$ sources, the combination of the $N_S$ BBAs $m_{1,j,k,l}^{\Omega*}, m_{2,j,k,l}^{\Omega*}, ..., m_{N_S,j,k,l}^{\Omega*}$ yields a new *BBA* $m_{j,k,l}^{\Omega}$ for each fact $F_{j,k,l}$.

$$m_{j,k,l}^{\Omega} = m_{1,j,k,l}^{\Omega*} \oplus m_{2,j,k,l}^{\Omega*} \oplus ... \oplus m_{N_S,j,k,l}^{\Omega*} \tag{9}$$

### 4.5 Source Reliability Estimation

To estimate the reliability of sources, we suppose that we already know the real correctness value of each fact $F_{j,k,l}$ about the attribute $A_{k,l}$. The basic idea here is that correct facts are often given accurate opinions by reliable sources. Using this idea, the reliability of each source can be estimated in 2 steps:

**Step 1:** source reliability calculation regarding each fact.

We define the *BBA* $m_{i,k,l}^{\Theta}$ over the FoD $\Theta$. This *BBA* represents the reliability degree of the source $S_i$ with respect to the fact $F_{j,k,l}$. Given the *BBA* $m_{j,k,l}^{\Omega}$ that quantifies the correctness of the fact, and the source belief degree $m_{i,j,k,l}^{\Omega}$ in the correctness of $F_{j,k,l}$, we can calculate $m_{i,j,k,l}^{\Theta}$ as follows:

$$\begin{cases} m_{i,k,l}^{\Theta}(R_i) = \frac{P}{P+N+C} \\ m_{i,k,l}^{\Theta}(\overline{R_i}) = \frac{N}{P+N+C} \\ m_{i,k,l}^{\Theta}(R_i \cup \overline{R_i}) = \frac{C}{P+N+C} \end{cases} \tag{10}$$

where $P$ and $N$ represent the true positive and the false negative values for the source $S_i$ about the attribute $A_{k,l}$ respectively:

$$\begin{cases} P = \sum_{i=1}^{|N_{F_{k,l}}|} m_{j,k,l}^{\Omega}(C_{j,k,l}) * m_{i,j,k,l}^{\Omega}(C_{j,k,l}) \\ N = \sum_{i=1}^{|N_{F_{k,l}}|} m_{j,k,l}^{\Omega}(\overline{C_{j,k,l}}) * m_{i,j,k,l}^{\Omega}(C_{j,k,l}) \\ C \in \mathbb{R}, C > 0 \end{cases} \tag{11}$$

**Step 2:** reliability aggregation for all attributes.

After calculating the reliability of each source with regard to each attribute, we can now introduce a new parameter $\alpha_{k,l}$ between 0 and 1 representing the importance of each attribute $A_{k,l}$ in the calculation of the sources reliability, where

$\alpha_{k,l} = 0$ means that the attribute is not important and it must not influence the reliability of the source. Whereas $\alpha_{k,l} = 1$ means that the attribute is highly important. We use the discount operation (4) to reduce the *BBA* $m_{i,k,l}^{\Theta}$. Next, the total reliability of each source $S_i$ is obtained by aggregating the discounted *BBAs*.

$$m_i^{\Theta} = m_{i,1,1}^{\Theta\alpha_{1,1}} \oplus m_{i,2,1}^{\Theta\alpha_{2,1}} \oplus ... \oplus m_{i,N_{A_1},1}^{\Theta\alpha_{N_{A_1},1}} \oplus m_{i,1,2}^{\Theta\alpha_{1,2}} \oplus ... \oplus m_{i,N_{A_{N_O}},N_O}^{\Theta\alpha_{N_{A_{N_O}},N_O}} \quad (12)$$

## 4.6   Algorithm Flow

So far, we have first described the evidential graph which is considered as the pillar of our evidential model. Next, we have explained how to compute the facts correctness given the sources reliability and their belief in the trustworthiness of each fact. After that, we have shown how to infer the sources reliability given a set of correct facts. Here, we recapitulate the overall process (Fig. 3) of the Evidence-based Conflict Resolution algorithm where both sources reliability and facts correctness are unknown. Formally, our EBCR-IS-IA proceeds as follows:

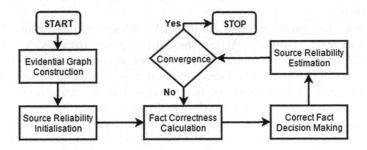

**Fig. 3.** Flow diagram for EBCR-IS-IA.

**Step 1:** given the imperfect information about each attribute, we first start by constructing the evidential graph i.e. determining the belief degree of each source in the correctness of each fact.

**Step 2:** once we have constructed our evidential graph, we initialize the reliability of each source $S_i$.

**Step 3:** we apply our facts correctness computation method explained in Sect. 4.4 so as to calculate the facts' correctness degree given the reliability of sources.

**Step 4:** with regard to each new *BBA* $m_{j,k,l}^{\Omega}$, the pignistic probability (5) is calculated so as to make decision (to select the correct facts). A correct fact

$F_{j,k,l}$ is the one that has the maximum pignistic probability $BetP(C_{j,k,l})$ among all facts that were proposed for the attribute $A_{k,l}$.

**Step 5:** in this step, we re-estimate the reliability of each source given the correct facts that have been determined in the previous step. Here, we follow the method explained in Sect. 4.5.

Note that **Steps 3**, **4** and **5** must proceed iteratively until convergence i.e. the mean of Jousselme distance (6) of the sources reliability between two successive iterations is less than a small positive number $\epsilon \in [0,1]$.

## 5    Experiments

In this section, we conduct preliminary experiments to evaluate the performance of the proposed EBCR-IS-IA method over synthetic perfect datasets (we let the case of imperfect datasets for future work). We also compare the EBCR-IS-IA method with baseline methods from the literature, including TruthFinder (TF) [16], 2-Estimates (2E) [6] and the naive conflict resolution method: vote (the correct facts are the ones that were given by the majority of sources). To evaluate the performance of these methods, we measure the Error Rate (ER) by computing the percentage of the methods' output that are mismatched with the ground truth.

To reach this purpose, The EBCR-IS-IA method and the other baselines were implemented in Matleb. We use the dataset generator provided by [14] to simulate different scenarios. Indeed, this dataset generator can control several characteristics of the generated dataset, namely the number of sources, the number of data items (a data item can be considered as an attribute of a specific object), the source coverage, the truth fact distribution per source, and the distinct fact distribution per data item. For evaluation, we run a set of experiments and we compare all methods by varying the number of sources. The number of data items was fixed at 1000, the number of facts was exponentially distributed

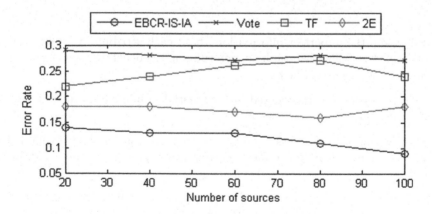

**Fig. 4.** The Error Rate of the methods versus the number of sources.

with maximum 10 distinct facts per data item, and the correct fact distribution per source was set at 80P, namely, 80 % of the sources provide 20 % of the correct facts, 20 % of the sources provide 80 % of the correct facts. The reported results are averaged over 100 experiments and shown in Fig. 4.

As can be seen in Fig. 4, the Error Rate of EBCR-IS-IA is the lowest among the baselines in the studied scenario. Besides the good performance of EBCR-IS-IA in the case of perfect information, it can also handle the case of imperfect information. Further evaluations of the EBCR-IS-IA over imperfect information are left for future work.

# 6    Conclusion and Future Work

In this paper, we have presented the evidence based conflict resolution model for independent sources and independent attributes. Our proposed model is based on the Demspter-Shafer theory, which is well-known for its usefulness to express the uncertain judgment of experts, and its efficiency to represent imperfect (uncertain, imprecise and incomplete) information. The initial experiments give preliminary validation about the performance of the proposed method and show that it can even perform better than baseline methods in the studied scenarios.

Several extensions are targeted for future work. First, we will carry through the validation of our proposal with synthetic and real-world imperfect data sets which will allow us to quantify the real benefit of the proposed method. Second, we plan to propose a new extension to our model by solving the problem of conflict resolution in the presence of multi-truths attributes.

# References

1. Aydin, B.I., Yilmaz, Y.S., Li, Y., Li, Q., Gao, J., Demirbas, M.: Crowdsourcing for multiple-choice question answering (2014)
2. Dempster, A.: Upper and lower probabilities induced by a multivalued mapping. Ann. Math. Stat. **38**(2), 325–339 (1967)
3. Dong, X.L., Gabrilovich, E., Heitz, G., Horn, W., Murphy, K., Sun, S., Zhang, W.: From data fusion to knowledge fusion. Proc. VLDB Endowment **7**(10), 881–892 (2014)
4. Dong, X.L., Saha, B., Srivastava, D.: Less is more: selecting sources wisely for integration. Proc. VLDB Endowment **6**(2), 37–48 (2012)
5. Elouedi, Z., Mellouli, K., Smets, P.: Assessing sensor reliability for multisensor data fusion within the transferable belief model. IEEE Trans. Syst. Man Cybern. Part B Cybern. **34**(1), 782–787 (2004)
6. Galland, A., Abiteboul, S., Marian, A., Senellart, P.: Corroborating information from disagreeing views. In: Proceedings of the Third ACM International Conference on Web Search and Data Mining, pp. 131–140. ACM (2010)
7. Jousselme, A.L., Grenier, D., Bossé, É.: A new distance between two bodies of evidence. Inf. Fusion **2**(2), 91–101 (2001)
8. Lee, S.K.: Imprecise and uncertain information in databases: an evidential approach. In: Proceedings of the Eighth International Conference on Data Engineering, 1992, pp. 614–621. IEEE (1992)

9. Li, Q., Li, Y., Gao, J., Zhao, B., Fan, W., Han, J.: Resolving conflicts in heterogeneous data by truth discovery and source reliability estimation. In: Proceedings of the 2014 ACM SIGMOD International Conference on Management of Data, pp. 1187–1198. ACM (2014)

10. Li, Y., Gao, J., Meng, C., Li, Q., Su, L., Zhao, B., Fan, W., Han, J.: A survey on truth discovery. arXiv preprint (2015). arXiv:1505.02463

11. Sentz, K., Ferson, S.: Combination of evidence in dempster-shafer theory. Technical report, Sandia National Labs., Albuquerque, NM (US); Sandia National Labs., Livermore, CA (US) (2002)

12. Shafer, G.: A Mathematical Theory of Evidence, vol. 1. Princeton University Press, Princeton (1976)

13. Smets, P.: Decision making in the TBM: the necessity of the pignistic transformation. Int. J. Approximate Reasoning 38(2), 133–147 (2005)

14. Waguih, D.A., Berti-Equille, L.: Truth discovery algorithms: an experimental evaluation. arXiv preprint (2014). arXiv:1409.6428

15. Wang, D., Amin, M.T., Li, S., Abdelzaher, T., Kaplan, L., Gu, S., Pan, C., Liu, H., Aggarwal, C.C., Ganti, R., et al.: Using humans as sensors: an estimation-theoretic perspective. In: Proceedings of the 13th International Symposium on Information Processing in Sensor Networks, pp. 35–46. IEEE Press (2014)

16. Yin, X., Han, J., Yu, P.S.: Truth discovery with multiple conflicting information providers on the web. IEEE Trans. Knowl. Data Eng. 20(6), 796–808 (2008)

# A New MGlaber Approach as an Example of Novel Artificial Acari Optimization

Jacek M. Czerniak$^{(\boxtimes)}$ and Dawid Ewald

AIRlab Artificial Intelligence and Robotics Laboratory, Institute of Technology,
Casimir the Great University in Bydgoszcz,
ul. Chodkiewicza 30, 85-064 Bydgoszcz, Poland
{jczerniak,dawidewald}@ukw.edu.pl

**Abstract.** The proposed MGlaber method is based on observation of
the behavior of mites called Macrocheles glaber (Muller, 1860). It opens
the series of optimization methods inspired by the behavior of mites,
which we have given a common name: Artificial Acari Optimization.
Acarologists observed three stages the ovoviviparity process consists of,
i.e.: preoviposition behaviour, oviposition behaviour (which is followed
by holding an egg below the gnathosoma) and hatching of the larva supported by the female. It seems that the ovoviviparity phenomenon in this
species is favoured by two factors, i.e.: poor feeding and poor quality of
substrate. Experimental tests on a genetic algorithm were carried out.
The MGlaber method was worked into a genetic algorithm by replacing
crossig and mutation methods. The obtained results indicate to significant increase in the algorithm convergence without side-effects in the
form of stopping of evolution at local extremes. The experiment was
carried out one hundred times on random starting populations. No significant deviations of the measured results were observed. The research
demonstrated significant increase in the algorithm operation speed. Convergence of evolution has increased about ten times. It should be noted
here that MGlaber method was not only or even not primarily created for
genetic algorithms. The authors perceive large potential for its application in all optimization methods where the decision about further future
of the solutions is taken as a result of the evaluation of the objective
function value. Therefore the authors treat this paper as the beginning
of a cycle on Artificial Acari Optimization, which will include a series of
methods inspired by behaviour of different species of mites.

**Keywords:** Acari · Nature-inspired methods of artificial intelligence ·
AAO · Artificial acari optimization · Genetics algorithms · MGlaber

## 1 Introduction

Dust mites are minor Arthropada classified to the class Arachnida. Second to
insects, mites (i.e. the subclass of the Arachnida named Acari or Acarina) are
the most diverse and complex group of arthropods found in the quarantine. Just

© Springer International Publishing Switzerland 2016
S. Kozielski et al. (Eds.): BDAS 2016, CCIS 613, pp. 545–557, 2016.
DOI: 10.1007/978-3-319-34099-9_42

like insects, but unlike their arachnid relatives (spiders, scorpions and the like), the feeding customs of mites go well beyond predation to include herbivory and parasitism. They differ from other arachnids by their morphological structure. There is high diversity of structures within the same order. Their common feature is a cephalothorax with abdomen as well as no signs of segmentation on the outside. Their name stems from that fact. The Acari name stems from the Greek language and means "a" "cari", i.e. "without the head". The branch of science studying mites is acarology. The body of mites includes gnathosoma and idiosoma. The mouth parts are adjusted to biting and sucking. Gnathosoma, i.e. the first, mouth-related, part of the mites body can (but does not have to) include: chelicerae, pedipalps, upper lip, labrum, over-pharynx fold and under-pharynx fold. Chelicerae are used to retrieve food; they may assume the form of pincers, cutting organ, suction organ or claw. The pedipalps are built very differently depending on the suborder: sometimes they are quite reduced, but sometimes extremely complex, and they may fulfill sensory functions. Most Acari are animals of microscopic size (0.1 1 mm). For some of the species the size of adult females overindulged with blood amounts up to 30 mm. There are about 50 thousand mite species. Many species have not been scientifically described yet. Acarologists estimate that the order of arachnids includes animals from microscopic ones to 3 cm large. Mites are saprophagous (such as e.g. Oribatida), but they also include parasites (mainly ecto-parasites) and predators [14,28]. Mites include not only human, plant and animal parasites, but they also spread pathogens (viruses, bacteria, protozoa, tapeworms, roundworms). Some predatory and parasitic mites are used in the biological fight against other mites, insects and weeds. Some of them may spread diseases or are classified to stored product pests. They live in all climatic zones, including polar areas. They have adapted to different environments: they live in the soil, home dust (the house dust mites), in the coastal zone of freshwater environments (the so-called water mites (Hydrachnidiae)) and even in the hot springs.

The Acari includes a host of plant parasites that can devastate crop by eating it or by spreading plant pathogens. Domestic and wild animals can also be infested by an often bewildering diversity of parasitic mites, including those that cause debilitating disease and deformity. Social insects like bees or ants and those that bore in timber are especially rich in associated mites and for most of these mites it is unknown what their potential effect may be if they are introduced into new environments.

A new method of MGlaber optimization is presented in the paper. It is based on the behaviour of a female mite Macrocheles glaber which was observed and documented by well-known acarologists Marquardt, Kaczmarek and Halliday (2014) [13,17]. This observation applies to the care for young by a female mite before and just after giving birth to them. Inspired by those studies, the authors of this paper presented modified classic Holland genetic algorithm. The proposed solution involves replacement of the crossing and mutation operators into the MGlaber operator. The research demonstrated significant increase of the algorithm operation speed. Convergence of evolution has increased about ten

times. It is also worth to notice, that stopping of evolution at local extremes
has not been observed [19–21]. It should be noted here that MGlaber method
was not only or even not primarily created for genetic algorithms. The authors
perceive large potential for its application in all optimization methods where the
decision about further future of the solutions is taken as a result of the evaluation of the objective function value. Therefore the authors treat this paper as
the beginning of a cycle on Artificial Acari Optimization, which will include a
series of methods inspired by behaviour of different species of mites.

## 2    Behavior of a Live Individual

Acarologists have known Macrocheles glaber (Muller, 1860) for a long time. It
is a species of mite widely spread over the world. Its presence is reported all
over Europe, in the Mediterranean region, in Russia as well as in North America
and in New Zealand. It is found in agricultural soils and on meadows as well as
in litter of coniferous and deciduous forests. Moreover, it is (female) most often
picked-up during harvest together with Coleoptera of Staphylinidae Scarabaei-
dae, and Silphidae. families. In addition, it also occurs on stored grass seeds and
in tree-covered areas on fields. Interestingly, M.glaber occurs as pioneer species
on mining wastelands and reclaimed areas such as e.g. abandoned military training grounds. However, customs and behaviour of those mites as regards their core
for young was not known until recently. The observation research was carried
out in a laboratory setting using IP camera system equipped with dedicated
optical system and peripheral fast data storage arrays by Marquardt and Kaczmarek [17]. The females of M. glaber are morphologically inseparable, so male
offspring were reared in the laboratory. The M. glaber was identified according to the criteria of Filipponi and Pegazzano (1962) [12] and Halliday (2000).
Behaviour of mites was observed in a 10 mm diameter chamber using the peat
substrate and methods of observation described by Marquardt and Kaczmarek
(2014) [17]. The results of these studies can be viewed in a video attached as a
supplement to their article and published on the publisher web site (ref. to sup.)
[18]. Marquardt, Kaczmarek and Halliday described three phases of M. glaber
female behavior, which they have observed using the above described instrumen-
tation. As noted above, that behaviour concerns taking care of young depending
on the external (ambient) conditions. Let these phases be illustrated in Fig. 1
inspirated by original figure from the article of the aforementioned authors [16].

The process of egg handling by female can be divided into three phases presented in Fig. 1. Those phases should be considered in conjunction with the "zero
phase". This term symbolically means a situation when a female bears offspring
(progeny) without any additional behavioral actions described in phases 1–3.
That note is justified, all the more that this natural behavior is emphasized in
the paragraph on the optimization method inspired by a M. glaber mite female
behavior. Different ways of female behaviour are related to environmental conditions in the place where she intends to place an egg. With a certain simplified
personalization, one can say, that a female assesses her offspring for its adjustment to the environment and, on the basis of this observation, she takes action

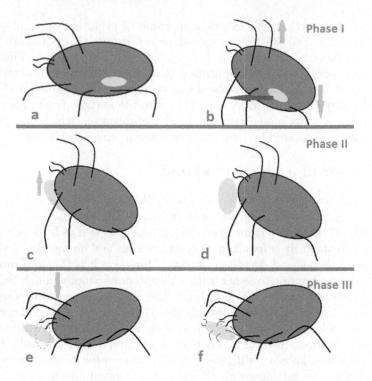

**Fig. 1.** Phases of egg handling [1]

aimed at the offspring adaptation to environmental conditions. When she deems that adaptation impossible, in other words when environmental conditions are too harsh a female performs an act of cannibalism. From her point of view it is the symptom of adaptation to environment, as she recovers to some extent the material and energy which she has lost while bearing her offspring.

**Phase I** behaviours before the egg is pushed out of the females genital opening. During this phase a female takes a characteristic body position. She lowers her abdomen and prepares for egg release. A female tries to compensate for adverse environmental conditions by extending eggs residence time inside of her. This phase is one of the longest ones. The duration of this phase usually ranges from 2 min and 33 s to 11 min and 33 s.

**Phase II** holding the egg under the gnathosoma and placing it in front of the body. Offspring protection in this phase consists in a female trying to provide more heat for an egg. The egg is already outside her body. This protection takes two forms, i.e. a female tries to keep an egg close to her, pressed against her body and covered from the top. The second method applied by a female is to place an egg in wanted and, so to say, deliberately selected slot in the ground and covering it with mobile elements of the ground. This phase typically lasts from 20 to 42 s.

**Phase III** changing the orientation of the egg and assisting the release of the larva. That phase seems to be particularly interesting. Watching it on the film recorded by acarologists during observation is very exciting. A female is holding an egg in front of her and manipulates it for a relatively long time trying to find an optimal position. It seems as if she is deliberately preparing for what she intends to do. Once she has determined the right position she starts to eat the egg. Acarologists do not provide the reason for those solicitous preparations. Perhaps she begins to eat it from the place close to the head of the offspring located inside the egg. Researchers refer to one of observations in which it seems that a female opens an egg helping the offspring to escape. In most of observations that phase ends with the act of cannibalism. This phase typically lasts from 27 s to 3 min.

## 3   New MGlaber Algorithm

Inspiration of M.glaber animal species female individual, which is referred to in the preceding paragraph has led to the development of a new MGlaber optimization method. This method belongs to the mainstream of Nature-Inspired Methods of artificial intelligence [5,8,15,22,23]. It should in particular be classified to the new subset in that mainstream i.e. Artificial Acari Optimization. MGlaber method consists in classification of the optimization result [1–4,6]. Assuming that $fit(x_i)$ function is given, which returns the result of a partial optimization specific for $x_i$ object one can observe the following solution [29]. It comes down to Heaviside staircase function $\theta(x)$ and its corresponding distribution. This function is defined in different ways. The following definition has been assumed for the purposes of this study:

**Definition 1. Heaviside staircase function for MGlaber**

$$\theta(fit(x)) \begin{cases} Phase_0 & if\ fit(x) \geq 75\,\%\ of\ max\_fit \\ Phase_1 & if\ fit(x) \geq 50\,\%\ of\ max\_fit \\ Phase_2 & if\ fit(x) \geq 25\,\%\ of\ max\_fit \\ Phase_3 & if\ fit(x) < 25\,\%\ of\ max\_fit \end{cases} \tag{1}$$

where $fit(x)$ are the objective function values, $max\_fit$ is a constant defining the maximum possible value of the function, while $\{Phase_0, Phase_1, Phase_2, Phase_3\}$ is a finite set of algorithm operation variants (Fig. 1).

- $Phase_0$ consists in the adoption of the solutions in its original (unchanged) form,
- $Phase_1$ consists in applying some form of the result amendment, e.g. mutation,
- $Phase_2$ consists in the assimilation ("cannibalism") of valuable solutions parts in order to obtain unique material,
- $Phase_3$ consists in total rejection of the solutions as a useless one.

Making further use of the Heaviside staircase function, one can propose a definition for the distribution function of a random variable. Therefore we obtain the notation of the discrete random variable distribution function using the following mathematical formula.

**Definition 2. Distribution function of a random variable**

$$F\xi(fit(x)) = \sum_k p_k \theta(fit(x)fit_k(x)) \tag{2}$$

Where $fit_k(x)$ are possible discrete random variable values, while $pk = Pr\{\xi = fit_k(x)\}$ is the probability that the random variable assumes $fit_k(x)$ value.

## 4 Materials and Methods

In order to verify the usefulness of the theoretically developed approach in practice, it was decided to modify the classical genetic algorithm of J.H.Holland [11,24–26]. This modification consisted in the fact that the classical crossover operator has been replaced by an additional operator of the MGlaber solution values assessment based on the mites counting. The applied modification enables valuation of the obtained local extreme.

### 4.1 Environment of Experiment

Prior to the crossover, the selection was made. As a selection operator the modified form of the classic roulette wheel method presented below was used. The modified roulette wheel method was used as a way of selection [27]. This will allow for further improvement of the algorithm efficiency in the future by using more optimum methods. Crossing and mutation operators are described in the paragraphs below. The algorithm is terminated when the population includes the individual with the fitness function value not lower than the assumed one. If there is no such an individual and the algorithm stops at a relative extreme, the algorithm will be automatically terminated when 10000 populations are exceeded. It is also worth mentioning that the proposed approach meets J.H.Holland's [5,7,9,10] postulate and represents classic algorithm without memory, characterized by overwriting (replacement) of the parent population with the population of children.

**Base Structures of the Algorithm.** The distribution of results obtained from the selection of individuals should correspond to actual distribution occurring in the reality. So overmuch determinism is not desirable here. Although the selection process is random, it is conducted so that individuals with the highest value of the fitness function were most likely to be selected for reproduction. This method has already been mentioned in the introduction to this paragraph, but as it constitutes the base for modification used in the this algorithm, it seems to be advisable to provide more details. Consecutive phases of that selection method are as follows: calculating the fitness function $eval(v_i)$ for each chromosome $v_i$, where $i \in [1, max\_pop]$ calculating total fitness of the population

$$F = \sum_{i=1}^{max\_pop} eval(v_i), \tag{3}$$

calculating the probability of selection $p_i$ for each chromosome $v_i$, where $i \in [1, max\_pop]$ calculating total fitness of the population

$$\forall_{i \in [1, max\_pop]}, p_i = \frac{eval(v_i)}{F} \tag{4}$$

calculating the cumulative distribution function $q_i$ for each chromosome $v_i$, where $i \in [1, max\_pop]$

$$q_i = \sum_{j=1}^{i} p_i \tag{5}$$

generating a random real number r from the range $[0,1]$ if $r < q_1$, then the chromosome vi should be selected; otherwise select the chromo-some $v_i$, where $i \in [2, max\_pop]$, for which $q_{i-1} < r \leq qi$. The Fig. 2 shows the visualisation of the roulette wheel method. As it may be seen, the name of the method seems to stem from the scheme described in the literature. Individuals are searched for in the roulette wheel calibrated proportionally to the fitness factors achieved by individual chromosomes.

Below we presented modified form of the classic roulette wheel method. calculating the fitness function strength(vi) (maximal values of strength were disks or drums will not be damaged) for each individual $v_i$, where $i \in [1, max\_pop]$, calculating the sum of all fitness function values

$$F = \sum_{i=1}^{max\_pop} strength(v_i), \tag{6}$$

improvement of the pseudo-code generator properties by rescaling F=F*100, sorting individuals in the ascending order with regard to fitness function

$$\exists_{Z(\Psi)} \left\{ \begin{array}{c} Z(\Psi) = \Phi, where(\Psi = (v_i, .., v_k, .., v_j)) \wedge \\ (\Phi = \Psi) \wedge \\ \left[ (strength(v_1 \leq .. \leq strength(v_i) \right] \end{array} \right\} \tag{7}$$

random selection of an integral number from the range $[0,F]$

$$\exists_{r \in C}[(0 \leq r) \wedge (r \leq F)] \tag{8}$$

selection of the first individual that fulfils the relationship below

$$\exists_{i \in [1, max\_pop]} \sum_{i=1}^{max\_pop} [(S = S + strength(v_i)), \tag{9}$$

$$((r \leq S * 100) \Rightarrow (i = max\_pop))]$$

Selection of chromosomes designed for reproduction is done according to modified method of the roulette wheel. It is also worth mentioning that the situation re-sembles the X-axis with a scale. Starting from zero in the direction of the sum, the system checks whether the previously randomly selected number is located within the range (Fig. 2). If it exceeds the range, the searching is continued. Otherwise it indicates that an individual for procreation has been selected. Visualisation of the roulette wheel method presented in the previous paragraph was based on the data set included in the Table 1. To give better picture of it, the visualisation of the modified roulette wheel method was based on the same data. This visualisation is shown in the last Fig. 2 of the paragraph.

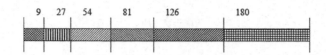

**Fig. 2.** Visualisation of the modified roulette wheel method

## 4.2   MGlaber in Genetic Algorithm

The first step of the MGlaber algorithm is classification of the offspring newly obtained as a result of the crossover operator operation. The classification is executed using the strength of an individual represented by the fitness function values as the main criterion. The following four classes are distinguished:

- the class of dominators - consisting of individuals, the strength of which is greater than or equal to 75
- the class of individuals for mutation - consisting of individuals, the strength of which is greater than or equal to 50
- the class of individuals for cannibalism - consisting of individuals "to be eaten", for transfer of chromosomes to a parent, if the strength is greater than or equal to 25
- the class of losers - consisting of individuals to be removed (deleted), if their strength is lower than 25

The next step of the algorithm is to allocate an offspring individual obtained as a result of crossing to specific class in order to modify it. If an individual is marked as a dominator, i.e. its usefulness for the population is high, then it is transferred to the new population table without any additional changes. If a child is classified as a member of the "to be mutated" class, this means that one of its genes will be randomly changed. The gene will be weakened for an odd child and strengthened for an even child. Such operation is justified, as increasing the convergence is not the main objective the modification. Indeed, the aim is a simultaneous improvement of the solution quality and the time to achieve a satisfactory result. If an individual is classified as a "cannibalism victim", there is a change of a genotype of one of its parents and it replaces an offspring individual in the next generation. Here, there is an experimentally unverified possibility to create a variant consisting in the enrichment of the parent chromosome with unique genes acquired from the "eaten" offspring. This is because eating offspring enables real mites to maintain positive energy balance which allows them to survive in adverse environment. Through this mechanism mites can ultimately survive longer and lay further eggs. The last category to which an offspring individual can be classified is a group of "individuals to be deleted". Offspring classified to this category is deleted from the evolution process for good. This is due to the fact that their strength is not sufficient enough to enhance the future population and processing so poor chromosome would only extend the evolution time contributing nothing to the solution. The pseudocode provided below reflects the essence of the procedure described above in a more formal way.

*Sample pseudocode of MGlaber*

```
switch (kids[0].fit)
  case 0.75*max_P:     //dominator
      C[i]= kids[0];
  case 0.50*max_P:     // to be mutated
          kids[0].chromosom[ch_m]='0';
          C[i]= kids[0];
  case 0.25*max_P:     // cannibalism victim
      do
      {
          if (P[kids[0].no_tab].chromosom[count] == '0')
            {
            P[kids[0].no_tab].chromosom[count] = '1';
            P[kids[0].no_tab].fit=P[kids[0].no_tab].fit+1;
          end = 1;
          }
      count++;
      } while(end!=1) ;
      end = 0;
      count=0;
      C[i]=kids[0];
  case 0.25*max_P:     // individuals to be deleted
      C[i]=P[kids[0].no_tab];
```

In order to perform experimental verification, the evolution process executed on a classic Holland algorithm was used with the population of 20 individuals, where each of them was described by a chromosome of 20 genes. In the simulation repeated tenfold the algorithm started from a randomly generated start population. Eight full evolution processes were executed for each such population. Four for the unmodified genetic algorithm and four for the MGlaber method which replaced the crossover operato. In the second case, there was also no additional evoking of the mutation operator, as it is integrated into MGlaber. Each of the four evolution processes was carried out with another type of the crossover operator. Thus, the following operations were executed, respectively:

– one-point crossover with a random crossover point,

**Table 1.** Results of $N = 10$ experiments

| No. | Crossing 1p. | | Crossing 2p. | | Crossing 3p. | | Crossing diagonal | |
|---|---|---|---|---|---|---|---|---|
| | GA | MGlaber | GA | MGlaber | GA | MGlaber | GA | MGlaber |
| 1 | 97 | 47 | 97 | 33 | 93 | 20 | 32 | 24 |
| 2 | 65 | 40 | 66 | 43 | 79 | 42 | 162 | 42 |
| 3 | 123 | 47 | 46 | 37 | 198 | 43 | 11 | 40 |
| 4 | 150 | 41 | 82 | 42 | 81 | 40 | 178 | 41 |
| 5 | 133 | 41 | 194 | 38 | 60 | 50 | 22 | 24 |
| 6 | 47 | 44 | 93 | 45 | 50 | 31 | 178 | 16 |
| 7 | 197 | 50 | 187 | 42 | 95 | 46 | 176 | 39 |
| 8 | 136 | 40 | 150 | 30 | 184 | 25 | 19 | 41 |
| 9 | 123 | 49 | 138 | 45 | 180 | 36 | 78 | 41 |
| 10 | 51 | 41 | 39 | 47 | 25 | 34 | 130 | 45 |
| Medium | 123 | 42,5 | 95 | 42 | 87 | 38 | 104 | 40,5 |

– double-point crossover with random crossover points,
– triple-point crossover with random crossover points,
– diagonal crossover with random positioning of individuals.

The results of the experiment are given in the Table 1. For clarification the Table 1 there is also a visualization of results in two versions (Figs. 3 and 4). The Fig. 4 additionally shows average percentage change of the evolution time. There was approximately 30 percent acceleration in each evolution regardless of the basic crossover method assumed. This means that with the MGlaber method it is possible to significantly increase convergence of evolutionary optimum search without loss of quality and without any recorded cases of stopping at local extremes.

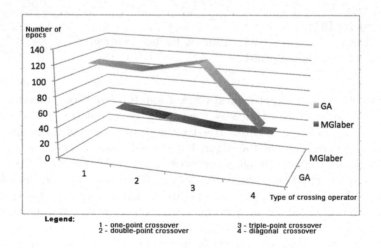

**Fig. 3.** Results of evolutions

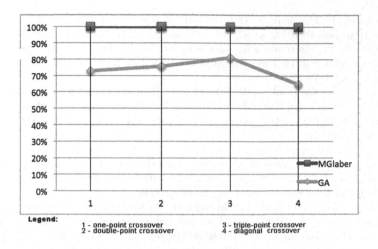

Fig. 4. Average percentage change of the evolution time

## 5    Conclusion

Application of the MGlaber method as a substitute of the crossover and mutation operator in the classical Holland genetic algorithm has changed significantly the algorithm operation. Significant increase in the evolution convergence has been observed. The number of evolution epochs decreased slightly more than three times. The results of individual simulations are specified in the following sections. An important feature of the applied method is the lack of the side effect, i.e. grounding (stopping) at local extremes. The quality of solutions has not deteriorated, and optimal results were, however, found already after several epochs. Further studies should focus on checking the aforementioned variant consisting in the enrichment of the parent chromosome with unique genes acquired from the "eaten" offspring. This mechanism, as indicated in non-published calculations of the authors, should help to strengthen convergence without the risk of falling into the schemata theorem. This is prevented by the possibility to acquire unique features, which characterize the "eaten" descendant.

**Acknowledgements.** The authors would like to thank those who have proved their goodwill and assistance during the research. We direct special thanks for inspiration and access to the lab to Tomasz Marquardt, Behavioral Research Laboratory, Department of Evolutionary Biology, Kazimierz Wielki University in Bydgoszcz.

## References

1. Angryk, R., Czerniak, J.: Heuristic algorithm for interpretation of multi-valued attributes in similarity-based fuzzy relational databases. Int. J. Approximate Reasoning **51**(8), 895–911 (2010)

2. Apiecionek, L., Czerniak, J., Dobrosielski, W.: Quality of services method as a ddos protection tool. IS'2014. AISC, vol. 323, pp. 225–234. Springer, New York (2015)
3. Apiecionek, L., Czerniak, J.M.: Qos solution for network resource protection. In: INFORMATICS 2013: Proceedings of the 12th International Conference on Informatics, pp. 73–76 (2013)
4. Apiecionek, L., Czerniak, J.M., Zarzycki, H.: Protection tool for distributed denial of services attack. In: Kozielski, S., Mrozek, D., Kasprowski, P., Małysiak-Mrozek, B., Kostrzewa, D. (eds.) BDAS 2014. CCIS, vol. 424, pp. 405–414. Springer, New York (2014)
5. Chan, F.T., Au, K., Chan, L., Lau, T.: Using genetic algorithms to solve quality-related bin packing problem. Robot. Comput.-Integr. Manuf. **23**, 71–81 (2007)
6. Czerniak, J.: Evolutionary approach to data discretization for rough sets theory. Fundamenta Informaticae **92**(1–2), 43–61 (2009)
7. Czerniak, J.M., Apiecionek, L., Zarzycki, H.: Application of ordered fuzzy numbers in a new ofnant algorithm based on ant colony optimization. In: Kozielski, S., Mrozek, D., Kasprowski, P., Małysiak-Mrozek, B., Kostrzewa, D. (eds.) BDAS 2014. CCIS, vol. 424, pp. 259–270. Springer, New York (2014)
8. Czerniak, J., Dobrosielski, W., Zarzycki, H., Apiecionek, L.: A proposal of the new owlant method for determining the distance between terms in ontology. In: Filev, D., et al. (eds.) is'2014. AISC, vol. 323, pp. 235–246. Springer, Heidelberg (2015)
9. De Jong, K., Spears, W.: Learning concept classification rules using genetic algorithms. In: Proceedings of the 12th International Conference on Artificial Intelligence, pp. 651–656 (1991)
10. Ewald, D., Czerniak, J., Zarzycki, H.: Approach to solve a criteria problem of the abc algorithm used to the wbdp multicriteria optimization. In: Angelov, P., et al. (eds.) IS'2014. AISC, vol. 322, pp. 129–137. Springer, Heidelberg (2015)
11. Farzanegan, A., Vahidipour, S.: Optimization of comminution circuit simulations based on genetic algorithms search method. Miner. Eng. **22**, 719–726 (2009)
12. Filipponi, A., Pegazzano, F.: Italian species of the glaber-group (acarina, mesostigmata, macrochelidae, macrocheles). Redia 47, pp. 211–238, ljubljana, Slovenia (1962)
13. Halliday, R.: The australian species of macrocheles (acarina: Macrochelidae). Invertebr. Syst. (formerly known as Invertebrate Taxonomy) **14**(2), 273–326 (2000)
14. Halliday, R., Holm, E.: Experimental taxonomy of australian mites in the macrocheles glaber group (acarina : Macrochelidae). Exp. Appl. Acarol. **1**, 277–286 (1985)
15. Kosiński, W., Prokopowicz, P., Ślezak, D.: On algebraic operations on fuzzy reals. In: Rutkowski, L., Kacprzyk, J. (eds.) Neural Networks and Soft Computing. Advances in Soft Computing, vol. 19, pp. 54–61. Springer, Heidelberg (2002)
16. Marquardt, T., Kaczmarek, S.: Continuous recording of soil mite behaviour using an internet protocol video system. Int. J. Acarol. **40**, 1–6 (2014)
17. Marquardt, T., Kaczmarek, S., Halliday, B.: Ovoviviparity in macrocheles glaber (müller) (acari: Macrochelidae), with notes on parental care and egg cannibalism. Int. J. Acarol. **41**(1), 71–76 (2015)
18. Marquardt, T., Kaczmarek, S., Halliday, B.: Video supplement [in] ovoviviparity in macrocheles glaber (müller) (acari: Macrochelidae), with notes on parental care and egg cannibalism. Int. J. Acar. **41**, 71–76 (2015)
19. Mikolajewska, E., Mikolajewski, D.: E-learning in the education of people with disabilities. Adv. Clin. Exp. Med. **20**(1), 103–109 (2011)
20. Mikolajewska, E., Mikolajewski, D.: Exoskeletons in neurological diseases - current and potential future applications. Adv. Clin. Exp. Med. **20**(2), 227–233 (2011)

21. Mikolajewska, E., Mikolajewski, D.: Non-invasive eeg-based brain-computer interfaces in patients with disorders of consciousness. Mil. Med. Res. **1**(14), 1 (2014)
22. Prokopowicz, P.: Methods based on the ordered fuzzy numbers used in fuzzy control. In: Proceedings of the Fifth International Workshop on Robot Motion and Control - RoMoCo 2005, pp. 349–354 (2005)
23. Prokopowicz, P.: Flexible and simple methods of calculations on fuzzy numbers with the ordered fuzzy numbers model. In: Rutkowski, L., Korytkowski, M., Scherer, R., Tadeusiewicz, R., Zadeh, L.A., Zurada, J.M. (eds.) ICAISC 2013, Part I. LNCS, vol. 7894, pp. 365–375. Springer, Heidelberg (2013)
24. Quagliarella, D.: Genetic Algorithms and Evolution Strategy in Engineering and Computer Science: Recent Advances and Industrial Applications. Wiley, Hoboken (1998)
25. Sadrai, S., Meech, J., Ghomshei, M., Sassani, F., Tromans, D.: Influence of impact velocity on fragmentation and the energy efficiency of comminution. Int. J. Impact Eng. **33**, 723–734 (2006)
26. Sameon, D., Shamsuddin, S.M., Sallehuddin, R., Zainal, A.: Compact classification of optimized boolean, reasoning with particle swarm optimization. Intell. Data Anal. **16**, 915–931 (2012). IOS Press
27. Shuiping, L., Hongzan, B., Zhichu, H., Jianzhong, W.: Nonlinear comminution process modeling based on ga-fnn in the computational commi-nution system. J. Mater. Process. Technol. **120**, 84–89 (2002)
28. Walter, D., Krantz, G.: A review of the glaber group (s. str.) species of the genus macrocheles (acari: Macrochelidae) and designation of species complexes. Acarologia **27**, 277–294 (2000)
29. Zolotová, I., Mihal', R., Hošák, R.: Objects for visualization of process data in supervisory control. In: Madarász, L., Živčák, J. (eds.) Aspects of Computational Intelligence. TIEI, vol. 2, pp. 51–61. Springer, Heidelberg (2013)

# Physical Knowledge Base Representation for Web Expert System Shell

Roman Simiński[(⊠)] and Tomasz Xięski

Institute of Computer Science, University of Silesia, Katowice, Poland
{roman.siminski,tomasz.xieski}@us.edu.pl

**Abstract.** Web applications have developed rapidly and have had a significant impact on the application of systems in many domains. The migration of information systems from classic desktop software to web applications can be seen as a permanent trend. This trend also applies to the knowledge based systems. This work is a part of the *KBExplorator* project – the main goal of this project is to provide a complete and easy to use web-based tool for the development of expert systems. The evaluation of the rules searching effectiveness in the proposed physical rule base model is the first experimental aim of this work. Experiments will be conducted to determine the duration of retrieving a single rule or group of rules in large rules sets. Decomposition of the rule knowledge base into the relational database is also a crucial issue of this work and therefore the presentation of the data model is the second goal of this work. The usage of a relational database in the web-based application is obvious, but its usage as the physical storage for the rule base is described in relatively small number of publications. Proposed decomposition conception and the model presented in this work has not been previously described. The positive results of experiments presented in this work allow us to continue the development of the system – in the next revision, the database interface layer will be implemented with the usage of a specialized API. This proposed software architecture allow us to transparently change the database engine as well as the programming language currently used in the application layer of the system.

**Keywords:** Knowledge base · Expert system shell · Web application

## 1  Introduction

Traditional expert system shells were developed as desktop applications. Meanwhile, web applications have grown rapidly and have had a significant influence on the application of such systems. The migration of information systems from classic desktop software to web applications can be observed as a permanent trend. This trend also applies to the knowledge based systems. Unfortunately the World Wide Web and some internet protocols (like the HTTP protocol) were originally introduced as a hypertext distribution infrastructure and were

© Springer International Publishing Switzerland 2016
S. Kozielski et al. (Eds.): BDAS 2016, CCIS 613, pp. 558–570, 2016.
DOI: 10.1007/978-3-319-34099-9_43

not created as the coherent and useful environment for development of applications such as expert systems [4]. In the literature we can find some attempts to build web-based expert systems [10,17], many of them are dedicated and tailored to particular applications [12,22]. The literature appears to offer contradictory views on the status and use of web-based expert systems [4], from "there are now a large number of expert systems available on the Internet" [8] to "there are not many ES on the web" [10]. The detailed analysis of these contradictory points of view goes beyond the scope of this study, but an interesting discussion can be found in the above mentioned reference [4].

The research conducted by the authors of this work was focused on domain independent tools for web expert systems development. The results of such analyses were described in the next section. The well-known and commercial expert system development tools have been extended to offer web-based development capabilities. In predominating number of cases, the proposed solutions are commercial and highly specialized or they are just software wrappers for existing tools adopted to work in the web environment. Therefore, authors claim that there is still a need to create easy to use, open source tools for developing web-based expert systems. A main goal of the *KBExplorator* project is to provide a complete and easy to use web-based solution for the development of expert systems. The user of the system has the possibility to create the knowledge bases using a dedicated (attributes, facts and rules) editor or to import rules from files in selected formats. The user can use different inference algorithms — classical as well as their modified versions (based on a new inference process implementation [20]). The methodological assumption and theoretical description of research realized within the *KBExplorator* project can be found in [13,14,18]. What is more, practical issues partially connected with the project are presented in [11,19], however proposed decomposition conception and the model presented in this work has not been previously published.

The main focus of this work is set on the effectiveness of rules searching. The process of retrieving the proper rules subset from the whole knowledge base is a crucial part of inference processes. The inference process can be divided into three main stages: matching, choosing and execution of rules. On each stage, inference algorithms select a single rule or a subset of rules from the whole rules set. Regardless of the considered inference algorithm, selecting of rules usually takes the biggest amount of processing time. The physical model proposed in this paper stores rules in a relational database. Each rules searching request requires the execution of a query, which retrieves data from proper data tables. Previous experiments have shown that the inference effectiveness for the proposed model is satisfactory [14,15,19], but these experiments were focused on the inference control strategies and were performed on rules sets of cardinality less than 1500.

The evaluation of the rules searching effectiveness in the proposed model is the first experimental goal of this work. The experiments will be done to determine the duration of retrieving a single rule or group of rules in large rules sets, counting up to 20 000 rules. The evaluation will be based on experiments on real rule sets, already stored in the database of a prototype version of the *KBExplorator* system and replicated to achieve the mentioned above cardinality.

The presentation of the decomposition of a rule knowledge base into the relational database is the second goal of this work. The usage of relational database in the web-based application is obvious, but its usage as the physical storage for a rule base is described in a relatively small number of publications [3,9,12,22]. Unfortunately those descriptions lack detailed information about an efficient (rule base) decomposition approach, meanwhile the expected increase in the number of web-based expert system applications is reported by these and other [4,8] publications.

## 2 Related Works

Several tools and languages are available for developing web-based expert systems – these tools use traditional expert system techniques and offer additionally the capacities for web-based development [8]. The Acquire system [1] provides an ability to develop web-based user interfaces through a clientserver development kit that supports Java and ActiveX controls. Unfortunately, detailed information on this topic is enigmatic, and it is impossible to get a precise description about the conceptual and physical representation of the knowledge base. The ExSys system provides the Corvid Servlet Runtime and implements the Exsys Corvid Inference Engine as a Java Servlet. Corvid uses *Logic* and *Action Blocks* to organize and structure the rules on a conceptual level [7]. The physical representation of a rule base has also been omitted in the information on the ExSys home page and tutorials. Jess is a popular rule engine for the Java platform. Rules written using Jess are saved in the form of an XML file which must contain a *rule-execution-set* element. The JESS engine cannot be used directly in a web-based application, but it is possible to use JESS within the JSP platform [2].

Another commercial expert system building tool is XpertRule, which offers a Knowledge Builder Rules Authoring Studio. Applications can be generated as Java Script/HTML files for deployment as web applications. Knowledge Builder supports the deployment of knowledge components and applications on the Windows platform – its applications are implemented via a .NET inference engine (Rules Server) with access to database, COM and .NET connectivity [21]. The Rules Server provides rules processing for other applications or for the transactions server. This mode of operation can also be exposed as a web service (via SOAP, REST etc.). With no doubt the rules are stored in a database, but there isn't any detailed information about the internal rule base representation available.

The eXpertise2Go's Rule-Based Expert System provides free building and delivery tools that implement expert systems as Java applets, Java applications and Android apps [6]. The eXpertise2Go offers an e2gRuleWriter module which allow to define the rules as well as their features and properties. The eXpertise2Go utilises knowledge bases stored in plain text files. Another expert system implementation is called Prolog, and some of its variants allow to run the rule interpreter within a web application [6]. It is possible to run inference by using CGI on the server, which builds the appropriate Prolog query and executes the interpreter. This approach works on the server side and is not tailored to the specifics of a web application. It requires the usage of relevant program which builds a Prolog query and executes the Prolog script, as well as generatates the HTML output. Prolog systems utilises knowledge bases stored in the program code [5].

The above described commercial expert system development tools have been extended to offer web-based development facilities. Unfortunately in predominating number of cases, the proposed solutions are commercial and highly specialized or they are just software wrappers for existing tools adopted to work in the web environment. Therefore, authors claim that there is still a need to create easy to use and open source tools for developing web-based expert systems.

The main subject of this work is the physical representation of rule knowledge bases. The majority of the analysed expert systems building tools store rules in text files, using particular formats (JESS rule language, XML, Prolog clauses). The Xpert Rule and ExSys systems are able to store rules in relational databases, but analysis of source materials does not allow us to conclude what is the physical structure of the system database. In the literature we can find some attempts to build web-based domain expert systems [3,9,12], but the presented information typically does not contain enough details about the rules storing method. Although in [22] the database diagram can be found, it does not concern rule base issues. The presentation of the decomposition of the rule knowledge base into the relational database is the main subject of the following section. Proposed proposition is the original, own authors' solution, applied in the *KBExplorator* system and it has not been presented yet.

## 3    Methods

This section presents three main issues – a conceptual model of a rule base (of the proposed web expert system shell), the architecture of such system and the decomposition of a conceptual model into a relational database.

### 3.1    Rule Knowledge Base — Conceptual Model

In the proposed web expert system shell the knowledge base is a pair $\mathcal{KB} = (\mathcal{R}, \mathcal{F})$ where $\mathcal{R}$ is a non-empty finite set of rules and $\mathcal{F}$ is a finite set of facts. $\mathcal{R} = \{r_1, r_2, \ldots, r_n\}$, each rule $r \in \mathcal{R}$ will have a form of Horn's clause: $r$ : $p_1 \wedge p_2 \wedge \cdots \wedge p_m \rightarrow c$, where $m$ — the number of literals in the conditional part of rule $r$, and $m \geq 0$, $p_i$ — $i$-th literal in the conditional part of rule $r$, $i = 1 \ldots m$, $c$ — literal of the decisional part of rule $r$.

For each rule $r \in \mathcal{R}$ we define the following functions: $concl(r)$ — the value of this function is the conclusion literal of rule $r$: $concl(r) = c$; $cond(r)$ — the value of this function is the set of conditional literals of rule $r$: $cond(r) = \{p_1, p_2, \ldots p_m\}$, $literals(r)$ — the value of this function is the set of all literals of rule $r$: $literals(r) = cond(r) \cup \{concl(r)\}$. We will also consider the *facts* as clauses without any conditional literals. The set of all such clauses $f$ will be called *set of facts* and will be denoted by $\mathcal{F}$: $\mathcal{F} = \{f : \forall_{f \in \mathcal{F}} \, cond(f) = \emptyset \wedge f = concl(f)\}$.

In this work, rule's literals will be denoted as pairs of attributes and their values. Let $A$ be a non-empty finite set of conditional and decision attributes[1]. For every symbolic attribute $a \in A$ the set $V_a$ will be denoted as the set of values of attribute $a$. Attribute $a \in A$ may be simultaneously a conditional and decision attribute. Also a conclusion of a particular rule $r_i$ can be a condition in an other rule $r_j$. It means that rules $r_i$ and $r_j$ are connected and it is possible that inference chains may occur. The literals of the rules from $\mathcal{R}$ are considered as attribute-value pair $(a, v)$, where $a \in A$ and $v \in V_a$.

For each rule set $\mathcal{R}$ with $n$ rules any arbitrarily created subset of rules $R \in 2^{\mathcal{R}}$ will be called *a group of rules*. In this work we will discuss specific subset $PR \subseteq 2^{\mathcal{R}}$ called *partition of rules*. Any partition $PR$ is created by *partitioning strategy*, denoted by $PS$, which defines specific content of groups of rules $R \in 2^{\mathcal{R}}$ creating a specific *partition of rules* $PR$. We may consider many partitioning strategies for a single rule base, each partitioning strategy $PS$ for rules set $\mathcal{R}$ generates the partition of rules $PR \subseteq 2^{\mathcal{R}}$: $PR = \{R_1, R_2, \ldots, R_k\}$, where: $k$ — the number of groups of rules creating the partition $PR$, $R_i$ — $i$-th group of rules, $R \in 2^{\mathcal{R}}$ and $i = 1, \ldots, k$. Rules partitions terminologically correspond to the mathematical definition of the partition as a division of a given set into the non-overlapping and non-empty subset. The partition strategies are described in the previous publication [14,18] and will not be described in details. In some cases we will assume that the particular expert system functions utilize the rule knowledge base decomposed into the group of rules, for example, modification of the classical inference algorithm is based on information extracted from the rules of groups generated by the particular rule base partition [15,20].

## 3.2 The Architecture of the Web Expert System Shell

The analysis of the requirements for the considered web expert system shell lead us to decompose the traditional rule base model into the relational database scheme. Detailed description of the decomposition process is described in the next subsection. The system is a client-server web application with clearly indicated client and server sides. In the server side three sub-layers are considered. *Database layer* is responsible for storing information about users and their (owned or shared) rules bases. *Rule base services layer* encapsulates data exchange with the database — all operations on rules are performed by the server-side services available via a specialized interface. *Application layer* offers services appropriate for the expert system shell — inference and explanations.

It is worth pointing out that actions typical for expert system shell should be divided into the client and server side. Client side initializes the inference, realizes the main event loop for the backward inference (forward inference can be fully implemented on the server side). Client side also realizes confirmation

---

[1] Decision attributes are attributes that are at least once included in a conclusion of any rule from $\mathcal{R}$.

of the rule's condition against a dynamically created fact set, acquisition of the facts from the application environment (usually from the user), initialization of recursive inference calls for sub-goals (backward inference). Client side utilizes JavaScript functions embedded in the HTML document, generated by the proper server side scripts from the application layer. The JavaScript functions use rules obtained from the server services via asynchronous AJAX requests. This distributed environment was used experimentally in the system described in [11,19]. We consider two main technologies for implementation of the server side of the system: W/LAMP and JaveEE. Works described in this paper has been done with the use of LAMP — MySQL working on the Linux server with Apache WWW server and PHP interpreter.

## 3.3   Rule Knowledge Base — Physical Model

In the proposed system the rule base is physically stored in the relational database. Any registered user of the system can create and manage multiple rules bases, each one containing a custom set of attributes, their values, facts and rules. Rule bases could be shared between users — it is possible to share particular rule bases for editing or only in the read-only mode. The proposed relational database model contains obvious items – for example an *user* entity containing typical data about users of the web application or *settings* entity with technical data associated with each user. For the problem considered in this work the most important are entities containing information about the rule base, which conceptual structure was described earlier. Any single user of the system can use many rule bases. The *knowledegeBase* entity contains essential information about the rule base — its name, description, creation date and a unique identifier as the primary key. As mentioned earlier the system allows knowledge base sharing, therefore the *knowledegeBase* entity is connected with a *kbShare* entity. This is an association entity, it connects entities named *knowledegeBase* and *user*.

Each particular knowledge base has its own attribute set – attributes definitions are stored in the *attribute* entity. An attribute is described by its name, type (symbolic, numeric continuous, numeric discrete), user specific description, knowledge engineers note and optional, additional information in the form of an HTML file. Every attribute (except numeric continuous attributes) has the values list stored in the *values* entity, each value is also described by its name (symbolic) or by a particular numeric value, user dedicated description and optional, additional information in the form of HTML file. The system allows to define facts permanently stored in a particular knowledge base – a fact is defined as an *attribute-relation-value* triple . The *fact* entity is designated for storing information about facts, this entity is connected with entities *attribute* and *value*. For numerical continuous facts their value is directly stored in the field named *continuousValue*. The relation between the attribute and value is stored in the field called *operator*. Figure 1 presents this part of the database structure as an ERD diagram.

The crucial part of the data model is presented on the Fig. 2. Information about each rule are represented in the *rule* entity, which contains all necessary

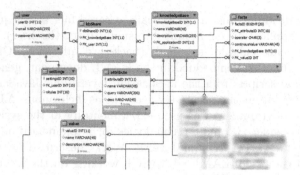

**Fig. 1.** First part of ERD diagram for rule base decomposition

information, including different kinds of descriptions. As was mentioned above, rules have a form of a Horn's clause. The literals of the rule are represented by the *attributeValue* entity, which stores the conditions and conclusion of each rule. The field *isConclusion* indicates that the particular literal is a conclusion of the rule. This entity associates *attribute* and *value* entities for symbolic and numerical discrete types, continuous numerical values and relation are stored directly in the proper fields.

The database model contains also other entities, dedicated for storing information about the scenarios for the inference processes, but this part of the model will be omitted because it is not directly connected with the main goals of this paper.

The above described conception of rule base decomposition allows to store multiple knowledge bases shared by the multiple users in the WWW environment. The prototype version of the system called *KBExplorator* was implemented and currently intensive tests and experiments are performed. An example

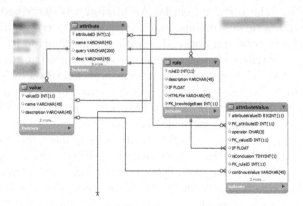

**Fig. 2.** Second part of ERD diagram for rule base decomposition

**Fig. 3.** The *KBExplorator* system — the rules list editor

of the system's user interface is presented on the Fig. 3. We are going to open free, public access to the system in spring of 2016.

## 4   Experiments and Discussion

The main problem of this work focuses on the effectiveness of rules searching. The process of retrieving the proper rule or subset of rules from the whole knowledge base is considered as the crucial part of inference processes. The proposed in this paper physical model stores rules in the database, each rule's data is divided into the two main entities: named *rule* and *attributeValue*, the latter is also connected with two others entities: *attribute* and *value*. It should be noted that the size of the *attributeValue* entity will grow when the user will add new rules to the his knowledge bases. Each rules searching request requires the execution of a SQL query, which retrieves data from proper data tables.

The authors decided to isolate the rules retrieval process for two reasons. Firstly, the conducted experiments allowed us to perform exhaustive tests focused on database services and enabled fine-tuning of particular database parameters. Secondly, in the *KBExplorator* system, the rules selection process is performed not only in the inference algorithms, but especially in the rules partitioning methods [16,18]. From this point of view experiments condiering the rules selection are important for further development of the *KBExplorator* system.

The aim of the first experiment described in this section was to evaluate the effectiveness of the rules' retrieval process, understood as the time of getting a result from a proposed SQL query. Finding relevant rules and presenting them to the user is a vital part of any inference, rules partitioning and explanation processes. That is why the information stored in multiple tables of the *KBExplorator* system must be joined together. A typical SQL query (except for the SQL_NO_CACHE part[2]) that is used in *KBExplorator* is shown on Algorithm 1. Its goal is to gather all information which forms a given rule in a previously

---

[2] The SQL_NO_CACHE option was added to the select query so that the server would not use the cache mechanism, which could falsify the experiment's result.

selected knowledge base. Thanks to the inner joins, key values get converted to human-readable data. It is also worth to note that the attributes in the premise and conclusion can be ordered differently for each rule by the user, so one should not omit the order by clause.

In order to test this query execution time, a simple PHP script was made (also presented on Algorithm 1), which for every rule identifier (previously gained form the *rule* table), sends the SQL query to the database engine and measures the response time using the `microtime()` function. The `microtime` function returns the current Unix timestamp with microseconds, and therefore the difference between the two points in time (registered before and after the query execution) allow us to determine the duration of such retrieval process. This procedure could be seen as a special case of forward inference, in which all of the rule premises are facts in the knowledge base. To rate the effectiveness of information retrieval, some descriptive statistics as the minimum, maximum or average as well as the median and standard deviation values (of the time duration of obtaining every rule from the knowledge base) were calculated.

---

**Algorithm 1.** Measuring of query execution time

---

```
foreach ($rule_ids as $r_id) {
  $rule_query = $pdo->prepare('SELECT SQL_NO_CACHE field_list
  FROM attribute INNER JOIN attributeValue ON
  attributeID = attributeValue.FK_attributeID INNER JOIN value
  ON valueID = attributeValue.FK_valueID
  WHERE FK_ruleID = ? AND FK_knowledgeBase = ?
  ORDER BY attributeOrder');

  //MEASURE TIME
  $msc=microtime(true);
  $rule_query->execute(array($r_id, $knowledgeBaseId));
  $msc=microtime(true)-$msc;

  //INSERT $msc to log table
}
```

---

The first experiment consisted of eight cases, concerning a real-world knowledge base. The knowledge base (created by domain experts from the building and construction industry branch) consisted of 4438 rules. The premises part of a given rules consisted of one to eighteen descriptors. The typical rule length was 2.66 descriptors. There were also 1720 unique decisions (when analysing the decisional part of rule). This rules set was duplicated five times (because gaining access to large, real-world knowledge base is very difficult) so that one can evaluate the rule retrieval process also on a larger base.

First four cases of the conducted experiment regarded the time of rule retrieval from the database, which did not use any column indexes (apart from primary keys). Results from cases five to eight regard the altered database structure – all columns with the *FK_* prefix (as well as the operator

**Table 1.** Analysis of rules' retrieval time (in seconds) based on their identifiers

| Case | Rule count | Order | Minimum | Maximum | Mean | Median | $\sigma$ |
|------|-----------|-------|---------|---------|------|--------|----------|
| 1 | 4438 | ascending | 0,22004 | 0,59827 | 0,22672 | 0,22237 | 0,02916 |
| 2 | 4438 | random | 0,22003 | 0,60644 | 0,22674 | 0,22376 | 0,02377 |
| 3 | 22190 | ascending | 0,21860 | 0,60182 | 0,22509 | 0,22229 | 0,02226 |
| 4 | 22190 | random | 0,21843 | 0,60169 | 0,22503 | 0,22232 | 0,02168 |
| 5 | 4438 | ascending | 0,00017 | 0,00063 | 0,00032 | 0,00031 | 0,00008 |
| 6 | 4438 | random | 0,00016 | 0,00091 | 0,00028 | 0,00024 | 0,00010 |
| 7 | 22190 | ascending | 0,00016 | 0,00105 | 0,00022 | 0,00018 | 0,00008 |
| 8 | 22190 | random | 0,00016 | 0,00335 | 0,00025 | 0,00020 | 0,00010 |

column) were rebuild to support database indexes. What is more, the process of obtaining rule identifiers were different. For cases number 1,3,5,7 the identifiers were sorted in ascending order whereas for even-numbered cases the identifiers where obtained randomly based on the following query: SELECT ruleID FROM rule WHERE FK_knowledgeBase = 1 ORDER BY RAND(). The results from the first experiment are presented in the Table 1.

When analysing the results for the first four cases one can observe that there is almost no noticeable time difference between them, regardless of the rules set size or the order of rules being retrieved. Cases from five to eight exhibit a significant improvement in information retrieval time, thanks to the usage of several column indexes. What is more, the rule order has a noticeable impact on maximal retrieval time – if rules are accessed sequentially (based on their identifiers) the maximal response time is lower than in cases which they are ordered randomly. The bigger the dataset is, the more noticeable difference can be observed. Fortunately the average and mean values differ only by few milliseconds, so generally speaking the rule order is not that important.

**Table 2.** Analysis of rules' retrieval time (in seconds) based on their conclusions

| Case | Rule count | Indexed | Minimum | Maximum | Mean | Median | $\sigma$ |
|------|-----------|---------|---------|---------|------|--------|----------|
| 1 | 4438 | no | 0,35344 | 37,97943 | 0,92294 | 0,57815 | 1,48545 |
| 2 | 4438 | yes | 0,00034 | 0,03927 | 0,00098 | 0,00061 | 0,00158 |
| 3 | 22190 | no | 1,23870 | 186,90694 | 4,05242 | 2,35636 | 7,36639 |
| 4 | 22190 | yes | 0,00136 | 0,20918 | 0,00489 | 0,00270 | 0,00890 |

The goal of the second experiment was to analyse the rules' retrieval time based on their conclusions. Therefore a list of 1720 rule conclusions was obtained from the database. Next, for each of the conclusions, the identifiers of relevant rules were gathered. Finally, the query from Algorithm 1 was send to the database multiple times (for each relevant rule) and the total execution time was measured. Therefore the results presented in Table 2 concern the time from gathering the relevant (to a given conclusion) rule identifiers to obtaining all information for those rules. Cases number one and three concern the database which did not use column indexes, whereas the even-numbered cases did.

The results prevented in the Table 2 clearly show that the non-optimal database structure (without indexes) may be the main factor of the poor performance of the database model of the *KBExplorator* system. Especially the maximum retrieval time of rule subsets above 37 and 186 seconds (for the 4438 and 22190 rulesets respectively) are considered as unacceptable. It is worth noticing that the discussed retreival process (of rules with a given conclusion) is only a small part of a backward inference. Therefore this phase should be as fast as possible. One should conclude that using database indexes is considered necessary in the inference process as well as other applications which require the fastest rules selection time.

## 5  Conclusions

This paper has two main goals. The presentation of a relational database model for the rule knowledge base of a web-based expert system shell was the first goal. Three main issues were discussed – a conceptual model of the rule base of the proposed web expert system shell, the architecture of such system and the decomposition of the conceptual model into the relational database. The evaluation of the rules searching effectiveness in proposed model was an experimental goal of this work. The experiments were conducted to determine the duration of retrieving a single rule or group of rules in large rules sets, counting up to 20 000 rules. The evaluation was based on experiments on real rule sets, already stored in the database of the *KBExplorator* system and replicated to achieve above cardinality.

In the authors opinion the experimental evaluation of rules retrieval effectiveness for large rule bases was an important part of *KBExplorator* development. Practical verification of the proposed relational model confirmed findings from previous works which were focused on inference control strategies. The selection of rules should be performed with time efficiency adequate to the needs and requirements of the algorithms implemented in such expert systems. The authors decided to isolate the rules retrieval process for two reasons. Firstly, the conducted experiments allowed us to perform exhaustive tests focused on database services and enabled fine-tuning of particular database parameters. Secondly, in the *KBExplorator* system, the rules selection process is performed not only in the inference algorithms, but especially in the rules partitioning methods [16,18].

When analysing the results of the experiments we observed that there was almost no noticeable time difference between the tested cases if the database did not use indexes – regardless of the rule set size or the order in which rules were being retrieved. The rule order had a noticeable impact on the maximal retrieval time only if indexes were a part of the database structure – the bigger the dataset was, the more noticeable difference was observed. Fortunately the average and mean values differed only by few milliseconds, so generally speaking the rule order was not that important. Database structure without indexes was regarded to be the main factor of the poor performance of the database model proposed for the *KBExplorator* system. One should also conclude that using database indexes is considered necessary in the inference process as well as other applications which require the fastest rules selection time. It is worth to notice, that results for rules sets counting maximum 22190 rules were acceptable from the users point of view, especially when backward inference was considered — rules retrieval time was negligible.

The positive results of performance tests presented in this work allow us to continue the development of the system. In the next stage we are going to encapsulate the database operation by implementating a specialized API. Other system components, working in the application and client layer will use the REST architecture to perform rules selection, and data will transferred using the XML and JSON notation. The proposed architecture allow us to transparently change the database engine as well as the programming language currently used in the application layer.

**Acknowledgements.** This work is a part of the project "Exploration of rule knowledge bases" founded by the Polish National Science Centre (NCN: 2011/03/D/ST6/03027).

# References

1. Acquired Intelligence: Acquired Intelligence Home Page. http://aiinc.ca (Accessed Oct 2015)
2. Canadas, J., Palma, J., Túnez, S.: A tool for MDD of rule-based web applications based on OWL and SWRL. Knowledge Engineering and Software Engineering (KESE6), p. 1 (2010)
3. Dokas, I.M.: Developing web sites for web based expert systems: a web engineering approach. In: ITEE, pp. 202–217 (2005)
4. Duan, Y., Edwards, J.S., Xu, M.: Web-based expert systems: benefits and challenges. Inf. Manage. **42**(6), 799–811 (2005)
5. Dunstan, N.: Generating domain-specific web-based expert systems. Expert Syst. Appl. **35**(3), 686–690 (2008)
6. eXpertise2Go: eXpertise2Go Home Page. http://expertise2go.com (Accessed Nov 2015)
7. Exsys: Exsys Home Page. http://www.exsys.com (Accessed Nov 2015)
8. Grove, R.: Internet-based expert systems. Expert syst. **17**(3), 129–135 (2000)
9. Grzenda, M., Niemczak, M.: Requirements and solutions for web-based expert system. In: Rutkowski, L., Siekmann, J.H., Tadeusiewicz, R., Zadeh, L.A. (eds.) ICAISC 2004. LNCS (LNAI), vol. 3070, pp. 866–871. Springer, Heidelberg (2004)
10. Huntington, D.: Web-based expert systems are on the way: Java-based web delivery. PC AI **14**(6), 34–36 (2000)

11. Jach, T., Xieski, T.: Inference in expert systems using natural language processing. In: Kozielski, S., Mrozek, D., Kasprowski, P., Małysiak-Mrozek, B., Kostrzewa, D. (eds.) Beyond Databases, Architectures and Structures. Communications in Computer and Information Science, vol. 521, pp. 288–298. Springer, Switzerland (2015)

12. Li, D., Fu, Z., Duan, Y.: Fish-expert: a web-based expert system for fish disease diagnosis. Expert Syst. Appl. **23**(3), 311–320 (2002)

13. Nowak-Brzezińska, A., Simiński, R.: Knowledge mining approach for optimization of inference processes in rule knowledge bases. In: Herrero, P., Panetto, H., Meersman, R., Dillon, T. (eds.) OTM-WS 2012. LNCS, vol. 7567, pp. 534–537. Springer, Heidelberg (2012)

14. Nowak-Brzezinska, A., Siminski, R.: New inference algorithms based on rulespartition. In: Proceedings of the 23th International Workshop on Concurrency, Specification and Programming, Chemnitz, Germany, 29 September - 1 October 2014, pp. 164–175 (2014). http://ceur-ws.org/Vol-1269/paper164.pdf

15. Nowak-Brzezińska, A., Simiński, R.: Goal-driven inference for web knowledge based system. In: Wilimowska, Z., Borzemski, L., Grzech, A. (eds.) Information Systems Architecture and Technology: Proceedings of 36th International Conference on Information Systems Architecture and Technology – ISAT 2015 – Part IV. Advances in Intelligent Systems and Computing, vol. 432, pp. 99–109. Springer, Switzerland (2015)

16. Nowak-Brzezinska, A., Wakulicz-Deja, A.: Exploration of knowledge bases inspired by rough set theory. In: Proceedings of the 24th International Workshop on Concurrency, Specification and Programming, Rzeszow, Poland, 28–30 September, 2015, vol. 1, pp. 64–75 (2015)

17. Riva, A., Bellazzi, R., Montani, S.: A knowledge-based web server as a development environment for web-based knowledge servers. In: IEE Colloquium on Web-Based Knowledge Servers (Digest No. 1998/307), pp. 5-1-5-5. IET (1998)

18. Simiński, R.: Multivariate approach to modularization of the rule knowledge bases. In: Gruca, A., Brachman, A., Kozielski, S., Czachórski, T. (eds.) Man–Machine Interactions 4. Advances in Intelligent Systems and Computing, vol. 391, pp. 473–483. Springer, Switzerland (2016)

19. Simiński, R., Manaj, M.: Implementation of expert subsystem in the web application-selected practical issues. Studia Informatica **36**(1), 131–143 (2015)

20. Siminski, R., Wakulicz-Deja, A.: Rough sets inspired extension of forward inference algorithm. In: Proceedings of the 24th International Workshop on Concurrency, Specification and Programming, Rzeszow, Poland, 28–30 September 2015, vol. 2, pp. 161–172 (2015)

21. Xpert Rule: Xpert Rule Home Page. http://www.xpertrule.com (Accessed Nov 2015)

22. Zetian, F., Feng, X., Yun, Z., XiaoShuan, Z.: Pig-vet: a web-based expert system for pig disease diagnosis. Expert Syst. Appl. **29**(1), 93–103 (2005)

# OSA Architecture

Ścibór Sobieski[1], Marek A. Kowalski[2]([✉]), Piotr Kruszyński[1], Maciej Sysak[1],
Bartosz Zieliński[1], and Paweł Maślanka[1]

[1] Department of Computer Science, Faculty of Physics and Applied Informatics,
University of Łódź, ul. Pomorska nr 149/153, 90-236 Łódź, Poland
`{scibor,piotr,maciej.sysak,bzielinski,pmaslan}@uni.lodz.pl`
[2] Faculty of Mathematics and Natural Sciences — College of Sciences,
Cardinal Stefan Wyszyński University in Warsaw,
ul. Dewajtis 5, 02-654 Warszawa, Poland
`kowalski@uksw.edu.pl`

**Abstract.** In this paper we present an in depth discussion of the architecture of a new plagiarism detection platform developed by a consortium of Polish universities. The algorithms used by the platform are briefly described in Sect. 3. The main goal of this paper is to present high level structures of services resulting from a very nontrivial attempt to strike an appropriate balance between locality and centralization, while working under strict constraint, both of technological and legal nature.

**Keywords:** Plagiarism detection · Vector space model · System architecture · Database applications

## 1 Introduction

The development of network technologies enabling a free exchange of information provided, along with priceless tools for honest research and learning, also enormous opportunities for plagiarism of all varieties of diploma works in Polish universities and other graduate schools. In order to better visualize the scale of potential plagiarism risk let us note that, according to the research by the Central Statistical Office of Poland (see [11]), at the beginning of academic year 2012/2013 there were 453 colleges and universities, including 153 public ones, and there were approximately 485.2 thousands of graduates at the beginning of academic year 2011/2012, where about 326.7 thousands were graduates of public schools. At this scale even the most devoted advisors can hardly be expected to recognize all instances of plagiarism, while being supported only by their own knowledge and contacts. Hence, there is an urgent need for an effective and efficient tool to assist supervisors and other academic employees.

The authors of this paper are members of the research team tasked with the development of the architecture of the new plagiarism detection system. Such a

This work is supported by MUCI (Międzyuniwersyteckie Centrum Informatyzacji — Interuniversity Centre for IT).

© Springer International Publishing Switzerland 2016
S. Kozielski et al. (Eds.): BDAS 2016, CCIS 613, pp. 571–584, 2016.
DOI: 10.1007/978-3-319-34099-9_44

system, in order to be effective, must not only compare the student work with generally available sources such as Wikipedia or Internet in general, but should also permit to create a shared database of diploma works from participating schools. This shared database would allow to check if a given thesis does not contain unattributed passages copied from a diploma work of a student from some other university, regardless of any contacts between advisors. The existence of a shared database turned out to be nontrivial request warranting a special system architecture which would guarantee the appropriate level of protection for copy and intellectual property rights, as well as their derivatives.

There are several problems which designers of plagiarism detection software must face (full discussion — see [8]). First of all, the notion of plagiarism is not very well defined. A common understanding of the word is presenting someone else's work or idea, in total or just part of it, as one's own. This understanding of plagiarism is well supported by copyright law and related codes but no precise definition is given.

Secondly, our academic experience indicates that non authentic texts are often originated in:

1. the available Internet resources. While the authors were unable to find any research on the verbatim copying of the Internet content, their experience as the diploma advisors (of all levels) indicates that students copy from web pages rather frequently.
2. purchasing a thesis — a huge problem in Polish (though not just Polish) academic life is the widespread practice of buying diploma theses or even purchasing custom written dissertations. It is easy to come, even unwillingly, across advertisements for services offering ready theses for sale or dissertation writing, worryingly open about their purpose. It suffices to enter in any search engine the keywords such as "dissertation writing help" to receive thousands of offers of "writing assistance". Many of these services emphasize that their offers are "plagiarism free", and it is only fortunate that not so many of them make good on this promise, as otherwise proving student's dishonesty might be nigh-on-impossible.

Finally, another important problem concerns legal status of student dissertations. According to Polish copyright and intellectual property regulations, all rights to a diploma (licentiate's, engineer's or master's) thesis belong to its author, that is, presumably, a student whose name is printed on the dissertation's title page. In particular, this implies that:

1. All personal and property rights belong to the student or students who authored the dissertation. The only exception is the right of the college or university to be the first publisher of the thesis.
2. The school is not the owner of the dissertation.
3. The college/university may use a third party service to detect plagiarism in student's dissertation, but it cannot permanently hand over the dissertation to any external plagiarism checking service.

4. The college/university cannot demand that students renounce their property rights to the dissertation (other rights are inalienable anyway).
5. Legal regulations permit creation of intra-university systems for checking and exchange of dissertations, as long as they serve exclusively the internal affairs of the group of the colleges/universities.

Interested reader can find the more thorough treatment of the problems mentioned above in [7,10,24,26].

In this paper we present an in depth discussion of the architecture of a new plagiarism detection platform developed by a consortium of Polish universities. The algorithms used by the platform are briefly described in Sect. 3. The main goal of this paper is to present high level structures of services resulting from a very nontrivial attempt to strike an appropriate balance between locality and centralization, while working under strict constraint, both of technological and legal nature.

According to legal limitations and postulates of the academic society the system architecture should meet the following requirements:

1. effective and scalable,
2. equipped with a central base of knowledge gathering information about reference texts,
3. based on partial information about texts, (i.e. not allowing to recover any of the original texts) and at the same time powerful enough to ensure effective detection of similarities between the originals,
4. able to detect similarities between a given Polish text and texts represented in the central base of knowledge and in the Internet,
5. able to detect similarities between texts in a given group (e.g. students' homeworks, projects, writings concerning a concrete task),
6. implemented in such a way that different similarity measures can be applied to text in different fields of sciences,
7. capable to store information on any successfully defended academic thesis.

## 2    Prior Work

There are several commonly used methods for plagiarism detection, which can be broadly divided into local and global similarity assessment methods. A comprehensive review can be found in [12]. Local similarity assessment is more direct and easy to interpret, as it matches chunks of reference and checked texts (see e.g., [5,20,22]).

Global similarity assessment methods use "global" features of larger fragments (or even whole documents). In OSA we use a variant of similarity measures based on vectors of word frequencies (see e.g., [2,21], cf. [19]). This is a popular choice, insensitive to the word reorderings, but which may lead to an increased false positive rate. Alternative approach, uses suffix data structures which preserve word order [13]. There are also more exotic approaches based on stylometry [6] and citation indices [4].

We were unable to find any system with architecture similarly balancing locality and centrality as our system.

# 3  Plagiarism Detection

In this section we describe a method implemented in the OSA project for detecting resemblance of input texts to the reference documents in local databases, owned and operated by the universities participating in the project.

Let $T$ be a text to be checked for plagiarism, or more generally, for similarities to documents in a given set of reference wordings. The departure point for possible similarities detection is transforming $T$ into a special digital form, called term frequency vector, which is well suited for comparisons to reference texts. A database $\mathbb{B}$ of reference texts and a dictionary $S$ are essential in this process (see [18]).

We begin with forming so-called skeleton of $T$, i.e., the lexically ordered set $T' = \{s_1, s_2, \ldots, s_k\}$ of all unique words in $T$ in their base forms taken from the dictionary $S$. It is accomplished in the following four steps.

1. Preliminary conversion which replaces all capital letters with the corresponding small letters and eliminates all non-alphabetic characters from the text.
2. Lemmatization, i.e., inflectional unification which brings all words to their base forms, called *lemmas*.
3. Synonymic unification which changes all substitutes to their basic forms.
4. Final conversion which eliminates redundant words and all expressions (single words and/or phrases interpreted as single words) belonging to the so-called list of exclusions, which will be described in what follows.

Given a fixed bijective function $f : S \to \{1, 2, \ldots, \text{card}(S)\}$ (called a numeration of $S$) we build the term frequency vector $\text{tfv}(T) = <Zs, Tf>$, where

$$Zs = \{f(s_1), f(s_2), \ldots, f(s_k)\}$$

and

$$Tf = \{a(T, s_1), a(T, s_2), \ldots, a(T, s_k)\},$$

with $\text{card}(S)$ being the number of elements in $S$ and $a(T, x)$ being the number of appearances of the word $x$ in the text $T^*$ resulting from the steps 1–3. In what follows we assume that $a(T, x) = 0$ if $x$ does not appear in $T^*$.

Since lemmas are not equally relevant in the lookout for similarities between texts, we define *ranks*. The rank of a particular lemma $w$ depends on how often $w$ appears in the skeletons of texts in $\mathbb{B}$ to which $T$ should be compared. Let $\boldsymbol{B}$ stand for the set of skeletons of the texts in $\mathbb{B}$, and let $W(\boldsymbol{B})$ consist of all unique words appearing in skeletons from $\boldsymbol{B}$. Assuming that $W(\boldsymbol{B})$ is lexically ordered we define the triple $\text{BoW}(\boldsymbol{B}) = <Bs, Bdf, Br>$ with the components $Bs, Bdf, Br$ being sequences with $\text{card}(W(\boldsymbol{B}))$ elements described below.

- $Bs_j$ (i.e., the $j$-th entry of $Bs$) is the numeration of the $j$-th word in $W(\boldsymbol{B})$.
- $Bdf_j$ is the number of the skeletons from $B$ in which the word $f^{-1}(Bs_j)$ appears.
- $Br_j = \log(IDF_j)$, where $IDF_j = \text{card}(S)/Bdf_j$.

The number $Br_j$ is said to be the rank of the word $w = f^{-1}(Bs_j)$. The more frequently $w$ appears in different skeletons, the less relevant it is. The use of logarithm smoothies differences between arguments $IDF_j$. The triple $\text{BoW}(\boldsymbol{B})$ is said to be *the bag of words*.

We use bags of words in two ways. Firstly, globally, in reference to whole database $\mathbb{B}$. Secondly, locally, in reference to its disjoint parts $\mathbb{B}_j$. More precisely, if $\boldsymbol{B}$ and $B_j$ are sets of the corresponding skeletons of texts in $\mathbb{B}$ and $\mathbb{B}_j$, respectively, we assume that

$$\boldsymbol{B} = \bigcup_{k=1}^{n} B_k, \ \forall_{i \neq j} Bi \cap Bj = \emptyset$$

and make use of $\text{BoW}(\boldsymbol{B}), \text{BoW}(B_1), \ldots, \text{BoW}(B_n)$.

Every word appearing in every skeleton of $\boldsymbol{B}$ has the $IDF$ factor 1, and consequently the word's rank is zero in $\boldsymbol{B}$. The set of these words is called *the global list of exclusions*. In reference to a given domain $B_j$ we build its local list of exclusions by gathering the lemmas which appear in every skeleton of $B_j$.

A lemma which appears in at least one skeleton of $B$ and does not belong to the global list of exclusions is said to be *a global keyword*. Similarly, a lemma which appears in at least one skeleton of $B_j$ and does not belong to its local list of exclusions is said to be *a local keyword* of $B_j$.

In the sequel we assume that the division of $\boldsymbol{B}$ into domains has already been done and skeletons in each $B_j$ have been obtained from text dealing with similar subjects. We also assume that the division is such that there is a need to compare $T$ to texts in only one part of $\boldsymbol{B}$. We now use the Rocchio algorithm, see [15], to point out the most appropriate one. We begin with the following definition. Given a word $x$ appearing in the skeleton $t \in B_k$ of the text $T$, the number

$$w_{k,t}(x) = a(T, x) * Br_k(x),$$

is said to be the weight of $x$ in $B_k$. Here we assume that $B_0 = \boldsymbol{B}$, $k = 0, 1, \ldots, n$ and we set $w_{k,t}(x)$ to 0, if $x$ does not appear in $t$. For each domain $B_i$, $i = 1, 2, \ldots, n$, we define its centroid $C^i = \{c_1^i, c_2^i, \ldots, c_{\text{card}(W(B_i))}^i\}$ by the formula

$$c_j^i = \alpha \frac{1}{\text{card}(B_i)} \sum_{t \in B_i} w_{0,t}(f^{-1}(B_i(s_j)) +$$

$$-\beta \frac{1}{\text{card}(\boldsymbol{B} \backslash B_i)} \sum_{t \in \boldsymbol{B} \backslash B_i} w_{0,t}(f^{-1}(B_i(s_j)),$$

where $j = 1, 2, \ldots, \text{card}(W(B_i))$ and the factors $\alpha = 16$ and $\beta = 4$ have been experimentally set, see [17].

For a given word $x$ let $j^i(x) \in \{0, 1, \ldots, \text{card}(W(B_i))\}$ be the value such that either $j^i(x) = 0$ or

$$c_{j^i(x)}^i = \alpha \frac{1}{\text{card}(B_i)} \sum_{t \in B_i} w_{0,t}(x) +$$

$$-\beta \frac{1}{\text{card}(\boldsymbol{B} \backslash B_i)} \sum_{t \in \boldsymbol{B} \backslash B_i} w_{0,t}(x).$$

According to Rocchio, the skeleton $t$ is classified to $B_k$, if the maximum

$$\max_i \frac{\sum_{s \in t} w_{0,t}(s) c^i_{j^i(s)}}{\sqrt{\sum_{s \in t} (w_{0,t}(s))^2} \sqrt{\sum_{s \in t} \left( c^i_{j^i(s)} \right)^2}}$$

is attained for $i = k$. Here, we assume that $0/0 = 0$. If the maximum is attained for more than one values of $k$, the skeleton $t$ can be classified to any, but only one, domain corresponding to these values. Fraction in the last formula can be interpreted as the cosine of the angle in $[0, \pi/2]$ between the vectors with the coordinates $w_{0,t}(s)$ and $c^i_{j^i(s)}$ , respectively. So, the smaller angle, the bigger its cosine.

When classification is done we search for similarities between $t$ and elements $y$ in $B_k$ by computing the quantities $I(t,y), C(t,y), R(t,y)$ motivated by those used in the SMART system, see [15, 16]. Let $t \cap y$ stand for the lexically ordered set of those local keywords which simultaneously appear in both skeletons $t$ and $y$. We set

$$I(t,y) = \frac{\sum_{s \in t \cap y} w_{k,t}(s) w_{k,y}(s)}{\sqrt{\sum_{s \in t \cap y} (w_{k,t}(s))^2 (w_{k,y}(s))^2}},$$

$$C(t,y) = \frac{\sum_{s \in t \cap y} \min(w_{k,t}(s), w_{k,y}(s))}{\min \left( \sum_{s \in t \cap y} w_{k,t}(s), \sum_{s \in t \cap y} w_{k,y}(s) \right)}.$$

In order to define $R(t,y)$ we assume that $t \cap y = \{s_1, s_2, \ldots, s_m\}$ and

$$ind(t,y) = \{i(s_1), i(s_2), \ldots, i(s_m)\}$$

is the text consisting of the words $0, 1, 2$ formed according to the rule

$$i(s) = \begin{cases} 0 \; if \; w_{k,t}(s) = w_{k,y}(s), \\ 1 \; if \; w_{k,t}(s) \neq w_{k,y}(s) \\ \quad and \; w_{k,t}(s) = \min(w_{k,t}(s), w_{k,y}(s)), \\ 2 \; if \; w_{k,t}(s) \neq w_{k,y}(s) \\ \quad and \; w_{k,y}(s) = \min(w_{k,t}(s), w_{k,y}(s)). \end{cases}$$

For $e \in \{1,2\}$ we consider the text $ind(t,y,e)$ created from $ind(t,y)$ by eliminating all appearances of $e$. We now set

$$R(t,y) = \frac{2 \max \{|ind(t,y,1)|, |ind(t,y,2)|\}}{\operatorname{card}(t \cap y)} - 1$$

Here, given a text $x$, $|x|$ denotes the number of words in $x$. To measure similarity between $t$ and $y$ we can use any mapping $\varphi : [0,1]^3 \to [0,1]$ which is an increasing function of each argument when two other arguments are fixed. The skeletons $t$ and $y$ (and the corresponding texts) are considered to be similar if

$$\varphi(I(t,y), C(t,y), R(t,y)) > p,$$

where $p$ is a fixed number from the interval $(0, 1)$. In extensive tests and simulations we obtained very good results for $p = 0, 5$ and

$$\varphi(x, y, z) = g(\max(x, yz)),$$

where

$$g(u) = 1 - \left(1 - \left(1 - 2 * \frac{\arccos(u)}{\pi}\right)^q\right)^{1/q}$$

with $q = 1.9$, see [23]. More detailed presentation of the material in this section is given in [9, 24].

## 4 Main Discussion

The method presented in this section is well-suited for reference databases of moderate size, say up to a few million texts. Its worst case cost linearly depends on the size. Searching for similarities in large-scale data requires distributed computing and special techniques like the cukoo hashing, see [14]. Therefore, for detecting similarities in the Internet the OSA system adopts the solution developed by the Institute of Computer Science at the Polish Academy of Science, see [25].

When designing a service, there are essentially two basic choices of architecture:

- **Fully centralized** — where the clients are thin and the actual work is done under the control of a central authority. Note that the central system itself might be distributed while being a centralized service from the point of view of clients.
- **Fully decentralized** — where the system consists of cooperating, but independent components of equivalent functionality.

Real systems usually employ the combination of this two architectures. Consider, for instance, the first iteration of GoogleFS [3] where the file system was decentralized but file metadata was stored on a single authoritative server.

Note that in case of plagiarism detection service, the plagiarism check can be made either at the central service or at a local server, in the second case, provided that all the local servers are supplied by the appropriate reference data. In our system we used the interesting combination of both approaches (see Fig. 1):

- Each university participating in the service will run its own local plagiarism check server. The local server will be supplied with the reference works (books, journal papers, Wikipedia, etc.) by the central server. Apart from it the local server will store the works of students from the university. Most of these works (preferably all, but there is the option of keeping some works strictly local for whatever reason like copyright, sensitive research and others) will be periodically sent to the central server. The local server performs the plagiarism check of the student works submitted to it, based only on the material it locally stores (reference works and local student works).

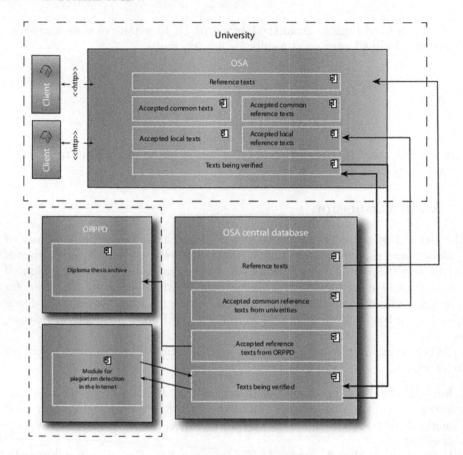

**Fig. 1.** Architecture of OSA.

– The central server has the authority over reference (non-student) works the universities can only submit requests for inclusion of some paper or book, but they cannot submit the works itself to the central repository. This is because of possible copyright issues but also to ensure that each reference text is not submitted many times under slightly different bibliographic data. In addition, as mentioned above, the central server periodically pulls a choice of student works submitted to the local server. The pulled student works are then checked for plagiarism (independently for the check executed at the local server). The reason for this additional check is that now the student work can be compared with student texts from other universities, which were unavailable for the local check. ORPPD, which stands for Ogólnopolskie Repozytorium Prac Dyplomowych (All-Polish Repository of Diploma Writings), is a diploma theses archive, government-owned and operated.

The advantages of this architectural choice are as follows:

- Each university participating in our plagiarism detection service has a local service, which is capable of performing its function even during temporary unavailability of the central service.
- Also, the local checks, which are often sufficient to uncover typical cases of plagiarism, can be performed quickly, giving immediate feedback to the instructor, who can then bear the waiting for the central system report, much more delayed due to the fact that the central server is under much heavier load than the local one.
- Finally, our system allows much more flexibility in the presence of delicate legal issues concerning copyright, industrial secrets or sensitive personal data. In the case of fully centralized system, it might have been impossible for such works to be submitted for checking, at least without severe security and privacy guarantees we are unwilling to take responsibility for. Here the university might decide not to send the work to the central repository precluding the possibility of comparing the work with the student texts from other universities but the checking can be at least performed locally.

The copyright law in Poland would put storing even the student works devoid of anything special about them in the service outside the control of the university in which those works originated in the legal grey area. Here, however, the algorithms used by the system help us. Consider the data flow diagram in Fig. 2. The submitted documents are first converted into txt format and then they are normalized: after tokenization, for each word we find its base form, a lemma in the parlance of linguists. Then we replace each word by its lemma (unknown words are left unchanged) obtaining a text skeleton ([]) of the original document. For further processing a text skeleton is then converted into the so called numeric

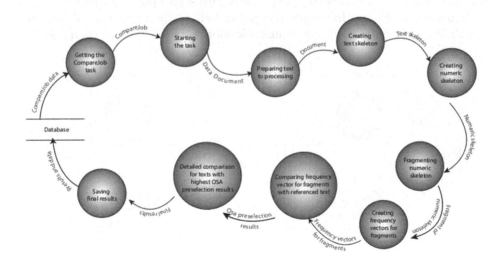

**Fig. 2.** Comparison job

skeleton [], in which each lemma is replaced by the id of this lemma in the base form database (or -1 in case of the unknown word). The plagiarism check algorithms we use require only numeric and text skeletons as input, hence it suffices for the central system to work that it stores only the skeletons. It turns out that according to Polish law that the text skeleton (and hence also numeric one) are sufficiently modified with respect to original that basic copyright limitations do not apply here and it is fully legal to submit the student work to the external (with respect to the originating university) repository.

## 5   System Architecture

The system is highly heterogeneous and designed with scalability in mind. As it utilizes a wide choice of technologies, it was necessary to apply some nonstandard techniques to enable reliable communication between decoupled components. For instance, in order to implement the asynchronous communication queues we use the relational database as a queue broker instead of relying on a specialized framework such as Java EE JMS. Hence, we only require from the subsystems to be able to connect to the relational database. Another advantage of using relational database is that the queue becomes transactional for free. Once the document is submitted (or consumed for analysis by the checking subsystem) the information about this fact can survive system crashes. The use of relational databases as data queue brokers has also other advantages which will be discussed further (see Fig. 3).

So far the local repository was completely implemented and is currently undergoing a testing phase. However the broad system architecture and system kernel adopted for the local repository is sufficiently versatile and scalable to be utilized for the central repository as well. As it was explained above the pivotal role is played by PostgreSQL database which serves both as a repository of skeletonized works and frequency vectors as well as the transactional request broker which facilitates all the communication between system components.

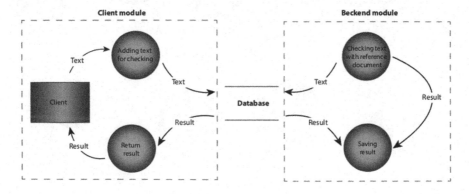

**Fig. 3.** Databases as data queue brokers

In particular it stores the requests for document checks as well as the outcomes of those checks. The requests are made from the client module, which is currently implemented as a web application using PHP and Yii framework. We emphasize that all the web applications communication with the rest of the system is mediated by the database. Apart from serving as a GUI, another task of the client module is to convert the submitted documents from word or PDF format into pure text files, which are then sent for further processing, which is performed by the execution module implemented in C++11. The execution module runs as a Linux daemon which polls periodically the database for new tasks. There can be many execution module processes running concurrently, either on the same or on different machines. The database takes care of race conditions and concurrency control, so that the different execution module processes do not interfere with one another. In this way the architecture of OSA system is very scalable. Note that the execution module task is much more computationally intensive than the database task, so that it makes sense to replicate the execution module, but it is unlikely that the database becomes the system bottleneck.

The tasks the execution module can run are as follows:

- **AddJob** — adds checked document skeleton to the repository;
- **ClearDatabaseJob** — clears the database;
- **RestartJob** — restarts the task, e.g., after system crash;
- **CompareJob** — does the actual comparison of submitted documents with the documents in the repository.
- **DecodeJob** — special job for decoding text with "cheated" charsets.

Figure 4 presents the fragment of database schema which concerns handling of the jobs queue. Each new task requested by the user through the (web) UI is added to the database (also see Fig. 3). The front-end inserts the appropriate entry into the `JobRequest` table, which contains information about job type and its present status (`NEW` for new jobs) and parameters necessary for its execution. The yet unfinished jobs are picked periodically, taking into account priorities, by execution module daemons, which then executes them, and, after successful completion, changes their status in the database into `COMPLETED`. In case of error, the status is also changed appropriately, so that information about the error can be passed to the user. When applicable, after successful completion the execution module also inserts appropriate data, such as job execution results or the outcome of comparison of a document with the reference database, into `JobResponse` table. The front-end periodically checks the status of added jobs and informs the user about the job completion, displaying when appropriate generated results. See [8] for a related material.

Note that tasks involving processing of documents rarely if ever process single documents. Instead they are meant to process the whole batches of documents. The most interesting and complex of tasks listed above is of course the CompareJob. The steps for each document are presented in Fig. 2. The first version of OSA system was implemented with C++11 because of the possibility of utilizing features specific to operating system used (like effective disc operations for

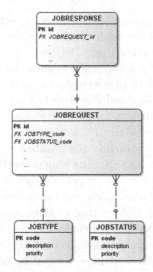

**Fig. 4.** Schema for jobs database

buffering), and, the ability to serialize effectively certain data structures by simply dumping/restoring memory region to disk. We also tried C++11 template metaprogramming, commonly used in the efficient implementations of numerical algorithms (see e.g. [1]). Finally, we utilized as much as possible the ability in C++ to use stack instead of heap and to define compact representations of data structures in order to improve cache locality.

The current version of the system utilizes Java instead of C++. This choice of programming language was influenced by the requirement of integration of OSA with the NEKST system (www.nekst.pl) and also the easiness of using with Java the standard frameworks for data distribution such as Hadoop. Another advantage of Java is the availability of standard web frameworks simplifying the design of the user interface. The implementation turned out to be sufficiently effective in practical use despite well known disadvantages of Java like non-deterministic resource management. This version is understood to be the base for future releases.

## 6    Conclusion

In the paper we presented the high level architecture of a new plagiarism detection platform developed by a consortium of Polish universities. Our design was influenced by the desire to strike a balance between locality and centralization. We also needed to satisfy strict functional and non-functional requirements, both of technological and legal nature. For instance, (because of copyright issues) it does not allow to recover any of the original texts. Our system is also scalable and should, eventually, store in the central repository information on any successfully defended academic thesis. Another interesting feature of our system is that it allows to use different similarity measures in different fields of science.

# 7    Final Notice

The interested reader is referred to http://osaweb.pl/ for up-to-date information about the systems and its features.

# References

1. Aragon, A.M.: A C++11 implementation of arbitrary-rank tensors for high-performance computing. Comput. Phys. Commun. **185**(11), 3065–3066 (2014)
2. Devi, S.L., Rao, P.R., Ram, V.S., Akilandeswari, A.: External plagiarism detection. Lab report for PAN at CLEF (2010)
3. Ghemawat, S., Gobioff, H., Leung, S.T.: The Google file system. In: Proceedings of the Nineteenth ACM Symposium on Operating Systems Principles, SOSP 2003, pp. 29–43. ACM, New York (2003). http://doi.acm.org/10.1145/945445.945450
4. Gipp, B., Beel, J.: Citation based plagiarism detection: a new approach to identify plagiarized work language independently. In: Proceedings of the 21st ACM Conference on Hypertext and Hypermedia, pp. 273–274. ACM (2010)
5. Hoad, T.C., Zobel, J.: Methods for identifying versioned and plagiarized documents. J. Am. Soc. Inf. Sci. Technol. **54**(3), 203–215 (2003)
6. Juola, P.: Authorship attribution. Found. Trends Inf. Retr. **1**(3), 233–334 (2006)
7. Kowalski, M.: Imitacja i ignorancja. Zeszyty Naukowe Politechniki Rzeszowskiej **15**, 69–74 (2008)
8. Kowalski, M., Kruszyński, P., Sobieski, S., Sysak, M.: Geneza, architekturai testy otwartego systemu antyplagiatowego. In: Hołyst, B., Pomykała, J., Potejko, P. (eds.) Nowe techniki badań kryminalistycznych a bezpieczeństwo informacji, pp. 257–273. PWN (2014)
9. Kowalski, M., Szczepański, M.: Identity of academic theses. In: Dobrzynska T., Kuncheva R. (eds.) Resemblance and Difference. The Problem of Identity, pp. 259-278. IBL PAN, IL BAN (2015)
10. Kowalski, M., Szczepański, M.: Akademicka przestępczość wcyberprzestrzeni. In: Hołyst, B., Pomykała, J. (eds.) Cyberprzestępczość i ochrona informacji, pp. 113–126. WydawnictwoWyższej Szkoły Menedżerskiej w Warszawie (2011)
11. Łysoń, P., Golaszewska, H., Maślankowski, J., Franecka, A., Jaworski, P., Kamińska, M., Rutkowska, M., Rybicka, K., Ulatowska, M., Wiktor, M.: Szkoły wyższe i ich finanse w 2012 r. Higher education institutions and their finances in 2012. In: Informacje i opracowania statystyczne, Statistical Information and Elaborations. Zakład Wydawnictw Statystycznych (2013)
12. Meuschke, N., Gipp, B.: State-of-the-art in detecting academic plagiarism. Int. J. for Educ. Integrity **9**(1) (2013)
13. Monostori, K., Zaslavsky, A., Schmidt, H.: Identifying overlapping documents in semi-structured text collections. In: Australasian Computer Science Conference (2000)
14. Pagh, R., Rodler, F.F.: Cuckoo hashing. J. Algorithms **51**(2), 122–144 (2004). http://dx.doi.org/10.1016/j.jalgor.2003.12.002
15. Rocchio, J.J.: Relevance feedback in information retrieval (1971)
16. Salton, G.: Developments in automatic text retrieval. Science (New York, N.Y.) **253**(5023), 974–980 (1991). http://dx.doi.org/10.1126/science.253.5023.974
17. Salton, G., Buckley, C.: Term-weighting approaches in automatic text retrieval. Inf. Process. Manage. **24**(5), 513–523 (1988). http://dx.doi.org/10.1016/0306-4573(88)90021-0

18. Salton, G., Wong, A., Yang, C.S.: A vector space model for automatic indexing. Commun. ACM **18**(11), 613–620 (1975). http://doi.acm.org/10.1145/361219.361220
19. Salton, G., Wong, A., Yang, C.S.: A vector space model for automatic indexing. Commun. ACM **18**(11), 613–620 (1975)
20. Schleimer, S., Wilkerson, D.S., Aiken, A.: Winnowing: local algorithms for document fingerprinting. In: Proceedings of the 2003 ACM SIGMOD International Conference on Management of Data, pp. 76–85. ACM (2003)
21. Si, A., Leong, H.V., Lau, R.W.: Check: a document plagiarism detection system. In: Proceedings of the 1997 ACM Symposium on Applied Computing, pp. 70–77. ACM (1997)
22. Sindhu, L., Thomas, B.B., Idicula, S.M.: Automated plagiarism detection system for malayalam text documents. Int. J. Comput. Appl. **106**(15), 13–16 (2014)
23. Szczepański, M.: Testy skuteczności algorytmu preselekcji otwartego systemu antyplagiatowego In: Holyst, B., Pomykala, J., Potejko, P. (eds.) Nowe techniki badan kryminalistycznych a bezpieczenstwo informacji, pp. 248–256. PWN (2014)
24. Szczepański, M.: Algorytmy klasyfikacji tekstów i ich wykorzystanie w systemie wykrywania plagiatów. Oficyna Wydawnicza Politechniki Warszawskiej (2002)
25. Szmit, R.: Fast plagiarism detection in large-scale data (submitted for publicaton)
26. Wu, H., Salton, G.: A comparison of search term weighting: term relevance vs.inverse document frequency. In: Proceedings of the 4th Annual International ACM SIGIR Conference on Information Storage and Retrieval: Theoretical Issues in Information Retrieval, SIGIR 1981, pp. 30–39. ACM, New York (1981). http://doi.acm.org/10.1145/511754.511759

# An Investigation of Face and Fingerprint Feature-Fusion Guidelines

Dane Brown[1,2(✉)] and Karen Bradshaw[1]

[1] Department of Computer Science, Rhodes University,
Grahamstown, South Africa
k.bradshaw@ru.ac.za
[2] Council for Scientific and Industrial Research, Modelling and Digital Sciences,
Pretoria, South Africa
p.dane49@gmail.com

**Abstract.** There are a lack of multi-modal biometric fusion guidelines at the feature-level. This paper investigates face and fingerprint features in the form of their strengths and weaknesses. This serves as a set of guidelines to authors that are planning face and fingerprint feature-fusion applications or aim to extend this into a general framework. The proposed guidelines were applied to the face and fingerprint to achieve a 91.11 % recognition accuracy when using only a single training sample. Furthermore, an accuracy of 99.69 % was achieved when using five training samples.

**Keywords:** Framework · Face · Fingerprint · Feature-level · Multi-modal biometrics

## 1 Introduction

Biometrics is defined as the measurement and analysis of unique biological and behavioural traits for human identification purposes [4]. Their widespread use have introduced security risks posed by forgers [13]. Furthermore, real-world conditions often result in degradation of the biometric data being modelled.

In a bid to counteract these real-world problems, multiple sources of biometric information have been used to improve the security, recognition accuracy and versatility of a biometric system. Multi-modal biometrics can also be used to solve non-universality and insufficient population coverage in well-planned applications [10].

Early use of multi-modal biometrics adopted the matching score level fusion approach. Later, the feature-level approach was shown to outperform the matching score level [18]. Feature-level fusion integrates feature sets corresponding to two or more biometric modalities. The widely used matching score level fusion does not utilize the rich discriminatory information available at the feature-level. The matching score level has been thoroughly reviewed and the results used to construct fusion frameworks [17]. These frameworks provide future research and

© Springer International Publishing Switzerland 2016
S. Kozielski et al. (Eds.): BDAS 2016, CCIS 613, pp. 585–599, 2016.
DOI: 10.1007/978-3-319-34099-9_45

applications a foundation on which to systematically implement multi-modal systems in the real-world. Feature-level fusion literature lack these frameworks, as it is a lesser studied problem. At present, the guideline used during feature-level fusion is limited to feature set compatibility – uncorrelated feature sets are to be used among different modalities and correlated feature sets among multiple samples of the same modality [22]. Biometric modalities, represented by an image, are independent and complementary. Based on the feature set compatibility guideline, the application of the same feature transformation method on different modalities can yield a very efficient multi-modal biometric system [17]. Fusion is often applied after transforming the feature space using linear or non-linear methods [13].

In this paper, different feature selection and transformation methods are applied to the face and fingerprint. The resulting feature sets are expected to produce an improved recognition performance compared with the two individual modalities. The scope includes the use of different sized datasets and varying the number of training samples from one to five during data modelling. The experimental results are subsequently used to determine feature-fusion guidelines, relevant to the face and fingerprint, based on the type of data acquired. The contribution of this paper are these guidelines, which serve as the foundation of a general feature-fusion framework that can be constructed in future.

The rest of the paper is organized as follows: Sects. 2 and 3 discuss quality enhancing and feature selection techniques, respectively. Three classifiers are explained in Sect. 4. Section 5 presents the related studies found in the literature. Sections 6 and 7 discuss the construction and application of the face and fingerprint feature-fusion guidelines, respectively. The experimental analyses and results are discussed in Sect. 8. Section 9 concludes the paper and discusses future work.

## 2    Quality Enhancement

Quality enhancement is used to recover the legibility of bad input data. This is particular to contours and pores in face images, and similarly the case for ridges and valleys in fingerprint images [16].

The biometric recognition process is often initiated by enhancing the quality of the input image [5,9,15,25]. This section discusses important image pre-processing techniques used on the biometric modalities in this paper.

### 2.1    Non-Local Denoising

Buades et al. [6] present an image denoising algorithm, called non-local means filtering (NL-means), described as neither local nor global. NL-means differs from typical neighbourhood filters as it compares the geometrical configuration in an entire neighborhood instead of a single greyscale pixel of one neighbourhood corresponding to another. NL-means preserves the edges of an image. This is important in fingerprint applications as ridges and valleys are key features. However, high filter strength and large neighbourhood size removes fine texture.

## 2.2   Pixel Normalization

Using this technique, pixel values are set to a constant mean and variance to reduce inconsistencies in lighting and contrast. This is essential for both face and fingerprint images as multiple samples are often captured under different conditions.

## 2.3   Histogram Equalization

Histogram Equalization effectively adjusts contrast intensities to an even amount based on the most frequent intensity values across an image histogram [5]. This uniform distribution is achieved by applying a non-linear transformation resulting in a minor side-effect on the histogram shape. This often produces better results than pixel normalization, but should be avoided in most histogram-based matching methods.

# 3   Feature Selection

Feature selection optimizes an objective function based on a requirement of specific features. The objective function reduces feature space by removing unwanted features. The remaining features are highly representative of the underlying image class [5]. The following subsections discuss the face and fingerprint features.

## 3.1   Local and Global Features

The texture pattern of a fingerprint contains richer information than singular points and minutiae [14]. The ridges and valleys that form the texture pattern are known as global features. Global features are effective in biometric fusion at the feature-level, but require registered points for alignment. Local features consist of these registration points known as minutiae and singular points.

Texture patterns, consisting of contours and pores can similarly be used as global features in face images. The local features are the coordinates of the eyes, nose and mouth. These local features are used to align global features in a similar way to fingerprints.

## 3.2   Core Detection and Region of Interest

The core point in a fingerprint image is often defined as the sharpest concave ridge curvature [11]. It is especially useful as a reference point during image registration. It can be used to define a regions of interest (ROI), which minimizes the discrepancy of stretch and alignment differences within same fingerprint classes.

Poincarè index is an orientation field based core detection algorithm [11]. It works well in good quality fingerprint images, but fails to correctly localize reference points in poor quality fingerprints with cracks, scars or poor ridge and valley contrast. However, since it does not rely on fine grain texture, NL-means can be used to enhance ridge and valley contrast before determining the orientation field.

Haar cascading is a popular method for detecting local facial features [1]. Similarly, local features in face images are used to create a border around the face, centred at the nose. Typical changes to the face such as hair, ears and neck are thus provided for.

### 3.3    Laplacian of Gaussian Filtering

A Laplacian of Gaussian (LOG) filter increases the dominant spectral components while attenuating the weak components [1,8]. However, the LOG filter can further degrade the recognition accuracy of badly registered images because the overlap between dominant spectral components of the training and testing images becomes sparse.

### 3.4    Gabor Filter

A Gabor wavelet is a commonly used method of frequency filtering. This filter is constructed using a special short-time Fourier transform by modulating a two-dimensional sine wave at a particular frequency and orientation with a Gaussian envelope.

The sine waves of the ridges in the fingerprint vary at a slow to medium rate in a local constant orientation. Therefore, it is tuned to specific orientations and frequencies in the bandpass range, isolating undesired noise while preserving the structure of the fingerprint. Similar effects can be achieve when applied to the face, based on the structure of contours. Thus, an effective bandpass filter is constructed when utilizing the frequency and orientation selective properties of a Gabor filter according to the modality [7].

## 4    Feature Transformation and Classification

Image classification algorithms aim to exploit highly discriminative features. These algorithms often transform a feature vector to another vector space. The following image classification algorithms are considered.

### 4.1    Eigen

Principal component analysis (PCA) is used in this Eigen classifier to maximize the total variance in data based on linear combination of features. Eigenvectors are the decomposition of features vectors into key components known as principal components, which can then be reconstructed into an approximation of the original image.

The largest variance in data is contained within the first few principal components. These are the key features that are modelled into classes. A training and testing model are compared based on the distances between eigenvalues during matching.

Given $N$ number of sample images $x_k$ the total scatter matrix is defined as [3]:

$$S_t = \sum_{k=1}^{N} (x_k - \mu)(x_k - \mu)^T ,$$

where $m \in \mathbb{R}^n$ is the mean image obtained from the samples.

### 4.2    Fisher

The total scatter matrix used in Eigen lacks some discriminative information at an inter-class level. On the other hand, linear discriminant analysis (LDA) performs extra class-specific dimensionality reduction by considering the between-class and within-class scatter matrix.

Fisher learns a class-specific transformation matrix, which can lead to inconsistent data in dynamic lighting conditions. Fisher generally requires more training data than Eigen in non-ideal conditions. However, an advantage of Fisher is lower training and testing time and reduced dimensionality compared with Eigen.

Given $C$ number of classes, the between-class scatter matrix is defined as [3]:

$$S_b = \sum_{i=1}^{C} N_k (\mu_k - \mu)(x_k - \mu)^T$$

and the within-class scatter matrix is defined as:

$$S_w = \sum_{i=1}^{C} \sum_{x_k \in \mathbb{X}_i} (x_k - \mu)(x_k - \mu)^T .$$

where $C - 1$ is the maximum number of non-zero generalized eigenvalues, which leads to extra dimensionality reduction.

### 4.3    Local Binary Patterns Histogram

The local binary patterns histogram (LBPH) is a robust texture feature descriptor. It uses a local binary pattern (LBP) operator that compares the centre pixel value to a set size of neighbouring pixels.

A special LBP operator called extended LBP (ELBP) is used in this work. The neighbourhood is extended to include interpolated pixels, based on a circular mask, allowing for fine grain texture to be captured. Spatially enhanced histogram matching is used to improve partial matching and automatic pixel normalization at a pixel level, circular neighbourhood level, and image level. This addresses the shortcomings of Eigen and Fisher in terms of illumination, scale and misalignment [2]. ELBP can also be used as a feature selector without the spatially enhanced histogram.

The spatially enhanced histogram trains a significantly smaller model and produces it faster than the former two classifiers. Furthermore, the training time is independent of the resolution of the images. Given $m$ circular neighbourhoods, their corresponding spatially enhanced histograms have size $m \times n$, where $n$ is the length of a single histogram.

## 5 Related Studies

Karki and Selvi [12] proposed a multi-modal biometric system designed to fuse the face, fingerprint and offline signature in parallel at the feature-level. A feature vector is concatenated and stored in a database by using parallel fusion. The biometric traits of an individual are recorded separately during data acquisition, but required feature selection and transformation techniques are applied to the modalities in parallel. The SVM classifier is used for matching.

Texture features are extracted by a Curvelet transform with a third level low-low subband from each trait. The low level coefficients from the subband of each trait are compatible and each form a feature vector. The three feature vectors are concatenated $(F_c)$. Five feature reduction methods are used to produce the reduced feature vectors, that is, feature averaging $(F_a)$, PCA $(F_{PCA})$, PCA on individual traits $(F_p)$, statistical moment features without fusion $(F_m)$ and feature concatenation by extracting significant coefficients only $(F_s)$. An SVM with a polynomial kernel of order 2 and parameter C set to 10 is the selected classifier.

Fingerprint, offline signature, and face samples of 100 users were captured to form a database, named ECMSRIT. PCA feature reduction performed on the concatenated feature vector of the ECMSRIT data produced the best equal error rate (EER) of 5.32 %. This result was followed by the following feature vectors: $F_a$, $F_p$, $F_c$, $F_m$ and $F_s$ with an EER of 12.00 %, 15.33 %, 19.31 %, 20.54 % and 23.54 %, respectively. The use of Curvelet transforms produced feature vectors that were robust to rotation of up to 10°.

Sharma and Kaur [20] designed a multi-modal system by integrating the face, fingerprint and palmprint at the feature-level, similar to Karki and Selvi's work, but without the use of the Curvelet transform. The feature vectors are extracted independently using PCA followed by their concatenation before classification using a multiclass SVM. The performance of the multiclass SVM classifier is compared to an artificial neural network (ANN).

PCA reduction is used to reduce the search space for the SVM. The SVM uses the radial basis function (RBF) kernel. The number of hyperparameters affecting the complexity of the trained model is less in the RBF kernel than the other kernels. This gives the RBF kernel an advantage over the other kernels because the dimensionality of the fused feature vector is often too high.

The fingerprint images were obtained from the DB3_(UPEK) database; face images collected by Markus Weber at California Institute of Technology and palm images were obtained from the CASIA palmprint database. A pseudo dataset was created containing 10 individuals by combining the separate datasets. Five training and testing images were used per individual.

The multi-modal system achieved a false acceptance rate (FAR) of 4 % and a false rejection rate (FRR) of 6 % on the dataset. The SVM significantly outperformed the ANN, but the details were not provided in the study.

Yao et al. [23] compared four PCA-based face and palmprint feature fusion algorithms. The proposed method filters EigenFaces and EigenPalms with a Gabor filter followed by weighted concatenation of the resulting feature vectors. The proposed system was designed to produce high accuracy with only a single training sample.

The AR face database and a palmprint database provided by Hong Kong Polytechnic University were used. The datasets consisted of 20 images per 189 individuals with a resolution of $60 \times 60$ in both cases. Fused datasets were created using parallel fusion. The highest genuine acceptance rate (GAR) of 95 % was achieved with six training samples, while 91 % GAR was achieved with only a single training sample.

## 6 Setting up the Guidelines

This section determines the feature-fusion guidelines that are relevant to the face and fingerprint. The face and fingerprint datasets consist of various scenarios as described in the following subsection.

### 6.1 Categories of Datasets

Pseudo multi-modal datasets, consisting of 40 individuals, were formed by pairing SDUMLA Fingerprint right index fingers [24] with ORL Face [19] and SDUMLA Fingerprint right middle fingers with Fei Face [21]. Fingerprint images organized into three groups, consisting of partials with absent core points, poorly-defined ridges and well-defined ridges. Face images were organized into two groups, consisting of standard faces and faces that consisted of poses and props. The interactions of image processing modules and classifiers, discussed in Sects. 2 to 4, were determined based on preliminary experiments conducted on the organized pseudo multi-modal datasets.

### 6.2 Preliminary Experiments

General results across all datasets indicated LBPH to be the classifier most robust to misalignment, dynamic lighting and scale. Eigen and Fisher achieved recognition accuracies similar to that of LBPH for face and fingerprint images consisting of standard faces and well-defined ridges, respectively. However, Eigen and Fisher performed poorly in the remaining datasets, which can be attributed to the high variance in data across multiple samples of face and fingerprint images contained within those datasets. Fisher, in particular, requires training and testing images with well-aligned texture and has a significantly lower dimensionality than Eigen.

Typical parameters of ELBP are the one pixel radius and eight neighbouring pixels. Consistent lighting and lower noise were achieved in the Eigen space by multiplying the parameters by four. The results were conclusive across all datasets. This reduced the variation in data across multiple samples of an individual. The ELBP operator outperformed the equalized histogram and pixel normalization under different lighting conditions. However, pixel normalization was applied to LBPH as histogram equalization caused negative effects on the spatially enhanced histogram and the ELBP operator was already part of the LBPH classifier. The best accuracies were achieved when applying the same feature transformation to the different modalities. This is based on the feature set compatibility guideline and confirms the assertion by Raghavendra et al. [17].

NL-means filtering improved core detection in fingerprint datasets consisting of poorly-defined ridges. The improved core detection resulted in a well-defined ROI. Partials were catered for by applying LBPH in a sliding window and selecting the ROI that produced the best confidence score. The ROI that best defined the pose and props face dataset made use of multiple Haar cascades to exclude props such as scarfs and hats.

The LOG filter significantly improved the recognition accuracy of all the fused datasets. It was particularly useful at lowering the data variance of multiple samples of face and fingerprint images consisting of poses and poorly-defined ridges, respectively.

The Gabor filter did not improve the accuracies of the fused datasets. Moreover, it reduced the accuracy of the Eigen and Fisher classifiers. It significantly improved the recognition accuracy of non-partial fingerprint datasets and slightly improved the recognition accuracy of all face datasets when using the LBPH classifier. On the other hand, the LOG filter lowered the accuracy of the LBPH classifier. Reducing the data variance by 1 % in the Eigen space, followed by reconstructing the image data, produced the best LBPH recognition accuracy across all datasets.

Table 1 provides a summary of the proposed guidelines, based on the results of these preliminary experiments. These guidelines are applied to a multi-modal database in the next section. The rest of the proposed methodology is based on the results of these preliminary tests. The best feature-fusion method, for the given datasets, on average, is illustrated in Fig. 1 and discussed as follows.

## 7   Applying the Guidelines

The following list refers to Fig. 1 and details the subsequent evaluation of the results.

1. The SDUMLA multi-modal database [24], consisting of 106 individuals, was used in the final experiments discussed in Sect. 8. The acquired data assumed the following form: Eight samples of the left thumbprint were selected from the fingerprint images consisting of partials with absent core points, poorly-defined ridges and well-defined ridges. The eight samples of frontal faces

**Table 1.** Feature-Fusion guidelines

| Stage | Name | Advantage | Disadvantage | Suggested Use |
|---|---|---|---|---|
| Quality Enhancement | Pixel Normalization | Reduces inconsistent lighting | Not very effective for big changes in lighting | All biometrics affected by lighting. The first step of quality enhancement |
| | Histogram Equalization | Reduces inconsistent lighting. | Minor histogram distortion. Introduces some noise. | Biometrics that are affected by lighting. Histogram-shape invariant classifiers such as Eigen and Fisher |
| | NL-means filter | Denoises and preserves edges | Can remove fine texture | Use before Poincar index |
| Feature Selection | LOG Filter | Improved feature discrimination, before transforming to the Eigen space | Requires consistent lighting | Remove noise in the upper and lower frequencies before fusion |
| | Gabor Filter | Improved feature discrimination, especially for fingerprints. | Requires tuning per application. | Adjust frequencies at a specific orientation and scale. Classifiers such as LBPH |
| | ELBP Operator | Minimizes inconsistent lighting. | Introduces noise. | Biometrics that are affected by lighting. Classifiers such as Eigen and Fisher |
| Feature Transformation and Classification | Eigen Classifier | LOG and ELBP supplement this classifier | Slow training. High dimensionality | Small image regions. Useful for fingerprints and fused datasets |
| | Fisher Classifier | LOG and ELBP supplement this classifier. Low dimensionality | Requires good training data | When Eigen dimensionality is too high |
| | LBPH Classifier | Works well on faces. Very robust. | Lowest accuracy after feature selection. | General purpose classifier. Face images. Applications with low storage requirements |

selected from the face images consisted of different poses and props – normal, smile, frown, surprise, look down, shut eyes, hat and glasses. The number of training samples were varied from one to five and the rest were used for testing. To the best of my knowledge there are no studies that fuse face and fingerprint data acquired from the SDUMLA multi-modal database.

2. The fingerprint and face datasets were automatically cropped to $75 \times 75$ using Poincarè index and multiple Haar cascades, respectively. NL-means was used to remove noise before the Poincarè index algorithm was performed. The fingerprint was cropped around the core point to reduce the amount of stretch caused by inconsistent fingerprint capturing. The Haar cascades detected the face, eyes, nose and mouth as outlined in Fig. 1. The outlining was used to remove or reduce partial occlusions that affect face recognition.

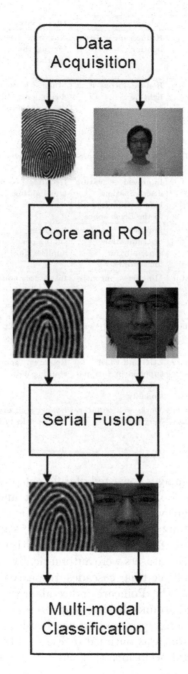

**Fig. 1.** Overview of proposed methodology.

3. The enhanced face and fingerprint feature vectors are combined using serial vector fusion, known as column concatenation.
4. The three classifiers divide the fused dataset into classes to create three baseline systems, as explained in Sect. 4. The baseline Eigen, Fisher and LBPH classifiers use histogram equalization and pixel normalization. The following feature selection techniques are applied to compare many multi-modal biometric recognition systems: Eigen and Fisher are used to classify combinations of the LOG and ELBP operator and PCA reduction is applied before using the LBPH classifier.

# 8    Experimental Analysis and Results

The following suffixes identify a feature-fusion scheme: Histogram equalization is a baseline system, henceforth, referred to as Eh; LOG is, henceforth, referred to as L; Extended LBP is, henceforth, referred to as LBP; Extended LBP followed by LOG is, henceforth, referred to as LBPL; LOG followed by Extended LBP is, henceforth, referred to as LLBP; and PCA reduction is, henceforth, referred to as PCA.

All recognition accuracies discussed in this section are measured at 0 % FRR.

The Eigen baseline fusion system always outperforms the face and fingerprint as illustrated in Fig. 2. LBPL performs the best for one training sample with an accuracy of 90.84 % and LLBP achieves an accuracy of 99.69 % with five training samples.

The Fisher classifier performs similarly to Eigen, but with a significantly reduced accuracy when using two training samples as shown in Fig. 3. This is attributed to a low overlap of the remaining principal components in the two samples caused by the huge reduction in dimensionality.

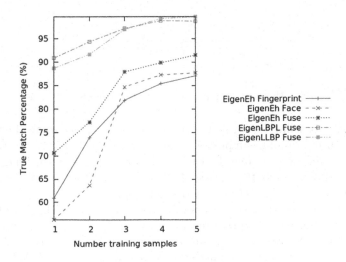

**Fig. 2.** Comparison of Eigen methods

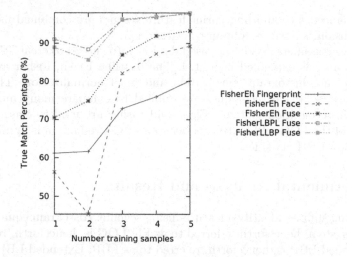

**Fig. 3.** Comparison of Fisher methods

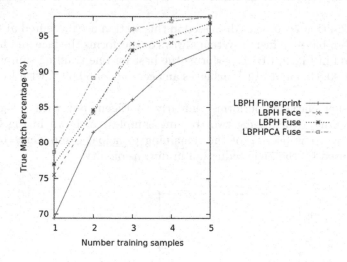

**Fig. 4.** Comparison of LBPH methods.

The LBPH baseline fusion system produces a lower accuracy than the face when using three training samples as illustrated in Fig. 4. Moreover, LBPH is a very good face texture classifier. LBPH has a poor response to the image processing modules described in the bottom half of Sect. 3. However, PCA reduction improves the recognition accuracy by 3 % on average.

The ELBP operator was successfully used together with the LOG filter to significantly improve feature discrimination in the Eigen space. EigenLLBP achieved the lowest EER of all the fusion schemes when using five training samples, at 0.31 % as seen in Fig. 5. Furthermore, it should be noted that this fusion scheme shows no increase in FAR after the EER point. EhL and LBP individually improved the accuracies over the baseline, but only the results of their combinations are included in this paper. LBPL achieved the best average recognition accuracy across the varied number of training samples. The results demonstrate the application of the proposed guidelines in this paper. There are no experiments that fuse the face and fingerprint datasets contained in the SDUMLA multi-modal database.

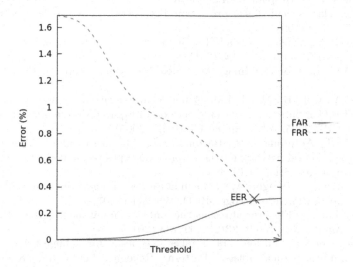

**Fig. 5.** Equal error rate.

## 9    Conclusion and Future Work

A comparison was performed on fingerprints, faces and their fused dataset using three baseline classifiers. The comparison was extended by combining a modified ELBP operator and a LOG filter. Additionally, principal components were removed from the LBPH training and testing images. The LBPH classifier achieved the best accuracy in the baseline systems and was robust to misalignment, dynamic lighting and scaling. The Eigen and Fisher classifiers yielded the best accuracies when combining the strengths of ELBP and LOG. Feature-level fusion research often makes use of well-known image processing and classification techniques without reasoning. Analyzing and testing many of these techniques to measure progress in the state-of-the-art is a non-trivial problem. Therefore, the guidelines introduced in this paper is the first step to solving the generalized feature-fusion framework problem.

In future, more combinations of biometric modalities and image processing modules will be investigated with additional experimentation toward a general multi-biometric feature-fusion framework. The framework will provide guidelines and measure progress in multi-modal biometric systems at the feature-level.

# References

1. Ahmadian, K., Gavrilova, M.: A multi-modal approach for high-dimensional feature recognition. Vis. Comput. **29**(2), 123–130 (2013)
2. Ahonen, T., Hadid, A., Pietikainen, M.: Face description with local binary patterns: application to face recognition. IEEE Trans. Pattern Anal. Mach. Intell. **28**(12), 2037–2041 (2006)
3. Belhumeur, P.N., Hespanha, J.P., Kriegman, D.: Eigenfaces vs. fisherfaces: recognition using class specific linear projection. IEEE Trans. Pattern Anal. Mach. Intell. **19**(7), 711–720 (1997)
4. Bharadwaj, S., Vatsa, M., Singh, R.: Biometric quality: from assessment to multi-biometrics. IIITD-TR-2015-003 (2015)
5. Bovik, A.C.: Handbook of Image and Video Processing. Academic Press, New York (2010)
6. Buades, A., Coll, B., Morel, J.M.: A non-local algorithm for image denoising. In: 2005 IEEE Computer Society Conference on Computer Vision and Pattern Recognition, CVPR 2005, vol. 2, pp. 60–65. IEEE (2005)
7. Budhi, G.S., Adipranata, R., Hartono, F.J.: The use of gabor filter and back-propagation neural network for the automobile types recognition. In: 2nd International Conference SIIT 2010 (2010)
8. Chikkerur, S., Cartwright, A.N., Govindaraju, V.: Fingerprint enhancement using STFT analysis. Pattern Recogn. **40**(1), 198–211 (2007)
9. Feng, J., Jain, A.: Fingerprint reconstruction: from minutiae to phase. IEEE Trans. Pattern Anal. Mach. Intell. **33**(2), 209–223 (2011)
10. Iloanusi, O.N.: Fusion of finger types for fingerprint indexing using minutiae quadruplets. Pattern Recogn. Lett. **38**, 8–14 (2014). http://www.sciencedirect.com/science/article/pii/S016786551300411X
11. Jain, A.K., Prabhakar, S., Hong, L., Pankanti, S.: Filterbank-based fingerprint matching. IEEE Trans. Image Process. **9**(5), 846–859 (2000)
12. Karki, M.V., Selvi, S.S.: Multimodal biometrics at feature level fusion using texture features. Int. J. Biometrics Bioinf. **7**(1), 58–73 (2013)
13. Kaur, D., Kaur, G.: Level of fusion in multimodal biometrics: a review. Int. J. Adv. Res. Comput. Sci. Softw. Eng. **3**(2), 242–246 (2013)
14. Maltoni, D., Maio, D., Jain, A.K., Prabhakar, S.: Handbook of Fingerprint Recognition. Springer Science & Business Media, Heidelberg (2009)
15. Peralta, D., Triguero, I., Sanchez-Reillo, R., Herrera, F., Benitez, J.: Fast fingerprint identification for large databases. Pattern Recogn. **47**(2), 588–602 (2014). http://dx.org/10.1016/j.patcog.2013.08.002
16. Porwik, P., Wrobel, K.: The new algorithm of fingerprint reference point location based on identification masks. In: Kurzyński, M., Puchała, E., Woźniak, M., żołnierek, A. (eds.) Computer Recognition Systems. Advances in Soft Computing, vol. 30, pp. 807–814. Springer, Heidelberg (2005)
17. Raghavendra, R., Dorizzi, B., Rao, A., Kumar, G.H.: Designing efficient fusion schemes for multimodal biometric systems using face and palmprint. Pattern Recogn. **44**(5), 1076–1088 (2011)
18. Rattani, A., Kisku, D.R., Bicego, M., Tistarelli, M.: Feature level fusion of face and fingerprint biometrics. In: Biometrics: Theory, Applications, and Systems, pp. 1–5 (2011)
19. Samaria, F.S., Harter, A.C.: Parameterisation of a stochastic model for human face identification. In: 1994 Proceedings of the Second IEEE Workshop on Applications of Computer Vision, pp. 138–142. IEEE (1994)

20. Sharma, P., Kaur, M.: Multimodal classification using feature level fusion and SVM. Int. J. Comput. Appl. **76**(4), 26–32 (2013)
21. Thomaz, C.E., Giraldi, G.A.: A new ranking method for principal components analysis and its application to face image analysis. Image Vis. Comput. **28**(6), 902–913 (2010)
22. Wang, Z., Liu, C., Shi, T., Ding, Q.: Face-palm identification system on feature level fusion based on CCA. J. Inf. Hiding Multimedia Signal Process. **4**(4), 272–279 (2013)
23. Yao, Y.F., Jing, X.Y., Wong, H.S.: Face and palmprint feature level fusion for single sample biometrics recognition. Neurocomputing **70**(7), 1582–1586 (2007)
24. Yin, Y., Liu, L., Sun, X.: SDUMLA-HMT: a multimodal biometric database. In: Sun, Z., Lai, J., Chen, X., Tan, T. (eds.) CCBR 2011. LNCS, vol. 7098, pp. 260–268. Springer, Heidelberg (2011)
25. Zou, J., Feng, J., Zhang, X., Ding, M.: Local orientation field based nonlocal means method for fingerprint image de-noising. J. Signal Inf. Process. **4**, 150 (2013)

# GISB: A Benchmark for Geographic Map Information Extraction

Pedro Martins[✉], José Cecílio, Maryam Abbasi, and Pedro Furtado

Department of Computer Sciences, University of Coimbra, Coimbra, Portugal
{pmom,jose,maryam,pnf}@dei.uc.pt

**Abstract.** The growing number of different models and approaches for Geographic Information Systems (GIS) brings high complexity when we want to develop new approaches and compare a new GIS algorithm. In order to test and compare different processing models and approaches, in a simple way, we identified the need of defining uniform testing methods, able to compare processing algorithms in terms of performance and accuracy regarding: large imaging processing, algorithms for GIS pattern-detection.

Taking into account, for instance, images collected during a drone flight or a satellite, it is important to know the processing cost to extract data when applying different processing models and approaches, as well as their accuracy (compare execution time vs. extracted data quality). In this work we propose a GIS Benchmark (GISB), a benchmark that allows to evaluate different approaches to detect/extract selected features from a GIS data-set. Considering a given data-set (or two data-sets, from different years, of the same region) it provides linear methods to compare different performance parameters regarding GIS information, making possible to access the most relevant information in terms of features and processing efficiency.

**Keywords:** Benchmark · GIS · Algorithms · Spatial-temporal databases · Bigdata · Performance · Experimentation · Pattern-detection

## 1 Introduction

With the communication technologies development and easier access to data, e.g., from drones or satellites, Geographic Information Systems (GIS) emerged as a new discipline. Nowadays GIS is applied in various domains to infer information with respect to location. Large amounts of data are generated in the form of images, flat files from sources, e.g., drones, satellites, sensors and other devices. Analyzing this information requires large computational resources (both processing and storage), different algorithms and processing techniques. Assess the best methods to apply to solve a certain problem requires to compare, processing time versus the extracted information and its quality/accuracy.

We propose a benchmarking tool to test and compare GIS data algorithms regarding: processing speed and extracted information accuracy. This objective

© Springer International Publishing Switzerland 2016
S. Kozielski et al. (Eds.): BDAS 2016, CCIS 613, pp. 600–609, 2016.
DOI: 10.1007/978-3-319-34099-9_46

raises many challenges regarding, the data sources heterogeneity, different data representations and file formats, classification algorithms and spatial data mining tools to extract data. GISB is designed to analyse geographic information from satellite imaging. This way we provide a uniform method to test and compare different GIS processing algorithms, techniques, the performance speed and quality/accuracy of the extracted information.

In this work we adopted a data-set based on satellite imaging, from two different years 1997 and 2014 (the exact same region), which is used in GISB to test detection algorithms. GISB evaluates the algorithms by comparing a set of features (e.g. areas, perimeters, counts, comparisons) and checkpoints that need to submitted using a specific API.

Section 2, describes the most relevant works in the related field. Section 3, shows GISB architecture. Section 4 provides a simple proof of GISB concept. Finally Sect. 5, shows the conclusions and proposed future work improvements.

## 2 Related Work

There are extensive works in the field of GIS regarding image processing such as: multiple algorithms for data extraction, data organization, storage and query languages, ontology's. However, there are no homogeneous method to compare each approach performance with each other. The following mentioned related-works are relevant in order to better understand the problem in hand.

In the work [7], the authors propose a framework to support high performance spatial queries and analytic, for spatial bigdata using Hadoop-GIS [3] to store information and process it with the aid of GPU based spatial operators. The GPUs are used to accelerate heavy duty geometric computation. We consider this work relevant because of the use of a MapReduce approach to capture/storage spatially oriented data. The data is captured from OpenStreetMap [5] imaging system and processed using GPUs. However the authors do not compare their approach with any other tool or benchmark in order to highlight the benefits of the proposed framework.

In [4], the authors propose a distributed system to handle big spatial data. At the same time they describe the four main components that are required for supporting big spatial data, they are: high-level language, spatial indexes, query processing engine, visualization component. This work resumes the main components for handling GIS imaging processing, but it does not mention any means of comparing them with each other.

In [2], the authors focus on building a framework based on a MapReduce architecture to create on-demand indexes for processing spatial queries and a partition-merge approach for building parallel spatial query pipelines. They focus on indexation, parallelization, and optimization for query efficiency. The authors test optimization by comparing their system with optimization and with extra optimization. Once again no possible comparison is possible to provide with other tools due to the non-existence of a benchmark for such purpose.

The works [1] and [6], together complement each other as a survey to introduce GIS processing phases, formats, existent data sources, tools and mining techniques.

By proposing GISB, we introduce a new opportunity for researches and industry to test and compare their tools, not only features and algorithms information extraction precision, but also processing performance.

## 3   Architecture

In this section, we start by describing the distinct parts which comprise GISB and its testing purpose.

Important to mention that GISB is a mainly designed for image mapping benchmark. However given a different data set, it is possible to reconfigure the benchmark parameters for other type of results comparison, e.g., medical imaging. With GISB we propose a set of uniform methods to compare the differences in performance and algorithms accuracy applied to each distinct research field. Ultimately, GISB allows comparing the different algorithms applied to different GIS research fields.

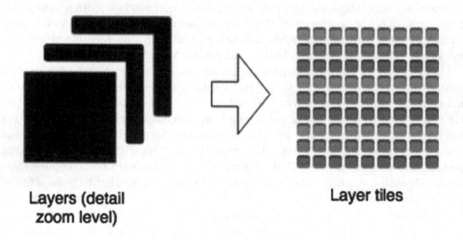

Layers (detail
zoom level)

Layer tiles

Fig. 1. GISB data-set

Given a GISB data-set example scenario, Figs. 1 and 2, based on satellite images, the data-set is based on 11 GB of satellite images related to a specif mapped area. The images are organized by different zoom scales (layers), each layer is organized in tiles (i.e., similar to a mosaic area). Finally each layer has a timestamp (i.e., year), referent to two different years. Each augmentation in the detail level, represents an increase in the image zoom and also an increase of the number of tiles to be processed. Globally there are 15 layers of detail each with a different number of tiles, each layer and respective tiles are identified by a unique ID.

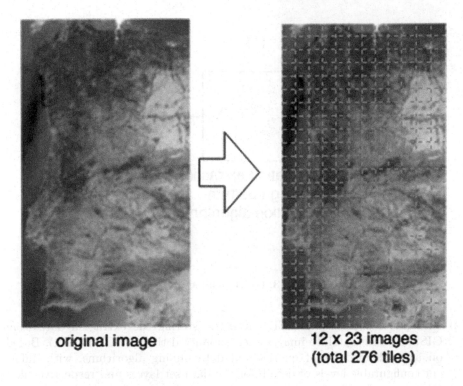

**original image**

**12 x 23 images
(total 276 tiles)**

**Fig. 2.** GISB example data-set (1 single layer) tiles

Note that, any data set can be applied to GISB in order to compare algorithms in terms of performance and accuracy. However, the data-set must be made available inside GISB so that they can be shared across users, allowing to compare algorithms in terms of quality, by GISB data mining algorithms, and performance.

Figure 3, shows the two main pipeline phases that are used to measure and compare each GISB processing stage.

(1) This phase is where from each layer the image tiles information is extracted using different algorithms depending on the desired objectives to be benchmarked. Each developer is required to write the processing algorithms to be benchmarked by GISB. By connecting to GISB APIs the developer algorithm extracts and submits data form/to GISB.

  Based on the extracted information detail from each layer (and each tile), a detailed (HTML format) benchmark report is automatically generated including performance measurements and algorithm completeness when detecting image mapped patterns, objects and measures.

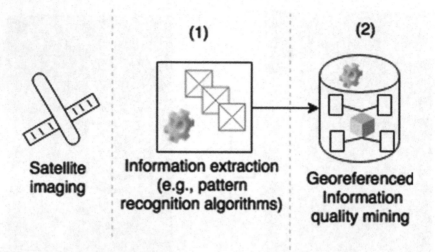

**Fig. 3.** GISB architecture

(2) This phase consists on GISB automatic storing the extracted data from GISB data-set satellite images (i.e., submitted thought GISB API). Based on a pre-defined set of queries and data mining algorithms, with different configurable levels of detail, for the data-set layers and respective tiles, GISB provides benchmark performance speed values, accuracy, statistical and visual information.

## 3.1   API

GISB provides different connection APIs methods to request the data-set and submit extracted information, allowing the developer to code the processing algorithms using any programming language and integrate GISB with any process (e.g., ETL). Figure 4 shows a high level illustration of how the developed algorithms interact with GISB API. The developer can connect to GISB API though three different methods:

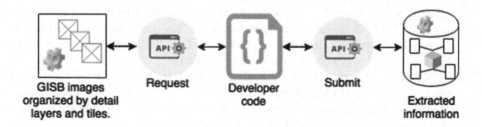

**Fig. 4.** GISB API

– By importing a Java library;
– Using web services to get and set data;
– Manually, by connecting to an IP:port (i.e., messages are passed in string format, images are locally stored as requested);

To each request there is an answer. Images can be transferred using GISB API or by requesting for a specific layer, specific images tiles, to be stored locally in a configured directory.

   **GISB "Request"**, Though GISB API the user can request map tiles from specific layers by specifying:

   The layer ID tile ID and data-set year, which the tile belong to (15 levels of layers detail available, and two distinct years for the same data-set);
– The number/ID of the tile. Together with the layer ID and data-set year, the developer can submit the tile extracted data back to GISB;
– Define/configure storage location and file format (i.e., where to store the tile and choose the file format), OR optionally, transfer the image tiles using an array format (e.g., ready to be use on MatLab);
– Number of existent layers and for each the respective number of tiles.

Other configurations are available though GISB API, for more details please direct to GISB documentation.

   **GISB "Submit"**, The developer can submit for each layer and tile information regarding:

– Forestall area;
– Cultivation area;
– Urban area;
– Green urban area;
– Count number of pools;
– Bridges count, and for each the respective length;
– Football field count;
– Traffic detection (i.e., true or false to detect traffic);
– Beach area and perimeter;
– Coast perimeter;
– Water reservoirs perimeter and area (when more than one, submit multiple times for the same tile ID);
– By comparing GISB data-set from two distinct years calculate the deforestation area. For this case the developer must choose the best detail level, or ultimately analyze all levels (all tiles).

Tiles processing speed are measured since the moment they are requested and made available to be processed, until the moment the tiles extracted information is submitted into GISB. This performance measure is automatically done by GISB, as well as reports regarding only the submitted information.

   Other features to compare algorithms performance vs. accuracy are proposed in the future work section.

## 3.2   Result Presentation

In this section we describe how it is possible to obtain using GISB. For more details and examples we redirect to GISB documentation. Figure 5 shows, the user interaction with GISB terminal to get benchmarked results. All results are outputted from GISB in HTML tables format.

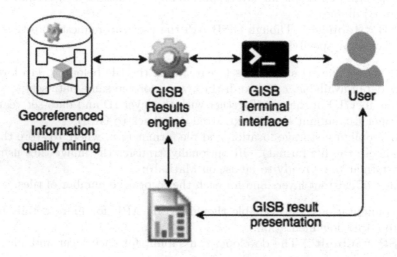

**Fig. 5.** GISB Result

The most complete report using the command line "*get all statistics;*". GISB generates a series of linked HTML pages/tables organized by layers and tiles, including images and charts comparing performance and algorithms accuracy.

With the purpose of evaluating the submitted algorithms completeness and precision, based on the information submitted to GISB regarding layers and tiles, GISB also provides base line values regarding accurate geographic information (e.g., sizes, areas) for the given data-set layer and tiles, allowing this way to evaluate the exact precision of tested algorithms.

Note that, it is also possible to extract only statistical GISB information from a certain layer and tile range.

## 4   GISB Demo Example

As GISB demonstration concept, we applied some fast algorithms called SLICK and DBSCAN to the first layer tiles, in order to estimate the desertification increase of the mapped data-set area. Our algorithm is very simple and fast, consisting only on comparing the color of the areas. Where green color range is considered non-desertification and brown color ranges is classified as desertification.

1997 GISB data-set                    2014 GISB data-set

**Fig. 6.** GISB desertification

**Table 1.** GISB demo result

| #Layer | 1 |
|---|---|
| #Tiles | 1 |
| Avg tile processing speed (sec) | 37 |
| Total processing time (sec) | 37 |

|  | Real | Benchmarked |
|---|---|---|
| Total desertification | 12% | 22% |
| Plantations desertification | 5% | NA |
| Meadow desertification | 4% | NA |
| Forest desertification | 7% | NA |
| Urban desertification | -4% | NA |

We used MatLab software, windows 8 64bits, 16GB RAM, intel i3 2.93 GHz. Figure 6 is provided by GISB html output. Based on the submitted information extracted by the benchmarked algorithm it is possible to visualize the detected desertification areas when comparing GISB two data-sets.

Some performance values extracted with GISB are presented in Table 1. Note that, the more information the tested processing algorithm submits for each layer and each tile, the more complete GISB output will be. For our demo we only submitted one value regarding the desertification area when comparing only the first full view map layer color differences between two distinct years. Table 1 shows, results extracted directly from GISB benchmark. The reduced information results from the limited extracted information provided by our algorithm. Also note that, the discrepancy between the real desertification area and the achieved by our algorithm (only based on the color ranges of the map) is very big and not precise. This is due to the superficial analysis made by our implementation. Moreover, we find that we have a good performance when analyzing the processing time for the entire map area (i.e., processing cost vs. accuracy).

## 5   Conclusions and Future Work

Due to the wide nature of GIS applications, GISB focus mainly on geographic satellite map processing.

GISB provides to each researcher the new opportunity to compare GIS based systems and algorithms in order to evolve and improve the related algorithms/work.

Based on a very simple test case we prove the benchmark concept.

GISB opens a new door to large scope of applications testing and improvement. Its principles are based on algorithms improvement, new processing techniques comparison, and most important the quality of the extracted GIS information versus the required processing time, using the defined algorithms.

This work received great attention from the GIS medical community regarding the comparison of different algorithms precision applied to medical imaging. The use of the same processing techniques proposed in GISB already allowed two different entities in two countries to compare and improve medical software for disease screening and detection.

Immediate future work includes the construction of a visual interface, available on the web, where each researcher can submit results and record information regarding different algorithms. Other future work direction, is the support for different GIS fields.

*Note, due to the sensitive nature of the original data-set used by this benchmark, medical brain ECT images, we adapted the benchmark to work with satellite imaging.*

**Acknowledgment.** This project is part of a larger software prototype, partially financed by CISUC research group from the University of Coimbra, and the Foundation for Science and Technology.

## References

1. Aissi, S., Gouider, M.S.: Spatial and spatio-temporal multidimensional data modelling: A survey. arXiv preprint (2012). arXiv:1208.0163
2. Aji, A., Wang, F., Saltz, J.H.: Towards building a high performance spatial query system for large scale medical imaging data. In: Proceedings of the 20th International Conference on Advances in Geographic Information Systems, pp. 309–318. ACM (2012)

3. Aji, A., Wang, F., Vo, H., Lee, R., Liu, Q., Zhang, X., Saltz, J.: Hadoop gis: a high performance spatial data warehousing system over mapreduce. Proc. VLDB Endowment **6**(11), 1009–1020 (2013)
4. Eldawy, A., Mokbel, M.F.: The era of big spatial data (2013)
5. Haklay, M., Weber, P.: Openstreetmap: User-generated street maps. IEEE Pervasive Comput. **7**(4), 12–18 (2008)
6. Perumal, M., Velumani, B., Sadhasivam, A., Ramaswamy, K.: Spatial Data Mining Approaches for GIS– A Brief Review. In: Satapathy, S.C., Govardhan, A., Raju, K.S., Mandal, J.K. (eds.) Emerging ICT for Bridging the Future - Volume 2. AISC, vol. 338, pp. 579–592. Springer, Heidelberg (2014)
7. Wang, F., Aji, A., Vo, H.: High performance spatial queries for spatial big data: from medical imaging to gis. SIGSPATIAL Spec. **6**(3), 11–18 (2015)

# SRsim: A Simulator for SSD-Based RAID

HooYoung Ahn[1(✉)], YoonJoon Lee[1(✉)], and Kyong-Ha Lee[2]

[1] School of Computing, Korea Advanced Institute of Science and Technology,
Daejeon, Korea
{hyahn,yjlee}@dbserver.kaist.ac.kr
[2] Scientific Data Research Center KISTI, Daejeon, Korea
bart7449@gmail.com

**Abstract.** RAID is a popular storage architecture devised to improve
both I/O performance and reliability of disk storages with disk arrays.
With the declining price of NAND flash-based solid state drives and their
performance gains, they have been widely deployed in systems ranging
from portables to Internet-scale enterprise systems. By virtue of the benefits, studies on developing RAID gears with solid state drives have been
performed in recent years. However, there are still some research issues on
realizing a reliable RAID system with arrays of solid state drives due to
the different characteristics of NAND flash memory. Moreover, the internal S/W architecture of the current commercial SSDs are not opened
to the public so that it is hard to test SSD-based RAID systems with
their devised algorithms. The fundamental algorithms in DBMS have
been optimized for the use of hard disk drives, which prefer sequential
data accesses. Recently, DBMS internals have been modified to make the
best use of solid state drives rather than simple hardware replacement.
To help the studies, we propose an open-source SSD-based RAID simulator named SRsim. SRSim helps researchers explore and experiment
their ideas on an array os solid state drives more easily and accurately.

**Keywords:** Storage systems · RAID (redundant array of independent
disks) · OLTP (online transaction processing) · SSD (solid state drives) ·
NAND flash memory

## 1 Introduction

Hard disk drives(HDDs) remain dominant in storage devices for several decades.
HDD manufacturers have met performance requirements by increasing RPMs
(rotations per minutes), and shrinking the platter size to spinning disks faster.
However, many studies indicated that the problem of above design direction had
reached the limit because of power consumption and thermal issues [10].

This work was partly supported by the R&D program of MSIP/COMPA
(2015K000260), IITP grant(R0126-15-1082), and MSIP grant(K-16-L03-C01) funded
by Korea government.

© Springer International Publishing Switzerland 2016
S. Kozielski et al. (Eds.): BDAS 2016, CCIS 613, pp. 610–620, 2016.
DOI: 10.1007/978-3-319-34099-9_47

Solid state drives(SSDs), which have no mechanical moving parts, have received attention as a new storage media. As SSDs have become more cost effective with the declining price of NAND flash memories, now they are used widely from embedded devices to storage servers. SSDs provide a lot of benefits over HDDs such as low power consumption, shock resilience, light weight, and significantly low latency. This enables to reduce the operating costs for enterprise data centers although the initial costs of adopting SSDs may still higher than HDDs. RAID was initially introduced to increase I/O performance and reliability of disk arrays. It stores files divided into fixed size blocks and distributes them to multiple disks which are combined into one large size reliable logical storage [7]. There are various RAID levels from the standard RAID level 0 to 6 and their variations. Each RAID level offers different I/O performance and fault tolerance for the storage systems. To store data in the proper RAID level, balance between cost, bandwidth, and reliability should be taken into consideration. Several researchers put their effort in improving DBMS performance in RAID environments [20, 22].

SSD arrays have started to get much attention in various research areas. Data centers discover the potential cost savings by deploying SSDs. Major web service providers, such as Ebay, Amazon, Apple, Dropbox, Google, and HP adopt SSDs in their data centers [4, 19]. Storage system manufacturers have also started to release their SSD-based solutions in large-scale database servers. Raid controller manufacturers begin to support SSDs such as Adaptec 8-series [21], HP SSD Smart Path [11]. This trend makes researchers look back the conventional algorithms and architectural designs of DBMS. The conventional DBMS is devised to well utilize sequential accesses and in-place updates on HDDs. This design is inefficient on SSDs since SSDs do not provide in-place updates and exhibit high update latencies due to erase operations. However, SSD internals are not standardized and manufacturers do not unveil their controller SW to the public. The limited research environment makes researchers hard to verify and experiment their ideas on SSDs. To overcome this problem, researchers have continued their effort to develop SSD simulators which behave like real SSDs [1, 8, 14]. However, although there are several widely known SSD simulators, none of them can simulate SSD array with RAID environment.

In this paper, we present SRsim (**S**SD-based **R**AID **Sim**ulator), an open source simulator framework for SSD based RAID. The simulator allows researchers to put little effort into their experiments on SSD array. Our simulator has implemented based on the extended FlashSim [3]. The remainder of this paper is organized as follows. In the next section, we present the basics of NAND flash memory and SSDs. Related work is summarized in Sect. 3. We present the design and implementation details in Sect. 4. Section 5 discusses our experimental results. Finally, we conclude our work and discuss future direction in Sect. 6.

## 2  Preliminaries

A SSD is composed of an array of NAND flash memories. Figure 1 shows an internal architecture of an off-the shelf 512 GB SSD. In the figure, each page size is 4 KB and the block size is 256 KB since each block is composed of 64 pages. A die is a 8 GB NAND flash chip and it is organized with 4 planes. The size of each plane is 2 GB and it is composed of 1024 blocks. Finally, the SSD becomes 512 GB which is composed of 8 64 GB packages having 8 dies. The package shares a bus channel in the SSD controller and it enables a SSD to interleave I/O requests and provide high I/O performance.

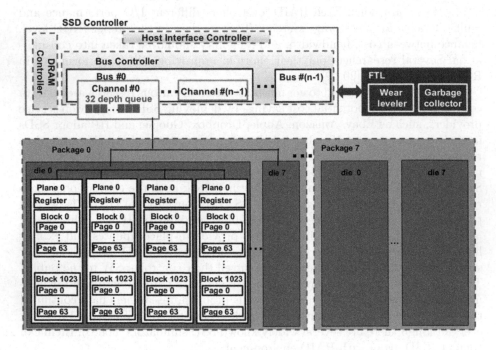

**Fig. 1.** 512 GB SSD internal architecture

NAND flash memory has unique properties different from HDDs. A major difference is that they cannot support in-place updates, which is updating data in overwriting manner. The basic operations of SSDs are read, write and erase. It is due to the different granularity of the unit of basic operations. Each read and write is performed with a page whereas an erase operation is performed with a block which is composed of dozens of pages.

Due to the out-of-place update operations, the garbage collector of FTL in Fig. 1 is triggered background GC (Garbage Collection) to get free blocks by erasing invalidated blocks. During the background GC process, incoming I/O requests are delayed until its over. In the FTL, there is another module which is called WL (Wear Leveler). It plays a role for making cells in the NAND flash memory wear out evenly since they have limited lifetime. The wear leveler stores wear statistics of all blocks in SSD and updates every time an erase operation occurs.

Although SSDs have lots of benefit, these background processes (GC and WL) make them cannot provide uniform performance. Researchers have devoted considerable efforts to find various reasons that lead the performance degradation of SSDs [6,18,23].

Kim *et al.* observed the performance degradation in SSD array due to the garbage collection of individual SSDs [13]. They attempt to improve the I/O performance of entire SSD array by coordinating the garbage collection time among SSDs.

# 3   Related Work

## 3.1   Flash-Based DBMS

With outstanding advantages over magnetic disks, researches on DBMS with flash-based SSDs have great attraction in a recent decade. In the early research stage, Lee *et al.* have shown that the I/O overhead of DBMS for transaction log, rollback and temporary data can be reduced by just adopting SSDs in DBMS [17].

However, existing algorithms and architectures in DBMS which are HDD-oriented designed cannot work efficiently on SSDs with the best performance. Many researchers have worked to solve this problem. Lee *et al.* show that the simple hardware replacement still makes I/O performance degradation in DBMS although they replace HDDs with SSDs [16]. Thus, they propose special SSD architectures which are optimized for OLTP workload processing. There are some researches propose new caching strategies for DBMS by using SSD [2,5,25].

Since the performance degradation in SSDs caused by random writes, several researchers especially focus on improving the performance of DBMS under OLTP workload which has large number of random writes to overcome the drawbacks of erase-before-write characteristic in Nand flash memory [12,15]. They reduce the number of block erase operations by logging updated logical pages into specific physical pages in a block of flash memory. However, all of these research approaches have no choice but to experiment their algorithms on single SSD simulators.

## 3.2   State-of-the Art SSD Simulators

Most previous work have focused on simulating internal hardware architecture of the SSD. Agrawal *et al.* propose the initial stage of research on SSD simulators [1]. It is an extended version of the well-known HDD simulator DiskSim developed by CMU [9]. Microsoft Research integrated basic operations of SSDs such as read, write, and erase into the DiskSim. Researchers in Pennsylvania State

University re-created the new version of above simulator in an object-oriented architecture [14]. They developed the simulator by adding more realistic components such as various FTLs and I/O queueing functionalities. Dayan *et al.* pointed out the flaw of previous simulators that they cannot estimate exact performance of overall systems using a SSD and it provided elaborate attempt to design the SSD controller, operating systems, and application modules which use the SSD more precisely [8]. In contrast to these previous approaches, SRsim simulates multiple SSDs with RAID architecture as well as a single SSD.

## 4    The SRsim

### 4.1    Overview

SRsim is an open source simulator. It simulates arrays of SSDs various RAID levels. In our simulator, a single SSD is implemented with well-designed FlashSim which is implemented with an object-oriented approach. SRsim is designed to support general disk traces and it is also written in C++ language like FlashSim. SRsim has the following features:

- simulate multiple SSD arrays
- support standard RAID levels with RAID 0, RAID 5, and RAID 6
- assign logical addresses in trace files to designated SSDs
- process I/O requests with corresponding parity updates
- process general storage traces which are enable to interoperable with DiskSim or FlashSim

### 4.2    Design and Implementation

Figure 2 presents an architecture of the SRsim. Overall, our simulator presents experimental results for general disk traces.

- RAID controller: plays a key role for parsing input trace files, managing stripe information and generate parities.
  - Addressing manager: The address range of each SSD is configured by the FlashSim settings. The address manager gets logical addresses from trace files and distributes data in a round robin manner to SSD arrays.
  - Striping manager: SRsim creates stripes with blocks stored in different SSDs. In the SSD-based RAID, updating data blocks in a stripe needs to update parities in different SSDs. The striping manager handles comprehensive stripe information in the SSD array and orchestrates read/write/update operations on data blocks and parities in SSD array.
  - Parity generator: The parity generator operates together with the striping manager. It is responsible for placing parities interleaved over all SSDs.

- SSD array: The SSD array in our simulator incorporates each SSD as a FlashSim simulator. The multiple independent FlashSim modules process I/O requests in parallel way.

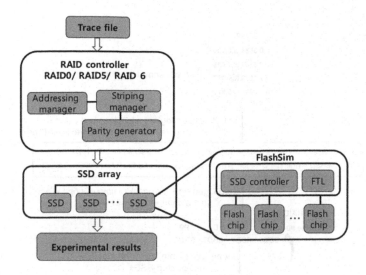

**Fig. 2.** SRsim architecture

### 4.3 Operational Process

Figure 3 shows the processing steps for incoming trace file. SRsim first scans the trace file (step 1) and then, converts logical addresses in the original trace files to newly calculated values. The reason for this address translation is to distribute data blocks of trace file to SSD array in ascending order and converge the original address into the address space of each SSD. The process of address translation is illustrated in detail in Fig. 4 (step 2). We first choose target SSDs to store data with modular operations of both original addresses in trace and the number of SSDs. Then, data blocks are distributed in a round-robin manner to the SSD array with large addresses which exceed the available address range of each FlashSim simulator. We cuts off the large address range into the available address space of the SSD array with Eq. (1). Finally, we rearrange the addresses in order to fit them in a single SSD with Eq. (2).

Then the striping manager generates stripes with data blocks stored in different SSDs and assign unique stripe IDs to each stripe. In this step, the number of parities is decided with the configured RAID level. In this figure, we illustrate a single stripe with 3 data blocks and 1 parity in RAID 5 to simplify the example (step 3). Finally, I/O requests for data blocks and parities are sent to multiple SSDs in parallel (step 4).

## 5   Experiments and Results

We conduct experiments to show the effectiveness and the precise operation of SRsim by deploying it on different RAID configurations such as the number of SSDs and various RAID levels.

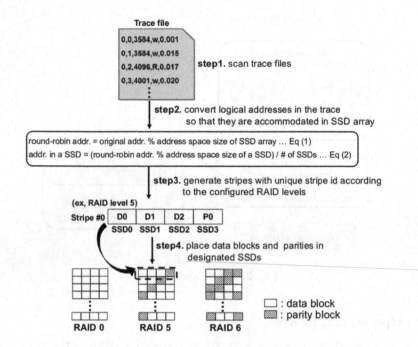

**Fig. 3.** Process of SRsim

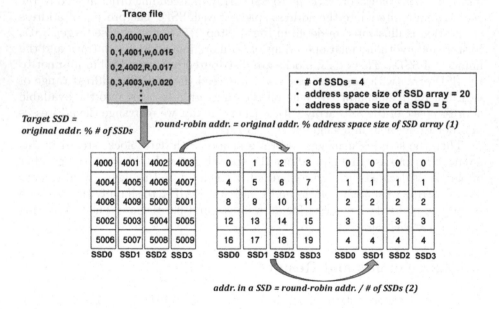

**Fig. 4.** Address translation in SRsim

## 5.1 Experimental Setup

All experiments are performed on a single server with 24 GB RAM and an Intel(R) Core(TM) i5-2500 K CPU @ 3.30 GHz running CentOS 6.6. Table 1 presents the specific configurations of our simulator. The workloads used in our experiments are two representative UMASS storage traces from OLTP applications that have different read/write ratio [24]. Financial 1 is write-dominant with 32.8 % read ratio and Financial 2 is read-dominant with 84.1 % read ratio.

**Table 1.** SSD-based RAID configuration

| Device configuration | |
|---|---|
| Total capacity | 256 GB |
| Reserved free blocks | 15 % |
| # of chips per device | 2 |
| # of dies per chip | 8 |
| # of planes per package | 4 |
| # of blocks per plane | 1024 |
| # of pages per block | 1024 |
| page size | 4 KB |
| FTL type | hybrid |
| **Flash operational latency** | |
| page read | 0.080 ms |
| page write | 0.150 ms |
| block erase | 1.5 ms |
| **RAID configuration** | |
| # of SSDs | 4, 6, 8 |
| Supported RAID levels | 0, 5, 6 |

## 5.2 Results for Realistic Workloads

We validated our simulator by running real workload traces. Figure 5 shows that SSD-based RAID 0 provides good performance with the larger number of SSDs. We use an $(n, k)$ erasure code which encodes $n$ data blocks into $k$ parity blocks. Figure 6 shows the normalized elapsed time for RAID 0 (6, 0), RAID 5 (5, 1), and RAID 6 (4, 2) configurations with six SSDs processing Financial 1 workload.

As expected, RAID 0 shows the best performance since RAID5 and 6 have additional I/Os for parity writes.

In Fig. 7, we evaluate the number of erase operations on 6 SSDs under two types of workloads. The erase operation is quietly related to both the lifetime and performance of SSDs. As can be expected, SSDs has fewer number of erases under Financial 2 than Financial 1 due to the less random write ratio of the workload.

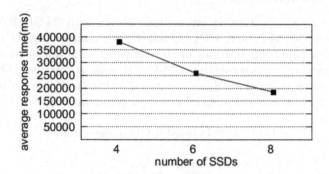

**Fig. 5.** Performance w.r.t the number of SSDs

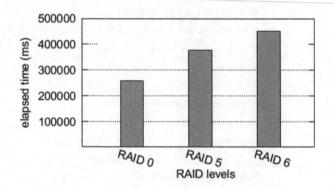

**Fig. 6.** Performance in various RAID levels

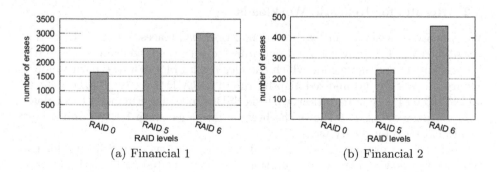

(a) Financial 1                    (b) Financial 2

**Fig. 7.** Erase counts with two workloads

The result shows that the number of erases is increasing from RAID 0 to RAID 6 due to the larger number of parity makes more additional random writes for parity updates. The main reason is that RAID 0 having no parities only needs to update data. On the other hand, RAID with more parities suffer from large amount of random write penalty from parity updates due to the out-of-place update scheme of SSDs.

## 6    Conclusion and Future Work

In this paper, we have presented the SRsim simulates multiple SSDs in RAID architecture. We have developed the SRsim as a general-purpose trace-driven simulator and validated it by performing experiments with real-world traces. Our research is a work in progress. We will continue our research on finding better ways to improve the performance of SSD-based RAID using parities. Specifically, we plan to prolong lifetime of each SSD in the SSD array by decreasing write amplification, which is a critical factor that limits the write performance and accelerates the aging of SSDs. We expect SRsim to be a useful environment for various researches related to SSD-based RAID. Our source-code is available for downloading at https://github.com/SRsim/simulator.

## References

1. Agrawal, N., Prabhakaran, V., Wobber, T., Davis, J.D., Manasse, M.S., Panigrahy, R.: Design tradeoffs for SSD performance. In: USENIX Annual Technical Conference, pp. 57–70 (2008)
2. Athanassoulis, M., Ailamaki, A., Chen, S., Gibbons, P., Stoica, R.: Flash in a DBMS: Where and how? IEEE Data Eng. Bull. **33**(EPFL–ARTICLE–161507), 28–34 (2010)
3. Bjørling, M.: Extended flashsim (2011). https://github.com/MatiasBjorling/flashsim
4. Bostoen, T., Mullender, S., Berbers, Y.: Power-reduction techniques for data-center storage systems. ACM Comput. Surv. (CSUR) **45**(3), 33 (2013)
5. Canim, M., Mihaila, G.A., Bhattacharjee, B., Ross, K.A., Lang, C.A.: SSD buffer-pool extensions for database systems. Proc. VLDB Endow. **3**(1–2), 1435–1446 (2010)
6. Chen, F., Koufaty, D.A., Zhang, X.: Understanding intrinsic characteristics and system implications of flash memory based solid state drives. In: ACM SIGMETRICS Performance Evaluation Review, vol. 37, pp. 181–192. ACM (2009)
7. Chen, P.M., Lee, E.K., Gibson, G.A., Katz, R.H., Patterson, D.A.: Raid: high-performance, reliable secondary storage. ACM Comput. Surv. (CSUR) **26**(2), 145–185 (1994)
8. Dayan, N., Svendsen, M.K., Bjørling, M., Bonnet, P., Bouganim, L.: Eagletree: exploring the design space of SSD-based algorithms. Proc. VLDB Endow. **6**(12), 1290–1293 (2013)
9. Ganger, G.R., Worthington, B.L., Patt, Y.N.: The disksim simulation environment version 2.0 reference manual (1999)

10. Gurumurthi, S., Sivasubramaniam, A., Natarajan, V.K.: Disk drive roadmap from the thermal perspective: A case for dynamic thermal management, vol. 33. IEEE Computer Society (2005)
11. HP: Optimized solid-state drive performance with hp SSD smart path, March 2014
12. Kim, Y.R., Whang, K.Y., Song, I.Y.: Page-differential logging: an efficient and DBMS-independent approach for storing data into flash memory. In: Proceedings of the 2010 ACM SIGMOD International Conference on Management of data, pp. 363–374. ACM (2010)
13. Kim, Y., Lee, J., Oral, S., Dillow, D., Wang, F., Shipman, G.M., et al.: Coordinating garbage collectionfor arrays of solid-state drives. IEEE Trans. Comput. **63**(4), 888–901 (2014)
14. Kim, Y., Tauras, B., Gupta, A., Urgaonkar, B.: Flashsim: A simulator for NAND flash-based solid-state drives. In: First International Conference on Advances in System Simulation, SIMUL2009, pp. 125–131. IEEE (2009)
15. Lee, S.W., Moon, B.: Design of flash-based DBMS: an in-page logging approach. In: Proceedings of the 2007 ACM SIGMOD International Conference on Management of Data, pp. 55–66. ACM (2007)
16. Lee, S.W., Moon, B., Park, C.: Advances in flash memory SSD technology for enterprise database applications. In: Proceedings of the 2009 ACM SIGMOD International Conference on Management of Data, pp. 863–870. ACM (2009)
17. Lee, S.W., Moon, B., Park, C., Kim, J.M., Kim, S.W.: A case for flash memory SSD in enterprise database applications. In: Proceedings of the 2008 ACM SIGMOD International Conference on Management of Data, pp. 1075–1086. ACM (2008)
18. Min, C., Kim, K., Cho, H., Lee, S.W., Eom, Y.I.: SFS: random write considered harmful in solid state drives. In: FAST, p. 12 (2012)
19. Narayanan, D., Thereska, E., Donnelly, A., Elnikety, S., Rowstron, A.: Migrating server storage to SSDs: analysis of tradeoffs. In: Proceedings of the 4th ACM European Conference on Computer Systems, pp. 145–158. ACM (2009)
20. Petrov, I., Gottstein, R., Hardock, S.: DBMS on modern storage hardware. In: 2015 IEEE 31st International Conference on Data Engineering (ICDE), pp. 1545–1548. IEEE (2015)
21. PMC: Adaptec series 8 raid, PMC-Sierra Inc. (2015)
22. Smolinski, M.: Efficient multidisk database storage configuration. In: Kozielski, S., Mrozek, D., Kasprowski, P., Małysiak-Mrozek, B., Kostrzewa, K. (eds.) BDAS 2015. CCIS, vol. 521, pp. 180–189. Springer, Heidelberg (2015)
23. Sun, H., Qin, X., Wu, F., Xie, C.: Measuring and analyzing write amplification characteristics of solid state disks. In: 2013 IEEE 21st International Symposium on Modeling, Analysis & Simulation of Computer and Telecommunication Systems (MASCOTS), pp. 212–221. IEEE (2013)
24. Umass Storage performance council (2007). http://traces.cs.umass.edu/index.php/
25. Wang, J., Guo, Z., Meng, X.: An efficient design and implementation of multi-level cache for database systems. In: Renz, M., Shahabi, C., Zhou, X., Cheema, M.A. (eds.) DASFAA 2015. LNCS, vol. 9049, pp. 160–174. Springer, Heidelberg (2015)

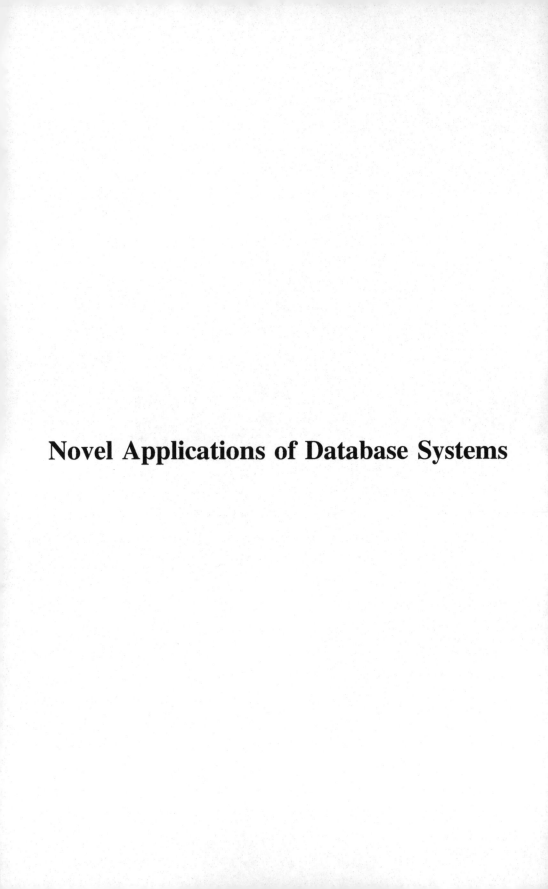

# Novel Applications of Database Systems

# Application of Reversible Denoising and Lifting Steps to LDgEb and RCT Color Space Transforms for Improved Lossless Compression

Roman Starosolski[✉]

Institute of Informatics, Silesian University of Technology,
Akademicka 16, 44-100 Gliwice, Poland
roman.starosolski@polsl.pl

**Abstract.** The lifting step of a reversible color space transform employed during image compression may increase the total amount of noise that has to be encoded. Previously, to alleviate this problem in the case of a simple color space transform RDgDb, we replaced transform lifting steps with reversible denoising and lifting steps (RDLS), which are lifting steps integrated with denoising filters. In this study, we apply RDLS to more complex color space transforms LDgEb and RCT and evaluate RDLS effects on bitrates of lossless JPEG-LS, JPEG 2000, and JPEG XR coding for a diverse image test-set. We find that RDLS effects differ among transforms, yet are similar for different algorithms; for the employed denoising filter selection method, on average the bitrate improvements of RDLS-modified LDgEb and RCT are not as high as of the simpler transform. The RDLS applicability reaches beyond image data storage; due to its general nature it may be exploited in other lifting-based transforms, e.g., during the image analysis for data mining.

**Keywords:** Image processing · Lossless image compression · Lifting technique · Denoising · Reversible denoising and lifting step · RDgDb · LDgEb · RCT · JPEG-LS · JPEG 2000 · JPEG XR

## 1 Introduction

Most color image compression algorithms independently compress the image components; since components in the RGB space are correlated, the compression is performed after transforming image data to a less correlated color space. For the lossless compression, the reversible color space transforms are employed, that are built using lifting steps [6]. In [9], we noticed that such step may increase the total amount of noise that must be encoded during compression. We replaced lifting steps with reversible denoising and lifting steps (RDLS), which are lifting steps integrated with denoising filters. We applied RDLS to a simple RDgDb [11] transform (also known as $A_{2,1}$ [14]) and found that RDLS improved bitrates of images in optical resolutions of acquisition devices. Experiments were performed for 3 significantly different standard image compression algorithms in lossless

© Springer International Publishing Switzerland 2016
S. Kozielski et al. (Eds.): BDAS 2016, CCIS 613, pp. 623–632, 2016.
DOI: 10.1007/978-3-319-34099-9_48

mode (LOCO-I/JPEG-LS[1] [16], JPEG 2000[2] [15], and HD-Photo/JPEG XR[3] [4,8]) and for a simple denoising filter. We found that the memoryless entropy of the component prediction error obtained with the nonlinear edge-detecting predictor MED [7,16] was a very efficient estimator of image component transform effects, suitable for selecting a filter for given image component. In this study, we apply RDLS to more complex color space transforms LDgEb [11] (denoted $A_{4,10}$ in [14]) and RCT (among others, used in JPEG 2000 standard) and evaluate RDLS effects using the same denoising filters, compression algorithms, and test images, as previously. Entropy estimation employing MED is used for selecting the denoising filter and deciding whether to exploit denoising.

The remainder of this paper is organized as follows. In Sect. 2 we briefly characterize RDLS, present the proposed RDLS-modified transforms along with the filter selection method and the denoising filters used, and describe the implementations and test data. Results are presented and discussed in Sect. 3; Sect. 4 summarizes the research.

## 2    Materials and Methods

### 2.1    RDLS

In [9] we proposed to replace color space transform lifting steps (Eq. 1) with RDLS (Eq. 2):

$$C_x \leftarrow C_x \oplus f(C_1, \ldots, C_{x-1}, C_{x+1}, \ldots, C_n) \qquad (1)$$

$$C_x \leftarrow C_x \oplus f(C_1^d, \ldots, C_{x-1}^d, C_{x+1}^d, \ldots, C_n^d) \qquad (2)$$

where $C_i$ is the $i$-th component of the pixel, $C_x$ is the component which is modified by the step, $C_i^d$ is the denoised $i$-th component of the pixel, $n$ is the number of components, and the operation $\oplus$ is reversible. Different denoising filters may be used for different components. For denoising of arguments of function f we may use any component of any pixel, but the $C_x$ of the pixel to which the RDLS is being applied; in this study for denoising of $C_i$ of specific pixel we use $C_i$ of this pixel and of its neighbors. RDLS, like the lifting step it originates from, is trivially and perfectly invertible and may be computed in-place. However note, that denoising exploited in RDLS is irreversible and it is not an in-place operation—computing the function f argument $C_i^d$ does not alter $C_i$. For more detailed characteristics of RDLS and examples of its application we refer the Reader to [9,12] and to the next subsection.

---

[1] Information technology–Lossless and near-lossless compression of continuous-tone still images–Baseline, ISO/IEC International Standard 14495-1 and ITU-T Recommendation T.87 (2006).

[2] Information technology–JPEG 2000 image coding system: Core coding system, ISO/IEC International Standard 15444-1 and ITU-T Recommendation T.800 (2004).

[3] Information technology–JPEG XR image coding system–Image coding specification, ISO/IEC International Standard 29199-2 and ITU-T Recommendation T.832 (2012).

## 2.2   RDLS-Modified Transforms

Here we show the application of RDLS to RDgDb and, for brevity, present only the RDLS-modified variants of LDgEb and RCT. Equation 3 shows the RDgDb transform, as it was presented in [11], where the definitions of other unmodified transforms may also be found. To obtain RDLS-modified RDgDb (RDLS-RDgDb) we first rewrite RDgDb using the notation as in Eqs. 1 and 2, where the same symbol denotes the pixel's component both before and after modifying its value by the lifting step or by the RDLS. Equation 4 persents RDgDb (both forward [left-hand side] and inverse) as a sequence of lifting steps; generally, the steps must be performed in a specified order. $C_1$, $C_2$, and $C_3$ denote $R$, $G$, and $B$ components of the untransformed image, respectively, and $R$, $Dg$, and $Db$ components of the transformed image, respectively. Next, we simply replace lifting steps (Eq. 1) in Eq. 4 with RDLS (Eq. 2) constructed based on them and obtain RDLS-RDgDb (Eq. 5). The employed denoising filters and method of selecting the filter for a given image component is described in the following subsection.

$$
\begin{array}{ll}
R = R & R = R \\
Dg = R - G \quad \Longleftrightarrow \quad & G = R - Dg \\
Db = G - B & B = G - Db
\end{array}
\tag{3}
$$

$$
\begin{array}{ll}
\text{step 1: } C_3 \leftarrow -C_3 + C_2 & \text{step 1: } C_1 \leftarrow C_1 \\
\text{step 2: } C_2 \leftarrow -C_2 + C_1 \quad \Longleftrightarrow \quad & \text{step 2: } C_2 \leftarrow -C_2 + C_1 \\
\text{step 3: } C_1 \leftarrow C_1 & \text{step 3: } C_3 \leftarrow -C_3 + C_2
\end{array}
\tag{4}
$$

$$
\begin{array}{ll}
\text{step 1: } C_3 \leftarrow -C_3 + C_2^d & \text{step 1: } C_1 \leftarrow C_1 \\
\text{step 2: } C_2 \leftarrow -C_2 + C_1^d \quad \Longleftrightarrow \quad & \text{step 2: } C_2 \leftarrow -C_2 + C_1^d \\
\text{step 3: } C_1 \leftarrow C_1 & \text{step 3: } C_3 \leftarrow -C_3 + C_2^d
\end{array}
\tag{5}
$$

The RDLS-RDgDb transform (Eq. 5) may be presented using component symbols like in standard definitions of color space transforms (Eq. 6). Note, that the same symbols denote components of regular transform and its RDLS-modified variant.

$$
\begin{array}{ll}
\text{step 1: } Db = -B + G^d & \text{step 1: } R = R \\
\text{step 2: } Dg = -G + R^d \quad \Longleftrightarrow \quad & \text{step 2: } G = -Dg + R^d \\
\text{step 3: } \quad R = R & \text{step 3: } B = -Db + G^d
\end{array}
\tag{6}
$$

In Eq. 7 we present the RDLS-LDgEb transform and in Eq. 8 the RDLS-RCT transform, that were obtained from LDgEb and RCT, respectively. The floor of division by integer power of 2 is computed using the arithmetic right shift.

$$
\begin{array}{ll}
\text{step 1: } Dg = -G + R^d & \text{step 1: } B = Eb + L^d \\
\text{step 2: } \quad L = R - \lfloor Dg^d/2 \rfloor \quad \Longleftrightarrow \quad & \text{step 2: } R = L + \lfloor Dg^d/2 \rfloor \\
\text{step 3: } Eb = B - L^d & \text{step 3: } G = -Dg + R^d
\end{array}
\tag{7}
$$

$$\begin{array}{ll} \text{step 1:} & Cv = R - G^d \\ \text{step 2:} & Cu = B - G^d \\ \text{step 3:} & Y = G + \lfloor (Cv^d + Cu^d)/4 \rfloor \end{array} \quad \Longleftrightarrow \quad \begin{array}{ll} \text{step 1:} & G = Y - \lfloor (Cv^d + Cu^d)/4 \rfloor \\ \text{step 2:} & B = Cu + G^d \\ \text{step 3:} & R = Cv + G^d \end{array}$$

$$(8)$$

It is worth noting, that components of RDLS-LDgEb and RDLS-RCT, as opposed to RDLS-RDgDb, may require greater bit depths, than their non-RDLS equivalents. In research reported herein, we used increased component depth only when component pixels actually exceeded original depth; in all such cases extending the depth by 1 bit was sufficient.

## 2.3   Denoising Filters and Filter Selection

For denoising we employed 11 low-pass linear averaging filters (smoothing filters) with $3 \times 3$ pixel windows. The filtered pixel component $C_i^d$ was calculated as a weighted arithmetic mean of the $C_i$ components of pixels from the window. The weight of the window center point was different for different filters—from 1 to 1024 (integer powers of 2 only), while its neighbors' weights were fixed to 1.

In each RDLS-modified forward transform step $s$ for all pixels of a given image a filter was selected individually for each component requiring denoising. E.g., in step 3 of forward RDLS-RCT (Eq. 8), 2 filters were selected for denoising of $Cv$ and $Cu$ and applied to all image pixels. All filter combinations were tested and we also allowed to not use the filtering. The combination resulting in the best estimated bitrate of the component being modified by the step $s$ was used for actual compression. Thus we performed an exhaustive search of filters in a given step, however only step 3 of RDLS-RCT requires denoising of 2 components. Filter(s) selection for a given step was not revised based on compression effects of components transformed in other steps, therefore even assuming the perfect estimation of compression algorithm bitrate, modifying this way transform with RDLS may result in bitrate worsening. Filter selection must be passed to the decoder along with the compressed data, but its cost is negligible.

The memoryless entropy of the component prediction error obtained using MED predictor was used as an estimator of compressed component bitrate, overall 3-component image bitrate was estimated as sum of computed this way entropies of 3 components. We denote this estimation method as H0_pMED. The compression ratio or bitrate $r$, expressed in bits per pixel (bpp), is calculated as $r = 8l/m$, where $m$ is the number of pixels in the image component, $l$ is the length in Bytes of the file containing the compressed component, including compressed file header; smaller $r$ denotes better compression. The memoryless entropy of the image component $H_0 = -\sum_{i=0}^{N-1} p_i \log_2 p_i$, where $N$ is the alphabet size and $p_i$ is the probability of occurrence of pixel component value $i$ in the image component, is also expressed in bpp.

## 2.4   Implementations and Test Data Used

We used the following sets of images:

- Waterloo—Set ("Colour set") of images from the University of Waterloo[4];
- Kodak—Image set from the Kodak corporation[5];
- EPFL—Image set from the École polytechnique fédérale de Lausanne[6] [3];
- A1, A2, and A3—Image sets from the Silesian University of Technology[7];
- A1-red.3, A2-red.3, and A3-red.3—reduced size (3×) sets A1, A2, and A3.

Sets A1, A2, and A3 contain unprocessed photographic images in optical resolutions of acquisition devices, or (A3) as close to such resolution, as possible without interpolation of all components. Except for Waterloo, all images may be characterized as continuous-tone photographic. The most widely-known Waterloo set contains both photographic and artificial images; some of them are dithered, sharpened, have sparse histograms of intensity levels [10], are computer-generated or composed of others. The same image sets were used for experiments in the previous study, their detailed characteristics may be found in [9].

We used the Signal Processing and Multimedia Group, Univ. of British Columbia JPEG-LS implementation, version 2.2[8], JasPer implementation of JPEG 2000 by M. Adams, version 1.900[9] [1], and JPEG XR standard reference software[10].

## 3   Results and Discussion

In Tables 1, 2, and 3, for RDLS-RDgDb, RDLS-LDgEb, and RDLS-RCT, respectively, we report average entropies obtained using H0_pMED estimator and average bitrates for JPEG-LS, JPEG 2000, and JPEG XR compression algorithms in lossless mode, summed for all 3 image components. Entropy and bitrate changes due to RDLS with respect to non-RDLS transform variants are also reported—in columns labeled $\Delta H_0$ for H0_pMED and $\Delta r$ for compression algorithms. Figure 1 for the examined transforms presents average entropy and bitrate changes due to RDLS of individual transformed components and of the 3-component image; in electronic version of the paper components are presented using colors of RGB components from which they were transformed.

We examined RDLS effects for several transforms, compression algorithms, and test image sets. In majority of cases, the RDLS-RDgDb obtains the best bitrates among all RDLS-modified transforms. For all these transforms, the

---

[4] http://links.uwaterloo.ca/Repository.html.
[5] http://www.cipr.rpi.edu/resource/stills/kodak.html.
[6] http://documents.epfl.ch/groups/g/gr/gr-eb-unit/www/IQA/Original.zip.
[7] http://sun.aci.polsl.pl/~rstaros/optres/.
[8] http://www.stat.columbia.edu/~jakulin/jpeg-ls/mirror.htm.
[9] http://www.ece.uvic.ca/~mdadams/jasper/.
[10] Information technology–JPEG XR image coding system–Reference software, ISO/IEC International Standard 29199-5 and ITU-T Recommendation T.835 (2012).

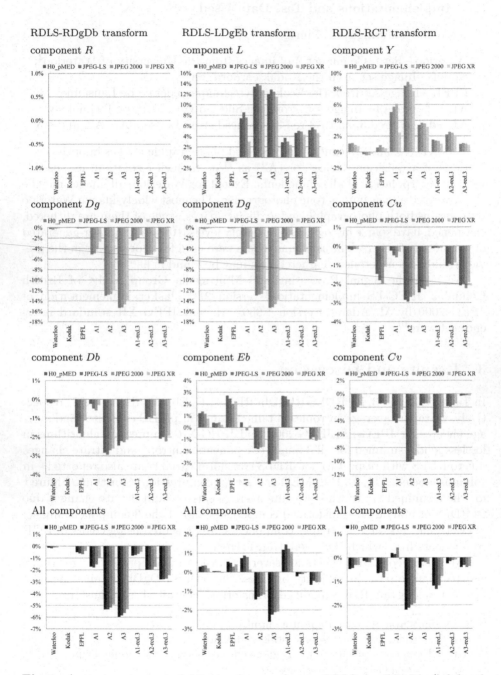

**Fig. 1.** Average entropy and bitrate changes due to RDLS for RDgDb (left-hand panels), LDgEb (middle panels), and RCT (right-hand panels) (Colour figure online).

**Table 1.** Effects of the RDLS-RDgDb transform and comparison to RDgDb

| Images | H0_pMED | | JPEG-LS | | JPEG 2000 | | JPEG XR | |
|---|---|---|---|---|---|---|---|---|
| | $H_0$ | $\Delta H_0$ | $r$ | $\Delta r$ | $r$ | $\Delta r$ | $r$ | $\Delta r$ |
| Waterloo | 8.9014 | −0.13% | 8.8498 | −0.18% | 11.1299 | −0.12% | 13.2834 | −0.11% |
| Kodak | 10.2876 | 0.00% | 9.5676 | 0.00% | 9.4755 | 0.00% | 10.8725 | 0.00% |
| EPFL | 11.1806 | −0.51% | 10.2746 | −0.62% | 10.6515 | −0.68% | 11.6264 | −0.39% |
| A1 | 6.7779 | −1.73% | 6.2722 | −1.80% | 6.2938 | −1.54% | 8.1445 | −0.99% |
| A2 | 14.6469 | −5.35% | 14.0353 | −5.34% | 14.2550 | −5.19% | 14.6526 | −4.94% |
| A3 | 15.1662 | −5.98% | 14.5854 | −5.85% | 14.7919 | −5.65% | 15.3186 | −5.34% |
| A1-red.3 | 9.0112 | −0.81% | 8.2770 | −0.76% | 8.4625 | −0.65% | 9.5931 | −0.49% |
| A2-red.3 | 14.0530 | −2.01% | 13.4235 | −1.99% | 13.8597 | −1.98% | 14.5830 | −1.77% |
| A3-red.3 | 13.3956 | −2.82% | 12.7130 | −2.80% | 13.0802 | −2.76% | 14.0082 | −2.46% |

**Table 2.** Effects of the RDLS-LDgEb transform and comparison to LDgEb

| Images | H0_pMED | | JPEG-LS | | JPEG 2000 | | JPEG XR | |
|---|---|---|---|---|---|---|---|---|
| | $H_0$ | $\Delta H_0$ | $r$ | $\Delta r$ | $r$ | $\Delta r$ | $r$ | $\Delta r$ |
| Waterloo | 9.0756 | 0.28% | 8.9888 | 0.33% | 11.2588 | 0.35% | 13.3249 | 0.21% |
| Kodak | 10.1512 | 0.05% | 9.4370 | 0.06% | 9.4282 | 0.06% | 10.8499 | −0.04% |
| EPFL | 11.3739 | 0.54% | 10.4858 | 0.49% | 10.8297 | 0.29% | 11.7821 | 0.41% |
| A1 | 6.9416 | 0.73% | 6.4348 | 0.87% | 6.4421 | 0.80% | 8.2329 | 0.13% |
| A2 | 14.6249 | −1.44% | 14.0454 | −1.30% | 14.2524 | −1.25% | 14.6502 | −1.19% |
| A3 | 15.3025 | −2.63% | 14.7217 | −2.24% | 14.9125 | −2.12% | 15.4212 | −2.08% |
| A1-red.3 | 9.2698 | 1.17% | 8.5435 | 1.44% | 8.7113 | 1.23% | 9.8088 | 1.05% |
| A2-red.3 | 14.1779 | −0.23% | 13.5386 | −0.10% | 13.9815 | −0.11% | 14.7211 | −0.05% |
| A3-red.3 | 13.6814 | −0.67% | 12.9811 | −0.46% | 13.3597 | −0.54% | 14.2952 | −0.53% |

**Table 3.** Effects of the RDLS-RCT transform and comparison to RCT

| Images | H0_pMED | | JPEG-LS | | JPEG 2000 | | JPEG XR | |
|---|---|---|---|---|---|---|---|---|
| | $H_0$ | $\Delta H_0$ | $r$ | $\Delta r$ | $r$ | $\Delta r$ | $r$ | $\Delta r$ |
| Waterloo | 8.9756 | −0.48% | 8.9236 | −0.43% | 11.1792 | −0.31% | 13.2790 | −0.31% |
| Kodak | 10.2758 | −0.14% | 9.5577 | −0.18% | 9.4893 | −0.19% | 10.9110 | −0.09% |
| EPFL | 11.2681 | −0.65% | 10.3995 | −0.63% | 10.7430 | −0.84% | 11.6942 | −0.57% |
| A1 | 6.9397 | 0.19% | 6.4536 | 0.15% | 6.4821 | 0.42% | 8.2503 | −0.07% |
| A2 | 14.6337 | −2.19% | 14.0611 | −2.10% | 14.2587 | −1.98% | 14.6532 | −1.93% |
| A3 | 15.2361 | −0.43% | 14.6042 | −0.17% | 14.7557 | −0.19% | 15.2918 | −0.27% |
| A1-red.3 | 9.1002 | −1.18% | 8.4063 | −1.33% | 8.5820 | −1.17% | 9.6624 | −0.78% |
| A2-red.3 | 14.0771 | −0.25% | 13.4646 | −0.15% | 13.8917 | −0.11% | 14.6175 | −0.08% |
| A3-red.3 | 13.2992 | −0.40% | 12.5899 | −0.32% | 12.9520 | −0.41% | 13.9179 | −0.37% |

obtained bitrates significantly differ among the compression algorithms. Constantly, except for the Kodak set, JPEG-LS obtains the best bitrates and JPEG-XR the worst; for Kodak the JPEG 2000 is the best. However, the bitrate improvements due to RDLS are similar for different compression algorithms and entropy estimated bitrate.

On average, RDLS improves bitrates of all the examined transforms. The bitrate improvements of RDLS-modified LDgEb and RCT are not as high as in the case of the simpler RDgDb transform and for the former transforms RDLS sometimes results in bitrate worsening of 3-component image and, to a greater extent, of individual components. For these transforms we did not find any objective image feature allowing to predict RDLS effectiveness—as opposed to RDgDb, that is the most effective for images in optical resolutions of acquisition devices.

LDgEb and RCT transforms contain steps, during which a given component $C_i$ is modified based on another one $C_a$ which has already been modified based on $C_i$ (see step 2 of LDgEb and step 3 of RCT). For example, let's look at steps 1 and 2 of forward LDgEb (its RDLS version is presented in Eq. 7). Step 1 modifies the $G$ component, that from this moment is denoted as $Dg$, step 2 modifies $R$, that is then denoted as $L$. Step 2 of the regular lifting transform may decrease in $L$ the amount of information which was originally present in $R$, that is of both noise and of the noise-free signal, and insert to $L$ a fraction of information originally present in $G$ (also consisting of noise and noise-free signal). Actually, as the transform is reversible, the information from $R$ is beforehand (in step 1) propagated to $Dg$, and then in step 2 a part of it (as a fraction of $Dg$) is subtracted from $L$. In step 1 components are subtracted—the use of the noise-free signal from $R$ is supposed to result in decreasing the correlation between $R$ and $Dg$ whereas noise from $R$ adds to noise in $Dg$. When we employ RDLS, step 2 applied to $R$ reduces the noise-free signal originally contained in it, but not the noise (noise from $R$ is not present in $Dg$ because filtering during step 1 prohibited propagating it); on the other hand RDLS avoids transferring to $L$ the noise originally present in $G$. In the case of our images the former effect, complemented by component distortions introduced by imperfect denoising we use, has greater impact on the component bitrate, than the later—see component $L$ (and $Y$ for RDLS-RCT) in Fig. 1. Also the $Eb$ component bitrate is worsened for some sets, which may be attributed to using for its calculation the denoised component $L$, bitrates of which are for some sets worsened by RDLS.

The filter selection method we use may be responsible for worse RDLS effects in the case of more complicated transforms. Since we select the best filters for a given step by analyzing filtering effects on bitrate of the component $C_i$ being modified by this step only, the selection may not be optimal if $C_i$ is then used in calculation of components modified in further steps. Better bitrates of such components and of overall 3-component image could be obtained by selecting filters based on overall 3-component image bitrate. Checking all filter combinations for all the steps would be too complex in practice, but employing a heuristics that would search for the best filters in a given step based on compression effects estimated for the entire image (similarly to [12]) seems worthwhile. We also note, that the selection process complexity might be substantially reduced by using only a certain number of image pixels for transform effect estimation—in [13,14] such optimization applied to a similar problem resulted in a close to optimum estimation.

## 4    Conclusions

RDLS effects differ among transforms, yet are similar for JPEG-LS, JPEG 2000, and JPEG XR algorithms as well as for entropy of the component prediction errors obtained using MED. On average, RDLS improves bitrates of all the examined transforms. The bitrate improvements of RDLS-modified LDgEb and RCT are not as high as in the case of the simpler RDgDb transform and for the former transforms RDLS sometimes results in bitrate worsening; both effects may be attributed to the employed method of selecting the denoising filters. As opposed to RDgDb, we did not identify an objective image feature to which the RDLS bitrate improvement could be linked. As the RDLS effects clearly depend on denoising filters used, we expect that the application of better filters may further improve bitrates of the RDLS-modified color space transforms. For some images the denoising filter parameters might be determined directly from the acquisition process parameters [2]. The bitrate improvements we obtained for some of the test-sets are useful from practical standpoint. In the ongoing research, we investigate other filters and filter selection methods as well as simplified compression effect estimators, that jointly are expected to result in greater bitrate improvements obtained at a significantly lower filter selection cost. RDLS recently was found effective for a much more complex, involving more interdependent steps, multi-level 2D DWT transform in lossless JPEG 2000 compression [12], however, its applicability reaches beyond image data storage. Due to its general nature it may be exploited in other lifting-based transforms, e.g., during the image analysis for data mining and skin segmentation [5].

**Acknowledgments.** This work was supported by BK-263/RAU2/2015 grant from the Institute of Informatics, Silesian University of Technology.

## References

1. Adams, M.D., Ward, R.K.: JasPer: a portable flexible open-source software tool kit for image coding/processing. In: 2004 Proceedings of the IEEE International Conference on Acoustics, Speech, and Signal Processing (ICASSP 2004), vol. 5, pp. 241–244 (2004). doi:10.1109/ICASSP.2004.1327092
2. Bernas, T., Starosolski, R., Robinson, J.P., Rajwa, B.: Application of detector precision characteristics and histogram packing for compression of biological fluorescence micrographs. Comput. Methods Programs Biomed. **108**(2), 511–523 (2012). doi:10.1016/j.cmpb.2011.03.012
3. De Simone, F., Goldmann, L., Baroncini, V., Ebrahimi, T.: Subjective evaluation of JPEG XR image compression. In: Proceedings of the SPIE, Applications of Digital Image Processing XXXII, vol. 7443, p. 74430L (2009). doi:10.1117/12.830714

4. Dufaux, F., Sullivan, G.J., Ebrahimi, T.: The JPEG XR image coding standard. IEEE Sig. Process. Mag. **26**(6), 195–199, 204 (2009). doi:10.1109/MSP.2009.934187
5. Kawulok, M., Kawulok, J., Nalepa, J.: Spatial-based skin detection using discriminative skin-presence features. Pattern Recogn. Lett. **41**, 3–13 (2014). doi:10.1016/j.patrec.2013.08.028
6. Malvar, H.S., Sullivan, G.J., Srinivasan, S.: Lifting-based reversible color transformations for image compression. In: Proceedings of the SPIE, Applications of Digital Image Processing XXXI, vol. 7073, p. 707307 (2008). doi:10.1117/12.797091
7. Martucci, S.A.: Reversible compression of HDTV images using median adaptive prediction and arithmetic coding. In: Proceedings of the IEEE International Symposium on Circuits and Systems, pp. 1310–1313 (1990)
8. Srinivasan, S., Tu, C., Regunathan, S.L., Sullivan, G.J.: HD Photo: a new image coding technology for digital photography. In: Proceedings of the SPIE, Applications of Digital Image Processing XXX, vol. 6696, p. 66960A (2007). doi:10.1117/12.767840
9. Starosolski, R.: Reversible denoising and lifting based color component transformation for lossless image compression (2015). arXiv:1508.06106 [cs.MM]
10. Starosolski, R.: Compressing high bit depth images of sparse histograms. In: Simos, T.E., Psihoyios, G. (eds.) International Electronic Conference on Computer Science. AIP Conference Proceedings, vol. 1060, pp. 269–272. American Institute of Physics, USA (2008). doi:10.1063/1.3037069
11. Starosolski, R.: New simple and efficient color space transformations for lossless image compression. J. Vis. Commun. Image Represent. **25**(5), 1056–1063 (2014). doi:10.1016/j.jvcir.2014.03.003
12. Starosolski, R.: Application of reversible denoising and lifting steps to DWT in lossless JPEG 2000 for improved bitrates. Sig. Process. Image Commun. **39**(A), 249–263 (2015). doi:10.1016/j.image.2015.09.013
13. Strutz, T.: Adaptive selection of colour transformations for reversible image compression. In: Proceedings of the 20th European Signal Processing Conference (EUSIPCO 2012), pp. 1204–1208 (2012)
14. Strutz, T.: Multiplierless reversible colour transforms and their automatic selection for image data compression. IEEE Trans. Circuits Syst. Video Technol. **23**(7), 1249–1259 (2013). doi:10.1109/TCSVT.2013.2242612
15. Taubman, D.S., Marcellin, M.W.: JPEG2000 Image Compression Fundamentals, Standards and Practice. The Springer International Series in Engineering and Computer Science, vol. 642. Springer, New York (2004). doi:10.1007/978-1-4615-0799-4
16. Weinberger, M.J., Seroussi, G., Sapiro, G.: The LOCO-I lossless image compression algorithm: principles and standardization into JPEG-LS. IEEE Trans. Image Process. **9**(8), 1309–1324 (2000). doi:10.1109/83.855427

# Daily Urban Water Demand Forecasting - Comparative Study

Wojciech Froelich[(⊠)]

Institute of Computer Science, University of Silesia, Sosnowiec, Poland
wojciech.froelich@us.edu.pl

**Abstract.** There are many existing, general purpose models for the forecasting of time series. However, until now, only a small number of experimental studies exist whose goal is to select the forecasting model for a daily, urban water demand series. Moreover, most of the existing studies assume off-line access to data. In this study, we are confronted with the task to select the best forecasting model for the given water demand time series gathered from the water distribution system of Sosnowiec, Poland. In comparison to the existing works, we assume on-line availability of water demand data. Such assumption enables day-by-day retraining of the predictive model. To select the best individual approach, a systematic comparison of numerous state-of-the-art predictive models is presented. For the first time in this paper, we evaluate the approach of averaging forecasts with respect to the on-line available daily water demand time series. In addition, we analyze the influence of missing data, outliers, and external variables on the accuracy of forecasting. The results of experiments provide evidence that the average forecasts outperform all considered individual models, however, the selection of the models used for averaging is not trivial and must be carefully done. The source code of the preformed experiments is available upon request.

**Keywords:** Forecasting water demand · Comparative study · Averaging forecasts

## 1 Introduction

Forecasting urban water demand is required for cost-effective, sustainable management and optimization of urban water supply systems. Numerous papers have been already devoted to the prediction of daily water demand. A set of rules based on a rough set methodology has been used to predict water demand one day ahead [3]. Linear regression and artificial neural network were reported in [32]. A non-linear regression was applied in [44]. A hybrid methodology combining feed forward artificial neural networks, fuzzy logic, and genetic algorithm has been investigated [31]. A comparison of multiple linear and nonlinear regression, auto-regressive integrated moving average, and two types of artificial neural network applied for urban water demand forecasting has been examined [1]. Dealing

© Springer International Publishing Switzerland 2016
S. Kozielski et al. (Eds.): BDAS 2016, CCIS 613, pp. 633–647, 2016.
DOI: 10.1007/978-3-319-34099-9_49

with seasonality in water demand data has been described in [15]. The application of Bayesian networks to the forecasting of water demand has been proposed in our recent studies [16,27]. The theory of chaos is an alternative approach to the forecasting of water demand [24,43]. An overview of existing water demand forecasting papers was made in [13,33]. The current status of time series analysis in hydrological sciences was presented in book form [25].

Besides the numerous investigations of individual forecasting methods, the literature suggests that forecast combinations often produce better results [40].

Linear programming [35], fuzzy approach [42], linear and logarithmic combinations [41] can be used for combining forecasts. A review of numerous empirical studies led to the recommendation to average at least 5 forecasting methods; however, it has also been concluded that increasing the number does not bring any benefit [4]. A review of methods for combining forecasts has been performed [9].

To the best of our knowledge, there are only two existing attempts to combine forecasts of water demand. The least squares method was applied to integrate the results of different forecasting models [45]. However, that approach is dedicated to long-term forecasting, i.e., analyzing the influence of climate, economic and social factors on water demand. In our paper, the short-term prediction that deals with different characteristics of the time series is investigated. The real data analyzed in [8] demonstrate similar auto-correlation and partial auto-correlation to those considered in this study. However, in [8] the off-line learning is assumed and the data are partitioned into two arbitrary periods for learning and testing respectively.

Many of the existing solutions assume off-line learning of the predictive model while neglecting the possibility offered by existing technology to deliver data in real time. The retraining of the predictive model as new data arrive has been rarely considered in [2]. In our study, on-line learning forecasting is assumed.

In spite of many studies, it is still not clear whether or not, and under what conditions, the use of additional explanatory variables related to weather brings any benefit. The acquisition and forecasting of meteorological data requires additional effort and cost, e.g., purchasing and maintaining at particular places or sensors for measuring rainfall. Many papers fail to examine whether such efforts are worth undertaking and to what extent they could improve the accuracy of predictions.

In this paper we address the aforementioned deficiencies and the requirements imposed by the undertaken project [26]. We perform a systematic comparative study of different forecasting methods by taking into account: statistical features of data, nonlinearity, influence of missing data and outliers on the accuracy of prediction, daily updating of the predictive model, and the influence of weather-related variables on the efficiency of prediction. In all of the performed experiments, scale-independent forecasting errors are calculated. In this way, our comparative study gives a clear reference point for further research.

The remainder of this paper is organized as follows. We present a brief theoretical background of time series forecasting in Sect. 2. Prepossessing and the analysis of source data is presented in Sect. 3. The results of comparative experiments on forecasting models are described in Sect. 4. Section 5 concludes this article.

## 2    Time Series and Forecasting Models

Let $x \in \Re$ be a real-valued variable and let $t \in [1, 2, \ldots, n]$ denotes a discrete time scale, where $n \in \aleph$ is its length. The values of $x(t)$ are observed over time. A time series is denoted as $\{x(t)\} = \{x(1), x(2), \ldots, x(n)\}$ [37]. Let us assume that $\{x(t)\}$ is known, constituting the available historical data. The goal of forecasting is to calculate $x'(t+1) = M(\{x(t)\})$, where: $x'(t+1)$ denotes the prediction and $M$ is the applied predictive model [7,38]. The main issue is to find the most effective forecasting model $M$ in terms of forecasting accuracy.

In this study, we assume that the predictive model may change with time, being retrained as the length of the collected time series grows. For that reason, the considered predictive model changes in time as $M(t)$; in fact, it is learned at every time step $t$ and then is exploited for the single-step ahead forecasting.

### 2.1    Temporal Partitioning of Data

The partitioning of time series to the learning and testing parts is an important issue determining finally obtained forecasting accuracy [7,37]. Let us denote $W$ as a period of time series starting at time $start(W)$ and finishing at time $end(W)$. The most straightforward possibility is to use all available data for learning. In such case, the idea of the growing window is applied for which the $start(W) = 1$, while $end(W)$ is always the last available step of time series. The length of the growing window $length(W) = end(W) - start(W) + 1$ increases as the amount of data increases. An exemplary growing window is shown in Fig. 1a. As the time flows, the growing window delivers an increasing amount of data to the learning process, enabling the capture of more knowledge by the model $M(t)$.

As the length of the time series grows, it becomes more difficult to fit the model to the entire data strand. For this reason, one may consider limiting the

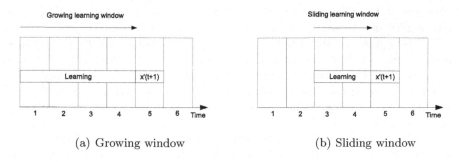

(a) Growing window          (b) Sliding window

**Fig. 1.** Partitioning of time series

learning data only to those data collected recently. The concept of the sliding window assumes that both $start(W)$ and $end(W)$ move as the time flows and the length of the sliding window $length(W)$ is constant. The example of the sliding window is shown in Fig. 1b.

## 2.2    Forecasting Errors

The error for any single forecast is calculated as $e(t) = x'(t) - x(t)$ [37]. There are numerous methods for measuring aggregated forecasting errors over a longer period of time. In this study, we decided to use only absolute percentage error (MAPE) as given by formula (1). We chose MAPE because it is a simple, easy to interpret, scale-independent error measure. The efficiency of forecasts is measured independent of the absolute values of the considered time series.

$$MAPE = \frac{1}{n} \sum_{t=1}^{n} |\frac{e_t}{x_t}| \times 100\,\% \tag{1}$$

## 2.3    An Overview of the State-of-the-art Predictive Models

For the purposes of this study, we selected a number of state-of-the art methods and checked their performance for the considered problem. A brief review of these methods is presented below [7,14,38]. The implementation of all methods has been made using the following tools: (1) the package related to programming language and software environment for statistical computing R [34], (2) the toolbox KNIME enabling to perform numerous data mining and time series forecasting experiments [22].

- **Naive** - The naive approach is used as a reference method for time series forecasting. It assumes that the forecast is the same as the previously observed value of time series, i.e., $x'(t+1) = x(t)$. In this case the model $M$ is obviously not trained.
- **Naive7** - During data prepossessing, it was discovered that the considered time series demonstrates weekly seasonality (the seasonality period is 7 days), e.g., the water demand during weekends is lower than during working days. As suggested in [37], we decided to check the accuracy of the so called naive seasonality, in our case the naive method with the lag 7, i.e., $x'(t+1) = x(t-6)$.
- **ES** - The standard exponential smoothing model is usually an effective method for stationary time series without trend and seasonality and can be applied for short-term forecasting [17]. In this study, the exponential smoothing model was learned using the function $ets$ from the 'forecast' package [18] with parameters indicating the traditional exponential additive model without trend and seasonality.
- **HW** - The extended version of the exponential smoothing is the Holt-Winters method. It takes both trend and seasonality into account within the data. The smoothing parameters $\alpha, \beta, \gamma$ refer to the estimated level, slope, and seasonal

effect, respectively. In this study, the HW model was learned using the function *ets* from the 'forecast' package [18] with the automatic adjustment of parameters.

- **ARIMA** - The Auto-regressive Integrated Moving Average (ARIMA) model is a linear model that can be used for both stationary and nonstationary time series [7]. In the case of nonstationarity detected in the data, the differentiation process is invoked and the differentiated data are predicted. After the prediction is made, an integration process occurs. To fit the ARIMA model to the data and adjust the parameters automatically, we used the function 'auto.arima' from the package 'forecast.' [18]. The applied 'auto.arima' also recognizes the seasonality in the data and implements the seasonal version of ARIMA. The extension version takes into account additional exogenous variables and is called **ARIMAX**.

- **ARFIMA** - The Auto-regressive Fractionally Integrated Moving Average model is dedicated to time series in which auto-correlation is recognized for far lags. It is a generalization of the ARIMA model and allows for long memory [14]. To the best of our knowledge, the method has been never used for the forecasting of water demand. For our purposes, the 'arfima' function from the R package 'forecast' was used.

- **GARCH** - Generalized Auto-regressive Conditional Heteroscedastic is a nonlinear model. It uses a Quasi-Newton optimizer to find the maximum likelihood estimates of the conditionally normal model. The details of this model are further described in the literature [28]. We used the 'garch' function from the R package for this implementation.

- **LR** - The Linear Regression model is used to relate a scalar dependent variable and one or more explanatory variables. Linear regression has been applied for forecasting [2]. As the linear regression model is not able to detect seasonality in time series, we implemented an extended version of linear regression with seasonal adjustment [17]. Seasonal adjustment is a method that relies on removing the seasonal component from the time series before the forecasting is made. The seasonal component is recognized and then subtracted from the original time series by the calculation $x_a(t) = x(t) - x(t-k)$, where $k \in \aleph$ is the time lag related to seasonality. The resulting seasonally adjusted series $x_a(t)$ are forecasted as series $x'_a(t)$. Afterward, the seasonal component is returned, resulting in the forecast of the original time series $x'(t) = x'_a(t) + x(t-k)$. The linear regression extended by seasonal adjustment is denoted as **LR-SA**. The Konstanz Information Miner (KNIME) was used for the implementation [22].

- **PR** - Polynomial Regression is a form of regression in which the relationship between the independent variable and the dependent variables is modeled as a polynomial. The use of polynomial regression for forecasting has been demonstrated [1,29,44]. The seasonally adjusted PR is denoted as **PR-SA**. The KNIME was used for the implementation.

- **ANN** - Artificial Neural Network is a system of interconnected units known as neurons which can compute output values from inputs. ANNs have already been used for the forecasting of water demand [32]. Although neural networks are able to represent any nonlinear function, it was not clear whether

the removing of seasonality from the time series helps to decrease forecasting errors. For that reason, we decided to verify that doubt experimentally using **ANN-SA** method. The KNIME was used for the implementation.
- **FR** - The Fuzzy Rule-based system can be used to map a set of input variables to the dependent variable. They have been used for forecasting in hydrology [5]. It was also decided to check the **FR-SA** method based on seasonal adjustment. The KNIME was used for the implementation in this case as well.

It is worth noting that the above selected forecasting models can be divided into two groups: (1) naive, naive7, ES, HW, ARIMA, ARFIMA, GARCH and (2) ARIMAX, LR, LR-SA, PR, PR-SA, ANN, ANN-SA, FR, FR-SA. The first group includes purely auto-regressive models dealing with univariate time series, using only the auto-regressive terms during forecasting. The second group includes multi-regressive models that are able to exploit additional explanatory variables (exogenous regressors) such as those related to weather. The goal is to improve the effectiveness of prediction by exploiting additional information.

## 3    Experimental Setup

The main problem with all experiments related to the forecasting of water demand is the lack of data that are not publicly available or licensed. In addition, those data are big [19]. Therefore all experiments performed in this study had to be performed using only a single set of real historical data gathered from the water distribution system of Sosnowiec (Poland). Being aware of that limitation we decided to perform numerous statistical experiments giving in this way the reference for what type of data the obtained results are valid. In spite of using only one data set, the results of experiments can be assumed as representative for all time series with the similar characteristics, i.e., data gathered from the highly industrialized urban regions.

For all univariate models and the ARIMAX method, the R software package was applied [34]. The parameters of these models were adjusted automatically. For the multi-regressive models with the exception of ARIMAX, the KNIME was used for the implementation. For the models marked by upper letters 'SA', we implemented the seasonal adjustment technique. For the LR, LR-SA, PR, and PR-SA models, the parameters were adjusted automatically by the processing nodes available in KNIME (further description of the details is available in the KNIME documentation). For the ANN, ANN-SA, FR, and FR-SA, we adjusted the parameters by trial-and-error. Particularly in the case of ANN, the experiments had to be repeated many times. We found that a single hidden layer with 20 neurons and the RProp training algorithm for multilayer feed-forward network led to the best results [36] with the maximum number of iterations for learning ANNs set to 100. For the FR model, the Mixed Fuzzy Rule Formation was the training algorithm [6] and the min/max norm were selected after numerous trials. The rounding for all calculations was assumed at $10^{-6}$.

## 3.1  Quality of Data

The considered water demand time series involved 2254 days, with the data missing for 78 of them. This was due to brakes in data transmission caused by discharged batteries or other hardware problems in the data transmission channels. This problem posed some difficulty because the missing values were irregularly dispersed across the entire time series. We decided to impute them by linear interpolation using function 'interpNA' from package R.

The second significant issue that we encountered was the occurrence of outliers that were also irregularly distributed across the time series. These were primarily due to pipe bursts and the subsequent extensive water leakage. To filter out the outliers, the commonly used Hampel filter was applied [30]. The parameters of the filter were tuned by trial-and-error to best exclude outliers. Comparative experiments using all three versions of the datasets (dropped missing values, imputed missing values, and filtered outliers) are described in Sect. 4.1.

## 3.2  Statistical Features of the Data

The stationarity of the time series is one of the main factors that determine the applicability of a particular predictive model. First, to test for unit root stationarity, the augmented Dickey Fuller test (ADF) was performed [12]. The resulting low p-value $= 0.05912$ suggested that the null hypothesis of nonstationarity should be rejected. As this result was puzzling, we decided to perform the Kwiatkowski-Phillips-Schmidt-Shin (KPSS) test [23]. The resulting p-value was very close to 0, rejecting the hypothesis of stationarity. To definitively resolve the issue, auto-correlation plots were performed. The pattern shown in Fig. 2a is typical for nonstationary time series with seasonality and trend [10]. As can be noticed in Figs. 2a, b the period of seasonality is 7, that is the interval of the observed peaks. The data were finally recognized as nonstationary.

In addition, we performed non-linearity tests. First, the ARIMA model was fitted to the entire time series and the residuals were checked using the Ljung-Box

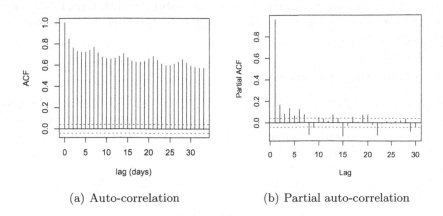

(a) Auto-correlation                 (b) Partial auto-correlation

**Fig. 2.** Correlation coefficients

**Table 1.** Forecasting accuracy

| Method | Missing values removed | Missing values interpolated | Outliers filtered out |
|---|---|---|---|
| Naive | 9.517191 | 8.271936 | 7.539984 |
| Naive 7 | 17.408470 | 14.762260 | 12.856880 |
| ES | 9.624532 | 8.315362 | 7.421575 |
| HW | 9.650111 | 8.371593 | 7.418528 |
| ARIMA | **8.125320** | **7.462123** | **7.352071** |
| ARFIMA | 10.023370 | 8.689852 | 7.410576 |
| GARCH | 52.592810 | 51.735601 | 47.982280 |
| LR | 10.300140 | 8.869507 | 8.026895 |
| LR-SA | 9.799139 | 9.753263 | 9.017286 |
| PR | 15.530600 | 14.259790 | 13.253680 |
| PR-SA | 13.680660 | 11.862580 | 16.797430 |
| ANN | 8.739082 | 8.783395 | 8.688714 |
| ANN-SA | 10.532720 | 10.348380 | 10.320770 |
| FR | 16.767290 | 14.374470 | 13.425140 |
| FR-SA | 15.427220 | 14.124800 | 12.173950 |

portmanteau test [21]. The p-value close to 0 rejected the hypothesis of linearity of the considered time series. Second, the test based on ANNs implemented as 'terasvirta.test' in R package was performed [39]. With a p-value close to 1, this confirmed that the considered time series are nonlinear.

# 4    Experimental Comparison of Forecasting Models

In this section, we present our testing of all forecasting methods, gradually selecting the best possible set up for the experiments and then selecting the best forecasting method.

## 4.1    Forecasting Accuracy

The results of MAPE using 1-step ahead forecasting and a growing window approach are shown in Table 1. The first column is related to the time series after a simple removal of missing data. The second column shows the results obtained after the interpolation of missing data. Finally, the third column shows the results after the filtering of outliers.

The experiments revealed that, independent of the considered column, the ARIMA method with MAPE = 7.352071 was the best individual forecasting method for the data set. It is worthy of note that data preprocessing leads to the decrease of forecasting errors for all models. For this reason, we chose to use only the fully preprocessed data for our further investigations. As shown by the third column of Table 1, the seasonal adjustment (performed for the methods marked by the 'SA' letters) did not help. It was beneficial only in the case of fuzzy rules (FR); however, the obtained MAPE was higher in comparison to the other methods. Surprisingly, the forecasting errors produced by the nonlinear GARCH model were very high; indeed, the highest among all considered models. However, after examining the literature, we would like to note that even when time series are not precisely linear, linear approximations are often adequate [11].

## 4.2  Influence of Weather Variables

In the next experiment, the water demand time series was associated with weather-related explanatory variables containing information on precipitation and daily mean temperature. Weather data were gathered and prepared at the professional weather station located in Sosnowiec at the University of Silesia. We constructed a multivariate time series consisting of the three considered variables: water demand, mean daily temperature, and precipitation. We checked whether such time series are co-integrated, sharing an underlying stochastic trend [11]. To test co-integration, the Johansen test [20] implemented as 'ca.jo' function in 'urca' library of the R package was used. The results given in Table 2 confirmed the null hypothesis of the existence of two co-integration vectors. The test value for the hypothesis $r <= 2$ was above the corresponding critical value at all significance levels: 10 %, 5 % and 1 %. The Johansen test suggested that the inclusion of weather related variables may improve the forecasting accuracy. Therefore, the mean daily temperature and precipitation were added as additional exogenous regressors to the previously considered multi-regressive models.

**Table 2.** Johansen test

|          | Test   | 10 %  | 5 %   | 1 %   |
|----------|--------|-------|-------|-------|
| $r <= 2$ | 18.77  | 10.49 | 12.25 | 16.26 |
| $r <= 1$ | 44.25  | 22.76 | 25.32 | 30.45 |
| $r = 0$  | 370.20 | 39.06 | 42.44 | 48.45 |

In the case of auto-regressive models, only the ARIMA was able to be extended to the so called ARIMAX model. The 'xreg' parameter of the 'auto.arima' function in R was used to include the additional two regressors. All previously selected multi-regressive models were appropriately complemented

The results of the experiment are shown in Table 3. In the first column, the previously obtained results are repeated. In the second column, the results generated by the models complemented with the exogenous variables are shown. Except for the ARIMA/ARIMAX models, there is a decrease of forecasting errors with the most dramatic observed for polynomial regression. However,

**Table 3.** Influence of weather on water demand

| Method | Demand | Demand + Weather |
|---|---|---|
| ARIMA/ARIMAX | **7.352071** | 7.896630 |
| LR | 8.026895 | **7.504034** |
| LR-SA | 9.017286 | **8.642562** |
| PR | 13.253680 | **7.852829** |
| PR-SA | 16.797430 | **11.235515** |
| ANN | 8.688714 | **7.790985** |
| ANN-SA | 10.320770 | **9.5854326** |
| FR | 13.425140 | **12.545179** |
| FR-SA | 12.173950 | **11.442812** |

despite the substantial decrease of errors produced by the complemented multi-regressive models, the errors were not lower than those for the ARIMA, ES, or HW models. The ARIMA model remained the best.

For further experiments, the weather related variables were included in multi-regressive models. To clearly distinguish the new models from those used previously, we complemented their names by the upper letter 'W'. The exception was the ARIMA model that turned out to be better then its extended version ARIMAX; therefore, we decided to investigate only the ARIMA model further.

### 4.3  Variable Length of Learning Period

In previous experiments, we used all of the available, continuously incoming data for learning, i.e., the concept of the growing window was used. This way, the real-world situation was simulated by assuming that the length of the learning period grows gradually as the data become available. This means that the calculation of MAPE reflected the mean errors, including those produced when the length of the learning period was very small and those produced after a long period of learning, e.g., after few years of gathering data from sensors. It was not clear how the actual length of the learning period influenced the forecasting accuracy.

In the next experiment, we evaluated forecasting errors by assuming different constant lengths of the training period. The concept of the sliding window was applied for this purpose. The objective of the experiment was to examine whether the use of shorter learning periods and the sliding window will lead to the decrease of forecasting errors.

It is also worth noting that by assuming different lengths of sliding window the testing period also changes. For example for the sliding window of 20 steps, the testing starts from step 21 of the historical time series, when the length of the sliding window is 100 the testing can start from the step 101. In consequence, the changing length of the testing period influenced slightly the mean value of MAPE produced by the nave model.

**Table 4.** Variable length of the learning period

| Method | 20 | 100 | 200 | 300 |
|--------|-----|------|------|------|
| Naive | 7.541293 | 7.541292 | 7.541288 | 7.541288 |
| Naive7 | 12.905700 | 12.905700 | 12.906101 | 12.906800 |
| ES | **7.371207** | **7.386846** | **7.406689** | **7.416100** |
| HW | 7.782162 | 7.780370 | 7.769285 | 7.653095 |
| ARIMA | 7.955182 | 7.878111 | 7.751845 | 7.600816 |
| ARFIMA | 7.683243 | 8.037972 | 8.08506 | 8.147527 |
| GARCH | 51.192145 | 48.168288 | 47.123126 | 46.742182 |
| LR-W | 10.228360 | 8.367324 | 8.286895 | 8.221234 |
| LR-SA-W | 13.430100 | 10.17988 | 9.981463 | 9.812342 |
| PR-W | 57.920000 | 12.45256 | 12.125321 | 11.923512 |
| PR-SA-W | 58.778430 | 16.77942 | 10.797430 | 10.252345 |
| ANN-W | 8.686030 | 7.85325 | 8.688714 | 8.764321 |
| ANN-SA-W | 10.771600 | 9.83916 | 10.320770 | 9.7843770 |
| FR-W | 10.915790 | 11.13910 | 11.245283 | 11.192943 |
| FR-SA-W | 10.852734 | 10.73221 | 10.836321 | 10.857395 |

As shown in Table 4, the exponential smoothing method overcame all other methods independent of the applied training period. The ES is the only model for which the errors grow as the learning period increases. The obtained results suggest that the locally working ES is better for shorter learning periods with increasing errors produced for longer learning horizons.

After further experiments with ES, we observed that decreasing the length of the sliding window below 20 leads to a further decrease of the produced errors; however, they begin to increase again when the learning period is shorter than 10. The lowest MAPE (7.367166) was obtained for the learning period of 13; however, this was still higher than that generated previously by the ARIMA using the growing window approach. For further experiments with averaging forecasts, we decided to use the ES-13 model instead of the standard ES.

Due to the substantially higher errors, we declined to further investigate the GARCH model. Similarly, the naive7 method was not used any more as it was not competitive to the simple naive model. As can be seen for multi-regressive models, the seasonal adjustment did not decrease errors below those generated by the basic versions of the methods. For that reason, we declined to further investigate the models LR-SA-W, PR-SA-W, and ANN-SA-W. Furthermore, due to higher errors (MAPE greater than 10), we declined to use the forecasting models based on fuzzy rules, i.e., FR-W and FR-SA-W.

For further experiments we selected the following models: naive, ES-13, HW, ARIMA, ARFIMA, LR-W, PR-W, and ANN-W.

## 4.4  Averaging Forecasts

The first experiment revealed that the averaging of forecasts produced by all considered models surprisingly led to very high MAPE (17.456027). Therefore, we decided to conduct further experiments by selecting different subsets of 5 forecasting models (suggested as the minimum number to average in the literature [4]).

Linear correlation was calculated for all pairs of the residual errors produced by the considered models; function 'cor' from the R package was applied. The objective was to find any negatively correlated series of errors that could be mutually compensated by averaging the corresponding forecasts. The results (rounded to $10^{-2}$) are presented in Table 5. To best present our results in Table 5, ARIMA and ARFIMA are shortened to AR and ARF, respectively. Unfortunately, as shown in Table 5, no negatively correlated errors were found to help select appropriate models to combine. Instead, we recognized two groups of models with highly positively correlated produced errors. The first group includes: naive, ES-13, HW, AR, and ARF. The second group contains LR-W, PR-W, and ANN-W. Further investigation of this phenomenon falls outside of the scope of this study. We performed the averaging of forecasts separately for the first and second groups of models. We also tried to combine models from both groups. After numerous trials it became clear that the best results were obtained by averaging forecasts produced only by the first group of models. The inclusion of any forecasts from the second group only increased the errors.

The best result ($MAPE = 7.181795$) was obtained by averaging forecasts generated by the following 5 models: naive, ES-13, HW, ARIMA, and ARFIMA. This result is better than the best individual model ARIMA ($MAPE = 7.352071$).

**Table 5.** Correlation of forecasting errors

|        | Naive | ES-13 | HW   | AR   | ARF  | LR-W | PR-W | ANN-W |
|--------|-------|-------|------|------|------|------|------|-------|
| Naive  | 1.00  | 0.95  | 0.98 | 0.91 | 0.92 | 0.01 | 0.02 | 0.01  |
| ES-13  | 0.95  | 1.00  | 0.98 | 0.95 | 0.95 | 0.06 | 0.06 | 0.05  |
| HW     | 0.98  | 0.98  | 1.00 | 0.94 | 0.94 | 0.04 | 0.04 | 0.03  |
| AR     | 0.91  | 0.95  | 0.94 | 1.00 | 0.97 | 0.03 | 0.04 | 0.03  |
| ARF    | 0.92  | 0.95  | 0.94 | 0.97 | 1.00 | 0.04 | 0.05 | 0.04  |
| LR-W   | 0.01  | 0.06  | 0.04 | 0.03 | 0.04 | 1.00 | 0.93 | 0.89  |
| PR-W   | 0.02  | 0.06  | 0.04 | 0.04 | 0.05 | 0.93 | 1.00 | 0.90  |
| ANN-W  | 0.01  | 0.05  | 0.03 | 0.02 | 0.04 | 0.89 | 0.90 | 1.00  |

# 5   Conclusions

In this paper we undertook an effort to obtain the best possible accuracy of 1-day-ahead forecasting of urban water demand time series. Daily updating of the forecasting model was assumed.

After the initially performed tests, it became clear that the daily water demand time series gathered from the urban area of the city are nonstationary and nonlinear. The raw time series includes both missing data and outliers that influence the effectiveness of forecasting. Therefore, we recommend the input of missing data and the filtering of outliers.

After numerous comparative experiments, it turned out that the ARIMA is the best individual, state-of-the art model for the considered time series. The weather related variables improve the forecasting accuracy, but only for those obtained by multi-regressive methods. Therefore, in the case of urban water demand gathered from an industrial city, the use of multi-regressive models with the exogenous weather variables does not lead to overall best results and is not recommended.

We confirm that the averaging of forecasts leads to improved forecasting accuracy; however, the selection of the models used for averaging is not trivial and must be carefully done.

**Acknowledgments.** The work was supported by ISS-EWATUS project which has received funding from the European Union's Seventh Framework Programme for research, technological development and demonstration under grant agreement No. 619228.

# References

1. Adamowski, J., Fung Chan, H., Prasher, S.O., Ozga-Zielinski, B., Sliusarieva, A.: Comparison of multiple linear and nonlinear regression, autoregressive integrated moving average, artificial neural network, and wavelet artificial neural network methods for urban water demand forecasting in montreal, canada. Water Resour. Res. **48**(1) (2012)
2. Alvisi, S., Franchini, M., Marinelli, A.: A short-term, pattern-based model for water-demand forecasting. J. Hydroinform. **9**(1), 39–50 (2007)
3. An, A., Chan, C.W., Shan, N., Cercone, N., Ziarko, W.: Applying knowledge discovery to predict water-supply consumption. IEEE Expert **12**(4), 72–78 (1997)
4. Armstrong, J.S.: Principles of Forecasting: A Handbook for Researchers and Practitioners, vol. 30. Springer, Heidelberg (2001)
5. Bardossy, A.: Fuzzy rule-based flood forecasting. In: Abrahart, R.J., See, L.M., Solomatine, D.P. (eds.) Practical Hydroinformatics. Water Science and Technology Library, vol. 68, pp. 177–187. Springer, Heidelberg (2008)
6. Berthold, M.R.: Mixed fuzzy rule formation. Int. J. Approx. Reason. **32**(23), 67–84 (2003)
7. Brockwell, P., Davis, R.: Introduction to Time Series and Forecasting. Springer, New York (2002)

8.  Caiado, J.: Performance of combined double seasonal univariate time series models for forecasting water demand. J. Hydrol. Eng. **15**(3), 215–222 (2010)
9.  Clemen, R.T.: Combining forecasts: a review and annotated bibliography. Int. J. Forecast. **5**(4), 559–583 (1989)
10. Cortez, P.C., Rocha, M., Neves, J.: Genetic and evolutionary algorithms for time series forecasting. In: Monostori, L., Váncza, J., Ali, M. (eds.) IEA/AIE 2001. LNCS (LNAI), vol. 2070, pp. 393–402. Springer, Heidelberg (2001)
11. Cowpertwait, P.S.P., Metcalfe, A.V.: Introductory Time Series with R, 1st edn. Springer Publishing Company, Incorporated, New York (2009)
12. Dickey, D.A., Fuller, W.A.: Distribution of the estimators for autoregressive time series with a unit root. J. Am. Stat. Assoc. **74**(366), 427–431 (1979)
13. Donkor, E., Mazzuchi, T., Soyer, R., Alan Roberson, J.: Urban water demand forecasting: review of methods and models. J. Water Resour. Plan. Manag. **140**(2), 146–159 (2014)
14. Ellis, C., Wilson, P.: Another look at the forecast performance of ARFIMA models. Int. Rev. Financ. Anal. **13**(1), 63–81 (2004)
15. Froelich, W.: Dealing with seasonality while forecasting urban water demand. In: Neves-Silva, R., Jain, L.C., Howlett, R.J. (eds.) Intelligent Decision Technologies, pp. 171–180. Springer International Publishing, Switzerland (2015)
16. Froelich, W.: Forecasting daily urban water demand using dynamic gaussian Bayesian network. In: Kozielski, S., Mrozek, D., Kasprowski, P., Malysiak-Mrozek, B., Kostrzewa, D. (eds.) Beyond Databases, Architectures and Structures, pp. 333–342. Springer International Publishing, Switzerland (2015)
17. Gardner, E.S., Mckenzie, E.: Forecasting trends in time series. Manag. Sci. **31**(10), 1237–1246 (1985)
18. Hyndman, R.J.: forecast: Forecasting functions for time series and linear models (2015). http://github.com/robjhyndman/forecast (r package version 6.2)
19. Jach, T., Magiera, E., Froelich, W.: Application of HADOOP to store and process big data gathered from an urban water distribution system. Procedia Eng. **119**, 1375–1380 (2015)
20. Johansen, S.: Estimation and hypothesis testing of cointegration vectors in gaussian vector autoregressive models. Econometrica **59**(6), 1551–1580 (1991)
21. Jung, L., Box, G.: On a measure of lack of fit in time series models. Biometrika **65**(2), 297–303 (1978)
22. KNIME: Professional open-source software. http://www.knime.org
23. Kwiatkowski, D., Phillips, P.C., Schmidt, P., Shin, Y.: Testing the null hypothesis of stationarity against the alternative of a unit root: how sure are we that economic time series have a unit root? J. Econom. **54**(13), 159–178 (1992)
24. Liu, J.Q., Zhang, T.Q., Yu, S.K.: Chaotic phenomenon and the maximum predictable time scale of observation series of urban hourly water consumption. J. Zhejiang Univ. Sci. **5**(9), 1053–1059 (2004)
25. Machiwal, D., Jha, M.K.: Hydrologic Time Series Analysis: Theory and Practice. Springer, The Netherlands (2012)
26. Magiera, E., Froelich, W.: Integrated support system for efficient water usage and resources management (ISS-EWATUS). Procedia Eng. **89**, 1066–1072 (2014)
27. Magiera, E., Froelich, W.: Application of Bayesian networks to the forecasting of daily water demand. In: Neves-Silva, R., Jain, L., Howlett, R. (eds.) Intelligent Decision Technologies, pp. 385–393. Springer International Publishing, Switzerland (2015)

28. MatíAs, J.M., Febrero-Bande, M., GonzáLez-Manteiga, W., Reboredo, J.C.: Boosting garch and neural networks for the prediction of heteroskedastic time series. Math. Comput. Model. **51**(3–4), 256–271 (2010)
29. Mavromatidis, L.E., Bykalyuk, A., Lequay, H.: Development of polynomial regression models for composite dynamic envelopes thermal performance forecasting. Appl. Energy **104**, 379–391 (2013)
30. Pearson, R.K.: Outliers in process modeling and identification. IEEE Trans. Control Syst. Technol. **10**(1), 55–63 (2002)
31. Pulido-Calvo, I., Gutirrez-Estrada, J.C.: Improvedfrigation water demand forecasting using a soft-computing hybrid model. Biosyst. Eng. **102**(2), 202–218 (2009)
32. Pulido-Calvo, I., Montesinos, P., Roldn, J., Ruiz-Navarro, F.: Linear regressions and neural approaches to water demand forecasting in irrigation districts with telemetry systems. Biosyst. Eng. **97**(2), 283–293 (2007)
33. Qi, C., Chang, N.B.: System dynamics modeling for municipal water demand estimation in an urban region under uncertain economic impacts. J. Environ. Manag. **92**(6), 1628–1641 (2011)
34. R: The R foundation for statistical computing. http://www.r-project.org
35. Reeves, G.R., Lawrence, K.D., Lawrence, S.M., Guerard Jr., J.B.: Combining earnings forecasts using multiple objective linear programming. Comput. Oper. Res. **15**(6), 551–559 (1988)
36. Riedmiller, M., Braun, H.: A direct adaptive method for faster backpropagation learning: the RPROP algorithm. In: Proceedings of the IEEE International Conference on Neural Networks (ICNN), pp. 586–591 (1993)
37. Shmueli, G.: Practical Time Series Forecasting, 2nd edn. LLC, New York (2011). Statistics.com
38. Shumway, R.H., Stoffer, D.S.: Time Series Analysis and Its Applications. Springer, New York (2000)
39. Tersvirta, T., Lin, C.F., Granger, C.W.J.: Power of the neural network linearity test. J. Time Ser. Anal. **14**(2), 209–220 (1993)
40. Timmermann, A., Codes, J.: Forecast combinations. In: Handbook of Economic Forecasting, pp. 135–196. Elsevier Press (2006)
41. Wallis, K.F.: Combining forecasts: forty years later. Appl. Financ. Econ. **21**(1–2), 33–41 (2011)
42. Xiao, Z., Gong, K., Zou, Y.: A combined forecasting approach based on fuzzy soft sets. J. Comput. Appl. Math. **228**(1), 326–333 (2009)
43. Xizhu, W.: Forecasting of urban water demand based on Chaos theory. In: Control Conference, CCC 2007, pp. 441–444, July 2007. (Chinese)
44. Yasar, A., Bilgili, M., Simsek, E.: Water demand forecasting based on stepwise multiple nonlinear regression analysis. Arab. J. Sci. Eng. **37**(8), 2333–2341 (2012)
45. Yin-shan, X., Ya-dong, M., Ting, Y.: Combined forecasting model of urban water demand under changing environment. In: 2011 International Conference on Electric Technology and Civil Engineering (ICETCE), pp. 1103–1107, April 2011

# Database Index Debug Techniques:
# A Case Study

Andrey Borodin$^{(\boxtimes)}$, Sergey Mirvoda, and Sergey Porshnev

Department of Radio Electronics for Information Systems,
Ural Federal University, Yekaterinburg, Russia
amborodin@acm.org, sergey@mirvoda.com, s.v.porhsnev@urfu.ru
http://urfu.ru

**Abstract.** The index corruption may lead to serious problems ranging from the temporary system outage to the loss of sensitive data. In
this article we discuss the techniques that we found helpful in assuring
the data index consistency during the development of specific indexing
algorithms for a multidimensional BI system featuring both OLAP and
OLTP aspects. The use of the techniques described in this article from
the very beginning of the project development helped to save sufficient
resources during the development and debugging.

**Keywords:** Data · Database · Index · Quality assurance · Durability ·
Debug

## 1 Introduction

Traditionally, database engines are divided into two classes: transactional
processing [10] and analytical processing [11]. The first class is intended to provide capabilities for the data collection and storage, while the second is intended
to transform, aggregate, and interpret data in various ways. Our team was tasked
with the creation of a database engine for the specific business intelligence calculations with multidimensional data [1,9]. Also, this data had a slowly changing
structure, but was rapidly updated in an unpredictable way. Moreover, calculations themselves were changing this data.

We had a fully functional system based on MS SQL Server database engine.
The only disadvantage of this system was performance. Deep profiling revealed
that performance was affected mostly by multidimensional data aggregation calculations. Grouping dimensions of these calculations we represented by hierarchical classifiers. We considered three strategies to solve the performance problem:

1. Use Microsoft SQL Server Analysis Services (SSAS). This option relies on
   similarity of data aggregation in SSAS with aggregation in our system.
2. Use PostgreSQL multidimensional GiST [15].
3. Implement middleware database engine as analytical and transactional component with MS SQL server as persistent backing storage.

© Springer International Publishing Switzerland 2016
S. Kozielski et al. (Eds.): BDAS 2016, CCIS 613, pp. 648–658, 2016.
DOI: 10.1007/978-3-319-34099-9_50

We found that the existing database engines did not satisfy our requirements. SSAS parent-child hierarchies were not recommended as main type in highly multidimensional cubes. Also transaction control and analytics on most recent data were required. PostgreSQL GiST performance was not good enough at the time (though it was improved eventually). Also GiST usage required development of our custom datatypes (elements of hierarchy) in Postgres. That is why we decided to implement the required features in our middleware database engine using specifically selected algorithms and data structures. We intentionally left transaction logging and recovery functionality to MS SQL Server, though we had to implement fully functional database index structures to deal with big amount of data (hundreds of gigabytes at the start of a project).

We revisited various multidimensional data access methods, adapted them accordingly, and proceeded to the production implementation. It turned out that the production implementation of these algorithms consume incomparably more resources than the selection of algorithms, evaluation of their performance, and verification of their applicability.

Most of resources were used during the process of quality assurance, specifically making sure that indices are not corrupted during the concurrent updates interleaved with aggregating query computations.

In this article we describe the techniques that proved to be of use during the debug process and helped us to ensure the high production quality of the database engine being developed.

## 2    Data Structures and Algorithms

The database test and quality assurance technologies are of certain interest for many developers [21,24]. But the database engines development, except for some famous open source projects (like Postgres), is performed by commercial companies. Thus, the published information usually represents the intersection of what developers want to publish and what can be published without potential risks for the intellectual property of the software companies. Nevertheless, most of the techniques can be deduced by the adaptation of common debug strategies, tricks, and ideologies [18,23,25] with adaptation to specifics of database algorithms.

### 2.1    Paged Memory Buffer

Database algorithms are designed to handle the amount of data much larger than the amount of RAM available. Thus, most of these algorithms operate in the specific memory buffer. The buffer works with the data chunks called pages. The data page is identified by the page address and can store the data in the external memory. The page can also be viewed and edited through the buffer. The page usually has a fixed size in bytes.

The buffe's interface has the functions allowing:

1. To allocate the page and produce the page address.
2. To fix the page in the buffer and return the pointer to accommodating frame.
3. To unfix the page from the buffer.
4. To deallocate the page from the storage.
5. To ensure the current page state is flushed to the disk storage.

The pages that are not fixed in the buffer frames are the candidates for eviction, if the frame is necessary to accommodate the unframed page. The page can be fixed in the buffer multiple times requiring multiple unfixed calls to mark it suitable for the frame eviction.

This strategy is coupling the concept of external storage and cache memory buffer. This is not always the case, *e.g.* in the PostgreSQL database, the shared memory buffer serves for multiple file storages. However, in the context of the debug techniques overview, we consider one memory buffer attached to the one and only storage.

## 2.2   Segregation List

One of the very common, though auxiliary, tasks in DBMS is the allocation of temporary records and maintaining temporary unordered record sets. These record sets support only the data appending and data traversal. As a rule, the data of these record sets is not persistent, but the disk flushes may be required anyway, due to the insufficient buffer memory. The common way to organize the record set is the segregation list data structure [13]. Usually segregation list is used for temporary data storage.

The segregation list is optimized for the fast data allocation with the compact used pages count. This is done by providing a way to quickly find the page with the sufficient storage for a new record.

The root page of the record set contains the starting page address of the chains of pages with the specific degree of the free storage. *E.g.*, one chain of pages with more than 127 free bytes, one chain of pages with more than 255 bytes, one chain of pages with more than 511 bytes, *etc.* Different strategies can be used for the degree calculation.

When a new record arrives, the structure finds the chain with the minimum suitable space free, extracts one page from the chain, allocates the record in it, enchains the modified page to a new corresponding chain, fixing the starts of both modified chains. The data structure can provide the efficient multithreaded access for appending the records of sufficiently different sizes, but it is a rare occasion.

## 2.3   B-Tree

A B-tree is a well-known, versatile, and ubiquitous database structure [2,13]. Technically, a database can operate solely on a B-tree. Usually B-trees are used to implement key-values associative containers, to implement join operations, to retrieve ranges of data and to store data in specific sorting order.

A traditional B-tree organizes data into a tree using a balanced tree data structure. Each node of a tree has a data page storing the whole information about the node. The leaf pages contain data, while the internal pages contain only keys and page addresses of the underlying pages. The main handle of the structure is the root page, which can be either internal or data page (if the stored data is small enough to fit in one page). A B-tree maintains one-dimensional order of keys and allows to efficiently find a record by key, extract the range of records with keys in a continuous range, merge indices, join an index with other indices, delete records from an index, bulk insert records into an index.

Our system uses a Multiversion B-tree [4] allowing to perform multiversion concurrency control. This means, that actual data can exist in a different state for different user sessions. If the data is changed in one session, each changed page is copied and marked as a private page. When the session commits data, the old root page is replaced by a new root page.

### 2.4   R-Tree Variants

An R-tree is a structure common for spatial databases. While a B-tree organizes the hierarchy of indexed data by arranging the continuous ranges of ordered data keys, an R-tree provides the structure of hierarchically arranged MBRs (minimum bounding rectangles) in a multidimensional space of independent keys. There are a lot of R-tree variants [12] differing mostly in the insertion algorithm, specifically in the way, in which a node (page) is split into two nodes, when it is overflowed. The R*-tree variant [5,16] is generally considered to be the most efficient one.

Usually R-trees are used to implement spatial search, data aggregations by window queries (multidimensional range queries) and nearest neighbor search.

Our database uses the R*-tree variant combined with the MVCC [13] approach.

Also, due to the frequent massive indices rebuild with a new data from an external source we had to implement the bulk-load algorithm. We used the VAM-split buffering algorithm [20]. We suppose it is more CPU-effective and has the same CPU efficiency as PostgreSQL buffered GiST build [15], though we have no actual data: PostgreSQL GiST did not exist at that time, and currently this part of a system is neither a bottleneck nor a bug source.

## 3   Sequential Test Script Techniques

Due to the test-driven development [3] methodology applied to the whole project, the main functionality of the indexing subsystem was covered by the unit tests. These tests were performing the basic strategies:

1. Various CRUD operations.
2. Bulk-load index construction.
3. Simultaneous multi-index CRUD.

4. Starts and commits of versioned indices for MVCC via session.
5. Complex scenarios based on the existing business-logic calculations.

Since our team had production implementation (based on MS SQL Server and Oracle RDMBS) deployed to the clients, we actively employed the 5th strategy applied to their data transformations and calculations. This allowed us to compare the expected and actual results of the test calculations.

The described tests turned out to be good at the detection of the bug existence. But we found it surprisingly difficult to spot the actual source of a bug. The index data structures could be corrupted but stay silent in memory for a serious length of time. This corruption could be frequently persisted to disk and brought back to the buffer, even read and overwritten in some cases. The actual detected malfunction of difficult bugs always stayed in the uncertain past of a data structure. Thus, we employed some techniques that allowed us to detect the bugs earlier, to trace the past of a structure, and to mitigate the bugs in the production environment.

### 3.1   .NET Environment Capabilities

Our database engine is implemented on the top of Microsoft .NET framework. This is the result of many preconditions and considerations, including the existence of the legacy code, performance comparisons, development cost tradeoffs, and business preferences.

The .NET runtime (CLR) prescribes the garbage-collected way of the memory management [22]. The use of this feature makes the programming significantly easier, but it is not suitable for the amounts of fragmented data bigger than the server's RAM. That is the case the databases have been dealing with for a long time. That is why, we effectively used C-style programming under C# language in the realization of low-level routines like the indexing or cursors. In general, this means the following:

1. Minimum managed allocations of memory. The managed memory allocations can serve as a way for the composition of temporary variables, but not as long-living allocations.
2. Long-living allocations are processed through the record sets (and, effectively, through the buffer memory, controllably swappable to the disk).

Unmanaged memory allocations have managed memory handles: *Page class, Record class, DataValue class, Index class, etc.* .NET provides the standard way to treat the objects with the controllable lifetime via the *IDisposable* interface. These objects also implement finalizers, suppressed by *Dispose()* call. A finalizer call is the first indication of a bug: an object with the controllable lifetime was not disposed properly. In the production environment this is the case to log for a possible investigation. In the development environment this is the case for marking the build as buggy. A finalizer call should not be performed in any case: a calculation termination request, a disk malfunction, a stack overflow (which is a bug itself), insufficient memory resources, and an internal error. Routine allocating resources has to dispose them, so that the system could proceed functioning.

## 3.2  Global Object Identification

Every managed handle of an unmanaged (in .NET terms) resource (a page, a record, a cursor, a session) has two global numeric incremental identifiers:

1. *Id* of the managed instance of a handle. Different *Ids* for different requests for a page with the same address, an event residing in same frame of the memory buffer.
2. *Id* of the persistent object within an object container. *Id* of the record set, *Id* of the record in the record set, *Id* of the page (the page address).

These identifiers should be read easily (a single decimal value), naturally indicate the order of objects creation, and be allocated deterministically. When the bug is found, the test scenario is reproduced until the corrupted object creation. Then manipulations with objects are inspected by the programmer.

## 3.3  Consistency Checks

The layout of the record with the data values is usually simple. The fixed-length data types are standardized by ABI. The only actual questions about storing the variable length data are the encoding/collation of strings and the reasons to use zero-terminated strings [14]. Both questions have adequate coverage in reliable scientific sources. The specific format of strings could possibly be used by the inverted indices like Postgres GIN [17].

The layout of an index page is, on the contrary, sophisticated and error prone. The R*-tree aims to minimize the amortized disk access count $DA$, so that:

$$DA_{search} \sim O(R + logN),\tag{1}$$

where $N$ is the size of a data set and $R$ is the size of a query result. But it can be shown that CPU costs are minimized, when the tree fan-out $f \to e$ [7]. This effectively means 3 records on the index page, and that is unsuitable. In order to optimize the CPU efficiency of searches, it is necessary to use the sqrt-decomposition or even maintain a small R-tree inside a page (an R-tree node). Both these improvements lead to the complex page layout.

This is why, page consistency checks are needed. The main points of a consistency check:

1. The page header and footer (if exist) are valid and correspond to the page type, location, and status.
2. Every offset on a page refers to an item inside the page. Offsets added to the record sizes fall within the page boundaries.
3. Every page address on a page refers to a non-deleted page.
4. Record keys fall within the local index ranges (if the information of the parent page is available).
5. Statistical information is consistent and sane (free memory offsets and sizes, all counts are positive and nonzero).

These checks can be performed only after the page modification, since they significantly slow down the system operations. However, in the debug build, it is reasonable to be able to switch them on and off depending on the operation performed.

This approach enables the programmer to reach the potentially corrupted object with almost-production system performance and then switch the system to the deep inspection mode in order to detect the earlier bug signs.

### 3.4   Dirty Pages Monitoring

When a buffer memory frame maps to the integral count of operational system's (OS) memory pages, the OS flag of 'dirty' page can be employed. Most OS APIs provide functionality to detect whether the memory was changed or not after a certain point (a function call).

When a frame is occupied by a memory page, the buffer memory manager marks the frame as clear. Index algorithms under the debug build inform the buffer, when (before and after) the page write occurs. Before that, the write buffer checks that a dirty flag is not set by anyone yet. This means that no routine used this frame detrimentally. After that, the write buffer checks that the write actually occurred and resets a dirty flag.

This technique creates a lot of conditionally-compiled code, but, when it is applicable, it guarantees that the corrupted data found on the pages of a particular data structure is produced by the code of that particular structure.

### 3.5   Unfixed Page Write Locks

Another way to isolate the memory from cross-writes (which occur mostly due to the similarity of the page layout and the corruption of the page addresses) is the read/write protection of OS memory pages. Most of index algorithms are designed in such a way that one user session keeps only a handful of pages fixed in the memory buffer. The OS allows to protect the non-fixed frames from reads and writes.

This approach also helps to ensure the correct order of the page fixations and unfixations during the development time.

### 3.6   Buffer Starvation

The similarity of closely allocated pages can be mitigated, for the test purposes, by restraining the size of a buffer.

The limited space purposed to accommodate the big count of pages causes the rotation of data through the persistent storage and the subsequent shuffling of pages in the buffer frames. Although this technique degrades the overall system performance, it does not require the specific software development efforts. Just a configuration change would be enough.

### 3.7    Managed Memory Index Implementation

We found out that the implementation of a managed memory (via garbage collector managed objects) index for the testing purposes requires incomparably less efforts than the implementation of an index in the buffered unmanaged memory.

For the testing purposes, we created the model of several indices in the .NET arrays and objects. It is still prone to errors of the multiversion conflicts resolution, but it is resistant to the memory management errors. The comparison of the state of the index tree pages to that of their managed node objects counterparts allows to detect the small erroneous changes in the tree node.

Actually, this technique is a boosted variation of the page consistency check with an important limitation: the test data have to be significantly smaller than RAM available. When the managed memory mirror-index starts to swap the heap to the disk, the performance degrades on many orders of magnitude.

## 4    Deterministic Parallel Test Scripting

Multithreaded operations bring a lot of entropy into the system. Hard concurrency bugs almost never manifest on the same objects. Nor they are stably exposed to consistency checks. Moreover, by design there is not much to do to make the concurrent bugs reproducible. The following is the description of the technique aimed to mitigate the indeterminism of concurrency bugs.

### 4.1    Script Run

Most of the multithreaded test scenarios are related to the manipulations with multiversion trees, since MVCC is a way of a system to deal with the concurrency. Given a defined multithreaded test scenario supplied with the test data, it is possible to create the check list of operations with the data manipulation objects.

Actually, this means logging every buffer call with its arguments to the interlocked sequential storage, preferably to RAM. But when the data is measured by gigabytes per second of the production build operations, it is viable to send the events flow to another system through the broadband network and apply the variations of RLE compression. It is noteworthy that the latency of the event fixation reduces the chance to reproduce a bug. Other lockable objects have to be logged too.

There is one more important thing to log. The functions spawning threads have to register them with the unique deterministic keys. This is necessary, because the permutations of the thread-spawn events may cause the deadlock during the debug process.

### 4.2    Debug Run

In most cases, when the bug is reproduced during the script run, it can be reproduced during the debug run.

After the log of events is formed, the debug run can lock the early-requested events to the point of the log, when they have to happen. It came out to be useful to employ the priority queue (heap) for awaiting the early-arrived events, when the active thread count peaks reach 64. This is a heuristic observation which does not have a theoretical ground. During the development process, we had to optimize the debug run algorithms, because the latency of the passing timely event also reduced the frequency of the bug reproduction.

### 4.3   Managing Debug Configuration

The employment of the technique requires the constant environment: the script run under x86 system would deadlock under x64 system, another size of the buffer may deadlock the system, heavy consistency checks reduce the frequency of the bug reproduction.

But the test script shall also be independent from the environment to the maximum extent possible. If the test calculations call to *now()* function and use its result as an operand of condition, undeterministic deadlock is possible. Other operations to avoid: reading the system-wide statistics, security checks, file system calls, network operations (except flushing the log buffer), any GUI operations, mutexes, and other IPC tools.

## 5   Analysis

Most described techniques solve two different tasks: justify system behavior as a bug and trace cause of this behavior. These techniques should be applied to collect bug description and conditions. The description must be enough detailed to justify system behavior as erroneous.

This article leaves aside topic of test case generation since it is very system-specific field. Typical set of test cases covers most frequent system use cases and provisioned edge cases.

The combination of all techniques allowed us to deploy the system to production. Currently, we experience on average 1 index corruption case on each server installation in 18 months. Each server is used by approximately 200 users, with concurrent peaks of 50 simultaneous calculations. No data was lost due to indexing the system errors, recovery redundancy handled all the events of errors. All servers are equipped with ECC-registered memory, the probability of the cosmic ray induction of bit flips [19] is mitigated.

This means, that the code still has bugs. But the frequency of their occurrence is covered by SLA.

We estimate that the usage of the debug technique from the beginning of the project development would save about 10 man-years. The total time of the project implementation is estimated to be about 50 man-years, including the requirements specification, research, development, quality assurance, deployment and post-production support.

# 6    Conclusion

There are no methods capable to guarantee the absence of bugs. Even the formal correctness proof only masks the problem and prevents the actual debug [6].

The described set of techniques can be generalized to other domains of the code quality assurance, but most of the techniques are related to the tree indices mapped to the memory buffer. The combination of all techniques would apparently ensure a better result.

This research is the introspective preparation for a future work on PostgreSQL GiST page layout improvements [8].

# References

1. Aksyonov, K., Bykov, E., Aksyonova, O., Antonova, A.: Development of real-time simulation models: integration with enterprise information systems. In: Proceedings of ICCGI, pp. 45–50 (2014)
2. Bayer, R.: Symmetric binary B-trees: data structure and maintenance algorithms. Acta informatica 1(4), 290–306 (1972)
3. Beck, K.: Test-Driven Development: by Example. Addison-Wesley Professional, Boston (2003)
4. Becker, B., Gschwind, S., Ohler, T., Seeger, B., Widmayer, P.: An asymptotically optimal multiversion B-tree. VLDB J. Int. J. Very Large Data Bases 5(4), 264–275 (1996)
5. Beckmann, N., Kriegel, H.P., Schneider, R., Seeger, B.: The R*-tree: an efficient and robust access method for points and rectangles, vol. 19, no. 2. ACM (1990)
6. Bloch, J.: Extra, extra-read all about it: nearly all binary searches and mergesorts are broken. Official Google Research Blog Date. Accessed 2 June 2006
7. Borodin, A.M., Mirvoda, S.G., Porshnev, S.V.: High dimensional data analysis: data access problems and possible solutions. St. Petersburg State Polytechnical University J. Comput. Sci. Telecommun. Control Syst. 6, 59–66 (2013)
8. Borodin, A.: [proposal] improvement of gist page layout. http://www.postgresql.org/message-id/CAJEAwVE0rrr+OBT-P0gDCtXbVDkBBG_WcXwCBK=GHo4fewu3Yg@mail.gmail.com
9. Borodin, A., Kiselev, Y., Mirvoda, S., Porshnev, S.: On design of domain-specific query language for the metallurgical industry. In: BeyondDatabases, Architectures and Structures, pp. 505–515. Springer (2015)
10. Codd, E.F.: The Relational Model for Database Management: Version 2. Addison-Wesley Longman Publishing Co. Inc, Boston (1990)
11. Codd, E.F., Codd, S.B., Salley, C.T.: Providing OLAP (On-Line Analytical Processing) to User-Analysts: an it Mandate, vol. 32. Codd and Date, Reading (1993)
12. Gaede, V., Günther, O.: Multidimensional access methods. ACM Comput. Surv. (CSUR) 30(2), 170–231 (1998)
13. Garcia-Molina, H., Ullman, J.D., Widom, J.: Database System Implementation, vol. 654. Prentice Hall, Upper Saddle River (2000)
14. Kamp, P.H.: The most expensive one-byte mistake. Commun. ACM 54(9), 42–44 (2011)
15. Korotkov, A.: Fast gist index build. https://wiki.postgresql.org/images/0/07/Fast_GiST_index_build.pdf

16. Korotkov, A.: A new double sorting-based node splitting algorithm for r-tree. In: Proceedings of Spring/Summer Young Researchers Colloquium Software Engineering, vol. 5 (2011)
17. Korotkov, A., at al.: Next generation of gin. http://www.sai.msu.su/~megera/postgres/talks/Next20GIN.pdf
18. LeBlanc, T.J., Mellor-Crummey, J.M.: Debugging parallel programs with instant replay. IEEE Trans. Comput. **100**(4), 471–482 (1987)
19. Leray, J.: Effects of atmospheric neutrons on devices, at sea level and in avionics embedded systems. Microelectron. Reliab. **47**(9), 1827–1835 (2007)
20. Manolopoulos, Y., Nanopoulos, A., Papadopoulos, A.N., Theodoridis, Y.: R-trees: Theory and Applications. Springer Science Business Media, Berlin (2010)
21. Mrozek, D., Małysiak-Mrozek, B., Mikołajczyk, J., Kozielski, S.: Database under pressure - testing performance of database systems using universal multi-agent platform. In: Gruca, A., Czachórski, T., Kozielski, S. (eds.) Man-Machine Interactions 3. AISC, vol. 242, pp. 637–648. Springer, Heidelberg (2014)
22. Rahman, M.: CLR memory model. C# Deconstructed, pp. 61–86. Springer, Heidelberg (2014)
23. Shapiro, E.Y.: Algorithmic Program Debugging. MIT Press, Cambridge (1983)
24. Smirnov, K., Chernishev, G., Fedotovsky, P., Erokhin, G., Cherednik, K.: R-tree re-evaluation effort: a report. Technical report (2014)
25. Zeller, A.: Why Programs Fail: a Guide to Systematic Debugging. Elsevier, Philadelphia (2009)

# AI Implementation in Military Combat Identification – A Practical Solution

Łukasz Apiecionek[1], Wojciech Makowski[2(✉)], and Mariusz Woźniak[2]

[1] Institute of Technology, Casimir the Great University in Bydgoszcz,
ul. Chodkiewicza 30, 85-064 Bydgoszcz, Poland
lapiecionek@ukw.edu.pl
[2] TELDAT Sp. z o. o.sp.k., Bydgoszcz, Poland
{wmakowski,mwozniak}@teldat.com.pl

**Abstract.** This paper presents the architecture of a communication system which was implemented in MiG-29 airplanes. This system provides a continuous on-line access to the situational awareness information which is necessary for the pilot. The interoperability of this system with other NATO systems allows to collect and transfer data between them. Artificial Intelligence methods are used to implement and improve this system. This modification enables the system to work faster and increases the situational awareness of the pilot on the battlefield.

**Keywords:** CID · Security · Network · Artificial intelligence

## 1 Introduction

The Polish Air Force has been undergoing intense changes since several years. During this time, the Air Force was modernized through investments in modern technologies, among others by acquiring brand new F-16 aircrafts and introducing numerous other changes which generate many powerful capabilities. Besides the investments in modernity, withdrawing older generation aircraft decreased the number of combat units. New machines, equipped with the systems compatible with NATO standards, allow the connection of the pilot and the aircraft to a network-centric system, present on todays battlefield. The use of these applications allows among others to exchange data with the tactical situation Link-16, which provides constant on-line access to information and increases the pilots situational awareness. This impacts directly the increase of the efficiency of the pilot, helps to finish the task and allows the pilot to return safely to the home airport.

The Polish Air Force still possesses aircrafts such as MiG-29, which are not interoperable with the aforementioned protocols due totheir level of technology and equipment. In order to restore the interoperability of this kind of aircraft, there is a possibility to perform complicated system changes.

Piloting of the combat airplane can be very stressful. Decisions need to be made quickly by the pilot during the flight. Possessing all the information which can help to complete his tasks is essential. The solution to this problem is to implement a Command Support System (CSS) to these aircrafts. This article features the elaborated system.

© Springer International Publishing Switzerland 2016
S. Kozielski et al. (Eds.): BDAS 2016, CCIS 613, pp. 659–667, 2016.
DOI: 10.1007/978-3-319-34099-9_51

# 2    CID Architecture

## 2.1    Template Selection

The aforementioned airplanes use older technology for military aviation that are used nowadays by the Polish Air Force. MiG-29 in the number of more than 30 units, can successfully fulfill provided tasks for a long time. These aircraft have a great combat value, and with the F-16 they form the core of the Polish combat aviation. For many years these aircraft were modernized and upgraded, but only half of them have recently undergone a comprehensive exchange of avionics, which moved them closer to the Western standards and allowed the cooperation in NATO structures [1]. The second half of the aircraft is still waiting for modernization. The planned upgrade includes increasing the pilots situational awareness on the modern battlefield. One of the improvements is the installation of liquid crystal displays, which, among other features also have the ability of viewing digital maps with the tactical situation. It is applied on the basis of the information provided by the new mission computer. However, at present, the data are entered into the computer before take-off and are not updated during the mission in the air. It is not possible for the pilot to build the current situational awareness.

## 2.2    System Requirements

Modern battlefield situation changes are extremely dynamical. These changes apply to the position of air and ground targets, as well as own troops. Transfer of such a large amount of dynamically changing data via the audio channel, as it is currently the case for the aircraft without implemented Combat Support System solutions, is archaic and does not meet the basic requirements of the modern battlefield. The pilot must receive updated information about the tactical situation regularly, in a transparent manner and in a way which does not distract him. This is very important for the effectiveness of eliminating the enemy targets, but also for preventing the accidents of own troops.

## 2.3    Information Source

The methods for obtaining the information about the situation from different sources on the battlefield make it necessary to implement many protocols used nowadays for data exchange standards. However, it is possible to use the COTS type finished product (commercial off-the-shelf - ready, working product off the shelf). One of these products that offers a very wide range of supported functionality is the Network Centric Data Communication Platform JASMINE, developed by TELDAT, a Polish company. This kind of system provides many unique internationally interoperable capabilities. It is a network-centric platform which enables the users to access a range of data, collected using different kinds of protocols. The JASMINE platform:

- supports (including automatization) processes of the command and management of troops at all levels, including a single soldier;
- enables achieving the information superiority and thereby creates the situational awareness of troops;
- provides possibility for building Common Operational Picture;
- substantially increases the security of military components and their elements, including soldiers and vehicles;
- creates modern, efficient, scalable, mobile and cheap multi-service ICT infrastructure, which enables construction of many independent networks for command center at operational and tactical level.

The system architecture proposed in this article is based largely on the solutions presented by the NATO STANAG ADatP-37 document (NATO Standard For Services To Forward Friendly Force Information To Weapon Delivery Assets). This standard sets guidelines for CID (Combat Identification) system class. High-level architecture of the system is presented in the Fig. 1.

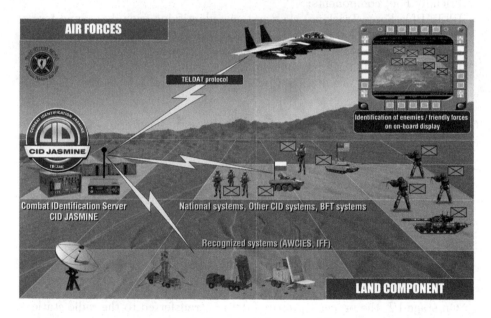

**Fig. 1.** System architecture [5]

The solution proposed by NATO was suitably modified considering the characteristics of the aircraft without implemented on-line protocols, among other things, the current restrictions on the transmission and presentation of data. The most important difference, compared to the original concept of NATO, is the introduction of a dedicated protocol and the data exchange medium between the aircraft and the CID JASMINE system [11]. Thanks to this protocol and medium, that aircrafts will receive the same information as the other objects in the air, supported by Link-16 data exchange protocol [12]. In addition, it will be possible to transfer the results from the analysis of information from reconnaissance systems. This functionality will be offered by JASMINE CID System improved under this project. The primary task of the system is to provide information about the current position of enemy and friendly forces, which is implemented as a series of the following stages:

- stage I - gathering information;
- stage II - information analysis;
- stage III - information passing to aircraft.

On stage I information is gathered in the system. The system uses the available protocols and data exchange means to acquire information about the current position of friendly forces and the location of the targets. Reports can be obtained from many sources, mainly from:

- sensors placed directly in the system, such as system IFF (Identification Friendly Foe) components;
- BFT (Blue Force Tracking) class systems, identifying friendly ground forces;
- systems of situational awareness and recognition of AWCIES class;
- command support systems of allied and own troops;
- joint MIP (Multilateral Interoperability Programme) database;
- JC3IEDM (Joint Consultation, Command and Control Information Exchange Data Model).

On stage II, the collected data must be processed with the aircraft characteristics and the earlier assumptions. The system is directed using the following rules:

- only the most important targets for the current run of the mission are transmitted to each aircraft;
- on each airplane the position of the allied troops that are in the immediate vicinity is presented in the first place;
- as the aircraft approaches it receives the first information about the position of allied forces in its vicinity;
- location reports are periodically supplemented with textual information provided to the pilot.

On stage III, the properly prepared data is transferred to the radio stations and received by the on-board radio telemetry channel.

Thus, after appropriate processing, the signals are transmitted to the mission computer, which after decoding it, sends the information to a display in the cockpit where it is presented on the background of a digital map display. Periodically sent text data is transmitted from the computer to the missions speech synthesizer and played there. The effect of the system work is the presentation of the current situation image on the pilots screen in real time.

## 2.4   System Security

One of the most important aspects of the proper operation of the presented system is the adequate protection of the whole process of data acquisition and transmission. The system can also operate on the interconnection of different classification levels. In that case it will need to use separating safety gates [4, 6, 7]. The process of the data transfer to the aircraft must also be protected. It is assumed to use the following mechanism [3]:

– authorization information based on a unique data type known only to the sender and the recipient;
– obfuscation of location information through the use of predetermined reference points;
– encryption of the transmitted information.

# 3   Data Acquiring Platform

The elements of the JASMINE System serve as a platform for data collection. This solution provides the standards and protocols, such as NFFI, FFI-XML-MTF, Link-16, VMF, including data acquired from IFF combat identification sensors, which, despite they seem to be useful both on land and in the air, do not enable the use of information from all these systems at the same time. NFFI, Link-16, VMF have been approved by NATO and are used, among others, during joint operations in Afghanistan. These protocols are very useful, but can be applied only in a few areas and scenarios. In order to improve, especially when the air-land operations are conducted, NATO's Combat Identification Server (CID Server) has been introduced, which collects information from various sources, such as:

– CID sensors such e.g. own enemy;
– BFT (Blue Force Tracking) position monitoring systems;
– situational awareness systems (SA).

CID is the process of obtaining a reliable, accurate view and the attributes of entities in the operations area in order to ensure the real use of tactical possibilities and weapons resources. JASMINE CID Server is an implementation of the NATOs conceptNATO in this area.

In order to provide a precise support, CID System has to be supplied with information from multiple of reliable sources. In this case, the data are delivered with the JASMINE system C3 components and database. This solution uses, among others:

– MIP database
– JC3IEDM model

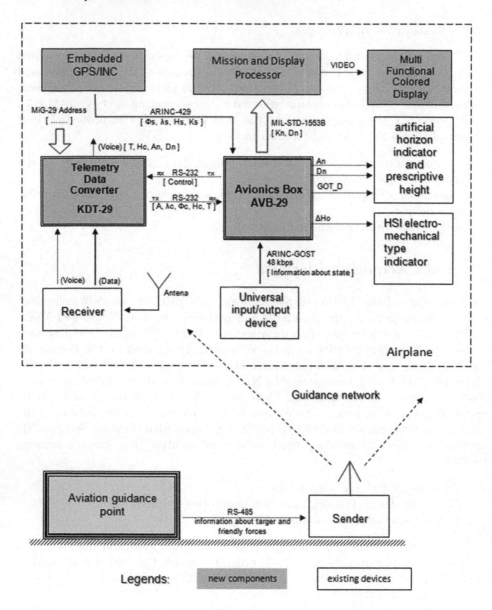

**Fig. 2.** System block diagram

## 4    The Use of AI Methods

In order to implement the Artificial Intelligence methods, the data flow of information which was presented has to be modified. The architecture solution diagram with AI method implemented is shown in the Fig. 3.

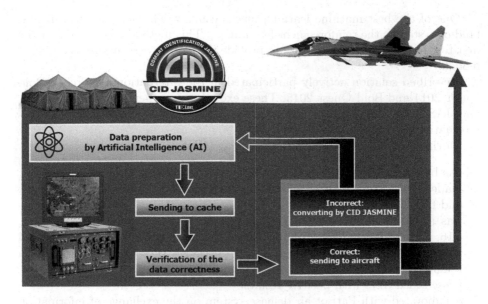

**Fig. 3.** Flow of information

According to this architecture, Artificial Intelligence is used for preparing the data which is to be placed in the cache. This data is selected based on the actual situation on the battlefield. The system prepares the package for the cache based on AI methods. This package contains a prepared prediction of the plane trajectory, using the collected data of the maneuvers which are performed most frequently.

The most effective AI method for this type of solution is one of the machine learning implementation [2,9,10]. Such a mechanism will improve significantly the dynamics of decision-making and eliminate the least delays [13]. This type of system would learn by examining its behavior patterns based on the pilots operations. After a sufficient number of inputs, AI will be able to predict how the pilot will behave. By this method the machine thinks with the pilot, which is a much faster solution.

For this purpose, a mechanism of Artificial Intelligence uses the following inputs:

- the position of own and foreign units;
- the type of foreign entities;
- the type of aircraft.

The output is a predicted concept of the future trajectory of the aircraft on the battlefield.

As a result, the prepared data is sent to cache. If the pilot decides to fly as provided by the AI method, the data is already given, otherwise it will be converted by CID JASMINE.

One of the best machine learning mechanisms which could be used in that kind of system is the reinforcement learning [8]. This method was chosen for the first test because of the possibility to provide the system description as an input model.

Described solution actively participated in the international exercises Bold Quest 2014 and Bold Quest 2015. These exercises were performed in the USA, and JASMINE CID was the only product from Poland in the history of this international exercise.

During Bold Quest exercises the CID JASMINE:

- has been accredited to connect to the training network;
- made positive tests with NATO military systems, including USA, Denmark and Italy;
- was an agent for the exchange of data, e.g. with airplanes A10, B1, F15, F16, F18, F18D and F18E-Super Hornet in the terms of: management and visualization of data availability from JASMINE CID server in plain text on remote terminal, and tracking the listed aircraft;
- passed load tests with positive results;
- collaborated with Patriot air defense system on the exchange of information on air situation

These exercises were performed with the MIP database, which contained the information about the aircrafts: A10, B1, F15, F16, F18, F18D and F18E-Super Hornet. By using the JC3IEDM, JASMINE CID had on-line access to this information.

## 5    Conclusion

The practical tests have already been performed. The implemented design concept is presented in this publication. CID JASMINE has already been tested on BQ 2014 and BQ 2015 exercises.

The first stage of the tests, the communication of the headquarters with the plane on the ground, has already been performed. The next step, air testing, will be performed in the near future. Implementation of the AI machine learning has been launched. It should be mentioned that the JASMINE System will be the first to use AI to increase the system performance for military use.

## References

1. Biuletyn konstrukcyjny p/o/r/u/5034/k/08
2. Angryk, R.A., Czerniak, J.: Heuristic algorithm for interpretation of multi-valued attributes in similarity-based fuzzy relational databases. Int. J. Approximate Reasoning 51(8), 895–911 (2010)
3. Apiecionek, L., Romantowski, M.: Secure IP network model. Comput. Method Sci. Technol. 4, 209–213 (2013)

4. Apiecionek, Ł., Romantowski, M., Śliwa, J., Jasiul, B., Goniacz, R.: Safe exchange of information for civil-military operations. In: Military Communications and Information Technology: A Comprehensive Approach Enabler, pp. 39–50 (2011)
5. Apiecionek, Ł., Biernat, D., Makowski, W., Lukasik, M.: Practical implementation of AI for military airplane battlefield support system. In: 2015 8th International Conference on Human System Interactions (HSI), pp. 249–253. IEEE (2015)
6. Apiecionek, Ł., Czerniak, J.M., Zarzycki, H.: Protection tool for distributed denial of services attack. In: Kozielski, S., Mrozek, D., Kasprowski, P., Małysiak-Mrozek, B. (eds.) BDAS 2014. CCIS, vol. 424, pp. 405–414. Springer, Heidelberg (2014)
7. Apiecionek, L., Romantowski, M.: Security solution for cloud computing (2014)
8. Bradtke, S.J., Barto, A.G.: Learning to predict by the method of temporal differences. Mach. Learn. **22**, 33–57 (1996). (Springer)
9. Kosinski, W., Prokopowicz, P., Slezak, D.: On algebraic operations on fuzzy reals. In: Rutkowski, L., Kacprzyk, J. (eds.) Neural Networks and Soft Computing. Advances in Soft Computing, vol. 19, pp. 54–61. Springer, Heidelberg (2003)
10. Kozielski, M., Skowron, A., Wróbel, Ł., Sikora, M.: Regression rulelearning for methane forecasting in coal mines. In: Kozielski, S., Mrozek, D., Kasprowski, P., Małysiak-Mrozek, B., Kostrzewa, D. (eds.) BDAS 2015. CCIS, vol. 521, pp. 495–504. Springer, Heidelberg (2015)
11. Kruszynski, H., Kosowski, T., Apiecionek, L.: CID server JASMINE. In: V Communications Conference in Sieradz (2014)
12. Lojka, T., Zolota, M., Zolotová, I., et al.: Communication engine in human-machine alarm interface system. In: Sincak, P., Hartono, P., Vircikova, M., Vascak, J., Jaksa, R. (eds.) Emergent Trends in Robotics and Intelligent Systems. Advances in Intelligent Systems and Computing, pp. 129–136. Springer, Heidelberg (2015)
13. Vidhate, D., Kulkarni, P.: Cooperative machine learning with information fusion for dynamic decision making in diagnostic applications. In: 2012 International Conference on Advances in Mobile Network, Communication and its Applications (MNCAPPS), pp. 70–74. IEEE (2012)

# Persistence Management
# in Digital Document Repository

Piotr Pałka[✉], Tomasz Śliwiński, Tomasz Traczyk, and Włodzimierz Ogryczak

Warsaw University of Technology, Nowowiejska 15/19, 00-665 Warsaw, Poland
{P.Palka,T.Sliwinski,T.Traczyk,W.Ogryczak}@ia.pw.edu.pl

**Abstract.** The CREDO Digital Document Repository enables short-and long-term archiving of large volumes of digital resources, ensuring bitstream preservation and providing most of the technical means to ensure content preservation of digital resources. The goal of the paper is to describe the design and implementation an innovative component of the CREDO Repository: the Persistence Management Subsystem (PMS). This subsystem sets guidelines for the file management system on replicas placement, and data relocation. The module responsible for scheduling access to the archive provides energy efficiency by setting suboptimal schedules. The module responsible for diagnose and exchange of data carriers calculates the probabilities of failure, and the information is used by the scheduling module to select appropriate storage areas for reading or writing of data, and for marking the areas as obsolete. Finally, the power management module is responsible for starting-up the storage areas only when necessary.

**Keywords:** Digital preservation · Digital archiving · Long-term preservation · Repositories management · OAIS

## 1 Introduction

The paper describes some concepts used in construction of digital repository named CREDO[1], designed for trustworthy storage of large volumes of digital information.

By design the repository is to act both as a secure file storage and as a long-term digital archive, providing metadata management, and storing digital resources packed together with their metadata in archival packages. When acting as an archive, repository meets requirements of the OAIS (Open Archival Information System [3], ISO 14721:2012) standard that provides recommended practices for implementation of digital archives.

One of the primary functions of the repository is the support for various currently available data carriers: hard drives, solid state drives, tapes. Reliability

---

[1] CREDO – the acronym of Polish name *Cyfrowe REpozytorium DOkumentów*, which means 'Digital Document Repository'. In Latin *credo* means 'I believe', which seems to be quite a good watchword for trustworthy digital repository.

© Springer International Publishing Switzerland 2016
S. Kozielski et al. (Eds.): BDAS 2016, CCIS 613, pp. 668–682, 2016.
DOI: 10.1007/978-3-319-34099-9_52

of information readouts is ensured by the data recording replication mechanisms in the used file system, as well as the distributed nature of the system that enables storing copies of the resources in remote locations.

The repository architecture is multi-tiered and consists of loosely-coupled subsystems, which enables, together with the emergence of new technologies, replacement and continuous upgrades of individual components.

This paper focuses on some problems related to long-term archive function of the CREDO repository. The purpose of the long-term archiving is to trustworthy store large amounts of digital data over a significantly long period, e.g. a few generations. This leads to many problems [8,10] such as obsolescence of technology, ensuring economic efficiency, etc.

The problems are quite specific due to the fact, that no known electronic or IT technology can be expected to survive such a long period, in particular no electronic digital media can guarantee data fidelity after dozens of years. Furthermore, many currently used file formats cannot ensure information readability after such a long time, because a software necessary to read them may be no longer available. And, in many cases, even if the digital data can be faithfully restored, this does not guarantee an intelligibility of the information, due to the possible lack of users that may understand that information.

The CREDO archive addresses most of the main long-term archiving problems.

– A technology obsolescence is addressed by possible utilization of various filesystems, based on various media, and by relatively easy replacement of system modules.
– Information readability is ensured by storing digital resources only in selected, sustained and non-proprietary formats, recognized as well-suited for digital archiving, together with technical metadata describing the resources, and documentation of formats used.
– Information intelligibility is partially facilitated by storing descriptive metadata of the archived resources.
– A storage uncertainty is addressed by:
  • regular multi-level monitoring of data correctness, with use of low-level (built in filesystem) and high-level CRC checks;
  • multi-level data replication: low-level replicas automatically maintained by filesystem, and high-level replicas maintained by CREDO archive;
  • automatic data dislocation between different locations,
  • limited support for dislocation into separate federated archives, including synchronization of activities on resource copies stored in separate archives;
  • continuously diagnosing and monitoring of data carriers and media replacement management.
– Economic efficiency is ensured by scheduling algorithms which appoint the schedule for turning the servers on, when all the data storing servers are normally in the off state.

The contributions of this paper are threefold: (i) the description of the Persistence Management Subsystem (PMS) implementation in CREDO long-term archive, and the detailed description of two important modules of the PMS: (ii) diagnose and replacement of data carriers module, and (iii) scheduling access to data carriers module.

## 2 State of the Art

Though digital data archiving becomes one of the most challenging problems of contemporary information technology, there are only a few fairly complete solutions.

Relatively many products are offered for a short-time storage of large amounts of data. Several hardware and software solutions are available, mainly suited to backup big data volumes in dedicated (e.g. MooseFS, ceph) or cloud (e.g. Amazon S3, Google Cloud) storage. Although these solutions are capable to store enormous data volumes, they do not address long-term archiving problems mentioned above.

The most promising solutions for long-term archiving, e.g. [5,17], are all based on OAIS standard [3], which addresses most of "logical" problems of digital archiving.

Most of existing long-term archiving solutions are based on tape storage. However, promising disk-based, energetically efficient solutions also exist, like our CREDO archive, or Archi-Clarin project [11] based on specialized hardware. There are also long-term cloud-based solutions, e.g. [17].

Many important problems of information archiving, especially on long-term horizon, cannot be solved without preserving and managing metadata. It is widely accepted, and confirmed by OAIS standard, that metadata should be stored packed together with the described information into so called archival packages [3,15]. Though metadata management is a key issue in context of archiving, the problem itself is very complex [16], and it is beyond the scope of this paper.

While in classical algorithms concerning the non real-time disk scheduling as SCAN [4], CSCAN [20], or C-LOOK [13] where the disks should be scheduled as quickly as possible and they are always on, in our problem, the carriers are mostly off, the energetic efficiency is the case, not a quickness. So, the long-term schedule for accessing the data carriers is taken into account. The problem of data centre energy-efficient network-aware scheduling is undertaken in the [7]. The paper describes the optimization of the trade-off between job consolidation and distribution of traffic patterns. Another solution for dealing with energy efficient resource management in virtualized cloud data centers is presented in [2]. A slightly different approach is to consider thermal-aware scheduling [22]. Other works on the energy efficient scheduling in data centers are [1,12,21].

An important issue of scheduling is strongly tied with a failure prediction. Papers [6,14] describes methods for prediction of disk failure. On the other hand, [18] analyses the data on disk replacement in a large system containing over

100,000 carriers. Authors of [19] present analysis of disk failure in the Internet archive. Paper [9] describes the head of the disk analysis, and the impact on operating temperature on the failure prediction.

## 3    Digital Document Repository Architecture

The CREDO Digital Document Repository is divided into several subsystems (see Fig. 1).

- Archive Management Subsystem (AMS) controls activities of the archive: ingest, search, and outgest sessions (as defined in OAIS standard [3]), internal administrative processes, etc., and interacts with end-user. The subsystem stores also in its database partial metadata of archived resources. This metadata enables fast search in the archive and is used to verify archive integrity and authenticity.
- Persistence Management Subsystem (PMS) monitors archive storage and works out recommendations for data allocation and relocation, ensuring low storage failure risk. It also optimizes the operation of the storage media to achieve energy savings. It co-operates with AMS, providing schedules of AMS activities.
- Security Management Subsystem (SMS) controls access to archives filesystems and the archive buffer. It interacts with AMS to grant access rights necessary for AMS activities but, for security reasons, it is separated from AMS. If necessary, it can be implemented in different technology and run on physically separate hardware.
- File Management Subsystems (FMS) consists of buffer filesystem, archive storage filesystem and some auxiliary modules. One repository can use many such subsystems, even implemented differently, but interacting with other subsystems of the repository with the same set of well-defined protocols. Each FMS can contain many filesystems, which may also be technologically diversified.

Due to the planned longevity, the repository cannot be dependent on any particular technology, as hardware and software. This is ensured by layer-component system architecture with loose coupling between subsystems, and clearly defined interfaces between them, which use standards-based communication protocols: RESTful Web Services containing self-describing XML messages, and SQL queries (which also may be replaced by REST/XML if necessary).

## 4    Persistence Management Subsystem

Persistence Management Subsystem (PMS) sets guidelines for the stream layer (file system) on replicas placement, relocation of data, data carriers diagnose and replacement, scheduling the access to the archive, and power management.

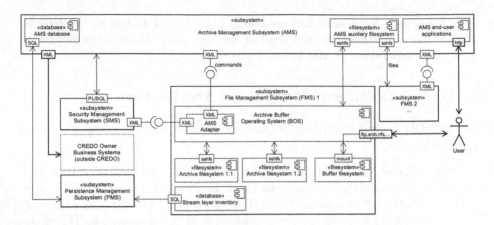

**Fig. 1.** CREDO Digital Document Repository architecture

*Replication.* Replication is the data migration method, that does not change the bit sequence on the data carrier. It assumes creating multiple copies of the same resource. Additionally, the replication assumes that the resource is copied onto another carriers. In CREDO, there are two levels of replication: low-level replication, that is assured by the file system; and high-level replication, provided with the PMS. Both replication methods serve to ensure data security. Moreover, the high-level replication can be assured on: (i) area level, where different copies of the same resource are kept under different paths, (ii) file system level, where a resource is stored on different file systems, (iii) archive level, where a resource is secured on a few archives.

The higher the level of the replication, the data is safer, as the mechanisms ensure: doubling the resource safety, diversification of a resource representation (when the copies are kept under different file systems), geographical dispersion, when resource is stored in different locations.

*Relocation.* The goal of the relocation is to prevent data aging by refreshing the resource on another: carrier, area, file system, or archive. PMS assumes package relocation using optimization methods, where the packages are relocated: periodically, on safer area, with sub-optimal areas usage, and with energy efficiency. Moreover, the relocation is coupled with the carriers replacement. The packages are automatically high-level replicated among the areas, the PMS schedules the plan for a long-period.

*Data carriers diagnose and replacement.* The goals of the diagnose and replacement of data carriers module are:

– Analysis of a risk concerning failures of data carriers and whole areas designed for storage of archival packages. The analysis consider single pieces, lots, models of carriers, and it goal is to predict the failure, and secure the data stored on the carrier by replication, relocation on the other carriers, renewing, or emptying it.

- Replacement of carriers decision support system.
- Failure prevention due to prediction of the failure.
- Interoperability with the data relocation, by automatic rewriting data from the carriers under threat.

To achieve the goals, the module for carriers, and area reliability assessment is used. It is common mechanism to provide the information about different: carriers, areas, batches of carries to PMS. Moreover it provides information independently of carrier type (hard disk, magnetic tape, CD, DVD, Blue-Ray, pendrive, etc.). Finally it solves the problem of a lack of technical solutions for monitoring, and managing carriers. The module allows to choose a source or destination for the archival packages during archive operation. Also, it specifies the moment for the migration. Finally, it determines the areas for powering-on during reading or writing the archival packages.

*Power management.* In CREDO the carriers are divided onto storage areas, that are subject to the PMS management. Single storage area has assigned a number of data carriers. The allocation of the packages onto storage areas is done by PMS, and a power management module manages areas starting-up, and shutting-down. Power management system is coupled with a scheduling module, and it have to start-up given area, when the schedule assumes operating on it. Similarly, when planned operations finish, the power management module shuts the area down.

## 5  Scheduling Access to Archive

As the repository lifetime is very long, the data carriers cannot be on the whole time. On contrary, the carriers will be off most of the time, and will be powered only on demand. In the repository using magnetic data tapes, the carriers should be loaded/unloaded into/from the reader if needed. Access to the data stored in the repository requires powering on the appropriate carrier, or loading the appropriate tape into the reader. Effective and secure usage of the archive requires separate algorithms to schedule access to its resources, and to manage powering on/off or loading/unloading data carriers. The needed data structures and algorithms constitute separate software module – the scheduling module.

### 5.1  Problem Description

There is given a set of operations that are to be scheduled. Operations belong to procedures. Single procedure consists of a sequence of sets of operations. The sequence defines the order in which the sets are executed. All operations within a set can be run in parallel. Each operation is given a time period within which it has to start and finish. The starting time of the operation can be given in advance or be undefined in which case it will be a result of the scheduling algorithm. Some of the operations require access to storage areas which constitute critical resources. In particular, the size, the number of parallel reads, and writes are limited. Additionally, some operations require exclusive access to the resource (area) either for write only, or for read and write.

Area is the smallest indivisible part of the archive of known capacity that can be turned on and off. Area can consist of a number of data carriers, but they are indistinguishable for the scheduling algorithm. However, their parameters can affect some aggregated parameters of the area.

Each operation has given a set of source areas (to read data from) and a set of target areas (to write to or to modify data). In the final schedule, only one area out of the given set can be chosen to read data from and/or one area to write data to.

The scheduling module should create the optimized schedule meeting the following multiple objectives (multiple criteria): ensuring data security and integrity, minimizing usage cost of the archive, load and I/O traffic leveling between different parts of the archive (areas).

*Notation.* Let $i$ denote operation ($i \in I$), and $p$ denote procedure ($p \in P$). Set of operations belonging to procedure $p$ will be denoted by $I_p$. Some operations utilize areas $o$ ($o \in O$), and each such action has defined separate set of possible areas for reading $U_i \subseteq O$ and writing data $S_i \subseteq O$. As noted earlier, for each operation $i$ time period $(T_i^e, T_i^l)$ is given within which the operation has to start and finish. The starting time of the operation will be denoted by $T_i^s$. The maximum number of parallel operations (read and write) in area $o$ is denoted by $R_o$, and parallel write operations by $W_o$.

Operations within a procedure are grouped to create sets of operations that can be run in parallel. There is a defined order of execution of such sets, and $G_p^k$ denotes the set of operations that can be executed simultaneously in position $k$, where $k = (1, \ldots, K)$.

Below is the list of all operation parameters relevant the scheduling algorithm.

**Predecessors.** Set of operations required to start/finish before given operation. The precedence type is defined later.

**Lead lag.** Required time period before given operation can start.

**Precedence type.** There are two precedence types relevant to the scheduling algorithm:

  **FS.** Operation can start only after all the operations in the Predecessors set have finished, plus given Lead lag.

  **SS.** Operation can start only after all the operations in the Predecessors set have started, plus given Lead lag.

$T^e$, $T^l$. Time period in which the operation has to be executed, i.e. cannot start earlier than $T^e$ and finish later than $T^l$.

**Source areas.** Set of areas to read data from. Only one will be chosen in the final schedule.

**Read times.** For each source area the estimated reading time.

**Target areas.** Set of target areas. Only one will be chosen to write to.

**Write times.** For each target area estimated write/modification time.

**Write size.** For each target area estimated size of the data written – due to the file system properties can be different for each area.

**Lock type.** Type of lock put on the target area to during execution of the operation. Possible lock types are:

**Full.** No other operation on the area allowed.

**Write.** Only read operations are allowed

**No.** No lock. All operations are allowed.

$D^{min}, D^{max}$. Minimum and maximum duration, defined for those operations which do not use resources (areas). The scheduling algorithm assigns the actual duration from this interval according to some strategy.

$T^s$. Starting time of the operation. This is input and output data for the algorithm. If it is not set, or beyond $(T^e, T^l)$, then the scheduling algorithm sets its value.

The list of area properties considered by the scheduling algorithm is as follows.

**Size.** The maximum number of bytes that can be stored in the area. In different file system, the actual size available for data can be different (smaller).

**Used space.** Number of bytes used by file system to store all the data.

$R_o$. Maximum number of I/O operations performed in parallel in the area $o$.

$W_o$. Maximum number of write I/O operations performed in parallel in the area $o$.

**Power off time.** Minimum time period with no operations, that results in area powering off in order to save energy.

**Power up cost.** Total cost of powering on the area

**Power on cost.** Cost of the area operation in the time unit.

**Power on time unit.** Time unit for which the area operation cost is defined.

**Reliability factor.** Some numerical quantification of area reliability.

The separate problem is acquiring information about the actual space used by the stored data on each area. This cannot be done during execution of write/modify operations. Therefore, a special write blocking operation will be provided only to read the used space of the area.

## 5.2   Basic Assumptions

During the development of the scheduling module following assumptions were made.

1. Multiple optimization criteria are to be considered.
2. Areas can be grouped in regions, each region is in possibly distinct geographical location.
3. To ensure data integrity, multiple (at least two) high level copies are created, each copy on separate region.
4. All the optimization criteria should be considered when choosing areas to perform given operation on. In particular, each operation can read data from one of the high level copies and write data to one of the areas belonging to the given region.

5. It is assumed, the execution time of each operation in a set of simultaneously executed operations can be estimated in advance. If it cannot be done, the scheduling of the following sets of operations in the procedure is suspended, i.e. will not be scheduled.

6. It is assumed that the execution time of each operation does not depend on the total number of operations executed in the same time period. Due to the physical properties of the file system this assumption obviously does not apply to the real systems in every usage pattern. In most systems, however, this assumption can be made if the number of parallel operations is limited.

7. For operations that use both, source and target resources, they are used in exactly the same time interval.

8. Previously created schedule in the presence of new operations (incremental scheduling), should stay unchanged as long as no constraints are violated.

## 5.3    Scheduling Algorithm

The algorithm is based on the construction heuristics. It is built using greedy method for operations sequenced according to some criteria.

Let $\tau$ be permutation of a set $I$ defining initial sequence of operations, i.e. if $i < j$, then operation $\tau(i)$ is placed before operation $\tau(j)$ in the sequence. Let $\tau(1)\ldots\tau(n-1)$ be operations used to create some partial, already existing schedule and the algorithm is about to schedule operation $\tau(n)$.

Let *event* be any change of the system state in a partial schedule, i.e. start or finish of any operation that requires resources (areas). All events in a partial schedule can be ordered according to their times: $t_1 < \ldots < t_m$.

The algorithm of scheduling operation $\tau(n)$ works as follows.

1. The operation that does not use resources is scheduled as soon as possible as long as it does not violate the precedence relations in a procedure.

2. Operation that uses resources is tentatively scheduled in each time $t_k \in \{t_1 \ldots t_m\}$, in a way that either operation starts or finishes in time $t_k$.

3. For each time $t_k \in \{t_1, \ldots, t_m\}$ all source areas and target areas are being considered for usage.

4. The algorithm selects time $\bar{t}$ (either as operation's start or finish), the single area from the set of source areas and/or single area from the set of target areas.

5. In the above selection the algorithm tries to minimize the increase of the objective value *marginal cost* resulting from execution of the operation. The objective here is an aggregation of the optimization criteria (see detailed description later).

6. Only those solutions are considered (times and resources/areas) for which no constraints are violated (precedence within procedure, area capacity, maximum number of parallel operations).

## 5.4   Initial Sequence of Operations

As a greedy construction heuristic is used and no backtracking is possible, proper initial sequence is crucial for obtaining good final schedules. The general principle here is to sequence first operations with lower degree of freedom (lower $T^l$, lower cardinality of source/target area sets, etc.) and later operations with higher degree of freedom. This way we maximize the likelihood for achieving good final schedules. The sequence should also meet the precedence requirements of operations within the procedures, that is operations that are sequenced later in a procedure should also be positioned later in the initial sequence.

The exact strategy of ordering the operations works as follows. Let $d_i$ be estimated duration of operation $i$ and $a_i$ the total cardinality of source areas set and target areas set. The duration of the operation can be deliberately undefined for some operations. Let operator $def$ return logical true if and only if its argument is defined and operator $undef$ return logical true if and only if its argument is undefined. Operation $i$ will be sequenced before operation $j$ only if

- $def(T_i^s)$ and $undef(T_j^s)$ or
- $def(d_i)$ and $undef(d_j)$ or
- $def(d_i)$ and $def(d_j)$ and $def(T_i^s)$ and $def(T_j^s)$ and $T_i^s + d_i < T_j^s + d_j$ or
- $(undef(d_i)$ or $undef(d_j)$ or $undef(T_i^s)$ or $undef(T_j^s))$ and
  - $def(T_i^l)$ and $undef(T_j^l)$ or
  - $def(T_i^l)$ and $def(T_j^l)$ and $T_i^l < T_j^l$ or
  - $def(T_i^e)$ and $undef(T_j^e)$ or
  - $def(T_i^e)$ and $def(T_j^e)$ and $T_i^e < T_j^e$ or
  - $a_i < a_j$ or
  - $def(d_i)$ and $def(d_j)$ and $d_i > d_j$

The above sequencing strategy will by further denoted by the operator $ord(i, j)$, which returns true only if operation $i$ should be sequenced before $j$.

To take into account the precedence of operations within the procedure, the operator $ord(i, j)$ is embedded into the following algorithm.

**Step 1.** $m := 0$, $S_0 := \emptyset$ (where $S$ is the set of sequenced operations).

**Step 2.** Let $C$ be the sum of sets $G_p^k$, for all procedures $p \in P$ and $k$ indicating the first set in a sequence of sets within procedure $p$ in which not all operations have yet been sequenced, i.e. $C = \bigcup_{p \in P, \min k : G_p^k \cup S_m \neq \emptyset} G_p^k$. The best operation $i \in C$ is selected, i.e. $i \in C : \nexists_{j \in C} ord(j, i) = \text{true}$. Add $i$ to set $S$: $S_{m+1} := S_m \cup \{i\}$. $m := m + 1$. Repeat Step 2.

## 5.5   Optimization Objective – Aggregation of Multiple Criteria

The following objectives are considered by the scheduling algorithm.

1. Total operation cost of the archive within the scheduled time period.
2. Reliability factor.
3. Relative area space usage.
4. Relative I/O operations usage.

All criteria, except the second one, are to be minimized.

*Total operation cost.* Each area has following cost related parameters.

- Power up cost – total cost of powering up the area.
- Power on cost – total cost of running the area in a defined unit of time.

For each operation depending on its placement in the schedule and used areas (source and/or target) the algorithm computes the total operation cost being the sum of the cost of powering it up (if the area is being actually powered up) and the running cost (if the running time increases when compared to the existing schedule). For each area additional 'Power off time' parameter is specified. It defines minimum time period without any operations which causes the area to be powered down to save energy. As data carriers can have different physical properties the reliability of an area can be partially affected by proper setting the three above mentioned parameters. For example, one can increase the Power up cost and Power off time to minimize the number of area's on/off operations.

*Reliability factor.* For each area some reliability factor is defined and areas with higher reliability are preferred, to increase data security and integrity.

*Relative area space usage.* Based on the area size, used space and the size change made by the operation the relative space usage of an area is determined. The algorithm tries to keep this value at similar level on all areas, preferring areas with lower values of this objective.

*Relative I/O operations usage.* Similarly, the number of actual I/O operations in relation to the maximum number of I/O operations is leveled between areas by the algorithm. This should prevent from the situation where some areas are extensively utilized leading to bottlenecks.

### 5.6   Scalarization with Preference Expressing Weights

The above mentioned criteria refer to different parameters of the schedule with different physical or logical interpretation. Furthermore, in the construction algorithm utilized here there is no need to know global objective, or even the absolute value of the marginal cost. It is sufficient to be able to evaluate relative differences among criteria for all examined potential solutions. Let $y = (y_1, y_2, y_3, y_4)$ and $z = (z_1, z_2, z_3, z_4)$ be values of the four criteria (objectives) under consideration for two solutions, respectively.

Let the vector $w = (w_1, w_2, w_3, w_4)$, $\sum_{i=1}^{4} w_i = 1$ of normalized weights express decision-maker preferences for individual criteria.

In the approach applied in the scheduling module the solution represented by vector $y$ is better than solution represented by vector $z$ if and only if

$$w_1 \frac{y_1 - z_1}{\max\{y_1, z_1\}} + w_2 \frac{z_2 - y_2}{\max\{y_2, z_2\}} + w_3 \frac{y_3 - z_3}{\max\{y_3, z_3\}} + w_4 \frac{y_4 - z_4}{\max\{y_4, z_4\}} < 0$$

Note the inverse of the fraction for criterion 2 which is maximized. If both solutions are equal (no one is better), then the solution with operation starting earlier is preferred.

# 6    Data Carriers Diagnose and Replacement Module

The goal of the diagnose and replacement of data carriers module is to create the generic mechanism handling different media, diagnosing them, and providing the failure predicting measures. The measures does not depend on specific media type. We assume their estimation basing on: (i) NARA (National Archives and Records Administration) guidelines, (ii) information provided by the media producer on the media lifetime (e.g. MTBF[2]), (iii) monitoring and diagnosing information embedded in the media (e.g. S.M.A.R.T information for disk drives).

To deal with the estimating the failure, we assume use of the reliability concept, that describes the ability of a system or component to function under stated conditions for a specified period of time. The reliability defines the probability of proper working defined by a function: $R(t) = P(T > t) = \int_t^\infty f(x)dx$, where $t$ is the reliable working time, $T$ is the assumed time for reliable working, and $R(t)$ is the reliability function. The reliability function is coupled with the MTBF parameter: $R(t) = e^{-t/MTBF}$.

We developed the method to predict the failure in situation where we have only basic information on a media (e.g. only NARA guidelines, and MTBF), or we have specific monitoring and diagnosing tool, as S.M.A.R.T.

1. With only the NARA guidelines available, we took the admissible operating media time as MTBF parameter.
2. With the MTBF available, we calculate the reliability function: $R(t) = e^{-t/MTBF}$, where $t$ is the time (in hours), that passed from the carrier manufacturing.
3. With the S.M.A.R.T parameter no 009 (Power On Hours) – POH, the reliability function is more accurate: $R(t) = e^{-POH(t)/MTBF}$, where $POH(t)$ is the value of POH in the current moment.
4. With the S.M.A.R.T parameter no 194 (Temperature Celcius) – TC, we can modify the working time of the disk, according to Arrhenius equation, that takes into account dependence of failure and working temperature. As the MTBF parameter is set in the reference temperature $TC_{ref}$ (mostly 30° C), having the information on working temperature, we can calculate the $AF$ factor, that is used to modify the working time.
5. Literature review gives us some clues, which S.M.A.R.T parameters warn us of impending failure. When one of the parameters: Read Error Rate (SMART 001), Reallocated Sectors Count (005), Seek Error Rate (007), Spin Retry Count (010), Calibration Retry Count (011), Reported Uncorrectable Errors (187), Command Timeout (188), Current Pending Sector Count (197), or Offline Uncorrectable (198) occurs, the probability of failure grows rapidly. Therefore, we modify the MTBF parameter as follows: $MTBF \leftarrow MTBF - \frac{\Delta P}{ADM\_P}$, where $\Delta P$ is the change of the S.M.A.R.T parameter, and $ADM\_P$ is admissible value of this parameter.

Having the reliability function, we can formulate the measures for reliability for each of the data carriers. We propose two measures:

---

[2] Mean Time Between Failures, parameter given by the media producers.

- 99th percentile of carrier failure probability: $CFP^{99} = -ln(0.99) * \frac{MTBF-t}{24}$
- Probability of carrier failure in one year: $CFY = 1 - e^{\frac{-365*24}{MTBF-t}}$

To aggregate the measures for areas of carriers, we propose the min and max operators ($A$ is a set of carriers $c$ belonging to an area):

- 99th percentile of an area failure probability: $AFP^{99} = \min_{c \in A} CFP_c^{99}$
- Probability of area failure in one year: $AFY = \max_{c \in A} CFY_a$

The module allows to choose a source or destination for the archival packages during archive operation. Also, it specifies the moment for the migration. Finally it determines the areas for powering-on during reading or writing the archival packages.

Having reliability measures calculated, we can choose a source or destination for the archival packages during archive operation, by selecting the destination areas with lower $AFY$ value, and source area with higher $AFY$ value. Also, the moment for migration is specified on the base of $AFP^{99}$ measure. Finally, the obsolete areas are determined on the base of $AFP^{99}$ measure.

## 7   Summary

Storing the data for dozens of years, introduces many issues, to mention only the most important: technology obsolescence, energetic efficiency, file format expiration, operating systems out-dating.

CREDO Persistence Management Subsystem provides a number of functionalities, that enables long-term data storing efficiency, assuming energy efficiency. By the design of replicas and relocation management, the data security is assured.

The scheduling and power management modules are responsible for a power saving, which is crucial for the long-term archival system.

Nevertheless, the long periods of shutting down the carriers may influence on their longevity, so monitoring and diagnosing of them is needed. The data carriers diagnose and replacement module serves as generic tool, ready to use for various kinds of carriers.

CREDO Repository is designed to trustworthy and cost-effectively store large amounts of digital resources (the instance now being built is expected to have over 2 PB capacity), with use of standard hardware and data carriers available now, and in the future. The solution addresses most of main problems of long-term digital archiving of digital resources, including possibility to adopt new technologies, and flexible metadata management, which should ensure information authenticity, readability and intelligibility.

The solution is designed for institutions that store large digital resources for long periods of time, e.g. institutions responsible for cultural heritage, mass media, state administration offices, health care institutions, etc.

**Acknowledgments.** The project entitled *Cyfrowe Repozytorium DOkumentów CREDO* (*Digital Document Repository CREDO*) is co-financed by the European Union through the European Regional Development Fund under the Operational Programme 'Innovative Economy' for the years 2007–2013, Priority Axis 1 – Research and development of modern technologies, Grant No. WND-DEM-1-385/00.

# References

1. Al-Fares, M., Radhakrishnan, S., Raghavan, B., Huang, N., Vahdat, A.: Hedera: dynamic flow scheduling for data center networks. In: NSDI, vol. 10, p. 19 (2010)
2. Beloglazov, A., Buyya, R.: Energy efficient resource management in virtualized cloud data centers. In: Proceedings of the 2010 10th IEEE/ACM International Conference on Cluster, Cloud and Grid Computing, pp. 826–831. IEEE Computer Society (2010)
3. Consultative Committee for Space Data Systems: Reference model for an open archival information system (OAIS). Recommended practice, June 2012. http://public.ccsds.org/publications/archive/650x0m2.pdf. Access: 01 Dec 2015
4. Denning, P.J.: Effects of scheduling on file memory operations. In: Proceedings of the Spring Joint Computer Conference, pp. 9–21. ACM, 18–20 April 1967
5. Giaretta, D.: Advanced Digital Preservation. Springer, Heidelberg (2011)
6. Hamerly, G., Elkan, C., et al.: Bayesian approaches to failure prediction for disk drives. In: ICML, pp. 202–209. Citeseer (2001)
7. Kliazovich, D., Bouvry, P., Khan, S.U.: Dens: data center energy-efficient network-aware scheduling. Cluster Comput. **16**(1), 65–75 (2013)
8. Lu, M., Chiueh, T.: Challenges of long-term digital archiving: a survey. Tech. rep., Experimental Computer Systems Lab, Department of Computer Science, State University of New York, October 2006. http://www.ecsl.cs.sunysb.edu/tr/rpe19.pdf
9. Mao, S., Chen, Y., Liu, F., Chen, X., Xu, B., Lu, P., Patwari, M., Xi, H., Chang, C., Miller, B., et al.: Commercial TMR heads for hard disk drives: characterization and extendibility at 300 gbit/in 2. IEEE Trans. Magn. **42**(2), 97–102 (2006)
10. Marasek, K., Walczak, J., Traczyk, T., Płoszajski, G., Kaźmierski, A.: Koncepcja elektronicznego archiwum wieczystego. Stud. Inform. **30**(2B), 275–307 (2009). http://zti.inf.polsl.pl/BDAS/2009/BDAS'09%20-%20KONCEPCJA%20ELEKTR ONICZNEGO%20ARCHIWUM%20WIECZYSTEGO.pdf?Id=646&val=1
11. Marasek, K., Walczak, J.: Long-term preservation of digital files in data network structures, 01 Dec 2015. http://www.ci.pw.edu.pl/content/download/1426/11818/file/KMJPW06102015-fin.pdf (in Polish)
12. Meng, X., Pappas, V., Zhang, L.: Improving the scalability of data center networks with traffic-aware virtual machine placement. In: 2010 Proceedings of the IEEE INFOCOM, pp. 1–9. IEEE (2010)
13. Merten, A.G.: Some quantitative techniques for file organization (1970)
14. Murray, J.F., Hughes, G.F., Kreutz-Delgado, K.: Hard drive failure prediction using non-parametric statistical methods. In: Proceedings of ICANN/ICONIP. Citeseer (2003)
15. Pater, K., Traczyk, T.: Opakowanie zasobów cyfrowych na potrzeby archiwizacji długoterminowej. Stud. Inform. **34**(2B(112)), 898–103 (2013). http://www.znsi.aei.polsl.pl/materialy/SI112/SI112_8.pdf

16. Płoszajski, G. (ed.): Standardy techniczne obiektów cyfrowych przy digitalizacji dziedzictwa kulturowego. Biblioteka Główna Politechniki Warszawskiej, Warszawa (2008). http://bcpw.bg.pw.edu.pl/dlibra/docmetadata?id=1262
17. Rabinovici-Cohen, S., Marberg, J., Nagin, K., Pease, D.: PDS cloud: Long term digital preservation in the cloud. In: 2013 IEEE International Conference on Cloud Engineering (IC2E), pp. 38–45, March 2013
18. Schroeder, B., Gibson, G.A.: Disk failures in the real world: What does an MTTF of 1, 000, 000 hours mean to you? In: FAST, vol. 7, pp. 1–16 (2007)
19. Schwarz, T., Baker, M., Bassi, S., Baumgart, B., Flagg, W., van Ingen, C., Joste, K., Manasse, M., Shah, M.: Disk failure investigations at the internet archive. Work-in-Progess Session, NASA/IEEE Conference on Mass Storage Systems and Technologies (MSST 2006) (2006)
20. Seaman, P.H., Lind, R.A., Wilson, T.L.: On teleprocessing system design: part iv an analysis of auxiliary-storage activity. IBM Syst. J. 5(3), 158–170 (1966)
21. Stage, A., Setzer, T.: Network-aware migration control and scheduling of differentiated virtual machine workloads. In: Proceedings of the 2009 ICSE Workshop on Software Engineering Challenges of Cloud Computing, pp. 9–14. IEEE Computer Society (2009)
22. Tang, Q., Gupta, S.K.S., Varsamopoulos, G.: Energy-efficient thermal-aware task scheduling for homogeneous high-performance computing data centers: A cyber-physical approach. IEEE Trans. Parallel Distrib. Syst. 19(11), 1458–1472 (2008)

# Intelligent FTBint Method for Server Resources Protection

Łukasz Apiecionek[1](✉) and Wojciech Makowski[2]

[1] Institute of Technology, Casimir the Great University in Bydgoszcz,
ul. Chodkiewicza 30, 85-064 Bydgoszcz, Poland
lapiecionek@ukw.edu.pl
[2] TELDAT Sp. z o. o.sp.k., Bydgoszcz, Poland
wmakowski@teldat.com.pl

**Abstract.** The subject of this article is the issue of security of network resources in computer networks. One of the main problems of computer networks are Distributed Denial of Service attacks, which can take all server resources and block them. The FTBint intelligent method can manage the amount of network traffic passed to a server and help the server to work during the attack. After the attack is recognized the number of connections provided to the server can be changed in time in an intelligent way. Such solution gives time to the server to dispose of the resources which were allocated incorrectly by the attacker. This new concept is different from the one used in the currently existing methods, as it enables the user to finish his work which had been started before the attack occured. Such user does not suffer from DDoS attacks when the FTBint method is used. The proposed method has already been tested.

**Keywords:** DDoS · Security · Network

## 1 Introduction

There are lots of useful servers operating in the Internet network nowadays. Users need a fast access to information provided by servers from every part of the network. One of the obstacles they can face are Denial of Service attacks, or rather Distributed Denial of Service nowadays. They cause network unavailability by blocking services via seizing system resources in computers in the network until they stop working. A user who has already started working in the system loses the connection and cannot even log out of the system, which has to do it for him after the connection timeout is reached or when a broken connection is detected. DDoS attacks are presently a serious obstacle for IT systems' efficient functioning and some new idea of dealing with them is necessary. There are only few common methods of fighting the DDoS attack problems [2–5,11]. The main idea behind them is to use the Intrusion Detection System and Intrusion Prevention System (IDS/IPS in short) solutions. Such systems are efficient provided that they have a description of well-known attacks or some kind of Artificial Intelligence solution which could learn the actions in some specific scenarios of attack.

© Springer International Publishing Switzerland 2016
S. Kozielski et al. (Eds.): BDAS 2016, CCIS 613, pp. 683–691, 2016.
DOI: 10.1007/978-3-319-34099-9_53

Other solutions suggest using a firewall mounted on the edge of the network. However, the systems based on this concept will only block the incoming traffic on specific ports or IP address ranges, which is not sufficient. During the attack the server will stop responding and become unavailable. This paper presents a new intelligent method called FTBint. This method is able to limit the traffic during the attack in an intelligent way and after the attack is over the network can smoothly return to its previous state. This solution was implemented and tested in a real environment. The structure of this paper is as follows. Section 2 shortly describes the issue of the DDoS attacks and introduces the proposed method for fighting them. Section 3 presents the results of the implementation of the described method. Section 4 provides a conclusion and discussion over the developed method.

## 2    FTB Intelligent Method Description

The DDoS attacks are widely described in the literature [5,11]. These attacks can be performed on various system resources: TCP/IP sockets [11,12] or DNS servers. Regardless of the method, the main principle is to simulate so many correct user connections that their number exceeds the actual system performance and drives it to abnormal operation. The transmission of the attackers' packets is done through the provider's network and if it cannot be blocked, it leads to data link saturation. Such saturation results in a lack of connection to the server. The proposed solutions designed to prevent such situations are not specific and their implementation is associated with many problems. The most common concern is the limited performance of the network devices. However, it is possible to limit the incoming traffic on a firewall and allow the servers to deal with the already established connection. This will let the users finish their work and the new users will be able to connect to the server. This is achieved via implementing intelligent network protection using the FTBint method.

The role of the input firewall is to control the incoming traffic on the edge of the network. When the network is about to give access to the server to the external users, a specific type of traffic has to be allowed by the incoming rules. For instance, in the case of an http server, usually the TCP port 80 has to be opened for the incoming connections. When an attack on the server occurs, this port is still open.

This situation leads in turn to the server overload. Thus, a special firewall FTBint module was developed, the role of which is to filter the traffic on the server's open port and to limit it according to the determined policy in an intelligent way.

During the server's regular work, all packets are passed through and the network is not under any attack. Recognizing the attack by the FTBint method is based on counting the opening network connections to the server during time slots $t1$. The attack is recognized when the opened connections counted in time slots exceed *packet_limit*. As a result, the intelligent FTBint filtration process is launched. At the beginning of the filtration a list of the IP addresses which communicate with the server correctly, i.e. which are not a part of the DDoS attack - *listIP* - is taken from the server. During the filtration each packet is checked whether it is on the list of the valid IP addresses *listIP*; if so, the packet is sent to the network, if not, a counter of the passed packets *packet_counter* is checked whether its value is greater than the allowable packet limit *packet_limit* in a time slot *t1*. If the limit of the packets is exceeded, the packet is dropped - *DROP*. In the following time slot the number of current packets *packet_counter* is zeroed and the above mentioned filtration process is restarted. Afterwards the FTBint method has to regulate the opening connection limit in an intelligent way. It is made in the following way:

- when the limit of the packets is exceeded in a given number of the subsequent time slots *subsequent_time_limit* the FTBint method recognizes that the network is facing a large attack on the server and in order to give the server some time to regain efficiency, the limit of packets *packet_limit* allowed to pass in time slots is decreased,
- if in the following time slots the packet limit is not exceeded, the FTBint method recognizes that the attack on the server started to lose its intensity and the *packet_limit* can be increased.

Changing the packet limit allows the server to handle the incoming connections which may be potentially correct or to release the resources used incorrectly by the attacker. Despite the attack, the server is still accessible to the users who were working on it when the attack was detected.

The process of decreasing the packet limit can depend on the server type, its needs and kind of work. Moreover, the limit values may require experimental determination or setting them basing on the server's resources, its operating system, the amount of memory and processor type.

A pseudocode of the main part of the algorithm responsible for passing the packets as well as narrowing the limits is shown below:

Algorithm pseudocode

```
packet_counter:=packet_counter + 1

if packet_counter < packet_limit then
    packet pass
else
    begin
    if IP address in listIP then
        packet pass;
    else
        packet drop;
    end;
if times_slots ends then
    begin
```

```
if packet_counter>packet_limit then
    overdrop_times=overdrop_times + 1;
    packet_counter=0;
if overdrop_times> subsequent_time_limit then
    packet_limit=packet_limit/2;
    overdrop_times=0;
else
    packet_limit=packet_limit*2;
    overdrop_times=0;
end;
```

The *packet_counter* variable contains the number of packets which are passed through in a certain time slot. When its value does not exceed the permissible limit *packet_limit*, the packet is sent, but when the limit is exceeded, further tests are performed. If the packet is present in the database of the known IP addresses *listIP*, the packet is passed through, otherwise it is dropped. When the *time_slots* timeout expires, a verification is performed whether the limit of packets was exceeded in this time slot. In this case, in the implemented method the counter of the limit was increased in the subsequent time slots. When the limit was exceeded in the following two time slots, the allowable limit of packets was decreased (*packet_limit/2*). If in the next time slots the limit was not exceeded, the limit of packets was increased (*packet_limit*2*).

## 3   Implementation Results

In order to verify if the server works continuously indeed, the method was implemented and tested. The implementation consisted of a module for the firewall's IPTables module on a Debian Linux system with 2.6.32 kernel. The tests simulating the most common types of attacks on the servers were performed on a simple network, which was built for this purpose according to well known structures and routing protocol requirements 789. A graphical interface to Asterix FreePBX distribution, working under CentOS with kernel 2.6.32 was used as an http server and an Apache 2.2.15 server, equipped with 1 GB of RAM, was used as a receiver. In order to perform the attack, a Sender machine was used - based on a Debian Linux with 2.6.32 kernel (equipped with 512 MB of RAM) and DDOSIM software (Layer 7 DDoS Simulator v0.2). The server was connected to a real network through the firewall with the FTBint method implemented, which was running on a Linux Debian with 2.6.32 kernel with 512 MB of RAM.

During the test, the memory usage and the http server response times were observed. The attack consisted of sending a large amount of HTTP Get requests to the server. Without the FTBint method, the server's memory usage rose up to 100 % and the server stopped responding. After launching the FTBint method, RAM memory usage was observed.

The FTBint method was configured with the following limits:

– time slot for analyzing the amount of transmitted data *time_slots* = 1 s;
– allowable limit of packets *packet_limit* = 30;
– minimal limit of packets = 10;

Five tests of the attack on the server were performed, each lasting one hour and under the following conditions:

– 100 HTTP GET requests sent every 30 s;
– 1000 HTTP GET requests sent every 30 s;
– 2000 HTTP GET requests sent every 10 s;
– 10000 HTTP GET requests sent every 10 s;
– 50000 HTTP GET requests sent every 30 s.

None of the cases resulted in a server overload. Prior to the attack, a connection with the server through the firewall was established and it remained active because it had been started before the attack and the computer was recognized as allowed to communicate with the server (Fig. 1).

During the test the amount of free RAM memory was observed on the http server and on the firewall. On the http server the memory remained mostly on a constant level (Fig. 2).

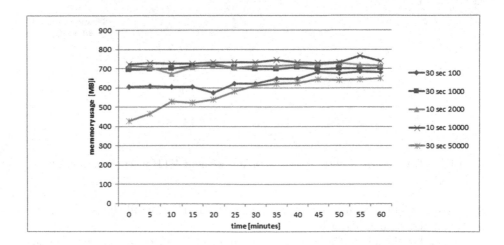

**Fig. 1.** RAM usage on http server

On the firewall the RAM usage remained on a constant level. During the test which consisted of sending 1000 packets every 10 s, the RAM memory usage was higher, which was probably a result of other operations of the device.

The RAM memory usage test was also performed in a network built without using the proposed FTBint method. The results are shown in Fig. 3. Without

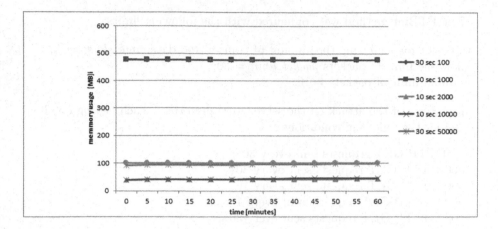

**Fig. 2.** RAM usage on firewall

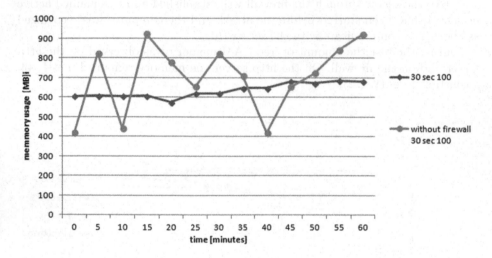

**Fig. 3.** RAM usage on http server with and without FTB method on firewall

the active method, the memory usage rose up to 100 % several times (the server was equipped with 1 GB of RAM).

In the description of the method a mechanism for intelligent narrowing the amount of packets in a single time slot was mentioned. Figure 4 shows a fragment of the operation of the implemented FTBint mechanism. It regulates the number of packets which can be transmitted in a time slot dynamically, depending on the recognized server load.

It is worth emphasizing that after the attack, the network regains its ordinary state. It means that using the FTBint method allows the network to work without requiring the administrator's action. The authors are planning to implement an improved solution for recognizing DDoS attacks, which will not be limited to counting the connections during the time slots. Some literature concerning using fuzzy sets for this purpose has already been published [1, 6, 8–10]. The authors are conducting research on this topic as well as in order to improve this element [7].

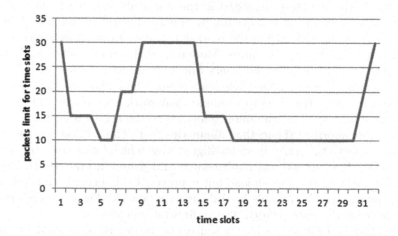

**Fig. 4.** Intelligent packet limit on time slots by FTBint method

**Table 1.** Method comparison

| Criteria | Existing methods | FTBint method |
|---|---|---|
| Server memory usage | Up to 100 % | Constant defined amount |
| Server availability during attack | No response | Response |
| Possibility to connect new user to the server during attack | Impossible | Possible, Network browsers tries to connect multiple time, so the user get the chance to connect |
| Possibility to finish the task with the server during attack | impossible | Possible |
| Connection limit during attack | Zero | Limited to defined amount by administrator according to proposed algorithm |
| Connection limit after the attack | Needs time and administrator's work to get back | No work required to back to previous state |
| Administrator work required after the attack | Required | Not required |

The Table 1 compares the FTBint method with other existing methods according to the server memory usage, possibility of initiating a connection during the attack, the connection limit after the attack stops and the required administrator's actions after the attack.

In this article the FTBint concept of eliminating DDoS attacks was introduced. While the methods suggested in the literature can block the access to the resources when the attack occurs, by using a firewall along with IDS/IPS mechanisms, during the time of the blockage no user from an external network can connect to the desired resources. Moreover, such solution does not allow to complete the work started by the users who were already connected. The users who worked with the server lose their connection. The FTBint method described in this article allows the users to continue their work even if the attack occurs. It is possible for the users who were connected to the server prior to the attack and the server informed the method about this fact. The FTBint method limits the connections to the server in an intelligent way, which lets a new user connect to the server. The method was implemented and tested in practice. It does not cause an increase of the firewall load but it prevents the server from overload by keeping the its load at a stable level during the attack. The proposed method may be successfully implemented on any firewall-type device.

The author is ready to provide the sources of the described method for further analysis.

# References

1. Angryk, R.A., Czerniak, J.: Heuristic algorithm for interpretation of multi-valued attributes in similarity-based fuzzy relational databases. Int. J. Approximate Reasoning **51**(8), 895–911 (2010)
2. Apiecionek, Ł., Czerniak, J.M., Zarzycki, H.: Protection tool for distributed denial of services attack. In: Kozielski, S., Mrozek, D., Kasprowski, P., Małysiak-Mrozek, B., Kostrzewa, D. (eds.) BDAS 2014. CCIS, vol. 424, pp. 405–414. Springer, Heidelberg (2014)
3. CA-1996-01, C.A.: UDP port denial-of-service attack. http://www.cert.org/advisories/CA-1996-01.html
4. CA-1996-21, C.A.: TCP syn flooding and ip spoofing attacks. http://www.cert.org/advisories/CA-1996-21.html
5. Chang, R.K.: Defending against flooding-based distributed denial-of-service attacks: a tutorial. IEEE Commun. Mag. **40**(10), 42–51 (2002)
6. Czerniak, J.: Evolutionary approach to data discretization for rough sets theory. Fund. Inform. **92**(1–2), 43–61 (2009)
7. Czerniak, J.M., Apiecionek, Ł., Zarzycki, H.: Application of ordered fuzzy numbers in a new OFNAnt algorithm based on ant colony optimization. In: Kozielski, S., Mrozek, D., Kasprowski, P., Małysiak-Mrozek, B., Kostrzewa, D. (eds.) BDAS 2014. CCIS, vol. 424, pp. 259–270. Springer, Heidelberg (2014)
8. Dobrosielski, W.T., Szczepański, J., Zarzycki, H.: A proposal for a method of defuzzification based on the golden ratio-GR. In: Atanassov, K.T., et al. (eds.) Novel Developments in Uncertainty Representation and Processing. AISC, vol. 401, pp. 75–84. Springer, Switzerland (2016). http://dx.doi.org/10.1007/978-3-319-26211-6_7

9. Ewald, D., Czerniak, J.M., Zarzycki, H.: Approach to solve a criteria problem of the ABC algorithm used to the WBDP multicriteria optimization. In: Angelov, P., et al. (eds.) IS 2014. AISC, vol. 322, pp. 129–137. Springer, Heidelberg (2015)
10. Kosiński, W., Prokopowicz, P., Ślęzak, D.: On algebraic operations on fuzzy reals. In: Rutkowski, L., Kacprzyk, J. (eds.) Neural Networks and Soft Computing. ASC, vol. 19, pp. 54–61. Springer, Heidelberg (2003)
11. Moore, D., Shannon, C., Brown, D.J., Voelker, G.M., Savage, S.: Inferring internet denial-of-service activity. ACM Trans. Comput. Syst. (TOCS) **24**(2), 115–139 (2006)
12. Schuba, C.L., Krsul, I.V., Kuhn, M.G., Spafford, E.H., Sundaram, A., Zamboni, D.: Analysis of a denial of service attack on TCP. In: 1997 IEEE Symposium on Security and Privacy, pp. 208–223. IEEE (1997)

# Lexicon-Based System for Drug Abuse Entity Extraction from Twitter

Ferdaous Jenhani[✉], Mohamed Salah Gouider, and Lamjed Ben Said

SOIE Laboratory, Institut Suprieur de Gestion de Tunis,
Rue de la Liberte. Bouchoucha, Tunis, Tunisia
jenhaniferdaous@gmail.com, ms.gouider@yahoo.fr, bensaid_lamjed@yahoo.fr

**Abstract.** Drug abuse and addiction is a serious healthcare problem and social phenomenon that has not received the interest deserved in scientific research due to the lack of information. Today, social media have become an ubiquitous source of information in this field since they are the environment on which addicted individuals rely to talk about their dependencies. However, extracting salient information from social media is a difficult task regarding their noisy, dynamic and unstructured character. In addition, natural language processing tools (NLP) are not conceived to manage social data and cannot extract semantic and domain-specific entities.

In this paper, we propose a framework for real time collection and analysis of Twitter data which heart is a personalized NLP process for the extraction of drug abuse information. We extend Stanford CoreNLP pipeline with a customized annotator based on fuzzy matching with drug abuse and addiction lexicons in a dictionary. Our system, ran on 86 041 tweets, achieved 82 % of accuracy.

**Keywords:** Information extraction · Twitter · NLP · Stanford CoreNLP

## 1 Introduction

Drug addiction is the dependency on chemical substance characterized with compulsive drug seeking and use despite harmful consequences on the addicted individual and his entourage. It destroys individuals brain cells and affects capacity to receive, analyze and manipulate information which changes the physiological and psychological behavior of addicted individual. Drug abuse and addiction is a disease like cancer and diabetes for which there are medications and behavioral therapy treatments. However, addicted individuals dont talk about their dependency to their entourage which hardens their help and limits medical discoveries in this field. However, they find in social media a new society where they can talk freely about their problems and feelings, share their experience, suffer their loneliness, pain, anxiety and depression, or even supply their substances. Thus, social media have become the unique and richer source of information in this field which drew our interest to exploit and analyze them in order to extract useful

© Springer International Publishing Switzerland 2016
S. Kozielski et al. (Eds.): BDAS 2016, CCIS 613, pp. 692–703, 2016.
DOI: 10.1007/978-3-319-34099-9_54

information such as drug types, routes of intake, units of doses, physical effects, psychoactive effects and medical conditions, useful for building interesting drug abuse and drug side effects patterns. We aim to demonstrate the richness of this external data source with information useful to push and improve scientific research in drug abuse and addiction epidemiological area.

Nevertheless, extracting salient information from social media is a challenging task compared to well-formed text because of their complex character since they are conversational free and short texts, unstructured, ungrammatical, informal, plenty with colloquialism, slang terms and non-standard abbreviations especially in drug abuse related communications which harden the extraction of trusted information useful for scientific research in this field. This raises challenges for existing NLP tools which are designed to handle grammatically-correct texts and fail with social data. Moreover, they are capable to detect efficiently common named entities such as Person, Location and Organization, and fail to detect domain-specific entities.

Stanford is the best in extracting common named entities from Twitter[1] data by achieving the highest average precision compared to Twitter-tailored NLP systems [15]. Therefore, we chose it to make extensions and enable it to detect domain-specific named entities and more semantic information. Indeed, we propose a customized NLP pipeline for Twitter analysis and information extraction. We extend Stanford NER capability with a customized annotator based on fuzzy lookup on prepared dictionary to overcome Twitter data informality. We implemented our system and test it on a series of tweets and got interesting results. The remainder of this paper is organized as follows; in Sect. 2, we investigate approaches of clinical information extraction from social media in general and drug abuse information in particular. Section 3 describes the proposed system as well as its different modules by explaining our choice to different techniques used. Section 4 is reserved to discuss experimental results, and, we conclude in Sect. 5.

## 2   Related Work

The interest in clinical information extraction has led to the development of sophisticated tools to extract information from regular, well-formed and grammatically correct documents. We distinguish MetaMap [6] and cTackes [22] which look for disease, signs, symptoms, drugs, etc. based on UMLS- Unified Medial Language System meta-thesaurus[2]. Nevertheless, their use for information extraction from social media especially conversational resources such as Facebook, Twitter, forums, etc., is not a rational choice regarding their informal and noisy character. To succeed information extraction from social media, some approaches used lookup matching with terms in dictionaries thanks to the availability of medical ontologies, thesaurus and online clinical data bases and

---

[1] www.Twitter.com.

[2] https://www.nlm.nih.gov/research/umls/.

resources. PREDOSE [12], a web semantic system, used Prefix Trie lookup technique on DAO,[3] which is a customized drug abuse ontology, to extract drug abuse semantic entities from forum posts in order to predict abuse behavior and detect adverse drug reactions. In [2] authors used string lookup on UMLS, CHV[4] and FAERS[5] to extract medical named entities and medical condition events in patient social media for diabetes prevention. In [7] authors used BioNER module of BeFree System to detect drug brand mentions from Twitter. Authors used BioNER including lengthy diseases collected from UMLS. Also, in [23], authors used Meaning-cloud for named entity recognition to identify drugs and effects from Spanish patient forums. This tool is based on dictionaries and authors built tailored dictionaries from CIMA[6] and MedDRA[7]. Some contributors rely on the use of learning techniques especially supervised algorithms. Supervised learning techniques require large and sufficient sets of manually annotated data for more accurate tags prediction of named entities. SVM, HMM, CRF, ME, etc., have shown good performance in the identification of disease names, symptoms and treatments for infectious disease detection and prediction tasks [1,4,5,8–10,13,14,18,20,21], the extraction of names of drugs and their side effects as well as medical conditions to uncover adverse drug reactions [2,3]. In [17], authors used CRF with bag of words, n-grams and word shape features. Despite the effectiveness of CRF, slang terms, misspellings and non standard abbreviations cannot be easily learnt. The idea of [24] is to integrate the use of gazetteers and Levenstein distance as feature types. With supervised learning, the choice of the classifier, the selection of features and building the training data set are major faced difficulties. Moreover, to learn misspelling and non-standard abbreviations, it is not evident to collect the sufficient training set.

## 3    Proposed Approach

Our work belongs to Named Entity Recognition (NER) which is one of the most important information extraction tasks. It addresses the problem of identification of named entities in free-form text and their classification on predefined types such as Organization, Person, Place, Date, Duration, Number, Currency, etc. It can additionally include extracting descriptive information from the text about the detected entities through filling of a small-scale template [19]. The task of NER was formally introduced in 1995 in the sixth Message Understanding Conference (MUC-6)[8] organized by DARPA. With the emergence of social media, new challenges raised for this task regarding the particularity of these data. In this work, we propose a customized framework for real-time Twitter data collection and analysis in order to extract drug abuse and addiction related

---

[3] http://knoesis-hpco.cs.wright.edu/predose/ontologies/DAO.owl.
[4] http://consumerhealthvocab.org/.
[5] http://www.fda.gov/.
[6] http://www.aemps.gob.es/cima.
[7] http://www.meddra.org/.
[8] http://cs.nyu.edu/faculty/grishman/muc6.

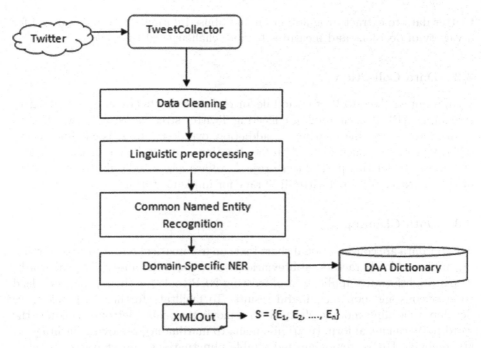

**Fig. 1.** Lexicon-based entity extraction framework

information. All the resources, modules and techniques employed are described in detail in subsequent sections and the corresponding framework is in Fig. 1.

### 3.1 Twitter Data

Twitter is the most popular social network producing millions of user generated contents about diverse topics among others the medical and healthcare one. Twitter gained its importance in public healthcare field since 2011 with the natural disasters that occurred in Japan and destroyed telephone lines and healthcare buildings. Doctors and patients exchange health care information using Twitter. It becomes an efficient bio-surveillance and risk factor assessment tool quicker and cheaper than the surveillance achieved by clinical centers still time-consuming processes. Moreover, it offers a geo-localization of diseases thanks to geographical information whatever explicit or implicit in text messages. It is a new public health capability thanks to the expressiveness of messages and coverage of topics especially those which people cannot discuss with medical staff such as drug addiction, sexual problems, psychological disorders, etc. Twitter is an epidemiological asset that created the web epidemiology field by participating in many medical applications such as influenza prediction, surveillance and monitoring of cancer, depression and other cardiovascular and infectious diseases, as well as health care quality of service enhancement [3]. In this work, we exploit twitter data to extract drug addiction and abuse information to be analyzed for a variety of decisions and measures to raise challenges of this field.

## 3.2  Data Collection

We developed TweetCollector module for real time tweets crawling from Twitter streaming API. It is charged of collecting Twitter streams using keywords such as drug abuse, cocaine, marijuana, addiction, overdose, etc. It is capable also to extract meta-data such as user profile, location, date and time of the message, etc. However, for this part of our project, we are interested only to the content of the message. We use Twitter4j library for implementation.

## 3.3  Data Cleaning

Twitter data are conversational short text, ungrammatical and informal containing undesirable characters, abbreviations, external links and colloquial words. The straightforward application of existing NLP tools for their processing lead to erroneous and non meaningful results. To facilitate further NLP tasks, we develop a cleaning module charged of: removing repeated letters to reduce the word to its canonical form (mariiiiijuanaaa to marijuana); removing hashtags #, @; removing URLs; removing undesirable punctuations and characters (?; !; ,; *; (; ); $, ]; [.&).

## 3.4  Linguistic Preprocessing

We perform word and sentence tokenization, stemming and lemmatization as well as POS-Part Of Speech tagging using Stanford CoreNLP. All the steps of the mentioned processing pipeline are necessary and indispensable for upcoming named entity recognition task.

## 3.5  Entity Extraction

Common named entities mainly Person, Location, Date, Organization, Number and Duration are detected using Stanford NER capabilities. However, to extract domain specific entities mainly:

Drug (DRUG): the substance used, it is mentioned explicitly in the post text or using a slang term;

Physical and Psychoactive effects (PSY-EFF, PHY-EFF): drug abuse causes physical pains, but major effects are psychoactive since they affect directly the behavior of the individual;

Medical condition (MED-COND): an abuse act is followed with a set of medical conditions such as overdose and addiction;

Unit (UNIT): it is necessary to know the doses taken by an individual which is detected thanks to numbers indentified by Stanford capability followed with the unit detected by our annotation system;

Route of Intake (ROI): It is necessary for some analysis to know the method used to take drugs. Each drug has one or more ways of administration.

We develop a customized annotator that extends Stanford CoreNLP pipeline [16] based on lexical lookup on Drug Abuse and Addiction Dictionary (hereafter DAA Dictionary) (Table 1).

**DAA Dictionary.** It is the set of tab separated triples (ET: Entity Type, RC: Real Concept, ST: Slang Term) manually created and fed with clinical information from DAO-Drug Abuse Ontology which is the first ontology dedicated to the drug abuse field created by domain experts namely CITAR researchers. We enriched it also with terms from online resources among others NIDA[9], WHO[10], DrugFacts[11], DrugBank[12]. Actually, social network discussions about healthcare issues are plenty with colloquialism. We used DrugSlang[13] to enrich our dictionary with drug slangs. Also side effects are described using slang terms, thus, we integrate terms from CHV-Consumer Health Vocabulary[14]. Substance prescription field is rich with lexicons and terms and DrugBank itself contains 7759 terms. However, we chose commonly used lexicons by digital abusers. Thus, DAA contains 318 entries where 225 are drug types, 25 physical effects, 33 psychological effects, 14 medical condition, 14 routes of intake and 7 units with their corresponding slang terms. Table 2 is an excerpt of DAA Dictionary:

**Table 1.** Excerpt of DAA dictionary

| Entity type | Real term | Slang term |
| --- | --- | --- |
| DRUG | marijuana, cocaine, heroin  tramadol, MDMA, LSD | puder, ganja, acid, junk  coke, bomb, hashish |
| PHY-EFF | Ciphalelgia, chronic-pain  Cardiovascular-problem | Headache, nausea, vomiting  constipation, stomache, pain |
| PSY-EFF | Depression, pleasure,anxiety | Dizzy, drowsy, bliss |
| MED-COND | Addiction, Overdose  Withdrawal | hook, Agonies  Dependent |
| ROI | Injection, smoking, oral  intranasal | Shot, Snort, blow, sniff |
| UNIT | Pill, tab, bag  milligram, cigarette | mg, ml, cigar |

---

[9] http://www.drugabuse.gov.
[10] http://www.who.int.
[11] http://www.drugabuse.gov/.
[12] http://www.drugbank.ca/.
[13] https://www.noslang.com/drugs/dictionary.
[14] http://consumerhealthvocab.org.

**Jaro Winkler Distance.** Jaro-Winkler distance [11] is the measure of similarity between strings originated from the Jaro distance with the Winkler modification. The higher the value of the distance is, more similar the strings are. The score is normalized so 1 for exact match and 0 for no similarity. It is best suited for short strings; therefore, for n-gram tokens we handle each unigram separately, then their concatenation.

**Annotation System.** Stanford offers the possibility to extend its pipeline using different methods. We propose a customized annotator which is invoked when Stanford NER, used for common entity extraction, fails and produces an outside 'O' tag. We check the corresponding lemma of the word encountered and measure its distance to each term in the dictionary by using Jaro-Winkler distance (JWD in the algorithm) and we keep the maximum of them. We fix a threshold value that allow us to capture possible correct semantic entities without being close neither limited to the dictionary. If kept distance is obove or equal to the threshold value, the entity type selected is kept as a tag. Our system is capable to:

- Capture the category of the word whatever its grammatical derivation thanks to lemmatization preprocess. For example, for the route of intake "inject", "injected", "injects", "injecting", etc. could be captured and recognized as ROI despite the dictionary contains only the term Injection.
- Solve the problem of spelling mistakes, non standard abbreviations as well as new slang terms which are derived from the real term. For example, if the user writes "marihuana", "mariwana", etc. instead of "marijuana",it is marked as DRUG entity type.
- Detect classes of n-gram tokens such as "heart attack" and "loss of appetite".

The following is the corresponding algorithm of the proposed annotator:

```
DAA _EntityAnnotation Algorithm
BEGIN
Input:      D: Twitter data set
            DAA {ET, RT, ST}: Dictionary
            th: distance threshold
Output:     S={E1,..., En}

for each S:sentence in D do
   for each Ti:token in S do
   while (T.NER == 'O') do
         Ch = T + ' ' + NextT
         Tag <-- GetTag(Ch)
         if (Tag != 'O')
         E <-- Assign(tag,NextT)
```

```
          S.Add(E)
      End if
      End if
 End while
          E <-- Assign(tag,Ti)
          S.Add(E)
   end for
end for

Function GetTag (token)
BEGIN
          While (not EOF)
              dr <-- JWD(tokens, RT)
              ds <-- JWD(tokens, ST)
              d <-- max(dr,ds)
              CurrTag <-- ET
            end while
        if(d>=th)
          Tag <-- CurrTag
        else
            Tag <-- 'O'
        end if
return Tag
END GetTag
END DAA _Entity_Annotation
```

For example, from the following sentence: "Jack had a heart attack after injecting 2 mg of heroin." Our system generates the output:
$< PERSON > Jack < /PERSON >$
$< O > had < /O >$
$< O > a < /O >$
$< MED - COND > heart < /MED - COND >$
$< MED - COND > attack < /MED - COND >$
$< O > after < /O >$
$< ROI > injecting < /ROI >$
$< NUMBER > 2 < /NUMBER >$
$< UNIT > mg < /UNIT >$
$< O > of < /O >$
$< DRUG > heroin < /DRUG >$

## 4   Experimental Results

The objective of this experimental study is to measure the performance and
accuracy of the proposed annotator in extracting correct semantic entities to be

used for more complex pattern identification tasks such as drug abuse events and drug side effects. The proposed framework was implemented using Java eclipse. We use Apache Commons package which implements Jaro-Winkler distance. We tested the system on 86 041 Tweets collected with our TweetCrawler for five days with a debit of two hours per day. The accuracy of our named identification system is affected by the value of the distance threshold selected. Thus, we have taken a subset of the data collected and analyze them each time with a slight variation of the threshold value, and calculate the number of positive entities, false positives and false negatives as well as the precision and recall. The precision is a measure of trust that instances marked as positives are really positives. Recall is a measure of trust that all positive instances are identified. These measures of trust are calculated as follows:

Precision (P) = P/P+FP

Recall(R) = P/P+FN

The results are depicted in Table 2.

We observe that the precision increases when we increase the threshold value.

**Table 2.** Statistics collected with a slight variation of the threshold distance value

| Threshold | #P | #FP | #FN | P | R |
|-----------|------|------|-----|------|------|
| 0.89 | 1874 | 1012 | 136 | 0.64 | 0.93 |
| 0.9 | 1886 | 612 | 180 | 0.77 | 0.91 |
| 0.92 | 1783 | 336 | 223 | 0.84 | 0.88 |
| 0.94 | 1765 | 272 | 318 | 0.87 | 0.84 |

This is obvious since more we are close to 1, more we are close to the terms of the dictionary and to the exact matching which is not our aim since we are handling free-form text generally informal. On the other side, the recall decreases gradually which means that we are losing the correct instances which are expressed using slang terms, abbreviations or misspellings. Actually, we are interested in high recall rather than high precision to guarantee that we are identifying all possible positive objects which are not all the time correctly written in social networks and conversational text such as Twitter. Therefore, we carry on our experiments using 0.89 threshold value. In Table 3 we add another measure which is a trade-off between precision and recall called F-measure calculated as follows: F-measure (F) = 2P/2P+FP+FN

According to results in Table 3 the proposed system is more efficient in detecting drugs, medical conditions and units. Thus, the proposed annotator is a good basis to detect efficiently drug abuse events rather than drug side effect event types. This is due, especially, to the frequent use of n-gram terms to describe physical and psychoactive effects. Indeed, the capacity of our annotator decreases when n increases in n-gram terms. In general, drug names, units and medical conditions are represented with unigrams which justifies the little number of false positives and false negatives.

**Table 3.** Number of positives, false positives and false negatives for each class category with a distance value fixed to 0.89

| Entity Type | #P | #FP | #FN | P% | R% | F% |
|---|---|---|---|---|---|---|
| DRUG | 50838 | 12632 | 1360 | 80 | 97 | 87 |
| PSY-EFF | 1606 | 3801 | 169 | 29 | 90 | 44 |
| PHY-EFF | 2250 | 2594 | 154 | 46 | 93 | 62 |
| MED-COND | 5450 | 2639 | 157 | 67 | 97 | 79 |
| UNIT | 1036 | 435 | 10 | 70 | 99 | 82 |
| ROI | 3706 | 3738 | 185 | 49 | 95 | 65 |
| TOTAL | 64886 | 25839 | 2035 | 71 | 96 | 82 |

In future work, we will ameliorate our annotator to be more efficient in detecting n-gram tokens. We are working on a solution to automatically update our dictionary with new terms (slangs, drugs, etc.) regarding that the drug abuse vocabulary used in social networks is continuously changing.

## 5 Conclusion

Social media have become a new capability in the healthcare field which analysis may uncover interesting clinical patterns. However, social media are complex compared to well-formed documents which harden information extraction. In this paper, we proposed a complete system for Twitter data crawling, cleaning and analysis which heart is a customized annotator that extends Stanford pipeline for drug abuse information extraction. The extracted inforation is useful for the identification of interesting drug abuse, addiction and drug side effects events which will be held in future work.

## References

1. Abboute, A., Boudjeriou, Y., Entringer, G., Azé, J., Bringay, S., Poncelet, P.: Mining twitter for suicide prevention. In: Métais, E., Roche, M., Teisseire, M. (eds.) NLDB 2014. LNCS, vol. 8455, pp. 250–253. Springer, Heidelberg (2014)
2. Abeed, S., Graciela, G.: Portable automatic text classification for adverse drug reaction detection via multi-corpus training. J. Biomed. Inf. **53**, 196–207 (2014)
3. Abeed, S., Rachel, G., Azadeh, N., Karen, O., Karen, S., Swetha, J., Tejaswi, U., Graciela, G.: Utilizing social media data for pharmacovigilance: a review. J. Biomed. Inf. **54**, 202–212 (2015)
4. Achrekar, H., Gandhe, A., Lazarus, R., Yu, S., Liu, B.: Twitter improves seasonal influenza prediction (2012)

5. Aramaki, E., Maskwa, S., Morita, M.: Twitter catches the flu: detecting influenza epidemics using twitter. In: Proceedings of 2011 Conference on Empirical Methods in Natural Language Processing, Edinburgh, Scotland, pp. 1568–1576 (2011)
6. Aronson, A.: Effective mapping of biomedical text to the umls metathesaurus: the metamap program. In: Proceedings of the AMIA (2001)
7. Carbonell, P., Mayer, M., Bravo, À.: Exploring brand-name drug mentions on twitter for pharmacovigilance. In: Digital Healthcare Empowering Europeans 2015 European Federation for Medical Informatics (EFMI), pp. 55–59 (2015)
8. Corley, C.D., Cook, D.J., Mikler, A.R., Singh, K.P.: Using web and social media for influenza surveillance. In: Arabnia, H.R. (ed.) Advances in Computational Biology. Advances in Experimental Medicine and Biology, vol. 680, pp. 559–564. Springer, New York (2010)
9. Culotta, A.: Toward detecting influenza epidemics by analyzing twitter messages. In: First Workshop on Social Media Analysis (SOMA 2010), Washington, USA (2010)
10. De Choudhury, M., Gamon, M., Counts, S., Horvitz, E.: Predicting depression via social media. In: Association for the Advancement of Artificial Intelligence (2013)
11. De Coster, X., De Groote, C., Destin, A., Deville, P.: Mahalanobis distance, jarowinkler distance and ndollar in usigesture (2012)
12. Delroy, C., Gary, A., Raminta, D., Amit, P., Drashti, D., Lu, C., Gaurish, A., Robert, C., Kera, Z., Russel, F.: PREDOSE: a semantic web platform for drug abuse epidemiology using social media. J. Biomed. Inf. **46**(6), 985–997 (2013)
13. Dredze, M.: How social media will change public health. IEEE Intell. Syst. **27**(4), 81–84 (2012). IEEE Computer Society
14. Lee, K., Agrawal, A., Choudhary, A.: Real time disease surveillance using twitter data: demonstration on flu and cancer. In: KDD 2013, Chicago Illinois, USA (2013)
15. Leon, D., Diana, M., Giuseppe, R., van Marieke, E., Genevieve, G., Raphal, T., Johann, P., Kalina, B.: Analysis of named entity recognition and linking for tweets. Inf. Process. Manage. **51**, 32–49 (2015)
16. Manning, C.D., Surdeanu, M., Bauer, J., Finkel, J., Bethard, S.J., McClosky, D.: The stanford corenlp natural language processing toolkit. In: Proceedings of 52nd Annual Meeting of the Association for Computational Linguistics: System Demonstrations, pp. 55–60 (2014)
17. Metke-Jimenez, A., Karimi, S.: Concept extraction to identify adverse drug reactions in medical forums: a comparaison of algorithms (2015)
18. Paul, M., Dredz, M.: You are what you tweet: analyzing twitter for public health. In: Proceedings of the Fifth International AAAI Conference on Weblogs and Social Media (2011)
19. Piskorski, J., Yangarber, R.: Information extraction: Past, present and future. In: Poibeau, T., Saggion, H., Piskorski, J., Yangarber, R. (eds.) Multi-source, Multilingual Information Extraction and Summarization. Theory and Applications of Natural Language Processing, pp. 23–49. Springer, Heidelberg (2013)
20. Sadilek, A., Kautz, H., Silenzio, V.: Modeling spread of disease from social interactions. In: Proceedings of the Sixth International AAAI Conference on Weblogs and Social Media (2012)
21. Sadilek, A., Kautz, H., Silenzio, V.: Predicting disease transmission from geo tagged micro blog data. In: Proceedings of the Twenty-Sixth AAAI Conference on Artificial Intelligence, pp. 136–142 (2012)

22. Savova, G., Bethard, S., Styler, W., Martin, J., Palmer, M., Masanz, J., Ward, W.: Towards temporal relation discovery from the clinical narrative. In: Proceedings of AMIA Annual Symposium (2009)
23. Segua-Bedmar, I., Martinez, P., Revert, R., Moreno-Shneider, J.: Exploring spanish health social media for detecting drug effects. Med. Inf. Decis. Making **15**, S6 (2015). From Louhi 2014: The Fifth International Workshop on Health Text Mining and Information Analysis. Gothenburg, Sweden
24. Zirikly, A., Diab, M.: Named entity recognition for arabic social media. In: Proceedings of NAACL-HLT, pp. 176–185, Denver, Colorado (2015)

# Manifold Learning for Hand Pose Recognition: Evaluation Framework

Maciej Papiez, Michal Kawulok$^{(\boxtimes)}$, and Jakub Nalepa

Institute of Informatics, Silesian University of Technology, Gliwice, Poland
michal.kawulok@polsl.pl

**Abstract.** Hand pose recognition from 2D still images is an important, yet very challenging problem of data analysis and pattern recognition. Among many approaches proposed, there have been some attempts to exploit manifold learning for recovering intrinsic hand pose features from the hand appearance. Although they were reported successful in solving particular problems related with recognizing a hand pose, there is a lack of a thorough study on how well these methods discover the intrinsic hand dimensionality. In this study, we introduce an evaluation framework to assess several state-of-the-art methods for manifold learning and we report the results obtained for a set of artificial images generated from a hand model. This will help in future deployments of manifold learning to hand pose estimation, but also to other multidimensional problems common to the big data scenarios.

**Keywords:** Hand pose estimation · Gesture recognition · Manifold learning

## 1 Introduction

Rapid increase in the amount of data collected and generated nowadays, poses new challenges concerned with extracting information from different sources. The real-world big data, such as digital photos and videos, recorded speech, medical images or word datasets, hide valuable knowledge behind a high number of dimensions [8,12,20,24]. If an $N$-dimensional image (which consists of $N$ pixels), presents a natural object, such as a human hand, then the observation space is much more constrained. In such cases, dimensionality reduction techniques facilitate classification, visualization and compression [15]. In a perfect scenario, these methods would output a representation that discovers the intrinsic dimensionality and structure of the data. In the work reported here, we focus on reducing the dimensionality of images that present a human hand. The hand is characterized with 26 degrees of freedom that define the bent angles of the joints, hence ideally the particular dimensions of the reduced feature space would correspond to the values of these angles.

There have been many attempts to model human hands and extract their features from still images, preceded with skin segmentation [9,10] and hand landmarks detection [5,6,17]. The majority of existing methods extract shape

© Springer International Publishing Switzerland 2016
S. Kozielski et al. (Eds.): BDAS 2016, CCIS 613, pp. 704–715, 2016.
DOI: 10.1007/978-3-319-34099-9_55

features from hand silhouettes, for example using shape context [1], Hausdorff distance [7], orientation histograms [3], or by combining multiple descriptors [18]. In such feature spaces, the similarity between two images presenting a hand can be established to recognize a hand pose given a set of reference images. Another, yet little investigated approach is to learn a feature space of reduced dimensionality from a set of images [4,27]. Such methods are based on the premise that the observed data lie on a low-dimensional manifold, which can be discovered given a sufficiently large training set. However, although these methods are effective in solving particular hand pose estimation problems, it is worth determining which manifold learning procedure offers the greatest coherence with the intrinsic dimensionality of a human hand.

In this paper, we propose how to evaluate the dimensionality reduction techniques, and we report the results obtained for a set of images generated from a hand model (Libhand [26]), encompassing two classes of the hand poses. We investigate several nonlinear dimensionality reduction methods, and we compare them with the linear one, namely principal components analysis (PCA). Although the intrinsic dimensions that correspond to the degrees of freedom of a hand are nonlinearly mapped onto the image space, the linear methods may be applicable for simple cases and this is actually demonstrated in this paper.

In Sect. 2, we present the state of the art concerning manifold learning methods in the context of hand pose modeling. In Sect. 3, we outline and justify the proposed evaluation framework. The obtained experimental results are reported and discussed in Sect. 4, and the paper is concluded in Sect. 5.

## 2 Related Work

### 2.1 Manifold Learning

Real-world, highly-dimensional data form certain structures in the input spaces, which may often be modeled with a manifold—a topological space, whose every point has a neighborhood equivalent to the neighborhood in a Euclidean space. There have been a number of techniques proposed to learn the manifold structures, and the most common ones are outlined in this section.

Isomap (*isometric mapping*) [25] is aimed at preserving pairwise geodesic distances between points in a high-dimensional space. Geodesic distances are measured along the manifold, rather than using simple Euclidean metrics. In such a way, two points lying on extreme ends of a Swiss roll may be close to each other in terms of Euclidean distance, but spread apart when the geodesic distance is considered. Isomap preserves these geodesic properties by constructing a neighborhood graph $G$, whose each point is connected to its $k$ nearest neighbors ($k$-NN). The geodesic distances are estimated as the shortest paths in $G$ between the data points [11], calculated with the Dijkstra's or Floyd's algorithm. Finally, the low-dimensional embedding is obtained using multidimensional scaling.

It is important to properly tune the value of $k$ in Isomap—if it is too large, then some paths may be "short-cut" in such a way that they no longer correspond to the geodesic distances, but represent the Euclidean paths, breaking the manifold shape. On the other hand, when $k$ is too small, the neighborhood graph may not approximate the high-dimensional manifold. If more than one connected component is determined with Isomap, then only the largest of them (in terms of the number of nodes) is chosen and processed. As a consequence, the output mapping may not contain all of the $N$ data points embedded. Another important weakness of Isomap is that it fails to recover the correct data structure when the input data is not uniformly sampled, e.g., in the case of a manifold with a "hole" [15]. This may be overcome by employing "manifold tearing" [13].

Locally linear embedding (LLE) [22] relies on neighborhood graph construction, similarly to Isomap. However, instead of preserving pairwise geodesic distances between all data points, LLE aims at retaining local properties of the manifold. Hence, short-circuiting may only defect the low-dimensional embedding locally—the errors will not propagate onto the rest of the manifold. This approach is underpinned with the assumption that a sufficiently close neighborhood of each data point lies on or close to a locally-linear patch of the underlying manifold. The local geometry of these patches is defined by linear coefficients that reconstruct each point from its $k$ neighbors. This method was later extended with an additional step to minimize the manifold "curviness" in the local tangent spaces—as it employs computing the Hessian in a local neighborhood, the method has been termed as Hessian locally linear embedding (HLLE) [2].

Another interesting method for dimensionality reduction is the t-distributed stochastic neighbor embedding (t-SNE) [14], which has been created to visualize high-dimensional data using two or three dimensions, while preserving the manifold structure at different scales. Contrarily to the previously described methods, t-SNE does not base its mapping on the Euclidean distances, but on the pairwise joint probabilities that model the similarity between the data points. Basically, each high-dimensional vector is mapped onto two- or three-dimensional subspace in such a way, that the similarity between the vectors in the input space is preserved in the low-dimensional embedding. This is achieved by modeling two probability distributions over pairs of vectors, namely (i) in the input and (ii) in the reduced space. Every pair of vectors is assigned with a probability of being picked, and these pairwise probabilities are obtained based on radial basis functions. The kernel width depends on the local data density, and it is controlled by the perplexity ($\rho$), which has a similar influence as $k$ in the aforementioned methods—the denser the data, the larger value of this parameter is required. Finally, the low-dimensional mapping is obtained by solving an optimization task that minimizes the Kullback-Leibler divergence between the two aforementioned distributions. Also, so as to reduce the time complexity, the input space dimensionality ($N$) may be reduced to $N'$ using PCA—it is recommended to reduce the number of dimensions to $N' = 30$.

## 2.2  Hand Pose Recognition Using Manifold Learning

One of the major challenges in estimating a hand pose is the high number of its degrees of freedom—depending on the model applied, the intrinsic hand dimensionality can reach up to 26 dimensions [21]. In an image of a hand, additional dimensions are present that are not related with the hand pose. They are concerned with many factors, including inter-personal variances or differences in the lighting conditions that may be reduced with color transforms [23]. Overall, the intrinsic hand dimensionality is still very small compared with the dimensionality of the input image space, however the relation between low-dimensional hand model and high-dimensional images is very difficult to discover—even a small variation of parameters, such as the camera view angle causes a substantial, nonlinear change of the hand appearance. Manifold learning may be helpful in recovering this relation, and this direction has been explored in the literature.

In 2008, Ge et al. [4] established a new method for recognizing hand gestures, termed distributed locally linear embedding (DLLE). It is based on LLE, and it takes into account additional information about the density of data points in the point neighborhood. By estimating the probability density function of the input data and using it in further calculations, DLLE captures more information from dense neighborhoods and less in the areas of high dispersion. In such a manner, more global manifold features are preserved. Next, the DLLE output is fed into a probabilistic neural network (PNN). This framework was applied to recognizing both static and dynamic hand gestures. In the former case, an input image is classified to a certain group with DLLE-PNN. This classification provides information about the presented gesture. For each gesture, a motion database of 30 images was set up. When a new image in the gesture sequence is processed, its relative position with respect to other images in that sequence is obtained by DLLE in the motion database. Hand model parameters are reconstructed as a linear interpolation of images from motion database, taking into account an extensive set of hand motion constraints defined in the paper.

In 2013, Wang et al. [27] developed a holistic approach to hand pose recognition employing another dimensionality reduction technique, namely locality preserving projections (LPP). The training data is artificially generated by rendering a 29 degree-of-freedom hand model from 29 viewpoints that are distributed on a sphere. In order to increase the accuracy, additional constraints (both static and dynamic) were added to the model. The data set consists of 1000 samples. Based on the artificial images, a 2D→3D mapping table is created, capturing the relation between projected image and hand model parameters. This framework has been tested with natural images registered by a web-camera that present the gestures of a frontally-oriented person. The manifold is learnt by extracting visual features from the image based on Hu's invariant moments. Having a reduced data representation, it is fed into subspace filtering algorithm (SFA), which is responsible for classification and prediction.

## 3  Proposed Evaluation Framework

Although there have been some approaches proposed towards using manifold learning for estimating the hand pose, there is a need for a comprehensive study to investigate and evaluate different techniques. This motivated us to create a validation framework which we apply to evaluate several manifold learning techniques. So as to focus on the intrinsic dimensions of the hand, we rely on artificial hand images generated using the Libhand model [26]. In this way, we eliminate the influence of the noise-related factors such as lighting conditions or inter-personal variations, as they are not helpful in determining the most appropriate embedding, which remains the main goal of our study. In the future, we plan to extend the validation, employing our HGR dataset [9] that contains real-life images of hand gestures[1].

The processing pipeline of our evaluation procedure is outlined in Fig. 1 (multiple blocks indicate a set of independent operations). First of all, we generate a set of $M$ vectors that define the hand model in the 26-dimensional parameter space. These vectors are grouped into $K$ ground-truth classes, each of which corresponds to the same hand pose (the details on how these vectors are obtained are reported later in Sect. 4.1). Subsequently, the images, each consisting of $N$ pixels, are generated and they are subject to manifold embedding and projected onto the $n$-dimensional feature space, where $n \ll N$. In the best scenario, the different classes of hand poses should be separated as much as possible, clearly indicating the differences in the images. So as to measure quantitatively the inter-class separability, we proceed with the cluster analysis in the embedded space and assess the clusters for the separability. Here, we apply $K$-means clustering. Ideally, each cluster contains the samples corresponding to the same hand pose—as the ground-truth labels are known, this may be evaluated quantitatively. Quality of the clustering reflects how well the manifold embedding discovers the intrinsic dimensions of the data, and it is evaluated based on the adjusted Rand index ($ARI$) [16].

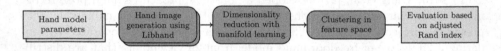

**Fig. 1.** The flowchart showing a high-level view of the validation procedure

So as to compute $ARI$, first a contingency table is created. The contingency table, given a data set $S$ of $M$ elements and two partitions of $S$ into subsets $\mathcal{G} = \{G_1, G_2, \ldots, G_K\}$ and $\mathcal{C} = \{C_1, C_2, \ldots, C_K\}$, summarizes the overlap between the ground-truth classes $\{G_i\}$ and the identified clusters $\{C_j\}$. The table is composed of $K \times K$ entries, where the $(i, j)$-th entry ($n_{ij}$) denotes the number of elements that are common for $G_i$ and $C_j$ subsets. All the numbers in each $i$-th row and $j$-th column are summed to $g_i = \sum_j n_{ij}$ and $c_j = \sum_i n_{ij}$, respectively. Naturally, all the partial sums should sum up to the number of observations $\left( \sum_j^K c_j = \sum_i^K g_i = M \right)$. Once the values in the contingency table are obtained, the adjusted Rand index is calculated as:

---

[1] HGR dataset is available at http://sun.aei.polsl.pl/~mkawulok/gestures.

$$ARI = \frac{\sum_i^K \sum_j^K \binom{n_{ij}}{2} - \sum_i^K \binom{g_i}{2} \sum_j^K \binom{c_j}{2} / \binom{M}{2}}{\left(\sum_i^K \binom{g_i}{2} + \sum_j^K \binom{c_j}{2}\right)/2 - \sum_i^K \binom{g_i}{2} \sum_j^K \binom{c_j}{2} / \binom{M}{2}}. \tag{1}$$

The index values range from $-1$ to $1$, where the maximal value indicates the best correspondence between the gold standard and the clustering being evaluated. The higher the index value is, the better clustering is with respect to the partition defined by the classes. It is important to notice that $ARI$ does not require matching between the labels—it relies only on the numbers of elements falling into particular $i$-th class and $j$-th cluster pairs. Basically, if the manifold embedding discovers the intrinsic dimensionality of a hand, then different images presenting the same hand pose should form concise groups in the embedded feature space, and they should be easily separated using simple $K$-means clustering. This is actually assessed using the proposed $ARI$-based measure.

## 4   Experimental Validation

In our study, we employed the Matlab framework[2] [15] for dimensionality reduction which contains implementation of various manifold embedding methods, and in our study we used Isomap [25], LLE [22], HLLE [2], t-SNE [14] and linear PCA. Section 4.1 provides the details on the dataset generation, and the obtained results are reported and commented in Sect. 4.2.

### 4.1   Dataset Generation

The artificial dataset of hand gesture images of $320 \times 240$ resolution was created using Libhand [26]—an open-source library which makes it possible to generate hand images, given the hand model parameters. The model used in Libhand consists of 18 joints, each of them possessing 3 degrees of freedom, namely bending, twisting and moving to a side. Although the hand model is characterized with 54 degrees of freedom, many of them should be rejected to obtain a visually correct result that corresponds to an allowable hand pose. Apart from varying hand model parameters, Libhand makes it possible to configure the camera point of view, modeled with three additional parameters (yaw, pitch and roll).

In the study reported here, a two-class hand pose data set was generated. It consists of 100 images divided into two equinumerous groups representing open (all fingers straight) and closed (all fingers bent significantly) hand gestures. At first, initial values of the Libhand parameters were set manually for both classes, and then the variations were imposed by adding uniformly distributed noise to: (i) camera's pitch, yaw and roll angles ($\pm 10\%$), (ii) hand joints' bending angle ($\pm 20\%$), and (iii) hand joints' side angle ($\pm 5\%$). In this way, we obtained a set of images presented in Fig. 2. It is worth noting that although this set appears to be a rather easy example, the hand shape variations within each class are significant.

---

[2] Available at http://lvdmaaten.github.io/drtoolbox.

Open hand (class $C_A$):

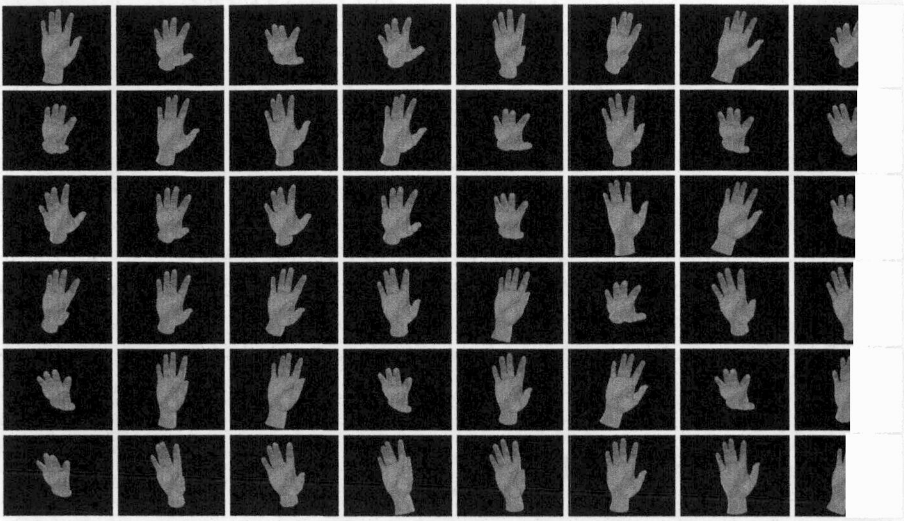

Closed hand (class $C_B$):

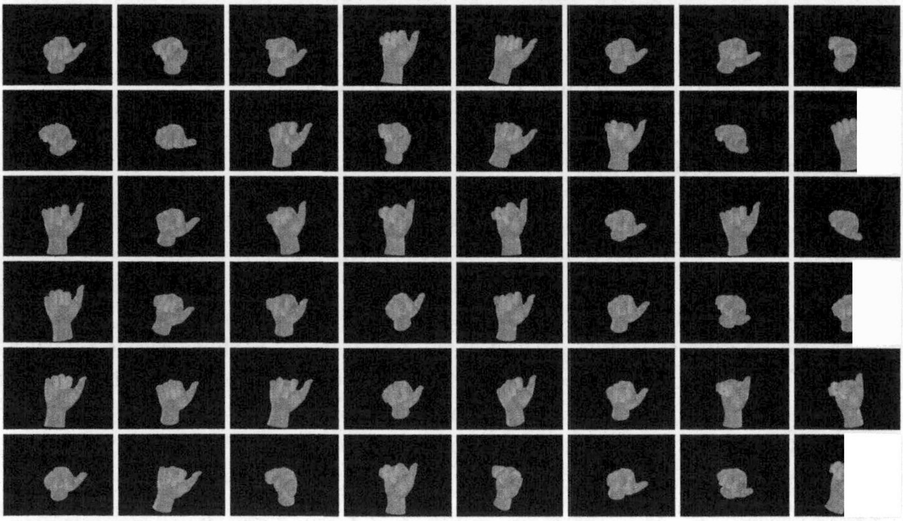

**Fig. 2.** The 2-class data set of hand gestures obtained with Libhand

## 4.2   Results and Discussion

First of all, in our experimental study we analyzed the sensitiveness of the investigated dimensionality reduction methods to their parameters. Basically, for all of the methods, the target dimensionality ($n$) of the embedded space is to be set. In addition, for Isomap, LLE and HLLE, the number of neighbors ($k$) used while building the manifold graph is an important parameter. Apart from the target dimensionality $n$, t-SNE is controlled using two additional parameters, namely perplexity $\rho$ and initial dimensionality $N'$ of the data, obtained from the input space using PCA. The obtained $ARI$ scores were generally very high for all the tested methods, and in some cases they reached 1, which means perfect separation using $K$-means clustering.

Figure 3 presents the obtained $ARI$ values for Isomap, LLE and HLLE, depending on the target dimensionality and the number of neighbors in the $k$-NN algorithm. It may be observed that it is Isomap, which is the least sensitive to the target dimensionality (for $n \geq 2$) and the results are quite stable for $k \notin (10, 20)$. Both LLE and HLLE are very sensitive to the target dimensionality (the values $d \in \{4, 5\}$ are the best), and high $ARI$ is obtained for $k \in (10, 20)$ and $k > 15$ for LLE and HLLE, respectively.

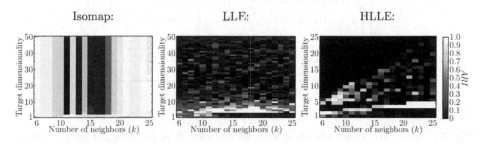

**Fig. 3.** Obtained $ARI$ values depending on the parameters of Isomap, LLE and HLLE

The recommended values of the t-SNE parameters are $\rho = 30$ and $N' = 30$, and the target dimensionality should be set to $n = 2$ or $n = 3$, as the method is primarily intended for visualization purposes. An important feature of t-SNE is that the optimization process applied for manifold learning involves the randomness—as a result, the optimization often falls into a local minimum, which may jeopardize the outcome. So as to deal with this problem, multiple runs are recommended. In the study reported here, we ran the learning 1000 times and distribution of the obtained $ARI$ values for $n = 2$ is presented in Fig. 4. It may be concluded that the outcome is generally good (the mean value is $ARI = 0.74$), however in ca. 10 % of cases the method fails completely to discover the intrinsic structure of the data. We have also verified the t-SNE behavior for larger $n$'s (see Fig. 5), and it occurred that the method presents definitely higher stability, achieving $ARI = 1$ in all cases. An important conclusion here is that the method can successfully be used for $n > 3$, and its applications may span beyond data visualization. Finally, we have investigated the method's sensitivity to the remaining parameters (i.e., perplexity $\rho$ and initial data dimensionality $N'$; the target dimensionality was set to $n = 2$). The results are reported in Fig. 6. It can be seen that initial dimensionality reduction is much of a help here, however this may have been attributed to the data simplicity—it can be seen in Fig. 5 that their structure is successfully discovered using PCA. On the other hand, t-SNE

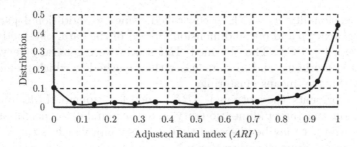

**Fig. 4.** Distribution of $ARI$ values for t-SNE with $d = 2$

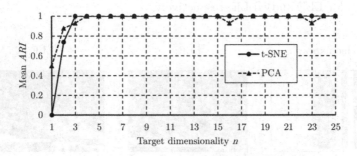

**Fig. 5.** Mean $ARI$ values obtained with t-SNE and PCA for different target dimensionality

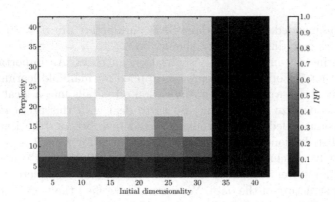

**Fig. 6.** Obtained $ARI$ values depending on the methods' parameters

fails completely for $N' > 30$, so the initial PCA-based dimensionality reduction seems to be a critical step here.

The times consumed by every method for learning the manifold structure are reported in Table 1. Not surprisingly, PCA offers the fastest computation, and taking into account the scores presented in Fig. 5, it would be the best choice

**Table 1.** Average times consumed to learn the manifold (in ms).

| Isomap | LLE | HLLE | t-SNE | PCA |
|--------|-----|------|-------|-----|
| 750 | 998 | 5670 | 820 | 110 |

for the analyzed case. Importantly, it would probably fail for more sophisticated cases that embrace many hand pose classes, but this example clearly shows that it should definitely be considered as a baseline approach. The remaining methods offer similar processing speed, apart from HLLE—taking into account its high sensitivity, this method does not seem to be very promising here.

In Fig. 7, we show several examples of how the data are separated after having been projected onto a two-dimensional feature space using different methods. It may be seen that *ARI* is higher, if the data form concise clusters, however low *ARI* values appear also for linearly separable cases, in which the data are distributed along the margin that separates the classes (e.g., for LLE). Also, the bottom-right t-SNE example could be easily separated (though nonlinearly), but the *ARI* value is very low here. This may explain low *ARI* stability observed in some cases—although the data could be easily separated in the feature space, they do not form concise groups that would result in high *ARI* scores. Potentially, the evaluation strategy could be extended with support vector machines (SVMs) [19] to identify linearly and nonlinearly separable cases as well, but taking into account that the ground-truth classes are obtained by imposing small

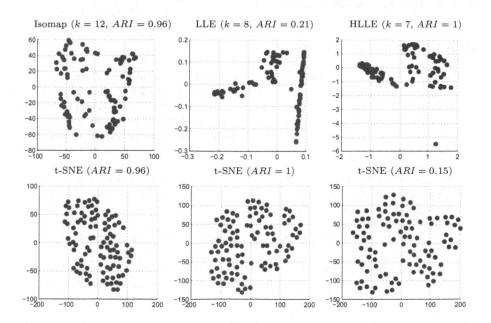

**Fig. 7.** Examples of data separation for target dimensionality $n = 2$

variance over the prototype vector of hand model parameters, it may be expected that these classes form concise groups in the feature space. Nevertheless, such an approach may be considered for more sophisticated, multi-class cases that we intend to focus in the future, especially if high *ARI* values cannot be achieved in such cases at all.

## 5    Conclusions and Future Work

In this paper, we report our preliminary attempt to evaluate different manifold learning techniques for recognizing a hand pose. The training data are clustered in the embedded subspace obtained with the manifold learning procedure, and the assessment consists in comparing the clustering outcome against the ground-truth labels. The elaborated procedure was used to evaluate manifold learning methods for a relatively simple dataset, which can be successfully clustered using unsupervised linear PCA. This allowed us to analyze the sensitiveness of the investigated methods, which is rather high for LLE and HLLE, and lower for Isomap and t-SNE.

Our ongoing work is concerned with preparing more demanding dataset that will make it possible to evaluate the manifold learning techniques not only in terms of their sensitiveness, but also on the basis how well they model the intrinsic dimensionality of a human hand. Furthermore, we plan to investigate the relation between the parameters of the hand model and particular dimensions of the embedded space. This will be an important step towards estimating the bent angles from real-life hand images.

## References

1. Belongie, S., Malik, J., Puzicha, J.: Shape matching and object recognition using shape contexts. IEEE Trans. Pattern Anal. Mach. Intell. **24**(4), 509–522 (2002)
2. Donoho, D.L., Grimes, C.: Hessian eigenmaps: locally linear embedding techniques for high-dimensional data. Proc. Nat. Acad. Sci. **100**(10), 5591–5596 (2003)
3. Freeman, W.T., Roth, M.: Orientation histograms for hand gesture recognition. In: Proceedings of the IEEE Conference on Automatic Face and Gesture Recognition, FG, pp. 296–301 (1995)
4. Ge, S.S., Yang, Y., Lee, T.H.: Hand gesture recognition and tracking based on distributed locally linear embedding. Image Vis. Comput. **26**(12), 1607–1620 (2008)
5. Grzejszczak, T., Gałuszka, A., Niezabitowski, M., Radlak, K.: Comparison of hand feature points detection methods. In: Camarinha-Matos, L.M., Barrento, N.S., Mendonça, R. (eds.) DoCEIS 2014. IFIP AICT, vol. 423, pp. 167–174. Springer, Heidelberg (2014)
6. Hachaj, T., Ogiela, M.R.: Human actions recognition on multimedia hardware using angle-based and coordinate-based features and multivariate continuous hidden markov model classifier. Multimedia Tools and Applications, pp. 1–21 (in press)
7. Huttenlocher, D., Klanderman, G., Rucklidge, W.: Comparing images using the hausdorff distance. IEEE TPAMI **15**(9), 850–863 (1993)

8. Kasprowski, P.: Mining of eye movement data to discover people intentions. In: Kozielski, S., Mrozek, D., Kasprowski, P., Malysiak-Mrozek, B., Kostrzewa, D. (eds.) BDAS 2014. CCIS, vol. 424, pp. 355–363. Springer, Heidelberg (2014)

9. Kawulok, M., Kawulok, J., Nalepa, J., Papiez, M.: Skin detection using spatial analysis with adaptive seed. In: Proceedings of the IEEE International Conference on Image Processing, ICIP 2013, pp. 3720–3724, September 2013

10. Kawulok, M., Kawulok, J., Nalepa, J., Smolka, B.: Self-adaptive algorithm for segmenting skin regions. EURASIP J. Adv. Sig. Process. **2014**(170), 1–22 (2014)

11. Kawulok, M., Smolka, B.: Competitive image colorization. In: Proceedings of the IEEE International Conference on Image Processing, ICIP 2010, pp. 405–408 (2010)

12. Kawulok, M., Wu, J., Hancock, E.R.: Supervised relevance maps for increasing the distinctiveness of facial images. Pattern Recogn. **44**(4), 929–939 (2011)

13. Lee, J.A., Verleysen, M.: Nonlinear dimensionality reduction of data manifolds with essential loops. Neurocomputing **67**, 29–53 (2005)

14. van der Maaten, L., Hinton, G.: Visualizing data using t-SNE. J. Mach. Learn. Res. **9**(2579–2605), 85 (2008)

15. van der Maaten, L., Postma, E.O., Herik, H.J.: Dimensionality reduction: a comparative review. J. Mach. Learn. Res. **10**(1–41), 66–71 (2009)

16. Milligan, G.W., Cooper, M.C.: A study of the comparability of external criteria for hierarchical cluster analysis. Multivar. Behav. Res. **21**(4), 441–458 (1986)

17. Nalepa, J., Grzejszczak, T., Kawulok, M.: Wrist localization in color images for hand gesture recognition. In: Gruca, A., Czachórski, T., Kozielski, S. (eds.) Man-Machine Interactions 3. AISC, vol. 242, pp. 81–90. Springer, Heidelberg (2014)

18. Nalepa, J., Kawulok, M.: Fast and accurate hand shape classification. In: Kozielski, S., Mrozek, D., Kasprowski, P., Malysiak-Mrozek, B., Kostrzewa, D. (eds.) BDAS 2014. CCIS, vol. 424, pp. 364–373. Springer, Heidelberg (2014)

19. Nalepa, J., Kawulok, M.: Adaptive memetic algorithm enhanced with data geometry analysis to select training data for SVMs. Neurocomputing **185**, 113–132 (2016). http://dx.doi.org/10.1016/j.neucom.2015.12.046

20. Nurzynska, K., Smolka, B.: PCA application in classification of smiling and neutral facial displays. In: Kozielski, S., Mrozek, D., Kasprowski, P., Malysiak-Mrozek, B., Kostrzewa, D. (eds.) BDAS 2015. CCIS, vol. 521, pp. 398–407. Springer, Heidelberg (2015)

21. Oikonomidis, I., Kyriazis, N., Argyros, A.A.: Efficient model-based 3D tracking of hand articulations using Kinect. In: BMVC, vol. 1(2), p. 3 (2011)

22. Roweis, S.T., Saul, L.K.: Nonlinear dimensionality reduction by locally linear embedding. Science **290**(5500), 2323–2326 (2000)

23. Starosolski, R.: New simple and efficient color space transformations for lossless image compression. J. Vis. Commun. Image Represent. **25**(5), 1056–1063 (2014)

24. Szwoch, M.: On facial expressions and emotions RGB-D database. In: Kozielski, S., Mrozek, D., Kasprowski, P., Malysiak-Mrozek, B., Kostrzewa, D. (eds.) BDAS 2014. CCIS, vol. 424, pp. 384–394. Springer, Heidelberg (2014)

25. Tenenbaum, J.B., Silva, Vd, Langford, J.C.: A global geometric framework for nonlinear dimensionality reduction. Science **290**(5500), 2319–2323 (2000)

26. Šarić, M.: Libhand: A library for hand articulation, version 0.9 (2011)

27. Wang, Y., Luo, Z., Liu, J., Fan, X., Li, H., Wu, Y.: Real-time estimation of hand gestures based on manifold learning from monocular videos. Multimedia Tools Appl. **71**(2), 555–574 (2013)

# A Meta-Learning Approach to Methane Concentration Value Prediction

Michał Kozielski[(✉)]

Institute of Electronics, Silesian University of Technology,
Ul. Akademicka 16, 44-100 Gliwice, Poland
michal.kozielski@polsl.pl
http://adaa.polsl.pl

**Abstract.** A meta-learning approach to stream data analysis is presented in this work. The analysis is based on prediction of methane concentration in a coal mine. The results of the analysis show that the chosen approach achieves relatively low error values. Additionally, the impact of a data window size on a learning speed and quality was verified. The analysis is performed on a stream of measurements that was generated on a basis of real values collected in a coal mine.

**Keywords:** Meta-learning · Algorithm selection · Stream data analysis · Prediction

## 1 Introduction

Coal mining is a heavy industry that plays an important role on an energy market and employs hundreds of thousands of people. Currently coal mines are well equipped with the monitoring, supervising and dispatching systems connected with machines, devices and transport facilities. Moreover, there are systems for monitoring natural hazards (methane, seismic and fire hazards) operating in the coal mines.

This work present a research that can be utilised within a safety system where data analysis can support coal miner security assurance but additionally, it can increase work efficiency. For example, due to the short-term prognoses about methane concentration, linked with the information about the location and work intensity of the cutter loader, it is possible to prevent emergency energy shutdowns and maintain continuity of mining. This will enable to increase the production volume and to reduce the wear of electrical elements whose exploitation time depends on the number of switch-ons and switch-offs.

Analysis of data collected from sensors and creating a data stream is restricted by memory and time constraints. Therefore, the methods dedicated to data stream analysis have to fulfil such requirements as: each data example (observation) has to be analysed only once and there is limited amount of time and memory for analysis. This type of methods, that perform online learning,

© Springer International Publishing Switzerland 2016
S. Kozielski et al. (Eds.): BDAS 2016, CCIS 613, pp. 716–726, 2016.
DOI: 10.1007/978-3-319-34099-9_56

has to be applied for a task defined in this work. There is a number of such methods however, according to a *No free lunch theorem* [17] there is no approach that could provide results of the best quality in each case of application. Different methods can perform better depending on a data set and its characteristics.

This work aims in application of meta-learning approach that performs algorithm selection to prediction task on a data stream. As the application of the analysis within a safety system is considered, the smallest possible error from the very beginning of analysis is desired. Therefore, the quality of the results in the initial phase of the prediction model creation is of interest of this work. Additionally, another goal of this work is to verify how the chosen parameter (size of a data window) of the approach impacts the prediction error.

The contribution of this work consists of the results showing that application of meta-learning approach to prediction task on data streams can result in a better prediction quality in the initial state of the model creation and in a quality similar to the best overall result. Additionally, it is shown in the work that analysis of the shorter data windows results in a better prediction quality in case of the chosen task.

This work is structured as follows. Section 2 presents the works related to the presented topic. A typical meta-learning approach to algorithm selection problem and the solution considered in this work are presented in Sect. 3. Section 4 presents the analysis that was performed and the results that were achieved. Section 5 presents the final conclusions.

## 2   Related Work

The domain of this paper covers several areas. The chosen studies related to these areas are presented in the following paragraphs.

Research in the field of natural hazards are mainly focused on seismic hazard analysis [7,8,15] and methane concentration prediction [6,11,19–21,25]. In this work we focus on the latter issue, however the analysis is performed on stream data instead of static hold-out data set.

Surveys of classification methods in data streams are presented in [3,13], whereas regression methods for data streams are presented e.g. in [1,4].

In this paper a meta-learning approach to algorithm selection for data stream analysis is considered. Meta-learning in general was surveyed in several papers, e.g. [5,14,23,24]. Whereas, meta-learning for algorithm selection issues were discussed, among others, in [22]. On the other hand, a survey of intelligent assistants for data analysis, which topic covers also meta-learning systems, was presented in [18]. A typical approach to meta-learning is based on analysis of data set characteristics and method performances. However, there are also more extensive approaches possible, that take, e.g., a method complexity and a data transformation process into account [5]. Knowledge about the machine learning processes that was collected in such approaches can be represented in the form of ontology [9,12].

From the perspective of the given work, an important approach was presented in [16], where algorithm selection on data streams was considered. In this work classification methods on data streams were analysed and justification of meta-learning approach was performed. However, the given work differs from the one presented in [16] as we take into consideration a certain task (methane concentration prediction) what imposes application of regression methods and a focus on initial learning phase. Additionally, in the given work a meta-learner is also a data stream dedicated method and different data stream window size is considered.

## 3   Methods

In this section a typical meta-learning architecture is presented at first. Next, an approach tested in the experiments is discussed in more details.

### 3.1   Algorithm Selection

Learning (basic-learning or simple-learning) system acquires knowledge about the analysed phenomenon in order to perform a defined task, e.g., classification or regression. Whereas, meta-learning system acquires knowledge about the multiple applications of a learning system [24]. In this way a meta-learning system should be able to point the approach (basic-learner) that will be the most successful in a given context (e.g. data set characteristics). A general scheme of such meta-learning system [18] is presented in Fig. 1. As an input for meta-learning process we take meta-data, that have to be generated on the basis source data sets and methods that were applied to them. As a result of meta-learning we receive a model delivering for a new data set an advice or ranking of methods that can be applied to this data set.

There is a number of characteristics that can be generated for a data set in order to create its meta-data. The following classes of meta-features can be defined [23]:

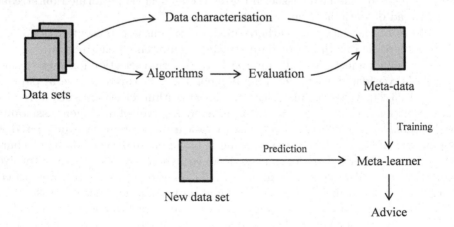

**Fig. 1.** General scheme of a meta-learning system

- simple, statistical and information-theoretic, e.g., number of examples, skewness, kurtosis or entropy,
- concept-based properties, e.g., uniformity of the class-label distribution throughout the feature space,
- case base properties, e.g., consistency - showing if there is any identical example belonging to a different class,
- model-based properties, where a simple model is built (e.g. decision tree) in order to deliver data characteristics, e.g., tree width and depth - illustrating complexity of a concept,
- landmarking, where several learning methods are applied to the data in order to evaluate their performance; the methods may be very simplified and fast (e.g. Decision stump or 1-Nearest Neighbour) or they may be applied to a small data sample,
- task-specific meta-features, where the approaches matching a chosen task are applied.

### 3.2   The Introduced Approach

It is assumed in this work, that the data have a form of a stream of measurements. Therefore, an appropriate solution is to apply online learning methods [2] because these methods do not perform learning on a whole data set or its part but they learn with each newly arriving observation.

The question is which of the available learning algorithms would be the best in the considered application. It can be very difficult to answer this question because each longwall in a coal mine can have slightly or even significantly different methane concentration characteristics. Another issue is that different learning methods can achieve different prediction quality during a learning process. Therefore, a meta-learning approach is worth considering in order to perform automatic selection of the best prediction method.

Considering meta-learning approach to data streams analysis it is possible to apply the system of the same structure as the one presented in Fig. 1. However, the definition of a meta-learning system can be paraphrased to be more specific, as it acquires knowledge about the applications of a learning system to previous data stream windows. Therefore, in this case the data sets illustrated in Fig. 1 are the consecutive data windows of the stream and the new data set is, thus, a new data window extracted from the stream. Knowing which methods achieved the best performance on previous data windows a meta-learner should be able to predict the best solution for the analysis of incoming data.

A window size is a parameter of the approach. Therefore, it is possible to take under consideration different values. In [16] a data window consisted of 1000 observations was utilised as it was assumed that too small window could miss an important structural information. However, in order to create a meta-learner model it is required to analyse several data windows. Therefore, we can expect that the larger the window is, the more time it will take to create a meta-learner of a good quality. Time is however, crucial in safety systems such as methane prediction in a coal mine and it is worth verifying if the shorter data windows are able to produce acceptable results.

The learning algorithms considered in this task have to be dedicated to stream analysis. Meta-learner can be implemented as a "classical" classification algorithm as it was presented in [16]. However, meta-data generated for consecutive data windows have also a form of a data stream. Therefore, a method dedicated to stream analysis can be also applied in this case.

In this work the analysis of methane concentration value stream is considered. This task imposes certain restrictions that impact meta-features that have to be generated. A stream delivers new data values constantly. Therefore, the set of meta-features should be created taking time restrictions into consideration. Additionally, landmarking is a very intuitive choice to meta-feature generation in case of data stream analysis. This is because base algorithms that are considered in a meta-learning system analyse a data stream online and their results are available for each data window.

## 4   Experiments

In order to verify how a meta-learning system can perform in case of a coal mine methane concentration value prediction task the experiments presented in this section were conducted.

The experiments were performed on a data set collected within the DISESOR project [10]. Originally the data set consisted of the measurements that were collected each second for 9 sensors:

- one anemometer,
- seven methanometers,
- binary indication if the longwall shearer is running.

This data set was aggregated in order to receive a single entry for each 30 s of measurements. The new data set was received applying the following aggregation operators within the given 30 s range:

- minimum operation for anemometer measurements,
- maximum operation for methanometer measurements,
- dominant operation for longwall shearer operation indication.

The task defined on this data set was to predict methane concentration value in 3 min horizon. However, the resulting data set consisted of 100547 observations, which quantity is not large enough for proper stream data analysis.

In order to receive a large number of observations a Bayesian network-based stream data generator introduced in [16] was utilised. This generator is implemented as a method available in the OpenML MOA package[1]. It creates a Bayesian network representing characteristics of the input data set, which is

---

[1] Available at http://www.openml.org.

methane data set in our case. Next, this Bayesian network is utilised to generation of a new data set, where the data set size is the method parameter. In this
way a new data set consisting of 10 millions of measurements was generated.

In order to verify how the data window size impacts the results of the analysis, data windows containing 1000, 100 and 10 observations were taken into
account. For each attribute of each window the following five meta-features were
generated: minimum, maximum, mean, median, standard deviation. The features that were chosen are very simple to make their calculation possibly fast.
Additionally to the simple features, landmarks were also generated in the form
of a ranking based on a method quality. The final set of meta-features consisted
of 50 attributes (there were 9 original features, times 5 simple meta-features,
plus 5 landmarks) plus decision attribute. In order to avoid time pressure the
meta-learner did not predict the best solution for the next window, but the shift
of two windows was chosen.

The analysis was performed utilising the methods implemented in MOA
tool [2]. The following five implementations of regression methods were
applied as base-learners: AMRules, FadingTaragetMean, Perceptron, Target-
Mean, FIMTDD. As a meta-learner an implementation of a classic Hoeffding
tree algorithm was chosen. The default parameters were set for all the methods.

### 4.1   Results

The performance of the analysed methods can be evaluated from the two perspectives. The first approach is to compare the total prediction quality, when the
whole data stream is taken under consideration. During the analysis a root mean
square error (RMSE) is calculated for each data stream window. Therefore, total
prediction quality is calculated as a mean value of RMSE for the processed data
windows. The RMSE values for different methods when different window size
was set are presented in Table 1.

Among the base-learners the best results were achieved by FIMTTD method.
This method performed significantly better than all the methods on average. The
meta-learning approach performed almost identically (slightly worse) as the best

**Table 1.** Total prediction quality (RMSE)

|                    | Window = 10 | Window = 100 | Window = 1000 |
|--------------------|-------------|--------------|---------------|
| AMRules            | 0.183101754 | 0.189378211  | 0.190092878   |
| FadingTaragetMean  | 0.263454507 | 0.268982924  | 0.269532804   |
| Perceptron         | 0.279492113 | 0.283744916  | 0.284210442   |
| TargetMean         | 0.262812313 | 0.268308288  | 0.268854005   |
| FIMTDD             | 0.13516606  | 0.142259622  | 0.143107532   |
| Average            | 0.224805349 | 0.230534792  | 0.231159532   |
| Meta-learner       | 0.13517236  | 0.142261527  | 0.143107388   |

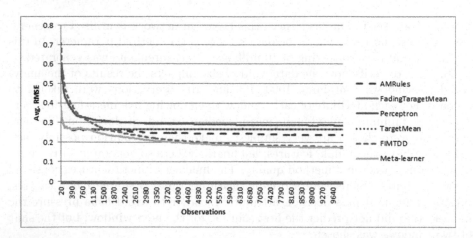

**Fig. 2.** Average RMSE of base-learners and meta-learner for initial 10000 observations, window size eq. 10

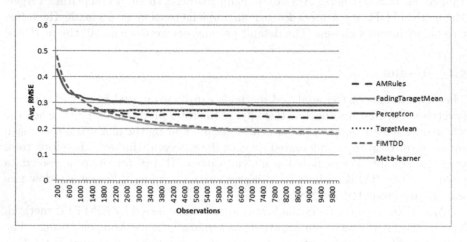

**Fig. 3.** Average RMSE of base-learners and meta-learner for initial 10000 observations, window size eq. 100

base-learner and hence, significantly better than all the base-learners on average. Additionally, the results from Table 1 show that for the given task the shorter data windows result in a better performance. However, the window size does not impact the ranking of the methods.

The second perspective considered in this work is to analyse the initial performance of the methods that are compared. Thus, the analysis of the first 10000 observations is presented in the following figures.

In order to compare how the methods behave at the initial learning phase a chart of an average RMSE from the beginning of a stream to the current window

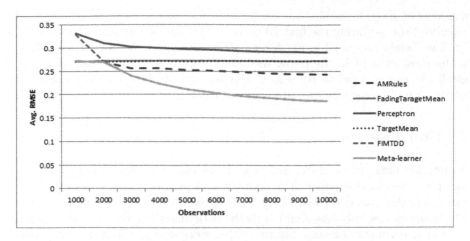

**Fig. 4.** Average RMSE of base-learners and meta-learner for initial 10000 observations, window size eq. 1000

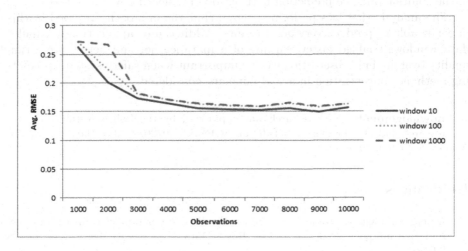

**Fig. 5.** Comparison of the meta-learning process during the first 10000 observations for different window sizes

was calculated for each method. Figures containing the results for windows of size 10, 100 and 1000 are presented in Figs. 2, 3 and 4 respectively.

It is noticeable in Figs. 2, 3 and 4 how meta-learner adjusts to faster learning base-learner at the begining and then how it adjusts to a base-learner that is better overall.

Finally, it is possible to compare how the window size impacts the quality of initial meta-learning phase. The comparison has to be performed at the granularity of the largest window. Therefore, the RMSE of the meta-learning approaches when window size was equal to 10 and 100 was averaged into the windows of 1000. It enabled to plot (see Fig. 5) and compare the quality of the learning process during the first 10000 observations for different window sizes.

The results presented in Fig. 5 show that the impact of different window size is the most evident in the initial meta-learning phase. When a window size is small (its size is equal to 10) a meta-learner achieves better results faster then in the case of larger window (size equal to 1000).

## 5    Conclusions

A meta-learning approach to prediction of methane concentration in a coal mine was presented in this paper. The analysis was performed on a stream of measurements that was generated on the basis of real values collected in a coal mine. The analysis was fully based on the methods dedicated to data stream learning.

The experiments presented in this paper were designed in a way that makes possible the implementation of the introduced solution in a real online system. Thus, very simple meta-features were chosen not to increase computational costs of the solution and the prediction horizon for meta-learner was increased.

The analysis that was performed shows that the approach proposed in this paper is able to produce very good results. Additionally, application of smaller data windows enabled faster learning of a meta-learner and better prediction quality overall. This observation can be important when safety systems predicting methane concentration in a coal mine are considered.

**Acknowledgements.** This research was supported by the Polish National Centre for Research and Development (NCBiR) grant PBS2/B9/20/2013 in the frame of the Applied Research Programme.

## References

1. Alberg, D., Last, M., Kandel, A.: Knowledge discovery in data streams with regression tree methods. Wiley Interdisc. Rev.: Data Min. Knowl. Disc. **2**(1), 69–78 (2012)
2. Bifet, A., Holmes, G., Kirkby, R., Pfahringer, B.: Moa: massive online analysis. J. Mach. Learn. Res. **11**, 1601–1604 (2010)
3. Gaber, M., Zaslavsky, A., Krishnaswamy, S.: A survey of classification methods in data streams. In: Aggarwal, C. (ed.) Data Streams. Advances in Database Systems, vol. 31, pp. 39–59. Springer, US (2007). http://dx.doi.org/10.1007/978-0-387-47534-9_3
4. Ikonomovska, E., Gama, J., Džeroski, S.: Learning model trees from evolving data streams. Data Min. Knowl. Disc. **23**(1), 128–168 (2011). http://dx.doi.org/10.1007/s10618-010-0201-y
5. Jankowski, N., Grąbczewski, K.: Universal meta-learning architecture and algorithms. In: Jankowski, N., Duch, W., Grbczewski, K. (eds.) Meta-Learning in Computational Intelligence. Studies in Computational Intelligence, vol. 358, pp. 1–76. Springer, Heidelberg (2011). http://dx.doi.org/10.1007/978-3-642-20980-2_1

6. Janusz, A., Sikora, M., Wróbel, U., Stawicki, Ł., Grzegorowski, M., Wojtas, P., Ślezak, D.: Mining data from coal mines: IJCRS'15 data challenge. In: Yao, Y., Hu, Q., Yu, H., Grzymala-Busse, J.W. (eds.) Rough Sets, Fuzzy Sets, Data Mining, and Granular Computing. LNCS, vol. 9437, pp. 429–438. Springer International Publishing, Heidelberg (2015)

7. Kabiesz, J.: Effect of the form of data on the quality of mine tremors hazard forecasting using neural networks. Geotech. Geol. Eng. 24(5), 1131–1147 (2006). http://dx.doi.org/10.1007/s10706-005-1136-8

8. Kabiesz, J., Sikora, B., Sikora, M., Wróbel, Ł.: Application of rule-based models for seismic hazard prediction in coal mines. Acta Montanist. Slovaca 18(4), 262–277 (2013)

9. Keet, C.M., Ławrynowicz, A., dAmato, C., Kalousis, A., Nguyen, P., Palma, R., Stevens, R., Hilario, M.: The data mining optimization ontology. Web Semant.: Sci. Serv. Agents World Wide Web 32, 43–53 (2015). http://www.sciencedirect.com/science/article/pii/S1570826815000025

10. Kozielski, M., Sikora, M., Wróbel, L.: DISESOR - decision support system for mining industry. In: 2015 Federated Conference on Computer Science and Information Systems (FedCSIS), pp. 67–74, September 2015

11. Kozielski, M., Skowron, A., Wróbel, Ł., Sikora, M.: Regression rule learning for methane forecasting in coal mines. In: Kozielski, S., Mrozek, D., Kasprowski, P., Małysiak-Mrozek, B., Kostrzewa, D. (eds.) Beyond Databases, Architectures and Structures. Communications in Computer and Information Science, vol. 521, pp. 495–504. Springer, Heidelberg (2015)

12. Ławrynowicz, A., Potoniec, J.: Pattern based feature construction in semantic data mining. Int. J. Semant. Web Inf. Syst. (IJSWIS) 10(1), 27–65 (2014)

13. Lemaire, V., Salperwyck, C., Bondu, A.: A survey on supervised classification on data streams. In: Zimányi, E., Kutsche, R.-D. (eds.) Business Intelligence. Lecture Notes in Business Information Processing, vol. 205, pp. 88–125. Springer, Heidelberg (2015). http://dx.doi.org/10.1007/978-3-319-17551-5_4

14. Lemke, C., Budka, M., Gabrys, B.: Metalearning: a survey of trends and technologies. Artif. Intell. Rev. 44(1), 117–130 (2015). http://dx.doi.org/10.1007/s10462-013-9406-y

15. Leśniak, A., Isakow, Z.: Space-time clustering of seismic events and hazard assessment in the Zabrze-Bielszowice coal mine, Poland. Int. J. Rock Mech. Min. Sci. 46(5), 918–928 (2009). http://dx.doi.org/10.1016/j.ijrmms.2008.12.003

16. van Rijn, J.N., Holmes, G., Pfahringer, B., Vanschoren, J.: Algorithm Selection on Data Streams. In: Džeroski, S., Panov, P., Kocev, D., Todorovski, L. (eds.) Discovery Science. LNCS, vol. 8777, pp. 325–336. Springer, Heidelberg (2014). http://dx.doi.org/10.1007/978-3-319-11812-3_28

17. Schaffer, C.: A conservation law for generalization performance. In: Proceedings of the 11th International Conference on Machine Learning, pp. 259–265 (1994)

18. Serban, F., Vanschoren, J., Kietz, J.U., Bernstein, A.: A survey of intelligent assistants for data analysis. ACM Comput. Surv. 45(3), 31:1–31:35 (2013). http://doi.acm.org/10.1145/2480741.2480748

19. Sikora, M., Sikora, B.: Improving prediction models applied in systems monitoring natural hazards and machinery. Int. J. Appl. Math. Comput. Sci. 22(2), 477–491 (2012). http://dx.doi.org/10.2478/v10006-012-0036-3

20. Sikora, M., Sikora, B.: Rough natural hazards monitoring. In: Peters, G., Lingras, P., Ślęzak, D., Yao, Y. (eds.) Rough Sets: Selected Methods and Applications in Management and Engineering. Advanced Information and Knowledge Processing, pp. 163–179. Springer, Heidelberg (2012). http://dx.doi.org/10.1007/978-1-4471-2760-4-10

21. Simiński, K.: Rough subspace neuro-fuzzy system. Fuzzy Sets Syst. **269**, 30–46 (2015)

22. Smith-Miles, K.A.: Cross-disciplinary perspectives on meta-learning for algorithm selection. ACM Comput. Surv. **41**(1), 6:1–6:25 (2009). http://doi.acm.org/10.1145/1456650.1456656

23. Vanschoren, J.: Understanding machine learning performance with experiment databases. Ph.D. dissertation, Katholieke Universiteit Leuven, Flanders, Belgium (2010)

24. Vilalta, R., Giraud-Carrier, C., Brazdil, P.: Meta-learning - concepts and techniques. In: Maimon, O., Rokach, L. (eds.) Data Mining and Knowledge Discovery Handbook, pp. 717–731. Springer, US (2010). http://dx.doi.org/10.1007/978-0-387-09823-4_36

25. Zagorecki, A.: Prediction of methane outbreaks in coal mines from multivariate time series using random forest. In: Yao, Y., Hu, Q., Yu, H., Grzymala-Busse, J.W. (eds.) Rough Sets, Fuzzy Sets, Data Mining, and Granular Computing. LNCS, pp. 494–500. Springer International Publishing, Heidelberg (2015). http://dx.doi.org/10.1007/978-3-319-25783-9_44

# Anomaly Detection in Data Streams: The Petrol Station Simulator

Anna Gorawska$^{(\boxtimes)}$ and Krzysztof Pasterak

Faculty of Automatic Control, Electronics, and Computer Science,
Institute of Informatics, Silesian University of Technology,
Akademicka 16, 44-100 Gliwice, Poland
{Anna.Gorawska,Krzysztof.Pasterak}@polsl.pl

**Abstract.** Developing anomaly detection systems requires diverse data for training and testing purposes. Real measurements are not necessarily reliable at this stage because it is almost impossible to find a diverse training set with exactly known characteristics. The petrol station simulator was designed to generate measurements that mimic real petrol station readings. The simulator produces datasets with exactly specified anomalies to be detected via anomaly detection system. The paper introduces foundations of the simulator with results. The discussion section presents future work in the area of stream data extraction and materialization in the Stream Data Warehouse.

**Keywords:** Petrol station simulator · Petrol station · Fuel leak detection · Sensor miscalibration · Statistical inventory reconciliation · Data stream · Stream data extraction · Stream Data Warehouse

## 1 Introduction

The biggest danger that comes with storage of petroleum or other hazardous substances is contamination of groundwater supplies caused by leaks and spills [2, 10]. The majority of petrol stations are not equipped with specialized appliances that prevent or detect leakages. Therefore, manual or automatic tank gauging [1,12], tank tightness testing [13,15], and inventory reconciliation techniques [7,14] are commonly used. In those methods the most important factors are frequency and detection accuracy, while as it was stated in [7] there is a variety of disturbing phenomena present in the inventory data that can negatively affect those factors. In reality problems described in [7] cannot be examined separately, as though they tend to overlap. As a consequence, detection accuracy may be extremely decreased or even a leakage may not be detected due to e.g. tank miscalibration [4].

The data gathered from a real petrol station usually does not contain hazardous anomalies, as though they appear very rarely. It is almost impossible to find a proper and diverse training set for anomaly detection purposes when relying solely on real data inventories. Thus, there was a need to implement a

© Springer International Publishing Switzerland 2016
S. Kozielski et al. (Eds.): BDAS 2016, CCIS 613, pp. 727–736, 2016.
DOI: 10.1007/978-3-319-34099-9_57

petrol station simulator to generate datasets with specified characteristics and anomalies. In the following paper, the petrol station simulator is understood as a software designed to reproduce the majority of phenomena that may occur on a petrol station in terms of fuel appliances operation. As a result, it can behave as a virtual petrol station which emulates the real one in terms of measurement data delivery.

## 2    Requirements and Prerequisites

The purpose of the petrol station simulator is not being plainly a mock data source for test purposes. Our main goal was to create a reliable data source which can reproduce uncommon and rare situations, such as compound anomalies occurring on a real petrol station.

Since the designed petrol station simulator is intended to be used as a source of interesting data for analysis and anomaly detection, it has to model a real petrol station accurately. The compatibility ought to be achieved on four levels: behaviour, configuration, data, and anomaly-level.

**Behaviour-Level Compatibility.** On a real petrol station fuel is stored in tanks and distributed to customers via nozzles that are installed in dispensers. A complex piping system connects tanks with specified nozzles, therefore the simulator is providing data according to determined schema of connections between petrol station's appliances, i.e. station infrastructure. Moreover, the petrol station simulator must generate fuel deliveries as cyclic events of refuelling the tank and simulate fuel purchase operations (*transactions*), that are consistent with dispensing fuel from the tank via nozzles.

All appliances in the petrol station infrastructure behave in a specific way that the petrol station simulator is mimicking. Behaviour-level compatibility ensures that output data are highly accordant to the real data and can be used as a substitute in a variety of applications.

**Configuration-Level Compatibility.** Sometimes there is a need to simulate operation of a certain petrol station, i.e. with a specified infrastructure and parameters (e.g. capacity of tanks, delivery cycle). Configuration-level compatibility ensures that virtual components of the simulated petrol station comply with their real counterparts and if needed they can be fully parametrized enabling simulation of an arbitrary real petrol station.

**Data-Level Compatibility.** Data acquired from a real petrol station has a specific format which can differ according to specific stations, just like applications that process the data can impose additional restrictions. Thus, the petrol station simulator has to produce data in a generic format that can be easily transformed to one required by an actual data recipient. By data-level compatibility we ensure that the petrol station simulator generates data that has exactly the same format as the real inventory dataset. Moreover, since data from real petrol stations is continuously delivered to the processing system, the simulator has to implement stream output data production [3,6,11] or at least batch (file) generation.

**Anomaly-Level Compatibility.** An anomaly can be understood as an abnormal and potentially dangerous situation, which sometimes occurs during normal operation of a petrol station. Anomaly-level compatibility ensures that all possible phenomena are implemented in the simulator, especially anomalies. Among the most common anomalies we can distinguish: fuel leak from a tank or piping, fuel surplus (increase in fuel volume in a tank not connected to natural fuel behaviour [7]), water influx in a tank, level probe hang [7] or density mismatch, tank or dispenser meter miscalibration [4].

# 3 The Petrol Station Simulator

For the anomaly detection purpose, three main types of test data can be distinguished: data from real petrols station with detected and identified anomalies, data from real petrol stations with artificially applied anomalies, artificial data obtained from the petrol station simulator. Since obtaining fully analysed real inventories with confirmed anomalies is generally difficult to accomplish, the two other are common sources of data.

The petrol station simulator is generating artificial measurements that can be configured so that they resemble measurements from a particular petrol station. The second solution is a simple application that applies anomalies to real data.

The main purpose of generating data was to obtain idealized measurements with anomalies included – which is helpful in the early stages of developing any anomaly detection system. Then real data with anomalies applied was used later, to test the behaviour of anomaly detection system when operating in more realistic environment.

Since output measurements from the simulator and the anomaly applier are compatible with real inventories, they can be used interchangeably, i.e. anomaly applying software can be used to process data generated by the simulator.

## 3.1 Dispensing Transactions

In the petrol station simulator, *transaction* is a single dispensing operation consistent with refuelling a car via nozzle. The amount of fuel per transaction ($V_{tr}$) is calculated according to Eq. 1, where $V_{tr_{avg}}$ represents *transaction average volume*, $V_{tr_{var}}$ *transaction volume variance*, and $r$ is a uniformly distributed random value, such as: $r \in \langle 0, 1 \rangle$.

$$V_{tr} = 2r \cdot V_{tr_{var}} + V_{tr_{avg}} - V_{tr_{var}} \tag{1}$$

For simulation purposes we have adopted a *queue model* to enable presentation of different traffic intensities on a petrol station. Similarly as in real petrol station, all available nozzles are organized in groups (dispensers). With each dispenser there is an associated waiting queue with newly incoming customers – fuel transaction precursors. The arrivals of customers are generated by a single source with a parametrized rate (i.e. *average number of transactions per hour, transaction average fuel volume, transaction volume variance*), knowing that customers usually prefer to select dispenser with the shortest waiting queue.

Let us assume that there are $m$ dispensers and waiting queue for $i$-th dispenser $(d_i)$ is denoted as $q_i$. The size of a single queue is $n_i$, the number of customers already queued – $c_i$, and the number of free slots in the $i$-th queue is $s_i = n_i - c_i$. The probability $p_i$ for a customer to select an $i$-th dispenser is as follows:

$$p_i = \frac{s_i}{\sum_{j=1}^{m} s_j} \tag{2}$$

In order to emphasize significant differences between daily and nightly rate of transactions we have decided to use the *sinusoidal model* of fuel transactions intensity to present daily cycle of transactions. We have assumed that its maximum (equal to twice the *average number of transactions*) is at 3:00 p.m. and its minimum (equal to 0) is at 3:00 a.m. Between aforementioned maximum and minimum values the intensity resembles a sine function.

The basic unit of time during simulation is one minute which refers to $\frac{2\pi}{60\cdot24}$, assuming that the whole day is equivalent to $2\pi$. Since the maximum intensity is intended to be at 3:00 p.m., the original time ought to be shifted by 3 h ($3\cdot60$ min). The resulting intensity of transactions $(T'_{avg})$ can be defined by the average number of transactions $(T_{avg})$ multiplied by the factor obtained from the sine function:

$$T'_{avg} = T_{avg} \cdot \left( 1 - \sin\left( \frac{t + 3\cdot60}{60\cdot24} \cdot 2\pi \right) \right) \tag{3}$$

## 3.2 Delivery

During normal operation the level of fuel stored in the tank is decreasing according to the amount of fuel that is dispensed from the tank during transactions. Therefore, to determine when deliveries must be triggered, it is necessary to define a *work area* for each tank, i.e. tank operating range between so called *low level* and *work level*. When the amount of fuel in the tank is smaller then the *low level* the delivery is triggered, as though the tank might destabilize with a small amount of fuel stored. The *work level* (WL) points to an upper limit of fuel stored in the tank. Filling the tank above this level may cause tank overfill, which might result in fuel spillage into the soil.

In the petrol station simulator the size of a delivery, i.e. the amount of fuel that will be poured into the tank, is calculated as follows:

$$V_d = r \cdot V_{WL_{var}} + V_{WL} - V_{WL_{var}} - V_s \tag{4}$$

In Eq. 4 $V_{WL}$ denotes volume consistent with the tank's *work level*, $V_{WL_{var}}$ its variance, and $V_s$ represents the amount of fuel currently stored in the tank.

In the petrol station simulator delivery is not an instantaneous event, i.e. the process of refuelling the tank takes time linearly proportional to the delivery volume from Eq. 4 and *refuel speed*. The *refuel speed* is a simulation global parameter that determines delivery throughput.

In the petrol station simulator deliveries are triggered when the amount of fuel stored in the tank is approaching the *low level* value. In reality deliveries are mainly scheduled according to tank's fuel consumption and other statistics.

### 3.3 The Simulation Configuration

The simulation configuration starts with determining global parameters, i.e. source of configuration (file or a local database), destination for generated data (e.g. database, text file, data stream), *refuel speed, transaction volume variance*, and time frame of the simulation (*start* and *end time*). Then every single tank can be configured separately in terms of basic functional parameters:

- *Capacity* – capacity of a fuel tank,
- *Low Level* and its *Variance* – the minimal secure amount of fuel in a tank,
- *Work Level* and its *Variance* – the maximal secure amount of fuel in a tank,
- *Average Transactions/Hour* – average number of fuel transactions per hour,
- *Transaction Average Volume* – average volume of fuel transactions,
- *Calibration Table/Curve* – function that transforms height measurements into volume measurements [4,7].

With tanks parametrized, anomalies of four types can be applied: tank or pipe leak, surplus, probe hang [7]. Anomalies may overlap and there is no upper limit on the number of anomalies applied on one tank. It is possible to induce an anomaly only in a specific time frame of the simulation by defining *start* and *end* time. All anomalies, except probe hang, require specifying *volume* or *speed*, i.e. intensity of the applied anomaly. It can be defined either as the total volume or speed (in litres per hour).

All parameters can be imported from a database or a file. Real petrol station's configuration, when saved in a database, can be directly transferred into the simulator. Moreover, some of the imported parameters can be further edited. The simulator allows also to export its configuration to a file.

## 4    Results

Usually on real petrol stations it can be observed that during nights the intensity of transactions is generally lower than during daytime.

Figure 1 presents fuel volume measurements from a 12-day simulation where petrol station configuration was extracted from a real station. The tank presented

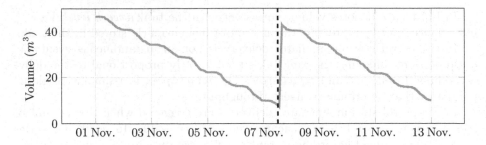

**Fig. 1.** Simulated fuel volume function

on Fig. 1 can store a maximum of 490000 l with a *work level* of 465550 l. Starting fuel volume in the simulation was set to *work level*. For the simulated tank the *lower level* was set to 50000 l – when the fuel level approached that value, delivery was triggered (marked with a vertical line on Fig. 1). The daily cycle of transactions was retained which is visible as undulations on the volume chart.

Obviously there are plenty of cases when the daily cycle of transactions may not be accurate. When a petrol station is located near highways, especially in border zones, the freight traffic is significant without regard to time of a day.

On Fig. 2 calibration curve from the same tank is presented in its ideal form (i.e. correctly recalculating height of the fuel level to corresponding volume in $m^3$) and altered in the upper half of the tank. While Fig. 1 shows measurements recalculated using the ideal calibration curve, Fig. 3 presents the very same tank but in the simulation when the altered calibration curve was used. Course of the volume function in the second case seems similar; however, maximum value is equal to 34 480 l, while *work level* was set to 465550 l. The changes to the calibration curve were applied from mentioned 34 480 l and according to them all the volumes above that level are lowered comparing to those calculated with ideal calibration curve.

In terms of anomaly detection, one of the crucial aspects to the simulator and associated anomaly applier is possibility of applying leakage. Figure 4 presents

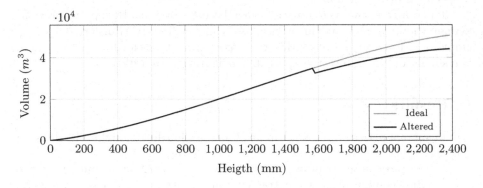

**Fig. 2.** Calibration curve – ideal and manually altered

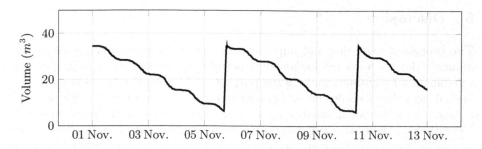

**Fig. 3.** Fuel volume in the tank affected by the alterations to the calibration curve

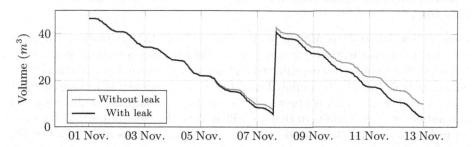

**Fig. 4.** Simulation no. 1 – data with and without a leak of 30 litres per hour

the same dataset as on Fig. 1 with and without a leak. Using the anomaly applier, the leak of 30 litres per hour was applied on the data from simulation no. 1 (Fig. 1) starting from the $5^{th}$ day. The change is clearly visible from the $6^{th}$ day where CV functions diverge.

Unfortunately, the comparison of leak applied to previously simulated data and the new simulation with leak configured seams pointless at this time. Figure 5 presents results of two separate simulations – with a leak applied on the simulated data (the same as on the Fig. 4) and another simulation with the same parameters including a leak of 30 litres per hour applied from the $5^{th}$ day. Simulations are always unique and it is not possible to produce the very same data.

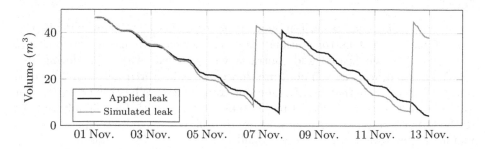

**Fig. 5.** Simulations no. 1 and no. 2 with leaks of 30 litres per hour

## 5    Discussion

The process of designing and implementing any software product, such as e.g. anomaly detection system for liquefied petroleum stations, requires proper and reliable data for training and testing purposes. The petrol station simulator has been designed and implemented in order to fulfil this requirement – it generates datasets with various anomalies applied, as well as without them (plain data). Nevertheless, there are still some areas in which the petrol station simulator has to be improved. They are not crucial for the simulator to comply with the requirements and prerequisites (i.e. four compatibility levels) described in Sect. 2; however, they can significantly ameliorate the detailed accordance with a real data source.

*Issue 1: Simulation Time Management – Stream Data.* In the simulator the internal time resolution is fixed – tempo of a simulation process cannot be arbitrary adjusted. We want to distinguish two distinct times: internal (simulation) time, which controls the whole simulated world (e.g. the frequency of measurements), and the real time, which is observable to the user and affects the duration of a simulation process. Thanks to that we will achieve the most important feature – mimicking a petrol station in real time by producing streams of data [3,6,8,11].

*Issue 2: Physical Modelling of Leaks From a Tank.* Currently, when a fuel leak from a tank is simulated, from each measurement of stored fuel volume the constant amount of fuel is subtracted. In reality the intensity of a leak directly depends on the fuel pressure exerted on the bottom of a tank. It means that the more fuel tank contains, the more intensively it leaks. Moreover, when a leaking point is located on the side wall of the tank on a certain height, it causes fuel to leak only when the fuel surface is above that point. Thus, according to [1] we propose the variable tank leak model, when the intensity of a leak is a function of the amount of fuel stored in the tank.

*Issue 3: Nozzle Miscalibration.* On a real petrol station fuel flow meters located in nozzles may miscalibrate over time, which leads to significant corruption of dispensed fuel measurements. This problem can be successfully modelled using the *linear model* of nozzle calibration: when a particular flow meter suffers from miscalibration, it increases or decreases its measured volume by a constant coefficient.

*Issue 4: Fuel Thermal Model.* Temperature changes affect significantly the behaviour of fuel on real petrol stations. Due to thermal expansion phenomenon, fuel expands when heated and contracts when cooled. Although, during seasonal changes in weather temperature slow changes in stored fuel volume can be noticed, it is delivery that may trigger more rapid effect. When relatively warmer or colder fuel is being delivered to the tank, it mixes with fuel already present in the tank causing the resulting mixture to rapidly change its temperature and volume. Some time after delivery (usually in few days) fuel tends to return to its normal temperature (accordingly to current weather conditions). Currently, the simulator generates data with a fixed fuel temperature $-15\,°C$. We propose the aforementioned fuel thermal model to be implemented.

*Issue 5: Petrol Station Infrastructure Modelling.* Currently, the simulator can simulate behaviour of a particular real petrol station by setting various parameter values, e.g.: number of tanks, tank volume or calibration curve. All these virtual devices are fixed, which means that currently the simulator cannot simulate e.g. manifolded tank systems [1]. All internal devices should be individually modelled and simulated, allowing the user to arbitrary build a custom virtual petrol station.

## 6   Conclusions

Issues 2–5 (presented in Sect. 5) address problems connected directly to the production of the most accurate datasets. However, it is issue no. 1 that refers to the most important future feature of the simulator. The main goal was to develop a reliable data source not only for the anomaly detection purposes. The future version of the simulator will be used in testing and development of the Stream Data Warehouse, especially the stream ETL process [5] and stream materialized aggregate list [9].

**Acknowledgments.** The authors would like to thank Professor Marcin Gorawski from Silesian University of Technology, Poland for support and mentoring, and undergraduate students Krzysztof Zagórski and Marek Bajorek for their collaboration during the implementation phase.

## References

1. EN 13160-5. Leak Detection Systems - Part 5: Tank Gauge Leak Detection Systems (2005)
2. Erkman, S.: Industrial ecology: an historical view. J. Cleaner Prod. **5**(1), 1–10 (1997)
3. Gorawski, M., Marks, P.: Towards reliability and fault-tolerance of distributed stream processing system. In: 2nd International Conference on Dependability of Computer Systems, DepCoS-RELCOMEX 2007, pp. 246–253, June 2007
4. Gorawski, M., Skrzewski, M., Gorawski, M., Gorawska, A.: Neural networks in petrol station objects calibration. In: Wang, G., Zomaya, A., Martinez Perez, G., Li, K. (eds.) ICA3PP 2015 Workshops. LNCS, vol. 9532, pp. 714–723. Springer, Heidelberg (2015). doi:10.1007/978-3-319-27161-3_65
5. Gorawski, M., Gorawska, A.: Research on the stream ETL process. In: Kozielski, S., Mrozek, D., Kasprowski, P., Małysiak-Mrozek, B. (eds.) BDAS 2014. CCIS, vol. 424, pp. 61–71. Springer, Heidelberg (2014)
6. Gorawski, M., Gorawska, A., Pasterak, K.: A survey of data stream processing tools. In: Czachórski, T., Gelenbe, E., Lent, R. (eds.) Information Sciences and Systems, pp. 295–303. Springer, Heidelberg (2014)

7. Gorawski, M., Gorawska, A., Pasterak, K.: Liquefied petroleum storage anddistribution problems and research thesis. In: Kozielski, S., Mrozek, D., Kasprowski, P., Małysiak-Mrozek, B., Kostrzewa, D. (eds.) BDAS 2015. CCIS, vol. 521, pp. 540–550. Springer, Heidelberg (2015)

8. Gorawski, M., Marks, P.: Towards automated analysis of connections network in distributed stream processing system. In: Haritsa, J.R., Kotagiri, R., Pudi, V. (eds.) DASFAA 2008. LNCS, vol. 4947, pp. 670–677. Springer, Heidelberg (2008)

9. Gorawski, M., Pasterak, K.: Research and analysis of the stream materialized aggregate list. In: Amine, A., Bellatreche, L., Elberrichi, Z., Neuhold, E.J., Wrembel, R. (eds.) CIIA 2015. IFIP AICT, vol. 456, pp. 269–278. Springer, Heidelberg (2015)

10. Sigut, M., Alayón, S., Hernández, E.: Applying pattern classification techniques to the early detection of fuel leaks in petrol stations. J. Cleaner Prod. **80**, 262–270 (2014)

11. Stonebraker, M., Çetintemel, U., Zdonik, S.: The 8 requirements of real-time stream processing. SIGMOD Rec. **34**(4), 42–47 (2005)

12. United States Environmental Protection Agency: Standard test procedures for evaluating leak detection methods: automatic tank gauging systems. Final report (1990)

13. United States Environmental Protection Agency: Standard test procedures for evaluating leak detection methods: non volumetric tank tightness testing methods. Final report (1990)

14. United States Environmental Protection Agency: Standard test procedures for evaluating leak detection methods: statistical inventory reconciliation methods. Final report (1990)

15. United States Environmental Protection Agency: Standard test procedures for evaluating leak detection methods: volumetric tank tightness testing methods. Final report (1990)

# Author Index

Printed in the United States
By Bookmasters